FUNDAMENTALS OF SURVEYING

THIRD EDITION

FUNDAMENTALS OF SURVEYING

THIRD EDITION

MILTON O. SCHMIDT

KAM W. WONG

University of Illinois at Urbana-Champaign

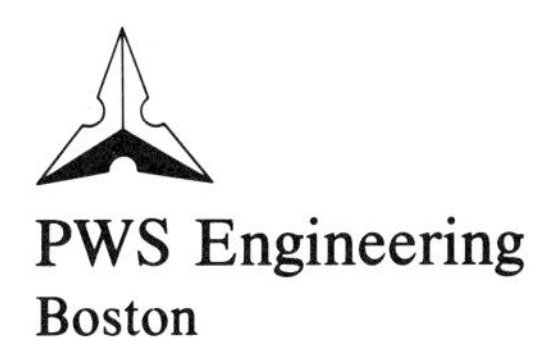

PWS Engineering

Boston

PWS PUBLISHERS

Prindle, Weber & Schmidt • Duxbury Press • PWS Engineering •
Statler Office Building • 20 Park Plaza • Boston, Massachusetts 02116

PWS Publishers is a division of Wadsworth, Inc.

Library of Congress Cataloging in Publication Data

Schmidt, Milton O. (Milton Otto), 1910–
Fundamentals of surveying.

Includes bibliographies and index.
1. Surveying. I. Wong, Kam W., 1940–
II. Title.
TA545.R3 1985 526.9 84-23687
ISBN 0-534-04161-2

ISBN 0-534-04161-2

Printed in the United States of America
85 86 87 88 89—10 9 8 7 6 5 4 3 2 1

Sponsoring Editor: *Ray Kingman*
Editorial Assistant: *Jane Parker*
Production: *Stacey C. Sawyer, San Francisco*
Manuscript Editor: *Sandra Spiker*
Interior Design: *John Edeen*
Cover Design: *John Edeen*
Interior Illustration: *Pat Rogondino & Mary Burkhardt*
Typesetting: *Bi-Comp, Inc., York, Pennsylvania*
Printing and Binding: *Maple-Vail Book Manufacturing Group*

PREFACE

Three factors have provided the impetus for this major revision of a traditional textbook. First, recent developments in surveying instrumentation and computer technology have resulted in fundamental changes in both field surveying procedures and computational methods. Second, changes in college curricula have resulted in significant reduction in the amount of instructional time allocated to surveying. Third, the increasing complexity of modern engineering projects, ranging from land subdivision to tunnelling, requires the use of the latest surveying technology.

Only three chapters from the previous edition of *Fundamentals of Surveying* remain substantially intact. More than twenty-five percent of the text materials included in this edition were not in the previous edition. In addition, there are about forty-five new example problems, one hundred and thirty new problems, and more than two hundred and thirty new illustrations. The important features of this new edition include:

1. a complete coverage of the principles and uses of electronic distance measuring (EDM) instruments and total stations;
2. a thorough coverage of the commonly accepted statistical methods used in error analysis;
3. emphasis of coordinate geometry in surveying computations, particularly in control surveys and traverse calculations;
4. a complete chapter devoted to the latest methods of horizontal control surveys, including the use of earth-orbiting satellites;
5. a completely updated chapter on the use of photogrammetry for reconnaissance measurements and topographic mapping;
6. relevant application examples throughout the book to stimulate student interest and to help bridge the gap between theory and practice; and
7. problems at the end of each chapter that supplement the textual materials as teaching and learning tools. In addition to the usual type of problems that drill students on methods and techniques, there are also problems designed to illustrate application environments as well as the decision-making aspects of surveying.

My goal is to present in one volume the basic fundamentals of modern surveying with extensive exposure to practical problems. Surveying is taught to students in different fields of specialization and on different levels of difficulty. It is not likely that the entire book can be covered in a one-semester course; the topics to be included and the thoroughness of coverage must depend on the objective of the course. I have organized the materials so that the chapters can be taught in different sequences and so that sections or chapters can be omitted in a course. My experience in teaching is that what matters most is not *how much* but *how well* a subject is taught. Having learned the basic material, a student should be able to learn the remaining material in the book by self-study when the need arises.

With the exception of few sections, Chapters 1 to 14 basically constitute the material that I have used in a one-semester course for civil engineering students. Some instructors may wish to substitute Chapter 15 (State Plane Coordinate Systems), Chapter 16 (Astronomical Determination of Direction), or Chapter 17 (Hydrographic Surveying) for Chapter 14 (Photogrammetry). In a less rigorous course, the subject of error analysis (Chapter 2 and relevant sections in other chapters) can also be omitted. Alternatively, the entire book can be covered in a two-semester course.

I wish to express my appreciation to Dr. Milton O. Schmidt, my co-author and colleague, and Mr. Ray Kingman, managing editor of PWS Engineering, PWS Publishers. Without their understanding and complete faith in me, such a major revision would not have been possible.

I gratefully acknowledge the assistance given by the many distributors and manufacturers of surveying instruments for having generously granted permission to use the many illustrative and photographic materials included in this edition. These organizations are recognized in the legend accompanying each figure.

This revision project has made me realize, more than ever before, how much I have learned from my former and present students and colleagues. This revised edition is a testimonial to their efforts as much as mine.

Any comment or suggestion for further improvement by users of this book will be gratefully received.

Kam W. Wong

CONTENTS

INTRODUCTION

1

1.1 SURVEYING DEFINED

Surveying may be simply defined as *earth measurements*. Surveying involves the measurement of elevations, distances, and angles and the processing of the measurements into positional information such as maps and coordinates. Survey activities are not limited to the surface of the earth but extend to the sea and deep underground, as well as extraterrestrial space. Surveying principles have also been applied to the measurement of other objects, such as human and animal bodies, machineries, artifacts, and plants.

1.2 BOUNDARY SURVEYS

One of the oldest uses of surveying is the demarcation of boundaries. Natural and man-made objects have historically been used to mark boundary corners and boundary lines. However, to determine the area of ownership and to provide written description of the location, size, and shape of the parcel of land, surveying measurements have always been needed. Today, boundary surveys are conducted not only for boundaries on the land's surfaces, but also in the air (such as condominiums and highrise buildings), at sea (such as off-shore oil leases), and underground (such as coal mines and rights-of-way for tunnels and shafts).

1.3 ENGINEERING SURVEYS

Surveying is also vital to the planning, design, construction, and operation of engineering facilities. Accurate topographic maps of the proper scales are needed for the planning and design of roads, dams, reservoirs, tall buildings, canals, pipelines, and so on. Surveying measurements are also needed for laying out designed engineering facilities, for computing surface drainage areas and volumes of earthwork, and for monitoring landslides, land subsidence, dam deformation, and structural settlements.

1.4 RESOURCES INVENTORY AND MANAGEMENT

Topographic maps and quantitative data are necessary for the proper inventory and management of natural resources such as land, forests, soils, water, and underground minerals. Since these functions are usually performed by national and state governments, topographic mapping and resource surveys are major responsibilities of these governmental units. Large utility companies (oil, gas, coal, and so on) also engage extensively in surveying activities.

1.5 SPACE EXPLORATION

The surveying and mapping profession has made major contributions to the exploration of space. To accurately track the orbits of the satellites and space vehicles, one must determine the relative positions of the tracking stations located throughout the world. The launch and landing sites as well as the associated facilities must be correctly located during construction. Even before the first man (Neil Armstrong) set foot on the lunar surface, detailed topographic maps were made of many parts of the moon for use in mission planning and scientific studies. Maps of Mars, Venus, and the Galilean satellites of Jupiter have also been made from television pictures returned from satellites (Reference 1.1). Although extraterrestrial mapping requires techniques more advanced than those normally used in conventional surveying and mapping projects, it does use many basic principles as well as instrumentations that are commonly used in surveying practices.

1.6 CHANGING TECHNOLOGY

The developments associated with space exploration have, in turn, introduced major changes to surveying technology. Instruments, field procedures, and computational methods that have been in use for decades are rapidly becoming obsolete. New instrumentations have imposed new measurement procedures and computational methods and have enabled surveyors to attain higher accuracy in a shorter time period and at lower cost. Accuracy that in the past could only be achieved through tedious field procedures and rigorous application of advanced mathematical skills can now be attained by the simple turning of a switch. Earth-orbiting satellites and electronic devices are becoming regular tools of the surveyors, even for such common tasks as boundary surveys. However, to make full efficient use of new instrumentations in the midst of this changing technology, a surveyor must be well-trained in the basic fundamentals of surveying.

Figure 1.1 shows a Zeiss Elta 2 electronic tacheometer used in engineering layout work. The instrument uses infrared light to measure distances. By pointing the telescope to a reflector target, both distance and angle can be measured simultaneously and automatically recorded for later processing. It has a built-in microcomputer that can provide real-time computation of positional coordinates.

Figure 1.2 shows a Wild "field to finish" surveying system, which includes a field unit (a) for measuring distances and angles with automatic data recording and an office data processing unit (b) that includes a microcomputer with a graphic terminal.

Figures 1.3a and b are computer-generated graphic models showing Mt. St. Helens before and after the eruption of May 18, 1980. The eruption removed the top 400 m of the mountain, leaving a crater 750 m deep. These graphic models were drawn using hundreds of spot elevations derived from aerial photographs of the mountain. Such aerial mapping and computer graphic techniques are also used extensively for the design and planning of transportation facilities.

Figure 1.4 depicts some of the earth-orbiting satellites included on the *Global Positioning System* (GPS) being developed by the U.S. Department of Defense. Development of the GPS began in 1973, and the first satellite was deployed in 1978. When the system is fully implemented, it will have 18 satellites, three in each of six orbital planes. By positioning a highly portable antenna over a survey station to track the radio signals emitted by the GPS satellites, the three-dimensional position of the survey station can be accurately determined. The suitability of the GPS for engineering surveying has already been clearly demonstrated. The GPS and other advanced positioning systems that are being developed will have a major impact on surveying projects of the future.

1.7 COMPLEX CHALLENGES

The complexity of surveying problems has grown at the same, if not a higher, pace as the improvement in surveying technology. Instead of surveying a largely rural landscape, boundary surveyors must now be con-

FIGURE 1.1 Zeiss Elta 2 electronic tacheometer used in engineering layout (courtesy of Carl Zeiss, Inc. Thornwood, N.Y.)

cerned with surveying the three-dimensional boundaries of condominiums located hundreds of feet above ground. In rebuilding the infrastructure of urban centers, surveyors are faced with the imposing task of surveying and locating vast mazes of tall buildings, overpasses, underpasses, power cables, sewage lines, water lines, gas mains, rapid transit rails, tunnels, and so on. Engineering projects are now more complex in scope, and surveying must be performed in a shorter period of time and at a much higher accuracy tolerance. One good example of the scope of surveying requirements of modern

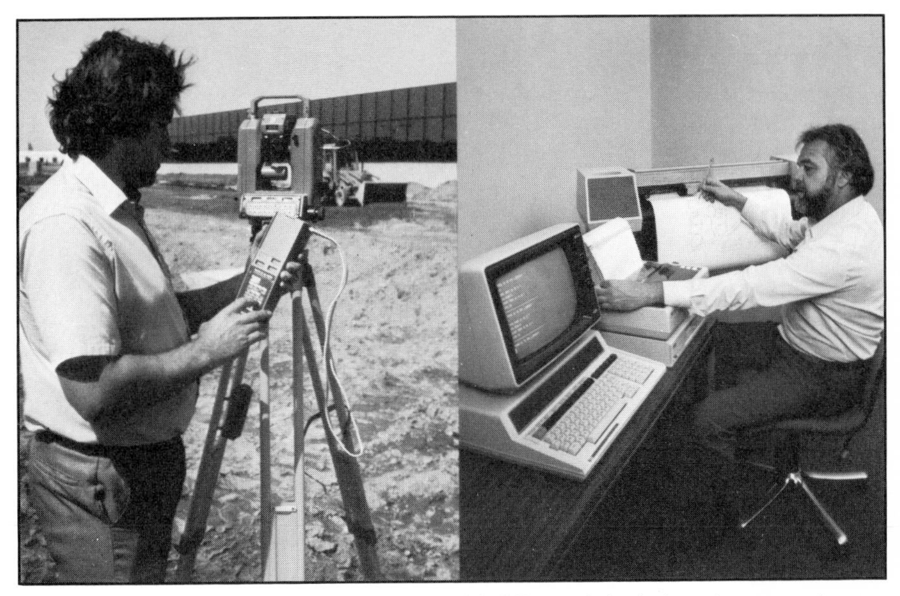

(a) Field unit

(b) Office unit includes microcomputer with graphic terminal

FIGURE 1.2 Wild "field to finish" surveying system (courtesy of Wild Heerbrugg Instruments, Inc.)

engineering projects is the Washington-Boston rail improvement project (Reference 1.3). Two thousand and four hundred (2,400) topographic maps at a scale of 1″ = 40′ and a contour interval of 1 ft were prepared for a corridor 62 miles wide and 456 miles long. The mapping project was completed in a period of only nine months!

1.8 SPHERICAL EARTH

One of the fundamental problems of surveying is the irregular shape of the physical surface of the earth, which defies mathematical description. To compound the problem, the density of the earth's material is not uniform and causes local as well as regional variations in the force of gravity. *Geodesy* is the science that deals with studies concerning the size and shape of the earth. To compute the relative positions of points located on or near the earth's surface, geodesists have traditionally approximated the earth with an ellipsoid. All position calculations are performed on the surface of the adopted ellipsoid. Spherical geometry and advanced mathematics must be used. Accurate surveys that extend over an area larger than about 10 miles in radius must take into consideration the spherical shape of the earth and the effects of gravity. This type of surveying is called *geodetic surveying*. Its

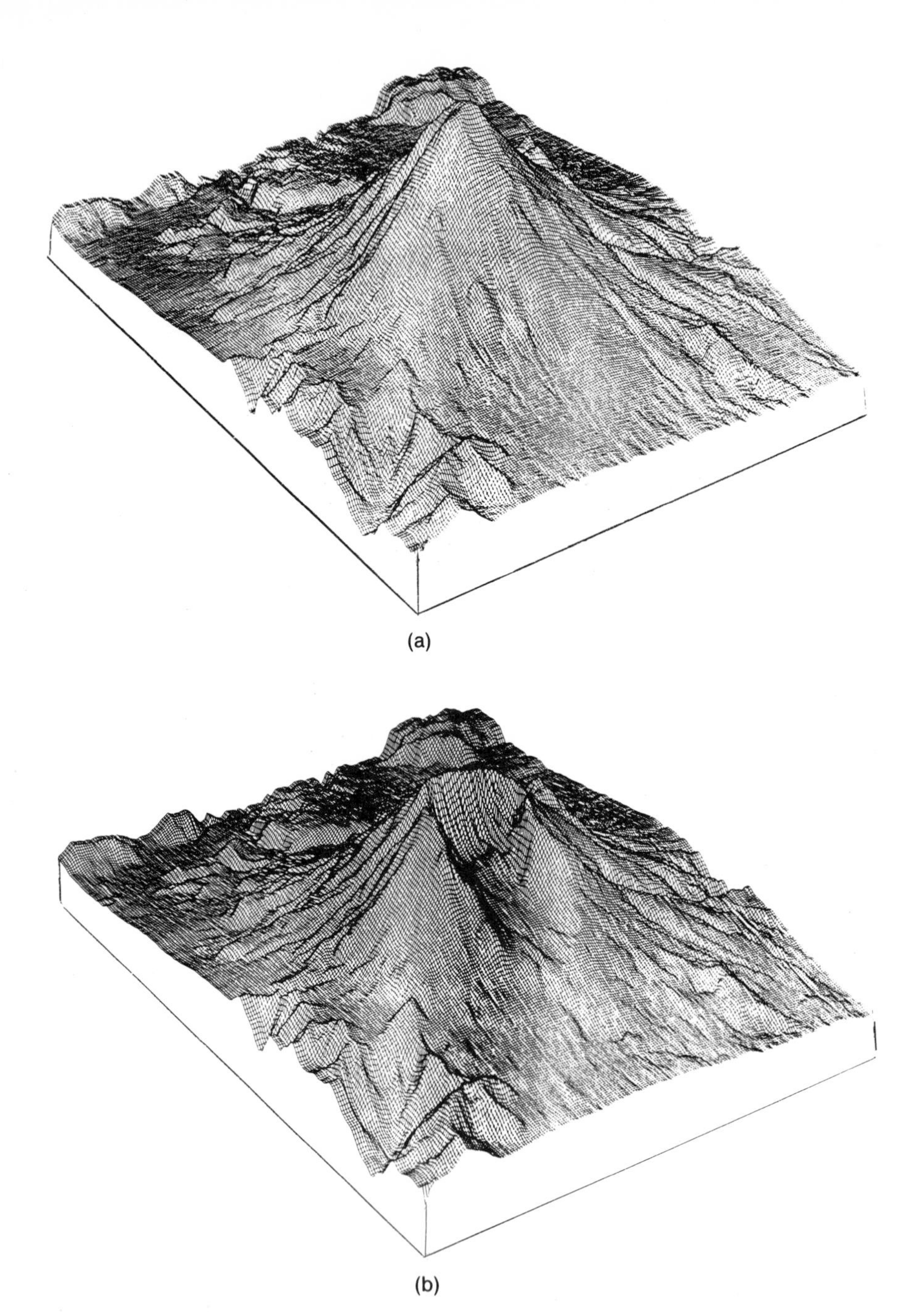

(a)

(b)

FIGURE 1.3 (a) Computer-generated graphic model of Mt. St. Helens before the eruption of May 18, 1980; (b) computer-generated graphic model of Mt. St. Helens after the eruption of May 18, 1980 (courtesy of U.S. Geological Survey)

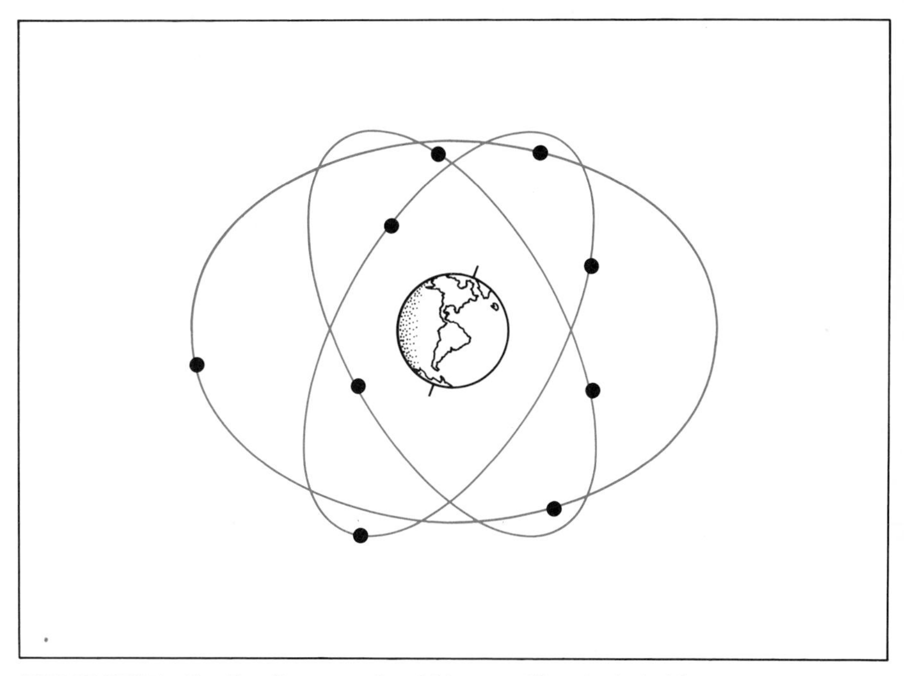

FIGURE 1.4 Some earth-orbiting satellites included in the Global Positioning System (GPS)

primary application is the surveying and mapping of an entire state or nation. It also finds applications in large engineering projects.

1.9 PLANE SURVEYING

Most common surveys require only a moderate level of accuracy and are confined to an area of less than a 10-mile radius. For such surveys it is usually sufficient to project points on the physical surface of the earth orthogonally onto a flat surface. In this manner, relative positions of points can be computed by plane geometry.

This book deals primarily with the fundamentals of *plane surveying*. However, both students and practitioners must be able to recognize situations where plane surveying methods can no longer satisfy the project requirements. For this reason the effects of the earth's curvature and the limitations of plane surveying methods will also be discussed throughout this book.

REFERENCES

1.1 American Society of Photogrammetry. *Manual of Photogrammetry,* 4th ed., Chapter XVII: "Satellite Photogrammetry." Falls Church, Va.: 1980. 883–977.
1.2 Beardslee, Daniel E. "The Kiket Island Nuclear Power Generation Site Survey," *Surveying and Mapping,* Journal of American Congress on Surveying and Mapping, Vol. 43, No. 2 (June 1983) 161–173.
1.3 Burns, Joseph P. "450 Miles of Rail Line Mapped in 9 Months," *Civil Engineering,* American Society of Civil Engineers, Vol. 48, No. 12 (December 1978) 51–53.
1.4 Chamard, Roger R. "Photogrammetric Mapping for Highways: Western U.S.," *Journal of Surveying Engineering,* American Society of Civil Engineers, Vol. 109, No. 1 (March 1983) 1–5.
1.5 Fisher, Irene K. "The Map as a Talisman—Different Ideas in Early Mapping," *Surveying and Mapping,* Journal of American Congress on Surveying and Mapping, Vol. 42, No. 2 (June 1982) 127–137.
1.6 Howard, John D. "Survey Control for I-205 Columbia River Bridge," *Surveying and Mapping,* Journal of the Surveying and Mapping Division, American Society of Civil Engineers, Vol. 107, No. SU1 (Nov. 1981) 33–44.
1.7 Richardson, Steven L. "Pioneers and Problems of Early American Photogrammetry," *Photogrammetric Engineering and Remote Sensing,* Vol. 50, No. 4 (April 1984) 433–450.
1.8 Sandilands, R. W. "Hydrographic Surveying in Great Lakes during the Nineteenth Century," *The Canadian Surveyor,* Vol. 36, No. 2 (June 1982) 139–163.

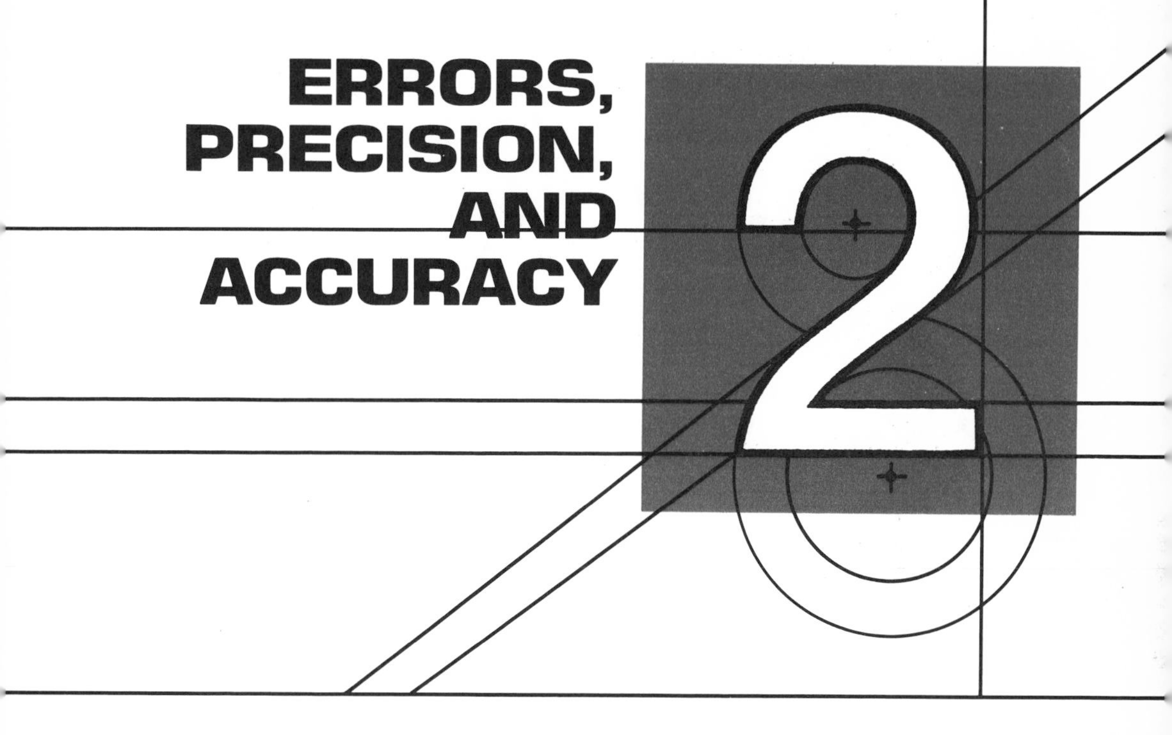

2 ERRORS, PRECISION, AND ACCURACY

2.1 INTRODUCTION

All surveying operations are subject to the imperfections of the instruments, the fallibility of the human operator, and the uncontrollable nature of the natural environment. In fact, no surveying measurement is exact, and the true value of the parameters being measured is never known. Therefore, a surveyor must thoroughly understand the sources of errors in various methods of surveying as well as the methodology for evaluating the achievable accuracy of a surveying program.

Two true stories illustrate the importance of the subject matter covered in this chapter. The first story concerns a tunneling project. A sanitary district was seeking rights-of-way to construct an 8-ft diameter underground tunnel. One section of the proposed tunnel passed underneath a power company's generating plant. Since that section of the tunnel would be located about 25 to 50 ft below the ground surface and the overburden materials consisted of sandy soils, the power company was concerned that surface settlement caused by the tunneling operation might result in serious damage to its buildings and generating facilities. Initially, the power company demanded guarantee from the sanitary district that there be absolutely no surface settlement. After some discussion, both sides finally agreed on the specification that surface settlement be less than "0.01 ft ± survey error." How the two parties arrived at this specification for settlement tolerance is unknown. It is safe to assume that the two sides had different conceptions of what that "± survey error" was. The power company might have been justified to expect that the survey error should be significantly less than 0.01 ft. However, the sanitary district might have intended to do the "best" that it could in conducting the survey without knowing exactly what the "best" was. Obviously, such a specification opens opportunities for abuses and later disputes. Fortunately, in this particular case, the construction was completed without causing any damage to properties of the power company.

The second story concerns a surveying program for monitoring ground subsidence associated with the solution mining of salt. In solution mining, fresh water is pumped underground where it dissolves the salt, and the resulting brine solution is pumped out to be processed for salt. As more and more salt is extracted, the underground cavity becomes larger and larger. When the overlying rock layer can no longer support itself, surface subsidence results. In this particular case, the salt field measured about 3/4 mile by 3/4 mile in area. There were 17 active wells located in the field. A registered civil engineer who was also a licensed land surveyor was hired to perform annual surveys to determine the elevations of the 17 active wells. The closest available elevation benchmark was located about 1 mile away. The engineer claimed that the maximum error in the measured elevations was ±0.01 ft. In fact, the elevation of one well was reported to be exactly 495.03 ft in seven consecutive surveys over a period of seven years. The elevation of another well was reported to be exactly 493.95 ft in six consecutive surveys. The engineer attributed this high accuracy to the dumpy level and the direct elevation rod that were used in the surveys. (A direct elevation rod has its scale marked on a ribbon that can be rotated to give direct elevation readings. It is never used in precise leveling.) The engineer, who was in his mid-fifties, further proudly explained that the surveys were done by his son and therefore must be trustworthy and reliable. Considering the types of instruments used in the surveys, the accuracy claimed by this engineer was not only incredible, but truly impossible.

All surveys are conducted to serve some specific purposes. The types of instruments, the field procedures, and the computational method to be used must be selected so that the survey results are accurate enough to serve the intended purposes. Moreover, the accuracy of the results must be verifiable and justifiable using commonly accepted statistical testing techniques. The accuracy of a survey should never be judged by the experience or reputation of the surveyor alone.

With the increasing availability of highly sophisticated survey instruments and methods, the specification that requires surveys to be done as accurately as possible could result in unnecessary expenditures amounting to thousands of dollars. However, because of the complexity of modern engineering, surveys that fail to provide the required accuracy could result in large financial losses as well as disastrous events.

This chapter will present the fundamental principles of measurement errors and the basic statistical techniques used for evaluating the accuracy of various methods of surveying and of survey results. These statistical techniques will be presented wherever appropriate throughout the remaining chapters of the book. These basic techniques are useful for evaluating any kind of measurement problem.

2.2 MEASUREMENT ERRORS

The *true error* in a measurement is the difference between the measured value of a parameter and its true value. Mathematically, the true error (e_i) of a measured value (x_i) is defined as follows:

$$e_i = x_i - x \tag{2.1}$$

where x is the true value of the measured parameter. In practice, except in the counting of discrete objects or events (such as counting the number of workers or the number of survey stations), the true value of a survey measurement is never known and can never be determined exactly. Consequently, the true error in a measurement also can never be exactly determined. The error in a measurement must be estimated by comparing it with another more accurately determined value of the same parameter. Let $\hat{x}$ represent such a value. Then, an estimate (v_i) of the true error (e_i) can be computed as follows:

$$v_i = x_i - \hat{x} \tag{2.2}$$

How close the estimate v_i is to the true error e_i will depend on the closeness of $\hat{x}$ to the true value x.

Errors in surveying measurements can be classified into the following three groups:

1. blunders,
2. systematic errors, and
3. random errors.

Blunders are simply mistakes caused by human carelessness, fatigue, and haste. A blunder can be of any sign (+ or −) and magnitude, and its occurrence is unpredictable. Blunders can be made by the most experienced surveyors. Therefore, blunders must be eliminated by careful work and by using field procedures that provide checks for blunders. Examples of blunders include transposition of digits in recording an observation, misreading the tape when a fractional distance is measured, and sighting on the wrong target when measuring an angle.

An interesting example of blundering involved an experienced surveyor who was doing a control survey for a construction project. One line of the angle he was measuring stretched across a green valley and was about 2 miles long. He was to sight on a specially designed white target and was using an instrument with a least count of 1 second. He measured the angle five times and found that the five measurements varied by as much as 1 minute of arc. He measured the angle five more times and found the same degree of inconsistency. After some perplexing moments, he finally discovered that he was sighting on the rear end of a grazing white sheep, rather than on the white survey target. (Had the sheep been sleeping and stayed immobile, the blunder would have been much more difficult to detect!)

Systematic errors are caused by some maladjustment of the surveying instruments and personal bias or inclination of the human operator, as well as by the natural environment. Both the signs and magnitudes of systematic errors behave according to some specific pattern or physical law of nature, which may or may not be known. When the law of occurrence is known, systematic errors can sometimes be modeled by a mathematical expression and the measurements can be corrected for the errors accordingly. One example of systematic error is the change in length of a steel measuring tape with temperature. When the coefficient of expansion of the steel is known and the correct length of the tape at some specific temperature has been previously determined, the temperature effect can be eliminated by measuring the field temperature and applying the proper correction to the field measurement.

Another example of systematic error is the maladjustment of a leveling instrument, which results in the line-of-sight not being perfectly horizontal when the level bubble is perfectly centered. This type of instrumental error can be minimized by adjusting the instruments properly and by checking them frequently. However, systematic errors are sometimes caused by poor instrument design and become an inherent part of the instrument. This type of error can only be detected by comparing the measurements made by one instrument with those obtained from a more accurate instrument or procedure. Such an operation is called *standardization,* or *calibration,* of the instrument.

A special type of systematic error is an error that always occurs with the same sign and magnitude and is therefore often referred to as a *constant error*. The most common source of constant error is the measuring instruments. For example, a 100-ft tape may in fact be only 99.99 ft long. Then, every tape length with this tape would contain a constant error of −0.01 ft. Constant errors of this type can be detected by proper calibration of the instruments.

Personal bias and human nature often cause a person to consistently center the cross-hair of a telescope slightly to the left or right of a target or to consistently either overestimate or underestimate the partial division of a scale. Such systematic errors are generally small and can be minimized by making conscious efforts to be careful and unbiased.

Random errors are caused by imperfection of the measuring instruments, inability of the human operator to make an exact measurement, and uncontrollable variations in the natural environment. Random errors exclude both blunders and systematic errors, although small undetected blunders and uncorrected systematic errors are often treated as random errors in surveying calculations. Theoretically speaking, the magnitude of the random errors should truly reflect the limitations of the instrument and its human operator. Random errors can be minimized by using better instruments and properly designed field procedures and by making repeated measurements.

2.3 CHARACTERISTICS OF RANDOM ERRORS

Random errors have the following characteristics:

1. Positive and negative errors of the same magnitude occur with equal frequency;
2. Small errors occur more frequently than large ones; and
3. Very large errors seldom occur.

For example, suppose that a distance is measured using the same instrument and the same care a large number of times, say 100,000 times. The mean, or average, of the 100,000 repeated measurements is computed, and the estimated error in each individual length measurement is computed by subtracting the mean value from the measured value—that is, by Eq. (2.2). The estimated error computed in this manner is called the *deviation from the mean*. Plotting the magnitude of the estimated errors against the frequency of occurrence may result in a *histogram* similar to that shown in Figure 2.1.

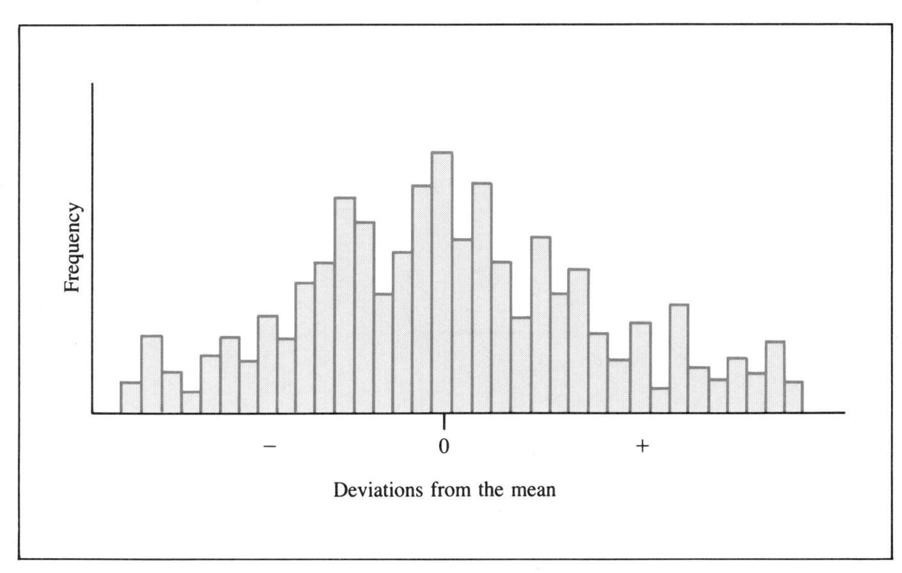

FIGURE 2.1 Distribution of random errors

In the 18th century, scientists recognized the distribution pattern of random measurement errors shown in Figure 2.1 and found that the pattern can be approximated by a continuous curve called the *normal curve of error*. This curve is now commonly known as the *normal distribution*, or *Gaussian distribution*—in honor of Carl Gauss (1777–1855). The normal distribution curve, as shown in Figure 2.2, is mathematically expressed by the following equation:

$$F(v) = \frac{1}{\sqrt{2\pi}\sigma} e^{-1/2(v/\sigma)^2} \tag{2.3}$$

where v is the random error, σ is a specific value for the random error (which will be defined later), and $F(v)$ is the normalized frequency. The curve is symmetrical about $v = 0$. The probability that the random error in a measurement takes on a value between a and b is equal to the area under the curve and bounded by the values of a and b as shown in Figure 2.3. In mathematical terms, if $P(a \leq v \leq b)$ represents this probability, then

$$P(a \leq v \leq b) = \int_a^b \frac{1}{\sqrt{2\pi}\sigma} e^{-1/2(v/\sigma)^2}\, dv \tag{2.4}$$

The curve is normalized so that the area under the entire curve is equal to 1. This is equivalent to saying that the probability that the random error in a measurement takes on a value between $-\infty$ and $+\infty$ is equal to 1, or 100%.

The value of σ is of particular interest. It can be computed from Eq. (2.4) that the probability that a random error takes on a value between $-\sigma$

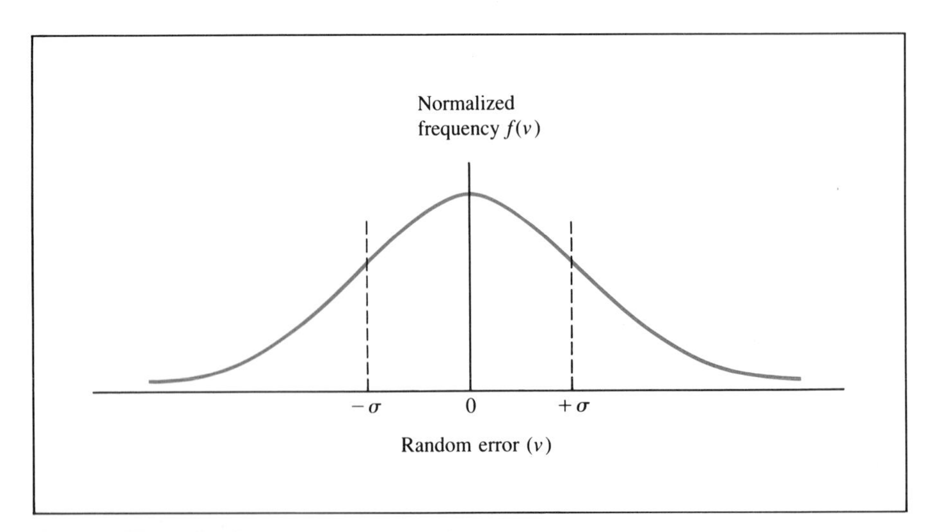

FIGURE 2.2 Normal curve of error

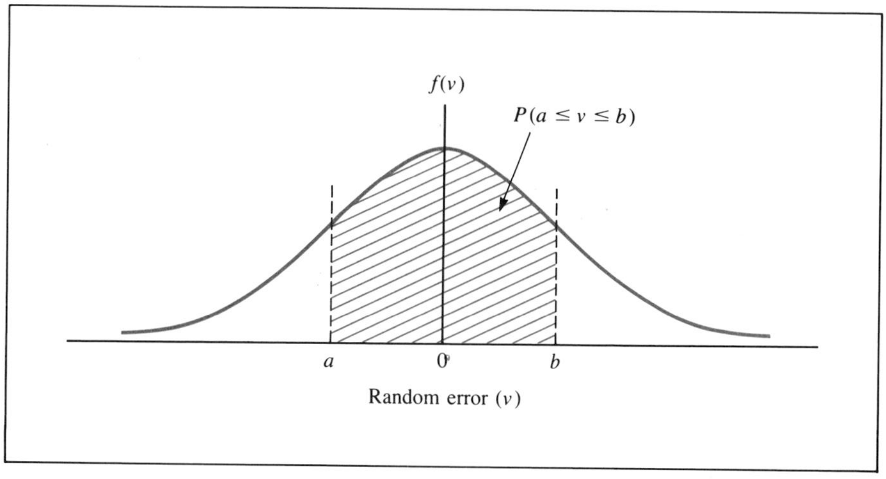

FIGURE 2.3 Probability of random errors

and $+\sigma$ is equal to 0.683 (or 68.3%). In terms of σ, some representative probabilities for selected error ranges are as follows:

Error Range	Probability (%)
$\pm 0.6745\sigma$	50.0
$\pm 1.00\sigma$	68.3
$\pm 1.6449\sigma$	90.0
$\pm 2.00\sigma$	95.4
$\pm 3.00\sigma$	99.7

The value σ is called the *standard error* and is often referred to simply as *sigma*.

For example, a measured distance is said to have a standard error of ±0.05 m if there is a 68.3%, 95.4%, and 99.7% probability that the random error in the measured value falls within the range of ±0.05 m (1σ), ±0.10 m (2σ), and ±0.15 m (3σ) respectively.

The standard error of a measurement can be determined if the exact distribution pattern, as shown in Figures 2.2 and 2.3, of the random errors is known. In practical terms this means that the measurement must be repeated almost an infinite number of times, and then a histogram (see Figure 2.1) of the deviations from the mean is plotted. In practice, since surveying measurements are usually repeated only a limited number of times—usually fewer than 10—only an estimated value can be computed for the standard error of a measurement. A method for computing the estimated values of standard errors will be discussed in the next section.

Although it has never been proven that random errors in measurements behave exactly according to the normal curve of error, it has been shown in practice that it represents the simplest approximation and has been found to be adequate in almost all surveying applications.

2.4 MEAN, STANDARD DEVIATION, AND STANDARD ERROR OF THE MEAN

Let $x_1, x_2, x_3, \ldots, x_n$ be n repeated measurements of the same quantity, and let it be assumed that all n measurements were made with the same instrument and same degree of care. The *mean*, denoted by $\bar{x}$, of the n measurements is computed as follows:

$$\bar{x} = \frac{\sum_{i=1}^{n} x_i}{n} \tag{2.5}$$

An unbiased estimate, to be denoted by $\hat{\sigma}_{xi}$, of the standard error of one measurement of the quantity can be computed by the following expression:

$$\hat{\sigma}_{xi} = \pm\sqrt{\frac{\sum_{i=1}^{n} (x_i - \bar{x})^2}{n - 1}} \tag{2.6}$$

The superscript (ˆ) is used to indicate that $\hat{\sigma}_{xi}$ is only an estimated, rather than an exact, value. *Unbiasedness* is a desirable quality for a statistical estimator. It is beyond the scope of this book to provide the statistical proof that Eq. (2.6) provides the best estimator for the standard error of one measurement of the quantity. The estimator $\hat{\sigma}_{xi}$ is often called the *standard deviation* of the set of n measurements or the *root-mean-square (RMS) error* of a single measurement.

An estimate for the standard error of the mean of the n measurements, to be denoted by $\hat{\sigma}_{\bar{x}}$, can be computed by the following expression:

$$\hat{\sigma}_{\bar{x}} = \pm\sqrt{\frac{\sum_{i=1}^{n} (x_i - \bar{x})^2}{n(n - 1)}}, \tag{2.7}$$

or

$$\hat{\sigma}_{\bar{x}} = \pm \frac{\hat{\sigma}_{xi}}{\sqrt{n}} \tag{2.8}$$

The quantity $\hat{\sigma}_{\bar{x}}$ is also often called the *RMS error of the mean*.

EXAMPLE 2.1

A distance was measured ten times yielding the following results: 574.536, 574.533, 574.530, 574.531, 574.532, 574.534, 574.535, 574.531, 574.531, and 574.533 m. Compute the mean ($\bar{x}$), standard deviation ($\hat{\sigma}_{xi}$), and an estimated standard error of the mean ($\hat{\sigma}_{\bar{x}}$). Also plot a histogram of the computed errors in the measurements.

SOLUTION

Measured Distances (m)	$v_i = x_i - \bar{x}$	v_i^2
574.536	0.0034	.00001156
.533	0.0004	.00000016
.530	−0.0026	.00000676
.531	−0.0016	.00000256
.532	−0.0006	.00000036
.534	0.0014	.00000196
.535	0.0024	.00000576
.531	−0.0016	.00000256
.531	−0.0016	.00000256
.533	0.0004	.00000016

$$\text{mean } (\bar{x}) = \frac{\sum_{i=1}^{n} x_i}{n} = 574.5326 \text{ m}$$

$$\text{standard deviation } (\hat{\sigma}_{xi}) = \pm\sqrt{\frac{\sum_{i=1}^{n} (x_i - \bar{x})^2}{n-1}}$$

$$= \pm\sqrt{\frac{0.00003440}{10-1}}$$

$$= \pm 0.00196 \text{ m}$$

rounding off: $\hat{\sigma}_{xi} = \pm 0.002$ m

$$\text{estimated standard error of the mean } (\hat{\sigma}_{\bar{x}}) = \pm\frac{\hat{\sigma}_{xi}}{\sqrt{n}}$$

$$= \pm\frac{0.00196}{\sqrt{10}}$$

$$= \pm 0.00062$$

rounding off $\hat{\sigma}_{\bar{x}} = \pm 0.0006$ m

Figure 2.4 shows a histogram of the computed errors in the ten measurements. An interval of 0.001 is used. Note that this histogram does not at all resemble

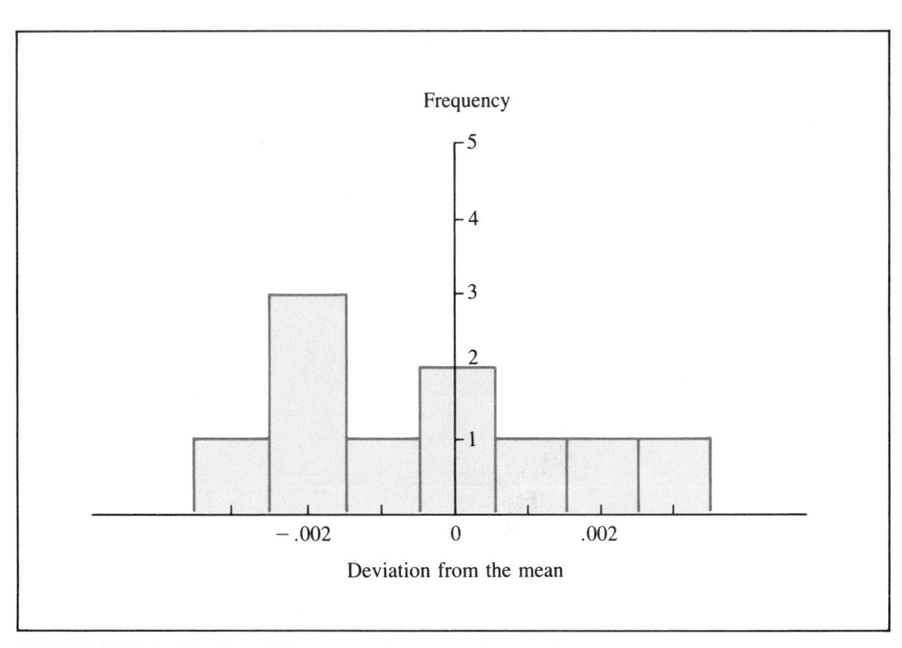

FIGURE 2.4 Histogram for a small sample

the pattern of the normal curve of error. This is usually the case when the number of repeated measurements is small. A small sample of measurement errors can rarely accurately display the true distribution pattern of the random errors.

It is a common practice to round off the final values of the actual or estimated standard error to one or two significant nonzero digits. For example, if $\sigma = \pm 43.21$, it is rounded off to $\sigma = \pm 40$ or ± 43; and if $\sigma = 0.00247$, it is rounded off to $\sigma = \pm 0.002$ or ± 0.0025. In Example 2.1, the value of $\hat{\sigma}_{xi}$ is rounded off to ± 0.002 m and $\hat{\sigma}_{\bar{x}}$ is rounded off to ± 0.0006 m. However, for intermediate calculations it is a good practice to carry two or three significant, nonzero numerical digits. For example, in computing for $\hat{\sigma}_{\bar{x}}$, $\hat{\sigma}_{xi} = \pm 0.00196$ m is used.

The value of a computed quantity should be rounded off to the same significant digit as its standard error. In Example 2.1, since $\hat{\sigma}_{\bar{x}} = \pm 0.0006$ m with the first significant, nonzero numerical digit in the fourth decimal place, the computed value for the mean ($\bar{x}$) should also be rounded off to the fourth decimal place—that is, $\bar{x} =$ 574.5326 m. Thus the mean should be reported as 574.5326 ± 0.0006 m. It should not be reported in any of the following forms: 574.5326 ± 0.00061824, 574.5326 ± 0.000618, 574.533 ± 0.0006, or 574.533 ± 0.001.

2.5 PROBABLE ERROR AND MAXIMUM ERROR

The *probable error* of a measurement is defined as being equal to 0.6745σ. There is a 50% probability that the actual error exceeds the probable error, as well as a 50% probability that it is less than the probable error. Thus if the standard error of an angle measurement is ±1.5 seconds, then the corresponding probable error would be ±(0.6745 × 1.5) seconds, or ±1.0 second. The term *probable error* was commonly used in surveying and engineering in the past. However, the term *standard error* is now used.

The *maximum error* in a measurement is most commonly defined as being equal to 3σ. There is 99.7% probability that the actual error falls within 3σ, and there is only a 0.3% probability that the actual error exceeds 3σ. For example, if the standard error of a distance measurement is ±0.05 ft, then the maximum error is ±(3 × 0.05 ft) or ±0.15 ft.

The maximum error is often used as the criterion for separating mistakes or blunders from random errors. For example, after the mean and standard deviation of n repeated measurements have been computed, the deviation of each measurement from the mean can be computed. If any measurement deviates from the mean by more than three times the standard deviation, that measurement is considered to contain a blunder. It is rejected, and a new mean and standard deviation is computed without this particular measurement.

EXAMPLE 2.2

Fifteen measurements of an angle are as follows: 27°30′41″, 32″, 30″, 38″, 31″, 34″, 39″, 37″, 32″, 03″, 40″, 42″, 30″, 37″, and 36″. Reject any measurement containing a blunder and compute the mean, standard deviation, and estimated standard error of the mean.

SOLUTION

Measurement	First Iteration $v_i = a_i - \bar{a}$	Second Iteration $v_i = a_i - \bar{a}$
27°30′41″	7.5″	5.4″
32″	−1.5″	−3.6″
30″	−3.5″	−5.6″
38″	4.5″	2.4″
31″	−2.5″	−4.6″
34″	0.5″	−1.6″
39″	5.5″	3.4″
37″	3.5″	1.4″
32″	−1.5″	−3.6″
03″	−30.5″ ← blunder	rejected
40″	6.5″	4.4″
42″	8.5″	6.4″
30″	−3.5″	−5.6″
37″	3.5″	1.4″
36″	2.5″	0.4″

First Iteration:

mean = 27°30′33.5″

standard deviation = ±9.3″

$$\text{estimated standard error of the mean} = \pm \frac{9.3}{\sqrt{15}} = \pm 2.4''$$

$$\text{maximum error of a single measurement} = \pm 3 \times 9.3'' = \pm 28''$$

reject measurement 27°30′03″

Second Iteration:

mean = 27°30′35.6″ round off to 27°30′36″

standard deviation = ±4.1″ round off to ±4″

$$\text{estimated standard error of the mean} = \pm \frac{4.1''}{\sqrt{14}} = \pm 1''$$

$$\text{maximum error of a single measurement} = \pm 3 \times 4.1'' = \pm 12''$$

$$\text{maximum error of the mean} = \pm 3 \times 1'' = \pm 3''$$

2.6 PRECISION AND ACCURACY

A measurement is said to have *high precision* if it has a small standard deviation. For example, suppose that Party A measured a distance with a standard deviation of ±0.05 m and Party B measured the same distance with a standard deviation of ±0.1 m. The measurement of Party A is said to be more precise than that of Party B. Figure 2.5 shows that a large standard deviation means a flatter distribution curve for the random errors.

A measurement is said to have *high accuracy* if it is close to the true value. High precision does not necessarily mean also high accuracy. A measurement that is highly precise is also highly accurate only if it contains little or no systematic error. For example, suppose that a distance has been measured ten times with an electronic distance measuring instrument, and that the mean was computed to be 2,579.34 m with an estimated standard error of the mean being ±0.03 m. Thus the small estimated standard error indicates that the mean is very precise. However, because the electronic components of the instrument were out of adjustment, the same mistake was repeated in each of the ten repetitions. When the distance was later measured with a more accurate instrument proven to be in good calibration, the distance was measured to be 2,579.863 ± 0.005 m. Thus the error in the first measurement could amount to as much as −0.52 m, indicating that it is a measurement of low accuracy.

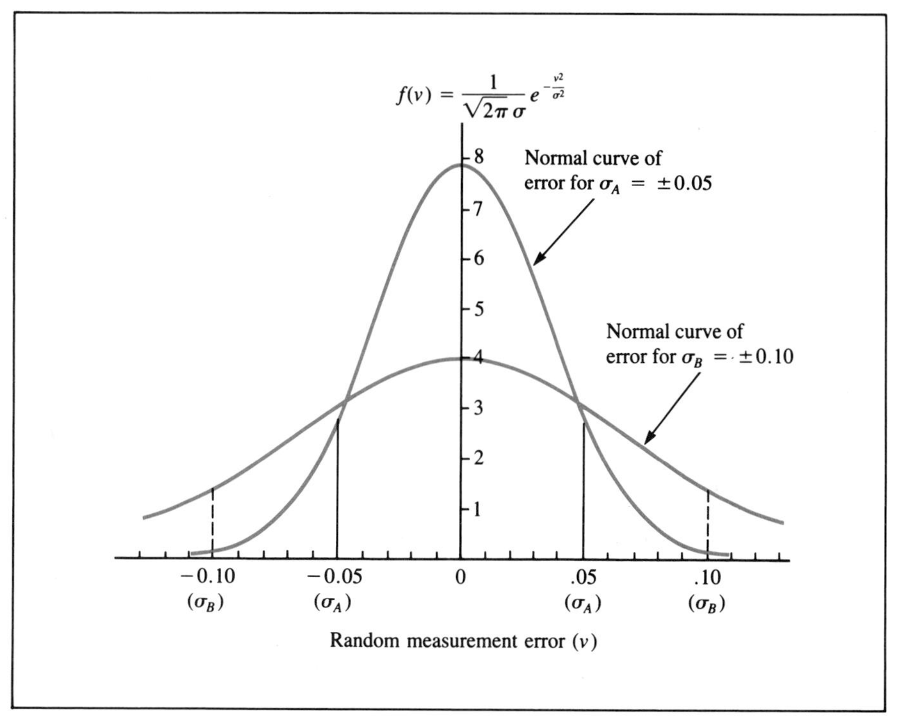

FIGURE 2.5 Standard error and the distribution of random errors

To obtain high precision and high accuracy in surveying, the following strategies must be followed:

1. Eliminate all blunders.
2. Eliminate or correct all systematic errors by frequent calibration and adjustment of the instruments.
3. Minimize the random errors by using good instruments and field procedures.

2.7 RELATIVE PRECISION

The term *relative precision* is commonly used to describe the precision of distance measurement in surveying. If a distance D is measured with a standard error σ_D, then

$$\text{relative precision of measured distance at } 1\sigma = \frac{1}{\dfrac{D}{\sigma_D}} \tag{2.9}$$

It is usually adequate to round off the denominator in the relative precision fraction to two nonzero digits. To avoid confusion and misunderstanding, the probability level at which the relative precision is computed should always be stated. For example, suppose that a distance was measured to be 3,673.24 ft with an estimated standard error of ± 0.03 ft.

$$\text{relative precision of measured distance at } 1\sigma = \frac{1}{\dfrac{3{,}673.24}{0.03}} = \frac{1}{120{,}000}$$

$$\text{relative precision of measured distance at } 3\sigma = \frac{1}{\dfrac{3{,}673.24}{3 \times 0.03}} = \frac{1}{41{,}000}$$

2.8 REPEATED MEASUREMENTS

Eq. (2.8) on page 16 shows the power of repeated measurements in surveying. If the estimated standard error of a single repetition is $\hat{\sigma}_{xi}$, Eq. (2.8) shows that the estimated standard error of the mean of n such repeated measurements would be $\hat{\sigma}_{xi}/\sqrt{n}$. Thus the precision of a measurement program can be increased by increasing the number of times the measurement is repeated. Suppose that an angle can be measured with an engineer's transit that has a least count of 1 minute, with an estimated standard error of ± 2 minutes in one repetition. The effect of the number of repetitions (n) on the estimated standard error of the mean is illustrated below:

No. of Repetitions (n)	$\hat{\sigma}_{\bar{x}}$ ($\pm$ minutes)
1	2
5	0.89
10	0.63
20	0.45
30	0.37
50	0.28
100	0.20
1,000	0.06
10,000	0.02 (or $\pm 1.2''$)

Thus theoretically speaking one can determine the angle with $\hat{\sigma}_{\bar{x}} = \pm 1''$ by using an engineer's transit, but it will take 10,000 repetitions! Even if a surveyor with the required patience could be found to perform the measurement, this could never be justified on economic grounds. To reduce the number of repetitions needed, a more precise instrument can be used.

The number of repetitions required to achieve a certain value of

$\hat{\sigma}_{\bar{x}}$ for a given value of $\hat{\sigma}_{xi}$ can be determined from Eq. (2.8) as follows:

$$\sqrt{n} = \pm \frac{\hat{\sigma}_{xi}}{\hat{\sigma}_{\bar{x}}}$$

and

$$n = \left(\frac{\hat{\sigma}_{xi}}{\hat{\sigma}_{\bar{x}}}\right)^2 \qquad (2.10)$$

Suppose that the same angle discussed above can be measured with an estimated standard error of $\pm 2''$ in one repetition by using a modern theodolite that has a least count of $1''$. The number of repetitions required to determine the angle with $\hat{\sigma}_{\bar{x}} = \pm 1.2''$ can be calculated using Eq. (2.10) as follows:

$$n = \left(\frac{2''}{1.2''}\right)^2 = 3$$

Thus, whereas it takes 10,000 repetitions to determine the angle with $\hat{\sigma}_{\bar{x}} = \pm 1.2''$ by using a 1-minute engineer's transit, it only takes three repetitions by using a 1-second theodolite. This example illustrates the advantage of using precise instruments.

2.9 LAW OF PROPAGATION OF RANDOM ERRORS

Suppose that the value of parameter Y can be calculated from the measured values of n other parameters, say $X_1, X_2, X_3, \ldots, X_n$. Let Y be related to the n parameters by a continuous function $F(\;)$; that is,

$$Y = F(X_1, X_2, X_3, \ldots, X_n) \qquad (2.11)$$

Furthermore, let $\hat{\sigma}_{X_i}$ be the estimated standard error of parameter X_i and $\hat{\sigma}_Y$ be the estimated standard error of Y. The *law of propagation of random errors* states that

$$\hat{\sigma}_Y^2 = \left(\frac{\partial F}{\partial X_1}\right)^2 \hat{\sigma}_{X_1}^2 + \left(\frac{\partial F}{\partial X_2}\right)^2 \hat{\sigma}_{X_2}^2 + \cdots + \left(\frac{\partial F}{\partial X_n}\right)^2 \hat{\sigma}_{X_n}^2 \qquad (2.12)$$

A derivation for Eq. (2.12) can be found in Section A.1 of Appendix A.

EXAMPLE 2.3

The radius (r) of a circular tract of land is measured to be 459.73 ft with an estimated standard error ($\hat{\sigma}_r$) of ± 0.01 ft. Compute the area (A) of the tract of land and the estimated standard error ($\hat{\sigma}_A$).

SOLUTION

$$A = \pi r^2 = 3.14159\ (459.73)^2$$
$$= 721{,}588 \text{ ft}^2$$

By the law of propagation of random errors,

$$\hat{\sigma}_A^2 = \left(\frac{\partial A}{\partial r}\right)^2 \hat{\sigma}_r^2$$

or

$$\hat{\sigma}_A = \left(\frac{\partial A}{\partial r}\right) \hat{\sigma}_r$$
$$= \pm(2\pi r)\hat{\sigma}_r$$
$$= \pm 2(3.14159 \times 459.73) \times (0.01)$$
$$= \pm 31 \text{ ft}^2$$

Expressing A and $\hat{\sigma}_A$ in acres,

$$A = 16.5654 \text{ acres}$$
$$\hat{\sigma}_A = \pm 0.0007 \text{ acre}$$

EXAMPLE 2.4

The area (A) of a rectangular field is to be determined by measuring its length and width. The length (l) of the field is measured to be 2,465.3 ft with an estimated standard error ($\hat{\sigma}_l$) of ± 0.2 ft, whereas the width (w) is measured to be ± 872.56 ft with an estimated standard error ($\hat{\sigma}_w$) of ± 0.08 ft. Compute the area and its estimated standard error ($\hat{\sigma}_A$).

SOLUTION

$$\text{area } (A) = l \times w$$
$$= 2{,}465.3 \times 872.56$$
$$= 2{,}151{,}122 \text{ ft}^2$$

By the law of propagation of random errors,

$$\hat{\sigma}_A^2 = \left(\frac{\partial A}{\partial l}\right)^2 \hat{\sigma}_l^2 + \left(\frac{\partial A}{\partial w}\right)^2 \hat{\sigma}_w^2$$

That is,

$$\hat{\sigma}_A^2 = w^2\hat{\sigma}_l^2 + l^2\hat{\sigma}_w^2$$
$$= (872.56)^2(0.2)^2 + (2{,}465.3)^2(0.08)^2$$
$$= 69{,}351.74 \text{ ft}^2$$

$$\therefore \qquad \hat{\sigma}_A = \pm 260 \text{ ft}^2$$

Converting feet to acres,

$$A = 2{,}151{,}122 \text{ ft}^2 = 49.383 \text{ acres}$$

$$\hat{\sigma}_A = \pm 260 \text{ ft}^2 = \pm 0.006 \text{ acre}$$

EXAMPLE 2.5

If the area of the rectangular tract of land in Example 2.3 above must be measured with a maximum allowable error of ±0.1 acre, what is the allowable relative precision in making the distance measurements at 1σ? Assume that both the length and width of the field are to be measured with the same relative precision.

SOLUTION

$$\text{maximum allowable error in area} = 3\hat{\sigma}_A$$

$$= \pm 0.1 \text{ acre}$$

$$\therefore \qquad \text{allowable } \hat{\sigma}_A = \pm 0.0333 \text{ acre}$$

$$= \pm 1{,}451 \text{ ft}^2$$

Let relative precision = $1/N$, then

$$\frac{1}{N} = \frac{1}{\dfrac{2465.3 \text{ ft}}{\hat{\sigma}_l}} = \frac{\hat{\sigma}_l}{2{,}465.3 \text{ ft}}$$

and

$$\frac{1}{N} = \frac{\hat{\sigma}_w}{872.56 \text{ ft}}$$

Therefore

$$\hat{\sigma}_l = \frac{2{,}465.3 \text{ ft}}{N}$$

and

$$\hat{\sigma}_w = \frac{872.56 \text{ ft}}{N}$$

From example 2.3

$$\hat{\sigma}_A^2 = w^2\hat{\sigma}_l^2 + l^2\hat{\sigma}_w^2.$$

Hence

$$(1{,}451 \text{ ft}^2)^2 = (872.56 \text{ ft})^2 \left(\frac{2{,}465.3 \text{ ft}}{N}\right)^2 + (2{,}465.3 \text{ ft})^2 \left(\frac{872.56 \text{ ft}}{N}\right)^2$$

Solving for N yields $N = 2{,}097$. Rounding off to two significant nonzero digits, $N = 2{,}100$. Therefore the allowable relative precision at 1σ is 1/2,100; that is, the allowable standard errors in measuring the length and width of the field are ± 1.2 ft and ± 0.4 ft respectively.

2.10 BASIC PRINCIPLES OF LEAST SQUARES

The method of *least squares* was developed in the 19th century by the French mathematician Adrien Legendre (1752–1833) for the purpose of curve fitting. The method is now extensively used in surveying to determine the most probable values of some unknown parameters from a set of measurements. For example, the simple survey network shown in Figure 2.6 has two known stations, 1 and 2, the rectangular coordinates of which have been determined from previous surveys. The coordinates of two new stations, 3 and 4, are to be determined by measuring eight angles and five distances. The coordinates of the two new stations can be determined by the method of least squares.

The method of least squares is based on the following assumptions:

1. There are more measurements than the minimum number needed to determine the unknown parameters.
2. The measurements contain only random errors.
3. The measurements are made independently from each other.

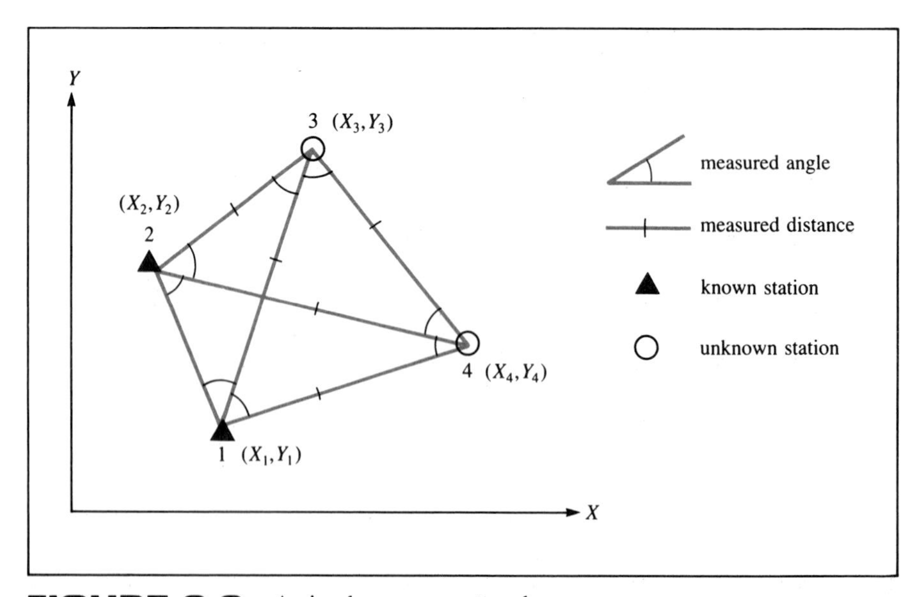

FIGURE 2.6 A simple survey network

Let $x_1, x_2, \ldots, x_n$ be the measured values of n parameters, and let $v_1, v_2, \ldots, v_n$ be the errors in these n measured values. Furthermore, let σ_i be the standard error of the measurement x_i. Then the *most probable solution* to the unknown survey parameters is that which results in satisfying the following condition:

$$\sum_{i=1}^{n} \left(\frac{\sigma_o v_i}{\sigma_i}\right)^2 = \text{minimum} \tag{2.13}$$

where σ_o is a constant that can be assigned any value. Hence the solution is called the *least-squares solution*. A derivation for this least-squares condition can be found in Section A.2 of Appendix A.

2.11 WEIGHTS AND WEIGHTED MEAN

Let

$$w_i = \frac{\sigma_o^2}{\sigma_i^2} \tag{2.14}$$

Then the least-squares condition given in Eq. (2.13) may be restated as follows:

$$\sum_{i=1}^{n} w_i v_i^2 = \text{minimum} \tag{2.15}$$

The parameter w_i is called the *weight* of measurement x_i; σ_o is called the *standard error of unit weight* because if the standard error (σ_i) of a measurement is equal to σ_o, then it has a weight of 1.

Note that the weight w_i is inversely proportional to the square of the standard error σ_i. Thus the more precise the measurement, the smaller will be its standard error and the larger will be its weight in a least squares solution.

Let $x_1, x_2, \ldots, x_n$ be n independent measurements of a quantity and let $\sigma_1, \sigma_2, \ldots, \sigma_n$ be the corresponding standard errors of the n measurements. Thus the measurements are assumed to be made with different precision. It can be shown that the most probable value ($\hat{x}$) of the quantity is given by the *weighted mean* of the n measurements—that is,

$$\hat{x} = \frac{w_1 x_1 + w_2 x_2 + \cdots + w_n x_n}{w_1 + w_2 + \cdots + w_n} \tag{2.16}$$

Furthermore, an *estimate of the standard error of the weighted mean* ($\hat{\sigma}_{\hat{x}}$)

can be computed from the following expression:

$$\hat{\sigma}_{\hat{x}} = \frac{\sigma_o}{\sqrt{\sum_{i=1}^{n} w_i}} \tag{2.17}$$

A derivation for Eq. (2.16) and Eq. (2.17) is given in Section A.3 of Appendix A.

EXAMPLE 2.6

Given the following three independent measurements of a distance:

$x_1 = 151.4$ ft $\quad \hat{\sigma}_1 = \pm 0.2$ ft
$x_2 = 151.3$ ft $\quad \hat{\sigma}_2 = \pm 0.1$ ft
$x_3 = 151.7$ ft $\quad \hat{\sigma}_3 = \pm 0.3$ ft

compute the weighted mean ($\hat{x}$) of the three measurements and an estimate of the standard error of the weighted mean ($\hat{\sigma}_{\hat{x}}$).

SOLUTION

It is usually convenient to let σ_o equal the value of the largest standard error. In this case, let $\sigma_o = \hat{\sigma}_3 = \pm 0.3$ ft. Then

$$w_1 = \frac{\sigma_0^2}{\hat{\sigma}_1^2} = \left(\frac{0.3}{0.2}\right)^2 = 2.25$$

$$w_2 = \frac{\sigma_0^2}{\hat{\sigma}_2^2} = \left(\frac{0.3}{0.1}\right)^2 = 9.00$$

$$w_3 = \frac{\sigma_0^2}{\hat{\sigma}_3^2} = \left(\frac{0.3}{0.3}\right)^2 = 1.00$$

Thus by letting σ_o equal the standard error of the least precise measurement, a weight of 1 is assigned to that measurement, whereas all other measurements will have a weight greater than or equal to 1.

Then, from Eq. (2.16),

$$\hat{x} = \frac{w_1x_1 + w_2x_2 + w_3x_3}{w_1 + w_2 + w_3}$$

$$= \frac{2.25 \times 151.4 + 9 \times 151.3 + 1 \times 151.7}{2.25 + 9 + 1}$$

$$= 151.35 \text{ ft}$$

and, from Eq. (2.17),

$$\hat{\sigma}_{\hat{x}} = \frac{\sigma_o}{\sqrt{\sum_{i=1}^{n} w_i}} = \pm \frac{0.3 \text{ ft}}{\sqrt{12.25}} = \pm 0.09 \text{ ft}$$

2.12 SIGNIFICANT FIGURES

The *significant figures* in a number are those digits with known values. They are identified by proceeding from left to right, beginning with the first nonzero digit and ending with the last digit of the number.

Some examples are as follows:

a. 541.6800 has seven significant figures.

b. 50.0006 has six significant figures.

c. 0.00058 has two significant figures.

d. 0.006200 has four significant figures.

e. 8.000050 has seven significant figures.

f. 51.0 has three significant figures.

For further clarification the following rules may be helpful:

1. All nonzero digits are significant.
2. Zeros at the beginning of a number merely indicate the position of the decimal point. They are not significant.
3. Zeros between digits are significant.
4. Zeros at the end of a decimal number are significant.

Note, however, that zeros at the end of a nondecimal number are not necessarily significant. The number 23,600 may have three, four, or five significant figures depending on whether its value is correct to the nearest hundred, ten, or unit. If this number is to be expressed to three significant figures, it can be written as 23.6×10^3.

2.13 ROUNDING OFF

The subject of significant figures is important in both field work and office computations. Since neither the measurements nor the quantities mathematically deduced from them can be exact, it is essential to use the appropriate number of significant figures to express a final result. Using more significant figures than the precision of the field measurement warrants is apt to convey a misleading impression about the quality of the end product of the survey.

When there are more significant figures in a quantity than are required, the number is rounded off to the number of places needed. When rounding off, it is advisable to adopt a definite system so that the laws of chance operate and equitable results are obtained. A specific rule also permits the checker to perform the task more speedily.

If the result is to be expressed to n significant figures, the nth figure should be retained as is if the figure following it is less than 5 in the $(n + 1)$th place. If the digit following the nth significant figure is greater than 5 in the $(n + 1)$th place, the nth figure should be increased by one unit. When the $(n + 1)$th digit is a 5, round off to the nearest even digit in the nth place.

The following examples illustrate the usual rules:

a. 6746.589 to five significant places is 6746.6.

b. 837848 to four significant places is 837800.

c. 468.767 to five significant places is 468.77.

d. 468.762 to five significant places is 468.76.

e. 468.755 to five significant places is 468.76.

f. 468.745 to five significant places is 468.74.

Whenever the standard error of a measured or computed value is known, the value should be rounded off according to the value of its standard error, as discussed in Section 2.4.

When performing addition (or subtraction) the sum cannot be more precise than the least precise number included in the addition. For example, consider the summation of the following three measured segments of a survey line:

$$\begin{array}{r} 24.217 \text{ ft} \\ 468.46\phantom{7 \text{ ft}} \\ 1{,}563.1\phantom{17 \text{ ft}} \\ \hline 2{,}055.777 \text{ ft} \end{array}$$

The sum cannot be properly expressed closer than to the nearest 0.1 ft, because one of the component quantities has been measured only to the nearest 0.1 ft. The sum should be expressed, therefore, as 2,055.8 ft.

When performing multiplication (or division) the number of significant figures in the product cannot be greater than the number of significant figures in the factor having the fewest significant figures.

For example, consider the combined operations of multiplication and division indicated by

$$\frac{5.27 \times 838 \times 51.3781}{5.2 \times 62.581076}$$

The result should be expressed to two significant figures, the same number of significant digits as in the term 5.2, which has the fewest significant digits.

Although the rates of change of the trigonometric functions depend on the function and the size of the angle, the following general guides, as applied to average size angles, will be helpful in determining how many places (significant figures) in the trigonometric function should be used in typical calculations.

1. For a 01′ error in angle, use five places.

2. For a 10″ error in angle, use seven places.

3. For a 01″ error in angle, use eight places.

Most electronic calculators give the trigonometric functions to eight or more significant digits.

Given the convenience of electronic calculators, always carry one or more extra significant digits during intermediate steps and perform the necessary rounding off only for the final results.

PROBLEMS

2.1 Listed below are six independent measurements of the length of a survey line:

827.461 ft
827.508 ft
827.512 ft
827.493 ft
827.468 ft
827.452 ft

Compute:

a. the mean,

b. the standard deviation of the six measurements,

c. the estimated standard error of the mean,

d. the maximum expected error (in ft) of the mean,

e. the relative precision of the computed mean at one sigma.

2.2 For each of the following sets of repeated measurements:

a. Compute the mean, standard deviation, and estimated standard error of the mean.

b. Check for the presence of any blunders. Reject any measurement containing a blunder and repeat the calculations in part (a).

c. Choose an appropriate interval and plot a histogram for the errors.

(I)	(II)	(III)
472.36 m	1,148.217 ft	147°33′51″
472.29	1,148.218	147°33′58″
472.31	1,148.215	147°33′55″
472.34	1,148.220	147°34′02″
472.35	1,148.251	147°33′57″
472.30	1,148.216	147°34′56″
472.33	1,148.219	147°33′54″
472.27	1,148.221	147°33′57″
472.34	1,148.216	147°33′55″
472.30	1,148.217	
	1,148.224	
	1,148.214	

2.3 Compute the mean, standard deviation, and estimated standard error of the mean for each of the following sets of repeated measurements:

(I)	(II)	(III)
54.32 ft	41°29′44″	614.1 m
54.31 ft	41°29′40″	614.3 m
54.30 ft	41°29′46″	614.2 m
54.35 ft	41°29′38″	614.5 m
54.31 ft		614.2 m
		614.7 m

2.4 A distance is measured to be 456.3 ft with an estimated standard error of ±0.1 ft. Compute for this measured distance:

a. the probable error,

b. the maximum error,

c. the relative precision at 1σ,

d. the relative precision at 3σ.

2.5 The area of a parking lot is calculated to be 5,474 ft^2 with an estimated standard error of ±2 ft^2.

a. What is the maximum survey error in the area?

b. What is the error range at the 50%, 90%, and 95% probability levels?

2.6 Compute the relative precision at 1σ for the following measurements:

a. l = 4,576.2 m $\quad \hat{\sigma}_l = \pm 0.3$ m

b. $l = 417.36$ m $\quad \hat{\sigma}_l = \pm 0.02$ m

c. $l = 1{,}729.1$ ft $\quad \hat{\sigma}_l = \pm 0.5$ ft

Round off the denominator of the relative precision to two significant nonzero digits.

2.7 In a surveying project, it is required that the distances be measured with a relative precision better than 1/100,000, or 1 part in 100,000. Does this language clearly indicate to you whether the stated relative precision is at 1σ, 2σ, or 3σ? How would you interpret this specification?

2.8 All the distances in a surveying project must be measured with a relative precision better than 1/10,000 at 1σ. Determine the maximum allowable survey error in measuring distances of the following lengths:

a. 46 ft

b. 4,253 ft

c. 140 m

d. 789.2 m

2.9 A long survey line was measured five times using a 100-ft steel tape. The average of the five measurements was found to be 742.56 ft with the standard error of the mean estimated to be ±0.01 ft. The same line was later measured with an electronic distance measuring instrument, and the distance was found to be 745.23 ft with an estimated standard error of ±0.02 ft. The electronic distance measuring instrument has been recently checked and found to be in good calibration.

a. Compute the relative precision of the tape measurement at 1σ.

b. Compute the relative precision of the electronic measurement at 1σ.

c. Which measurement is more precise?

d. Which measurement is more accurate?

e. How can you explain the large difference in the two measured values?

2.10 Based on previous experience using the same instruments, a surveyor has estimated that a single measurement of an angle has a standard error of ±2″. How many repetitions should be made if the standard error of the mean must not exceed ±1″?

2.11 A single measurement of an angle using a particular instrument is estimated to have a standard error of ±5 minutes. Determine the minimum number of repetitions needed if the maximum allowable error in the mean angle is to be less than the following amount:

a. ±2′

b. ±10′

c. ±30″

2.12 A single measurement of a distance is estimated to have a relative precision of 1/10,000 at 1σ. Compute the relative precision (at 1σ) of the mean of five such independent measurements.

2.13 The length and width of a rectangular field were measured to be as follows:

$$\text{length} = 1{,}357.4 \pm 0.2 \text{ ft } (\hat{\sigma})$$
$$\text{width} = 589.3 \pm 0.1 \text{ ft}$$

Compute the area and its estimated standard error.

2.14 The length and width of a rectangular field are approximately 5,420 ft and 1,510 ft respectively. If the area of the field must be determined with a standard error of ±0.1 acre, determine the relative precision at 1σ with which the length and width of the field must be measured.

2.15 A long survey line was divided into four sections. Each section was measured several times, and the average of the repeated measurements computed together with the estimated standard error of the mean. The results are listed below:

Section	Mean Length	Estimated Standard Error of the Mean
1	125.45 m	±0.02 m
2	278.32	±0.04
3	183.21	±0.03
4	110.35	±0.02

Compute:

a. the total length L of the baseline,

b. the estimated standard error of the total length,

c. the relative precision of length L at 1σ.

2.16 A long survey line is measured in three separate sections with the following results:

$$\text{Section 1: } l_1 = 472.34 \text{ ft} \qquad \hat{\sigma}_1 = \pm 0.05 \text{ ft}$$
$$\text{Section 2: } l_2 = 973.27 \text{ ft} \qquad \hat{\sigma}_2 = \pm 0.07 \text{ ft}$$
$$\text{Section 3: } l_3 = 87.42 \text{ ft} \qquad \hat{\sigma}_3 = \pm 0.01 \text{ ft}$$

Compute the total length of the line and its estimated standard error.

2.17 A tract of land is trapezoidal in shape. The two parallel sides l_1 and l_2 have been measured to be as follows:

$$l_1 = 472.3 \text{ ft} \qquad \hat{\sigma}_1 = \pm 0.1 \text{ ft}$$
$$l_2 = 583.7 \text{ ft} \qquad \hat{\sigma}_2 = \pm 0.3 \text{ ft}$$

The perpendicular distance between the two sides is measured to be 241.8 ft with an estimated standard error of ±0.2 ft. Compute the area of the tract and its estimated standard error.

2.18 The diameter (d) and height (h) of a cylindrical storage tank have been measured to be as follows:

$$d = 54.86 \text{ ft} \qquad \hat{\sigma}_d = \pm 0.04 \text{ ft}$$
$$h = 32.15 \text{ ft} \qquad \hat{\sigma}_h = \pm 0.01 \text{ ft}$$

Compute:

a. the volume of the storage tank;

b. the estimated standard error of the volume.

2.19 Prove, using the basic principle of least squares, that the mean of n repeated measurements is the most probable value of the measured quantity. Assume that all n repetitions are made with the same precision.

2.20 An angle was measured ten times with an optical theodolite by observers A and B on two separate days. The calculated results are as follows:

Observer A	Observer B
mean = 42°16′25.2″ $\hat{\sigma}_{\bar{x}} = \pm 3.2''$	mean = 42°16′20.4″ $\hat{\sigma}_{\bar{x}} = \pm 1.6''$

Compute:

a. the weighted mean of the two observers' results;

b. the estimated standard error of the weighted mean.

2.21 Compute the weighted mean and the estimated standard error of the weighted mean for the following four independent measurements of a distance:

$$l_1 = 2{,}746.34 \text{ ft} \qquad \hat{\sigma}_1 = \pm 0.02 \text{ ft}$$
$$l_2 = 2{,}746.38 \text{ ft} \qquad \hat{\sigma}_2 = \pm 0.06 \text{ ft}$$
$$l_3 = 2{,}746.26 \text{ ft} \qquad \hat{\sigma}_3 = \pm 0.05 \text{ ft}$$
$$l_4 = 2{,}746.31 \text{ ft} \qquad \hat{\sigma}_4 = \pm 0.04 \text{ ft}$$

2.22 A line was carefully measured ten times on three different days. The mean and estimated standard error of the mean for each day's measurement were computed to be as follows:

Day	Mean	Estimated Standard Error of the Mean
1	2,815.4 ft	±0.5 ft
2	2,816.7 ft	±0.3 ft
3	2,816.3 ft	±0.2 ft

Compute:

a. the weighted mean of the three measurements;

b. the estimated standard error of the weighted mean.

REFERENCES

2.1 Ang, Alfredo H-S., and Tang, Wilson H. *Probability Concepts in Engineering Planning and Design, Volume 1—Basic Principles,* Wiley, New York: 1975.

2.2 Barry, B. Austin. *Engineering Measurements,* Wiley, New York: 1966.

2.3 Freund, John E. *Mathematical Statistics,* 3rd ed. Prentice-Hall, Englewood Cliffs, N.J.: 1980.

2.4 Mikhail, E. M., and Gracie, G. *Analysis and Adjustment of Survey Measurements.* Van Nostrand Reinhold, New York: 1980.

2.5 Mikhail, E. M. *Observations and Least Squares.* Harper & Row, New York: 1976.

TAPE MEASUREMENT

3.1 INTRODUCTION

One of the fundamental operations in surveying is the measurement of distances. Some of the most common methods for measuring distances include pacing, taping, stadia, and the use of electronic distance-measuring equipment.

The approximate distance between two points can be measured by pacing. The length of a step is usually quite regular for each person. Thus by counting the number of steps a person takes to walk from one point to the other and multiplying the number by the average length of that person's step, the distance can be determined. An experienced person can obtain an accuracy (1σ) of 1 part in 100 (that is, 1/100) of the distance in pacing. An instrument called a *pedometer* operates on the principle of pacing. When it is attached to a person's boot, it automatically records the number of steps the person takes while traveling between two points. Such an instrument is commonly used by hikers as well as professionals such as engineers, geologists, surveyors, and foresters performing reconnaissance missions.

However, distances ranging from a few feet to several tens of miles can be measured almost instantaneously with extremely high accuracy by using electronic distance measuring (EDM) equipment. Such instruments measure distances using electromagnetic waves and will be the subject of discussion in Chapter 6.

Distances can also be measured by a method called *stadia,* which requires an angle-measuring instrument and a graduated rod. The stadia method combines distance measurement with the measurement of elevation differences. Therefore it is uniquely suited for topographic mapping. This method will be discussed in Chapter 9.

This chapter is devoted to the discussion of measuring distances by graduated tapes in a process called *taping*. (The term *chaining* is also frequently used to describe the taping operation.) It has its origin from the common use in the past of the link chain, which was devised by the Englishman Edmund Gunter (ca. 1620). The *chain* used in the early surveys was 33 ft long and comprised 50 links. All recorded distances, however, were expressed in units of a 66-ft chain (see Figure 3.1). It was made of wire and the many links exposed many wearing surfaces so that the length of the chain increased as much as a half-foot in a season. A means of adjusting the length of the last link was provided and frequent comparisons with a "standard" chain were required. The link chain is now obsolete, but steel tapes graduated in units of links and chains (1 chain = 100 links) are still in use.

In spite of the increasing availability of EDM equipment, measuring tapes remain one of the most versatile surveying tools. A good quality, portable steel tape can be purchased for less than $100 (1984). Distances up to 200 or 300 feet long can be easily measured with a steel tape to an accuracy (1σ) of 1/3,000. By using proper care and field procedure, small areas can be mapped using a measuring tape alone with an accuracy that is adequate for many engineering projects. The steel tape is also an indispensable tool in construction layout. As in all surveying operations, the successful

FIGURE 3.1 A 66-ft chain

use of tapes for distance measurement requires a thorough understanding of measurement principles, equipment, field procedure, and the sources of systematic and random errors.

3.2 UNITS OF LENGTH AND AREA

The basic units of length used within the United States are the *foot* and the *meter*. The foot (ft) is of Anglo Saxon origin and is universally used in English-speaking countries. The meter (m) is of French origin and has become the adopted unit for international and scientific usage. On December 23, 1975, the provisions of the Metric Conversion Act of 1975 became law in the United States. The purposes of this legislation were to declare a national policy of coordinating the increasing use of the metric system in the United States and to establish a U.S. Metric Board to assist in the voluntary conversion to the new system. The term *metric system* specifically means Le Systeme International d'Unités (abbreviated SI) or the International System of Units established by an International Conference of Weights and Measures in 1960. Agencies of the Federal Government are gradually converting to the SI units in the national surveying and mapping programs. It is likely that in time the meter will replace the foot in all areas of engineering.

In the English system of length and area measurements, the following equivalent relationships exist:

1 foot (ft) = 12 inches (in.)
1 yard (yd) = 3 feet
1 mile (mi) = 5,280 feet
1 Nautical mile = 6,076.10 feet
1 chain (ch) = 100 links (lk)
1 chain = 66 feet
1 chain = 4 rods
1 rod = 16.5 feet
1 acre (ac) = 43,560 square feet (ft^2)
1 acre = 10 square chains

A decimal system of linear measurements based on the foot unit has been widely used for practically all surveying work. In engineering practice, the decimal system is extended to measurements of less than one foot, the foot unit being subdivided into tenths and hundredths. The building trades still use the older English units in which the foot is divided into inches, quarter-inches, and so on; hence architects and engineers frequently have to convert units from one system to the other.

The *rod* (sometimes called a *perch*, or a *pole*), which is a unit of 16 1/2 feet, has had considerable use in land measure.

The unit of area is the *acre*, which has been standardized at 1/8 mile in length and 1/80 mile in width. It is, therefore, 660 feet long and 66 feet wide. Because of these dimensions, Gunter made use of a 66-foot chain. This made the acre 10 chains long and 1 chain wide, or 10 square chains in area, and thus reduced land measure to a decimal system.

The following relationships exist in the SI system of length and area measurements:

1 meter (m)	= 1,000,000 micrometers (μm)
1 meter	= 1,000 millimeters (mm)
1 meter	= 100 centimeters (cm)
1 meter	= 10 decimeters (dm)
1 kilometer (km)	= 1,000 meters
1 hectare	= 10,000 square meters (m^2)

There is no officially sanctioned unit of area in the SI system, although the *hectare* (10,000 m^2) is in common use.

In the metric system of weights and measures certain prefixes are so widely used as to deserve special mention here. *Kilo* means 1,000 and *milli* indicates one-thousandth. To provide for multiples larger than 1,000 and for subdivisions smaller than one-thousandth, the prefixes *mega* meaning 1,000,000 and *micro* meaning one-millionth have been generally recognized. A special term is *micron,* which means one-millionth of a meter or one-thousandth of a millimeter and is indicated by the abbreviation μm, in which μ is the symbol for the Greek letter *mu*. Table XVI in Appendix B lists the letters in the Greek alphabet, which are commonly used as mathematical symbols in engineering and surveying.

In southwestern United States, including Texas and California, the old Spanish unit *vara* was used in some old surveys. One vara is equal to 33 inches in California, and 33 1/3 inches in Texas.

3.3 CONVERSION OF LENGTH UNITS

The meter was originally defined as being 1/10,000,000 of a meridional quadrant of the earth. Following the execution of surveys of geodetic accuracy and the deliberations of outstanding geodesists, an international treaty in 1875 provided for the creation of an International Bureau of Weights and Measures. At the first conference in 1889, new standards for the metric system were adopted. The meter was redefined in terms of the distance between certain markings on a platinum-iridium bar at 0°C. This is known as the International Prototype Meter and is kept at the International Bureau of Weights and Measures at Sèvres, France.

The United States Congress legalized the use of the Metric system on July 28, 1866. This legislation was, however, only permissive. It did not make mandatory the use of the meter. Effective on April 5, 1893, all legal units of measure used in the United States were defined as exact numerical multiples of metric units. This action, known as the Mendenhall Order, established the relationship between the yard and meter as follows:

$$1 \text{ U.S. yard} = \frac{3,600}{3,937} \text{ meter} = 0.9144018288 \text{ meter}$$

or

$$39.37 \text{ inches} = 1 \text{ meter}$$

In 1960 the International Meter was officially defined in a supplementary manner as 1,650,763.73 wavelengths of the orange-red light of krypton 86, a rare gas extracted from the atmosphere. The U.S. inch thus becomes equal to 41,929.483 wavelengths of the krypton light. The new definition of the meter relates it to a constant of nature, the wavelength of a specified kind of light, which is believed to be immutable and can be reproduced with great accuracy in any well-equipped laboratory.

The precision requirements in length measurements increased greatly following World War II, and the difference between the U.S. inch and the British inch became especially important in gage-block standardization. As a result of several years of discussion the directors of the national standards laboratories of the United Kingdom and the United States entered into an agreement effective July 1, 1959, whereby uniformity was established for use in the scientific and technical fields. The new relationship between the yard and meter became

$$1 \text{ yard} = 0.9144 \text{ meter (exactly)}$$

or

$$1 \text{ foot} = 0.3048 \text{ meter}$$

and

$$1 \text{ inch} = 25.4 \text{ millimeter}$$

Thus the new value for the yard is smaller by two parts in one million than the 1893 yard. However, it is emphasized that any data expressed in feet derived from and published as a result of geodetic surveys within the United States will continue to bear the original relationship as defined in 1893; namely,

$$1 \text{ U.S. survey foot} = \frac{1,200}{3,937} \text{ meter} = 0.3048006096 \text{ meter}$$

The foot unit defined by this equation is referred to as the *U.S. Survey Foot.*

In all survey work except that of geodetic accuracy, the distinction between the two definitions of the foot can be disregarded.

For the conversion of area units, the following relationship is used:

1 hectare = 2.47104 acres

3.4 TAPES

The most common surveying tapes are made of a steel ribbon of constant cross-section bearing graduations at regular intervals. Others are made of a steel alloy or of a metallic or nonmetallic cloth. There is great diversity of tapes with respect to lengths and widths and manner of graduation.

The 100-ft steel tape is the surveyor's favorite device for measuring distance (see Figure 3.2). Its width varies from 1/4 to 1/2 in. and its thickness from 0.020 to 0.025 in. The graduations and identifying foot numbers are either stamped on soft (babbit) metal previously embossed on the tape at the foot divisions or etched into the metal of the tape. Loops are riveted to each end of the tape and usually rawhide thongs are attached to the loops. When not in use, the tape should be stored on its reel. If not kept on a reel, it should be gathered up in 5-ft lengths to form a figure eight and then thrown into a single small circle (see Figure 3.2b and c).

Nickel-steel alloy tapes, known as *Invar* or *Lovar,* have a coefficient of thermal expansion about one-thirtieth that of steel and are used when the taping specifications prescribe a high order of accuracy. The relative insensitivity of these tapes to temperature changes and the fact that they do not readily tarnish or rust on exposure to the elements make them particularly suitable for important survey tasks. However, the alloy metal is relatively soft and special care must be taken to prevent these tapes from becoming kinked or broken.

Woven tapes are 5/8 to 7/8 in. wide and most commonly 50 or 100 ft long (see Figure 3.3). The nonmetallic type is made of synthetic yarns without metallic threads. It offers excellent wearing properties, high tensile strength, and dimensional stability. The metallic type contains very fine, noncorrosive metallic strands woven in with the yarn. The woven tape is impregnated with a paintlike material for protection and graduations are applied to the surface. A leather-covered metal case with a built-in reel is used to wind up the tape when not in use. Because the materials in this kind of tape are susceptible to temperature and moisture changes, woven tapes should be used only when relatively low accuracies can be tolerated. A relatively new measuring product is the fiberglass tape, which is flexible, strong, nonconductive, and will not stretch or shrink because of temperature or moisture. Available in 50- and 100-ft lengths, it is well suited for all construction work.

(a) Mounted on reel

FIGURE 3.2 100-ft steel tape (*continued on p. 44*)

Steel tapes, whether foot or metric, are graduated in various ways and care must be taken to read the graduations properly in order to prevent making a blunder. Of special significance is the location of the end marks, particularly the zero mark. It may be located in different places on the various tapes. Various kinds of end arrangements for foot tapes are shown in Figure 3.4. On some tapes, such as that shown in Figure 3.4a, zero is at the end of the tape ribbon and the attached loop is not included in the graduated portion. On others, such as that shown in 3.4b, the zero is at the very end of the loop. On still others, such as those shown in c and d, the zero is marked at some distance from the end of the ribbon and there is a blank piece of tape between the zero and the end loop. Some tapes, such as those shown in a, b,

(b) Steel tape gathered on 5-ft lengths

(c) Steel tape after being thrown into a small loop

FIGURE 3.2 *(continued)*

FIGURE 3.3 Woven tapes (courtesy of Keuffel & Esser Company)

c, and d, have the first foot or meter subdivided into tenths or hundredths; others, such as those shown in e and f, have an extra unit graduated from the zero toward the end loop. Figure 3.5 shows two common graduation styles in metric tapes. Metric tapes in lengths of 30 m and 50 m are now commonly available. Metric tapes are commonly graduated in meters, decimeters, and centimeters.

3.5 TAPING ACCESSORIES

Various accessory equipment is usually used with tapes. Some of the most common items will be briefly described, and a few are depicted in Figure 3.6.

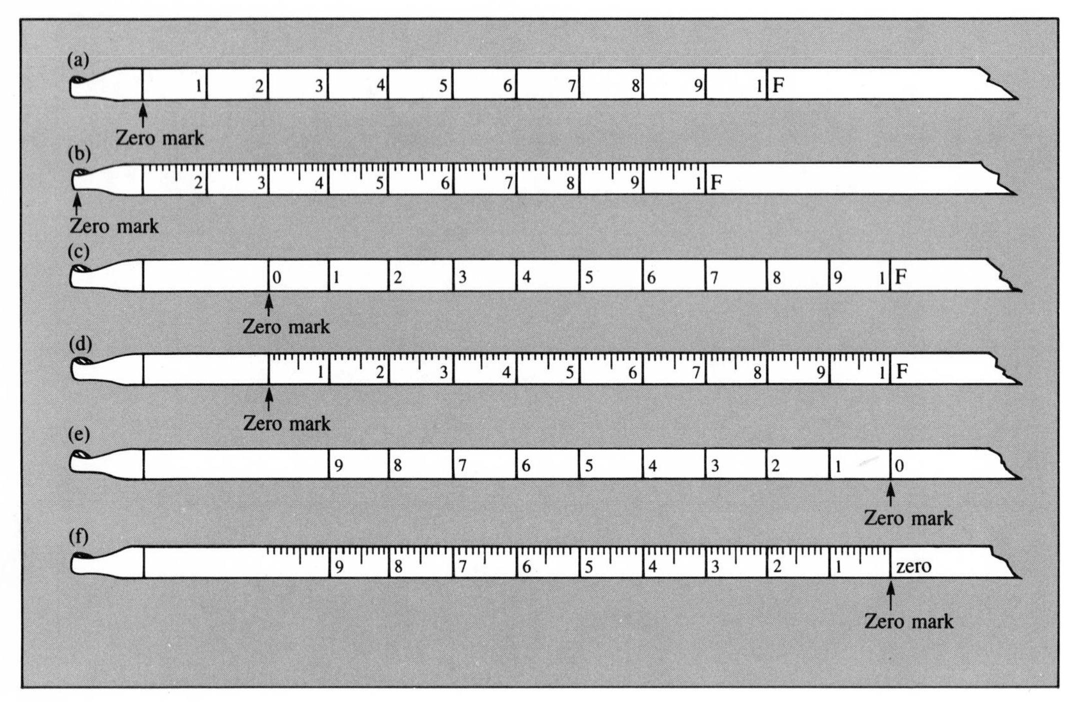

FIGURE 3.4 End arrangements in foot tapes

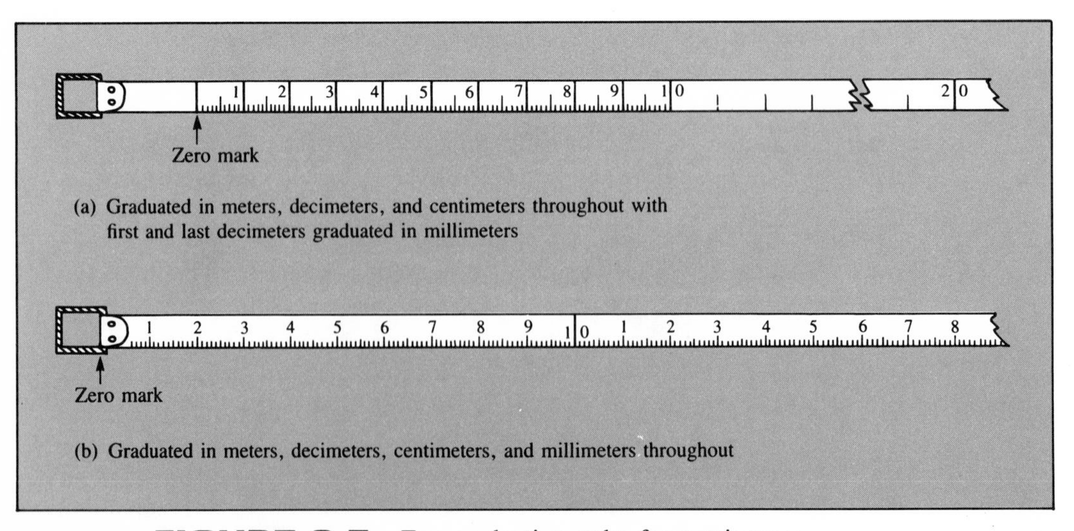

FIGURE 3.5 Two graduation styles for metric tapes

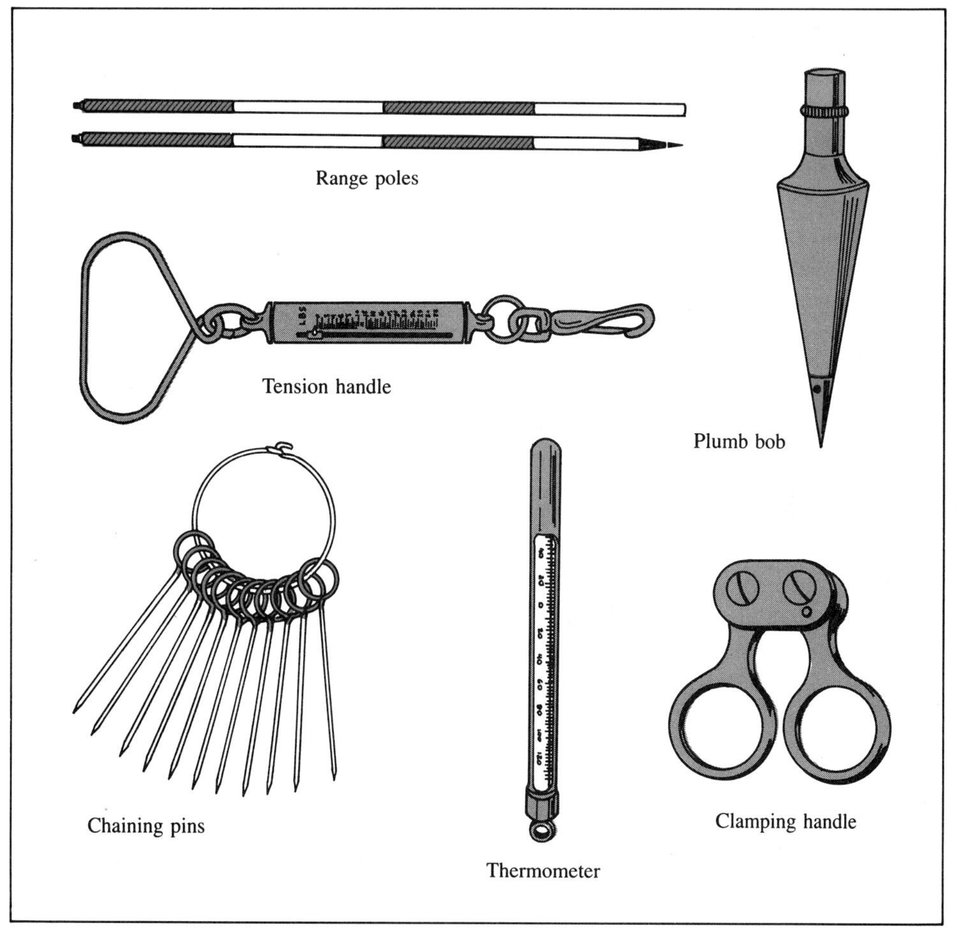

FIGURE 3.6 Taping accessories (courtesy of Keuffel & Esser Company)

Steel pins with a ring at one end and pointed on the other are called *chaining pins* or *taping arrows*. They are used for marking tape ends on the ground and for tallying the number of tape lengths in a given line.

The *tension handle* is used to apply the appropriate pull to the tape when fairly careful measurements are to be made.

The *clamping handle* is employed to grip the flat ribbon of steel tape without kinking it.

In addition to such special equipment as the transit and the hand level, which will be described in subsequent chapters and which are used to convert taped slope distances to their horizontal equivalents, other accessories include *range poles* for aligning the taping, plumb bobs, thermometers, and cutting tools like hatchets or machetes for clearing the line of vegetation.

3.6 CARE OF EQUIPMENT

Although the steel tape is relatively strong, it will readily break if it is subjected to a hard pull when it is kinked. Therefore the tape should be kept straight when in use. It should never be jerked, pulled around corners, or bent in order to obtain a better hand grip on it. Vehicles of any kind should never be permitted to run over it, even though the tape may remain flat on the pavement surface. When a steel tape is unwound hastily, it may uncoil itself suddenly and twists will develop. These must be carefully removed before any tension is applied. Steel tapes rust readily and should be wiped dry after use.

Range poles should not be used to loosen stakes or stones or as javelins. Such practice will blunt the steel point, bend the shaft, and render the pole useless for accurate use as a target for transit observations. To avoid losing pins, a piece of brightly colored cloth can be tied to the ring of each.

The zero end of a woven tape receives the roughest treatment and is most subject to wear. In order to decrease the chances for breakage here, most woven tapes are specially reinforced with a plastic-rubber laminated to the first 6 inches. However, violent or hard pulls should not be applied to a woven tape.

3.7 THE TAPING PARTY

The minimum taping party consists of two persons, one at each end of the tape. The person holding the zero end of the tape proceeds in front of the line and is called the *front tapeperson*. The person holding the rear end of the tape is called the *rear tapeperson*. The front tapeperson reads the graduation on the tape, sets the chaining pin, and records the measurement. The rear tapeperson is responsible for keeping the tape on line by lining up the front end of the tape with a range pole set at the other end of the line. When the party has three persons, two should be assigned to the front end of the tape. One of the two should be responsible for holding the tension handle and applying the proper amount of tension, while the other person sets the chaining pin, reads the graduation, and records the measurements. The most experienced member of the party usually reads the graduation.

3.8 TAPING HORIZONTAL DISTANCE OVER LEVEL GROUND

The simplest taping operation is measuring over level ground where the tape can be stretched out while fully supported by the ground, as shown in Figure 3.7a. The basic procedure for measuring the horizontal distance between points *A* and *B* in Figure 3.7a will be described in the following paragraphs. It is assumed that the party consists of two persons, and that the equipment includes a 100-ft steel tape, a set of 11 chaining pins, and two range poles. It is assumed that the tape has the first foot subdivided.

A range pole is set at each end of the line, and the tape is unwound and laid out on the ground with the zero end forward. The front tapeperson then takes the zero end of the tape, hands one pin to the rear tapeperson, and moves forward along the line. When the 100-ft end of the tape comes up even with the beginning point (*A*), the rear tapeperson calls out "halt." At this signal the front tapeperson halts and holds out a chaining pin along the general direction of the line. By using the two range poles for reference, the rear tapeperson directs the front tapeperson to place the chaining pin on line by the use of right or left signals.

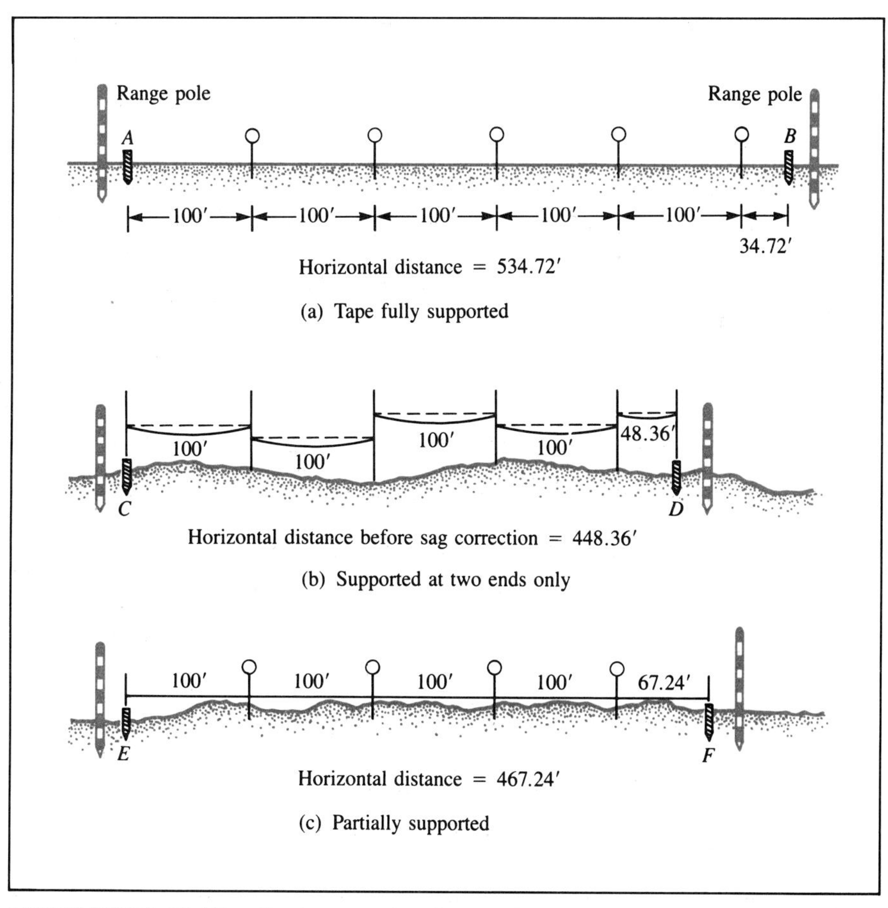

FIGURE 3.7 Taping over level ground

As soon as the tape has been placed on line, the rear tapeperson holds the rear end of the tape exactly even with the beginning point. The front tapeperson takes up position to one side of the line (not on the line), kneels, and applies tension (about 10 lb) to the tape with the left arm bearing against the left leg. The right hand is then free to place the pin on line and at the zero mark of the tape. The pin may be set vertically, but more often it is given a slant at a right angle to the tape, by which it can be placed more conveniently and accurately in position (see Figure 3.8).

When the front tapeperson sets the pin, the rear tapeperson must hold the 100-ft mark of the tape precisely over the beginning point (*A*). Hence, before setting the pin, the front tapeperson waits for the signal "right here" from the rear tapeperson. As soon as the pin is set, the front tapeperson calls out "right here," which is the signal for the rear tapeperson to move ahead.

At the beginning point (*A*), the rear tapeperson holds one pin and the front tapeperson begins with ten pins on a ring. As soon as the front tapeperson sets the first pin, the pin that the rear tapeperson holds indicates the fact that one tape length has been measured. After the next pin has been set, the rear tapeperson pulls the pin that marks the end of the previous tape length and now has two pins, indicating that two tape lengths have been measured.

FIGURE 3.8 Setting taping pin with tape supported by ground

Accordingly, the number of pins that the rear tapeperson holds, not counting the pin set in the ground, indicate the number of full tape lengths that have been measured. When the tenth or last pin is set, the front tapeperson calls out "tally." The rear tapeperson now has ten pins to bring forward, and the taping proceeds. Thus the number of *tallies* indicates the number of thousands of feet that have been measured.

Instead of using pins to record the number of full tape lengths, a distance equal to the full tape length can be recorded in the field book each time the front tapeperson sets a pin.

When the end of the line is reached, the front tapeperson halts and the rear tapeperson comes up to the last pin set. The tape is adjusted so that a full foot mark is opposite the pin and the terminus falls within the end foot length, which is subdivided into tenths. Proper tension is applied, and the front tapeperson reads the graduation opposite the survey mark at point B. Suppose that the front tapeperson obtains a reading of 0.28 ft while the rear tapeperson holds the tape at the 35-ft mark. If the first foot of the tape is graduated in tenths starting from the zero mark, as shown in Figure 3.4c, then the correct reading for this section of the line is $35.00 - 0.28$ ft $=$ 34.72 ft. The front tapeperson should call out "Cut point two eight," and the rear tapeperson calls out "thirty-five." They both make the subtraction mentally, and check each other on the result, 34.72 ft. If the rear tapeperson has seven pins in hand, not counting the one on the ground, the total distance is 734.72 ft.

If the tape has an extra foot graduated in tenths starting from the zero mark, as shown in Figure 3.4e, then the front tapeperson's reading must be added. In this case, the front tapeperson should call out "Add zero point two eight," and the correct length of this section of line would be 35.28 ft. The total length of the line would then be 735.28 ft.

If the taping is done on a hard surface, such as a sidewalk, steel rail, or pavement, the position of the end of the tape is marked with a colored lumber crayon, called *keel*. In this case the number of the tape length is recorded beside the mark as a means of keeping the count of tape lengths measured. To avoid mistakes the rear tapeperson calls out the number of his mark just before the front tapeperson records the next number.

When the ground is rough and uneven or when there are obstructions such as bushes and fences along the line, a horizontal distance can be measured by supporting the tape at its two ends only, as illustrated in Figure 3.7b. A clamping handle can be used to provide a better hold on the tape, and a plumb bob is used to vertically transfer a graduation mark from the tape to the ground or to line up a graduation mark on the tape with the survey mark on the ground (see Figure 3.9). In this method of taping, one should apply proper tension to the tape and hold the two ends so that the straight line joining the two ends is approximately level. Furthermore, because of the sag of the tape, it should be obvious that the straight line distance between the two ends is slightly less than the actual length of the tape. Therefore, in precise measurement, correction for sag must be applied to the measured

FIGURE 3.9 Dropping point with plumb bob

distance. The method for applying sag correction will be discussed in Section 3.11.

To minimize the errors caused by sag and by point transfer using plumb bob, it is often preferable to have the tape only partially supported when the ground is only slightly uneven or when the obstacle is relatively small, as illustrated in Figure 3.7c.

3.9 TAPING HORIZONTAL DISTANCE OVER SLOPING GROUND

Consider two points *A* and *B*, which are situated several tape lengths apart along a slope. The horizontal distance between the two points can be measured by supporting the tape at the two ends only, as shown in Figure 3.10.

Assuming that the distance is to be measured downhill, the front tapeperson holds one end of the tape at a comfortable height above the

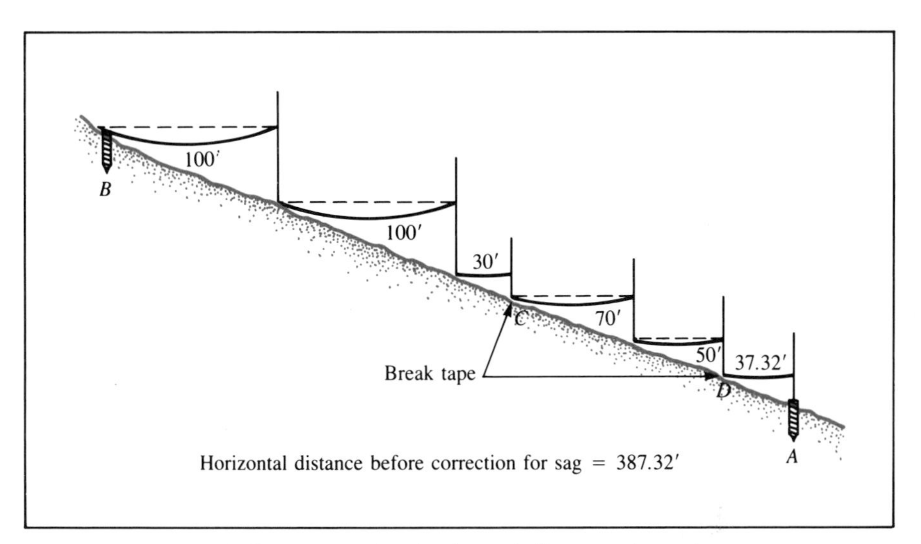

FIGURE 3.10 Measuring horizontal distance along slope

ground so that the straight line joining the two ends are approximately level. Tension is applied and the position of the end of the tape is transferred by means of a plumb bob to the ground, where a point is set or, on a hard surface, a keel mark is made. If accurate work is needed, the tape is stretched a second time and the mean of the two measurements taken.

It is hardly feasible or good practice to hold the tape more than 5 ft above the ground; hence, if the slope is more than 5 ft per tape length it is necessary to *break* tape, as the process is called. In this case the front tapeperson pulls the tape forward until the rear end comes up to the rear tapeperson. The front tapeperson then goes back to a point where the height difference is not more than 5 ft and here, at some full foot mark, plumbs down to the ground. The rear tapeperson then comes up and holds the foot mark at the ground point; the front tapeperson goes forward until another point is found for which the height difference is approximately 5 ft and again plumbs down and fixes a new ground point. This process is repeated until the full tape length has been measured. It is immaterial what foot marks on the tape are used, and no record of them is kept.

On less accurate surveys, a range pole may be used to plumb from the end of the tape to the ground. In going uphill, of course, the rear tapeperson must hold the end of the tape above the ground, and its elevated position is likewise found by plumbing from the ground point.

3.10 TAPING SLOPE DISTANCE

When a distance lies along an uniform slope where a tape can be either fully or partially supported by the ground, and when the difference in height between the two end points has already been or is to be determined in a separate survey, it may be preferable to measure the slope distance directly, as shown in Figure 3.11a. The corresponding horizontal distance between the two end points can be computed from the following simple formula:

$$d = \sqrt{s^2 - \Delta h^2} \tag{3.1}$$

where d is the horizontal distance, s is the slope distance, and Δh is the difference in height between the two end points. Note that the accuracy of the computed horizontal distance d depends not only on the accuracy of the measured slope distance s but also on the accuracy of the height difference Δh. By the error propagation formula given in Eq. (2.12),

$$\hat{\sigma}_d^2 = \left(\frac{\partial d}{\partial s}\right)^2 \hat{\sigma}_s^2 + \left(\frac{\partial d}{\partial \Delta h}\right)^2 \hat{\sigma}_{\Delta h}^2 \tag{3.2}$$

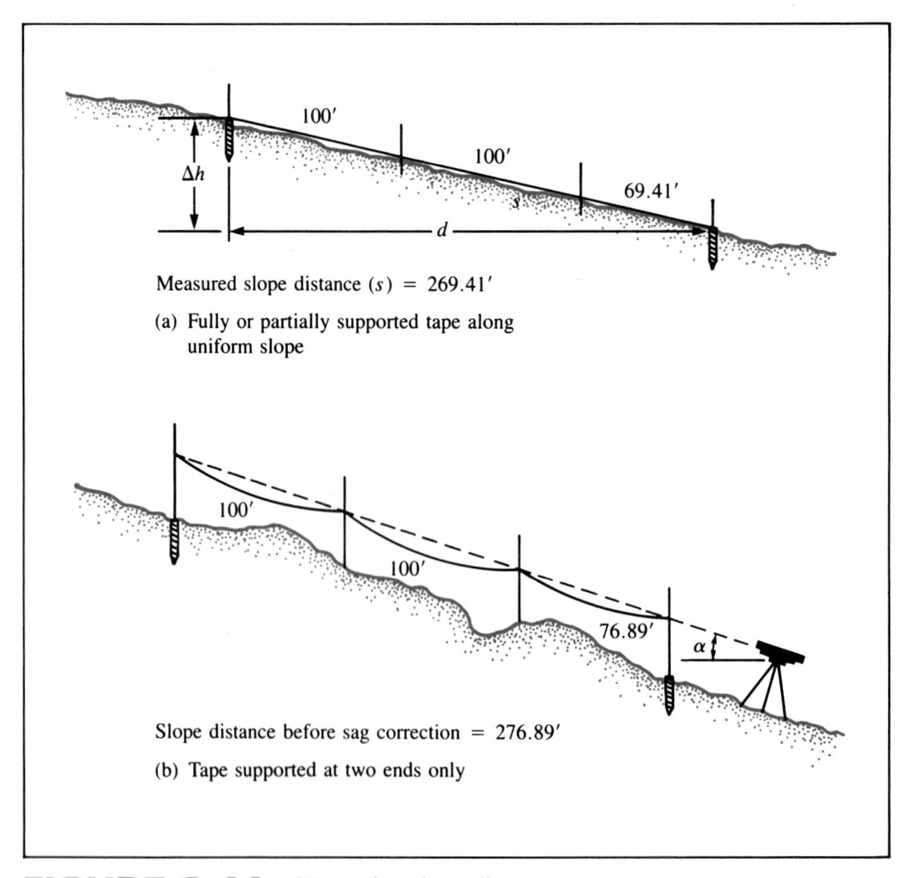

FIGURE 3.11 Measuring slope distance

where $\hat{\sigma}_d$, $\hat{\sigma}_s$, and $\hat{\sigma}_{\Delta h}$ are the estimated standard errors for the computed horizontal distance (d), the measured slope distance (s), and the measured height difference (Δh), respectively. Taking the partial derivatives of d with respect to s and Δh in Eq. (3.1) and substituting into Eq. (3.2) yields the following expression:

$$\hat{\sigma}_d^2 = \left(\frac{s^2}{s^2 - \Delta h^2}\right) \hat{\sigma}_s^2 + \left(\frac{h^2}{s^2 - \Delta h^2}\right) \hat{\sigma}_{\Delta h}^2 \tag{3.3}$$

EXAMPLE 3.1

Given the following field measurements and their estimated standard errors:

$$s = 472.36 \text{ ft} \qquad \hat{\sigma}_s = \pm 0.08 \text{ ft}$$
$$\Delta h = 21.6 \text{ ft} \qquad \hat{\sigma}_{\Delta h} = \pm 0.1 \text{ ft}$$

compute the horizontal distance (d) and its estimated standard error ($\hat{\sigma}_d$).

SOLUTION

$$d = \sqrt{s^2 - \Delta h^2} = \sqrt{(472.36)^2 - (21.6)^2}$$
$$= 471.87$$

$$(\hat{\sigma}_d)^2 = \frac{(472.36)^2}{(472.36)^2 - (21.6)^2} (0.08)^2 + \frac{(21.6)^2}{(472.36)^2 - (21.6)^2} (0.1)^2$$
$$= 0.006413 + 0.000021$$
$$= 0.006434$$
$$\hat{\sigma}_d = \pm 0.08 \text{ ft}$$

In Example 3.1, the net effect of the measurement error in the height difference (Δh) is negligible. However, if $\hat{\sigma}_{\Delta h} = \pm 1$ ft, then the corresponding value of $\hat{\sigma}_d$ would be increased to ± 0.09 ft; and if $\hat{\sigma}_{\Delta h} = \pm 2$ ft, $\hat{\sigma}_d$ would be increased to ± 0.12 ft.

Before electronic pocket calculators became commonly available, it was tedious to find the square root of a large number by hand calculation. Therefore, in the past, instead of using Eq. (3.1) to compute the horizontal distance, a slope correction (C_g) was usually applied to the measured slope distance to obtain the corresponding horizontal distance. From Eq. (3.1),

$$\Delta h^2 = s^2 - d^2$$
$$= (s - d)(s + d)$$

Therefore

$$s - d = \frac{\Delta h^2}{s + d} = C_g$$

When the slope is small, d can be approximated by s, and the slope correction can be approximated by the following expression:

$$C_g = \frac{\Delta h^2}{2s} \tag{3.4}$$

For instance, in Example 3.1, $C_g = 21.6^2/(2 \times 472.36) = 0.49'$. Therefore, $d = s - C_g = 472.36 - 0.49 = 471.87$ ft, which is the same as previously obtained using Eq. 3.1. The slope correction needed to reduce a slope measurement to the horizontal is always negative.

Over rough terrain a slope distance can also be measured by supporting the tape at the two ends only, as shown in Figure 3.11b. In this method, a transit is usually needed to measure the slope angle (α) and to align the ends of the tape along the same slope. The calculations needed for sag correction are significantly more involved than the horizontal case. It is very difficult to measure distances accurately by this method. When slope distances must be measured over rough terrain, considerations should be first given to the use of either electronic distance measuring equipment (Chapter 6) or method of stadia (Chapter 9).

3.11 CORRECTIONS FOR SYSTEMATIC ERRORS IN TAPING

The distance between the two end points of a fully stretched tape varies slightly with changes in temperature, tension, and mode of support. The errors caused by these sources behave according to some physical laws of nature and can be modelled with simple mathematical expressions. They are, therefore, systematic errors.

All steel survey tapes are carefully graduated by the manufacturer under controlled conditions of temperature, tension, and mode of support. A steel tape purchased from reputable makers usually will have a length not different from its nominal length (for example, 100 ft for a 100-ft tape) by more than 0.01 ft. But a tape that has been in use may have become kinked or been repaired so that its length has been appreciably altered. Such tapes are frequently in error as much as 0.02 ft or more. Hence all survey tapes should be frequently calibrated against a standard tape or a specially constructed calibration range. Figure 3.12 shows an outdoor calibration range that consists of two concrete pillars. A metal disk is imbedded into the concrete at both the bottom and the top of the pillar. A mark is scribed onto each metal disk to represent the end mark of a 100-ft distance. The two bottom marks are used to calibrate tapes when fully supported by the ground, while the two marks on the top of the pillars are used to calibrate tapes when supported at the two ends only. A similar calibration range can

(a) Tape fully supported

(b) Tape supported at two ends

FIGURE 3.12 Outdoor calibration range

be located inside a building. The temperature, tension, and mode of support during calibration should be recorded for future reference. Although the field temperature cannot be controlled, whenever possible the same tension and mode of support should be used in the field measurement as in calibration. Otherwise, corrections must be applied.

Tapes can be calibrated by the National Bureau of Standards for a fee.

LENGTH CORRECTION

The difference between the nominal length of a tape and its actual length under the conditions of calibration is known as the *length correction,* C_l. When measuring with a 100-ft tape, its nominal length is 100 ft and every full tape length will be recorded as 100 ft in field measurement. If the tape is actually 100.02 ft long when calibrated (or standardized) under a specific set of conditions, then the tape is said to be 0.02 ft too long.

For example, if a distance is measured with the tape just mentioned and found to be 705.76 ft, the resultant error would be $7.05 \times 0.02 = 0.14$ ft, and the corrected length, therefore, would be $705.76 + 0.14 = 705.90$ ft.

Likewise, if the tape is too short, the correction must be subtracted from the recorded length.

Note that these corrections are made when a distance is being measured between fixed points. If a distance is to be established, the sign of the correction would be reversed. For example, if in staking out a city subdivision it is necessary to set two iron pins exactly 600 ft apart, and if the true length of the tape is 100.02 ft, the measured distance with this tape will be $600.00 - 0.12$ ft $= 599.88$ ft.

TEMPERATURE CORRECTION

The correction C_t to be applied to the observed length of a survey line because of the effect of temperature on the steel tape can be computed from the following expression:

$$C_t = 0.00000645(T_1 - T_0)L \tag{3.5}$$

where 0.00000645 is the coefficient of thermal expansion of steel per 1°F, T_1 is the field temperature, T_0 is the temperature under which the tape is calibrated, and L is the length of the line. For example, if $T_0 = 68$°F and $T_1 = 83$°F, the temperature correction for a 100-ft steel tape would be

$$C_t = 0.00000645(83 - 68)100 = 0.0097 \text{ ft}$$

Thus note that a 100-ft tape will have its length changed very nearly 0.01 ft for each change of 15°F in the temperature. For small ranges of temperature and on ordinary work, this error may not be important, but the inexperienced surveyor is apt to underestimate the importance of this source of error when the ordinary temperatures of winter and summer measurements are encountered. For example, a summer temperature of 100°F and a winter temperature of 25°F are not uncommon. This difference in temperature of 75°F causes a change in the length of the tape of 0.05 ft, which is equivalent to a discrepancy of 2.6 ft in a mile. This error is greater than that permitted from all sources combined on many surveys, and yet it is frequently disregarded entirely. The coefficient of expansion of steel is 0.0000116 per 1°C.

EXAMPLE 3.2

The measured (observed) length of a traverse line was 876.42 m. The field temperature was 24°C and the 30-m steel tape was exactly 30 m at 20°C. Find the corrected length of the line.

SOLUTION

$$C_t = 0.0000116(24 - 20)876.42 = +0.04 \text{ m}$$

$$\text{corrected length} = 876.42 + 0.04 = 876.46 \text{ m}$$

SAG CORRECTION

A tape supported only at the ends will sag in the center by an amount that is related to its weight and the pull. The shortening effect of the sag is essentially the difference between the axial length of the tape hanging in a catenary and the chord distance between the ends. Sag causes the recorded distance to be greater than the actual length being measured. When the tape is supported at its midpoint, the effect of sag in the two spans is considerably less than when it is supported at the ends only. Furthermore, as the number of equally spaced intermediate supports is increased, the distance between the end marks will closely approach the length of the tape when supported through its length.

The sag correction C_s can be calculated by either one of the following equations:

$$C_s = \frac{W^2L}{24P^2} \tag{3.6a}$$

or

$$C_s = \frac{w^2L^3}{24P^2} \tag{3.6b}$$

in which W = the total weight of the section of tape located between supports,
L = the interval between supports,
P = the tension on the tape,
w = weight per foot of tape.

The units of weight and tension must be compatible (that is, both either pounds or kilograms). Furthermore, the total sag correction for a tape resting on multiple supports would be the sum of the sag corrections for the separate intervals. Hence the total sag effect for a 30-m tape supported at its midpoint and the ends would be twice the calculated sag for a 15-m span.

EXAMPLE 3.3

A 100-ft steel tape weights 2 lb and is supported at the ends only with a pull of 12 lb. Find the sag correction.

SOLUTION

$$C_s = \frac{2^2 \times 100}{24 \times 12^2} = -0.12 \text{ ft}$$

If the tension were increased to 20 lb, the shortening is reduced from 0.12 ft to 0.04 ft, which shows the desirability of using a higher tension on the tape when unsupported and also the fact that the error in any case is considerable. It is a good practice to avoid the effects of sag by taping on the ground whenever conditions permit.

The most practicable way of dealing with sag is to calibrate the tape with a specific pull (say 20 lb) when supported at the ends and to use the tape in the field in this manner whenever it has to be elevated above the ground. Very seldom should it be necessary to evaluate the sag effect through the use of Eq. (3.6a) or (b).

EXAMPLE 3.4

A 30-m steel tape weighs 0.336 kg and is supported at the 0, 15, and 30-m points under a tension of 5 kg. What is the sag correction?

SOLUTION

$$C_s = \left[\frac{(0.168)^2 15}{24(5)^2}\right] 2 = -0.001 \text{ m}$$

TENSION CORRECTION

Since a steel tape is elastic to a small extent, its length is changed by variations in the tension applied. This change in length is not to be associated with the effect of the sag on the tape due to variations in tension but rather with the elastic deformation of the tape.

It can be calculated from the expression

$$C_P = \frac{(P_1 - P_0)L}{AE} \tag{3.7}$$

in which C_P = the elongation of the tape of length L in feet,
P_1 = the applied tension in pounds,
P_0 = the calibration tension in pounds,
A = the cross-sectional area of the tape in square inches,
E = the modulus of elasticity of the tape material (for steel 29,000,000) in pounds per square inch (psi).

An ordinary 100-ft steel tape will stretch only about 0.01 ft for an increase of 15 lb in tension. The need for tension correction can be completely avoided by always using the same tension in both field measurements and tape calibration. Ordinarily, tensions of 10 lb and 20 lb are used for conditions of full support and end support, respectively.

EXAMPLE 3.5

A 100-ft steel tape having a cross-sectional area of 0.0046 in.^{2} is correct length under a pull of 12 lb. Calculate the elongation (nearest 0.001 ft) due to a tension of 20 lb.

SOLUTION

$$C_P = \frac{(20 - 12)100}{(0.0046)(29{,}000{,}000)} = +0.006 \text{ ft}$$

A convenient method of ascertaining the cross-sectional area of a tape is to perform the following calculation after weighing the tape and making a suitable weight correction for those portions that extend beyond the end marks. The specific weight of steel can be taken as 490 lb per ft^3. Assume the net weight of the tape to be 1.70 lb.

$$\frac{\text{length (in.)} \times \text{area (in.}^2\text{)}}{1{,}728} \times 490 = \text{weight (lb)}$$

Hence,

$$\text{area} = \frac{1{,}728 \times 1.70}{1{,}200 \times 490} = 0.0050 \text{ in.}^2$$

For a given tape supported at the ends, there is a specific tension that will stretch (elongate) the tape by an amount exactly equal to the shortening effect of sag. If this pull is employed in the field, the tension and sag corrections cancel each other. This particular pull is commonly termed the *normal tension*. It can be mathematically evaluated but direct laboratory determination is both more instructive and reliable.

EXAMPLE 3.6

A distance was measured using a 100-ft steel tape in five sections: 100′ + 100′ + 100′ + 100′ + 100′ + 89.32′ = 589.32′. The tape was supported at the two ends only during the measurements. The field temperature was 80°F and a tension of 20 lb was used. The tape was previously calibrated using a tension of 15 lb at a temperature of 68°F with the tape fully supported. The tape was found to measure actually 100.02 ft. The tape had a total weight of 2.80 lb and a cross-sectional area of 0.003 in.^{2}.

Compute the corrected distance.

SOLUTION

$$\text{length correction } (C_l) = (+0.02) \times \frac{589.32}{100} = +0.12 \text{ ft}$$

$$\text{temperature correction } (C_t) = (0.00000645)(80 - 68) \times 589.32 = +0.05 \text{ ft}$$

$$\text{sag correction for each full tape length} = \frac{(2.80)^2 \times 100}{24 \times (20)^2} = -0.08 \text{ ft}$$

$$\text{sag correction for the last partial tape length} = \frac{(0.028)^2 \times (89.32)^3}{24 \times (20)^2} = -0.06 \text{ ft}$$

$$\text{total sag correction } (C_s) = -(0.08 \times 5 + 0.06) \text{ ft} = -0.46 \text{ ft}$$

$$\text{tension correction for each full tape length} = \frac{(20 - 15) \times 100}{(0.003)(29 \times 10^6)} = +0.0057 \text{ ft}$$

$$\text{tension correction for last partial tape length} = \frac{(20 - 15) \times 89.32}{(0.003)(29 \times 10^6)} = +0.0051 \text{ ft}$$

$$\text{total tension correction } (C_P) = +(0.0057 \times 5 + 0.0051) \text{ ft} = +0.03 \text{ ft}$$

$$\text{total correction} = C_l + C_t + C_s + C_P = 0.12 + 0.05 - 0.46 + 0.03 \text{ ft} = -0.26 \text{ ft}$$

$$\text{corrected distance} = 589.32 - 0.26 = \underline{589.06 \text{ ft}}$$

EXAMPLE 3.7

A 100-ft tape actually measures 99.963 ft when it is supported at the two ends only, with the temperature at 68°F and a tension of 15 lb. The tape weighs 2.80 lb and has a cross-sectional area of 0.0065 sq. in. If the field temperature is also 68°F, what tension can be applied so that the tape measures exactly 100.00 ft when supported at the two ends only?

Modulus of elasticity of steel is 29,000,000 psi.

SOLUTION

Let x be the required tension.

$$\text{sag correction for 15-lb tension} = \frac{(2.80)^2 100}{24(15)^2} = +0.145 \text{ ft}$$

$$\text{sag correction for } x\text{-lb tension} = \frac{(2.80)^2 100}{24x^2} = +\frac{32.667}{x^2} \text{ ft}$$

$$\text{decrease in sag correction due to increase in tension} = 0.145 - \frac{32.667}{x^2}$$

Setting

$$0.145 - \frac{32.667}{x^2} = 100.00 - 99.963$$

$$\frac{32.667}{x^2} = 0.108$$

$$x^2 = \frac{32.667}{0.108}$$

and

$$x = \underline{17.4 \text{ lb}}$$

3.12 SLOPE AND ALIGNMENT ERRORS

When the straight line joining the two ends of a tape is not perfectly horizontal, its slope distance is always longer than the projected horizontal distance. As shown in Figure 3.13a and b, when one end of a 100-ft tape is either too high or too low by 1′, the slope distance is actually longer than the horizontal projection by +0.005′. Therefore, if the horizontal distance is recorded as 100 ft, the measurement would have an error of +0.005 ft. If a 500-ft line is measured in five sections, and the tape is inclined by 1% (1-ft drop or rise per 100 ft horizontal distance) at each section, the total error due to slope alone would amount to $+0.005 \times 5$ ft = +0.025 ft. Due to the cumulative nature of the slope error, special care should be taken to level the tape.

The error caused by a tape not being kept on line is identical to that caused by a tape not being held horizontal. Figure 3.14 shows the error caused by an intermediate point C being off-line by 1 ft. The actual straight line distance between points A and B is 199.990 ft. Thus, if the distance is treated as equal to 200 ft, the error would amount to +0.01 ft. However, a 1-ft misalignment is hardly excusable even when the alignment is done by eye only. Visual alignment using a range pole at the end of the line should

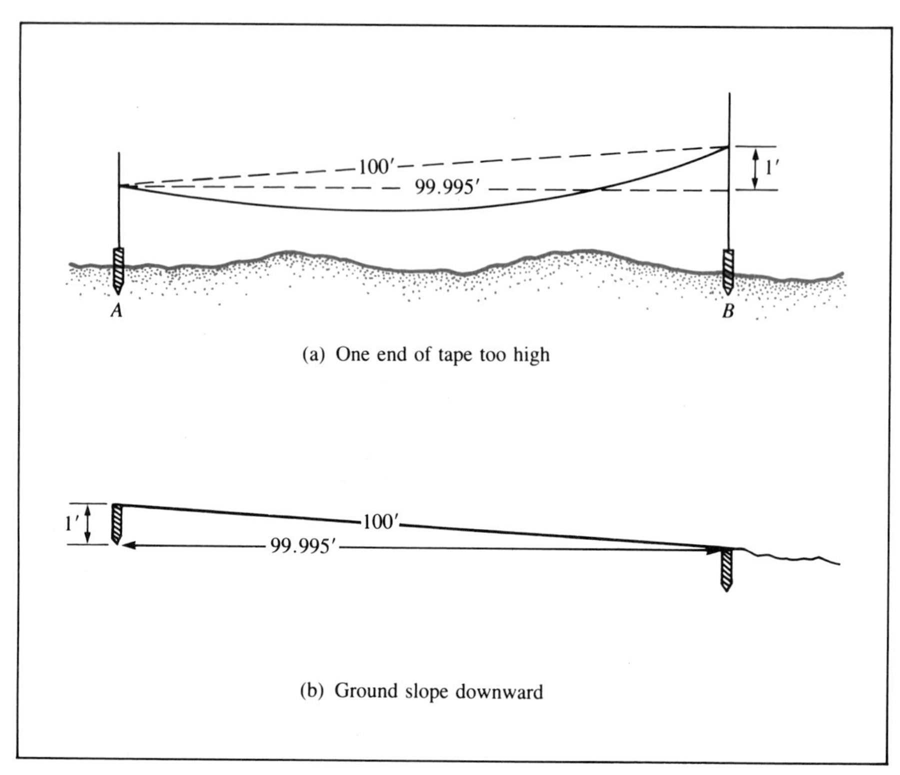

FIGURE 3.13 Slope error

keep a tape on line within 3 to 4 inches, which would correspond to a distance error of only +0.0003 ft, a negligible error for almost all surveys involving steel tape.

The magnitude of both slope and alignment error can be computed using Eq. (3.4).

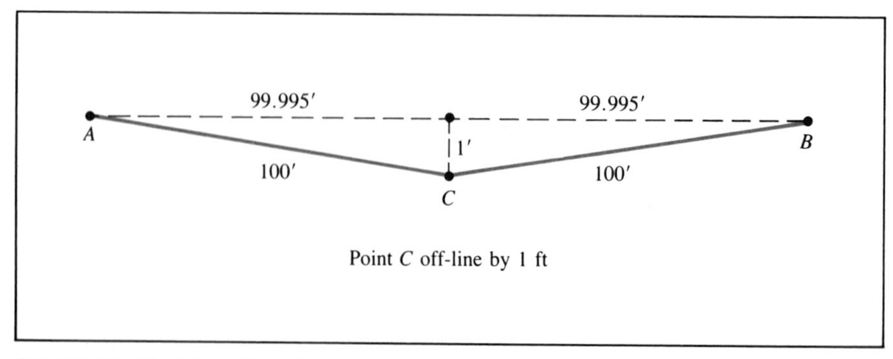

FIGURE 3.14 Alignment error

EXAMPLE 3.8

What error results from having one end of a 30-m tape

a. off-line by 0.1 m?

b. too low by 0.8 m?

SOLUTION

a. $\text{error} = \dfrac{\Delta h^2}{2s} = \dfrac{(0.1)^2}{2 \times 30} = +0.0002 \text{ m (negligible)}$

b. $\text{error} = \dfrac{\Delta h^2}{2s} = \dfrac{(0.8)^2}{2 \times 30} = +0.011 \text{ m}$

3.13 SOURCES OF RANDOM ERRORS IN TAPING

Random errors can occur at every step of a surveying operation. In taping the most serious sources of random errors include pin setting, tape reading, plumbing, incorrect tension, and temperature reading. Random errors cannot be totally eliminated, but can be minimized by using proper field procedures.

3.14 MISTAKES

Some of the common mistakes made in taping and recording are as follows:

Faulty tallying. Failure to correctly count the number of full tape lengths is not unusual.

Misreading the tape. A frequent mistake is that of misreading the tape, as when 6 is read for an inverted 9. Thus 86 is read for 89, or vice versa. Also, when the numbers on a tape become worn, an 8 may be read as a 3, and so on. Mistakes of this kind are prevented if the tapeperson will form the habit of looking at the flanking graduations before calling off the distance.

Calling and recording numbers. Numbers are easily reversed or misunderstood when they are called out to be recorded. The zero digit and the decimal point are most likely to cause mistakes. Thus the number 40.4 should be called as ''forty, point four.'' Otherwise, it may be misunderstood as ''forty-four'' and recorded as 44.0. The recorder should always repeat such numbers as are called, before recording them.

One-foot mistakes. It is easy to make a mistake of one foot when measuring a fractional tape length by subtracting incorrectly.

Mistaking the end mark. The end marks are differently arranged on different tapes. Hence the surveyors should always assure themselves of the position of the end marks of each tape before it is used.

3.15 CHECKS

It is a common practice to measure each distance first forward, say from point A to point B, and then backward, from point B to point A. However, close agreement between the two measurements does not necessarily mean high accuracy. The same systematic errors may be present in both measurements. For example, suppose that a 100-ft tape is used to measure a distance that is 1,000 ft long and that the forward measurement differs from the backward measurement by only 0.05 ft. If the tape were in fact too long by 0.05 ft, the systematic error due to incorrect tape length alone would amount to $0.05 \times 5 = 0.25$ ft. This error could not be detected by comparison of the two repeated measurements, because the same tape is used in both measurements. Therefore, it is important that the sources of systematic errors be recognized and their effects corrected for in the measurements.

3.16 ACCURACY

From the discussions presented in the previous sections, it is obvious that the major sources of errors in taping are systematic in nature. The accuracy of a tape measurement can be evaluated by estimating the magnitude of the error introduced by each of these sources. Under normal field conditions, following good field procedures, and using a good quality tape the length of which has been calibrated to ± 0.02 ft (1σ), a standard error of ± 1 part in 3,000 (that is, 1/3,000) of the measured distance can be achieved without applying corrections for temperature, tension, and sag; that is, the maximum survey error (3σ) for a measured distance of 3,000 ft would not exceed ± 3 ft. This level of accuracy is adequate for most boundary surveys of rural land and for many topographic mapping projects. For this reason one might hear a surveyor proudly proclaim that he or she has never had to apply corrections for systematic errors in taping, one obvious explanation being that this person has never been required to perform surveys with accuracy exceeding 1/3,000.

By applying corrections for all the known sources of systematic errors, a standard error of ± 1 part in 15,000 (1/15,000) of the measured distance can be commonly achieved. To achieve this level of accuracy, the following requirements must be fulfilled:

1. The length of the tape is calibrated with the standard error not to exceed ± 0.006 ft.

2. The field tension is measured with a standard error of ±2 lb.
3. The field temperature is measured with a standard error of ±5°F.
4. The slope is determined with a standard error of ±0.5%, which is equivalent to an error of ±0.5 ft in height difference for a distance of 100 ft.
5. The standard error of pin marking is not to exceed ±0.01 ft.

This level of accuracy is often required for surveying land boundaries in urban areas.

By using *invar tape* to reduce the temperature effect and by using extreme precautions in field measurements, a standard error of ±1 part in 700,000 (that is, 1/700,000) of the measured distance is achievable. To obtain such high accuracy, it is essential that the actual error caused by any of the possible sources (including calibration, alignment, point marking, slope, tension, temperature, sag, and tape reading) be limited to less than 1/500,000. The tape must be aligned so that the end of a 50-m tape would be off-line by no more than 1 inch. Field temperature should be taken with a thermometer accurate to within 0.3°C and attached to one end of the tape by a light metal holder. Tension is measured with a high-grade commercial balance, and tape stretchers should be used to hold the ends of the tape. A tape stretcher consists of a round staff, pointed at the lower end for driving into the ground, and an attachment for fastening the tape, which can then be moved up and down on the staff to adjust for proper height. A typical survey party for such tape measurements may consist of six to seven persons, and usually three to four tapes are used. Thus accurate distance measurement using tapes, although technically feasible and commonly performed in the past, is a tedious and time-consuming task. With the availability of electronic distance measuring equipment, which can achieve the same or higher accuracy in a few minutes of measuring time, there is now rarely any need for this type of accurate tape measurement.

The Kern Distometer ISETH (see Figure 3.15) is an instrument designed for making accurate measurement of structural deformations and terrain displacements using invar wire. The device has a measuring range between 1 and 50 m and can measure displacements up to 100 mm. For distances shorter than 20 m, displacement or deformation can be measured with a standard error of ±0.02 mm. For distances greater than 20 m, a standard error of ±1 part in 1,000,000 of the distance is achievable.

3.17 TAPE SURVEYS

The dimensions of a structure, the floor plan of a building, the layout of a street intersection, and the landscape of a small residential lot can all be easily mapped by using only tape and taping accessories. The basic geometric principles most frequently employed in tape surveys are illustrated in

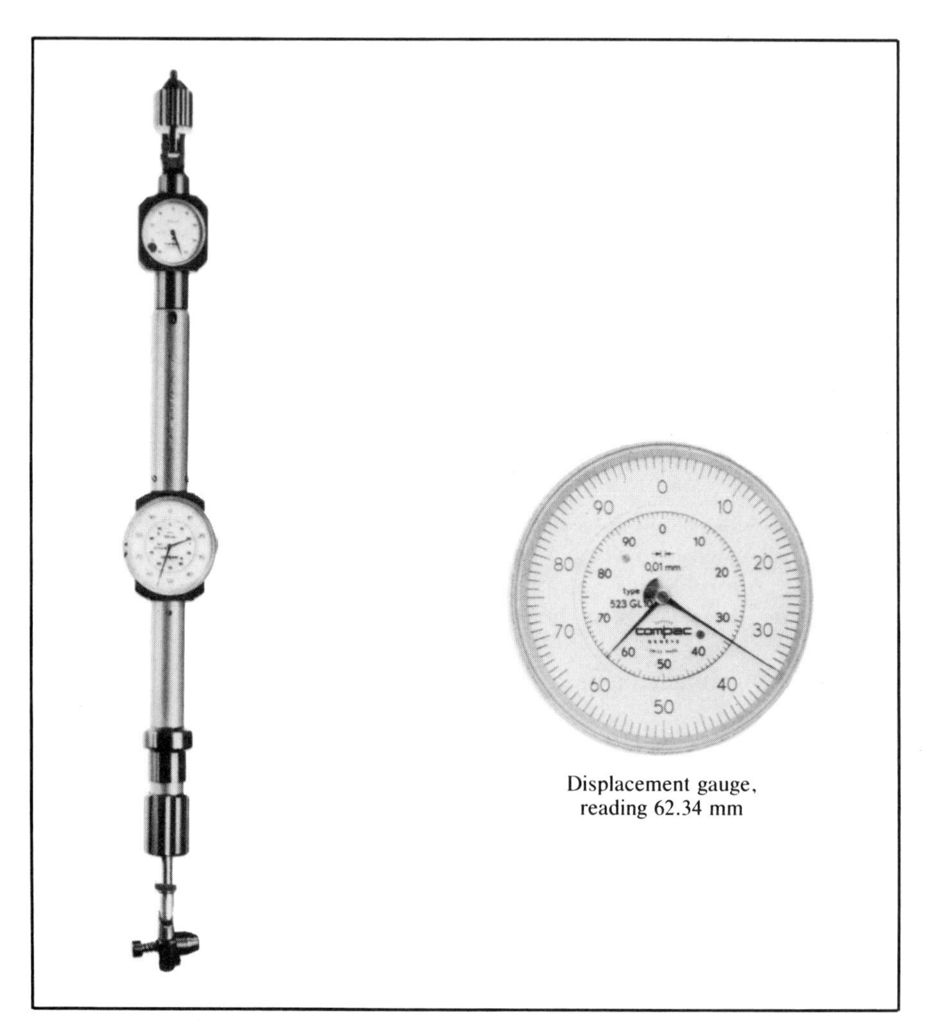

FIGURE 3.15 Kern Distometer ISETH (courtesy of Kern Instruments, Inc.)

Figure 3.16 and explained as follows:

a. The relative position of any three points can be defined by measuring the three sides of the triangle formed by the three points.

b. Once the relative position of two points, *A* and *B*, is known, the position of a third point *C* can be located by measuring the distances from *B* to *C* and from *A* to *C*. This method is called intersection by distances.

c. Given that the relative position of points *A* and *B* is known, the positions of new points can be located by measuring offset distances from the line joining *A* and *B*.

d. A right angle (that is, a 90° angle) can be established by using the proportion of 3-4-5 for the three sides.

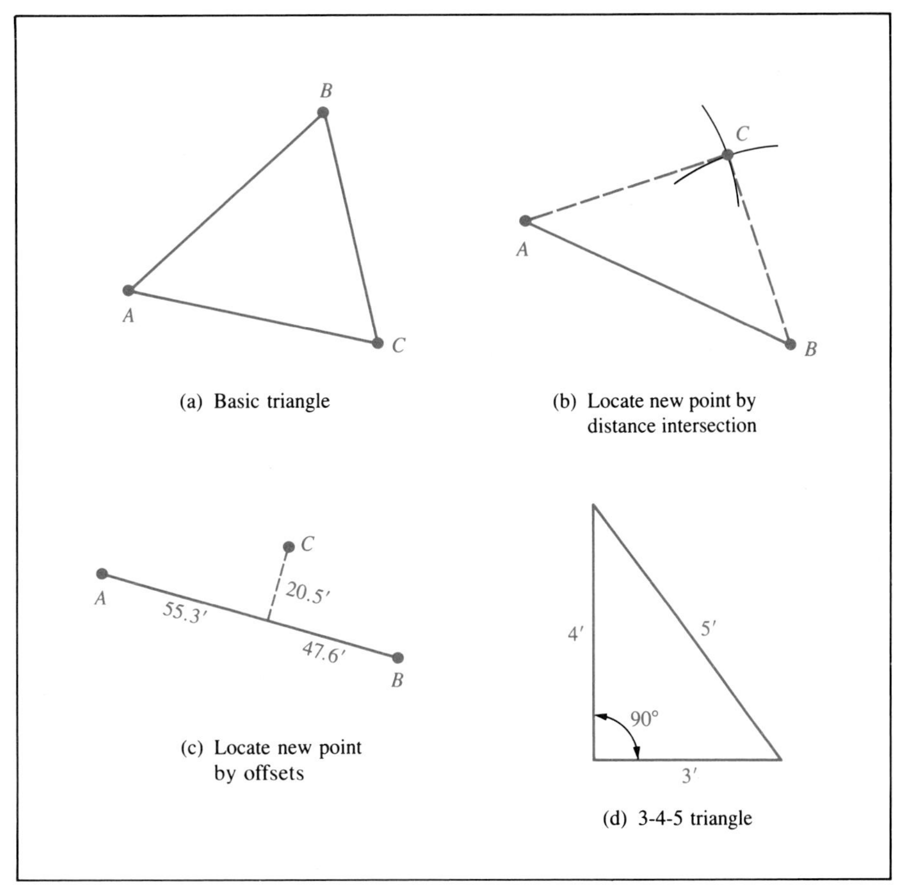

FIGURE 3.16 Basic geometric principles used in tape surveys

Figure 3.17 illustrates the application of these principles in the tape survey of a small lot. Important distances, such as lines *AB*, *BC*, *CD*, and *DA* in Figure 3.17, should be measured two times (once forward and once backward) to provide a check. Sufficient measurements must be made so that a drawing can be prepared from the field measurements. To avoid the need for a return to the site for additional measurements and to provide checks for the survey, it is always a good policy to take more measurements than absolutely needed for preparing the drawing. For example, in Figure 3.17 the outside dimensions of the building are measured and the locations of the corners are tied to the base lines *AB*, *BC*, and *AD* by offsets.

Tape surveys are particularly suited for the mapping of small areas that are relatively flat, such as street intersections, parking lots, and small playgrounds. For the mapping of large areas or hilly and mountainous terrain, other methods such as stadia (Chapter 9) and photogrammetry (Chapter 14) become more practicable.

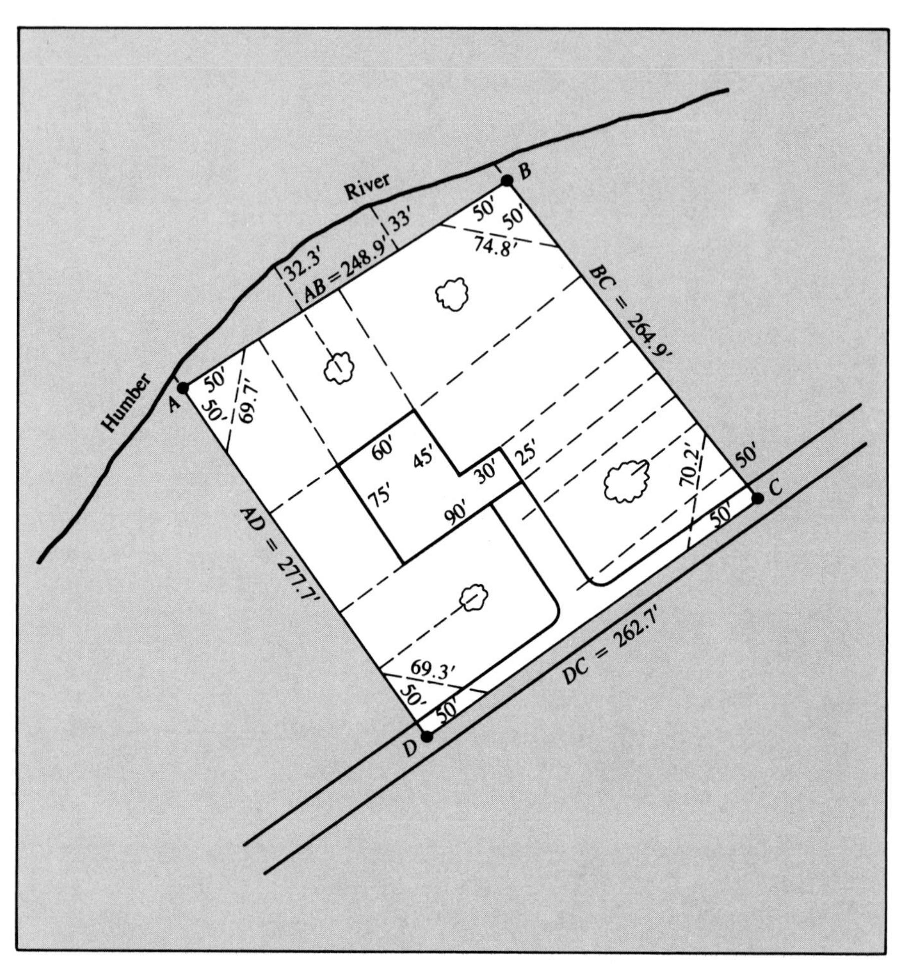

FIGURE 3.17 Tape survey of a small area

Tape measurements can also play a significant role in the detection and monitoring of ground movements. Figure 3.18 illustrates the use of offset measurements to measure the rate of movement in a landslide. The baseline is defined by two points, *A* and *B*, located outside of the landslide. The rate of movement is monitored by measuring the offset distances to slide points (*a*, *b*, *c*, *d*, and *e*) at regular time intervals.

3.18 DOUBLE RIGHT ANGLE PRISM

In conducting tape surveys, the measurement of offset distances can be greatly facilitated by the use of a *double right angle prism*. As shown in

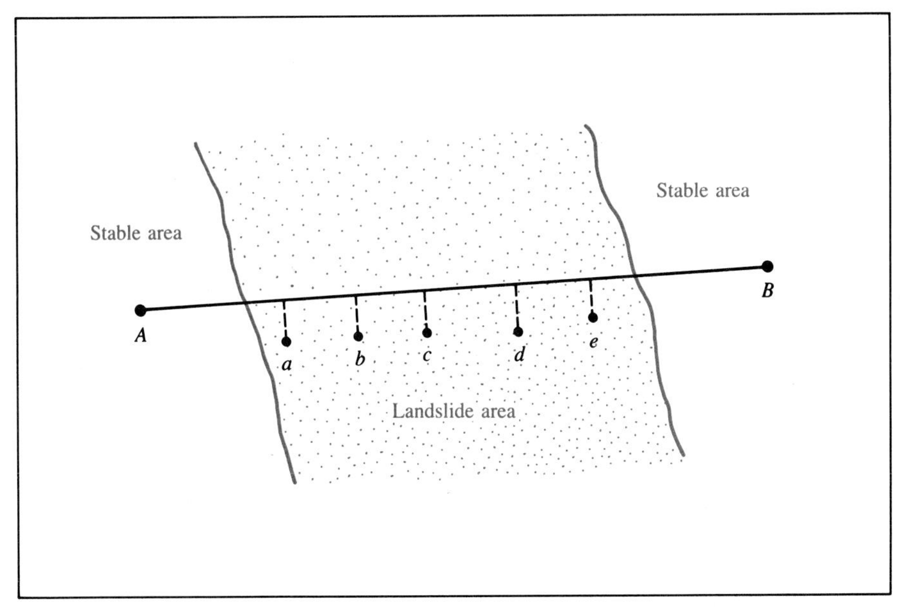

FIGURE 3.18 Landslide movement by offset measurements

Figure 3.19, the device consists of two right angle prisms, a piece of flat viewing glass, and a lightweight plummet. The upper prism provides a view to the right, and the lower prism provides a view to the left. The view in front of the prism is seen through the flat viewing glass mounted between the two right angle prisms. Figure 3.20 illustrates the use of this device to establish the offset direction from a line *AB* to the corner of a building. It is assumed that a range pole is being held at points *A* and *B*. The double right angle prism is first located on the line *AB* by holding it over a point, say P_1, so that the images of the range poles are properly aligned along a vertical line. It is then moved either to the left or to the right until the image of the building corner is aligned with the images of the two range poles. The correct position of the offset point, say P_2, is then indicated by the lightweight plummet. In windy conditions the lightweight plummet can be replaced by a metal rod called *Jacob's staff,* which can be inserted into a bottom opening.

3.19 FIELD NOTES

The field notes for tape measurements must contain all the information needed to convert the raw field data into correct horizontal lengths. Figure 3.21 shows an example of taping notes. The note records the conditions under which the tape was calibrated, as well as the field conditions. The names of the members of the party and the task performed by each

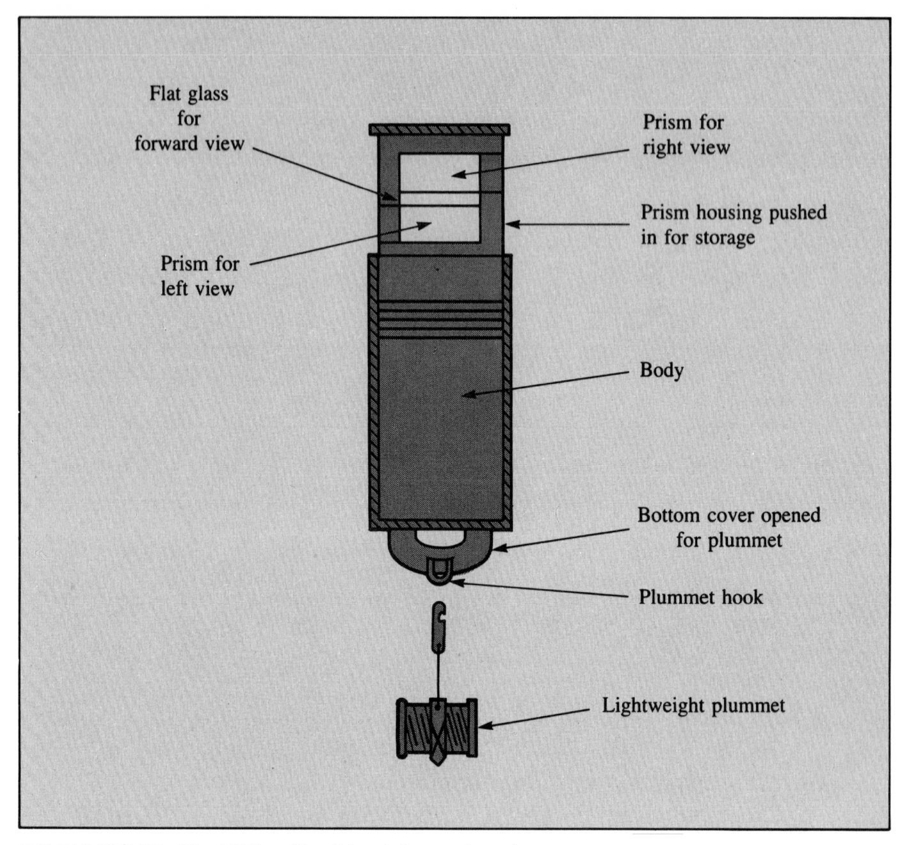

FIGURE 3.19 Double right angle prism

member are also recorded. For each taped distance, both the number of full tape lengths and the fractional tape are recorded. For example, for distance 1-2, a full tape length of 50 m was first measured. Then the rear tapeperson held the tape at the 20.50-m mark (as denoted by H in the note), while the front tapeperson read the tape at 0.006 m from the zero mark, denoted as cut (C) in the note. Hence the total distance for line 1-2 was 50.00 + 20.50 − 0.006 m = 70.494 m.

In the case of tape surveys, a sketch such as that shown in Figure 3.17 should be drawn in the field note to show the distances measured and relative locations of the relevant features.

Recording the measurements of a surveying project in a field book is a most important task. Despite the meticulous refinement with which a survey may have been executed, its objectives will not be attained if the documentation is deficient in quantity, defective in quality, illegible, or ambiguous. The field notes of a survey represent the original record of what was found and what was done. After the stakes have rotted and more durable markers have become obliterated, the field notes may become the sole evi-

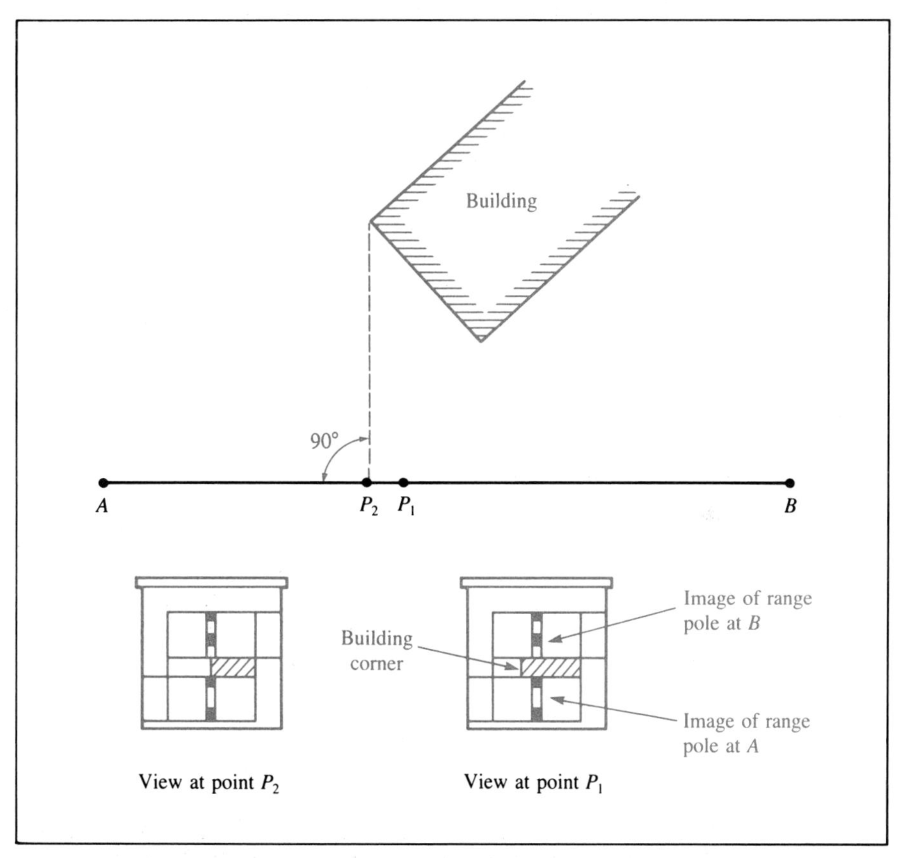

FIGURE 3.20 Locating an offset point with a double right angle prism

dence of a survey. When the notekeeping is closely supervised or performed by the party chief, its quality may be a significant indication about the diligence, thoroughness, and care with which the survey was conducted.

The field notebook record is particularly important in certain kinds of preliminary engineering investigations, such as for highways and major structures. Initial decisions affecting planning and design are based to a large extent on boundary, topographic, and other surveys. Even if such surveys are complete and accurate, poor notekeeping will portray an incorrect picture of the project situation and influence the deduction of erroneous conclusions.

The engineer's field book is generally used for recording surveying data. The most common page size is $4\frac{5}{8}$ in. $\times$ $7\frac{1}{2}$ in. Bound and loose-leaf types are available. They are ruled horizontally with six vertical columns on the left page, and cross lines and a single vertical red line on the right page. A single page number is placed in the upper right-hand corner of each double page.

Course	Distance in full tape	Fraction	Distance	Average distance	Corrected distance
1-2	50	N 20.50	70.494		
		C 0.006			
				70.491	70.503
2-1	50	H 20.50	70.488		
		C 0.012			
2-3	50	H 14.30	64.218		
		C 0.082			
				64.215	64.226
3-2	50	H 14.30	64.212		
		C 0.088			
3-4	50	H 9.30	59.250		
		C 0.050			
				59.255	59.265
4-3	50	H 9.30	59.259		
		C 0.041			

Sept. 9, 1989
Sunny, warm, 80°F
Locker No. 21

Party:
J. Jones – head tape
S. Smith – rear tape
R. Brown – note

50 meter steel tape
Calibrated length = 50.006 m
Calibration condition: Tension = 10 lb
Temp. = 72°F
Fully supported

Field condition: All measurements fully supported
Tension = 10 lb
Temp. = 80° F

FIGURE 3.21 Taping notes

The selection of a field book should be based primarily on considerations of permanency of the written record. The book must be able to withstand repeated handling and the aging effects of time for the many years it may be in a filing repository. The book should be strongly bound so that rough field usage will not break the binding, and the lines on opposite pages should be in correct register. A leather-bound field book best satisfies these requirements.

Loose-leaf books are convenient when performing work involving the progressive transfer of notes from field to office. However, the sheets are easily lost or misplaced, or the recorder may fail to place an identifying caption on each one.

The duplicating notebook provides for detaching the original leaf while the carbon copy remains in the book.

Ownership information should be written on both the front cover and the first page of the book. The next two or three pages should be reserved for an index, and the entire book should be paginated. Some important practices in notekeeping are:

1. Make clear and legible entries with a well-sharpened, medium-hard (3H or 4H) black pencil.

2. Letter all entries.
3. Make no erasure of observed data. Erroneous entries should be struck out by a single line and the corrected value entered above the original value. Erasures may destroy the credibility of the field record.
4. Make liberal use of sketches but draw lines with a straight edge.
5. Indicate the general orientation of field drawings with a north-pointing arrow.
6. Use abbreviations and symbols to promote conciseness, but be certain their meaning is known to the office engineer.
7. Keep the index up to date.
8. Do not hesitate to use narrative to explain more clearly some significant aspect of a surveying project.
9. Acknowledge the original source of any data used in initiating or closing a survey. Cross-reference important information.
10. Record data in its original observed form. For example, if the distances are measured in feet, do not record their equivalents in meters.

Uniformity in the format and recording of field survey notes is important, particularly on a given project or within an organization. Some of the fundamental requirements that all good field notes should satisfy are:

1. *Neatness*. Neatness and orderliness will promote clarity.

2. *Completeness*. Every pertinent item of information should be recorded. In addition to the basic record of survey measurements, certain supplementary information should be documented. This will include date, party chief, recorder, instrument operator, location, equipment, and comments on weather conditions that could affect measurements. It is particularly important that each page of a set of loose-leaf notes display its own number as well as the total number of pages in the set, such as "page 2 of 23 pages."

3. *Clarity*. All entries and statements must admit only one interpretation—the correct one.

4. *Legibility*. Illegible entries reflect discredit on the notekeeper and are costly to the engineering firm because the data are numerically uncertain. A return trip to the project area may be necessary.

5. *Accuracy*. It is absolutely essential that the field records correctly portray the data acquired by the various measurements. Placing data on the covers of book matches, on memo pads, or in improvised fieldbooks with the intention of subsequently transcribing such information into the official notebook is bad practice. Relying on the memory is both hazardous and costly because it may be impracticable to return to the project site for a single item of overlooked information.

6. *Integrity*. The notes should reflect the personal integrity of the chief of party and recorder. The data should be honestly acquired and not be molded or altered in any way in order to satisfy stated accuracy requirements. If field note entries are to earn validity, they must be recorded at the time the measurements are made.

PROBLEMS

3.1 To determine the length of her pace, a young engineer taped off a distance of 300 ft over flat ground. She then walked naturally over the distance five times and recorded the following number of paces: 123, 121, 123, 122, and 123. Determine (a) the average length of her pace and (b) the estimated standard error of the average pace.

3.2 The recorded (raw, observed) length of a line was 2,418.62 ft. Then, the tape was calibrated under the same conditions that prevailed during the field measurement and found to be 99.97 ft long. Find the corrected length of the line.

3.3 The observed slope distance between two survey points situated on a 4.8% grade was 837.56 ft. What is the horizontal distance?

3.4 The distance on a slope making an angle of 4°35′ with the horizontal was found to be 1,265.02 m. Calculate the horizontal distance.

3.5 Points *A* and *B* are situated on a uniform slope along which the measured ground distance was 568.36 m. If the difference in elevation between the two points is 22.55 m, what is the horizontal distance?

3.6 The slope distance between two points was measured to be 463.21 ft with an estimated standard error of ±0.05 ft. The difference in elevation between the two ends of the line was found to be 35.6 ft. Compute (a) the horizontal distance, (b) the estimated standard error of the computed horizontal distance if the measured elevation difference is known to have a standard error of ±0.5 ft.

3.7 A horizontal distance of 475.00 ft is to be established by measuring with the tape lying on the ground, which has a uniform grade of 3.70%. What nominal measurement should be made to effect this layout?

3.8 A steel tape was standardized while fully supported at 68°F under a 15-lb pull and found to be 100.003 ft long. The tape was then employed to measure the distance between two boundary corners at a temperature of −4°F with the same tension and mode of support. The observed distance was 5,287.65 ft. Find the corrected distance.

3.9 A 100-m track course is to be laid out using a 30-m steel tape that is known to be 30.012 m long at 20°C. If the field temperature is 14.5°C, find the nominal measurement that will establish the desired distance.

3.10 In staking out a subdivision it is necessary to effect the layout of a design dimension of 620.00 ft. The 100-ft steel tape is known to be 0.007 ft short at 68°F and the field temperature is 68°F. Find the nominal length that will establish the requisite distance.

3.11 Because of temporary irregularities in the ground surface between the monumented ends of a 100.000-ft tape base, a 100-ft steel tape weighing 1.50 lb was standardized while suspended by the ends only and under a tension of 10 lb. The temperature was 98°F and the observed length of the tape was 99.890 ft. Find the length of the tape while fully supported at a temperature of 68°F and under a pull of 10 lb.

3.12 The observed distance between two steel pins was found to be 328.41 m. The measurement was made at 27°C with a 30-m steel tape, which was immediately thereafter compared with a nearby outside tape base and found to be 29.986 m long. Find the corrected distance.

3.13 A distance of 348.65 ft is to be staked in the field using a 50-ft steel tape, which actually measures 50.03 ft at a temperature of 60°F. The field temperature is 23°F. Calculate the nominal length that should be used to establish the desired distance.

3.14 The recorded length of a traverse line was 892.75 ft. It was measured at 5°F with the steel tape supported throughout on a 5.5% slope. If the tape is 100.015 ft long at 68°F while supported in the same manner, determine the corrected length of the line.

3.15 In the course of a forest survey the recorded distance between two survey points was 80.98 chains. The steel chain is correct length at 60°F, but the field measurement was conducted at 92°F. What is the corrected distance to the nearest 0.1 ft?

3.16 What error (nearest 0.01 ft) results from having one end of a 200-ft tape (a) too low by 3.8 ft? (b) off-line by 1.0 ft?

3.17 Compute the shortening effect of sag (nearest 0.01 ft) for a 100-ft steel tape weighing 1.45 lb when supported at its ends under a pull of (a) 10 lb; (b) 15 lb.

3.18 The average weight per meter of a 30-m steel tape is 0.0146 kg. Calculate the sag effect under a tension of 5 kg when supported only at the ends.

3.19 How much error (nearest 0.01 ft) is introduced in effecting the layout of the 850.00-ft side of a large mill building if the nominal length of the 100-ft tape is considered as its true length? The field temperature is −10°F and the tape was actually 99.96 ft long at 68°F.

3.20 Determine the elongation of the tape of Problem 3.18 if the tension is increased to 7.5 kg. The average *AE* value is 38,560 kg.

3.21 The distance between two fixed points in a desert region was measured at a temperature of 108°F with a 100-ft steel tape that was true length at 68°F and was observed to be 2640.56 ft.

a. What is the corrected distance?

b. What would probably be the observed distance if measured at 22°F?

3.22 A 100-ft steel tape weighs 1.25 lb and has a cross-sectional area of 0.0032 sq in. At 68°F, when supported at the ends only under a pull of 15 lb, it is 100.003 ft long. Calculate its length under a tension of 22 lb at a temperature of 60°F with the manner of support remaining the same.

3.23 A 100-ft steel tape was calibrated under a tension of 10 lb, and at a temperature of 68°F while fully supported. It was found to be 99.99 ft long. In the field a tension of 10 lb was also used. But the tape was supported at the two ends only, and a field temperature of 85°F was recorded. A line was measured in the following sections: 100′, 100′, 100′, and 85.32′. Compute the correct length of the line. The tape weighed 2 lb and had a cross-sectional area of 0.0035 sq in.

3.24 A 100-ft steel tape was found to measure 99.97 ft when supported at the two ends only with the temperature at 68°F and a tension of 10 lb. If the field temperature actually measures 80°F, what tension can be applied so that the tape measures exactly 100.00 ft when supported at the two ends only on the field? The tape weighs 2.0 lb and has a cross-sectional area of 0.0050 sq in. Modulus of elasticity of steel is 29×10^6 psi.

3.25 A 30-m tape was calibrated under a tension of 5 kg and a temperature of 20°C while fully supported. It was found to be actually 30.005 m long. In the field it was used under a tension of 5 kg, a temperature of 35°C, and supported at the two ends only. A line was measured in five sections with the following results: 30 m, 30 m, 30 m, 30 m, and 28.641 m. Determine the correct length of the line. The tape weighed 0.35 kg.

3.26 Show by means of a sketch the distance measurements needed to map the details shown in Figures P3.1 and P3.2.

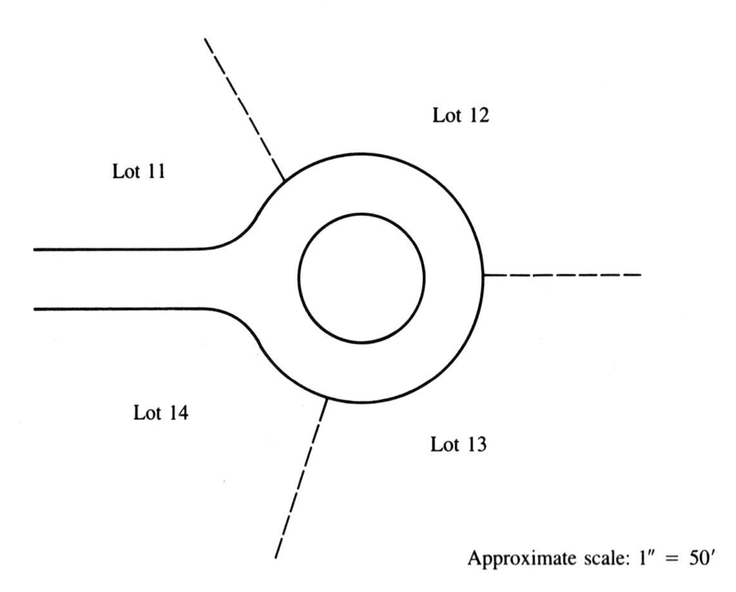

FIGURE P3.1 A cul-de-sac

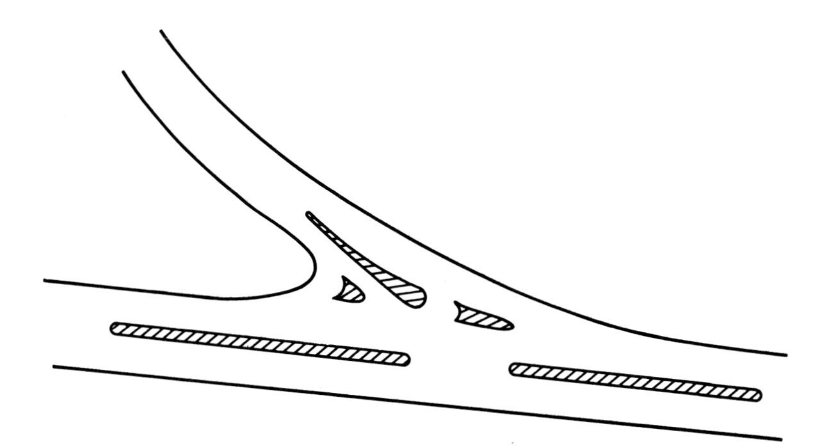

Approximate scale: 1″ = 50′

FIGURE P3.2 A Y-intersection

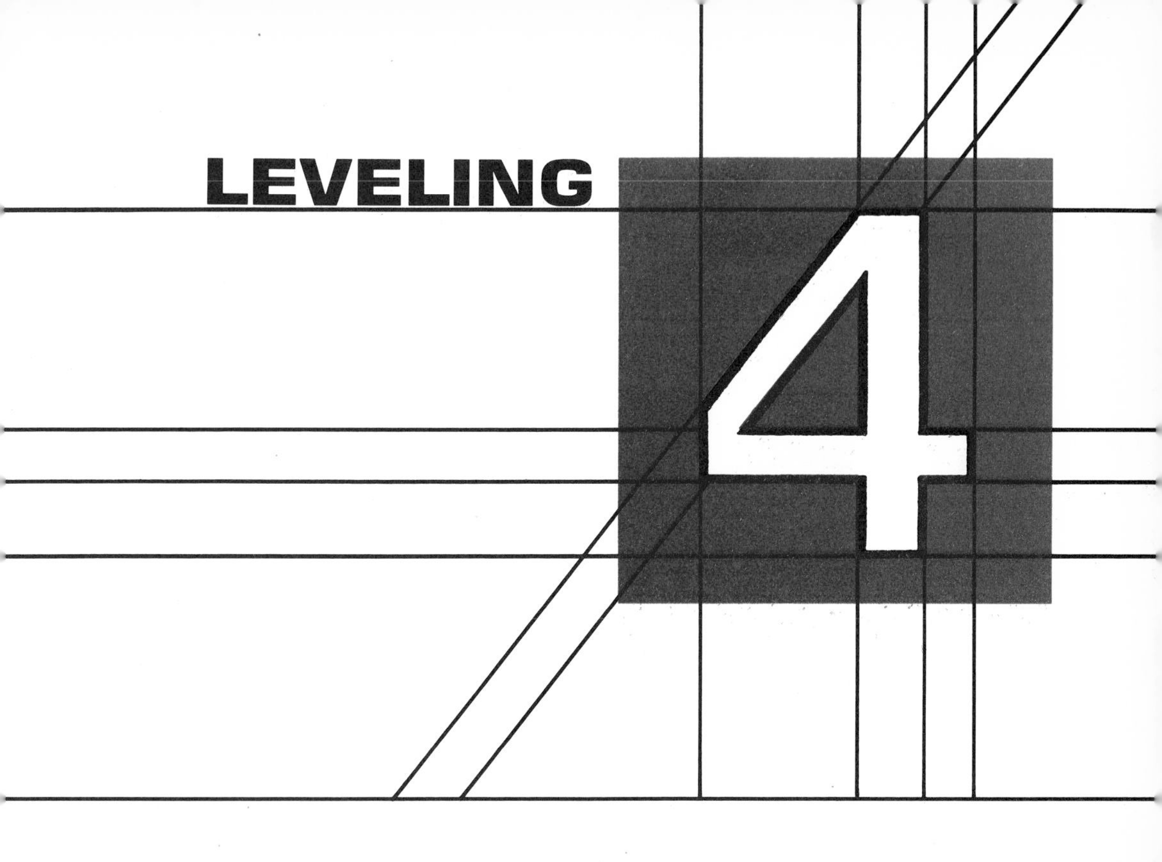

4.1 INTRODUCTION

The *elevation* of a point in space is its vertical distance above or below a surface of reference, called *datum*. The most commonly used datum is the sea level. However, a level surface at any specified elevation can be used to serve as a datum. The elevation of a point can be considered as its vertical coordinate. It can be plus (+) or minus (−) depending on whether the point is above or below the datum.

The operation of determining the difference in elevation between points is called *leveling*. The horizontal distance between points can be measured by taping, as discussed in Chapter 3, whereas the vertical distance (more commonly referred to as elevation or height difference) between points is measured by leveling.

The word *height* is often used interchangeably with the word *elevation,* although the height of an object should mean the vertical distance between the top and bottom of the object. The word *altitude* also has similar meaning as elevation. However, the altitude of a point always means the vertical distance of the point above sea level. Thus the altitude of an aircraft means the height of the aircraft above sea level.

Leveling is an important operation in almost all engineering and construction projects. Leveling is needed to determine the undulation of the land surface in topographic mapping. The elevations at various points along a highway or a sanitary sewer must be carefully determined during construction to achieve the designed slope. Before the placement of concrete or asphalt in street construction, the manhole frame must be set at the correct elevation so that its top will be flush with the surface of the finished roadway. Excavation for a building must be carried out to a prescribed depth or elevation. Masonry footings to be used as foundations for a structure must be in the correct vertical as well as horizontal position. Timber piling must be cut off at the proper elevation, bridge piers must be constructed to a designed elevation, and railroad bridges over highways must have the requisite minimum vertical clearance. The elevations of the foundations of buildings adjacent to a large excavation can be continuously monitored to determine and control the amount of settlement.

Various methods of leveling will be discussed in this chapter.

4.2 ELEVATION DATUM

The most commonly used datum for measuring elevation differences is the *mean sea level (MSL)*. The mean sea level at a location is the average height of the surface of the sea for all stages of the tide at that particular location. Accurate determination of the MSL requires the averaging of hourly tide readings at the particular location for a 19-year period.

The MSL differs from one location to another. The elevation of a point determined from an Atlantic coast mean sea level datum and the elevation of the same point determined from a Pacific coast datum will not agree. The elevation of the Pacific coast is approximately 2 feet higher than the coast lines of the Atlantic and Gulf states. Even within a distance of a few

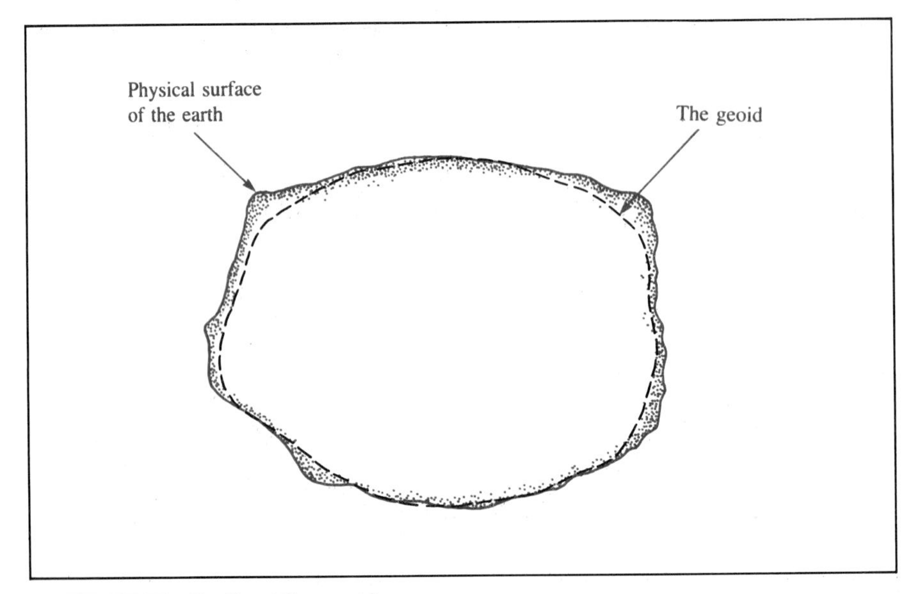

FIGURE 4.1 The geoid

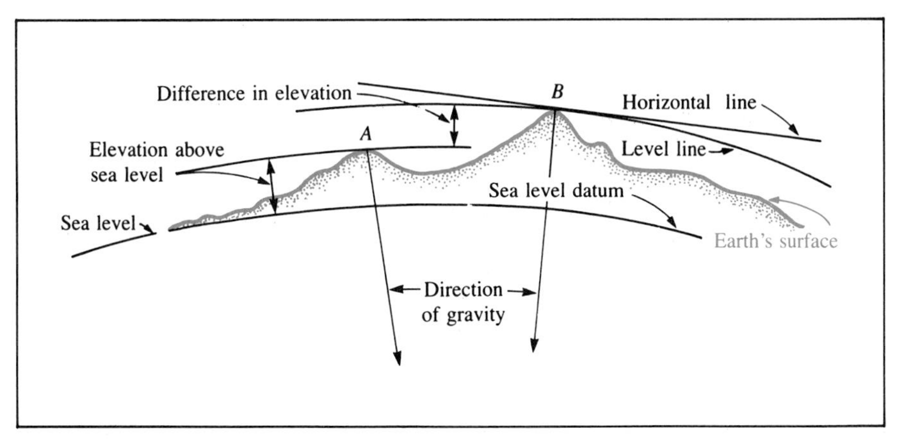

FIGURE 4.2 Elevation and elevation difference

miles along the same coast line, the MSL determined from tide gauge readings may differ slightly from one location to another.

The mean sea level at a particular location on earth, if extended in all directions over the entire earth, would be an undulating spheroid as shown in Figure 4.1. This spheroid is called the *geoid,* which has the following two characteristics: (1) every point on its surface has the same gravity potential, and (2) the direction of gravitational pull is perpendicular to the geoidal surface at all points.

In the United States, the National Geodetic Survey used as reference surface the "Sea Level Datum of 1929," which was redesignated in 1973 as the *National Geodetic Vertical Datum of 1929*. This datum is based on the mean sea level observed at 26 tide stations along the Atlantic and Pacific Coasts and the Gulf of Mexico. About 60,000 miles of level lines in both the United States and Canada were adjusted to this reference datum.

Figure 4.2 illustrates the meaning of elevation and elevation difference. The elevation of summit *A* is measured by the vertical distance along the direction of gravity between the sea level datum and point *A* or between the sea level datum and a parallel curved surface through *A*. Likewise the elevation of summit *B* is equal to its vertical distance above the sea level datum. The difference in elevation between summits *A* and *B* is equal to elevation *B* minus elevation *A*. It is equal to the vertical distance between the imaginary curved surfaces parallel to the sea level datum, one passing through *A* and the other through *B*.

4.3 CURVATURE AND REFRACTION

It is essential to understand the nature of the earth's curvature and atmospheric refraction as they affect leveling operations.

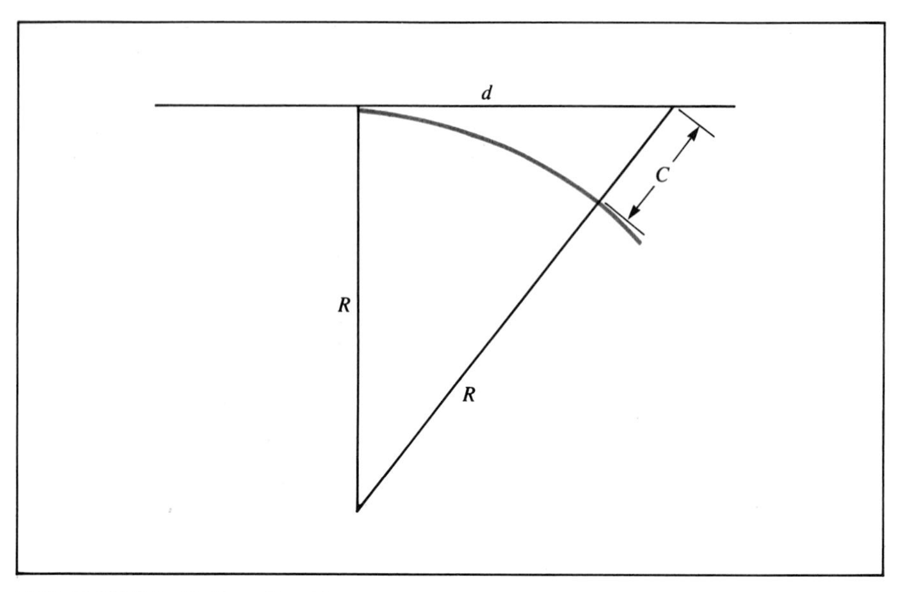

FIGURE 4.3 Earth's curvature

The definition of a level surface indicates that it parallels the curvature of the earth. A line of constant elevation, termed a *level line,* is likewise a curved line and is everywhere normal to the plumb line. However, a *horizontal line of sight* through a surveyor's telescope is perpendicular to the plumb line only at the point of observation. Hence it should be carefully distinguished from a level line.

Figure 4.3 shows a section passing through the earth's center. The magnitude of the curvature C in relation to the radius R of the earth and the tangent distance d can be deduced at once in the same manner as Eq. (3.4) for the grade correction in slope measurements. Hence

$$C = \frac{d^2}{2R} \tag{4.1}$$

Because of atmospheric refraction, rays of light transmitted along the earth's surface are refracted or bent downward so that the actual line of sight is along a curve that is concave downward. Unlike the earth's curvature, which is directly calculable from geometrical principles, refraction depends on the state of the atmosphere and is variable.

The effects of the earth's curvature and atmospheric refraction are shown in Figure 4.4. For a sight distance AB, a level line is deflected from the horizontal line the distance BF. This deflection, according to Eq. (4.1), is proportional to the square of the distance. For a sight distance of 1 mile and a value of 3,959 miles as the average radius of the earth (R), the curvature amounts to 0.667 ft.

Refraction will cause the line of sight to be deflected downward by a small angle θ. The magnitude of this angle θ is directly proportional to the angle γ subtended by the line of sight at the center of the earth. The coefficient of refraction (m) is defined as a constant so that

$$\theta = m\gamma \tag{4.2}$$

The coefficient of refraction m has an average value of 0.071.

The refracted distance BE shown in Figure 4.4 is approximated by the following relationship:

$$BE \approx d \cdot m\gamma \tag{4.3}$$

where d is the tangent distance AB shown in Figure 4.4.
Since

$$\gamma \approx \frac{d}{R}$$

then

$$BE \approx m\frac{d^2}{R} \tag{4.4}$$

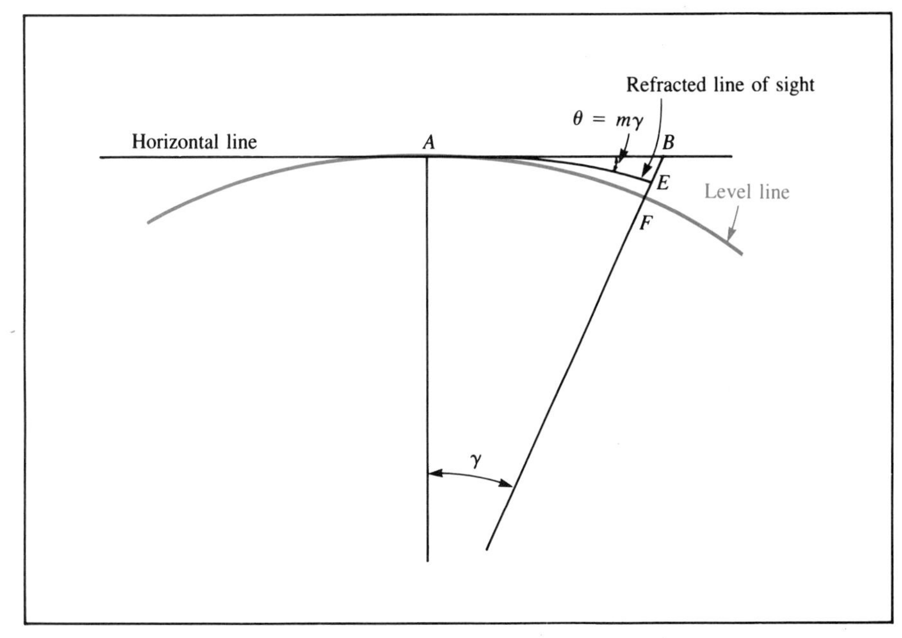

FIGURE 4.4 Earth's curvature and atmosphere refraction

From Eq. (4.1)

$$C = \frac{d^2}{2R}$$

Hence

$$BE \approx 2mC \approx 0.14\,C \tag{4.5}$$

From Eq. (4.4), note that the magnitude of refraction is proportional to the square of the distance d. Although it varies with atmospheric conditions, the magnitude of refraction is frequently approximated to be 14% of the effect of curvature, as shown in Eq. (4.5). Refraction can be considered to reduce the effect of curvature by this percentage. The combined effect of earth's curvature and atmospheric refraction in feet can be closely approximated by the expression

$$C + R = 0.574M^2 \tag{4.6}$$

where M is the length of sight in statute miles, or by

$$C + R = 0.021F^2 \tag{4.7}$$

where F is the length of the sight in thousands of feet. For sight distances of 100, 200, 300, and 500 ft, $C + R$ amounts to 0.0002, 0.0008, 0.0019, and 0.0052 ft respectively.

For metric units a convenient equivalent equation with $(C + R)$ in meters and the sight distance K in kilometers is as follows:

$$C + R = 0.0675\,K^2 \tag{4.8}$$

4.4 BASIC PRINCIPLE OF A LEVEL

The basic instrument used to measure differences of elevation is the *level*. Although of many types and designs, a level consists essentially of a telescope for sighting and a leveling device for maintaining the line of sight in a horizontal position. A level is constructed so that the line of sight of the telescope is perpendicular to its vertical axis, as shown in Figure 4.5. After a level has been properly set up on a tripod, its vertical axis should coincide with the direction of gravity and the line of sight should be horizontal. As the telescope is rotated about its vertical axis, the line of sight moves in a horizontal plane.

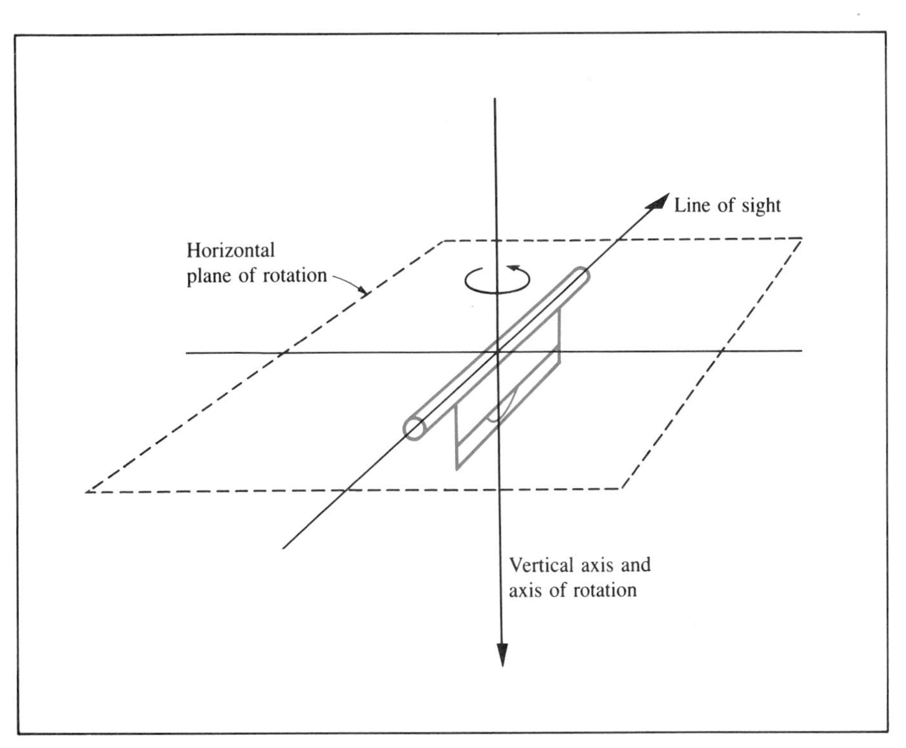

FIGURE 4.5 Basic principle of a level

4.5 BUBBLE TUBE

A *bubble tube,* as shown in Figure 4.6, is used in many levels to establish a horizontal line. The bubble tube is a glass vial of uniform cross-section. Its inside, upper surface is accurately ground in longitudinal section to the arc of a circle of specific radius. The tube is nearly filled with ether or some nonfreezing liquid, the remaining volume being a vapor space called the *bubble*. The buoyancy of the liquid lifts the bubble to a position symmetrical with the highest point in the tube. Since this highest point is on the arc of a circle that lies in a vertical plane, the tangent at that point will be truly horizontal. This tangent is called the *axis of the bubble tube*. The location of the highest point in the tube is indicated by divisions marked symmetrically on both sides of the point. Thus when the bubble is centered on the division marks, as shown in Figure 4.6a, the axis of the bubble tube is in a horizontal position.

The *sensitivity* of the bubble tube is determined by the radius of the circular arc and is expressed by the seconds of arc subtended by one division of the bubble tube. For example, suppose that the division marks on a bubble tube are spaced 2 mm apart and that for each division mark the bubble is off-centered, the axis of the bubble tube is inclined from the horizontal by 30

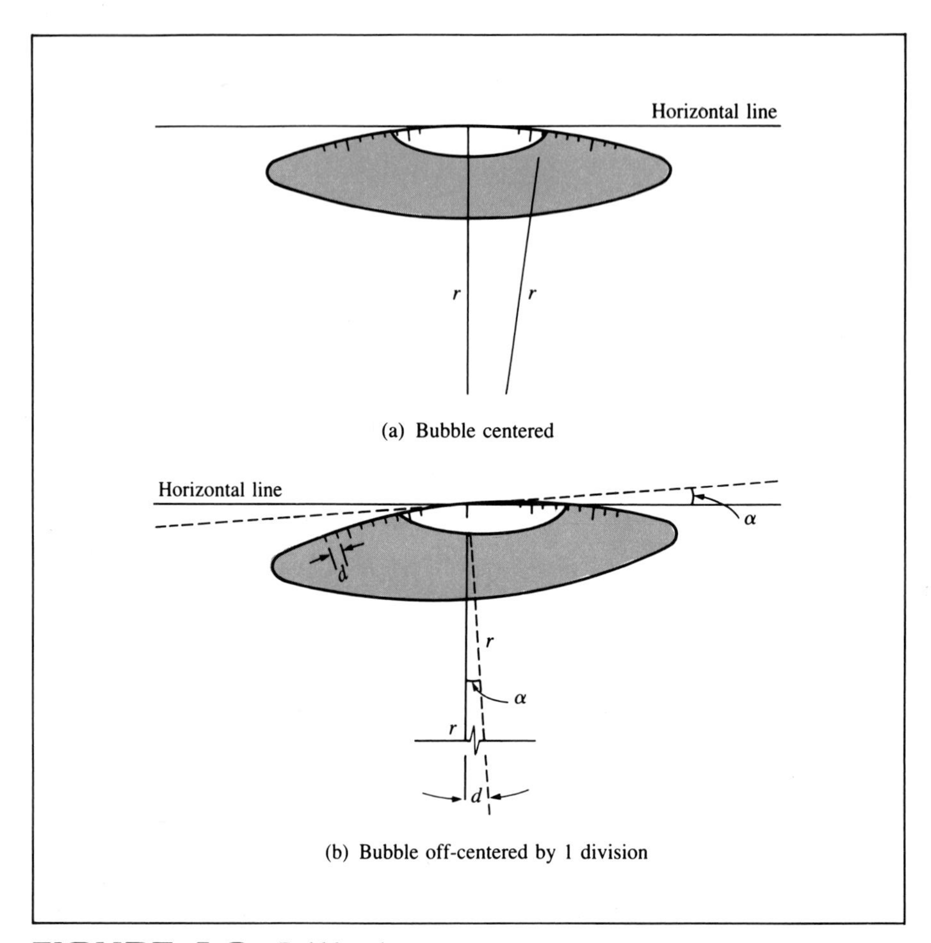

FIGURE 4.6 Bubble tube

seconds. Such a bubble tube is said to have a sensitivity of 30 seconds per 2-mm division. Figure 4.6b shows that the following relationship exists:

$$r = \frac{d}{(0.4848) \times 10^{-5}\alpha''} \tag{4.9}$$

where d is the spacing between division on the bubble tube expressed in a linear unit, r is the radius of curvature of the bubble tube in the same linear unit, and α'' is the angular displacement per division expressed in seconds of arc. For example, for a bubble tube with a sensitivity of 30″ per 2-mm division, its radius of curvature amounts to 13,751 mm, or 13.751 m.

Equation 4.9 indicates that if the radius of the bubble tube is large, a small angular movement of the tube will be accompanied by a large linear displacement of the bubble. For a given length of bubble division d, the radius of the level vial will vary inversely with the sensitivity expressed in seconds. Furthermore, a 10″ bubble tube is twice as sensitive as a 20″ bubble tube.

4.6 CIRCULAR BUBBLE VIAL

A *circular bubble vial,* as shown in Figure 4.7, is used in many modern levels to approximately establish a horizontal plane. When the circu-

(a) Bubble off-centered

(b) Bubble centered

FIGURE 4.7 Circular bubble vial

lar bubble vial is mounted on an inclined surface, the bubble is displaced from a circular mark at the center of the bubble vial, as shown in Figure 4.7a. By centering the bubble within the circular mark, as shown in Figure 4.7b, the surface is approximately leveled. Such a bubble vial is often referred to as a *bull's-eye level*.

4.7 TRIPODS

A *tripod* is a three-legged stand used to support a level or other surveying instrument during field measurements. It consists of a head and three legs that are fitted with pointed metal shoes. A survey instrument is usually secured to the tripod head by a threaded bolt.

Figure 4.8 shows two models of tripods. The extension leg tripod (a) provides more flexibility in setting up and can be more conveniently trans-

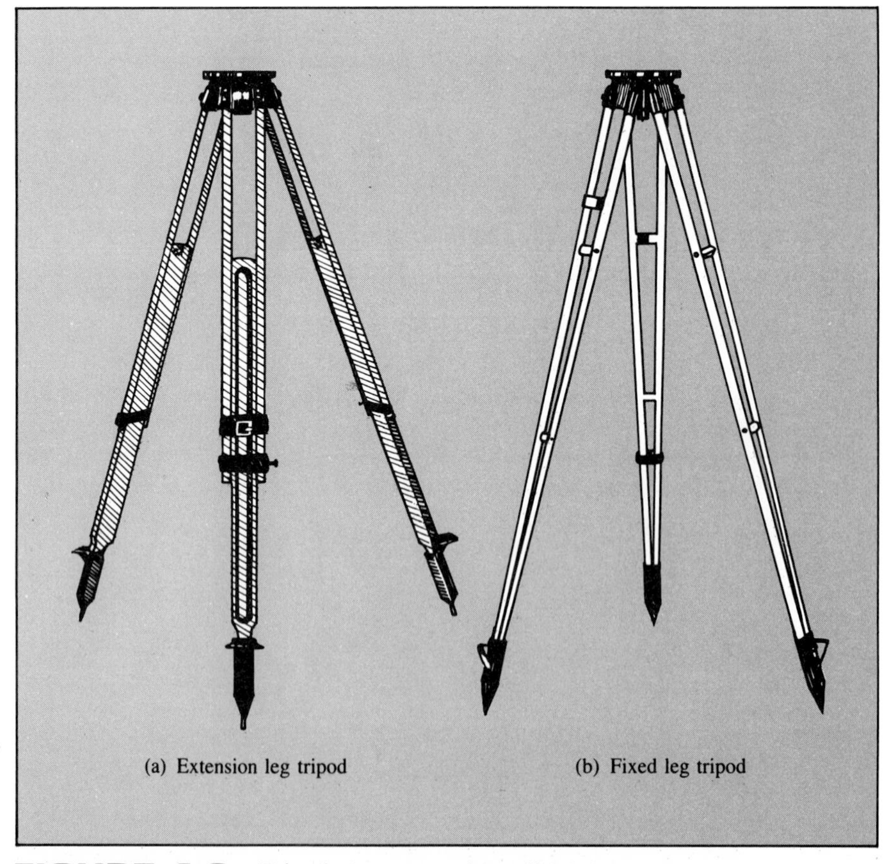

FIGURE 4.8 Tripods (courtesy of Keuffel & Esser Company)

ported. The fixed leg tripod (b) provides more rigidity and therefore stability during measurement. Both types are available in a special design called *wide frame,* which provides even greater stability. Tripods are usually made of either wood or aluminum.

Great care should always be taken to secure a stable setup of the tripod to assure the safety of the instrument mounted on it. On level terrain this is accomplished by having each leg form an angle of approximately 60° with the ground. When pressing on the tripod shoes, especially when the setup is over hard soil, caution must be exercised to apply the pressure parallel to the legs and not vertically to avoid breakage of the legs. If the setup is over a hard smooth surface, such as pavement or sidewalk, the tripod leg hinges should be more snugly tightened and the tripod shoes should be placed, if at all possible, in cracks to prevent the tripod from collapsing. When setting up on sloping ground, stability will be increased by having one leg pointed uphill and two legs downhill. If the tripod is being transported or is in storage, its head should always be covered with a plastic or metal cap to avoid accidental damage to either the head or the threads on the lock bolt.

4.8 DUMPY LEVEL

For many years the American-made *dumpy level* was popularly employed in leveling operations. Its spindle revolves in the socket of a leveling head controlled in position by four leveling screws. The leveling head is screwed to a tripod. At the lower end of the spindle is a ball-and-socket joint, which permits a flexible connection between the instrument proper and the foot plate. Hence when the foot screws are manipulated, the level is moved about this point as a center. Figure 4.9 shows a dumpy level with an internal focusing telescope providing an erect image. The telescope is rigidly attached to a horizontal bar that houses the level tube. The line of sight of the instrument is the line fixed by the intersection of the cross-wires and the center of the objective lens. When the instrument is in proper adjustment, the line of sight is parallel to the axis of the level tube. For the dumpy level shown in Figure 4.9, the sensitivity of the level tube is 20″ of arc per 2 mm graduation and the magnifying power of the telescope is 32 diameters. The meaning of the term *dumpy level* is associated with the dumpy appearance of older models of this type, which had inverting eyepieces and relatively short telescopes.

In setting up the dumpy level, the tripod legs are adjusted so that the leveling head base is approximately level, and then the legs are firmly pushed into the ground. On a steep hillside, two legs are placed on the downhill side and one uphill. The foot screws are operated in diagonal pairs, and in each pair one screw is turned inward and the other outward, the proper direction being determined by the fact that the bubble moves in the same direction as

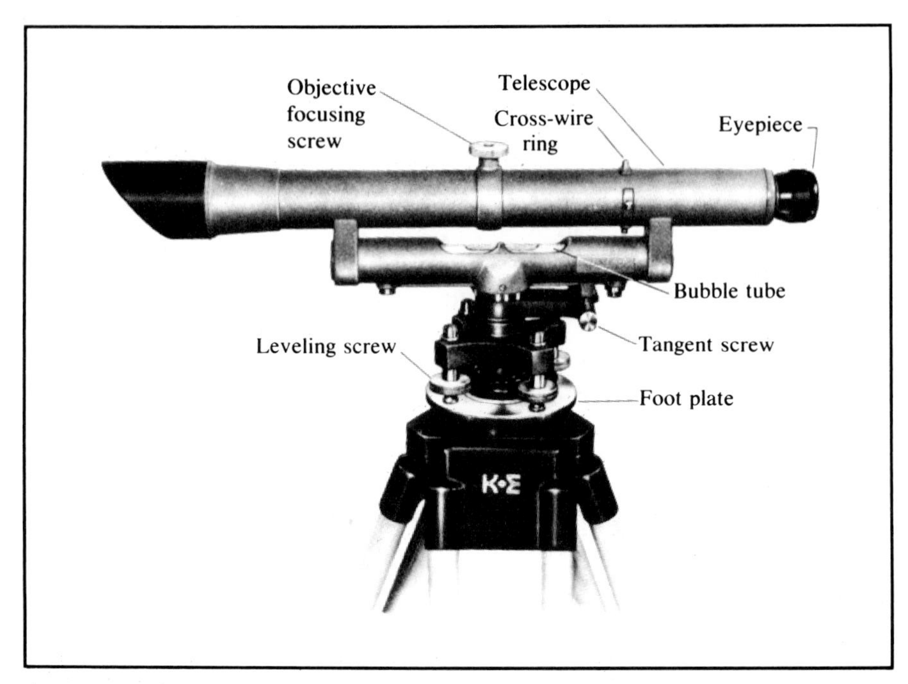

FIGURE 4.9 A dumpy level (courtesy of Keuffel & Esser Company)

the left thumb. If one screw is turned more rapidly than the other, either the pair will bind or the leveling head will become unstable and may wobble from side to side. Either condition is corrected by turning one screw only until the proper bearing of both has been re-established.

Figure 4.10 illustrates the procedure for leveling an instrument with four foot screws. The telescope is rotated so that the bubble tube is lined up with a diagonal pair of foot screws—1 and 3 as shown in (a). The level bubble is centered by turning foot screws 1 and 3 simultaneously. The level bubble moves in the same direction as the turning motion of the left screw (1). The telescope is then rotated 180° so that the bubble tube is again aligned with foot screws 1 and 3 as shown in (b). If the bubble tube is properly adjusted, the bubble should resume the center position without any operator assistance. If the bubble is off-centered by a small amount, it should be brought back toward the center by one-half the distance. If the bubble is off-centered by more than 1 division in this second position, the bubble needs adjustment. The adjustment is effected by bringing the bubble back halfway with the adjusting screw at one end of the bubble tube. The bubble is then brought to center with the foot screws. Such an adjustment, however, should be accompanied by a peg test, which is described in Section 4.15.

After the bubble has been centered using foot screws 1 and 3, the procedure described above is repeated with the bubble tube aligned with the second pair of foot screws, 2 and 4, as shown in (c) and (d).

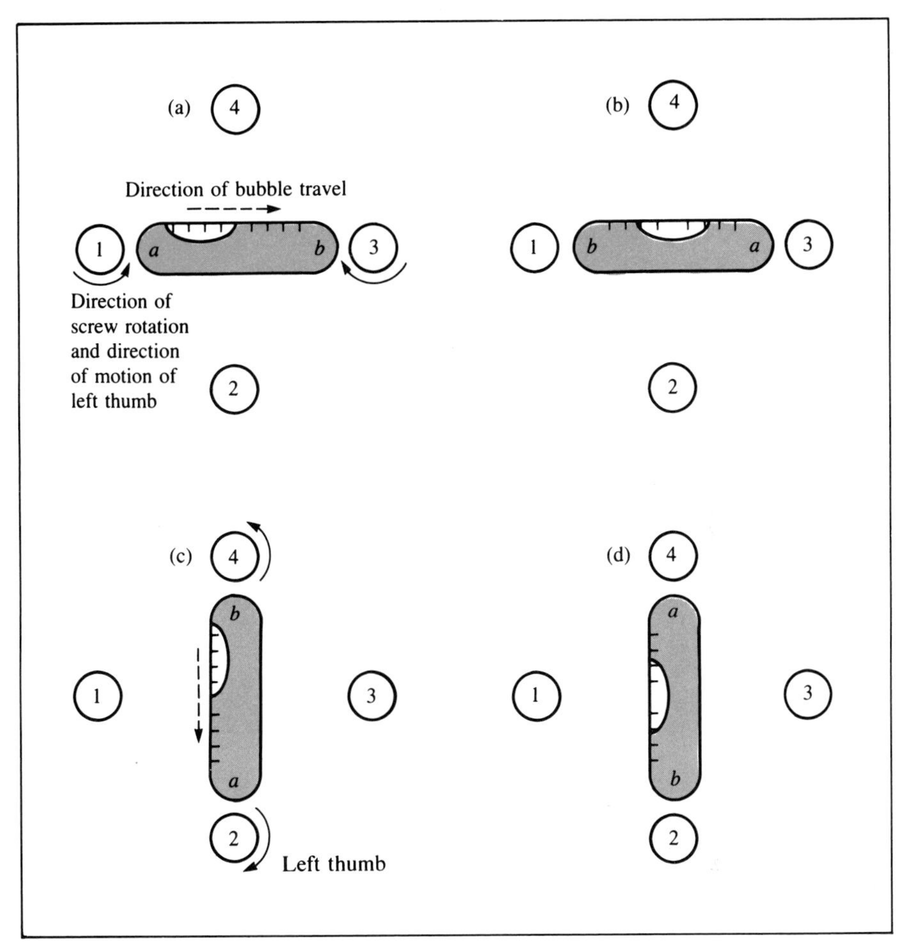

FIGURE 4.10 Leveling an instrument with four foot screws

The entire procedure using both sets of foot screws should be repeated until the bubble remains centered in both alignments. Once this is achieved, the instrument is properly leveled and the bubble should remain centered when the telescope is rotated about its vertical axis.

4.9 TILTING LEVEL

A *tilting level* has a telescope that can be tilted about a horizontal axis. This design enables the operator to quickly and accurately center the bubble and thus bring the line of sight into a horizontal plane. Figure 4.11 shows a level instrument of this type. A tilting screw is provided to raise or

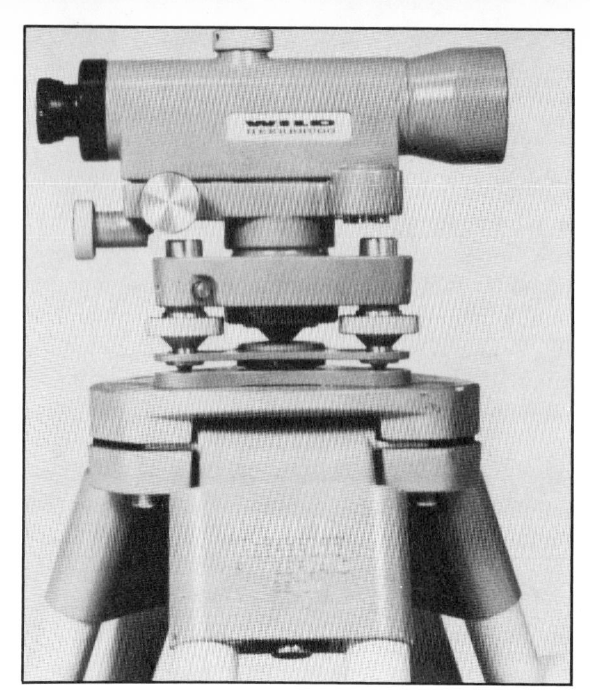

FIGURE 4.11 Wild N10 Tilting Level (courtesy of Wild Heerbrugg Ltd.)

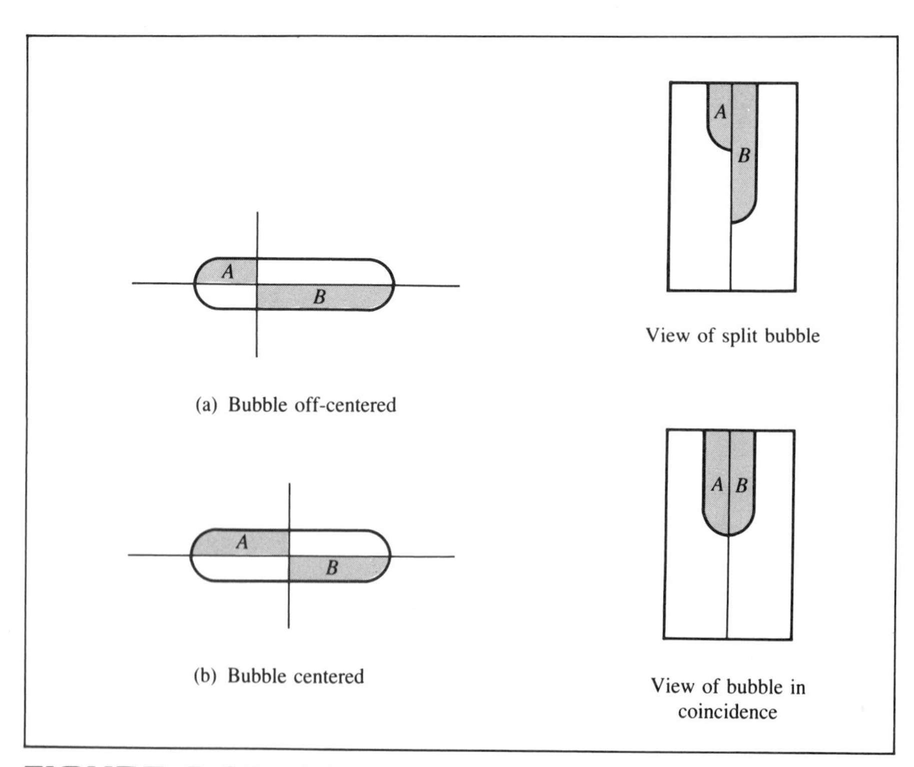

FIGURE 4.12 Coincidence-type bubble

lower the eyepiece end of the telescope. The bubble tube is enclosed in a metal cover. A split image of the two opposite ends of the bubble is viewed optically through a small microscope situated next to the eyepiece of the telescope. When the bubble is off-centered, a split image of the two ends of the bubble is seen through the viewing microscope, as shown in Figure 4.12a. As the bubble moves in its tube, the images of the two parts of the bubble move in opposite directions. When the bubble is correctly centered, the two images coincide to form a continuous U-shaped curve, as shown in Figure 4.12b. This arrangement permits the accurate centering of the bubble

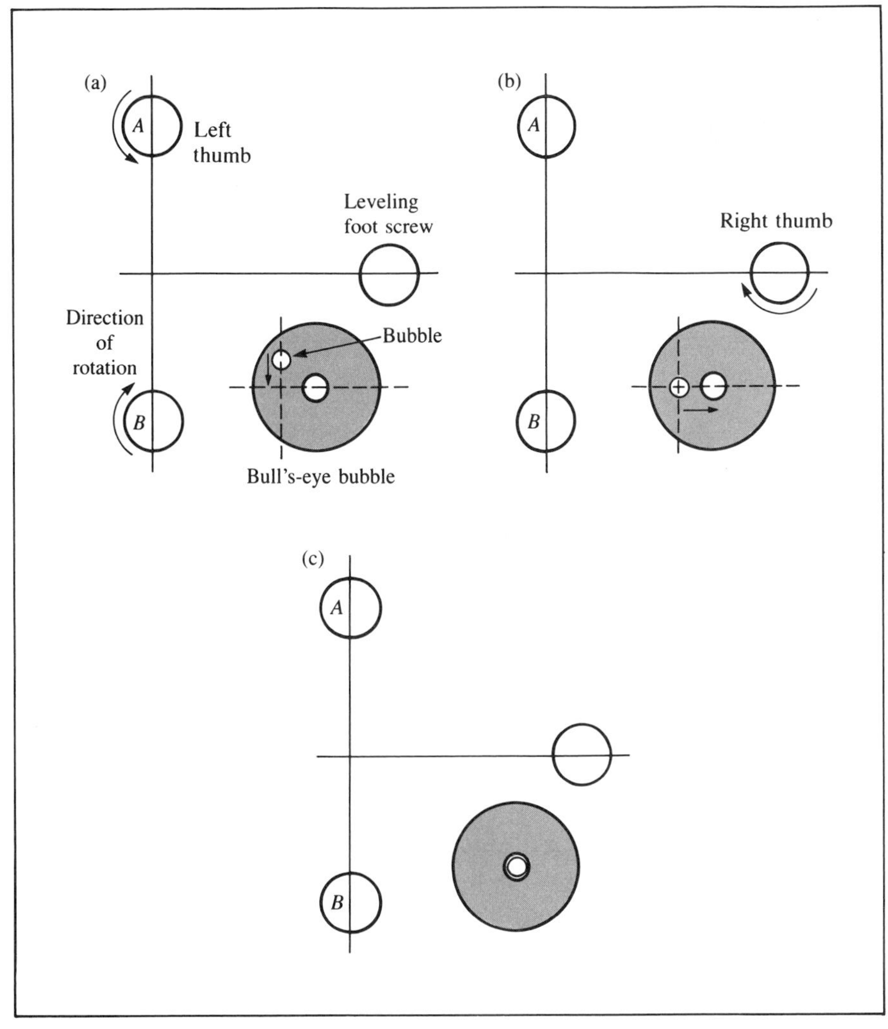

FIGURE 4.13 Leveling an instrument with three foot screws and a bull's-eye level

from the normal observing position of the operator. There is no need to walk about the level in order to view the bubble.

Such a level instrument is first approximately leveled by means of a bull's-eye level. The split bubble is then brought into coincidence by turning the tilting screw. Once coincidence is achieved, the line of sight should be in a horizontal position.

The instrument shown in Figure 4.11 is supported by three foot screws, as is the case with most modern survey instruments. Figure 4.13 illustrates the procedure used to level an instrument with three foot screws and a bull's-eye level. The movement of the level bubble can be divided into two mutually perpendicular directions. One direction is parallel to an imaginary line joining any two of the foot screws, say *A* and *B*. The second direction is then parallel to a perpendicular line from the third foot screw to the line joining the two other foot screws. The bubble is first moved using the two screws *A* and *B*, as shown in (a). It is then centered by using the third foot screw as shown in (b). After acquiring some experience, an operator can center the bubble by turning all three foot screws simultaneously.

4.10 AUTOMATIC LEVEL

One of the most significant improvements in leveling instrumentation has been the *automatic,* or *self-leveling, level.* The distinctive feature of this type of level is an internal compensator that automatically makes horizontal the line of sight and maintains it in that position through the application of the force of gravity. As soon as the instrument is leveled by means of a circular bubble, the movable component of the compensator swings free to a position that makes the line of sight horizontal. For the best performance of the compensator, which operates within a limited angular range from the horizontal, it is always essential to accurately center the circular bubble and to maintain it in good adjustment.

Automatic or self-leveling level instruments are accurate, fast, and easy to use. Their employment substantially increases the productivity of a leveling party, and they have been found to be acceptable for the highest grades of leveling work.

The Lietz B-2C Precision Automatic Level shown in Figure 4.14 has a compensator that operates within a range of ±10 minutes of arc from the horizontal. It keeps the line of sight horizontal within an accuracy of ±0.3 seconds. The circular level vial used for preliminary leveling has a sensitivity of 10 arc minutes for 2-mm movement. It also has a tangent screw that can be operated from either side of the instrument for rotating the telescope.

A wedge-shaped cross-hair (also called *reticule*) is commonly used in precision levels. The wedge shape on the left side of the reticule is intended for more precise centering of the horizontal cross-hair on a graduation line, as shown in Figure 4.15.

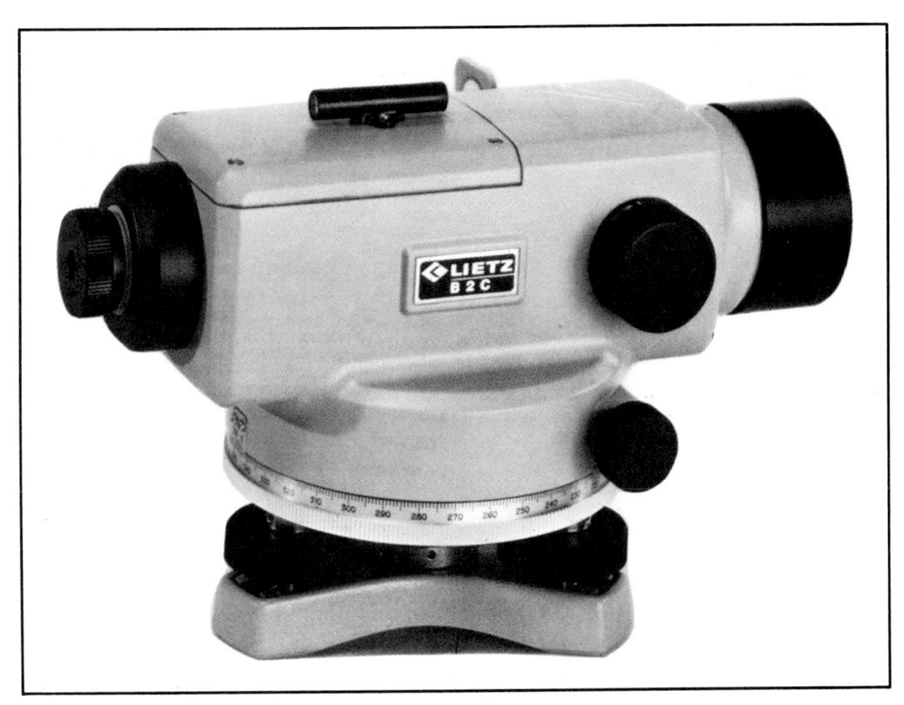

FIGURE 4.14 Lietz B-2C Precision Automatic Level (courtesy of The Lietz Company)

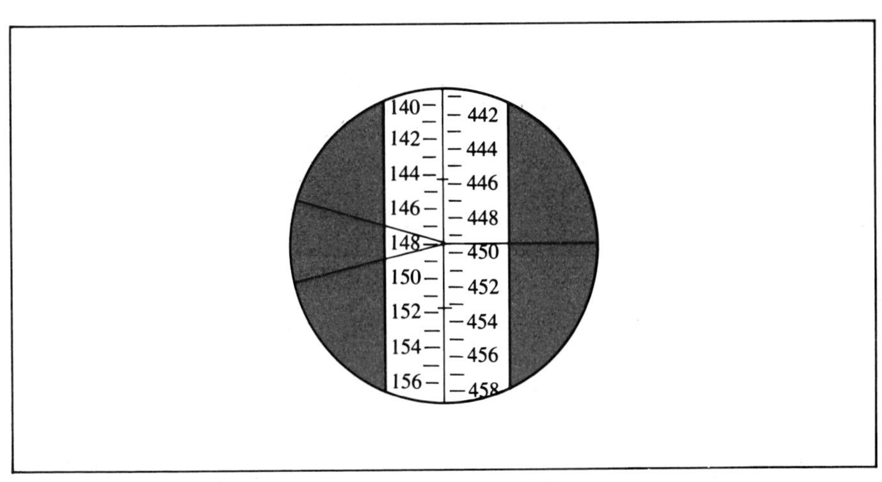

FIGURE 4.15 Use of wedge-shaped reticule (courtesy of Wild Heerbrugg Instruments, Inc.)

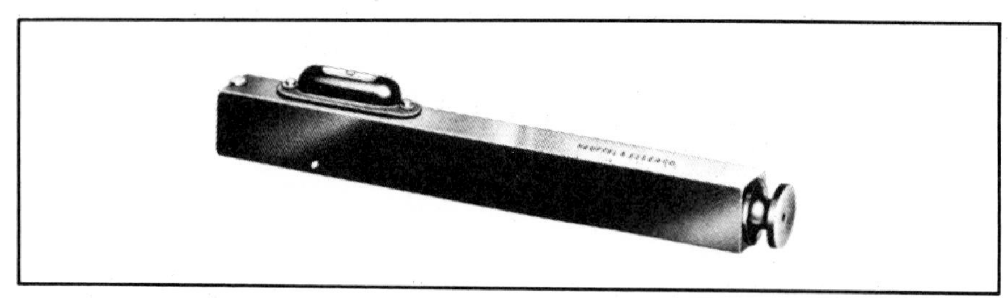

FIGURE 4.16 Hand level (courtesy of Keuffel & Esser Company)

4.11 HAND LEVEL

For rough leveling and for sight distances of not more than 50 ft, a hand level has much utility (see Figure 4.16). A *hand level* usually consists of a metal sighting tube about 5 to 10 inches long. It is equipped with a small level vial and the level bubble can be viewed through the eyepiece with the reticule. Hand levels are extremely useful in setting grade stakes to rough tolerances and for leveling the two ends of a tape in taping.

4.12 LEVEL RODS

Level rods are used to measure the vertical distance between a line of sight and a survey point, as shown in Figure 4.17. There are many different styles of level rods; among the most commonly used is the so-called *Philadelphia rod* shown in Figures 4.18 and 4.19. The Philadelphia rod is usually constructed of wood with a graduated metal scale mounted on the

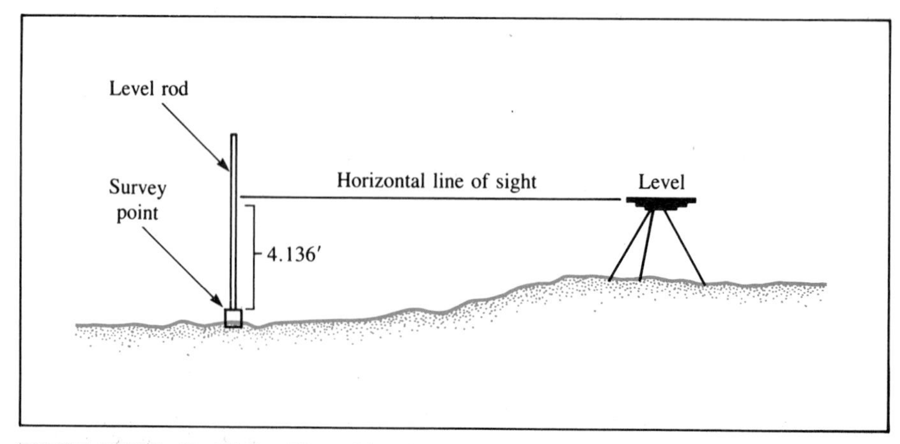

FIGURE 4.17 Use of level rod

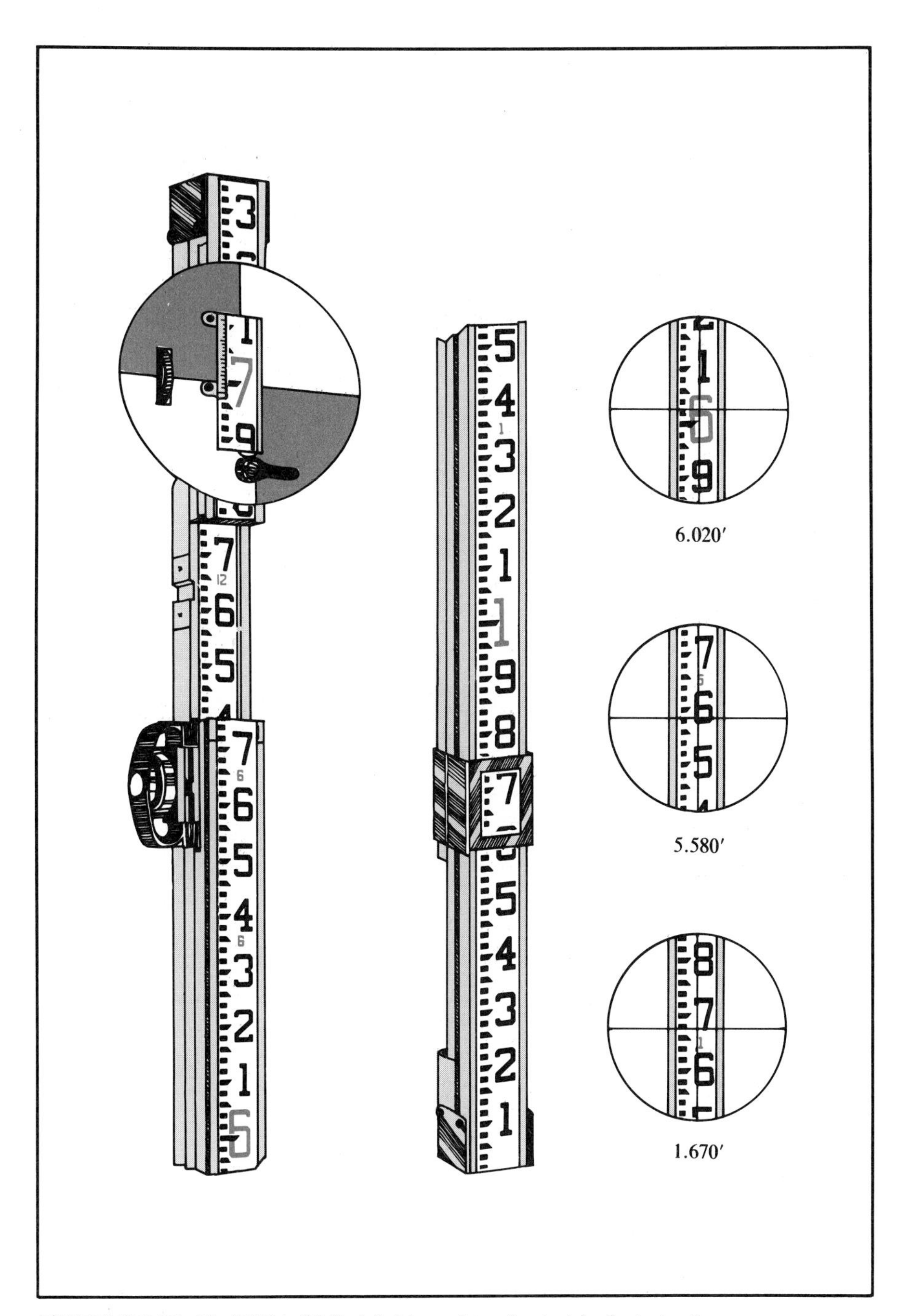

FIGURE 4.18 Philadelphia rod graduated in feet, tenths, and hundredths (courtesy of Keuffel & Esser Company)

FIGURE 4.19 Philadelphia rod graduated in meters, decimeters, and centimeters

rod face. It consists of two sections and usually measures 12 or 13 ft when fully extended and 6 or 7 ft when the two sections are closed. Foot rods are graduated in feet, tenths, and hundredths, with the smallest unit being 0.01 ft, as shown in Figure 4.18. Metric rods are graduated in meters, decimeters, and centimeters, with the smallest unit being 0.01 m, as shown in Figures 4.19 and 4.20. Some level rods, such as that shown in Figure 4.21, are equipped with a calibrating spring that automatically compensates for climatic variations on the wood while the metal scale maintains its accuracy.

During long distance sighting, when it is difficult to read the rod graduations accurately, a sighting target such as that shown in Figure 4.22 is fitted over the leveling rod. In this arrangement the person holding the rod would move the target up or down the rod according to signals from the

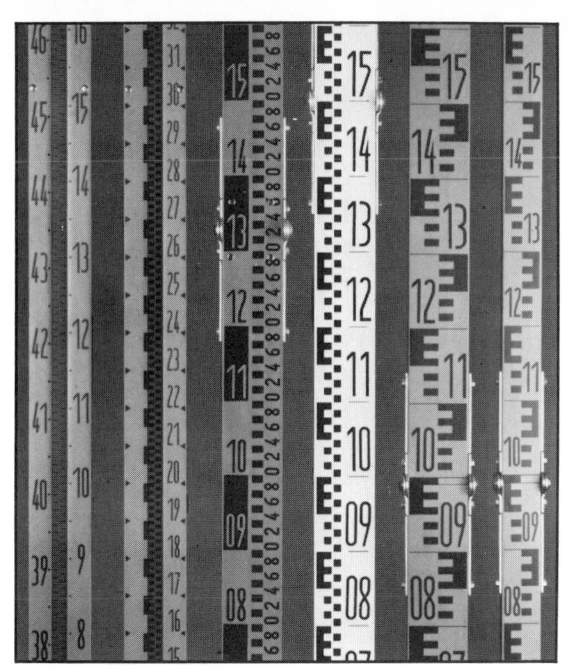

FIGURE 4.20 Metric rod graduations (courtesy of Carl Zeiss, Inc., Thornwood, N.Y.)

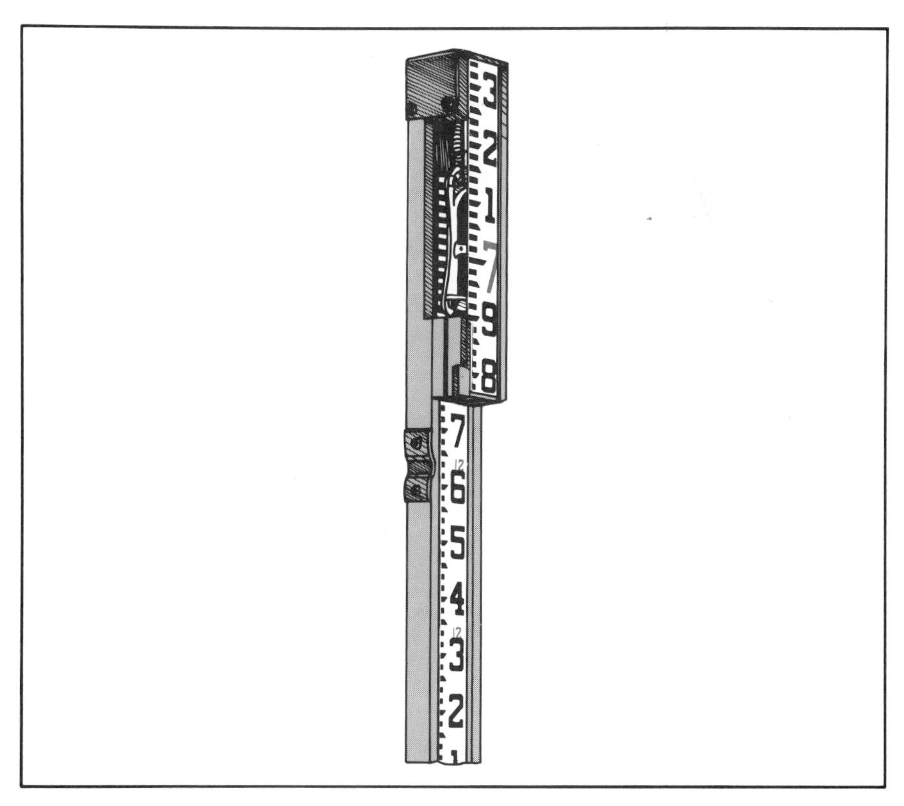

FIGURE 4.21 Level rod with calibrating spring (courtesy of Keuffel & Esser Company)

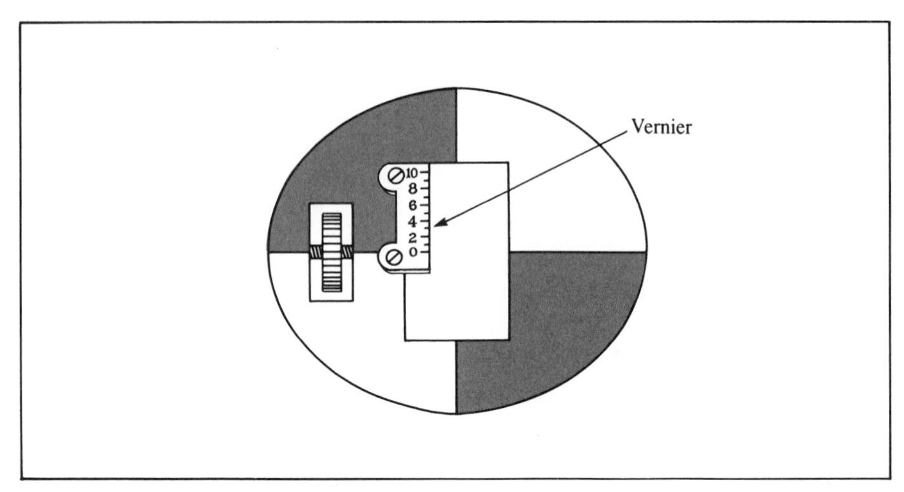

FIGURE 4.22 Sighting target for level rod

person operating the level instrument. When the target is positioned so that the horizontal cross-wire of the telescope coincides with the cross-mark on the target, the person holding the rod then reads the rod. Such a target is usually fitted with a vernier. The smallest whole unit on the rod graduation is read with the zero index on the vernier, and the fractional unit is read from

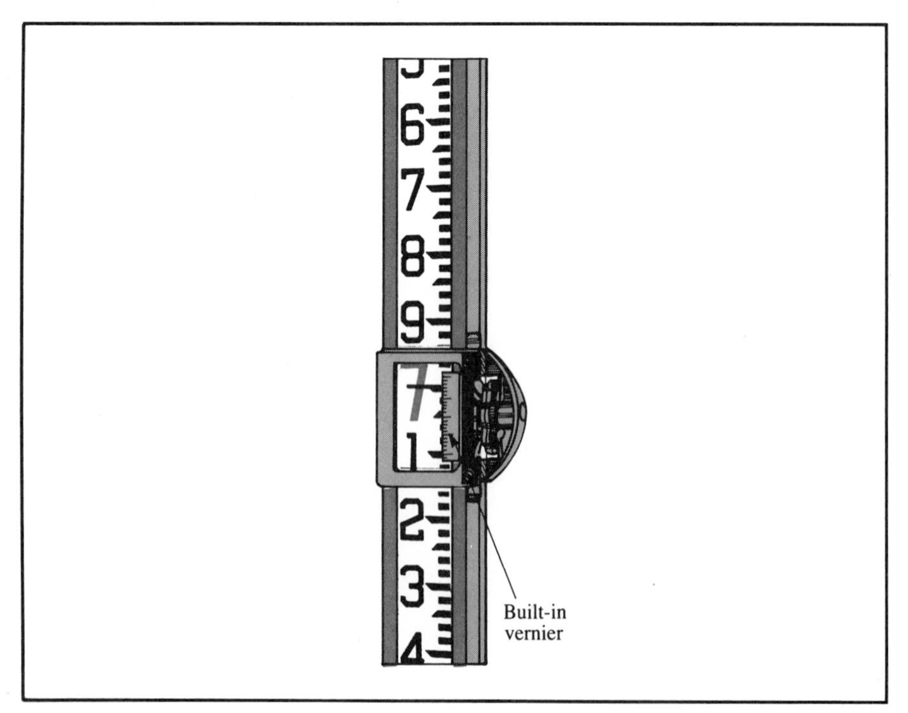

FIGURE 4.23 Scale on backside of a Philadelphia rod (courtesy of The Lietz Company)

the graduation mark on the vernier that coincides with a graduation mark on the rod (see Figure 4.23). The operational principle of a vernier will be discussed in the next section.

If the target must be positioned on the upper section of the rod as shown in Figure 4.18, it is clamped exactly on the 7-ft mark. The upper section is then lowered or raised till the target is lined up with the horizontal cross-hair. The rod must then be read using the vernier located on the backside of the rod clamp and the graduation scale mounted on the backside of the upper rod section. The scale on the backside is graduated so that the reading increases downward as shown in Figure 4.23. If the target is exactly 7 ft above the ground, the zero index on the clamp vernier should read 7 ft exactly. As the target is raised, the reading increases correspondingly, thus giving directly the height of the target above ground.

4.13 VERNIER

A *vernier* is a device commonly used to provide accurate reading of fractional units on a scale. Figure 4.24 illustrates the principle of a vernier with 10 divisions. Each division on the vernier is 9/10 of the dimension of the smallest unit on the main scale. Thus when the first graduation mark on the vernier coincides with 6.1 mark on the main scale, the zero index of the vernier is exactly at the 6.01 position of the main scale.

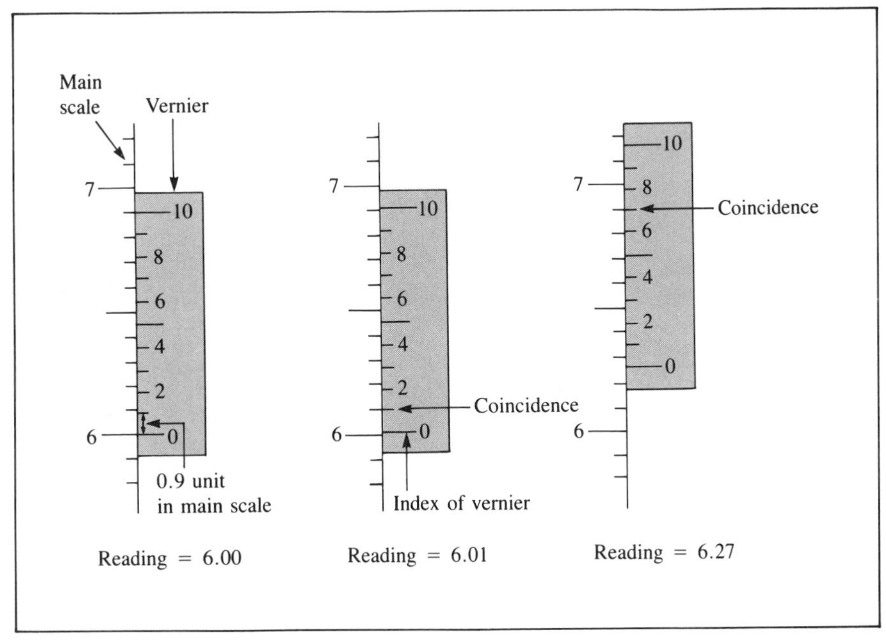

FIGURE 4.24 Principle of vernier

When using a vernier, the zero index of the vernier is used to read the main scale to the smallest graduation unit. The fractional unit is then read from the vernier graduation mark that is in coincidence with a graduation mark on the main scale.

Some verniers are graduated with more than 10 divisions. In general, let n be the number of divisions on the vernier. Then the vernier gives direct reading to $1/n$ of the smallest unit on the main scale.

4.14 DIFFERENTIAL LEVELING

Differential leveling is the most widely used method for determining difference of elevation. It is often referred to as *spirit leveling* because the bubble tubes of many older levels were filled with alcohol. Figure 4.25 illustrates the method of differential leveling for determining the elevation of a survey point P from an existing benchmark, called BM 1. A benchmark (BM) is a permanent object of known elevation.

Suppose that the distance between BM 1 and point P is so long that several intermediate setups are required. The length of sight is usually dictated by the accuracy with which the rod can be read and by the topographic relief. It may range from a few tens of feet to a few hundred feet.

Figure 4.26 shows the corresponding field notes. BM 1 is assumed to have a known elevation of 913.22 ft above a mean sea level datum. With the level set up at point A and the level rod held vertical at BM 1, a backsight (BS) reading of 9.42 ft is read on the rod. This reading is recorded in the backsight (BS) column in the field book and on the line that pertains to the point where the rod is located (that is, BM 1). The height of the instrument (HI) is then computed as the sum of the elevation of the BS station and the BS reading—that is, $HI = 913.22 + 9.42 = 922.64$ ft.

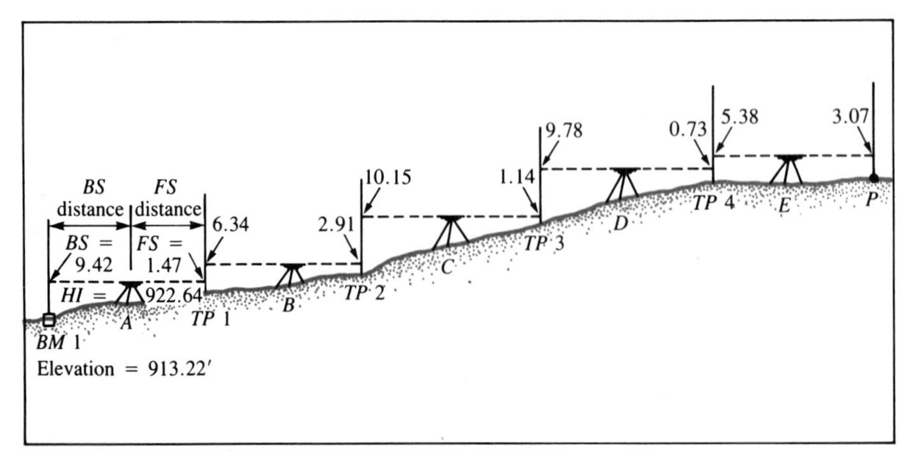

FIGURE 4.25 Differential leveling

Differential leveling
Turkey Run Creek to Route 24

From Points	BS	HI	FS	Elv.
BM	9.42	922.44		913.22
TP 1	6.34	927.51	1.47	921.17
TP 2	10.15	934.75	2.91	924.60
TP 3	9.78	943.39	1.14	933.61
TP4	5.38	948.04	0.73	942.66
P			3.07	944.97
ΣBS =	41.07	ΣBS =	9.32	−913.22
	−9.32			31.75
	31.75			
			check	

Locker No. 47

⊼ J.E. Slattery
Rod W.D. Fooks
June 1, 1983

BM 1 – Brass plug set on NW abutement of highway bridge over Turkey Run Creek on Illinois Route 45

FIGURE 4.26 Field notes for differential leveling

Next the rod is moved to a point located ahead of the level so that the foresight (*FS*) distance to this point from the level is approximately equal to the *BS* distance. This intermediate point is called a turning point (*TP*). It should be a stable object with a well-defined elevation point, such as a pointed object or the flat level surface of a sidewalk. The turning points are numbered sequentially. With the rod held at *TP* 1, a foresight reading of 1.47 ft is obtained. The *FS* reading is then subtracted from the *HI* to give the elevation of *TP* 1—that is, elevation of *TP* 1 = 922.64 − 1.47 = 921.17 ft.

This procedure is then repeated with the level set up at point *B*. Here a *BS* reading of 6.34 ft is first obtained with the rod held at *TP* 1, and then a *FS* reading of 2.91 is obtained with the rod held at *TP* 2. The *HI* at point *B* is 921.17 + 6.34 = 927.51 ft, and the elevation of *TP* 2 is 927.51 − 2.91 = 924.60 ft.

With the level set up at point *C*, a *BS* reading of 10.15 on *TP* 2 and a *FS* reading of 1.14 on *TP* 3 are obtained. The *HI* is computed as 924.60 + 10.15 = 934.75 ft, and the elevation of *TP* 3 is 934.75 − 1.14 = 933.61 ft.

With the level set up at point *D*, a *BS* reading of 9.78 on *TP* 3 and a *FS* reading of 0.73 on *TP* 4 are obtained. The *HI* is computed as 933.61 + 9.78 = 943.39 ft, and the elevation of *TP* 4 is 943.39 − 0.73 = 942.66 ft.

Finally, with the level at point *E*, a *BS* reading of 5.38 on *TP* 4 and a *FS* reading of 3.07 on point *P* are obtained. The *HI* is computed as 942.66 + 5.38 = 948.04 ft, and the elevation of point *P* is 948.04 − 3.07 = 944.97 ft.

As soon as a line of levels is completed, the correctness of the field calculations as shown in the field notes in Figure 4.26 should be checked. The difference in elevation between the beginning and end points can be computed in two ways as follows:

1. Elevation difference = elevation of point *P* − elevation of *BM* 1
= 944.97 − 913.22
= 31.75 ft

2. Elevation difference = Σ *BS* readings − Σ *FS* readings
= 41.07 − 9.32
= 31.75 ft

This arithmetical check is called a *page check.* It merely confirms the correctness of calculating heights of the instrument and the elevations of turning points. Errors or mistakes in reading the rod can not be disclosed by a page check.

4.15 SYSTEMATIC ERRORS

There are four major sources of systematic errors in differential leveling. These are:

1. Inclination of line of sight due to curvature of the earth and atmospheric refraction.
2. Inclination of line of sight due to maladjustment of the level.
3. Changes on the dimension of the graduated scale due to temperature.
4. The rod not being held plumb.

EARTH CURVATURE AND REFRACTION

The problems of earth curvature and refraction and the combined effects on the line of sight have already been discussed in detail in Section 4.3. At a sight distance of 300 ft, this source of error amounts to only 0.002 ft. Moreover, it can be rendered completely negligible by keeping the *BS* distance approximately equal to *FS* distance for each set up. Figure 4.27, though grossly exaggerating the curvature effect, shows that when the *BS* distance equals the *FS* distance, the error introduced by earth curvature and refraction does not affect the measured difference in elevation. Let Δh rep-

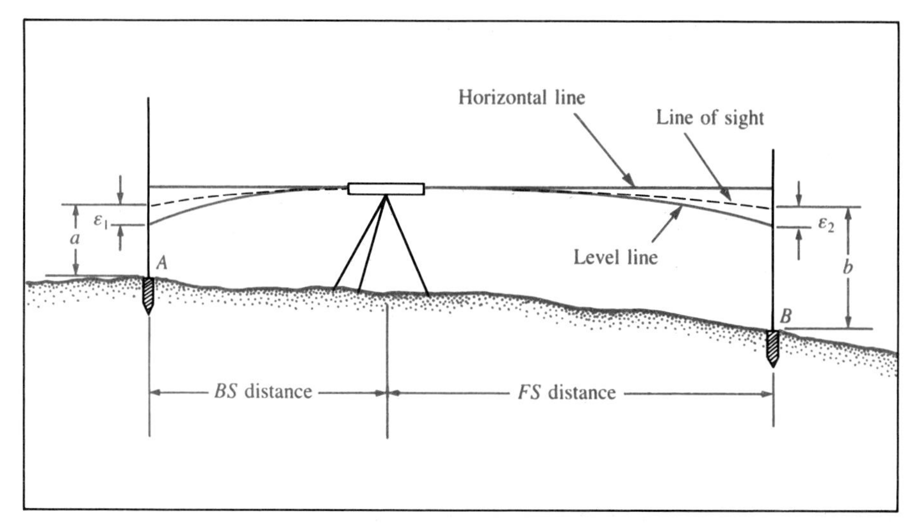

FIGURE 4.27 Inclination of line of sight by curvature and refraction

resent the difference in elevation between points A and B in Figure 4.27. Let a be the BS reading with the rod at point A, and ε_1 be the reading error caused by earth curvature and refraction. Similarly, let b represent the FS reading with the rod at point B, and ε_2 be the corresponding error. Then

$$\Delta h = (a - \varepsilon_1) - (b - \varepsilon_2)$$

That is

$$\Delta h = (a - b) - (\varepsilon_1 - \varepsilon_2) \tag{4.10}$$

But $\varepsilon_1 = \varepsilon_2$ when the BS distance equals the FS distance. Therefore, the second term in Equation 4.10 becomes zero and the correct difference in elevation is obtained by the following expression:

$$\Delta h = a - b \tag{4.11}$$

MALADJUSTMENT OF THE LEVEL

When the line of sight of a level is not perfectly parallel to the axis of the level bubble, the line of sight is actually inclined even though the level bubble is perfectly centered (see Figure 4.28). The same condition exists when the compensator of an automatic level consistently causes the line of sight to be either above or below the horizontal position. Again, the error introduced by this source can be completely eliminated by balancing the BS and FS distances.

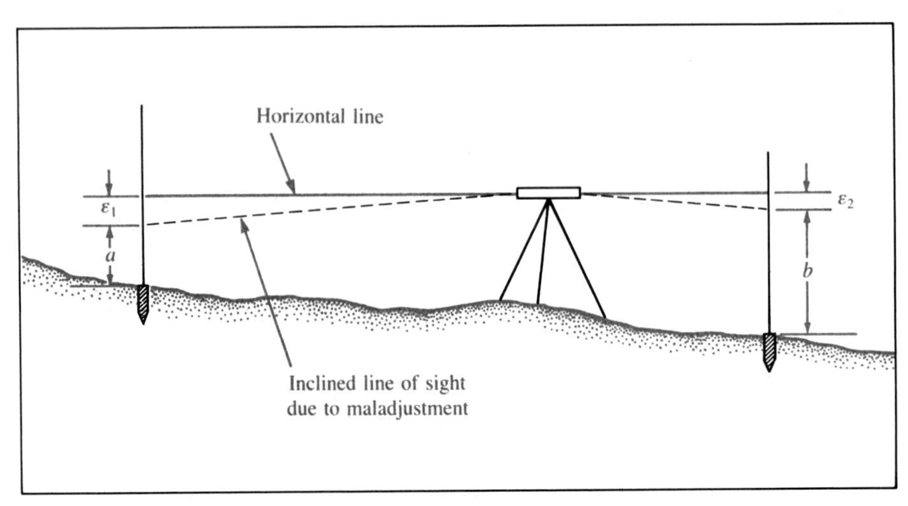

FIGURE 4.28 Inclination of line of sight due to maladjustment

Even though the errors introduced by the inclination of line of sight may be relatively small for short sight distances, the combined effect can be appreciable when all the *FS* distances are consistently longer than the *BS* distances, or vice versa.

EXAMPLE 4.1

Calculate the error in the preliminary elevation of a point resulting from leveling a distance of 24,000 ft up a uniform slope with backsights of 300 ft and foresights of 100 ft. Assume that when the bubble is centered, the line of sight is consistently inclined upward 0.004 ft per 100 ft. Disregard all other sources of error.

SOLUTION

The total error in the backsight readings is (60)(3)(0.004) or +0.72 ft. The total error in the foresight readings is (60)(1)(0.004) or +0.24 ft. Hence the elevation of the end point would be too high by 0.48 ft.

EXAMPLE 4.2

Include $C + R$ in the solution of Example 4.1.

SOLUTION

The constant error due to curvature and refraction alone per setup would be (see Section 4.3) 0.0019 − 0.0002 or 0.0017 ft. For 60 setups the accumulative effect would be 0.10 ft. The combined error from these two sources (maladjustment and $C + R$) is +0.58 ft.

The proper alignment of the line of sight of a level can be checked by a procedure called *peg test* (see Figure 4.29). Two wooden construction stakes are driven into the ground at a distance of approximately 300 ft apart. The topography of the area should be relatively flat. With the level set up at the mid-point between the stakes, a *BS* reading *a* is obtained with the rod at stake *A* and a *FS* reading *b* is obtained at stake *B*. Assuming that the error due to the inclined line of sight is ε_1, the correct difference in elevation (Δh) between the two stakes is computed as follows:

$$\begin{aligned} \Delta h &= (a + \varepsilon_1) - (b + \varepsilon_1) \\ &= (a - b) \end{aligned}$$

Next the level is moved to stake *A* so that its eyepiece is just in front of the rod held at stake *A*. The rod at *A* is then read by looking backward through the objective lens. In this manner the cross-hair would not be visible. The rod can be accurately read by holding a pencil as pointer in front of

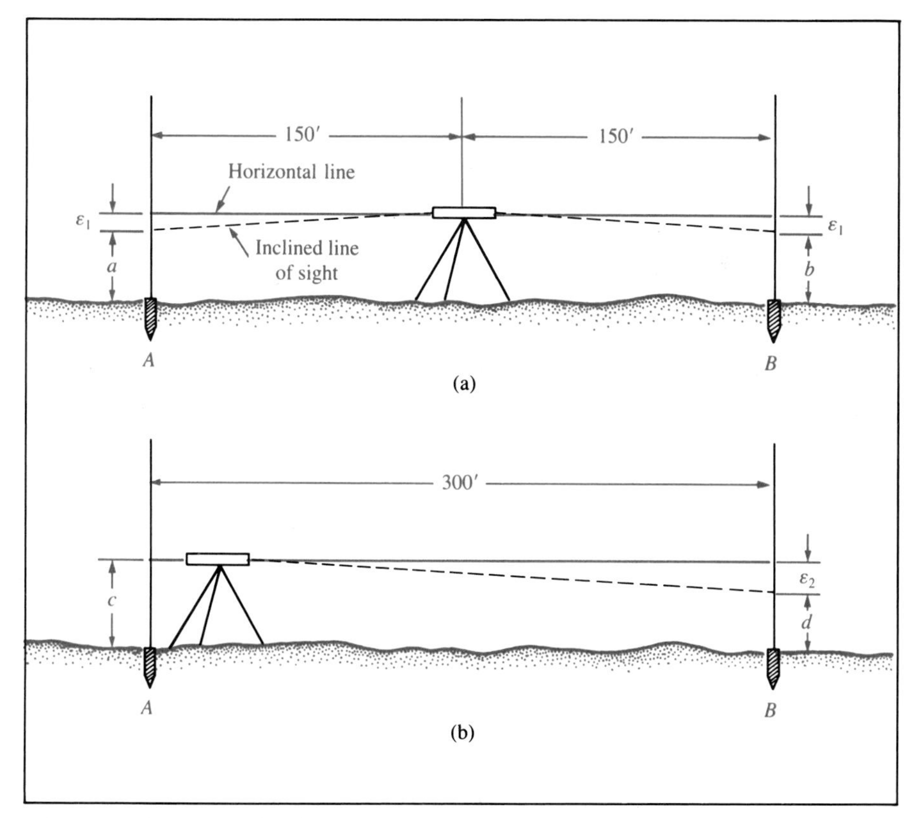

FIGURE 4.29 Peg test

the graduation. Suppose the rod reading is c. Then a FS reading d is obtained with the rod held at stake B. Let ε_2 represent the error in the FS reading due to inclination of the line of sight. Then ε_2 can be computed from the following relationship:

$$\Delta h = c - (d + \varepsilon_2) = a - b$$

and

$$\varepsilon_2 = (c - d) - (a - b) \tag{4.12}$$

The line of sight is inclined downward if ε_2 is positive, and upward if negative.

For example, let the four rod readings be as follows: $a = 2.04$ ft, $b = 5.16$ ft, $c = 4.73$ ft, and $d = 7.81$ ft. Then from Eq. (4.12)

$$\begin{aligned}\varepsilon_2 &= (4.73 - 7.81) - (2.04 - 5.16) \\ &= 0.04 \text{ ft}\end{aligned}$$

Therefore the line of sight is inclined downward 0.04 ft over a distance of 300 ft, or $0.04/3 = 0.013$ ft per 100 ft.

TEMPERATURE EFFECT

Temperature affects the dimension of the metal scale on a level rod in the same manner that it does on a steel tape (see Chapter 3). When the metal scale is bonded to a wooden rod, as is the case with rods commonly used in engineering and boundary surveys, the dimension of the scale is further affected by the expansion or contraction of the wood due to temperature.

The temperature range encountered during the time required for running a level line is relatively small. The effect of temperature can usually be ignored except when very high precision is required. However, this source of error must be recognized when surveys conducted during different periods of the year are compared with each other. For example, suppose that the elevation of a building foundation is surveyed once during the summer and once during the winter to measure the amount of settlement. Suppose that the average field temperature is 90°F during the summer survey and 20°F during the winter survey. The difference in elevation between the foundation and the reference benchmark is 50 ft. The change in the measured elevations due to temperature alone can be computed using Eq. (3.5):

$$\begin{aligned}\text{change in elevation} &= 0.00000645(T_1 - T_0)L \\ &= 0.00000645(90° - 20°) \times 50 \text{ ft} \\ &= 0.023 \text{ ft}\end{aligned}$$

Error due to temperature can be reduced by the use of invar rods. The graduated scale of this type of rod is made of invar steel, which has a coefficient of expansion about 1/30 that of regular carbon steel.

THE ROD NOT PLUMB

If the rod is not plumb when a reading is taken, a positive systematic error results. The magnitude depends on the size of the rod reading; that is, it is greater near the top than near the bottom of the rod. Its value may be estimated by use of Eq. (3.4).

EXAMPLE 4.3

Calculate the error in an observed rod reading of 10.00 ft if the rod is out of plumb (in the plane joining rod and instrument) by 6 in. at that height.

SOLUTION

$$\text{error} = \frac{\Delta h^2}{2s} = \frac{(0.5)^2}{2 \times 10} = +0.01 \text{ ft}$$

This source of error is minimized by carefully plumbing the rod for all readings. This is done by the rodholder standing squarely behind the rod and balancing it between the fingertips of both hands (see Figure 4.30). When the wind is blowing, this is more difficult to do. For windy conditions and especially when readings are taken near the top of the rod, the rodholder should "rock the rod" slowly toward and away from the level, and the level operator should take the *lowest* reading. However, the rod—especially a flat bottom rod—should not be rocked when its bottom is resting on a flat surface. For more careful work a rod level is used to plumb the rod (see Figure 4.31).

4.16 RANDOM ERRORS

The principal sources of random errors that affect the accuracy of leveling results are discussed below.

THE BUBBLE NOT CENTERED

If the bubble is not centered when the rod is read, an accidental error in the reading results. The magnitude of the error depends on the sensitivity of the bubble tube. Thus for the usual bubble tube, which has a sensitivity of 15″ per 2 mm, the line of sight is displaced 0.02 ft at a distance of 300 ft when

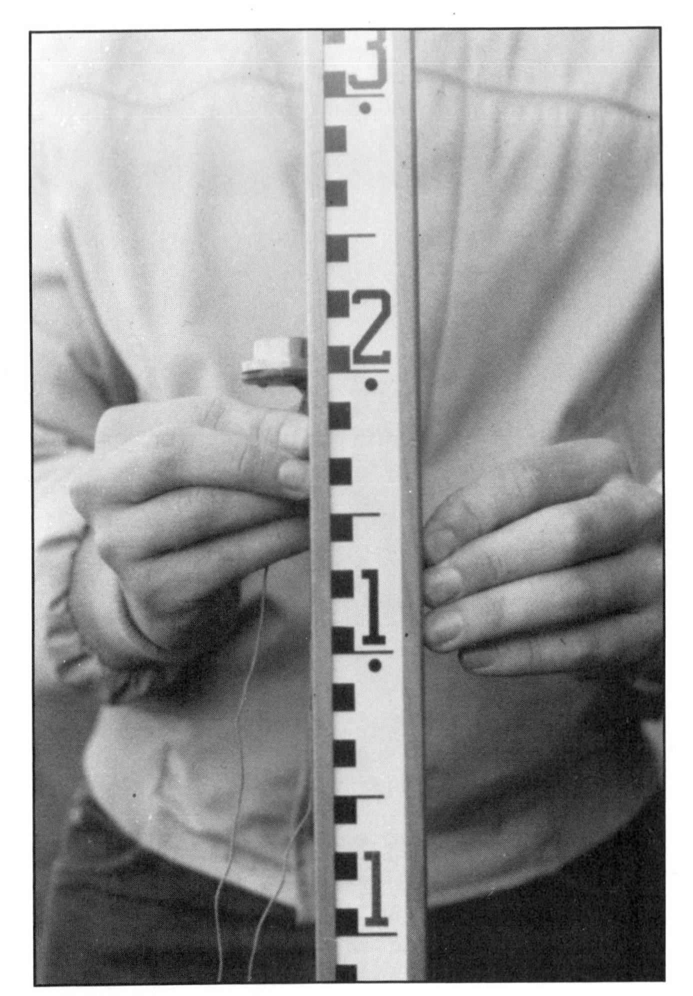

FIGURE 4.30 Holding a level rod plumb

the bubble is one division off center. This source of error is minimized by carefully centering the bubble just before the rod reading is taken. On sunny days direct sunlight on the tripod may cause uneven expansion resulting in the level bubble gradually drifting away from the centered position.

INCORRECT READING OF THE ROD

This source of error is due to the fact that the eye is not able to judge exactly where the horizontal cross-wire apparently cuts the rod. It is an accidental error and its magnitude depends on the distance to the rod, the quality of the telescope, the weather conditions, and so on. The error should not exceed 0.005 ft at a distance of 300 ft, and this may be taken as about the maximum distance for good results of ordinary precision. If

FIGURE 4.31 Rod level

weather conditions are adverse, the length of sight should be shortened accordingly.

Reading errors due to estimating a fractional part of a division on a level rod can be minimized by using a *parallel plate micrometer*. Figure 4.32 shows such a device mounted in front of the telescope of an automatic level. A micrometer screw is used to turn a parallel plate prism, which deflects the line of sight. After the level instrument has been set up and properly leveled, the micrometer screw is turned so that the horizontal cross-wire coincides with a division mark on the rod. The rod is then read to the smallest full division in the usual manner, and the fractional distance that the horizontal wire is above the full division is read from the micrometer screw. Usually the micrometer can be read directly to 0.01 of the smallest unit on the rod and estimated to 0.001 unit. For example, for a foot rod with 0.01 ft as the smallest unit, the micrometer can give direct reading to 0.0001 ft. For a metric rod that has 0.01 m as the smallest unit, the micrometer can give direct reading to 0.0001 m.

POOR TURNING POINTS

A turning point should be a stable object with a well-defined elevation point. Usually a short time elapses between the *FS* and *BS* readings on a turning point so that the level instrument may be moved ahead to the next setup. Furthermore, the level rod must also be turned to a different direction for the *FS* reading. The turning point should be so well-defined that both the *BS* and *FS* readings can be taken with the rod resting on the same elevation

FIGURE 4.32 A Lietz B-2C automatic level with parallel plate micrometer attachment (courtesy of The Lietz Company)

surface. Examples of good turning points include the top of a fire hydrant, a flat level surface on a sidewalk or curb, top of a construction stake driven into the ground, a spike driven into tree root, and high point on a boulder. Even for leveling of ordinary precision, a turning point is never taken on bare ground or turf. A metal turning pin, such as that shown in Figure 4.33, may be used to serve as turning point when appropriate natural objects are not available. If firmly pushed into the ground, it provides an excellent turning point.

PARALLAX

When any distant object is viewed through the telescope, the cross-wires will at the same time appear to be sharp only when they lie in the common focal plane of the objective and eyepiece lenses. If the plane of the cross-wires is very close to but not coincident with the common focal plane of the lenses, the cross-wires may appear to be quite distinct, but they will seem to move about on the object with the slightest movement of the eye of the observer. A similar phenomenon is the apparent movement of the window sash with respect to any out-of-door object if the observer moves his or her head slightly. This condition in the telescope is called *parallax* and is to be prevented by careful focusing of the eyepiece on the cross-wires before viewing any distant objects.

FIGURE 4.33 Turning pin

HEAT WAVES

On hot sunny days when the line of sight is close to hot objects such as pavement and metal tanks, heat waves can significantly degrade the accuracy of leveling. Errors from this source can be minimized by limiting the length of sights and keeping the line of sight high above ground and away from heat sources. Under extreme conditions leveling operations should be halted until the heat waves have subsided.

WIND

Wind may vibrate the instrument and make it difficult to keep the bubble centered and to read the rod correctly. Errors from this source can be minimized by decreasing the lengths of sights and increasing the stability of the tripod by forcing the tripod shoes deeper into the soil and placing the legs farther apart.

In summary, good field procedures in leveling include the following:

1. maintaining the instrument in proper adjustment;
2. keeping the bubble centered when reading the rod;
3. plumbing the rod;
4. balancing the backsight and foresight distances;

5. using good turning points; and
6. minimizing the effects from the natural environment including temperature, heat waves, and wind.

4.17 MISTAKES

Mistakes commonly made in leveling work include the following:

1. Misreading the rod. A particularly common blunder is misreading the rod by a full foot. This happens most often when the marks on the rod are obscured by leaves on trees, on the ground, fences, and so on. This kind of mistake is to be prevented by habitually looking at the markings and numbers on both sides of the horizontal cross-wire. Another common reading mistake is omitting one or more zeros after the decimal point. For example, 5.004 may be misread as 5.04.

2. Not setting the rod on the same point for a foresight and the subsequent backsight.

3. Recording and computing. As in other kinds of work, mistakes may be made in calling numbers and in recording them. To avoid mistakes, such numbers should always be called back. Conscious efforts should be made to avoid recording *BS* readings under the *FS* column or vice versa. Page check should always be made before leaving the field.

4.18 CARE OF LEVEL AND ROD

Care of the level, as of any other instrument, begins with a general inspection of the instrument when it is removed from its carrying case at the start of a workday. Try all clamps, motions, and screws to make certain there is no binding or malfunctioning. Attach the head to the tripod securely but not so tightly that difficulty will be experienced in removing it. Always use the carrying case to transport the level over considerable distances or rough terrain during operations. Carry the level and tripod under the arm, not over the shoulder, when walking beneath trees, over rough ground, and in confined areas. The tripod should always be set up in a stable position so that there is no possibility of the tripod shoes slipping on smooth surface, such as on concrete sidewalks. The instrument must never be left unattended.

Care of the rod consists of keeping it clean, unwarped, and readable. Dragging it through the brush or sounding water depths with it will soon foul the face of the rod and render the graduations illegible. The Philadelphia rod should not be carried over the shoulder while fully extended because of the extreme flexibility of the rod in this condition. The rod, particularly in the extended condition, should not be left leaning against a wall or a tree because of the possible damage that may ensue if it falls. It should be placed flat on the ground with the graduations facing upward. From time to time the continuity of the graduations from the top of the lower section to the bottom of the top section should be checked for accuracy. The section clamp should also be periodically checked to ensure that the sections clamp together correctly when "high rod" is used.

4.19 PROPAGATION OF RANDOM ERRORS

Let $\hat{\sigma}_r$ denote the estimated standard error of a rod reading. It represents the combined effect of all sources of random errors. Suppose that n setups of the level are required to run a level line from a benchmark to a new survey monument. Furthermore, let Δh denote the measured difference in elevation between the benchmark and the monument. Then

$$\Delta h = \Sigma\, BS \text{ readings} - \Sigma\, FS \text{ readings}$$

By applying the law of propagation of random errors (see Section 2.9), the following expression can be derived for the estimated standard error ($\hat{\sigma}_{\Delta h}$) of Δh:

$$\hat{\sigma}_{\Delta h} = \sqrt{2n}\hat{\sigma}_r \qquad (4.13)$$

Moreover, let h_1 denote the elevation of the benchmark with a corresponding estimated standard error $\hat{\sigma}_{h_1}$. Since the elevation (h_2) of the new survey monument is computed from h_1 by

$$h_2 = h_1 + \Delta h \qquad (4.14)$$

the corresponding estimated standard error ($\hat{\sigma}_{h_2}$) of the derived elevation can be computed from the following expression:

$$\hat{\sigma}_{h_2} = \pm\sqrt{\hat{\sigma}^2_{h_1} + \hat{\sigma}^2_{\Delta h}} \qquad (4.15)$$

Eq. (4.13) and Eq. (4.15) are extremely useful for evaluating the accuracy of a leveling project.

EXAMPLE 4.4

The elevation of a survey monument is to be determined by differential leveling from an existing benchmark located approximately 3,000 ft away. All the *BS* and *FS* distances will be approximately 150 ft long. It is estimated that the standard error of a single rod reading ($\hat{\sigma}_r$) is about ±0.005 ft. Determine the accuracy with which the difference in elevation can be established.

SOLUTION

Number of setups required:

$$(n) = \frac{3{,}000 \text{ ft}}{2 \times 150 \text{ ft}} = 10$$

From Eq. (4.13),

$$\begin{aligned} \hat{\sigma}_{\Delta h} &= \sqrt{2n}\hat{\sigma}_r \\ &= \pm\sqrt{2 \times 10} \times 0.005 \text{ ft} \\ \therefore \hat{\sigma}_{\Delta h} &= \pm 0.02 \text{ ft} \end{aligned}$$

EXAMPLE 4.5

The elevation of the existing benchmark in Example 4.4 is 763.25 ft with an estimated standard error of ±0.03 ft. Determine the estimated standard error of the elevation derived for the new survey monument.

SOLUTION

From Eq. (4.15),

$$\begin{aligned} \hat{\sigma}_{h_2} &= \sqrt{\hat{\sigma}^2_{h_1} + \hat{\sigma}^2_{\Delta h}} \\ &= \pm\sqrt{(0.03)^2 + (0.02)^2} \\ &= \pm 0.04 \text{ ft} \end{aligned}$$

EXAMPLE 4.6

Suppose that the difference in elevation discussed in Example 4.4 must be established with an estimated standard error of ±0.01 ft or smaller. What is the maximum allowable estimated standard error on a single rod reading?

SOLUTION

From Eq. (4.13),

$$\begin{aligned} \hat{\sigma}_r &= \frac{1}{\sqrt{2n}}\hat{\sigma}_{\Delta h} \\ &= \pm\frac{1}{\sqrt{20}} \times .01 \text{ ft} \\ \therefore \hat{\sigma}_r &= \pm 0.002 \text{ ft} \end{aligned}$$

4.20 CLOSURE ERROR

Whenever possible a level line should be closed either on the starting benchmark or on a second benchmark of known elevation. In Figure 4.34 a level line is run from an existing National Geodetic Survey (NGS) benchmark (*BM* 159) to three new survey benchmarks and then is closed on

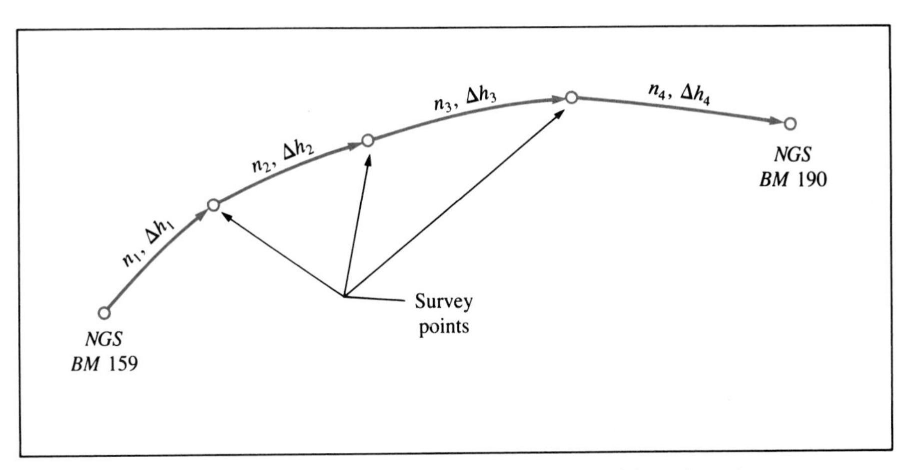

FIGURE 4.34 A level line closing on a second benchmark

a second NGS benchmark, *BM* 190. This is the situation most commonly encountered in a highway or a pipeline survey. In Figure 4.34 the arrows indicate the directions of the level lines, and Δh_i and n_i represent the measured difference in elevation and number of setups for line section i respectively. In Figure 4.35 a level line starts and ends on the same U.S. Geological

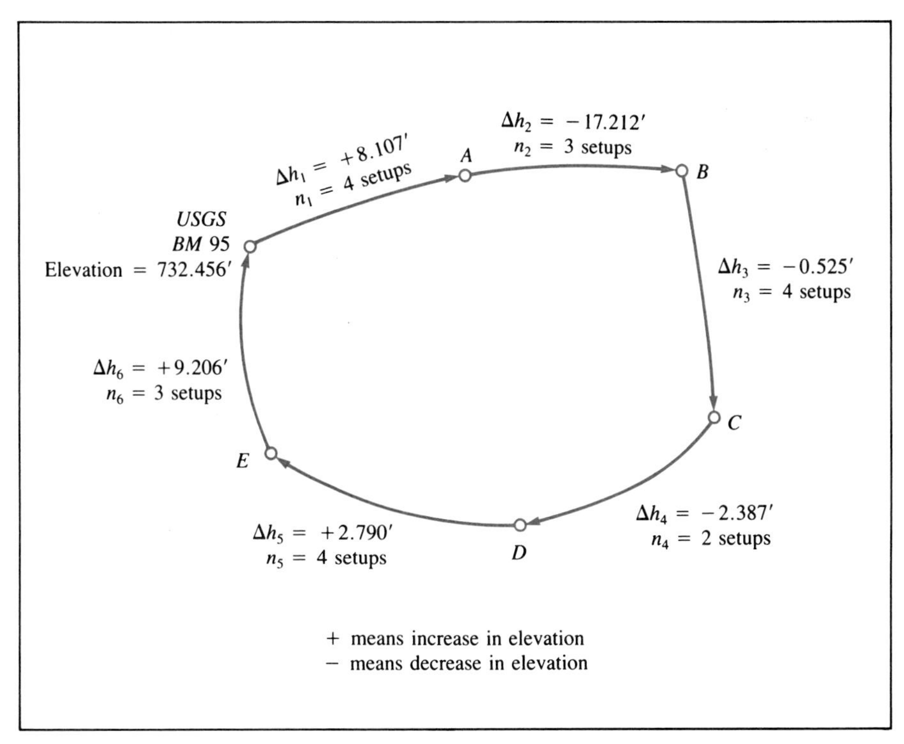

FIGURE 4.35 A level loop

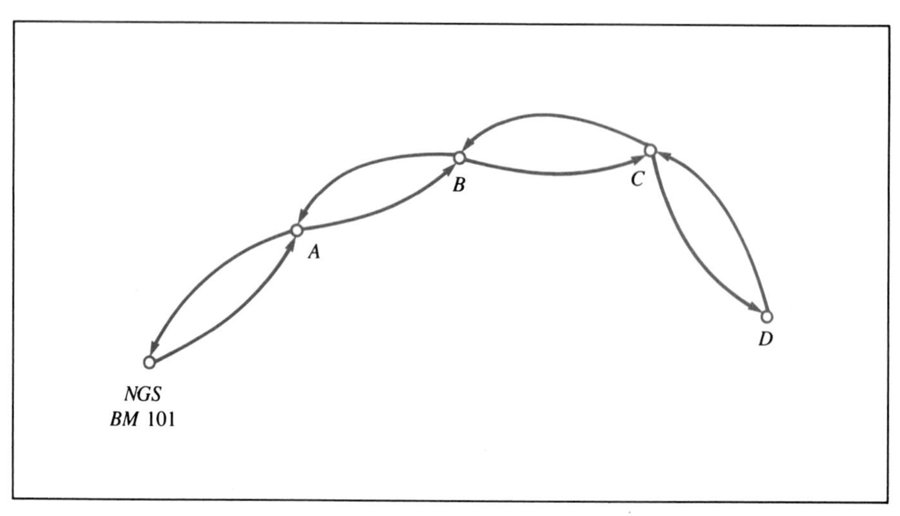

FIGURE 4.36 A double-run level line

Survey (USGS) benchmark, *BM* 95. Such a level line is called a *level loop*. Figure 4.36 shows a special kind of loop, in which the level line connects the same survey points, once forward and once backward. Such a level line is said to be *double-run*.

The closure error of a level line or loop is computed as the difference between the measured and known elevation of the closing benchmark—that is,

$$\text{closure error } (\varepsilon) = \text{measured elevation} - \text{known elevation} \quad (4.16)$$

In the case of a level loop, the closure error may also be computed as the algebraic sum of the elevation differences for the line sections in the loop.

Corrections for closure error can be distributed to the line sections by proportion. Assuming that all the rod readings along a level line are made with the same care and accuracy, the number of setups needed for each of the line sections can be used for proportioning. Therefore

$$\text{closure correction for } \Delta h_i = -\frac{n_i}{\sum n_i}(\varepsilon) \quad (4.17)$$

Table 4.1 shows the calculation performed for distributing the closure error for the level loop shown in Figure 4.35 and computing the final elevations for points A, B, C, D, and E.

Figure 4.37 shows the field notes and the calculation for distributing closure error along a level line that closes on a second benchmark of known elevation. Note that the number of instrument setups for a level line section is equal to the number of turning points + 1. Thus between stations 201 and 202, there are three turning points but four setups. The closure correction is

TABLE 4.1 Adjustment of a Level Loop

Points	Measured Δh (ft)	Correction (ft)	Corrected Δh (ft)	Elevation (ft)
BM 95				732.456
	+8.107	+0.004	+8.111	
A				740.567
	−17.212	+.003	−17.209	
B				723.358
	−0.525	+.004	−0.521	
C				722.837
	−2.387	+.002	−2.385	
D				720.452
	+2.790	+.004	+2.794	
E				723.246
	+9.206	+.003	+9.209	
BM 95				732.455✓

Closure error = −0.021 ft
Total no. of setups = 4 + 3 + 4 + 2 + 4 + 3 = 20

BM	BS	HI	FS	Elv.	Closure correction	Adjusted elevation	
NGS BM 100	4.321	235.777		231.456	0.0	231.456	
TP 1	5.410	237.516	3.671	232.106			
TP 2	4.893	235.682	6.727	230.789			
sta. 201	6.011	239.475	4.218	231.464	− 0.004	231.460	$\frac{3}{10}$ X(−0.012) = −0.004
TP 3	4.892	239.615	2.752	234.723			
TP 4	5.398	240.649	4.344	235.271			
TP 5	2.116	237.402	5.363	235.286			
sta. 202	5.673	238.177	4.898	232.504	− 0.008	232.496	$\frac{7}{10}$ X (−0.012) = −0.008
sta. 203	4.059	238.575	3.661	234.516	− 0.010	234.506	$\frac{8}{10}$ x (−0.012) = −0.010
TP 6	4.180	237.545	5.210	233.365			
NGS BM 100			4.330	233.215	− 0.012	233.203	$\frac{10}{10}$ x (−0.012) = −0.012
			−	231.456			
				1.759			
Σ BS =	46.933						
Σ FS =	45.174						
	1.759		check				

Closure error = 233.215 − 233.203 = +0.012

Total No. of setups = 3 + 4 + 1 + 2 = 10

FIGURE 4.37 Adjustment of a level line

computed only for the survey stations. The correction for station 202, for example, is computed as follows:

$$\begin{array}{l}\text{closure correction}\\\text{for Station 202}\end{array} = -\frac{\begin{array}{c}\text{no. of setups between } BM \text{ 100}\\\text{and Station 202}\end{array}}{\begin{array}{c}\text{total no. of setups along}\\\text{the level line}\end{array}} \times \text{closure error}$$

$$= -\frac{3 + 4}{3 + 4 + 1 + 2}(0.012)$$

$$= -0.008 \text{ ft}$$

4.21 ADJUSTMENT OF LEVEL NETWORK BY SUCCESSIVE ITERATION

When elevations are to be determined for a large number of survey stations distributed throughout a construction site, the level lines should be run in short closed loops. The adjoining loops then form a level network, as shown in Figure 4.38. The loops within the network can be adjusted for

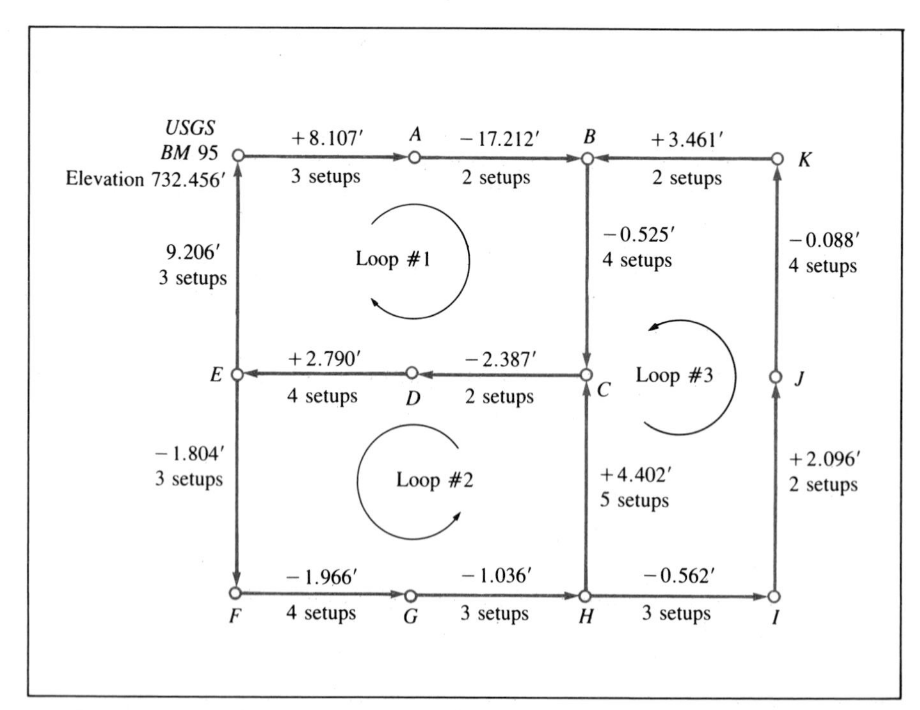

FIGURE 4.38 A level network

closure error using the procedure described in Section 4.20. For example, the network in Figure 4.38 can be adjusted as follows:

1. Adjust loop 1.
2. Adjust loop 2. Use the adjusted elevation differences from step 1 for line sections common to loops 1 and 2.
3. Adjust loop 3. Use the latest adjusted elevation differences from steps 1 and 2 for line sections common to loops 1 or 2.
4. Repeat steps 1 to 3 until the closure error is negligible for all three loops.

Usually two or three iterations are sufficient to reach the final solution.

4.22 LEAST-SQUARES ADJUSTMENT OF LEVEL NETWORK

Let Δh_{ij} be the measured elevation difference from points i to j in a level network. Let v_{ij} be the most probable error in the measured quantity Δh_{ij}. Furthermore, let h_i and h_j be the most probable elevations for points i and j respectively. Then the following relationship exists among these parameters:

$$v_{ij} + \Delta h_{ij} = h_j - h_i$$

Rearranging terms yields

$$v_{ij} + h_i - h_j = -\Delta h_{ij} \tag{4.18}$$

One such equation, which is called an *observation equation,* can be written for each measured elevation difference in the network. The system of equations can then be solved by the method of least squares (see Section 2.10) to determine the most probable values of the elevations of the new stations. Such an adjustment is most efficiently performed with an electronic computer. One major advantage of a least-squares solution is that, as a by-product of the adjustment, estimated standard errors can be derived for all the computed elevations. Figure 4.39 shows a computer printout for a least-squares adjustment of the network in Figure 4.38.

It is beyond the scope of this book to deal with the computational details of a least-squares adjustment. Interested readers are referred to Reference 4.5.

BS STATION	FS STATION	ELV. DIFF.	RMS ERROR	MOST PROBABLE ERROR
1	2	-17.212	.005	-.002
2	3	-.525	.006	-.007
3	4	-2.387	.005	-.002
4	5	2.790	.006	-.004
5	6	-1.804	.006	.000
6	7	-1.966	.006	.000
7	8	-1.036	.006	.000
8	9	-.562	.006	-.002
9	10	2.096	.005	-.002
10	11	-.088	.006	-.003
11	2	3.461	.005	-.002
8	3	4.402	.007	.004
5	95	9.206	.006	-.003
95	1	8.107	.006	-.003

NO. OF DEGREES OF FREEDOM - 3

BM	ADJ. ELEVATIONS	RMS ERROR
1 (A)	740.566	.005
2 (B)	723.356	.007
3 (C)	722.838	.007
4 (D)	720.453	.007
5 (E)	723.247	.005
6 (F)	721.443	.007
7 (G)	719.476	.008
8 (H)	718.440	.008
9 (I)	717.881	.009
10 (J)	719.978	.009
11 (K)	719.894	.008
95 (BM 95)	732.456	.000

FIGURE 4.39 Least-squares solution for a level network

4.23 ORDERS OF ACCURACY

The terms *first-, second-,* and *third-order accuracy* for leveling projects have been in common use since they were officially adopted by federal government agencies in 1957. This classification was replaced in 1975 by one that consists of the following five grades of accuracy:

1. First-Order, Class I
2. First-Order, Class II
3. Second-Order, Class I
4. Second-Order, Class II
5. Third-Order

This new classification was developed by the *Federal Geodetic Control Committee* (FGCC) in 1975 and described in detail in two publications (see References 4.2 and 4.3). The committee consisted of representatives from different agencies in the federal government. All federal agencies are mandated to comply with these standards. Engineers and surveyors in private practice traditionally follow the same set of standards.

The accuracy standards and specifications developed by FGCC are designed primarily for the purpose of developing a national network of geodetic control monuments distributed throughout the United States. Some of the guidelines and specifications are not applicable for local surveys confined to an area of few square miles. Table 4.2 summarizes the standards and specifications prepared by FGCC and includes only those items relevant to surveying in private practice. Complete details on the FGCC standards and specifications can be found in References 4.2 and 4.3.

In Table 4.2 the maximum allowable closure error is expressed as a function of the length of the level line or loop in kilometers. Thus if a level loop measures about 3 km long, the maximum allowable closure error is $\pm 12 \text{ mm} \sqrt{3} = \pm 21 \text{ mm}$ for third-order leveling.

4.24 PRECISE LEVELING

The accuracy obtainable in leveling operations depends largely on the quality of the instrumentation and on the care that is taken in field operations to minimize random errors from all sources.

Table 4.3 lists the accuracy capabilities and other characteristics of several levels. Figures 4.40 and 4.41 show two levels suitable for first-order leveling.

Modern levels such as those listed in Table 4.3 are well-constructed, precision instruments that should remain in good adjustment through long

TABLE 4.2 Summary of Standards of Accuracy and General Specifications for Leveling (Summarized from Reference 4.2)

Classification	First-Order		Second-Order		Third-Order
	Class I	Class II	Class I	Class II	
Principal uses Minimum standards; higher accuracies may be used for special purposes	Basic framework of the National Network and of metropolitan area control Extensive engineering projects Regional crustal movement investigations Determining geopotential values		Secondary control of the National Network and of metropolitan area control Large engineering projects Local crustal movement and subsidence investigations Support for lower-order control	Control densification; usually adjusted to the National Network. Local engineering projects Topographic mapping Studies of rapid subsidence Support for local surveys	Miscellaneous local control: may not be adjusted to the National Network. Small engineering projects Small-scale topographic mapping Drainage studies and gradient establishment in mountainous areas
Instrument standards	Automatic or tilting levels with parallel plate micrometers; invar scale rods		Automatic or tilting levels with optical micrometers or three-wire levels; invar scale rods	Geodetic levels and invar scale rods	Geodetic levels and rods

Field procedures	Double-run; forward and backward		Double-run; forward and backward	Double- or single-run	Double- or single-run
Maximum allowable closure error for level line or loop (K = total distance in kilometer)	4 mm $\sqrt{K}$	5 mm $\sqrt{K}$	6 mm $\sqrt{K}$	8 mm $\sqrt{K}$	12 mm $\sqrt{K}$
Maximum length of sight	50 m	60 m	60 m	70 m	90 m
Maximum difference between *BS* and *FS* distances per set up	2 m	5 m	5 m	10 m	10 m
Maximum difference between sums of *BS* and *FS* distances per level line or loop	4 m	10 m	10 m	10 m	10 m

TABLE 4.3 Characteristics of Several Levels (All Data from Manufacturer's Literature)

Model No. (Manufacturer)	Accuracy	Level Mechanism	Telescope
Kern GK1-A Automatic Engineer's Level (Kerns Instruments, Inc)	Mean error, 1 km, double run level ±2.5 mm	Automatic level, by magnetic suspension; setting accuracy ±0.5 to ±1.0 second of arc	Erect image; magnification 25×; minimum focus distance 7.5 ft
Eagle 2 Automatic Level (Keuffel & Esser Company)	Mean square error ±0.003 ft $\sqrt{S}$ miles	Automatic level; air damped compensator; working range ±5 min; ±1 sec nominal accuracy	Erect image; magnification 32×; minimum focus 11 ft
B-2C Precision Automatic Level (The Lietz Company)	Standard deviation for 1.6 km double run leveling ±1.0 mm; ±0.5 mm with parallel plate micrometer attachment	Automatic level; magnetic self-damping; compensating range ±10 min; setting accuracy ±0.3 second	Erect image; magnification 32×; minimum focus 1.3 m

Wild N2 Engineer's Level (Wild Heerbrugg Instruments, Inc.)	Standard deviation of 1 km double run leveling ±2 mm; ±1 mm with parallel plate micrometer	Tilting screw setting; setting accuracy of level ±0.8″	Erect image; magnification 30×, shortest focus 1.6 m
Wild N3 Engineer's Level (Wild Heerbrugg Instruments, Inc.)	Standard deviation of 1 km double run leveling, ±0.2 mm	Tilting screw setting; setting accuracy ±0.2″	Built-in parallel plate micrometer; erect image; magnification 42×, shortest focus 2.15 m
Wild N10 Engineer's Compact Level (Wild Heerbrugg Instruments, Inc.)	Standard deviation of 1 km double run leveling ±2.5 mm	Tilting screw and split-bubble; setting accuracy ±1.5″	Erect image; magnification 20×; minimum focus 1.35 m
Zeiss Ni 1 Self-Leveling Level (Carl Zeiss, Inc.)	Nominal ±0.2 mm $\sqrt{K\ (\text{km})}$	Automatic level; mirror suspended by wires; working range ±5′; setting accuracy ±0.2″	With parallel plate micrometer; minimum focus 1.4 m; magnification 40×; erect image
Zeiss Ni 2 Self-Leveling Level (Carl Zeiss, Inc.)	Nominal ±0.7 mm $\sqrt{K\ (\text{km})}$	Automatic level; mirror suspended by wires	Erect image; minimum focus 11 ft; magnification 32×

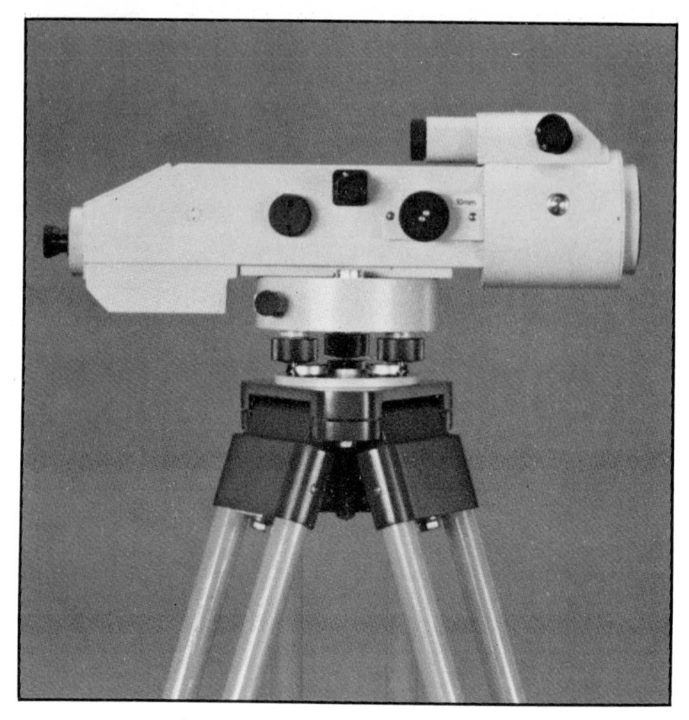

FIGURE 4.40 A Zeiss Ni 1 precise automatic level (courtesy of Carl Zeiss, Inc., Thornwood, N.Y.)

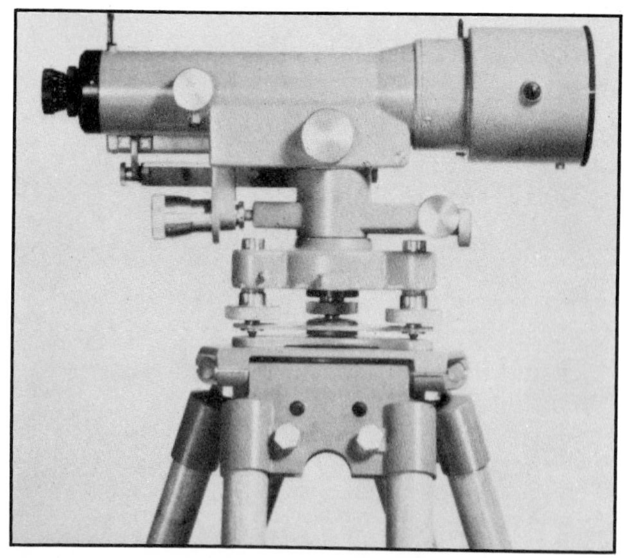

FIGURE 4.41 A Wild N3 Precision Level (courtesy of Wild Heerbrugg Instruments, Inc.)

periods of careful use. Nevertheless, correct alignment of the horizontal line of sight should be checked periodically using the peg test, which was discussed in Section 4.15. Except for simple adjustment of the bull's-eye level, adjustment of modern levels is best left to a qualified instrument technician.

Figure 4.42 shows a level rod used in first- and second-order leveling. It is equipped with a rod level, a thermometer, and a metal foot piece, called *rod shoe,* which is designed for more accurate setting of the rod over a benchmark or turning point. The metal scale is made of invar steel and is held at constant tension by a spring at the top of the wooden staff. It is 12 ft long and constructed in one piece. Figure 4.43 shows a precise metric rod equipped with two supporting legs for accurate and stable plumbing of the rod.

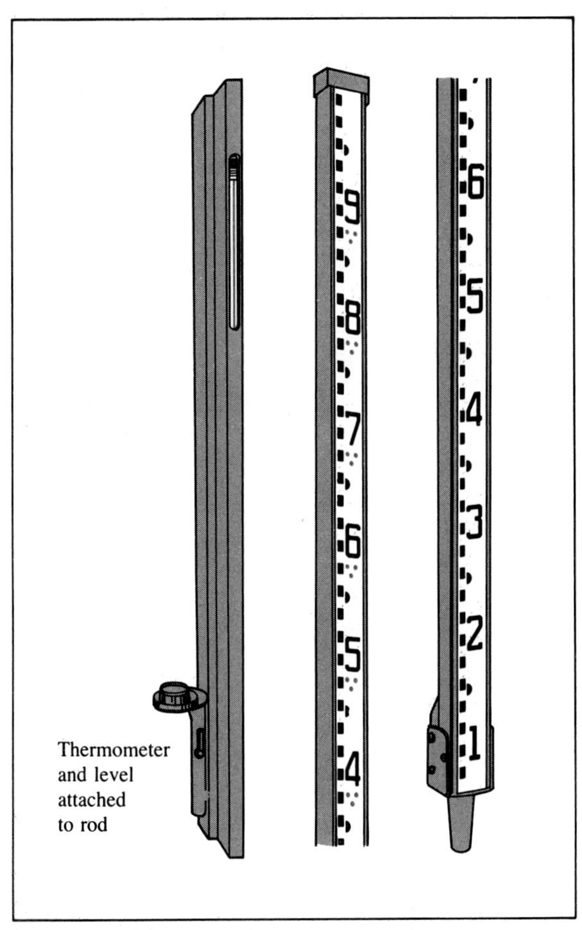

FIGURE 4.42 Precise level rod (courtesy of Keuffel & Esser Company)

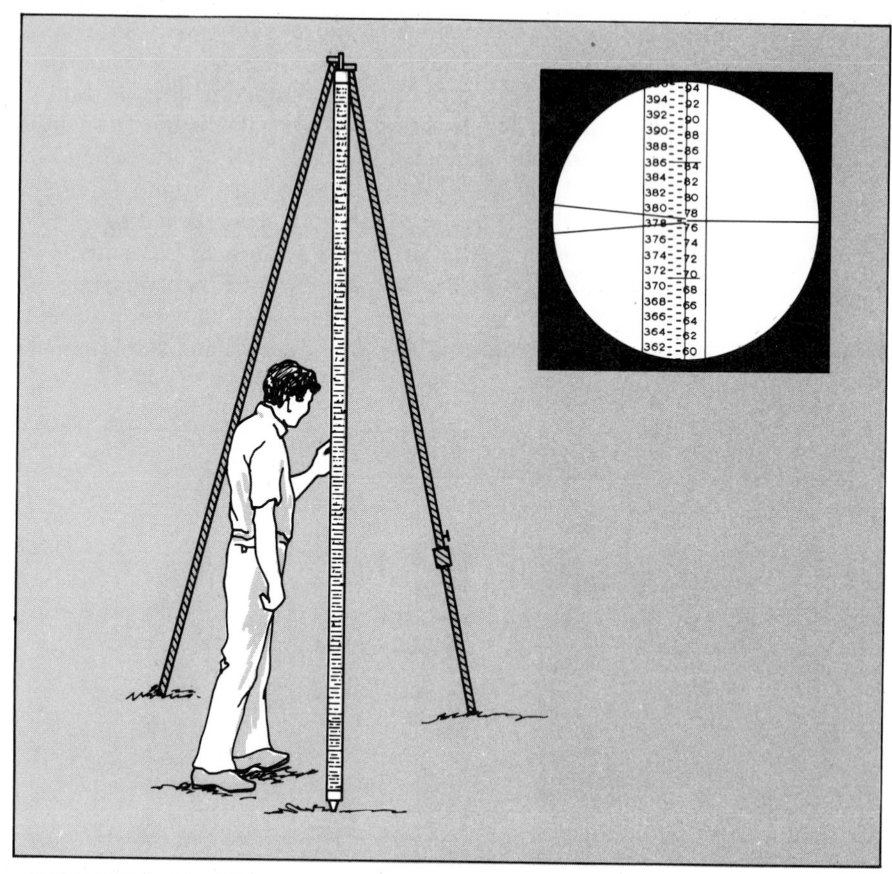

FIGURE 4.43 Invar staff with supporting legs (courtesy of Wild Heerbrugg Instruments, Inc.)

The accuracy of rod reading can be significantly increased by the use of an optical micrometer, as discussed in Section 4.16. The use of an optical micrometer is mandatory for first-order leveling and is also recommended for second-order, Class I leveling (see Table 4.2). When an optical micrometer is not available, the accuracy of rod reading can be improved by reading the rod with all three horizontal wires in the reticule and taking the average of the three readings. Such a procedure is called *three-wire leveling* and was used extensively in the past. However, three-wire leveling is time consuming and can no longer be justified on economic grounds when optical micrometers of relatively low cost are now available.

Figure 4.44 shows an example of field notes for three-wire leveling. One distinction from the regular field notes is that there is no need to compute each *HI* and turning-point elevation. The desired difference in elevation between stations is computed as the difference between total backsight and foresight readings.

Three - wire leveling from USGS "G-1" to USGS "6 LES 1948"					
	BS			FS	
Threads	Mean	Stadia	Threads	Mean	Stadia
1.984		0.646	4.719		0.692
1.338	1.337	0.650	4.027	4.027	0.692
0.688			3.335		
5.647		0.714	4.410		0.746
4.933	4.933	0.715	3.664	3.663	0.748
4.218			2.916		
5.802		0.781	5.822		0.739
5.021	5.022	0.779	5.083	5.084	0.735
4.242			4.348		
6.007		0.768	5.018		0.728
5.239	5.239	0.767	4.290	4.291	0.725
4.472			3.565		
ΣBS	16.531		ΣFS	17.065	
	− 17.065				
Δh =	− 0.534′				

K&E level # 104326	August 27, 1983
Rods No. 10	Level – Weers, A.J.
Temp. 72° F	Book – McCarthy, C.
Cloudy	
USGS "G-1" on south	side of east entrance
to Engineering Hall,	University of Illinois at
Urbana-Champaign,	18" above ground
in wall, aluminum	Tablet stamped "G-1"

FIGURE 4.44 Field notes for three-wire leveling

4.25 U.S. NATIONAL VERTICAL CONTROL SYSTEM

The basic vertical control system of the United States consists of an extensive network of lines of first- and second-order differential levels executed by the National Geodetic Survey, which was formerly known as the Coast and Geodetic Survey, or often simply called the Coast Survey. This network contains approximately 500,000 benchmarks along 70,000 kilometers of level lines. The elevations of these points with respect to a sea level datum was determined by differential leveling.

The federal leveling program in the United States began in 1856 when the Coast Survey initiated work along the Hudson River for the control of tide gauges. The U.S. Lake Survey of the Corps of Engineers undertook in 1875 the task of determining elevations for the water levels in the Great Lakes and for benchmarks in their harbor areas. Then in 1877 the Coast Survey was authorized to begin the transcontinental leveling that stretched generally along the 39th parallel of latitude from the Atlantic to the Pacific. This project was completed in the early 1900s. In 1929 the existing first-order leveling nets in the United States and Canada were adjusted to conform to mean sea level recorded at 26 tide stations along the Atlantic and Pacific coasts and the Gulf of Mexico (21 in the United States and 5 in Canada). The resulting datum was initially called the "Sea Level Datum of 1929," which was changed to "National Geodetic Vertical Datum of 1929 (NGVD29)" in 1973. A readjustment of the NGVD is now in progress at the National Geodetic Survey and is planned for completion by the end of 1988.

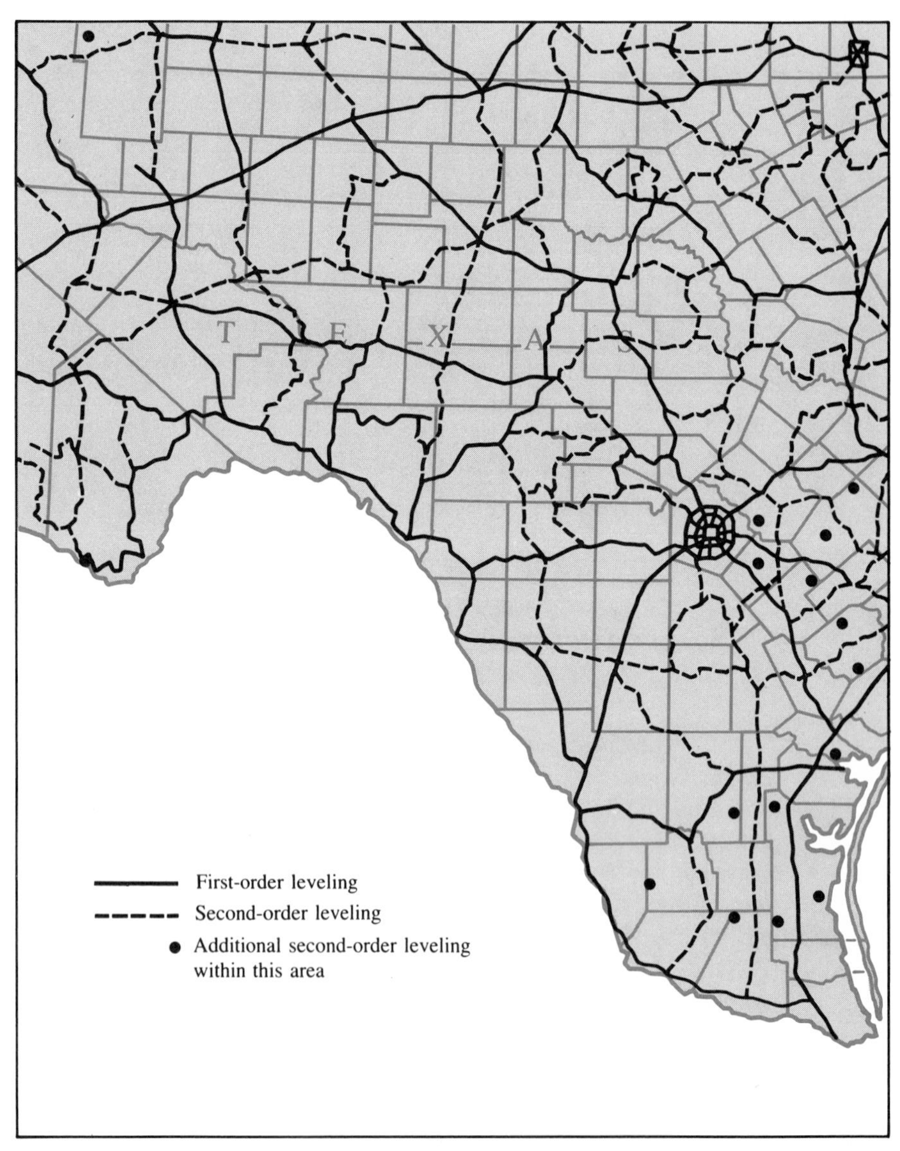

FIGURE 4.45 NGS level network (courtesy of National Geodetic Survey)

The new datum is to be called the North American Vertical Datum of 1988 (NAVD 88) (see Reference 4.4).

A small portion of the present network of NGS first- and second-order level lines is shown in Figure 4.45. Not included are the lines of third-order levels that have been executed by the Topographic Division of the U.S. Geological Survey and that total many thousands of miles.

Information concerning the locations, descriptions, and elevations of federal benchmarks can be obtained from the National Geodetic Survey, Rockville, Maryland 20852. The distribution and general locations of benchmarks within a region can best be obtained from a NGS Geodetic Control Diagram. These diagrams portray with distinctive color and symbolization both horizontal and vertical control established by the NGS, the U.S. Geological Survey, and, where available, other governmental agencies in a format having latitudinal and longitudinal dimensions of 1° and 2° respectively. Each diagram is subdivided into two 1° quadrangles as shown in Figure 4.46a. Then a six-digit numbering system is employed to further

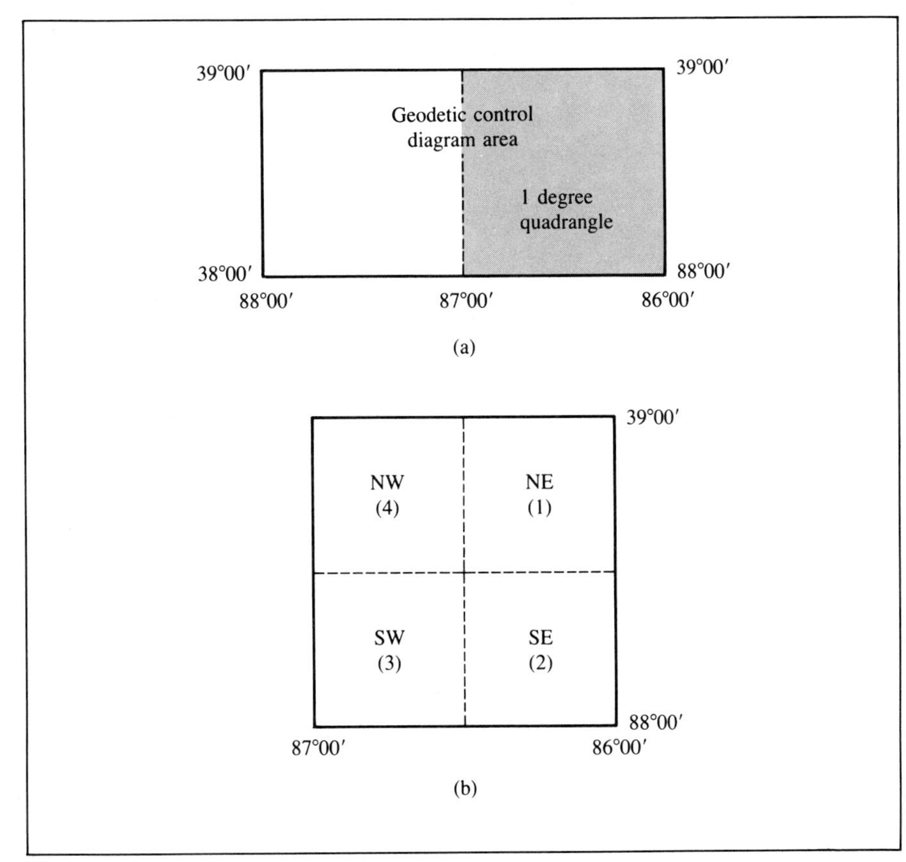

FIGURE 4.46 Geodetic control diagram (courtesy of National Geodetic Survey)

subdivide the 1° quadrangle into four 30′ quadrangles. As shown in Figure 4.46b, the southeast 30′ quadrangle would be identified as 380862. The 38 and the following three digits refer to the latitude and longitude, respectively, of the lower right-hand corner of the 1° quadrangle. The last digit, 2, refers to the southeast 30′ quadrangle. These 30′ quadrangles are numbered clockwise with number one being that in the northeast. Figure 4.47 shows locations of NGS level lines in a particular 30′ quadrangle.

Figure 4.48 shows the manner in which the NGS publishes benchmark elevations. It should be noted that the elevations are expressed in meters as well as feet and refer to the National Geodetic Vertical Datum of 1929. The words "Sea Level" with respect to such elevations are no longer

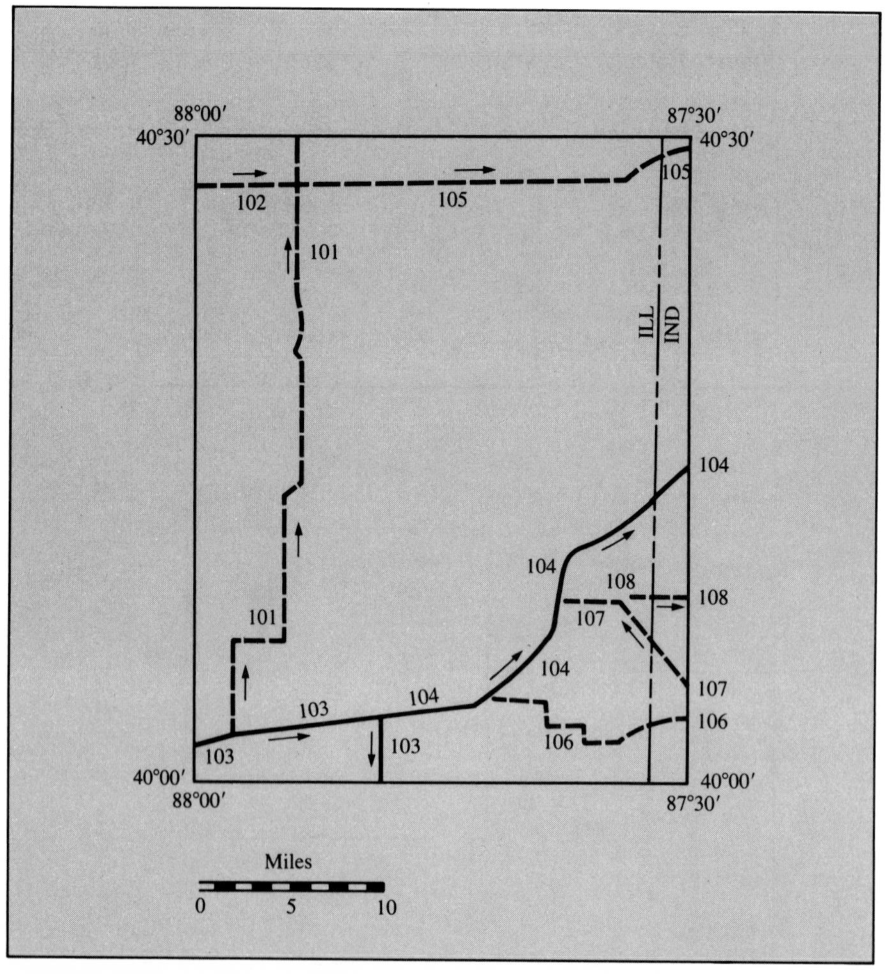

FIGURE 4.47 NGS level lines (courtesy of National Geodetic Survey)

LINE 101
(Second-order)

The original field work (L-5383) was done in the winter of 1934–35 by a party supervised by E. E. Stohsner. The line was recovered in November 1954 by a party supervised by Clarence Symns. Releveling (L-15995) was done in the vicinity of Homer in the summer of 1956 by a party supervised by Clarence Symns. In February 1966, the portion of the line in the vicinity of Rankin was readjusted.

Elevations for the original field work are based on a supplementary adjustment of 1939. Elevations for the releveling are based on a supplementary adjustment of 1959 and the readjusted portion is based on a supplementary adjustment of 1966.

Benchmark	Adjusted elevation (Meters)	Adjusted elevation (Feet)
M 167	203.894	668.942
**Z* 23	199.816	655.563
M 167 *X*	204.230	670.045
A 24	204.004	669.303
B 24 reset (1958)		671.744
D 24	203.467	667.541
E 24	202.999	666.006
G 24	208.759	684.903
H 24	227.139	745.205
J 24	237.810	780.215
K 24	228.162	748.561
L 24	219.753	720.973
M 24	212.495	697.161
N 24 reset (1959)		695.620
P 24	238.544	782.623
Q 24 reset (1959)		749.834
R 24	225.746	740.635
S 24	231.233	758.637
T 24	224.082	735.176
U 24	219.259	719.352
V 24	212.396	696.836

*Changed elevation

FIGURE 4.48 Benchmark elevations (courtesy of National Geodetic Survey)

officially used. Obviously the description that accompanies the survey data is important because a benchmark that cannot be recovered is of no value.

Some older government benchmarks were stamped with the elevations obtained from the original surveys. Such elevation data should be verified with the original surveying agency before use. Elevations of benchmarks of the national networks are subject to deterioration or change due to vandalism or natural causes such as ground subsidence, fault movement, and change in underground water table. Only the latest elevation on a benchmark should be used.

4.26 MONUMENTATION

Monumentation refers to the process of marking in an enduring manner the physical position of survey points. Proper monumentation together with supporting records and reports is the sole means for documenting surveys and providing the necessary relationship between field surveys and maps and charts.

A surveying monument should be so constructed that it remains unmoved during the period of intended use. It should not be easily bent, moved, or destroyed. It should not be affected by seasonal frost heave and should be properly protected so that opportunity for damage by motor vehicles, snow removal equipment, or vandalism can be minimized.

Before 1978 both the National Geodetic Survey and the U.S. Geological Survey used extensively a bronze disc embedded in cement or concrete as an elevation benchmark. The metal disc bears the name of the organization that established the benchmark and measures 4 inches in diameter (see Figure 4.49). The disc is commonly set with cement in holes that are drilled into flat surfaces of bridge abutments, foundations of large buildings, and concrete steps in front of federal buildings such as court houses and post offices. When a benchmark had to be located in an open field, a concrete post extending from the ground surface to several inches below the frost line was often used as monument. Figure 4.50 shows a standard design that was commonly used for USGS benchmarks.

In the early 1970s, both the USGS and the then Coast and Geodetic Survey started using a rod-type monument as a benchmark. It is constructed of 5/8-inch copper-coated rods, which are driven in successive coupled sections into the ground until strong resistance is met at a depth greater than 10 ft. A brass benchmark tablet is crimped to the top end of the rod and is set either at or below ground surface. By using a hand-held power hammer for

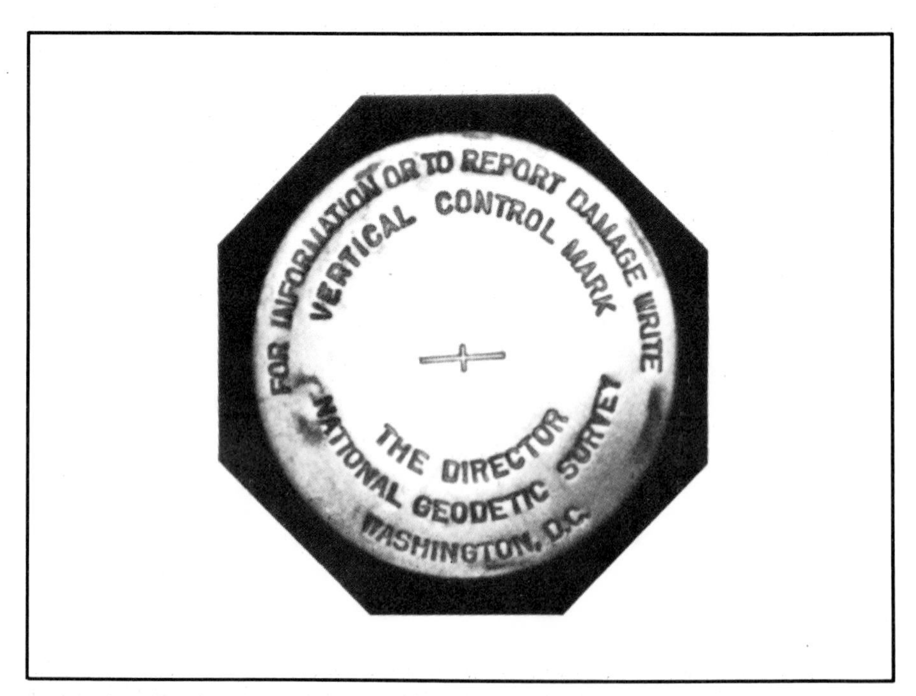

FIGURE 4.49 Bronze tablet as benchmark (courtesy of National Geodetic Survey)

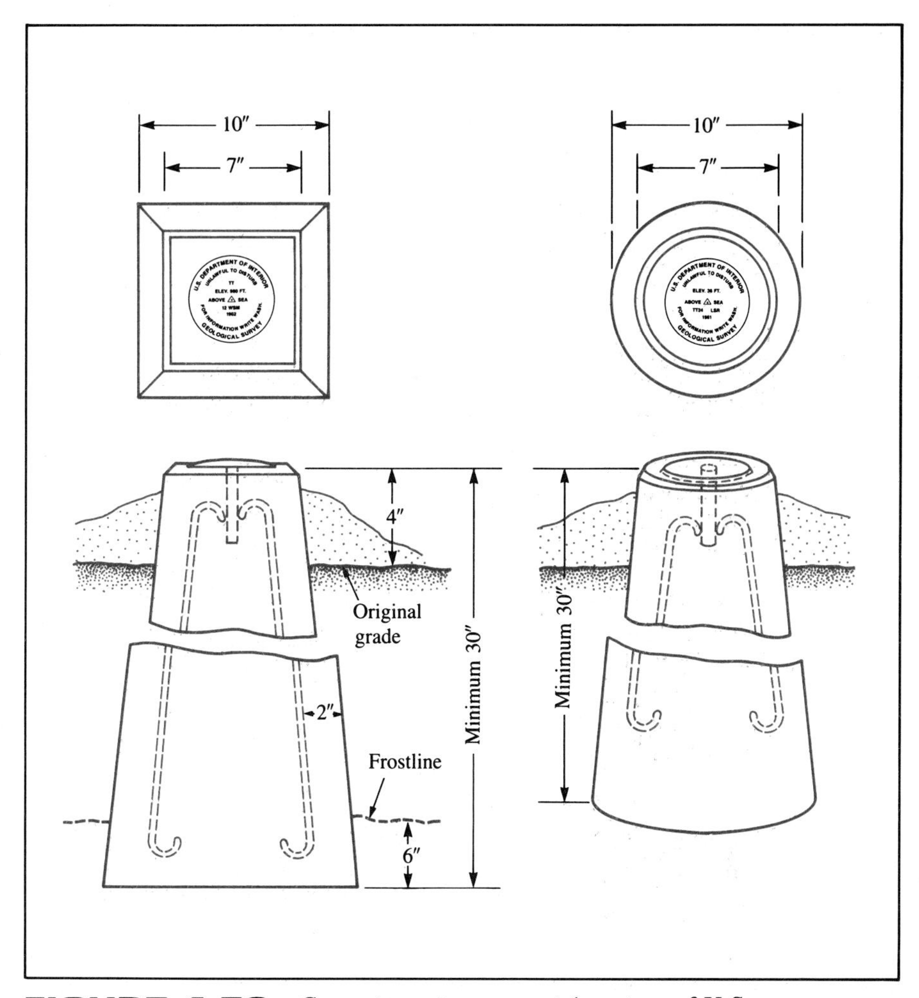

FIGURE 4.50 Concrete post monument (courtesy of U.S. Geological Survey)

driving the rod into the ground, such a monument can be easily and quickly installed. This type of monument was found to be just as stable as the concrete-post type.

In 1978 the National Geodetic Survey introduced a new rod-type monument, the design of which is shown in Figure 4.51. It consists of a stainless steel rod driven to a depth of 1.5 to 15 m. It is encased in a lubricant inside a PVC pipe sleeve. The sleeve extends to a depth beyond the influence of any soil movements due to frost, shrinkage, and swelling. The point of reference for elevation is the top of the stainless steel rod, which is set about 10 in. below the ground surface and is protected by an outside PVC pipe and an aluminum access cover. The National Geodetic Survey logo and the benchmark designation are stamped on the flange of the aluminum cover.

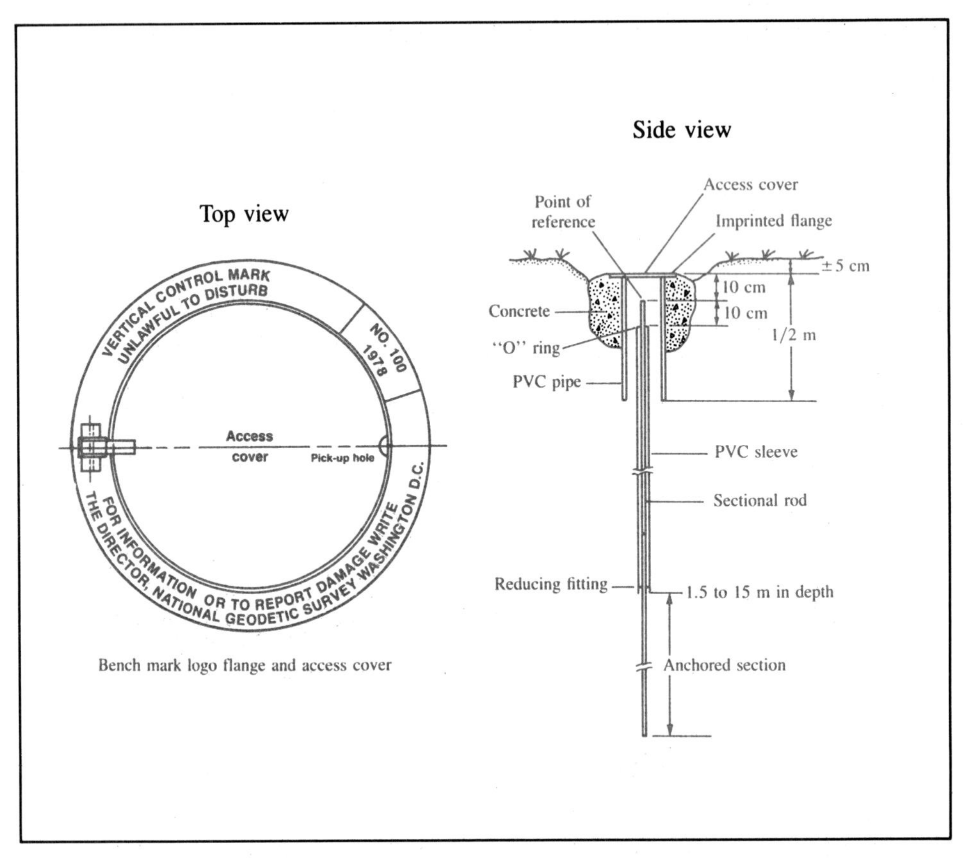

FIGURE 4.51 Stainless steel rod benchmark (courtesy of National Geodetic Survey)

The location of a federal benchmark is usually marked by a nearby witness post, as shown in Figure 4.52. A detailed description on the location as well as the latest elevation of a federal benchmark can be obtained by writing directly to the federal agency that established the benchmark.

The following is a typical National Geodetic Survey benchmark description:

> L 9—About 3.8 miles southwest along the Northern Pacific Railroad from the station at Blackduck, Beltrami County, at Hines; 84 feet northwest of the northeast corner of the station, 33.8 feet west of the west rail, and 39.9 feet east of the center line of U.S. Highway 71. A standard disk, stamped "L 9 1931" and set in the top of a concrete post. Elevation 1,404.174 ft, 427.993 m.

Figure 4.53 shows an elevation marker that is particularly suited for monitoring the settlement of buildings and foundations. It consists of a

FIGURE 4.52 Witness post (courtesy of National Geodetic Survey)

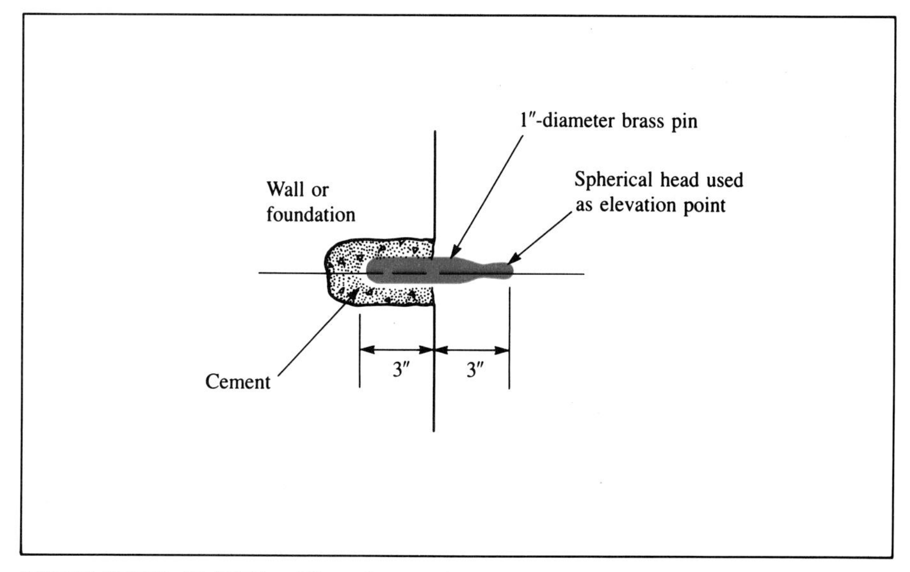

FIGURE 4.53 Elevation marker in wall or foundation

1-inch diameter brass rod, one end of which is firmly embedded in the foundation or outside wall of a building. The spherical head of the exposed end of the rod is used as elevation point.

4.27 RECIPROCAL LEVELING

When a line of levels crosses a broad body of water or a ravine making it impossible to balance the backsight and foresight distances, it is necessary to take sights much longer than are ordinarily permissible. Under such circumstances, errors due to rod reading, curvature, and refraction as well as the inclination of the line of sight become particularly significant. A special procedure termed *reciprocal leveling* can be used to obtain the best results.

In Figure 4.54 the elevation of survey point *A* is to be determined by leveling from *BM* 36. At a setup near *BM* 36, a backsight is taken on *BM* 36 and a foresight on *A*. The difference in elevation is computed as (*BS* reading − *FS* reading). Next the level is set up near point *A*. Again a *BS* reading is

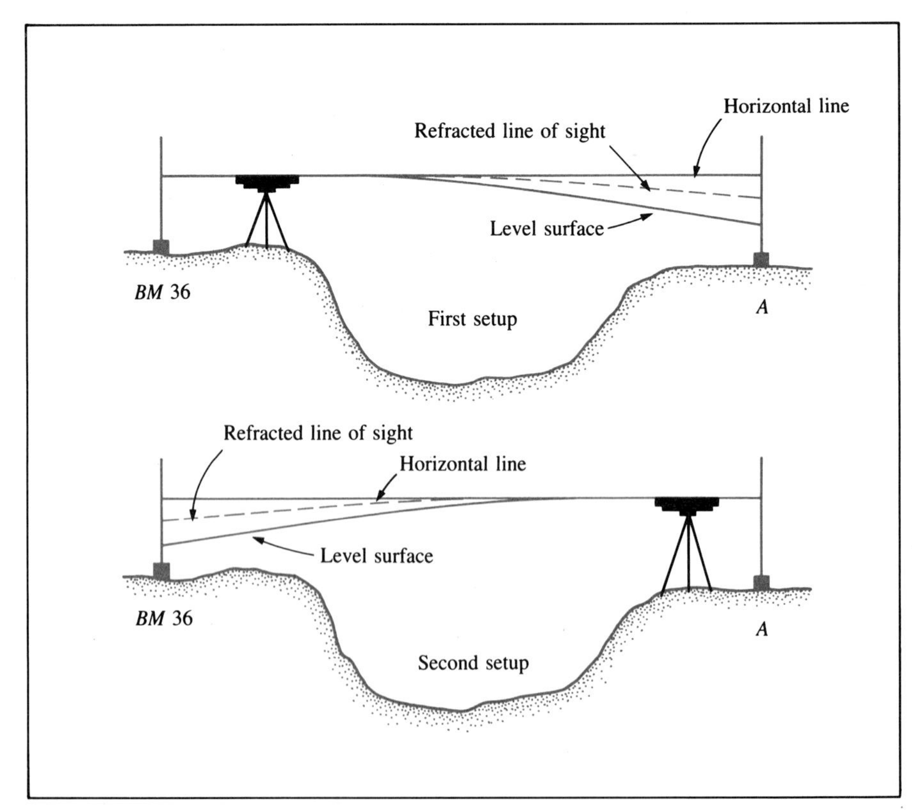

FIGURE 4.54 Reciprocal leveling

taken on *BM* 36 and an *FS* reading on point *A*. The difference between these two readings yields a second measurement on the elevation difference. Assuming that atmospheric refraction remains constant during the time period between the two setups, the correct difference in elevation is computed as the mean of the two measured differences.

Rod reading errors on the long sights can be minimized by using a rod target and taking the mean of several repeated readings.

EXAMPLE 4.7

The elevation of *BM* 36 in Figure 4.54 is 742.87 ft. From a setup on the left bank, the *BS* reading was 4.73 ft and the *FS* was 6.27 ft. At the second setup (on right bank) of the level, the *BS* reading was 5.89 ft and the *FS* was 7.63 ft. Find the elevation of point *A*.

SOLUTION

Mean difference in elevation is

$$\frac{(+4.73 - 6.27) + (5.89 - 7.63)}{2} = -1.64 \text{ ft}$$

Hence elevation of point $A = 742.87 - 1.64 = 741.23$ ft

4.28 PROFILE LEVELING

Profile leveling is the operation of determining the elevations of points at regular intervals along a fixed line. Before sewers, highways, railroads, and similar projects are designed and constructed, stakes are set at intervals, usually 100 ft, along the centerline. These 100-ft points are called *stations*. Points intermediate between full stations are termed *plus stations*. Hence a stake situated 2,400 ft from the point of beginning of the project would be identified as Station "24 + 00." A stake marking a stream crossing 75 ft in advance of station 24 + 00 would be designated as Station "24 + 75." It is usually advisable to assign a stationing of, say, 15 + 00 to the beginning point of a route engineering project. This will prevent the generation of negative stationing if the centerline is extended backward.

For engineering purposes a ground profile is the trace of the intersection of an imaginary vertical plane with the ground surface. This profile is usually plotted on specially prepared profile paper, on which the vertical scale is much larger than the horizontal; on this plotted profile various studies relating to the fixing of grades and the estimating of costs are made. The field work of profile leveling provides the data for this work.

Assuming that the alignment has been fixed on the ground by setting stakes at 100-ft stations, the level party first determines by the usual procedure of differential leveling the height of the instrument, which has been set up conveniently near the line of stakes. Foresight readings, called *ground*

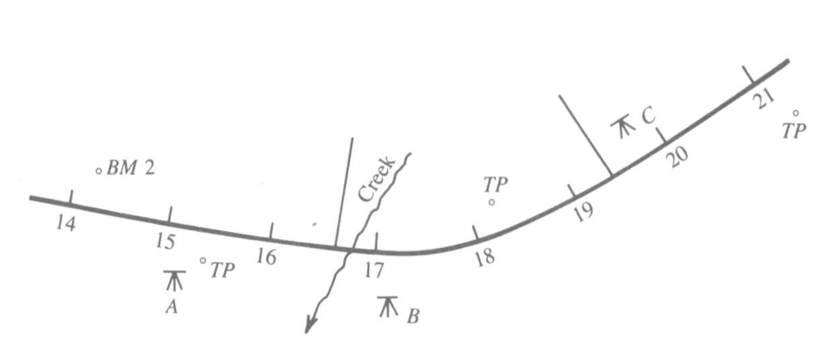

(a) Plan

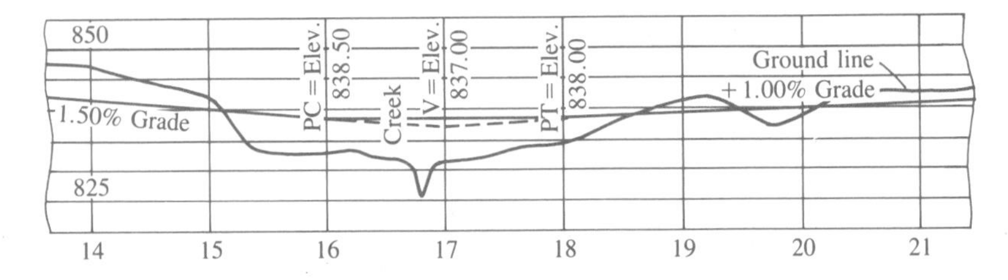

(b) Profile

Highway location-Fontana to Walworth						Dietzgen level No.22 June 15, 1968	F.R. Grant ⊼ M. Hughes Rod 10
Sta.	BS	HI	FS	GR	Elev.		
BM2	1.76	849.63			847.87	USGS BM2	Sunny, calm
14				2.2	847.4		80° F
15				7.0	842.6		
TP	0.63	838.66	11.60		838.03		
+40				4.9	833.8		
16				6.3	832.4		
+70				6.8	831.9	Spring Creek	
+75				13.2	825.5		
+80				6.6	832.1		
17				6.2	832.5		
18				4.5	834.2		
TP	9.76	846.32	2.10		836.56		
19				5.1	841.2		
+40				6.2	840.1		
+60				10.1	836.2		
20				8.5	837.8		
+40				3.5	842.8		
21				3.7	842.6		
BM3			3.24		843.08	USGS BM3	
	12.15		16.94		847.87	Correct elev. =	843.11
			12.15	check	4.79		
			4.79				

(c) Field notes

FIGURE 4.55 Profile leveling

rod readings, are then taken on the ground at each stake and at intermediate "plus-station" points where there is a marked change in the ground slope.

Since these ground-rod readings are used only for plotting the profile and have no relation to the determination of benchmark elevations, they are taken usually to the nearest tenth of a foot only. Accordingly, all elevations of ground stations are computed to the nearest tenth of a foot only.

Figure 4.55 illustrates the principle of profile leveling. A plan view of the centerline is shown in (a) and the field notes are shown in (c). The level is first set up near *A* and backsighted to *BM* 2 to establish the *HI*. Ground rod readings are then made at stations 14 and 15. Since station 16 is too low to be sighted from this setup, a turning point (*TP* 1) is selected and an *FS* reading is made on it. Since this *FS* reading on *TP* 1 is to be used for propagating the level line, it must be read to 0.01 ft. The level is then moved to position *B*. Here a *BS* reading on *TP* 1 is made to establish the *HI*, and then *GR* readings are made at stations 15 + 40, 16 + 00, 16 + 70, 16 + 75, 16 + 80, 17 + 00, and 18 + 00. It is obvious from the profile in (b) why it is necessary to take plus-station readings to show the stream crossing and the changes in slope. To avoid excessively long lines of sight, a second turning point (*TP* 2) is selected.

The level is moved to position *C*, and the profile leveling is continued. The level line is closed on a second known benchmark (*BM* 3). Adjustment for small closure errors is usually not needed in this type of application, but the page check should always be performed.

4.29 CROSS-SECTION LEVELING

Cross-sections are profiles taken transverse to the centerline of a project. They provide the data for estimating quantities of earthwork and for other purposes. The procedure consists of taking profile readings at right angles to the centerline at each station along the route and at any plus-stations where a cross-section is necessary to represent correctly the ground surface. For a roadway the cross-section profile extends both to the right and to the left of the centerline, usually to the right-of-way line or at least as far as any possible earthwork will be constructed.

There are many different forms in which cross-sectioning data can be recorded in the field books. Figure 4.56 shows a form of field notes that is well suited for surveys along an existing highway, where uniformity of the roadway cross-section makes it satisfactory to obtain rod readings at fixed distances to the left and right of the centerline. Each ground-rod reading is recorded on the right-hand page as the numerator of a fraction; the denominator is the elevation of the point, found by subtracting the numerator from the *HI*. The differential levels for determining the heights of instrument are recorded as usual on the left-hand page. Figure 4.57 shows the manner in which the cross-section data of Figure 4.55 are plotted and also the relation-

	Cross-section notes					Route 47			November 4, 1977	
Sta.	BS	HI	FS	Elev.		Berger level No. 4 Fair, cool			Williamson, Inst. Lessler, Rod	
BM	6.15	757.58		751.43		10 in. Oak, 40′L. Sta. 412			Rogers, Tape	
	7.32	761.74	3.16	754.42		Left		℄	Right	
						30 ft	12 ft		12 ft	30 ft
415						4.6 / 57.1	5.0 / 56.7	5.8 / 55.9	6.2 / 55.5	6.8 / 54.9
+50						6.5 / 55.2	7.1 / 54.6	7.2 / 54.5	7.6 / 54.1	8.2 / 53.5
416						4.0 / 57.7	4.2 / 57.5	4.5 / 57.2	4.8 / 56.9	5.4 / 56.3
417						2.8 / 58.9	3.0 / 58.7	3.1 / 58.6	3.6 / 58.1	4.0 / 57.7
TP			4.10	757.64						

FIGURE 4.56 Field notes for cross-section leveling over uniform terrain

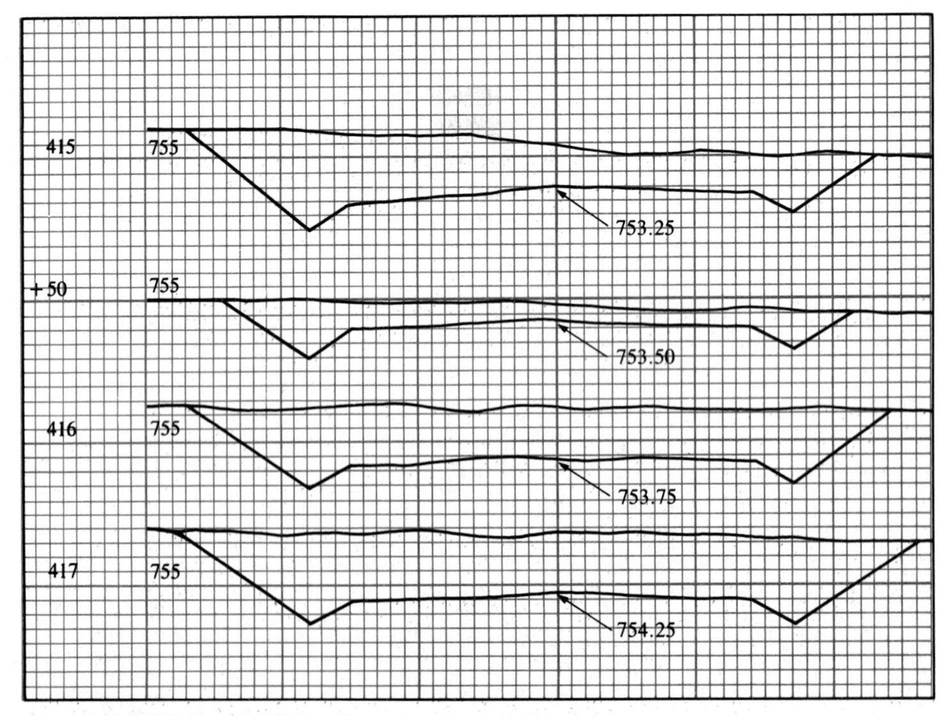

FIGURE 4.57 Cross-section profiles

ship of the proposed roadway to the ground line. The same scales are usually used both vertically and horizontally. Methods for computing volumes of earthwork from such cross-sections will be presented in Chapter 10.

Figure 4.58 shows a form of field notes for cross-sectioning over rough terrain. The ground rod readings are recorded as numerators, whereas the distances of the *GR* readings from the centerline are recorded as the denominator. Figure 4.59 shows a cross-sectioning operation being conducted over rough terrain.

4.30 MONITORING VERTICAL MOVEMENT BY LEVELING

Ground subsidence can be caused by the mining of underground minerals, the withdrawal of underground water, or by natural erosion processes. Vertical settlements of buildings can be caused by underground tunneling and excavation on adjacent sites, as well as by the weight of the building itself. Crustal movements associated with landslides and fault movements also have measurable vertical components. These vertical movements can all be detected and monitored by a regular program of leveling in which elevations of benchmarks are determined at regular time intervals. In this type of application, it is particularly important to have a good estimate on the accuracy of the benchmark elevations. It should be possible to determine whether the change in elevation at a benchmark is possibly due to survey error or actual vertical movement.

For example, let the elevation of a benchmark during a previous survey be h_0 with an estimated standard error of $\hat{\sigma}_0$. Suppose a new survey yields an elevation of h_1 with an estimated standard error of $\hat{\sigma}_1$ for the same benchmark. The change in elevation (Δh) is then computed as the difference in the two elevations—that is,

$$\Delta h = h_1 - h_0 \tag{4.19}$$

By applying the law of propagation of random error to Eq. (4.19), the following expression can be derived:

$$\hat{\sigma}_{\Delta h} = \sqrt{\sigma_0^2 + \sigma_1^2} \tag{4.20}$$

where $\hat{\sigma}_{\Delta h}$ is an estimated standard error of the change in elevation. There is a 68%, 95%, and 99.7% chance that the survey error can amount to $\pm\hat{\sigma}_{\Delta h}$, $\pm 2\hat{\sigma}_{\Delta h}$, and $\pm 3\hat{\sigma}_{\Delta h}$ respectively. Thus if the change in elevation exceeds $3\hat{\sigma}_{\Delta h}$, it is most likely that vertical movement has actually occurred.

However, suppose that a subsidence detection program is required to detect any vertical movement in excess of a specified magnitude, say M—that is,

$$3\hat{\sigma}_{\Delta h} \leq M \tag{4.21}$$

Cross section leveling along Illinois Route 150

Oct. 10, 1983
Warm, sunny, 71°F
Kern GK1-A level #2

⊼ A. J. Brown
▯ B. Rogers
◫ C. Yates
Tape G. Jones

USGS benchmark at NW corner of intersection with Route 123

Sta.	BS	HI	FS	Elev.								
BM 101	4.32	647.53		643.21								
0 + 00					10.3/30	4.1/25	5.3/10	8.4	6.9/8	1.2/2.5	7.3/40	
0 + 45					12.6/35	5.3/25	1.3/10	9.6	5.4/10	6.3/20	9.4/45	
1 + 00					10.6/35	8.9/25	2.3/10	6.3	2.3/7	10.3/25	3.5/40	
2 + 00					8.7/30	4.9/20	3.2/8	4.1	6.3/10	8.9/25	10.6/45	
2 + 50					11.6/35	6.3/25	1.7/10	3.1	1.3/10	4.9/15	10.7/30	
3 + 00					11.0/40	8.3/25	4.9/10	5.3	4.6/10	7.9/25	11.3/35	
TP1	4.89	645.69	6.73	640.80								
4 + 00					6.1/40	7.3/30	6.3/10	4.9	3.6/10	8.2/20	9.3/25	11.5/35
5 + 00					3.1/30	7.3/25	4.5/10	3.6	2.0/10	1.0/25	5.7/35	
5 + 75					4.2/35	8.3/20	6.0/10	3.3	4.7/10	5.9/20	10.3/40	

FIGURE 4.58 Field notes for cross-section leveling over rough terrain

FIGURE 4.59 Cross-sectioning (courtesy of California Division of Highways)

For the sake of simplicity, it can be assumed that the elevation of the same benchmark is determined with the same estimated standard error (say $\hat{\sigma}_h$) during each survey. Then from Equation 4.20

$$\hat{\sigma}_{\Delta h} = \sqrt{2}\hat{\sigma}_h \tag{4.22}$$

Substituting Eq. (4.22) into Eq. (4.21) yields the following requirement for $\hat{\sigma}_h$:

$$\hat{\sigma}_h \leq \pm \frac{M}{4.24} \tag{4.23}$$

EXAMPLE 4.8

The elevation of a benchmark located on a building foundation was found to be 463.265 ft with an estimated standard error of ±0.005 ft in May 1984. In September of the same year the elevation was found to be 463.247 ft with an estimated standard error of ±0.008 ft. Compute the change in elevation Δh and its estimated standard error $\hat{\sigma}_{\Delta h}$.

SOLUTION

$$\Delta h = 463.247 - 463.265 = -0.018 \text{ ft}$$

$$\hat{\sigma}_{\Delta h} = \pm\sqrt{(.005)^2 + (.008)^2} = \pm 0.009 \text{ ft}$$

$$2\hat{\sigma}_{\Delta h} = \pm 0.018 \text{ ft}$$

$$3\hat{\sigma}_{\Delta h} = \pm 0.027 \text{ ft}$$

Thus the change in elevation is less than the maximum expected survey error ($3\hat{\sigma}_{\Delta h}$). It is possible that the change in elevation was due to survey error.

EXAMPLE 4.9

A subsidence detection program is required to detect any vertical movement in excess of 0.01 ft. How accurately must the elevation of a benchmark be determined during each survey?

SOLUTION

From Eq. (4.18)

$$\hat{\sigma}_h \leq \pm \frac{0.01 \text{ ft}}{4.24} = \pm 0.0024 \text{ ft}$$

Therefore the benchmark elevation must be determined with an estimated standard error less than or equal to ±0.0024 ft.

4.31 TRIGONOMETRIC LEVELING

Trigonometric leveling is the process of determining elevation differences by measuring distances and vertical angles. It is particularly useful for mountainous areas, off-shore construction, and other situations where it is difficult to perform precise differential leveling. The topic of trigonometric leveling will be discussed in Section 6.14 after sufficient background information on angle and electronic distance measurements have been presented.

4.32 BAROMETRIC LEVELING

Barometric leveling is the process of determining elevation by measurement of atmospheric pressure. It is based on the principle that atmospheric pressure decreases with increases in elevation. The precise surveying *altimeter* shown in Figure 4.60 is an improved version of the old aneroid

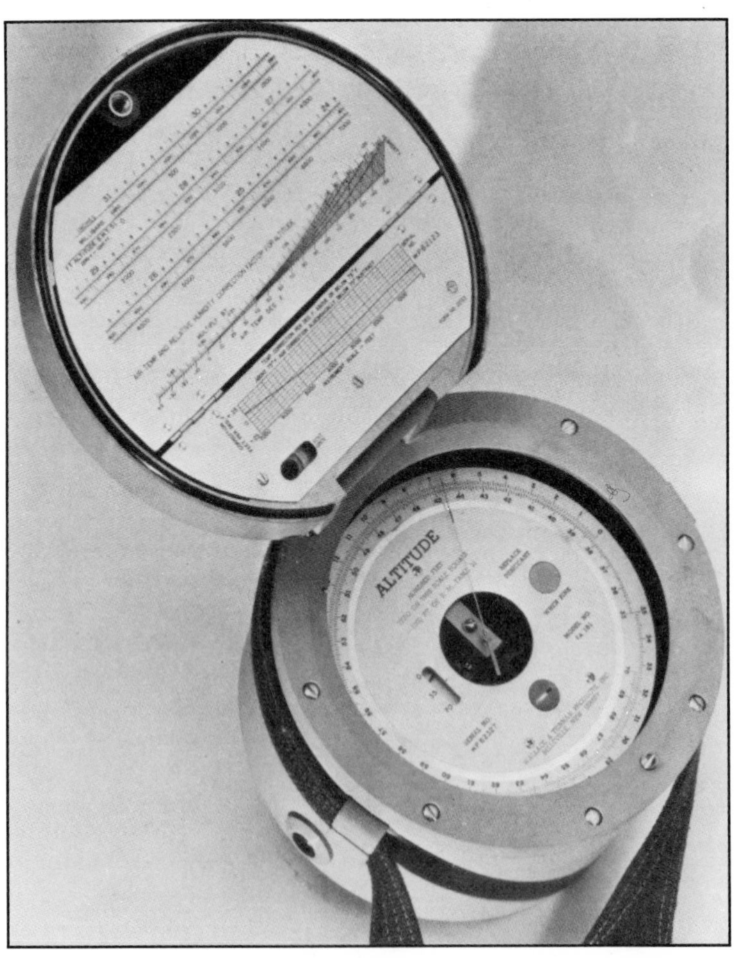

FIGURE 4.60 Surveying altimeter (courtesy of Wallace & Tiernan, Inc.)

barometer. It is remarkably sensitive to changes in atmospheric pressure and its operation is simple. The graduation of the dial into feet or meters makes possible direct readings of altitude.

The surveying altimeter is used primarily to determine differences in elevation between survey stations. In a procedure called the *single-base method,* two altimeters are used. One altimeter remains at a benchmark of known elevation where readings of the altimeter and a thermometer are made at regular time intervals, such as every 15 min. The second altimeter, called the *roving altimeter,* is transported to those points where elevations are desired. Readings are made of the altimeter and a thermometer. The time of readings are also recorded. The elevation difference between a field station and the reference benchmark is computed as the difference in the readings of the two barometers at the same point in time. Correction for differences in temperature (and sometimes humidity) can be applied to the measured difference in elevation.

Barometric leveling is used primarily in reconnaissance missions to obtain approximate elevation data. With proper field procedure and good equipment, barometric leveling in a stable atmosphere may yield differences in elevation with errors not in excess of ± 1 m.

PROBLEMS

4.1 The elevation of a benchmark is known to be 437.083 ft above the Cairo Datum, which is actually 20.434 ft below the National Geodetic Vertical Datum of 1929. Determine the elevation of the benchmark above the NGVD29 datum.

4.2 The sea level datum at San Francisco corresponds to a reading of 8.60 ft on the tide staff there. The City of Oakland datum intersects the staff at 11.59 ft. If the sea level elevation of a point is 1.89 ft, what is its elevation relative to the Oakland datum?

4.3 A determination of the sensitiveness of the bubble tube of a dumpy level was made. With the rod at a distance of 300 ft, a reading of 5.320 was obtained. After moving the bubble through five divisions, the rod reading was 5.155. Find (a) the sensitivity of the level tube per 2-mm graduation and (b) its radius of curvature.

4.4 Calculate the uncertainty in the reading of a level rod at a distance of 240 ft if the bubble is off center by one-fourth of a division in a 16″ level vial.

4.5 What are the level rod readings for each of the following diagrams?

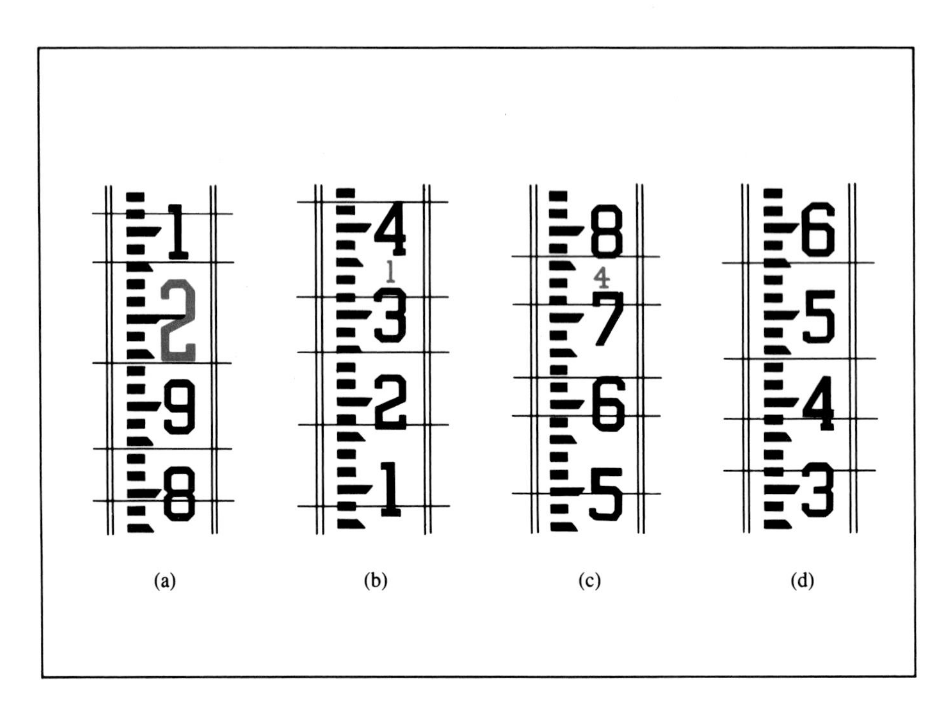

4.6 Complete the following record of differential levels, determine the elevation of point *A*, and make a page check.

Sta.	*BS*	*HI*	*FS*	Elev.
BM 36	3.37			507.37
TP 1	6.14		8.21	
TP 2	5.08		4.17	
TP 3	11.10		6.52	
TP 4	4.32		5.91	
TP 5	8.89		0.32	
Point *A*			1.15	

4.7 A peg test of a tilting level was made over a distance of 200 ft. With the instrument midway between stakes *A* and *B*, the reading on *A* was 6.47 and on *B*, 5.14. With the instrument beside *B*, the backsight on *B* was 4.82 and the foresight on *A* was 6.14. Is the line of sight inclined upward or downward and how much per 100 ft?

4.8 An automatic level was tested over a distance of 250 ft to determine the inclination of the line of sight when the circular bubble was centered. With the level midway between two points *A* and *B*, the reading on *A* was 5.205 and on *B*,

4.365. With the instrument beside *A*, the backsight on *A* was 4.190 and the foresight on *B* was 3.365. Is the line of sight inclined upward or downward and how much per 100 ft?

4.9 What is the error in the reading of a level rod which is 8 in. out of plumb at a height of 11.2 ft when:

a. a backsight of 10.50 is observed?

b. a foresight of 1.05 is observed?

4.10 A line of differential levels was executed from *BM* 205 (elev. 527.468 m) to Station *B* in Death Valley, California. The preliminary elevation of Station *B* was determined to be −71.872 m. The invar rod was known to be 1.0 mm too short (distributed uniformly over the 3.75 m rod) at 20°C, and its thermal coefficient of expansion was 0.00000036 per 1°C. The average field temperature was 33°C. Find the revised elevation of *BM* 237.

4.11 A line of differential levels was run from *BM* 105 (elev. 481.23 ft) to survey station *A*. The preliminary elevation of station *A* was determined to be 941.86 ft. It was then discovered that the 11.0-ft rod was 0.015 ft too long with the error uniformly distributed over the entire length of the rod. Calculate the revised elevation of station *A*.

4.12 Assume now that the rod of Problem 4.11 was not long but was short by 0.010 ft because of wear of the shoe or foot piece. What change will be effected in the elevation of station *A*?

4.13 A line of differential levels was executed with an automatic level down a long slope for a distance of 8.45 miles. Commonly accepted rules for leveling were violated by carelessly permitting uniform backsights of 100 ft and foresights of 300 ft. Following completion of the levels, a peg test revealed the line of sight was inclined downward 0.008 ft per 100 ft. If the preliminary or observed difference of elevation was −802.41 ft, find the corrected difference of elevation. Do not neglect curvature and refraction effects.

4.14 A line of levels was run through a tunnel beneath the Mississippi River at New Orleans. All benchmarks and turning points were located in the crown or roof of the tunnel and the level rod was inverted when readings were obtained. Complete the following record. Datum is mean sea level.

Sta.	*BS*	*HI*	*FS*	Elev.
BM 57	10.02			8.42 ft
TP 1	11.43		9.16	
TP 2	9.78		8.42	
TP 3	10.14		7.20	
TP 4	4.29		5.89	
Sta. *A*			11.03	

4.15 A level line is run to determine the difference in elevation between *BM* 102 and Station *B*, which are located approximately 3,000 ft apart. All the *BS* and *FS* distances are kept approximately equal to 150 ft. An automatic level is used.

a. The line-of-sight of the automatic level has a setting accuracy of ±1.0 sec ($\hat{\sigma}_s$). Determine the estimated standard error of the measured difference in elevation due to this source of error.

b. Suppose that the level rod can be read with a standard error ($\hat{\sigma}_r$) of ±0.003 ft. Determine the estimated standard error of the measured difference in elevation due to this source of error.

c. Determine the total estimated standard error ($\hat{\sigma}$) of the difference in elevation due to the two sources of error.

4.16 A level line was run between points *A* and *B* to determine the elevation difference between the two points. The distance between the two points is estimated to be about 6,000 ft. The *BS* and *FS* distances were kept approximately equal to 150 ft each. It is estimated that the root-mean-square (RMS) error of a single rod reading is ±0.005 ft.

Determine the root-mean-square (RMS) error of the elevation difference between *A* and *B* if:

a. the level line is run only once from *A* to *B*;

b. the level line is run twice, once from *A* to *B* and once from *B* to *A*.

4.17 In Problem 4.16 above, what is the maximum acceptable RMS error of a single *BS* or *FS* rod reading if the elevation difference between *A* and *B* must be determined with an RMS error of ±0.01 ft or better? Assume that the level line is run only once.

4.18 Complete the following pages of field notes and distribute the closure errors to determine the adjusted elevations for the survey stations. Also perform a page check on your calculation.

(a)

Sta.	*BS*	*HI*	*FS*	Elev.	
BM 20	2.440			1,123.223	
Sta. 33	4.259		3.910		
TP 1	3.418		3.490		
TP 2	5.529		3.515		
Sta. 34	2.694		2.391		
TP 3	3.674		0.795		
TP 4	7.677		2.028		
TP 5	11.314		0.441		
Sta. 36	0.624		8.750		
Sta. 37	2.841		8.977		
TP 6	2.644		4.330		
BM 30			7.268		
Known elevation of *BM* 30 = 1,124.495 ft					

(b)

Sta.	*BS*	*HI*	*FS*	Elev.	
BM 103	5.996			1,131.565	
TP 1	2.595		2.024		
TP 2	2.739		1.545		
Sta. 50	3.082		1.337		
TP 3	2.890		2.080		
Sta. 51	2.800		2.033		
BM 122			2.334		
Known elevation of *BM* 122 = 1,140.286 ft					

4.19 For each of the level lines or loops, distribute the closure errors and compute the elevations of the survey stations.

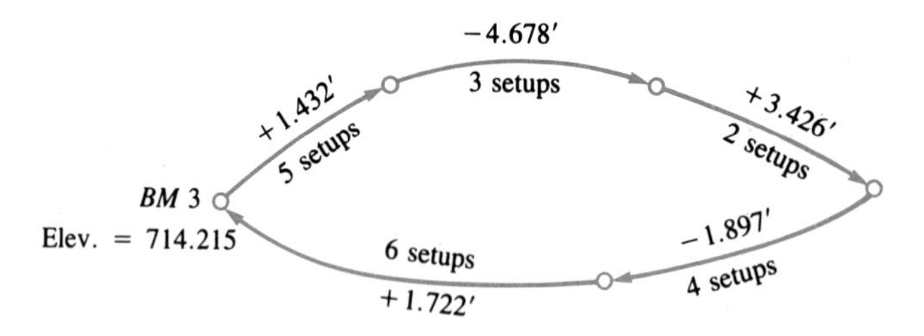

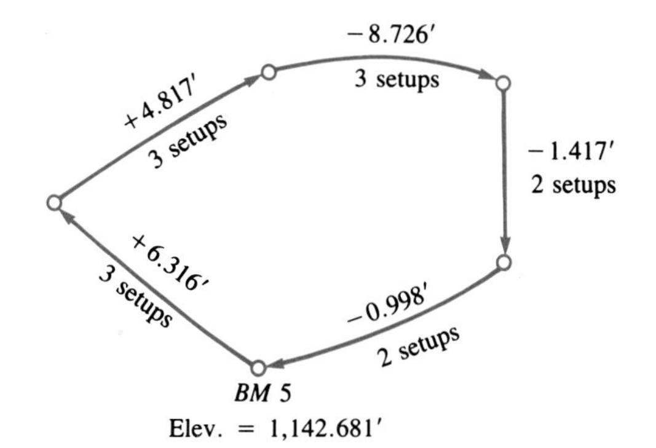

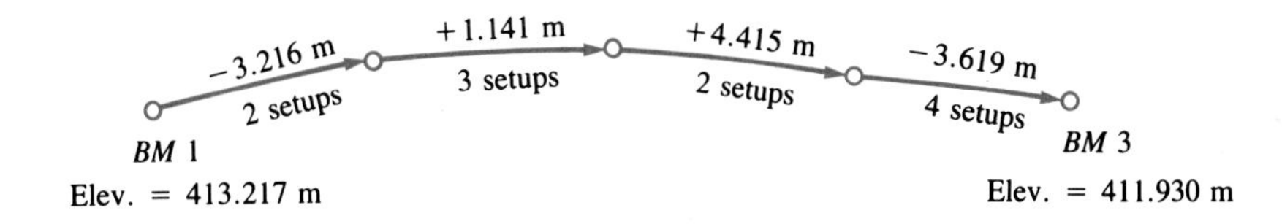

4.20 Adjust the level network shown in Figure 4.38 by successive iteration, and determine the elevations of stations *A*, *B*, *C*, *D*, *E*, *F*, *G*, *H*, *I*, *J*, and *K*.

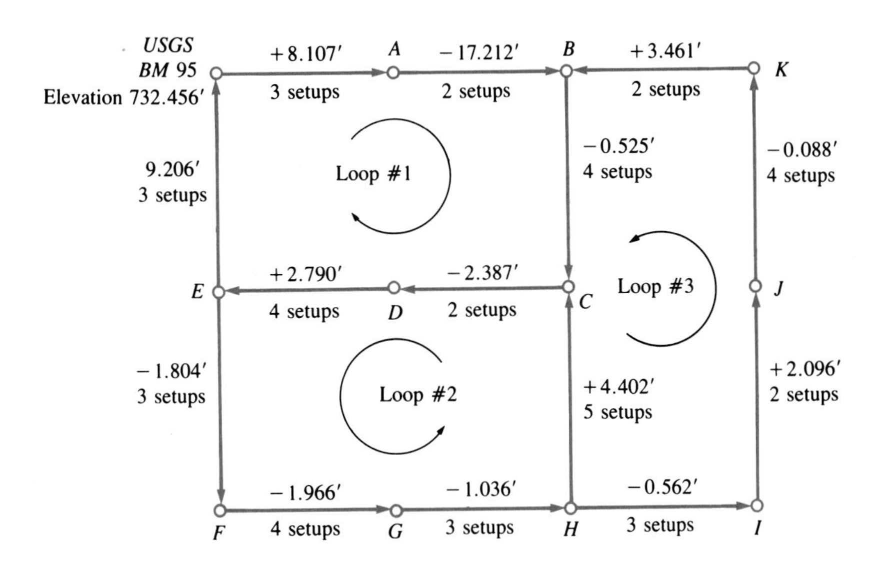

4.21 Adjust the following network by successive iteration, and determine the elevations of survey stations A, B, C, D, E, F, G, H, and I.

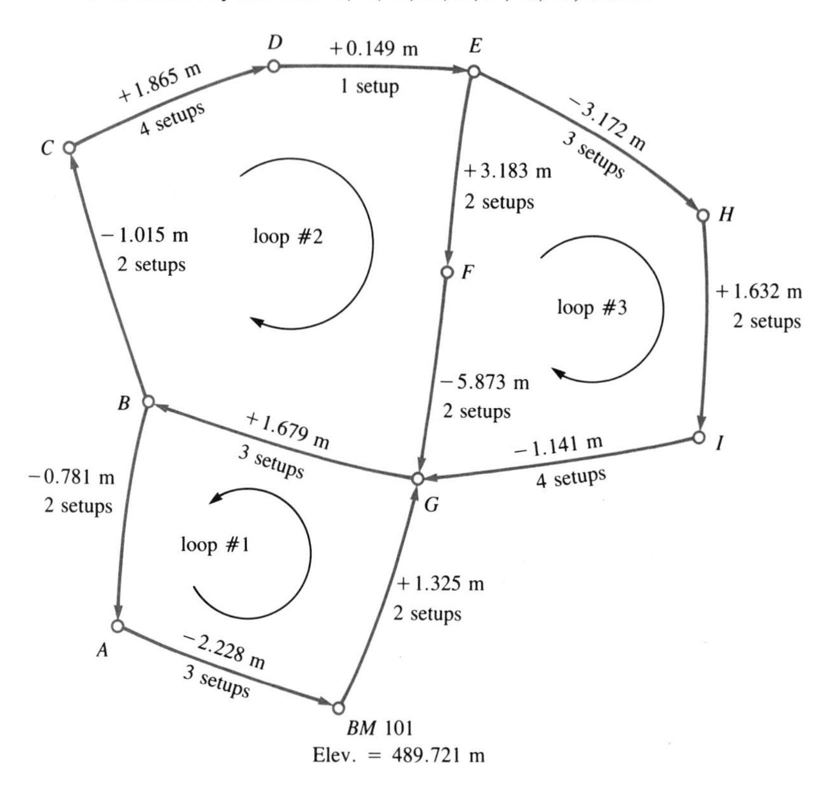

4.22 Complete the following page of field notes from three-wire leveling, and compute the elevation difference between NGS BM 1432 and station 43.

From NGS *BM* 1432 to Station 43					
Thread Readings	Mean	Thread Interval	Thread Readings	Mean	Thread Interval
1.435			3.221		
1.175			2.753		
0.915			2.286		
2.532			2.779		
2.275			2.538		
2.020			2.295		
2.569			2.825		
2.324			2.550		
2.080			2.275		
1.771			2.924		
1.535			2.677		
1.300			2.432		

4.23 A rod-type elevation monument is located in an area where ground subsidence is suspected. The elevation of this monument is to be determined every six months by differential leveling from a nearby NGS benchmark. This benchmark is located at an approximate distance of 2,000 ft. The leveling is to be performed with an accuracy sufficient to detect any vertical movement in excess of ± 0.02 ft. The NGS benchmark is known to be stable, and its elevation can be assumed to be error free for this purpose.

a. How accurately (that is, maximum allowable estimated standard error) must the elevation of the subsidence monument be determined during each survey?

b. If a sight distance of about 100 ft is to be used for both backsights and foresights, what is the maximum allowable estimated standard error for each rod reading?

c. Name two level instruments that are appropriate for this job.

d. Is an invar rod needed for this job? The elevation difference between the NGS benchmark and the subsidence monument is about 30 ft. The maximum difference in temperature between surveys is about 80°F.

4.24 Given below are the elevations and the RMS errors measured from two surveys for three subsidence monitoring points.

a. Compute for each point the change in elevation, the RMS error of the change, and the maximum expected survey error in the change.

	Elevations			
Point	June 1974	RMS Error	June 1984	RMS Error
101	563.14	±0.03	563.01	±0.06
102	579.26	±0.04	579.11	±0.05
103	533.19	±0.06	533.15	±0.07

b. Which point has elevation change exceeding maximum expected survey errors?

4.25 A line of levels is to be carried over the Mississippi River at Cairo using reciprocal leveling procedures. At *BM* 105 (elev. 427.080 ft) the backsight reading from the nearby instrumental position was 8.415 ft and the foresight reading across the river to station *B* was 10.015 ft. Shortly thereafter with the level set up close to station *B*, the backsight reading on *BM* was 5.428 ft and the foresight reading on station *B* was 6.852 ft. Find the elevation of station *B*.

4.26 Complete the following record of profile leveling and show the page check.

Sta.	*BS*	*HI*	*FS*	*GR*	Elev.
BM 41	3.26				827.37
5 + 00				5.2	
6 + 00				4.9	
TP	0.56		5.89		
7 + 00				8.5	
7 + 60				9.2	
8 + 00				11.7	
9 + 00				7.5	
BM 42			6.81		

REFERENCES

4.1 Berry, Ralph Moore. "History of Geodetic Leveling in the United States," *Surveying and Mapping,* American Congress on Surveying and Mapping, Vol. 36, No. 2 (June, 1976) 137–151.

4.2 Federal Geodetic Control Committee. *Classification, Standards of Accuracy, and General Specifications of Geodetic Control Surveys,* National Oceanic and Atmospheric Administration, 1974. Available from National Geodetic Information Center, NOS/NOAA, Rockville, MD 20852.

4.3 Federal Geodetic Control Committee. *Specifications to Support Classification, Standards of Accuracy, and General Specifications of Geodetic Control Surveys,* National Oceanic and Atmospheric Administration, 1975, Revised 1980. For Sale by Superintendent of Documents, U.S. Government Printing Office, Washington, D. C. 20402.

4.4 Lippold, Jr., H. R. "Readjustment of the National Geodetic Vertical Datum," *Surveying and Mapping,* American Congress on Surveying and Mapping, Vol. XL, No. 2 (June 1980) 155–164.

4.5 Mikhail, Edward M. *Observations and Least Squares.* Harper & Row, New York: 1976.

4.6 Ropes, Gilbert E. "Vertical Control of the Great Lakes," *Journal of the Surveying and Mapping Division,* American Society of Civil Engineers, Vol. 91, No. SU1 (April 1965) 35–49.

4.7 U.S. Geological Survey, "Leveling," *Topographic Instruction of the United States Geological Survey,* Book 2, Chapters 2E1–2E5, 1966. For Sale by Superintendent of Documents, U.S. Government Printing Office, Washington, D.C. 20402.

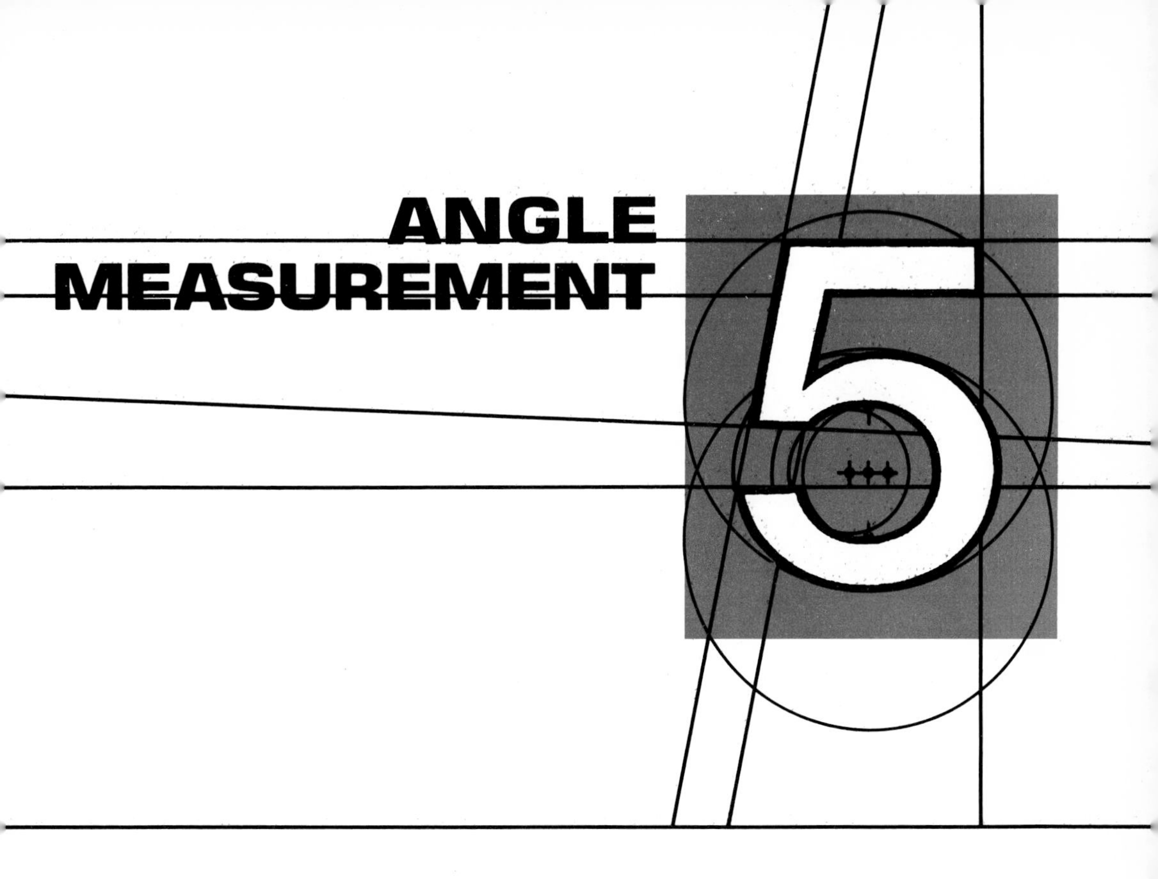

5.1 INTRODUCTION

Fundamentally, the purpose of surveying is to determine the relative locations of points on or near the surface of the earth. To fix the position of a point, both distance and angular measurements are usually required. Such angular measurements are either horizontal or vertical, and they are most commonly accomplished with instruments called *transits* or *theodolites*.

Although the ancients developed devices for angular measurements, it was not until 1571 that the concept of the modern transit was evolved. In that year Thomas Digges, an English mathematician and surveyor, published one of the earliest treatises on surveying in which he described his "Topographical Instrument," which was a forerunner of today's engineer's transit.

Before 1800 practically all surveying instruments were brought to America from France and England. As the frontier moved westward, the demand for surveying services greatly increased, and equipment soon was in short supply. The high prices of European equipment and the long delay in delivery were prime factors stimulating American manufacture of such instruments. David Rittenhouse, who was born in Germantown, Pennsylvania, in 1732 and who became an accomplished astronomer and surveyor at an early age, is reputed to have made the first surveying telescope in the United States and, independent of European practice, was the first to use lines of spider web as cross-hairs in the focal plane of the telescope. A compass bearing his name made for George Washington can be found in the Smithsonian Institute.

Various claims have been made about the origin of the first American transit, but both William J. Young and Edmund Draper should be credited, since available records indicate each independently produced a transit instrument around 1830.

The scope of this chapter includes an introduction to the basic concepts of angles and directions and a detailed treatment of the four most commonly used types of angle-measuring instruments: *vernier transits, scale-reading theodolites, digital theodolites,* and *electronic theodolites.*

5.2 UNITS OF ANGLE MEASUREMENT

An angle between two lines at a point is given by the difference in the directions of the lines. Only plane angles are considered here. The magnitude of an angle can be expressed in different units, most of which are basically derived from the division of the circumference of a circle in various ways. The principal systems of units are as follows:

Sexagesimal System. The circumference is divided into 360 parts called *degrees.* A degree is subdivided into 60 *minutes* (60′), and the minute is subdivided into 60 *seconds* (60″). This system is used almost exclusively in surveying practice in the United States. Long-established usage, the correlation of time and arc units in engineering astronomy (1 hr = 15°), and other considerations favor the continued use of the sexagesimal system in surveying.

Centesimal System. The use of decimalized degrees in various engineering computations has certain inherent advantages. This led to the creation of the centesimal system in which the circumference of a circle is divided into 400 parts called *grads* (or grades). Hence $100^g = 90°$. The grad is divided into 100 centesimal minutes (100^c), and the centesimal minute is divided into 100 centesimal seconds (100^{cc}). Thus an angle can be expressed as 236.4268^g with the first pair of digits after the decimal point representing centesimal minutes and the second pair centesimal seconds. However, the distinction between the minutes and seconds need not be explicitly denoted, as shown in the following addition of angular quantities where the simplicity of the grad system is readily apparent. It has wide usage in Europe.

EXAMPLE 5.1

Find the sum of the three tabulated angles on the next page.

SOLUTION

Centesimal System		Sexagesimal System	
	100.4527		75°51′23″
	251.7590		207°18′41″
	312.0314		340°39′57″
Sum =	664.2431	Sum =	623°50′01″
or	264.2431^g	or	263°50′01″

An angle expressed in grads can be converted to degrees by multiplying the grad value by 0.9. Conversely, an angle expressed in degrees can be converted to grads by dividing the degree value by 0.9. The conversion calculations are illustrated by Examples 5.2 and 5.3 below.

EXAMPLE 5.2

What is the sexagesimal equivalent of 264.2431^g?

SOLUTION

$$264.2431 \times 0.9 = 237.81879^\circ$$
$$0.81879^\circ \times 60 = 49.1274'$$
$$0.1274' \times 60 = 7.644''$$

Hence the sexagesimal value is 237°49′07.644″.

EXAMPLE 5.3

What is the grad equivalent of 263°50′01″?

SOLUTION

$$50' = \frac{50}{60} = 0.83333^\circ$$

$$01'' = \frac{1}{60 \times 60} = 0.00028^\circ$$

$$\frac{263.83361^\circ}{0.9} = 293.1485^g$$

The Radian (Rad). One *radian* is defined as the angle at the center of a circle that is subtended by an arc having exactly the same length as the radius. Let γ be the angle, expressed in radians, which is subtended by an arc of length s along the circumference of a circle of radius r. Then, the following relationship exists:

$$\gamma = \frac{s}{r} \text{ radians}$$

One radian equals $360°/2\pi$, or approximately 57.30°. Equivalent values for 1 radian are as follows:

Centesimal System	Sexagesimal System
1 rad = 63.6619772^g	1 rad = 57.2957795°
1 rad = $6{,}366.19772^c$	1 rad = 3,437.74677′
1 rad = $636{,}619.7724^{cc}$	1 rad = 206,264.806″
also, 1^{cc} = 0.000001571 rad	
and 1″ = 0.000004848 rad	

The radian, in contrast with the other units of angular measure previously mentioned, is sometimes referred to as the natural unit of angle because there is no arbitrary number, like 360, in its definition. It is the official unit for the plane angle in the SI system.

5.3 HORIZONTAL, VERTICAL, AND ZENITH ANGLES

In surveying angles are measured either in a horizontal plane, yielding horizontal angles, or in a vertical plane, yielding vertical angles. In Figure 5.1, points A, B, and C are three points located on the earth's surface. Points A', B', and C' are the projections of points A, B, and C, respectively, onto a horizontal plane. Angles $A'\hat{B}'C'$, $B'\hat{C}'A'$, and $C'\hat{A}'B'$ are the horizontal angles.

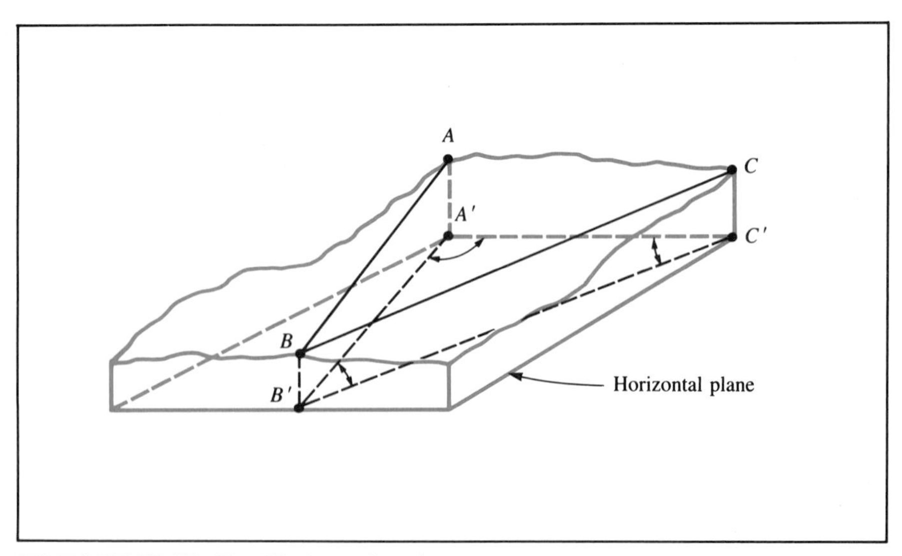

FIGURE 5.1 Horizontal angles

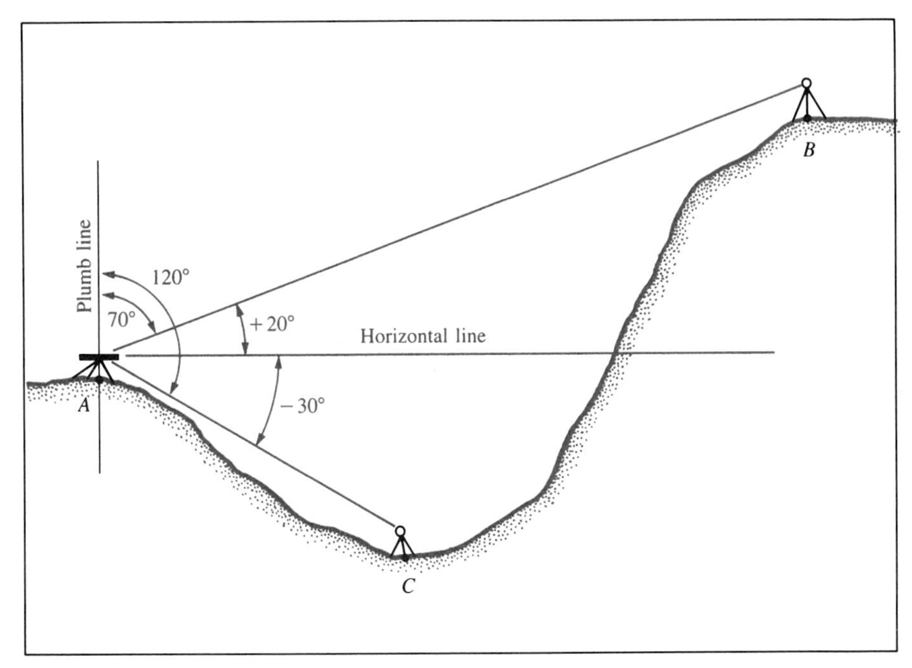

FIGURE 5.2 Vertical and zenith angles

A *vertical* angle is measured in a vertical plane using the horizontal plane as reference plane. When the point being sighted on is above the horizontal plane, the vertical angle is called an *angle of elevation* and is considered a positive angle. When the point being sighted on is below the horizontal plane, the angle is termed an *angle of depression* and is considered a negative angle. The value of a vertical angle can range from −90° to +90°.

A *zenith* angle is also measured in a vertical plane but uses the overhead extension of the plumb line as reference line. Its value ranges from 0° to 180°. In Figure 5.2, the vertical angles measured at station *A* to targets at stations *B* and *C* are 20° and −30°, respectively. The corresponding zenith angles are 70° and 120° to stations *B* and *C*, respectively. Zenith angles are also frequently referred to as *zenith distances*.

5.4 TRUE BEARINGS

It is convenient to choose or fix a reference line to which directions of all the lines of a survey are referred. Such a reference line is called a *meridian*. The *true meridian* passing through a point on or near the earth's surface is the great circle that passes through that point and the north and south poles of the earth.

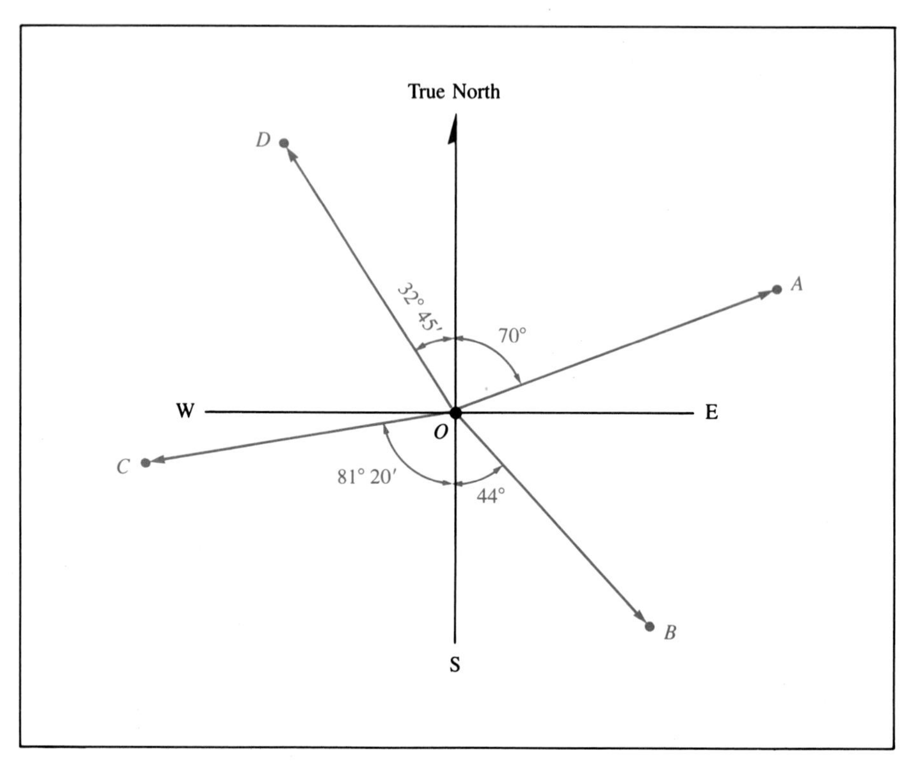

FIGURE 5.3 True bearings

The *true bearing* of a line is the acute angle that the line makes with the true meridian. The true bearing of a line is expressed as ''North (or South) so many degrees to the East (or West).'' Thus, in Figure 5.3, the true bearing of line *OA* is North 70° to the East or simply expressed, N 70° E. The true bearing of a line never exceeds 90° and is never referenced to the east or west line. For example, it is incorrect to express the bearing of line *OA* in Figure 5.3 as either E 20° N or S 110° E. The correct true bearings for lines *OB*, *OC*, and *OD* in Figure 5.3 are: S 44° E, S 81°20′ W, and N 32°45′ W, respectively.

5.5 MAGNETIC BEARINGS AND DECLINATION

A magnetic needle, when allowed to come to rest in the earth's magnetic field, will point in the direction of the magnetic north. The direction in which the magnetic needle rests is called the *magnetic meridian*.

The magnetic bearing of a line is the acute angle that the line makes with the magnetic meridian. In Figure 5.4, the magnetic bearings of lines *OE*,

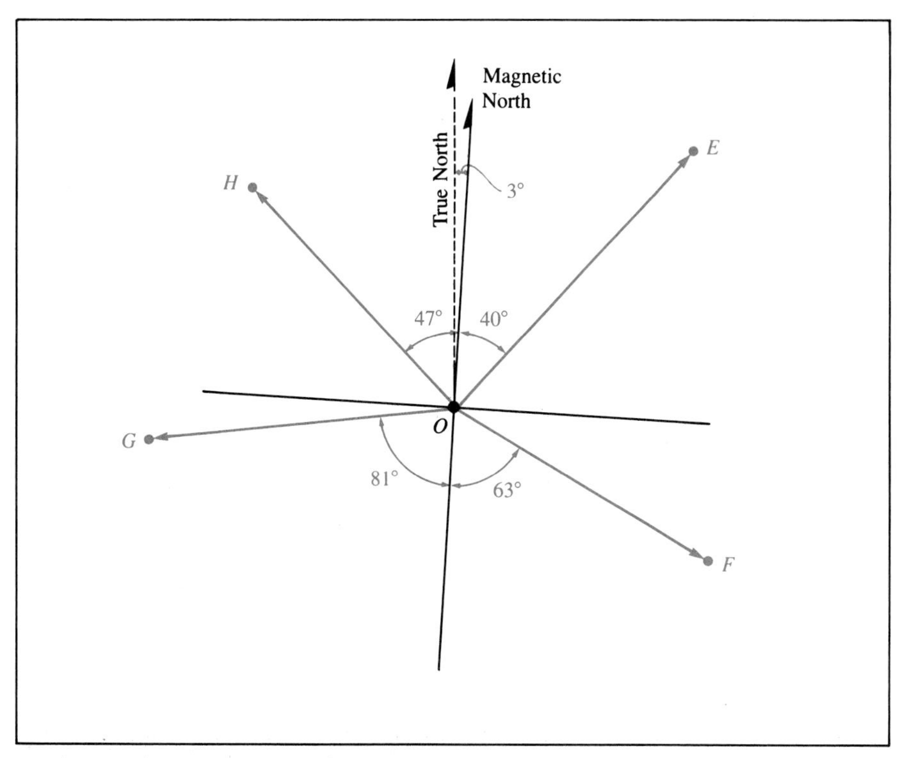

FIGURE 5.4 Magnetic bearings

OF, *OG*, and *OH* are: N 40° E, S 63° E, S 81° W, and N 47° W, respectively. Since the magnetic meridian is 3° east of the true meridian, the corresponding true bearings for lines *OE*, *OF*, *OG*, and *OH* are: N 43° E, S 60° E, S 84° W, and N 44° W, respectively.

The angle between the magnetic and true meridians is called the *magnetic declination*. The declination is expressed as the angular distance east (or west) of the true meridian. Thus a declination of 3°E means that the magnetic meridian is 3° east of the true meridian, as shown in Figure 5.4.

The magnitude of magnetic declination varies from one location to another, and at a given location, also changes with time. Figure 5.5 shows the distribution of magnetic declination in the United States for the epoch 1980. Depicted are the lines of equal magnetic declination. The position of the agonic line or line of 0° declination is to be noted. To the west of it the declination is east, and to the east of it the declination is west.

In general, the average value of the magnetic declination changes from one year to the next, and the change usually continues in one direction for many years. This long-period variation is termed the *secular change*. The amount in one year is known as the *annual change*. Some isogonic charts, such as that shown in Figure 5.5, display a set of lines showing places of equal annual change. Such lines are called *isopors*.

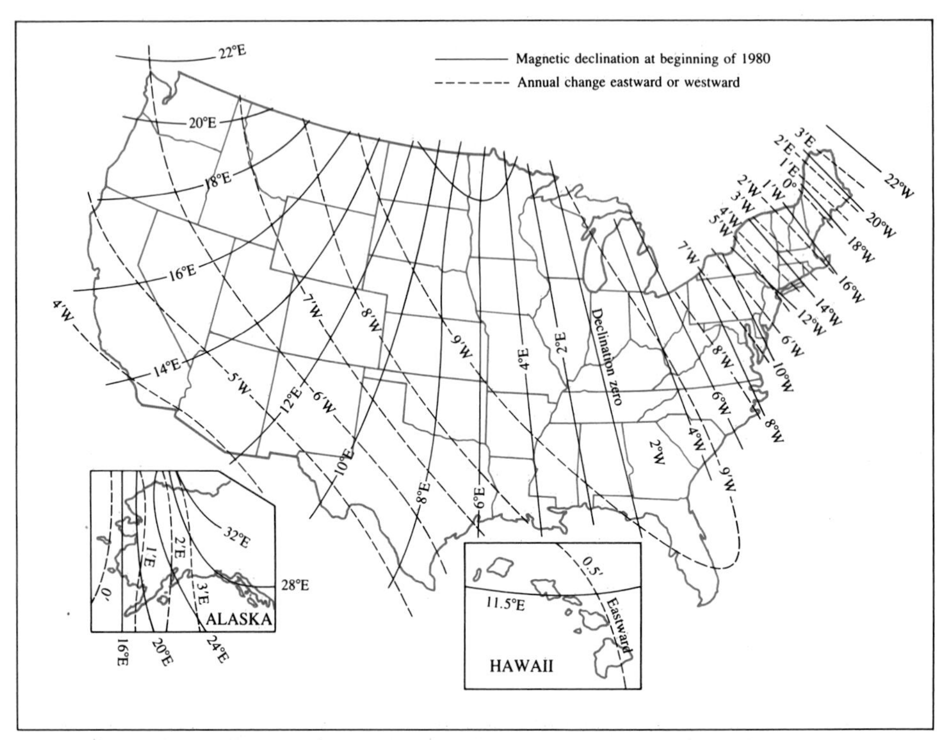

FIGURE 5.5 Magnetic declination in the United States—epoch 1980 (from a chart of the same title published by the U.S. Geological Survey)

The small seasonal effect on the declination is called the *annual variation*. For example, at a given location the declination may be slightly greater in winter than in summer.

There is also a daily variation—that is, a slight swing—amounting to only a few minutes of arc between morning and afternoon observations. Superimposed on this diurnal variation may be irregular changes associated with magnetic storms which may last many hours or even several days.

5.6 AZIMUTHS

The direction of a line may also be expressed by its azimuth. The *azimuth* of a line is the clockwise angle that the line makes with the north end of the reference meridian. The magnitude of an azimuth may range from 0° to 360°. For example, the azimuths of lines *OA*, *OB*, *OC*, and *OD* in Figure 5.6 are: 70°, 136°, 261°20′, and 327°15′, respectively, with respect to the true meridian.

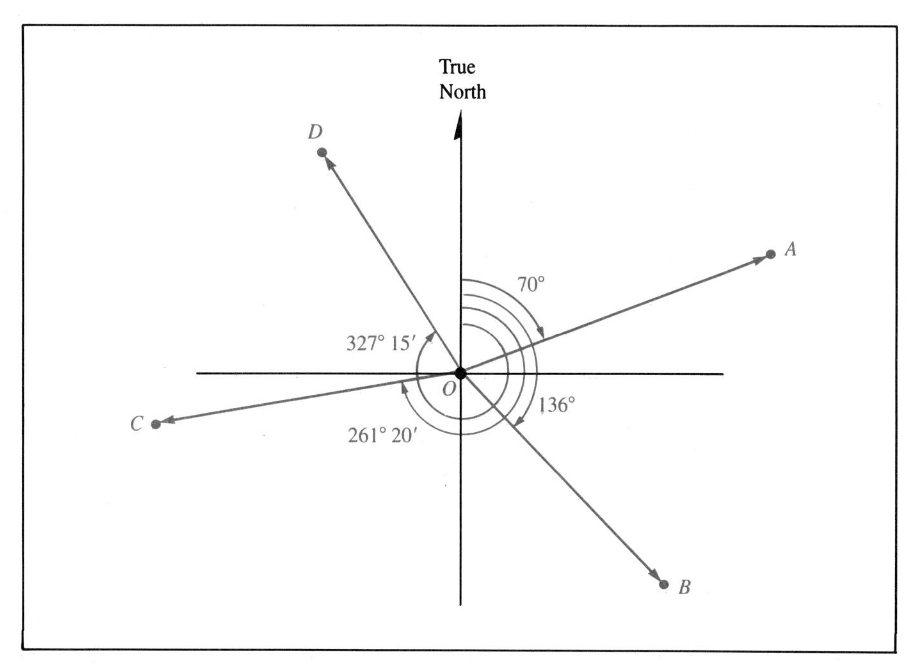

FIGURE 5.6 Azimuth of a line

Sometimes azimuths are referred to the south end of the meridian, in which case the azimuths of lines *OA*, *OB*, *OC*, and *OD* in Figure 5.6 are 250°, 316°, 81°20′ and 147°15′, respectively. All the azimuths in a surveying project should refer to the same end of the meridian. In this book, unless otherwise stated, the term azimuth will refer to the north end of the meridian.

5.7 BACK BEARING AND BACK AZIMUTH

The back bearing of a line *OA* is the bearing of the same line going from *A* to *O*. In Figure 5.7, the true bearing of line *OA*, going from *O* to *A*, is N 30° E. The *back bearing* of the same line *OA* (but going from *A* to *O*) is S $(30° + \theta)$ W, where θ is an angle due to the convergence of the meridians. The magnitude of θ depends on the east-west distance between the two end stations (*A* and *O*) and the average latitude along the line. For example, for an east-west distance of 1 mile and an average latitude of N 40°, θ amounts to about 44 seconds. Equations for computing the convergence of meridians will be given in Section 8.16.

Similarly, the *back azimuth* of line *OA* in Figure 5.7 is $30° + 180° + \theta = 210° + \theta$.

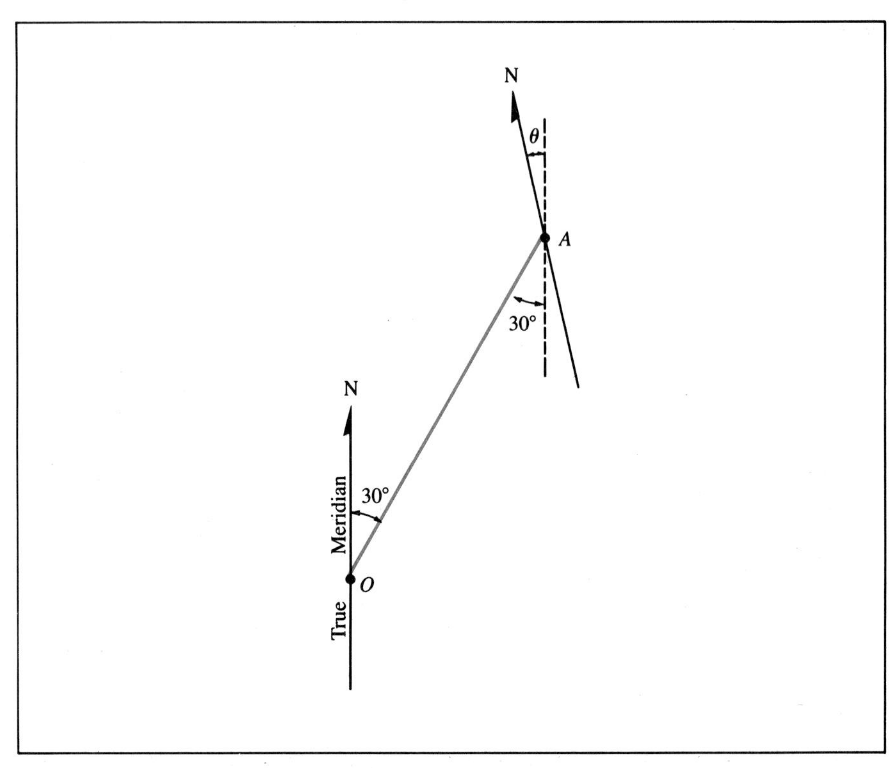

FIGURE 5.7 Back bearing

5.8 MAGNETIC COMPASS

The earliest literary reference to what later became known as the magnetic compass established the fact that crude versions of this device appeared in Northern Europe around the year 1200. It seems probable, however, that the Chinese at least 100 years earlier had used magnetized needles suspended from threads for the purpose of defining directions.

The essential features of the compass are shown in Figure 5.8a and b. These include a circle graduated in quadrants from 0° to 90° both east and west from both the north and south points of the compass, a magnetized steel needle supported on a steel pivot and jeweled bearing and counterbalanced against the dip of the needle, and also a line of sight fixed with respect to the north-south points of the compass box. Since the needle dips down to the north, the counterbalance is placed south of the pivot and serves to indicate the north and the south ends of the needle. Note that, in sighting any points, since the graduated circle turns while the needle remains stationary on its pivot, the west and east points of the circle are interchanged to give the correct bearing if the north end of the needle is read. For example, the bearing of the line of sight in Figure 5.8a is N 25° W.

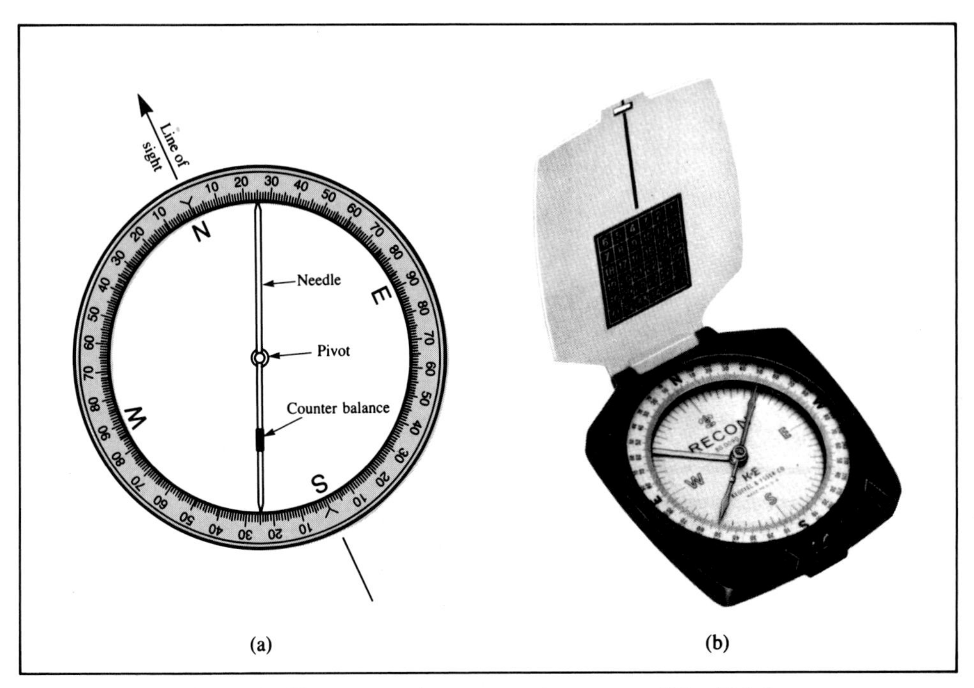

FIGURE 5.8 Magnetic compass (courtesy of Keuffel & Esser Company)

A clamp is provided to lift the needle from its pivot when not in use. It is important that this be done; otherwise the needle will jar on its bearing and soon become sluggish and insensitive.

Since the magnetic needle is affected by any nearby iron, steel, or other electromagnetic influence, a compass should not be used in close proximity to power lines, vehicles, hydrants, and like objects.

5.9 GYRO COMPASS

Most writers cite the famous French physicist Foucault (1819–1868) as the inventor of the first gyroscopic instrument for determining true direction. Although the earliest successful gyro compasses were used in ships, their large and heavy motors and low limits of accuracy precluded acceptance by surveyors. In the early 1950s, the requirements for a rapid and reliable method of determining azimuth in deep mines stimulated the development of what has been variously termed the gyro compass, gyro attachment, or gyro azimuth surveying instrument.

Figure 5.9 shows the Wild GAK 1 gyro attachment set up on the top of an optical transit modified with a bridge that supports the attachment and rests on the telescope standards. Three centering pins on the bridge insure

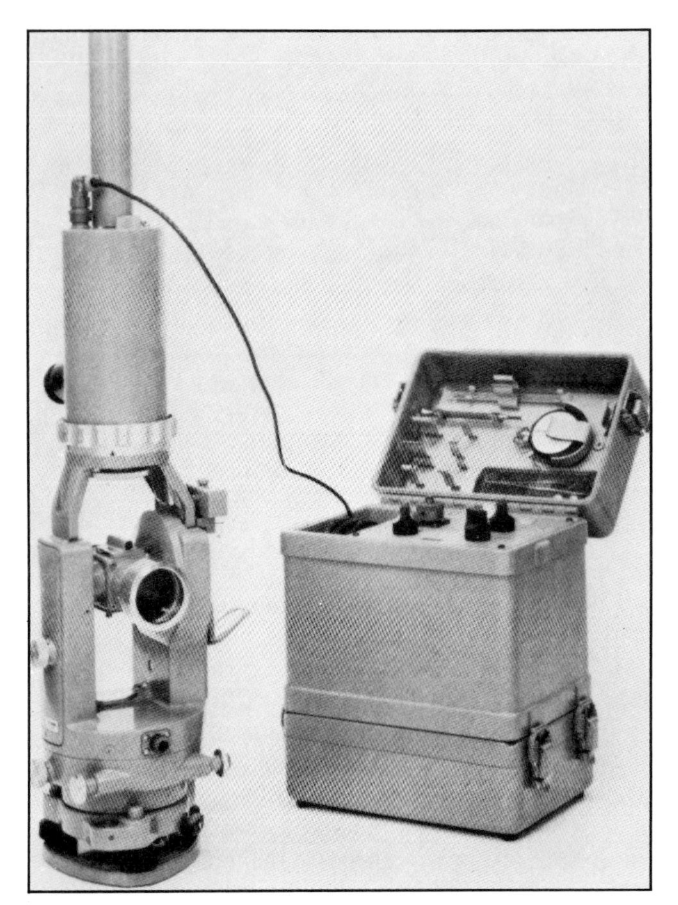

FIGURE 5.9 GAK1 Gyro Compass on Wild T1b theodolite (courtesy of Wild Heerbrugg Instruments, Inc.)

that the gyro always has the same position relative to the line of sight of the transit's telescope. The gyro motor is suspended on a thin metal tape and rotates at a speed of about 22,000 r.p.m. Under the influence of the earth's spinning motion, the spin axis of the motor seeks the meridian plane and oscillates about it over a small angular range. By measuring the size of these oscillations the north direction is taken to be at their midposition.

The gyro attachment can be useful whenever the accuracy requirements for orienting a survey lie within the capabilities of the instrument. The manufacturer states that the GAK 1 will provide a true azimuth with a standard error of not more than $\pm 20''$ in a total operating time of 20 minutes.

The gyro compass is particularly useful in the transfer of direction between the several levels of deep subsurface works such as mines and tunnels. It has also found application in certain surface surveying operations where the continued prevalence of cloudy skies has made astronomic determination of direction impracticable. Astronomic procedures for determining directions by sighting to the sun or stars will be discussed in Chapter 16.

The latest and most accurate version of a gyro compass is a lightweight gyro azimuth surveying instrument that was developed by the military for use by field artillery units. It consists of gyroscopic unit and theodolite integrally combined. Under rigorous test conditions this instrument was able to establish direction with uncertainties of only a few seconds.

5.10 PRINCIPAL ELEMENTS OF AN ANGLE-MEASURING INSTRUMENT

There are four common types of angle-measuring instruments: vernier transits, which are also often referred to as engineer's transits; scale-reading theodolites; digital theodolites; and electronic theodolites. However, all angle-measuring instruments have the following basic elements: (1) a line of sight, (2) a horizontal axis about which the line of sight revolves, (3)

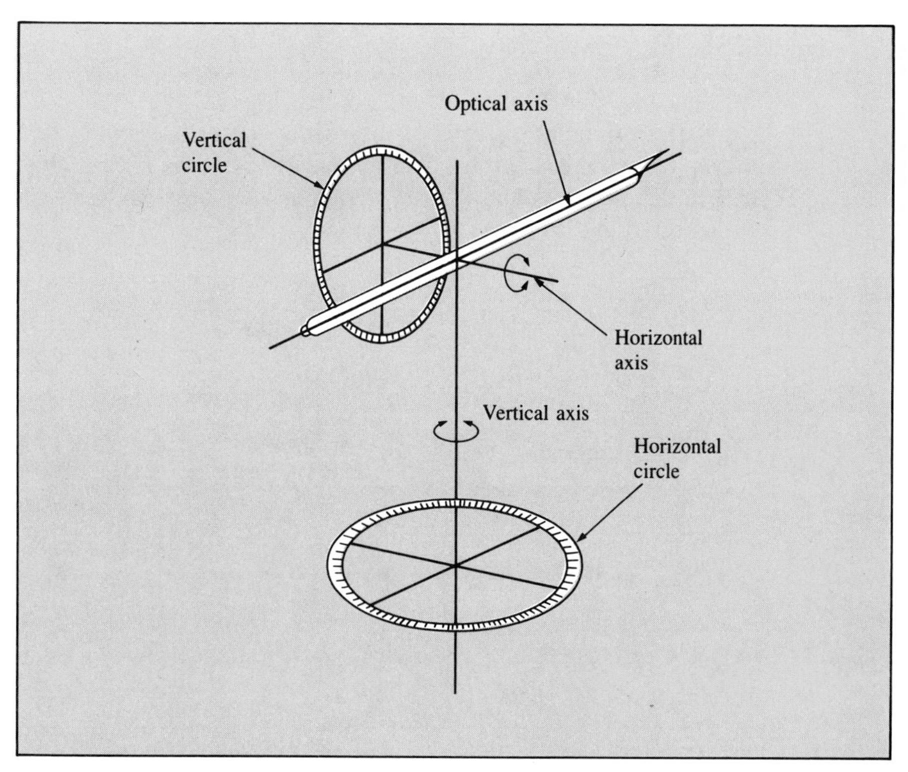

FIGURE 5.10 Principal elements of an angle-measuring instrument

a vertical axis about which the line of sight can be rotated, (4) a graduated vertical circle for measuring vertical angles, and (5) a graduated horizontal circle for measuring horizontal angles. These five basic elements are illustrated in Figure 5.10 on the previous page. When an instrument is in perfect adjustment, the following conditions exist:

1. The line of sight, the horizontal axis, and the vertical axis are mutually perpendicular;
2. The horizontal axis is perpendicular to the vertical circle; and
3. The vertical axis is perpendicular to the horizontal circle.

The characteristics and operation of the four types of angle-measuring instruments will be discussed in detail in the remaining parts of this chapter.

5.11 SURVEYING TELESCOPE

The surveying telescope serves the two purposes of fixing the direction of the line of sight and of magnifying the apparent size of objects in the field of view.

The *line of sight* is the line fixed by the intersection of the cross-wires and the optical center of the objective lens.

The principal features of the surveying telescope are the objective lens, the cross-wires that are also often referred to as the reticule, and the eyepiece. These parts and their relations to each other for a vernier transit are illustrated in Figure 5.11.

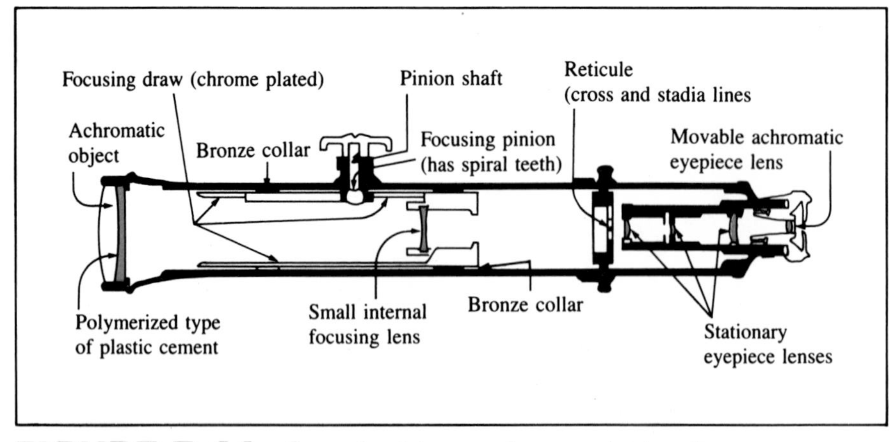

FIGURE 5.11 Surveying telescope for a vernier transit (courtesy of Keuffel & Esser Company)

THE OBJECTIVE LENS

The objective lens forms an image of any object within its field of view. This image is a real image and lies in a plane within the telescope, at a distance from the lens that depends on the curvature of the lens surfaces and the distance to the object. The distance from the lens to the image is called the *focal distance* (f), for that particular object. If the object is at a great distance, the image will be formed at a distance from the lens called the *principal focal length* (F). These focal lengths will, of course, be different for different telescopes. Thus the objective lens may be compared with the lens of a camera, which forms an image at a definite distance from the lens—that is, on the photographic plate or film. Since this distance varies for different objects, a means of changing the focal distance (that is, of focusing the lens) is provided on the telescope by the focusing screw shown in the illustration.

The use of an achromatic objective and eyepiece will correct lens aberrations that cause imperfect images with edges fringed with rainbow colors.

THE EYEPIECE

The image that is formed by the objective lens is inverted and small in size. A system of eyepiece lenses is used, therefore, to magnify the image and, in many telescopes, to reinvert or to yield an erect image of the object. The eyepiece, then, may be thought of as a microscope with which to view the image formed by the objective lens. An image may be viewed through the eyepiece only when it lies in the focal plane of the eyepiece; accordingly, a small focusing adjustment is also provided for the eyepiece. An image that has thus been brought into the common focal plane of both the objective and eyepiece lenses will appear to be magnified and distinct.

Figure 5.11 is a typical surveying telescope used in a vernier transit. It is of the internal focusing type with erecting eyepiece. The objective is fixed at the far end of the telescope, and focusing is accomplished through the use of the small internal focusing lens, which moves in response to rotation of the knob on the pinion shaft. Two major advantages of the internal-focusing telescope are (1) the interior of the telescope is virtually free from moisture and dust because both ends of the telescope are closed, and (2) an instrumental constant, which is relevant to stadia observations (Chapter 9), is practically eliminated.

THE CROSS-WIRES

When the image of any object is seen plainly, it lies in the common focal plane of both the objective and the eyepiece lenses. The position of this common plane can be altered by moving (focusing) the eyepiece. If cross-wires are placed in this common focal plane, they will appear to be projected on the object viewed and will serve to fix the line of sight on any point of the object. This condition will be effected if the cross-wires are fixed in a stationary position and if the eyepiece is then focused upon them and, finally, if the image of the object to be viewed is brought into the common focal plane by focusing the objective lens.

For proper use of the telescope, the eyepiece must first be focused on the cross-wires and then the objective lens focused on the object to be observed. Since the position of the cross-wires is fixed, it will be necessary to focus the eyepiece only once for the day's work, unless the position of the eyepiece somehow becomes altered.

The manner of mounting the cross-wires in a vernier transit is illustrated in Figure 5.12. The cross-wires consist of finely drawn platinum wire glued in position on a heavy brass ring. For modern theodolites, the cross-wires are formed by lines etched onto a glass diaphragm. Four threaded holes are drilled into the edge of the ring to receive capstan-headed screws that are inserted through slots in the barrel of the telescope tube. The heads of these screws bear against curved washers. By this arrangement the cross-hair ring is held suspended by the capstan screws and within the telescope tube. It is thus held securely and firmly in place, but is subject, as occasion arises, to a small amount of lateral movement by turning the capstan screws. This movement is used for purposes of adjustment, to be described later, and for this reason the capstan screws are commonly called *adjusting screws*.

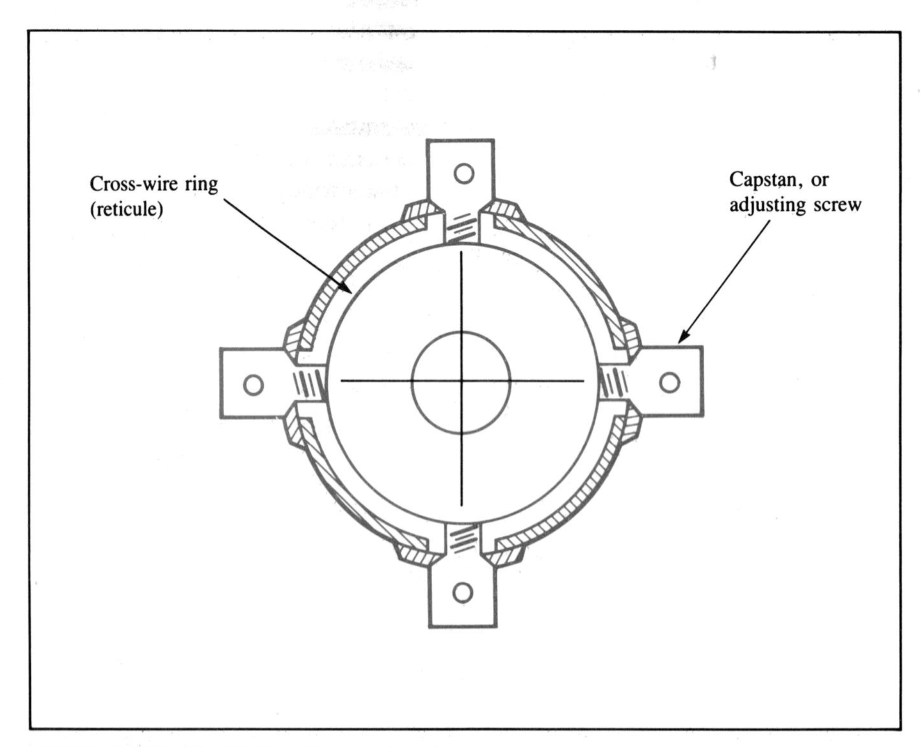

FIGURE 5.12 Cross-wire ring

Some reticule designs incorporate two additional horizontal cross-wires situated equal distances above and below the main horizontal cross-wire. They are called *stadia lines*.

MAGNIFICATION

The magnification of a telescope is fixed by the ratio of the focal lengths of the objective and the eyepiece lenses. It can be determined closely for any telescope by viewing, at a close range, a graduated rod with one eye looking through the telescope and with the other naked eye. Thus two images are seen, one being the magnified and the other the natural size of the rod. By a suitable adjustment of these images, the observer may count the number of divisions on the unmagnified rod, which is covered by one of the magnified divisions. This number is the measure of the magnification of the telescope, expressed as diameters. The magnification for telescopes for transits varies from 15 to 30 diameters, and for levels from 25 to 40 diameters.

High magnification, beyond proper limits, is a disadvantage, because it limits the field of view and reduces the illumination or brightness of the objects viewed. Accordingly, the size of the aperture of the objective lens and the quality of the lens system are considerations as important in a good telescope as the magnification.

The field of view of a telescope refers to the angle at the eye of the observer subtended by the arc whose magnitude is the width of the field viewed through the telescope. The field of view varies from approximately 1°30′ for a magnifying power of 20 to 45′ for a magnification of 40.

The illumination of the image depends on the magnifying power, the quality and number of the lenses, and the effective size of the objective. The use of coated lenses substantially increases the amount of light transmitted through the telescope. The advantages are (1) a sharper and brighter image, and (2) the elimination of stray light produced by internal reflections. Coated lenses are easily recognized by their purple color.

CARE OF TELESCOPE

The optical system of a telescope is particularly vulnerable to severe shocks that may disturb the lens alignment and to dust and dirt that may not only accumulate on the exposed metal surfaces but also penetrate the interior of the tube. Ordinarily, the only parts of the telescope visible to the person operating the instrument will be the faces of the objective and eyepiece lenses, the focusing devices such as knobs and knurled rings, and the capstan heads of the reticule adjusting screws, because the telescope is never dismantled in the field.

The proper care of the lenses is especially important. Optical glass is not particularly hard and may easily be scratched. Dust should be carefully brushed away with a clean camel's-hair brush, holding the lens downward when brushing so that any loosened dirt will fall away. Fingerprints on lens surfaces should be avoided because the oil transmitted from the human skin to the glass will tend to retain dust. If further cleaning is necessary, use special lens-cleaning tissue, which can be lightly wadded and gently passed over the lens surface. Rubbing must be avoided. Lens-cleaning fluid, obtainable from an optician, can be used with lens tissue for removal of dust not yielding to the first treatment. Usually, breathing on the lens will provide sufficient moisture for ordinary cleaning. Solvents other than ether or alcohol must not be used. When dust has worked its way into the interior of the telescope, it is necessary to have the instrument serviced by a qualified technician.

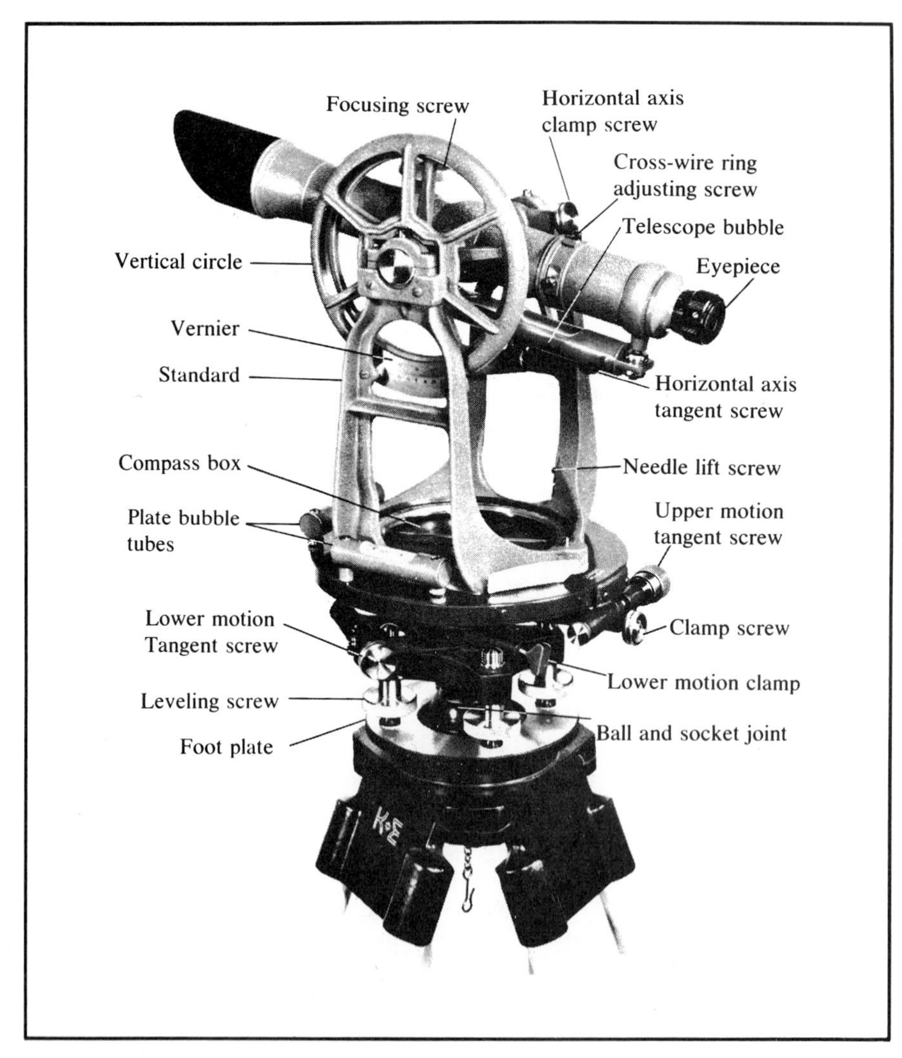

FIGURE 5.13 Engineer's transit (courtesy of Keuffel & Esser Company)

5.12 PARTS OF A VERNIER TRANSIT

A typical vernier transit is shown in Figure 5.13. It consists of three major parts: the leveling head, the lower plate, and the upper plate. These parts are illustrated in Figure 5.14.

1. *Leveling head.* The leveling head is the assembly that supports the instrument on the tripod and provides a means for leveling the instrument by the use of

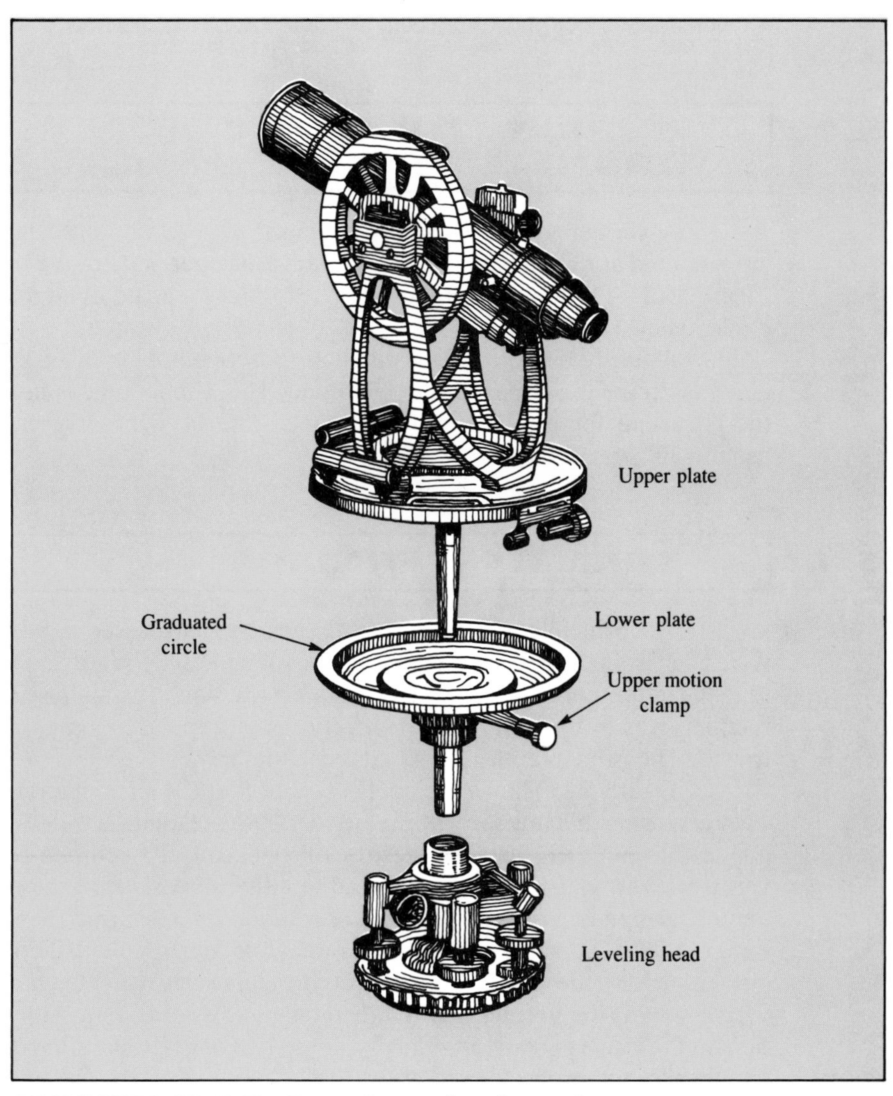

FIGURE 5.14 Parts of an engineer's transit

four foot screws. It is so constructed that the instrument can be shifted laterally on the foot plate in order to accomplish centering over a specific point on the ground.

2. *Lower plate.* The lower plate consists of a hollow spindle that is accurately fitted to a socket in the leveling head and carries the graduated horizontal circle. The rotation of the lower plate is controlled by a clamp screw, which provides a means for locking it in place. A slow-motion tangent screw permits the lower plate to be rotated a small amount relative to the leveling head.

3. *Upper plate.* The upper plate consists of a spindle attached to a circular plate bearing the verniers, the standards that support the telescope, the plate bubble tubes, and usually a magnetic compass. This spindle coincides with the socket in the lower plate spindle and the two are held together by a clamp screw that may be loosened to permit movement of the upper plate relative to the lower. A small rotation of the upper plate is effected by the use of a tangent screw.

5.13 BUBBLE TUBES ON A TRANSIT

A vernier transit is usually equipped with two bubble tubes, which are mounted at right angle to each other on the upper plate of the transit (see Figure 5.13). These bubble tubes are used to level the graduated horizontal circle. Their sensitivity is commonly 70″ per 2-mm division.

A third bubble tube, labelled as telescope bubble in Figure 5.13, is attached to the lower part of the telescope. This bubble tube is used to level the telescope for establishing a horizontal line of sight. Its sensitivity is usually 30″ per 2-mm division.

5.14 TRANSIT VERNIERS

The principle of verniers has already been discussed in Section 4.13. A vernier is an auxiliary scale that slides along a main scale and is used to read the latter more closely than its smallest division. The main scale may be rectilinear as in the case of some level rods or circular as on an engineer's transit. The principle of design and use is the same.

Figure 5.15 shows the application of the vernier to a circular scale. Two arrangements are shown, the vernier at (a) reading to minutes and the one at (b) reading to half minutes, or thirty seconds. Each one is a double vernier—that is, it is arranged to read in either direction, to the right or to the left, depending on the direction of rotation in measuring the angle (see Figure 5.16). The value of an angle is found as follows: first, the horizontal circle is read to the smallest division on the main scale using the index on the vernier; then, the vernier line that is most nearly coincident with a graduation on the circle is read; and, finally, these two angular quantities are added together to obtain the total angle.

Thus, in Figure 5.15a, the clockwise angle reading (inner row) is 342°30′ + 05′ = 342°35′. The counterclockwise angle reading (outer row) is

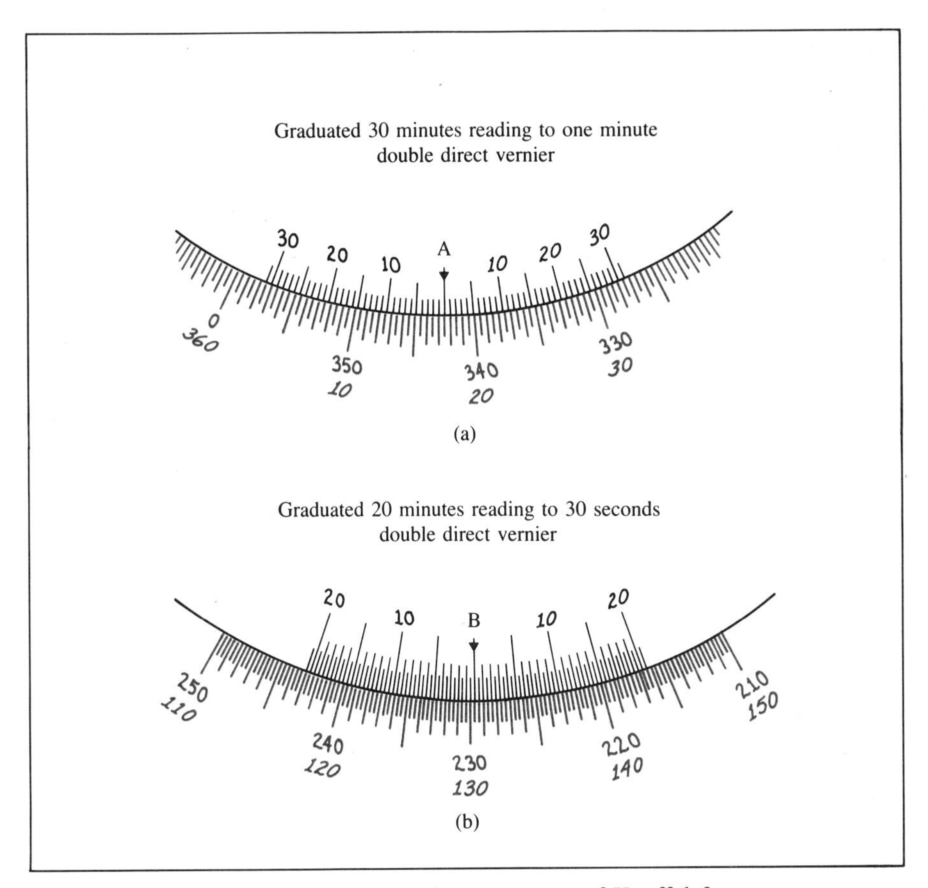

FIGURE 5.15 Transit verniers (courtesy of Keuffel & Esser Company)

17° + 25′ = 17°25′. Similarly, in Figure 5.15b the clockwise angle reading (inner row) is 229°40′ + 10′30″ = 229°50′30″. The counterclockwise angle (outer row) is 130°00′ + 9′30″ = 130°09′30″. Note that the vernier is always read in the same direction as the scale. This relationship is indicated by the slope of the numbers, both on the vernier and on the scale. Thus the numbers that slope to the left on the scale and on the vernier are to be taken together.

Various arrangements of verniers will be found on the different instruments. Care should be exercised to determine correctly the characteristics of each vernier before it is used.

While reading a vernier, one eye should be positioned directly over the line of coincidence. A hand-held magnifying glass is often helpful in reading the vernier. The line of coincidence can also be identified from the positions of the two flanking lines on the vernier, as illustrated in Figure 5.17. These flanking lines will differ from coincidence by equal amounts but in opposite directions. Blunders in reading the vernier can also be minimized

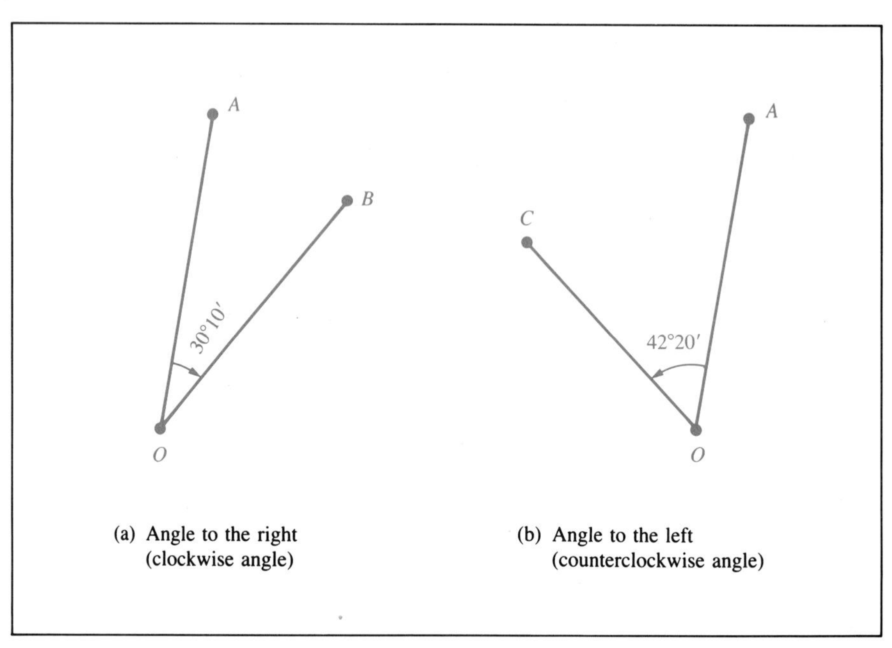

FIGURE 5.16 Angles to the right or left

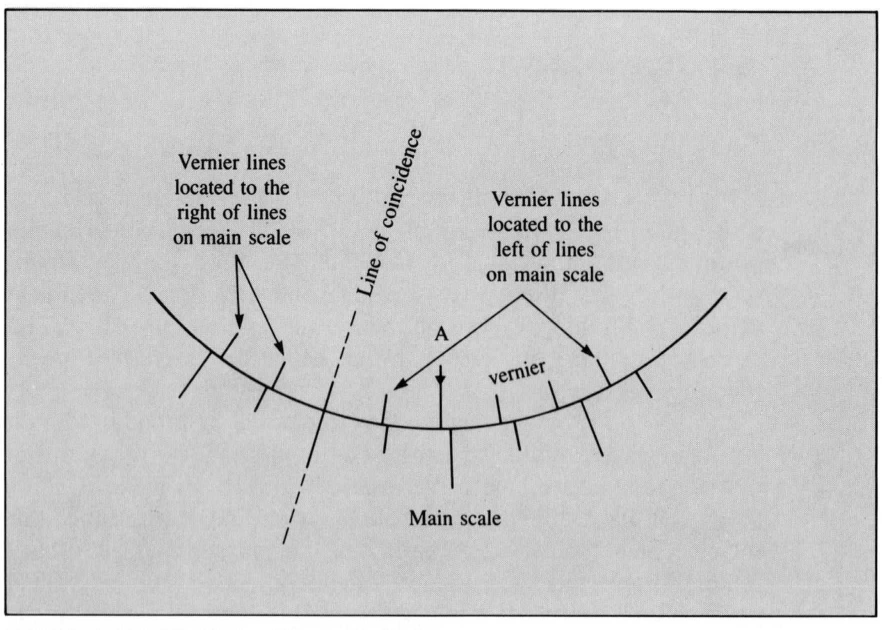

FIGURE 5.17 Line of coincidence

by making a rapid preliminary reading of the angle by estimating the fractional part of the circle division.

A vernier transit is usually equipped with two verniers, labelled *A* and *B* verniers, which are located exactly 180° apart. Thus, at any particular position of the telescope, the readings from the *A* and the *B* verniers should be exactly 180° apart. When the telescope is in the direct position—that is, the telescope bubble is below the telescope—the eyepiece should be on the side of the *A* vernier. In measuring angles, both *A* and *B* verniers are sometimes read for each pointing of the telescope to increase the accuracy of the reading and to reduce the opportunity of making undetected blunders. For example, if the *A* vernier reads 87°34′ and the *B* vernier reads 267°33′, then the minutes from the two readings are averaged to yield the final reading as 87°33′30″. The degree reading is always obtained from the *A* vernier.

5.15 UPPER AND LOWER MOTIONS

A vernier transit can be rotated about its vertical axis by using either the upper motion or the lower motion. The part of the instrument that rotates on the inner spindle and includes the upper plate, the standards, and the telescope is commonly called the *upper motion*. It is controlled by the upper-motion clamp and tangent screw. When the transit is rotated by using the upper motion, the graduated horizontal circle is moved with respect to the two stationary verniers. Thus a different angle reading is obtained for each pointing of the telescope.

The part of the instrument that rotates on the outer spindle and includes the graduated circle is commonly called the *lower motion*. It is controlled by the lower-motion clamp and tangent screw. When the transit is rotated using the lower motion, the two verniers move together with the horizontal circle as a single unit. Thus the angle reading on the verniers remains unchanged as the transit is rotated about the vertical axis. The lower motion is thus used to point the telescope along a desired direction without disturbing the angle reading on the vernier.

An angle-measuring instrument that has both an upper and lower motion is called a *repetition instrument*. An instrument that has only an upper motion is called a *direction instrument*.

5.16 SETTING UP A TRANSIT

The vernier transit is first securely mounted on a tripod by screwing the foot plate onto a threaded tripod head. The wing nuts on the top of the tripod legs should be tightened snugly at the beginning of the transit operations and left in that position throughout the measurement. A typical tripod head used for vernier transits is shown in Figure 5.18.

FIGURE 5.18 Tripod head used for vernier transit

To help in setting up the transit over a survey point, a plumb bob is attached to a hook hanging below the foot plate. With the transit approximately centered on the foot plate and the four foot screws equalized, the tripod legs are so positioned that the foot plate is nearly horizontal and the plumb bob is close to the survey point. Then the departure from the point is noted, and the tripod is bodily lifted and shifted laterally until it is within $\frac{1}{2}$ inch of the point. Pressing on the spur of each tripod leg will serve to make the setup stable as well as aid in bringing the plumb bob closer to the point. If the setup is on a slope, one leg of the tripod should be positioned on the uphill side, while the two other legs are positioned on the downhill side. Final centering of the transit over the survey point is achieved by shifting the leveling head of the transit. The leveling head is freed by loosening two adjacent foot screws. It is then shifted until the plumb bob is directly over the survey point. The head is then secured in place by turning the same two foot screws, but in the opposite direction, until they bear snugly on the foot plate.

The transit is next leveled by first rotating it about its vertical axis until each plate bubble tube is approximately parallel to the line joining a pair of diagonally opposite leveling screws (see Figure 5.19a). Each plate bubble is separately centered by turning uniformly in opposite directions the foot screws that control it. In Figure 5.19a, bubble tube *ab* is centered using foot screws 2 and 4, while bubble tube *cd* is centered using foot screws 1 and 3. The bubble will move in the same direction as the left thumb. After both bubbles are centered, the transit is rotated 180° about its vertical axis, so that

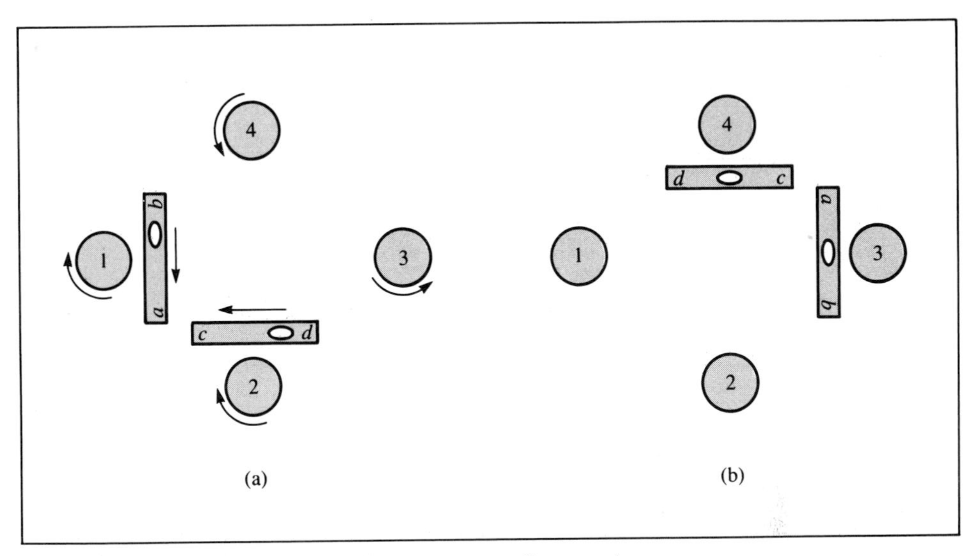

FIGURE 5.19 Leveling a vernier transit

the bubble tubes assume the positions shown in Figure 5.19b. If the bubbles have moved out of center, they are brought back halfway by using the foot screws. If the bubbles are off-centered by more than one division, the bubble tubes need to be adjusted. If at any time during the leveling operation the foot screws become so tight that an opposite pair cannot be turned simultaneously, the entire assembly can be loosened by loosening one screw only. Normal leveling procedure can then be continued.

5.17 MEASUREMENT OF ANGLES BY REPETITION

The lower motion of an angle-measuring instrument can be used to accumulate angles in repeated measurements, thus eliminating the need for reading the circle at each and every pointing of the telescope. This procedure is called the *repetition method*. Figure 5.20 shows the field notes obtained from measuring a horizontal angle $P\hat{O}Q$ by the repetition method with a vernier transit. The measurement procedure is as follows:

Step 1. Set up the transit over the survey point using the procedure described in Section 5.16.

Step 2. Clamp the lower motion. Release the upper motion, and set the *A* vernier approximately on zero. Clamp the upper motion, and then set the *A* vernier on zero using the upper-motion tangent screw.

Step 3. Release the lower-motion clamp and vertical circle clamp. Point the telescope so that station *P* is in field of view. Clamp the lower motion and vertical circle. The vertical cross-hair is set precisely on the target using the lower-motion tangent screw. The vertical tangent screw is used to move the center of the cross-

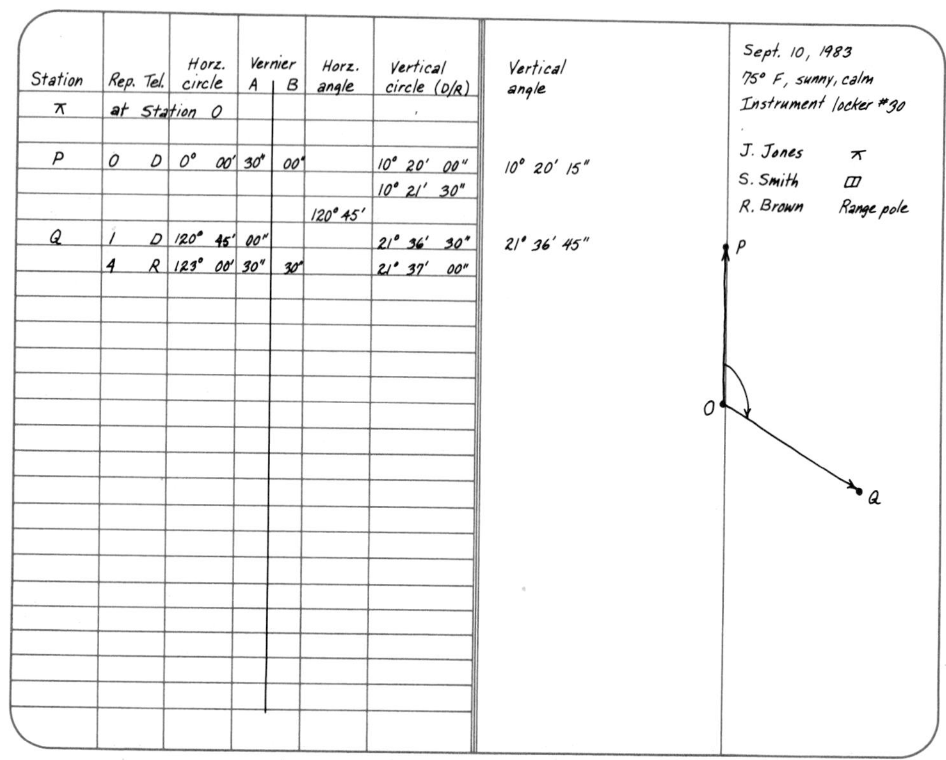

Station	Rep.	Tel.	Horz. circle	Vernier A	Vernier B	Horz. angle	Vertical circle (D/R)	Vertical angle
⊼	at Station O							
P	0	D	0° 00'	30"	00"		10° 20' 00"	10° 20' 15"
							10° 21' 30"	
						120° 45'		
Q	1	D	120° 45'	00"			21° 36' 30"	21° 36' 45"
	4	R	123° 00'	30"	30"		21° 37' 00"	

Sept. 10, 1983
75° F, sunny, calm
Instrument locker #30

J. Jones ⊼
S. Smith
R. Brown Range pole

FIGURE 5.20 Measurement of angles by repetition

hairs directly over the target. Since station *P* is sighted on first, it is referred to as the backsight station. Read both the *A* and *B* verniers and record the readings. It should be recorded in the note that the telescope is in the direct (*D*) position.

Step 4. If vertical angle is to be measured at the station, set the middle cross-hair exactly on the target using the vertical tangent screw. Read and record the vertical angle.

Step 5. Release the upper-motion and vertical-circle clamps. Turn telescope to Station *Q*, called the *foresight station,* so that the target is in the field of view. Clamp both the upper motion and the vertical motion. Set the cross-hairs precisely on the target using the upper-motion and vertical tangent screws. Read and record the angle reading on vernier *A*.

Step 6. If vertical angle is to be measured, read and record the vertical angle.

Step 7. Release the lower-motion clamp and rotate the telescope about its horizontal axis by 180°. This rotation of the telescope is often referred to as *reversing* (or *inverting,* or *plunging*) the telescope. The telescope is now in a reversed (or inverted, or plunged) position.

Step 8. Point the telescope to the backsight station using the lower motion. Set the cross-hairs precisely on the target using tangent screws. The verniers need not be read at this pointing.

Step 9. Read and record the vertical angle if needed. This reading should be recorded as being taken with the telescope in the reversed (*R*) position.

Step 10. Point the telescope to the foresight station using the upper motion. Set the cross-hairs precisely on the target using the tangent screws. Since two repetitions of the angle have been made, the reading on the *A* vernier should be equal to two times the actual value of the angle. If only two repetitions are needed, both the *A* and *B* verniers are read and recorded.

Step 11. If more than two repetitions are desired, the telescope is again reversed and thus returned back to the direct position. Steps 8 to 10 are then repeated for as many repetitions as necessary. The *A* and *B* verniers are read only at the last pointing to the foresight station. In the field notes shown in Figure 5.20, four repetitions (two direct and two reverse) are made. It is usually sufficient to read the vertical angles during the first direct and reverse pointing, as shown in the field notes in Figure 5.20.

Step 12. If vertical angles are measured only once during the above steps, the index error on the vertical circle should now be recorded. The telescope is brought to the horizontal position, and the telescope bubble is carefully centered by using the vertical tangent screw. The vernier on the vertical circle is then read. This reading, termed the *index reading,* will ordinarily be zero in the case of a properly adjusted transit. The index reading, as is the case with any vertical angle reading, should be recorded with the proper sign indicating above (+) or below (−) the horizontal.

To illustrate the application of the index reading to an observed vertical angle, assume that an angle of +4°21′ was recorded and that the index reading was found to be +0°01′. The corrected vertical angle is +4°20′. Also, suppose that a depression angle of 3°49′ was recorded and the index reading was +0°02′. The corrected vertical angle is −3°51′.

If the vertical arc of the transit is a full circle and if the object is observed first with the telescope direct and then with the telescope inverted, the mean of the two readings will be the correct value of the vertical angle, the index error having thus been eliminated.

This is the case for the example shown in Figure 5.20.

Step 13. The averaged value of the horizontal angle is then computed using the following procedure:

a. *Find number of complete revolutions*

Reading at 1-*D*	=	120°45′00″
− average initial reading (average of *A* and *B* verniers)		−15″
1 repetition of angle	=	120°44′45″
		× 4
4 repetitions of angle	≅	482°59′00″
No. of completed revolutions	=	1

b. *Compute average angle*

Average final reading (average of *A* and *B* verniers)	=	123°00′30″
+ 1 revolution		+360°
Correct final reading	=	483°00′30″
Average initial reading (average of *A* and *B* verniers)	=	−15″
4 × angle	=	483°00′15″
Average angle	=	120°45′04″

5.18 GENERAL CHARACTERISTICS OF THEODOLITES

Theodolite is a European term for angle-measuring instruments. Through common usage, the term may now be applied to any angle-measuring instruments that possess the following general characteristics:

1. Compared with the vernier transit, the head or alidade is compact, lightweight, and of reduced height. The weight (without tripod) of the vernier transit is approximately 16 to 18 lb. A typical theodolite weighs about 10 lb.
2. The telescope is short, has a large objective, and can focus at relatively short distances. Its cross-hairs are lines etched on glass, and it may have rifle sights mounted on its top for preliminary pointing. It is of the internal-focusing type so that in stadia work (Chapter 9) the stadia constant ($F + c$) is zero and the multiplying factor is 100.
3. Three leveling screws support the instrument.
4. It is equipped with an optical plummet.
5. The horizontal and vertical circles are made of glass with etched graduations and numerals.
6. The circles and all optical and mechanical systems are completely enclosed so that the instrument is moistureproof and dustproof.
7. The instrument has a circular level, a single tubular level in the plane of the horizontal circle, and either an index level attached to the vertical circle or an automatic vertical indexing device. This latter device usually consists of a compensator that eliminates the influence of any inclination of the vertical circle.
8. All readings, both of horizontal and vertical circles, and observations of the bubbles can be made from the eye end of the telescope. There is no need for the operator to walk about the transit in order to obtain the readings. Some models provide digital readings of the horizontal and vertical circles, which may be graduated in the centesimal as well as the sexagesimal modes.
9. Usually the upper part of the head can be separated and lifted out from the centering and leveling base, which is called the *tribrach*. This permits easy interchange of theodolite with targets or electronic distance measuring (EDM) instruments on certain operations.
10. A wide range of accessories can be used with the theodolite and thus increase its usefulness and versatility.
11. Most instruments are housed in light, streamlined metal cases into which the theodolite can be fastened very securely and adequately protected when not in use.
12. The reticule and reading systems are provided with internal illumination for night operations.
13. The instrument may be capable of complete integration with an electronic distance measuring unit.

There is a wide variety of theodolites, differing from each other slightly in design and with least counts ranging from 1 minute to 0.5 second. Before the mid-1970s, the basic designs of the theodolites remained unchanged for over two decades. Then the miniaturization of EDM instruments stimulated some design changes to facilitate the mounting of EDM equipment on the theodolites. Recent advances in electronics have further resulted in the development of theodolites with digital readouts or displays of the horizontal and vertical angles as well as total stations, which can measure both angles and distances.

Several common types of theodolites will be discussed in the following sections. The operating characteristics of total stations will be discussed in Chapter 6 with the EDM instruments.

5.19 SCALE-READING THEODOLITE

Figure 5.21 shows a WILD T16D scale-reading theodolite. This type of theodolite is particularly suited for engineering layout work. The horizontal circle is graduated both clockwise and counterclockwise so that horizontal angles can be measured to either the left or right of the backsight station. Both the horizontal and vertical circles are read directly with an optical microscope to the nearest minute and by interpolation to 0.1′. The vertical circle reads 0° when the telescope is pointed at the zenith or directly overhead and 90° when the telescope is horizontal. Thus these angular values are zenith angles.

An air-damped compensator provides automatic indexing of the vertical circle; that is, the vertical circle is automatically set to read 90°00′ at the horizontal position. If the instrument is out of level beyond the working range (±6′) of the compensator, a red warning screen is seen in the vertical circle image. The theodolite should then be releveled before taking any zenith angle measurements. The horizontal circle appears yellow while the vertical circle is white. This reduces the probability of reading the wrong circle.

The Wild T16D has only one set of horizontal motion clamp and tangent screws. However, it has a horizontal circle clamp that permits the instrument to be operated as a repetition instrument. When the horizontal circle clamp is in the lock (down) position, the horizontal circle is locked and the circle reading does not change when the theodolite is rotated about its vertical axis. When the circle clamp is unlocked (up position), the circle reading changes with the rotation of the theodolite.

The telescope is reversible and produces an erect image. Its magnification is 30 power. The field of view at 1,000 ft is 27 ft, and the shortest focusing distance is 5.6 ft. The reticule (Figure 5.21) has a set of stadia ticks on both the horizontal and vertical crosslines and two parallel vertical lines

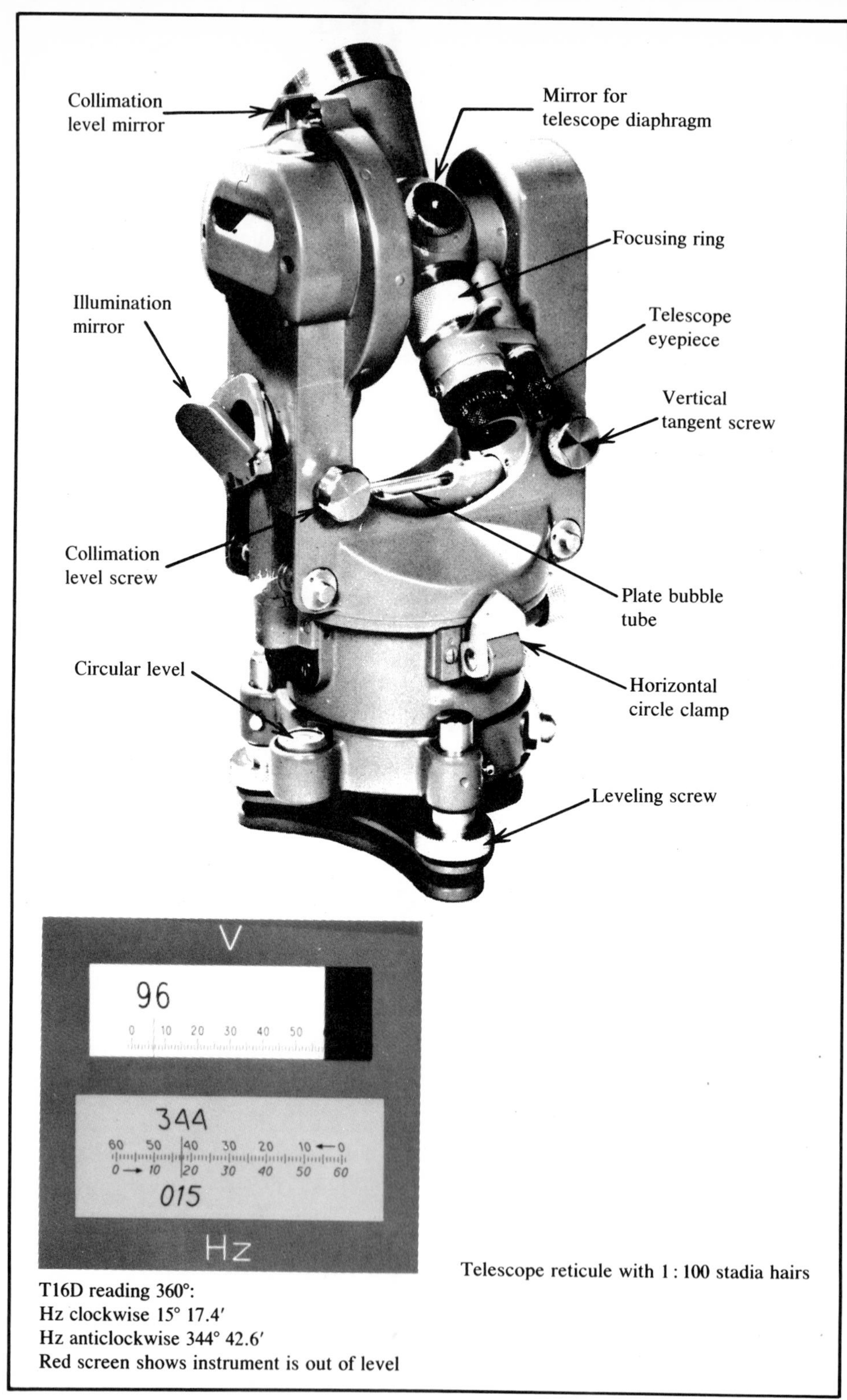

FIGURE 5.21 Wild T16D scale-reading theodolite (courtesy of Wild Heerbrugg Instruments, Inc.)

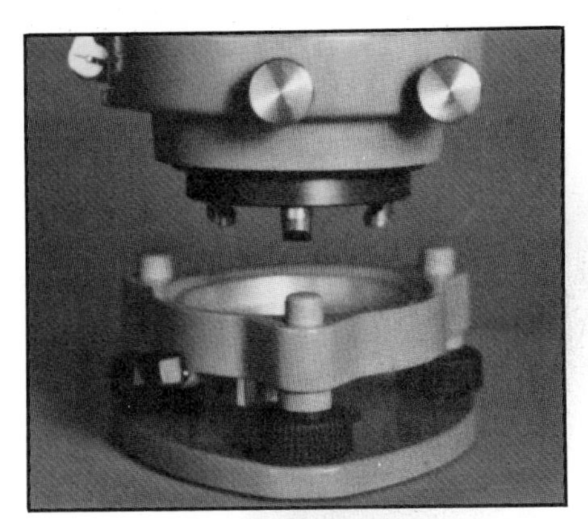

FIGURE 5.22 Tribrach (courtesy of Wild Heerbrugg Instruments, Inc.)

in the lower half of the field. These are used for precise pointings on distant range poles and plumb lines that would be otherwise obscured by the use of a single vertical sight line. For daytime use tilting mirrors provide illumination for both circles.

The tribrach or base (see Figure 5.22) is a detachable part of the theodolite and contains the three leveling screws and the circular level. These screws are completely enclosed and dustproof. A locking device holds the instrument head and tribrach together.

The instrument has a single plate level bubble, which has a sensitivity of 30 seconds per 2 mm graduation. An optical plummet facilitates the centering of the instrument over a survey point.

Care must be exercised in reading horizontal circles that are graduated for both clockwise and counterclockwise angles, such as that shown in Figure 5.21. When angles are measured to the right (clockwise) of the backsight station, the lower scale indicated by an arrow pointing to the right (→) should be used. The graduation numerals on this scale slant to the right. When angles are measured to the left of the backsight station (counterclockwise), the upper scale should be used. This scale has an arrow pointing to the left (←), and the graduation numerals slant to the left.

Confusion often occurs in setting the circle to read 0°00′. The correct setting for 0°00′ is shown in Figure 5.23.

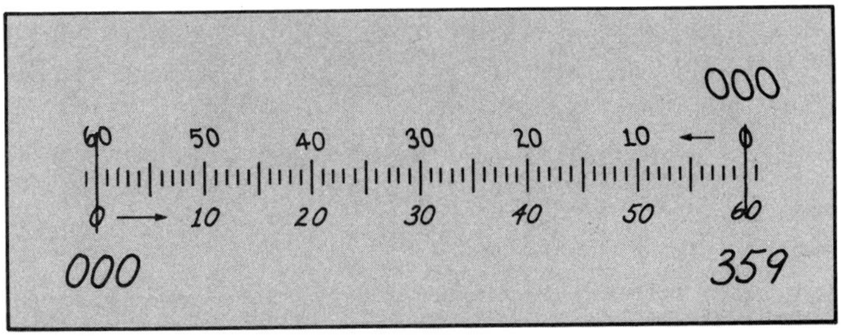

FIGURE 5.23 0° setting

5.20 DIGITAL THEODOLITE

A digital theodolite provides direct reading of both the horizontal and vertical angles in numerical digits. Figure 5.24 shows a Lietz TM-20H Digital Theodolite, and Figure 5.25a shows the scale readings as viewed

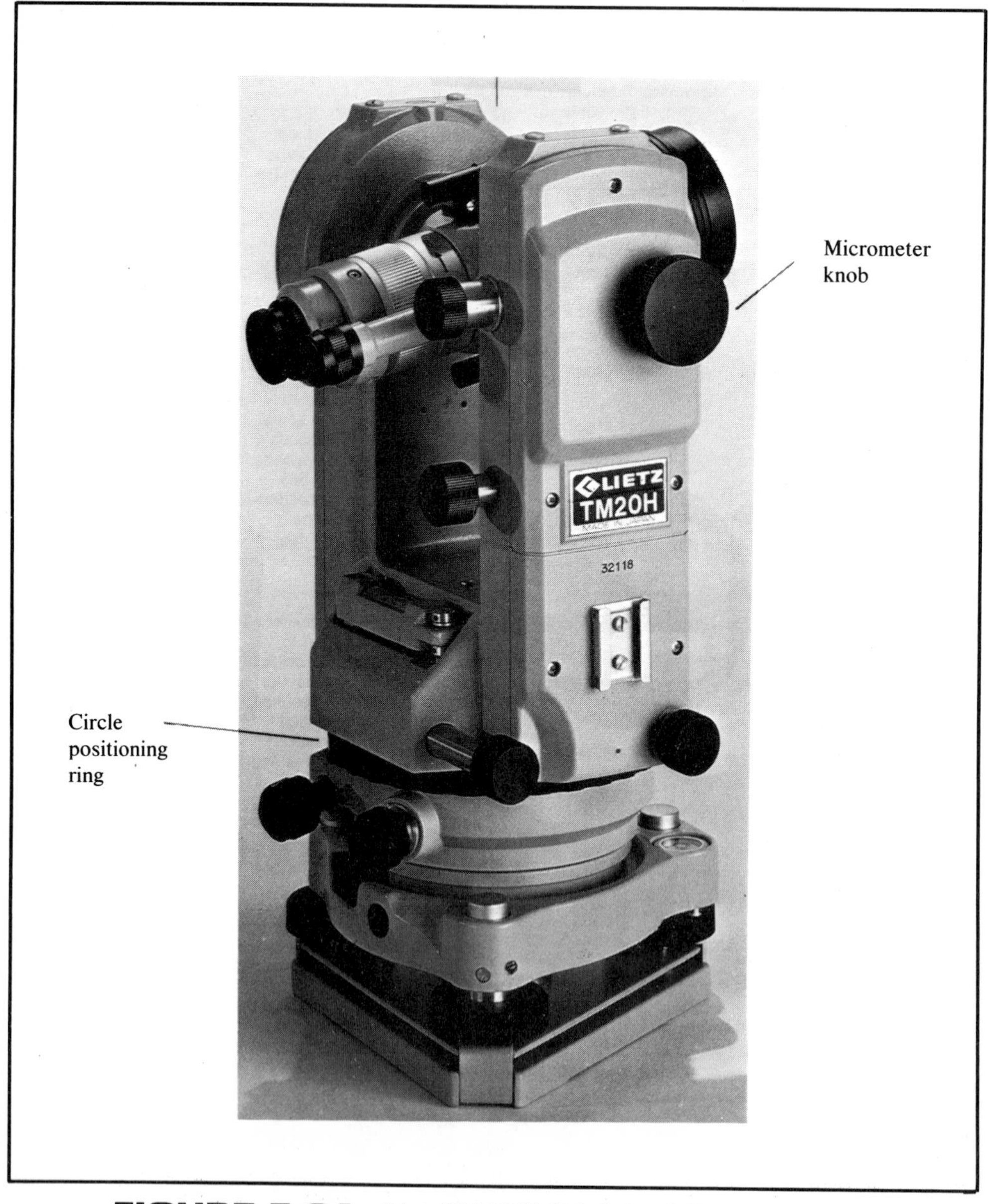

FIGURE 5.24 Lietz TM-2OH digital theodolite (courtesy of The Lietz Company)

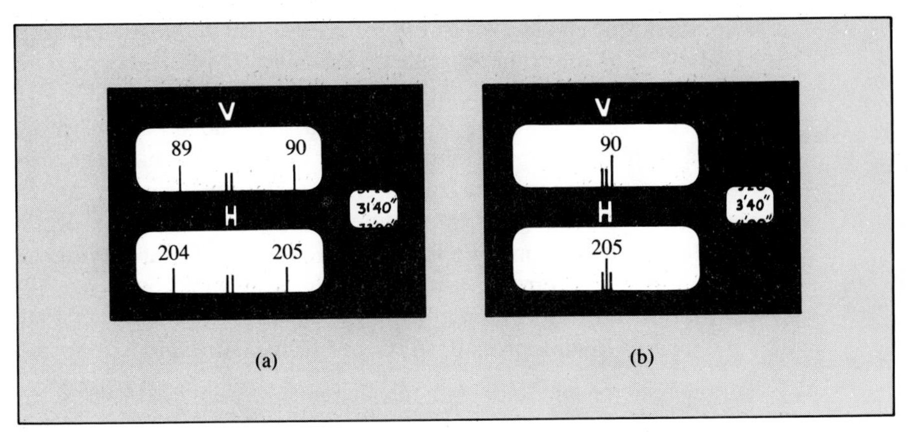

FIGURE 5.25 Reading the horizontal circle on the Lietz TM-2OH digital theodolite (courtesy of The Lietz Company)

through the optical-reading microscope. The upper window (marked by the letter *V*) shows the full degrees in the vertical circle, and the lower window (*H*) shows the full degrees in the horizontal circle. The micrometer scale on the right gives the minutes and seconds for the horizontal and vertical circles.

Suppose that Figure 5.25a shows the readings on the scale windows after the cross-hairs of the theodolite have been set on a target. To read the horizontal circle, the micrometer knob is turned until a full graduation mark bisects the double index line, as shown in Figure 5.25b. In this setting the horizontal angle reads 205°3′40″. Similar procedure is used for reading the vertical circle, as illustrated in Figure 5.26. Starting from the readings shown in Figure 5.26a, the same micrometer knob used for reading the horizontal circle is turned until a whole degree mark on the vertical circle bisects the

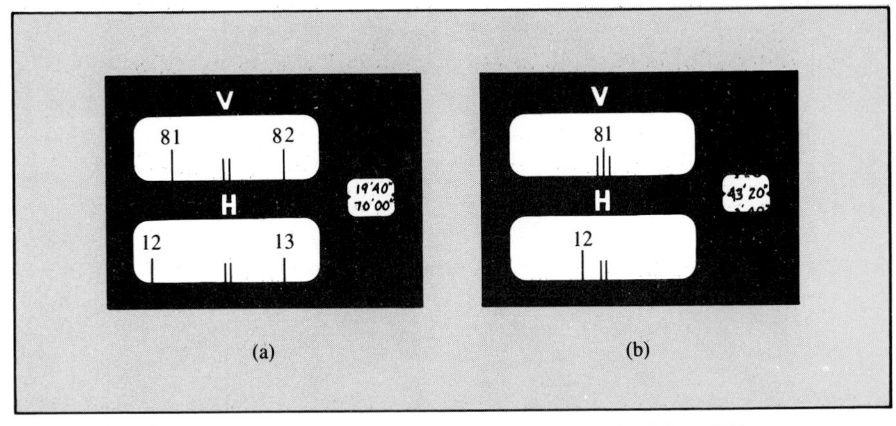

FIGURE 5.26 Reading the vertical circle on the Lietz TM-2OH digital theodolite (courtesy of The Lietz Company)

double index line, as shown in Figure 5.26b. The vertical circle is then read as 81°43′20″. The micrometer scale on the Lietz TM-20H can be read directly to 20 seconds and with estimation to 10 seconds.

The Lietz TM-20H has both a lower and an upper motion. It is equipped with a circle-positioning ring for quick positioning of the horizontal circle to some approximate reading. By rotating the ring, the horizontal circle reading is changed. It provides a convenient means of setting the horizontal circle to some approximate reading without having to use the upper motion. The circle-positioning ring is automatically fixed when the lower-motion screw is clamped.

The following procedure is used to measure an angle from 0°00′00″.

1. Turn the micrometer knob until the micrometer scale reads 0′00″.
2. Loosen both the upper- and lower-motion clamps.

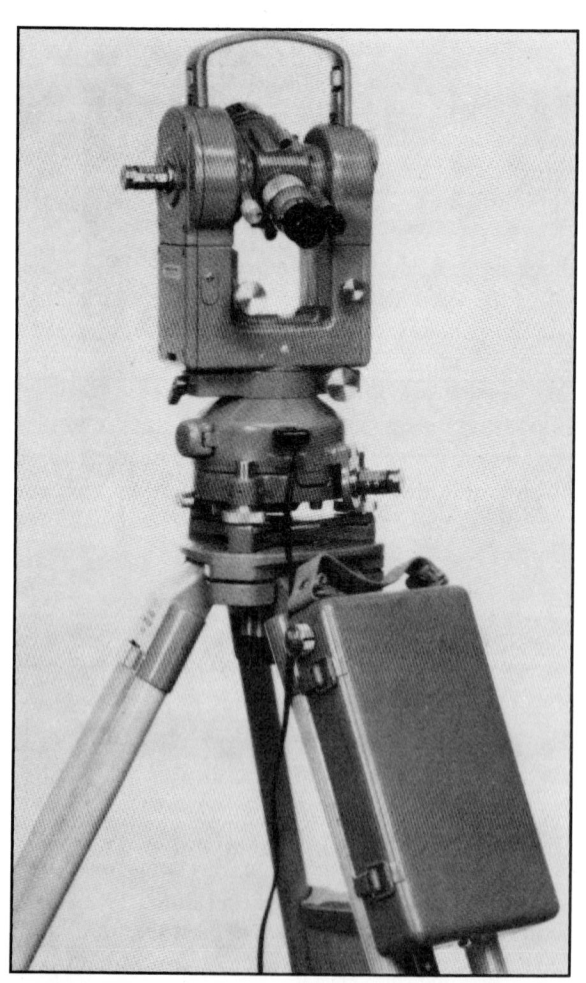

FIGURE 5.27 Wild T2 universal theodolite (courtesy of Wild Heerbrugg Instruments, Inc.)

3. Look through the micrometer eyepiece and turn the circle-positioning ring to set the 0° mark between the double index line for the horizontal circle.
4. Clamp the upper-motion screw, and use the upper-motion tangent screw to set the 0° mark exactly between the double index line.
5. Set the cross-hairs on the backsight station using the lower-motion screw.
6. Clamp the lower-motion screw.
7. Set the cross-hairs on the foresight station using the upper-motion screw.
8. Turn the micrometer knob until a full degree mark bisects the double index line for the horizontal circle. Read the horizontal circle.

Figure 5.27 shows a Wild T2 Universal Theodolite, and Figure 5.28a illustrates the display windows for the vertical circle as viewed through the reading microscope. To read the vertical circle, a micrometer knob is turned

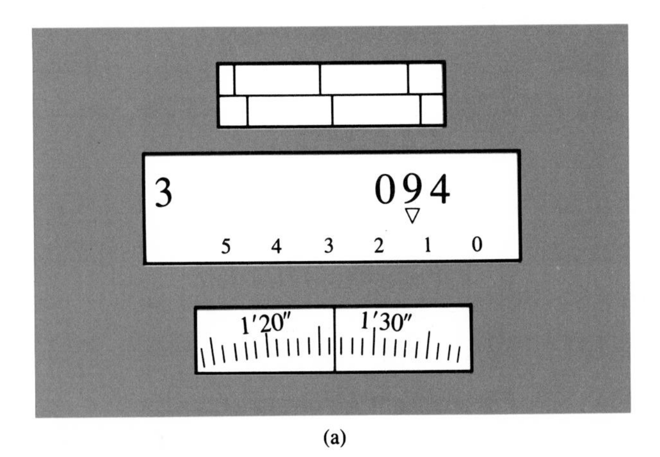

(a)

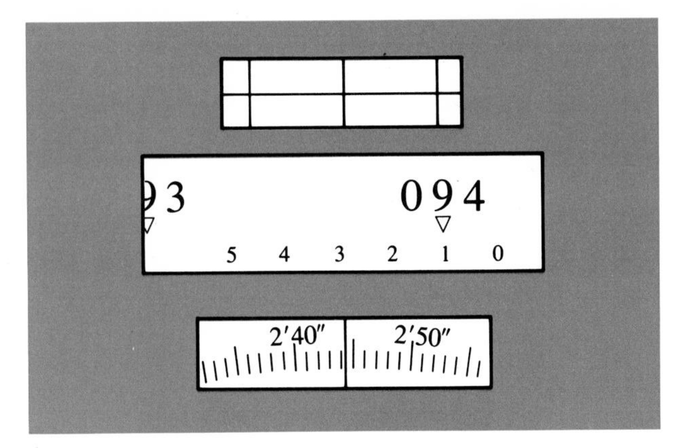

(b)

FIGURE 5.28 Reading the Wild T2 universal theodolite (courtesy of Wild Heerbrugg Instruments, Inc.)

until the vertical lines displayed in the top window are in coincidence as shown in Figure 5.28b. The number of degrees and tens of minutes are then read from the middle window, and the minutes and seconds are read from the bottom window. Thus the vertical circle reading shown in Figure 5.28b should be 94°12′44.4″. The horizontal circle is read in a similar manner. A selector knob is used to select a view of either the horizontal or vertical scale through the reading microscope. Both the horizontal and vertical circles can be read directly to 1 second and estimated to 0.1 second. The Wild T2 has no lower motion and is therefore a direction instrument. It has a circle setting knob for setting the horizontal circle to a desired reading.

Figure 5.29 shows the Kern DKM3 Theodolite, which can be read directly to 0.5″ and estimated to 0.1″. The vertical circle reading shown in Figure 5.29 is 82°53′01.8″.

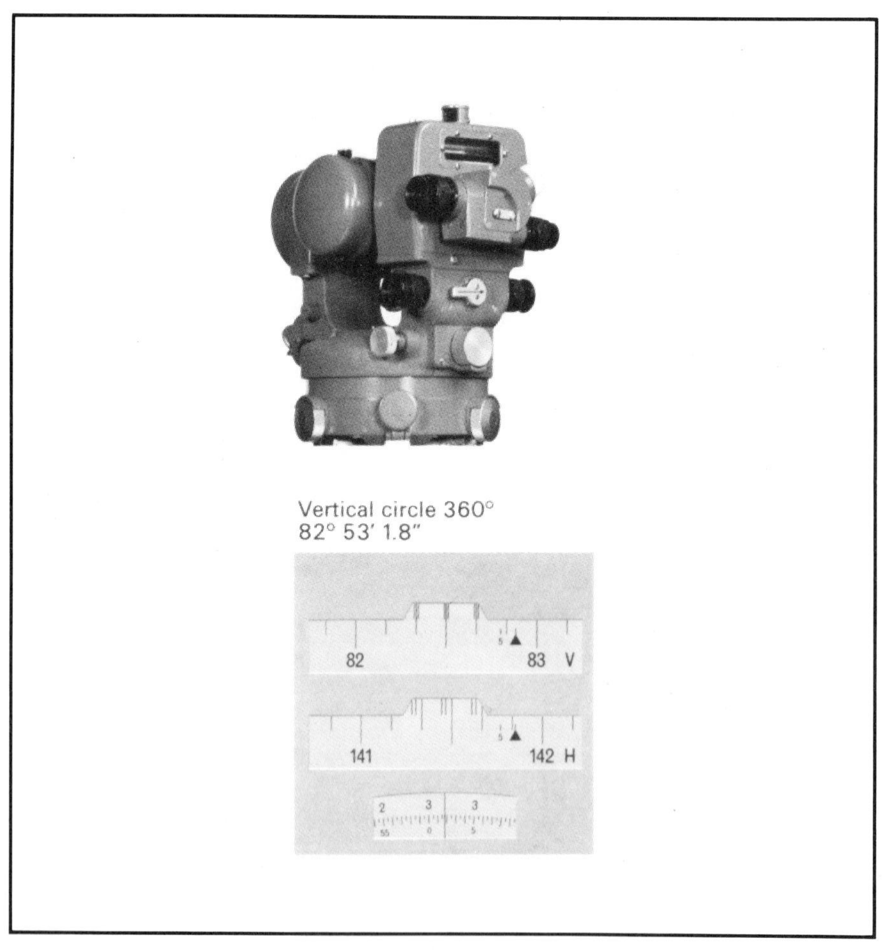

FIGURE 5.29 Kern DKM3 theodolite (courtesy of Kern Instruments, Inc.)

5.21 ELECTRONIC THEODOLITE

An electronic theodolite provides a visual digital display of the circle readings, thus eliminating the need for a reading microscope. A digital display of either the horizontal or vertical angle is selected by the push of a button. Either circle reading can be set to 0°00′00″ by the touch of a button, and either clockwise or counterclockwise angles can be displayed. Usually, two identical displays are located on opposite sides of the upper plate so that the display can be read with the telescope in either the direct or reversed position.

Figure 5.30 shows a TOPCON DT-20 electronic digital theodolite, which gives a direct display to 20 seconds of arc. Figure 5.31 illustrates a

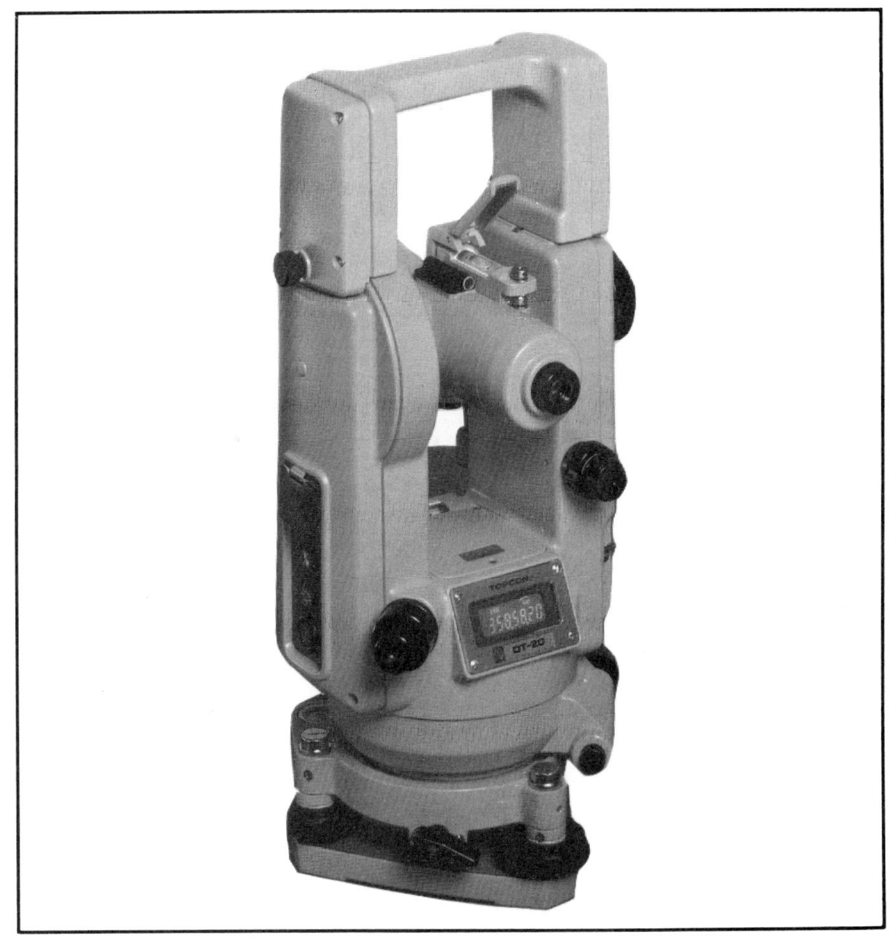

FIGURE 5.30 TOPCON DT-20 electronic digital theodolite (courtesy of TOPCON Instrument Corporation of America)

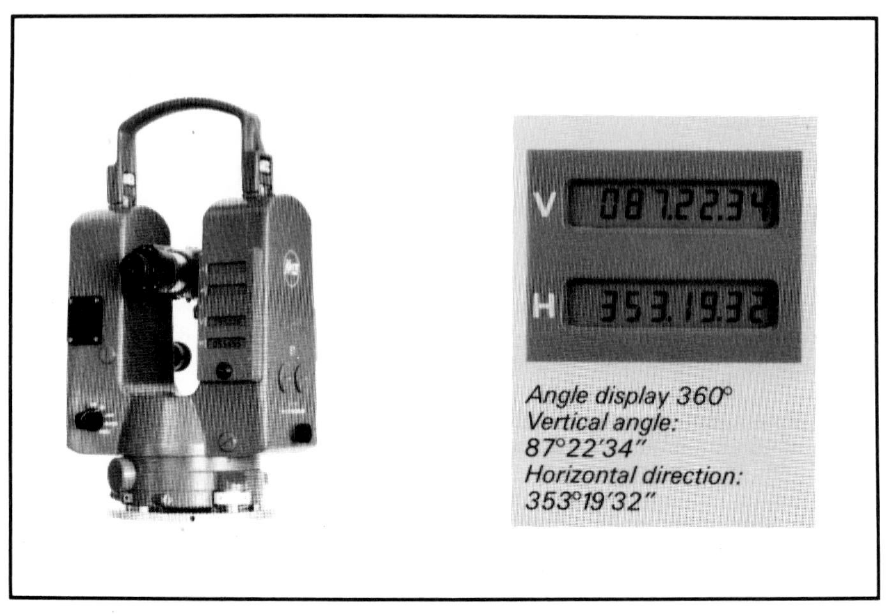

FIGURE 5.31 Kern E1 electronic theodolite (courtesy of Kern Instruments, Inc.)

Kern E1 Electronic Theodolite, which displays circle readings directly to 1 second.

5.22 SETTING UP A THEODOLITE

A theodolite is usually equipped with an optical plummet for centering over a survey point, a bull's-eye bubble level for preliminary leveling, and one plate bubble for precise leveling of the instrument. The following procedure may be used to set up a theodolite equipped with these features and an extension leg tripod.

1. Set the extension legs of the tripod to the proper length for the instrument operator.
2. Attach a plumb bob to the tripod head. Although an optical plummet is available with the instrument, it is usually more convenient to perform the preliminary centering over the survey point using a plumb bob.
3. Keeping the tripod head level, set the tripod approximately over the survey point. Step on the spur of each tripod leg to drive it firmly into the ground.
4. Center the plumb bob directly over the survey point by adjusting the length of the tripod legs. If the tripod head looks excessively off level, lift tripod from ground and repeat steps 3 to 4.
5. Mount theodolite on the tripod, and level it with the bull's-eye bubble.

6. If steps 3 and 4 have been performed properly, the survey point should be within view of the optical plummet. The theodolite is then exactly centered over the survey point by loosening the fastening screw on the tripod head and moving the theodolite around slowly. During this operation the theodolite should always be held firmly by its standard with one hand. Once it is centered properly, the fastening screw is tightened firmly.

7. Check to see if the bull's-eye bubble is still centered. If it has moved off center, repeat steps 6 and 7.

8. Rotate the theodolite so that its plate bubble is parallel to a line joining any two of the foot screws, say screws *A* and *B* as shown in Figure 5.32a. The level bubble is centered using the same two foot screws.

9. The theodolite is rotated 180° so that the plate bubble assumes the position shown in Figure 5.32b. If the plate level is in perfect adjustment, the bubble should remain centered. If the bubble has moved off center, it is brought halfway back towards the center using the same two foot screws (*A* and *B*).

10. Rotate the theodolite so that the plate bubble is now perpendicular to the line joining the two previous foot screws (*A* and *B*) (see Figure 5.32c). Center the bubble with the third foot screw (*C*) only.

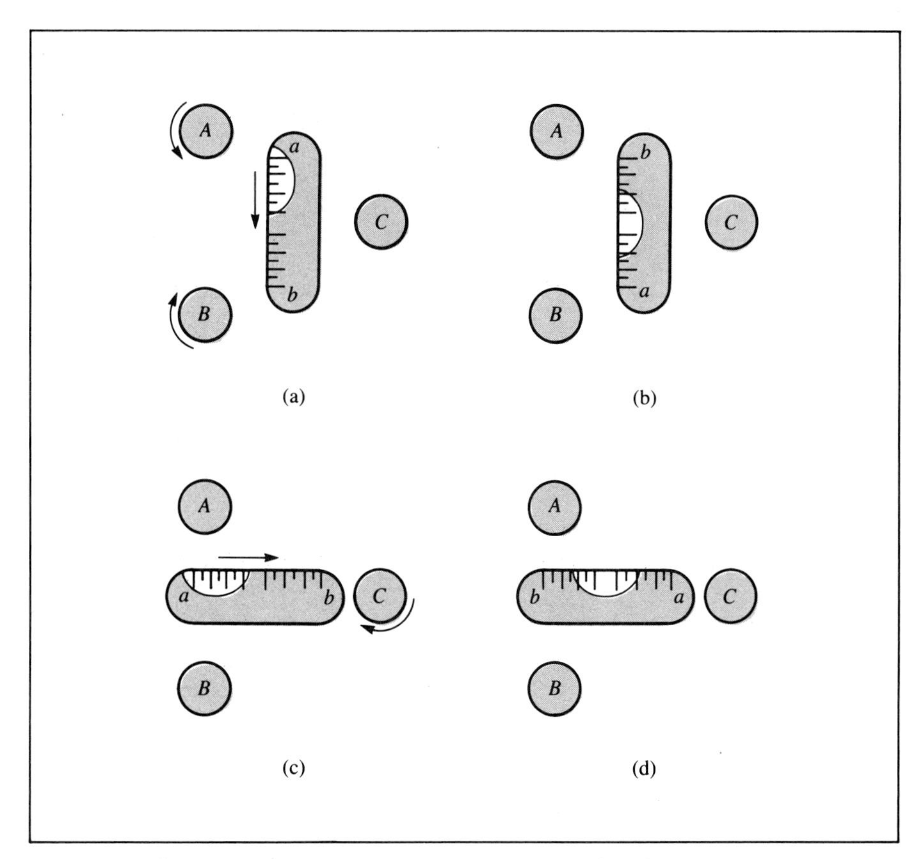

FIGURE 5.32 Leveling a three foot screw instrument

11. Rotate the theodolite 180° so that the bubble assumes the position shown in Figure 5.32d. If the bubble has moved off center, bring it back halfway using the third foot screw (C).

12. Check the optical plummet to make sure that the instrument is still properly centered over the survey point. If not, repeat steps 6 to 12.

13. The theodolite is now ready for making angle measurements.

5.23 MEASUREMENT OF A HORIZONTAL ANGLE BY THE DIRECTION METHOD

The procedure for measuring a horizontal angle ($A\hat{B}C$) by the *direction method* is outlined below. The corresponding field notes are shown in Figure 5.33.

1. Set up the theodolite over station B.

2. With the telescope in the direct (D) position, backsight to station A and set the horizontal circle to read 0°00′00″.

Station	Tel.	Direction	Average direction	Reduced direction
	⊼ at	Station B		
A	D	0° 00′ 00″	0° 00′ 2.5″	0° 00′ 00″
	R	180° 00′ 05″		
C	D	75° 09′ 30″	75° 09′ 34″	75° 09′ 31.5″
	R	255° 09′ 38″		
A	D	90° 00′ 00″	89° 59′ 55.5″	0° 00′ 00″
	R	269° 59′ 51″		
C	D	165° 09′ 32″	165° 09′ 30.5″	75° 09′ 35.0″
	R	345° 09′ 29″		
A	D	180° 00′ 00″	180° 00′ 01″	0° 00′ 00″
	R	0° 00′ 02″		
C	D	255° 09′ 33″	255° 09′ 34.5″	75° 09′ 33.5″
	R	75° 09′ 36″		
A	D	270° 00′ 00″	269° 59′ 59″	0° 00′ 00″
	R	89° 59′ 58″		
C	D	345° 09′ 28″	345° 09′ 29.5″	75° 09′ 30.5″
	R	165° 09′ 31″		

Sept. 21, 1983
78° F, windy, sunny
Instrument locker #5

J. Jones
S. Smith
R. Brown

Targets
⊼
◫

A
B
C
B

Mean $A\hat{B}C$ = 75° 09′ 32.6″

Std. deviation = ± 2″

Estimated standard error of the mean = ± 1″

FIGURE 5.33 Field notes for measurement of horizontal angle by direction

3. Foresight to station C with the upper motion. Read and record the horizontal circle (75°09′30″).
4. Reverse the telescope and use the upper motion to sight on station C again. The horizontal circle is read and recorded as a reverse (R) reading (255°09′38″). This reading should differ from the reading obtained from step 3 by 180° plus or minus a few seconds for an instrument that has a least count of 1 second.
5. Use the upper motion and backsight to station A. Read and record the horizontal circle reading (180°00′05″).
6. Average the minutes and seconds of the direct (D) and reverse (R) readings to each station.
7. Subtract the averaged direction to the backsight station (in this case, station A) from that of the foresight station to obtain the reduced direction.

Steps 1 to 7 above yield one set of angle measurement for the angle $A\hat{B}C$. It actually consists of one direct and one reverse measurement of the angle. By averaging a direct and a reverse reading, any small error due to imperfect leveling of the instrument is minimized. Usually several sets are measured to increase the precision of the averaged value. If N sets are measured, then the initial backsight readings of the N sets should be successively increased by 360°/N. For example, in the field notes shown in Figure 5.33, four sets of the angle ABC are measured. The first set has an initial backsight reading of 0°. The second set has an initial reading of 360°/4 = 90°. The purpose of this change in the initial reading is to use different parts of the horizontal circle to minimize the effect of any systematic error due to small graduation errors in the horizontal circle.

After the required number of sets have been measured, the mean and standard deviation of the measured angle should be computed. If the measurement from any set exceeds three times the value of the computed standard deviation, then that measurement is likely to contain a blunder error and should be replaced with a new set of measurements. In the example shown in Figure 5.33, the standard deviation of the four sets is computed to be ±2″. Since all the measured angles from all four sets are within ±6″ of the computed mean, the measurements are all acceptable.

An estimated standard error of the mean angle can be computed using Eq. (2.8). In this example, estimated standard error of the mean is $\pm 2''/\sqrt{4} = \pm 1''$. Thus it can be recorded that $A\hat{B}C = 75°09'33''$ with an estimated standard error of ±1″.

5.24 TESTS AND ADJUSTMENTS

When a transit or theodolite is in perfect adjustment, its three principal axes should be mutually perpendicular to each other. These axes include the vertical and horizontal axes and the line of sight. In addition, the plate bubble should be perfectly centered when the upper plate of the instrument is in a horizontal position. If the instrument is equipped with a telescope bubble tube, that bubble should also be perfectly centered when the line of sight is horizontal. These conditions should be frequently field tested, and the instrument should be properly adjusted whenever necessary.

The test procedures described here apply equally well to transits and theodolites. However, most theodolites are so constructed that only field adjustments of the level bubbles are advisable. Adjustment of the line of sight or the standards should best be performed by trained technicians.

The following testing and adjustment program should be undertaken under favorable atmospheric conditions, over terrain permitting stable set-ups, and preferably with the instrument in the shade. Adjusting pins that properly fit the adjusting screws, which are commonly called the capstan screws, should be used.

ADJUSTMENT OF PLATE-BUBBLE TUBE

Relation. The axis of a plate-bubble tube should lie in a plane perpendicular to the vertical axis.

Test. The instrument is set up and carefully leveled with the bubble tube parallel with a pair of leveling screws. The plate is then rotated on its vertical axis until the tube is turned end for end over its pair of leveling screws. If the correct relation exists, the bubble will remain centered; if not, the bubble will be displaced and the amount of the displacement will be double the error of adjustment because of the reversal of conditions.

Adjustment. If on reversal the bubble is displaced, say by four divisions, the adjustment is made by bringing it back two divisions by means of the capstan adjusting screw at the end of the bubble tube. The bubble is then centered with the foot screws and the test repeated for verification.

ADJUSTMENT OF THE CROSS-WIRE RING

Relation. The vertical cross-wire should lie in a plane perpendicular to the horizontal axis.

Test. The instrument is set up and the vertical cross-wire sighted on a definite point in the field of view. The telescope is then rotated slowly about its horizontal axis by using the vertical tangent screw. If the correct relation exists, the cross-wire will apparently remain on the point. If not, the point will appear to move off the cross-wire as the telescope is rotated.

Adjustment. This adjustment procedure applies only to transits. To adjust the cross-wire ring, both pairs of capstan screws that hold the ring in position are loosened slightly so that it may be rotated by means of pressure of the fingers on the screws, or by tapping with a pencil, until the correct position has been obtained.

ADJUSTMENT OF THE LINE OF SIGHT

Relation. The line of sight should be perpendicular to the horizontal axis.

Test. Set the instrument up and level it carefully. Take a backsight on a point as *A*, as in Figure 5.34a, with the telescope in the normal position. Invert the telescope and set a point at *D*. Reverse the horizontal axis end for end by turning the plate about the vertical axis, and take a second backsight on *A* with the telescope in the inverted position (Figure 5.34b). Reinvert the telescope and set a point at *E*.

The lack of perpendicularity between the line of sight and the horizontal axis is represented by angle α in the illustration; it is evident that, on inverting, the line of sight is deflected from the true prolongation of line *AB* by an angle equal to 2α. Accordingly, after reversal of the horizontal axis and inversion of the telescope the second time, the angle between the two foresights *D* and *E* is equal to 4α.

Adjustment. This adjustment procedure applies only to transits. The adjustment is made by sighting point *E*; then, by loosening one capstan screw of the horizontal pair and tightening the other, the vertical cross-wire is fixed on a point *F* that is set $\frac{1}{4}$ of the distance from *E* toward *D*.

After this adjustment the test is repeated; if the adjustment is perfect, the line of sight will fall on point *C* both before and after reversal of the horizontal axis.

ADJUSTMENT OF THE STANDARDS

Relation. The horizontal axis should be perpendicular to the vertical axis.

Test. Set the transit near a building where a definite point can be sighted that requires the telescope to be elevated through a large vertical angle. Level the plate carefully and sight the elevated point. Depress the telescope and set point *A* near the ground. Reverse the horizontal axis end for end by turning the plate about the vertical axis, invert the telescope, and sight the elevated point again. Depress the telescope a second time; if the adjustment is perfect, the line of sight will fall on point *A*, as previously set. If not, a second point *B* is set near the ground.

Adjustment. This adjustment procedure applies only to transits. The adjustment is made by raising or lowering one end of the horizontal axis until, after repeated reversals, the line of sight falls on the same point near the ground. The horizontal axis rests on journals, and provision is made for raising or lowering one of them, usually by first loosening setscrews on top of the standards and then by turning a capstan screw under the journal and between the two legs of the standards.

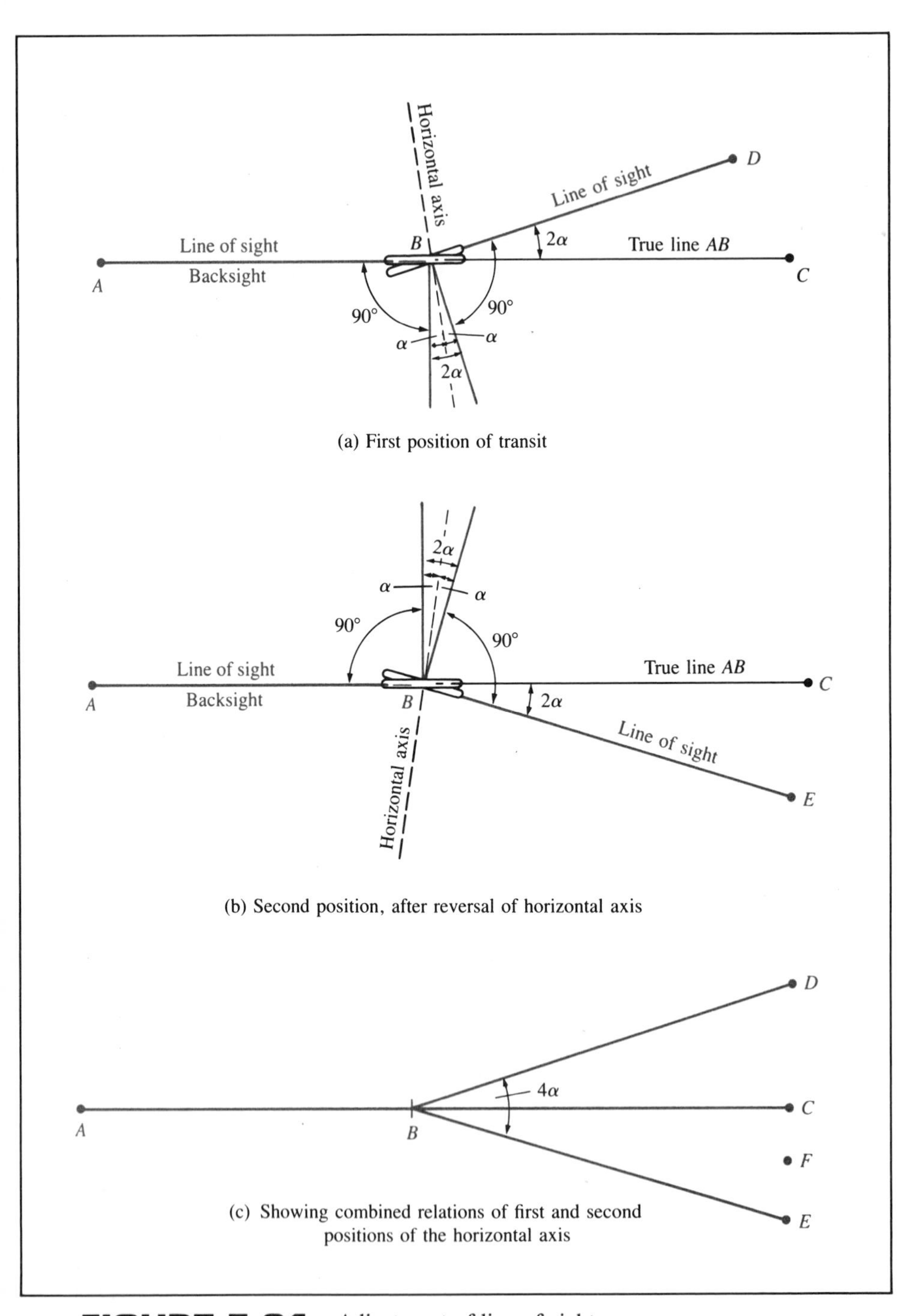

FIGURE 5.34 Adjustment of line of sight

ADJUSTMENT OF THE TELESCOPE BUBBLE TUBE

Relation. The axis of the bubble tube attached to the telescope should be parallel to the line of sight.

Test. This is accomplished essentially in the same manner as the "peg test" for the engineer's level explained in Section 4.15. An alternate procedure would begin with setting up the instrument, leveling it with the plate bubbles, and driving a stake at a taped distance of 150 ft. With the bubble of the level tube attached to the telescope carefully centered, a reading to the nearest 0.01 ft would be obtained on a level rod held on the stake. Another stake would be set in the opposite direction and at the same distance from the instrument and be progressively hammered into the ground until the reading on the rod now placed on it is exactly the same. Here again the bubble of the attached level tube must be carefully centered when the reading is obtained. It is now obvious that the tops of both stakes are at the same elevation even though the axis of the telescope bubble tube was inclined with the horizontal.

The instrument is now set up close to either stake (minimum focusing distance is usually 5 to 10 ft), and a reading is obtained on the rod held on that stake. The rod is removed to the distant stake where the same reading must be obtained in order to define a horizontal line.

Adjustment. The adjustment is made by placing the middle horizontal wire at the correct reading on the distant rod by using the vertical clamp and tangent screw and then centering the bubble of the attached level tube by means of its capstan adjusting screws.

ADJUSTMENT OF THE VERTICAL ARC

This adjustment procedure applies only to transits. When the plate bubbles and the bubble on the telescope are centered, the vertical arc should read zero. If adjustment is necessary, provision is made for moving the vernier until the index reads zero on the vertical circle.

5.25 CARE OF INSTRUMENTS

Transits and theodolites are precision instruments and must be handled with care if quality results are to be obtained and costly repairs avoided. Listed below are precautions that should always be exercised:

1. When removing an instrument from its carrying case, observe how it was secured in the case so that it can be correctly returned.
2. Make certain the instrument is properly fastened to the tripod but avoid overtightening.

3. When carrying the instrument while it is mounted on a tripod, hold the tripod waist high in a horizontal position with the transit in front. The clamp screws should be loosened so that shock due to impact with any solid object is lessened.
4. Never leave the instrument unattended when it is set up.
5. Turn snugly all clamp and leveling screws to a firm bearing. Do not check them for tightness by giving them an extra twist. Capstan and other adjusting screws are particularly susceptible to damage due to overstress.
6. Have a waterproof hood available in case of sudden rain.
7. Always use the sunshade. If it is to be removed or replaced, do so with a clockwise rotary action so that the objective lens is not loosened.
8. Between regular maintenance performed by qualified technicians, the threads of leveling, clamp, and tangent screws should be cleaned and lightly lubricated.
9. The exposed vertical circle and vernier of transits should be cleaned with a chamois or very soft cloth. Do not touch the graduations with the fingers.
10. When returning an instrument to its carrying case, place the dust cap over the objective, center the head on the foot plate, and equalize the foot screws. If the carrying case does not close readily, find out what is wrong. Never use force to close it.
11. Because of the greater use of glass in theodolites, problems can arise under hot and humid conditions, which favor the growth of fungus on the surface of optical glass. This growth can become so pronounced on lenses, prisms, and graduated circles as to make the instrument inoperable. At the end of the day's operations, the instrument should be exposed to air currents in a well-ventilated room. If the transit is to be shipped and will remain in its container for an extended period, a suitable dessicant should be packed with it.

5.26 CONSTRUCTION LAYOUT TECHNIQUES

In performing construction layout using transits and theodolites, one should use procedures that minimize the effects of errors due to imperfect adjustment and leveling of the instrument. Some of the basic construction layout techniques are discussed in the following paragraphs.

PROLONGING A STRAIGHT LINE

Suppose that the two points *A* and *B* fix the direction of the line that is to be prolonged. The instrument is set up over *B* and a backsight is taken on *A* with the telescope in the direct position. The telescope is inverted and a temporary point *C* is marked on a stake set on the line of sight and at a distance as far as can conveniently be seen. The instrument is then reversed, and a second sight is taken on *A* with the telescope remaining in the inverted position. The telescope is then made direct and a second temporary point *D* is marked on the foresight stake. A point midway between *C* and *D* is now found and fixed as a permanent point on the true prolongation of line *AB*.

If the points *C* and *D* should happen to coincide, that is the permanent point sought.

Possibly the two temporary points will not fall within the width of one stake. Then two or more stakes may be required to complete the process, with care taken not to disturb a point that has once been set and to ensure that the final point is properly fixed.

WIGGLING-IN

Occasionally it may be necessary to establish a point on a line that is defined by termini that cannot be occupied by a transit and are not intervisible because of some obstruction such as a hill. The general procedure for placing the intermediate point *B* on line is indicated in Figure 5.35. Initially, an estimate is made of the position of the line and the instrument is set up at point *B′* and a backsight taken to point *A*. Then, the telescope is inverted, a sight taken toward point *C*, the distance *CC′* measured, and the approximate distance the instrument must be moved laterally to place it on line is estimated by the instrument operator. This trial-and-error procedure continues until the final movement of the instrument is effected by shifting it on the tripod head. Finally, the location of *B* is checked by using the procedure discussed above for prolonging a straight line. Backsight can be made to either point *A* or *C*.

RANDOM LINE

When it is desired to connect with a straight line two distant points that are not intervisible, it is frequently accomplished by projecting from one point a straight line that is estimated to fall nearly on the other point. When the line thus projected has been run out, its position relative to the distant point is measured, usually by a swing offset, and from these data the direc-

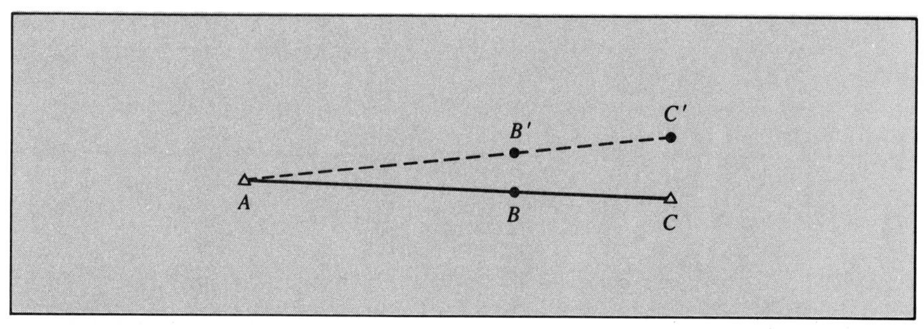

FIGURE 5.35 Wiggling-in

tion of a true line between the two points is calculated and established on the ground. The projected or trial line is called a *random line*.

INTERSECTION OF TWO LINES

A problem frequently encountered, especially on route surveys, is that of finding the point of intersection of two lines. Figure 5.36 depicts the situation. The intersection of lines *AB* and *CD* is to be found. An instrument is set up on point *C*, a foresight taken on *D*, and stakes *E* and *F* are located in positions estimated to flank within a few feet the location of the point of intersection *G*. A cord is stretched tightly between points *E* and *F*. The instrument is next set up at point *A*. The foresight through *B* will intersect the stringline *EF* at point *G*. Provided two transits are available, an obvious alternative to this procedure is to occupy points *A* and *C*, take foresights to *B* and *D* respectively, and progressively direct a person holding a range pole to each line until the point *G* is reached.

LAYING OUT AN ANGLE

Suppose that, in Figure 5.37, the direction of the line *OD* is to be established by turning a specified angle from an existing line *OC*. The instrument is first set up over point *O*. With the telescope in the direct position and the circle set to read 0°00′00″, a backsight is made to station *C*. The specified angle is then turned off using the upper motion and a point D_1 is set on the ground along the line of sight. Next, the telescope is inverted, and a back-

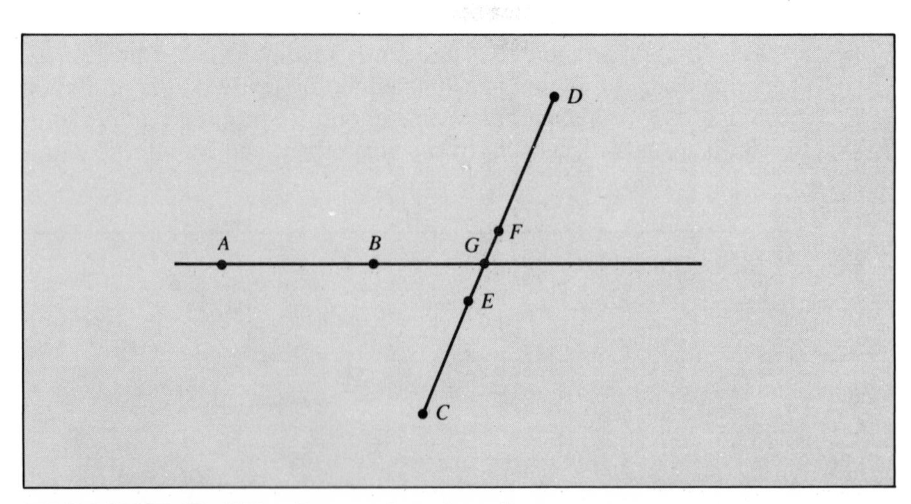

FIGURE 5.36 Intersection of two lines

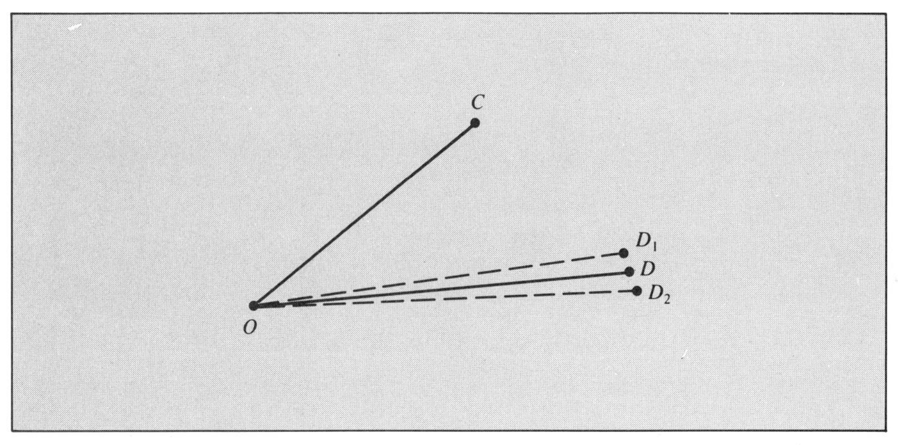

FIGURE 5.37 Laying out an angle

sight is again made on point C with the circle set at 0°00′00″. The specified angle is then turned off using the upper motion, and a second point D_2 is located on the ground. The mid-point between points D_1 and D_2 is then the location for point D.

After point D has been located, the angle $C\hat{O}D$ may be measured several times. If the average value of the several repeated measurements differ from the specified angle, the length of the perpendicular offset from point D is computed and used to establish the correct direction of the line OD.

5.27 SIGHT TARGETS

For ordinary surveying that involves relatively short sight distances, some of the commonly used targets for angle measurements include range poles, chaining pins, and pencils held plumb over the survey point. In calm weather conditions, a paper target suspended by a plumb bob, such as that shown in Figure 5.38a, may be used. A self-adhesive paper target with red and white quadrants, such as that shown in Figure 5.38b, is particularly useful as a target point on walls of buildings or other structures. The self-adhesive target has a paper backing that can be peeled off before applying the target to a surface.

Figure 5.39 shows a target system that can be used for more accurate surveys. It consists of a tribrach, a target carrier, a short-distance target, and a long-distance target. The tribrach has a bull's-eye bubble. The target carrier has an optical plummet for centering over a survey point, and the short-distance target can be illuminated with batteries for night observations. The long-distance target is suitable for sights up to 4 miles long.

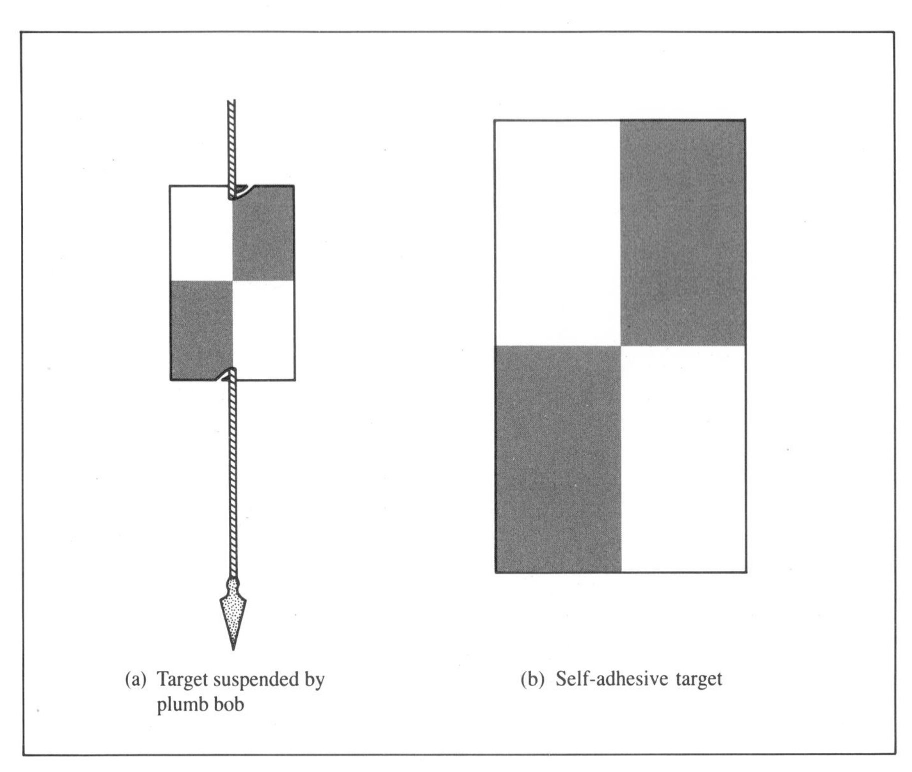

FIGURE 5.38 Paper sight targets

5.28 SOURCES OF ERRORS

Most errors in angle measurements behave in a random manner over the course of a survey project. Some of the most common sources of errors are described below.

1. *Imperfect Adjustment of Instrument.* All surveying transits and theodolites should be frequently checked and properly adjusted before use. In the field, errors from this source can be minimized by always making direct and reverse readings in both measurement and layout works.

2. *Eccentricity in Setting Up.* Both the measuring instrument and the target must be carefully centered over the respective survey point. In windy conditions, an optical plummet should be used. Error due to improper centering, called eccentricity, is particularly serious for short lines. A rule of thumb worth remembering is that a chord 1 inch long will subtend an angle of nearly one minute of arc at a distance of 300 ft. Thus, for short distances, a small eccentric error can result in a large error in the measured angle.

3. *Faulty Pointing to Target.* To minimize any error due to imperfect adjustment of the cross-hair reticule, the center part of the cross-hairs should always be used in horizontal angle measurement. Use of the upper or lower part of the vertical

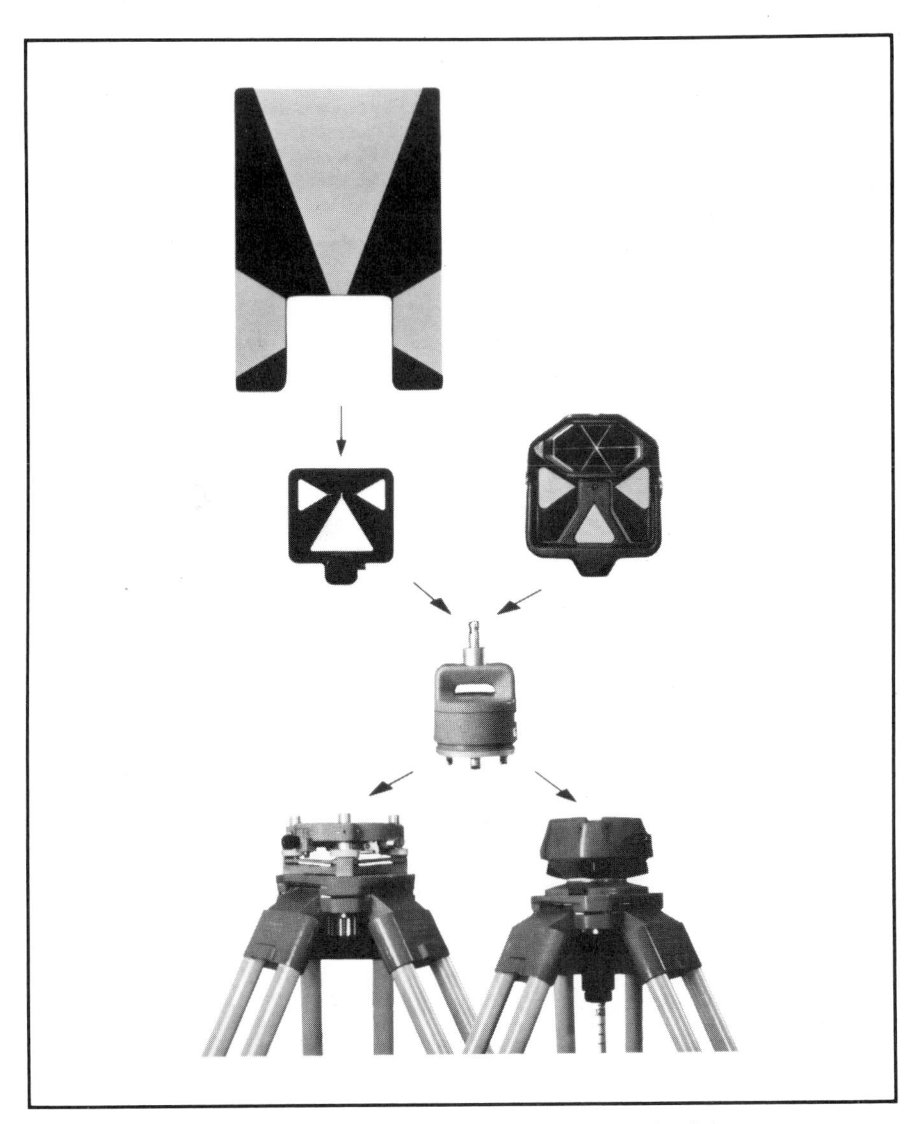

FIGURE 5.39 A sight target system (courtesy of Wild Heerbrugg Instruments, Inc.)

cross-hair should be avoided. When sighting on a supposedly vertical object such as a range pole or chaining pin, the lowest part of the object should be used for pointing to minimize any error due to incorrect plumbing. Special care should be exercised to recognize any uneven lighting of the target. Figure 5.40 shows the uneven lighting effect on a round object. The left part of the object is shaded. As the observer centers the cross-hairs only on the part that shows up by the lighting, a sighting error occurs. This problem can be minimized by using targets of suitable size and shape.

4. *Unstable Setups.* Swampy or thawing ground will make it impossible to obtain good results with a transit or theodolite unless special instrumental supports

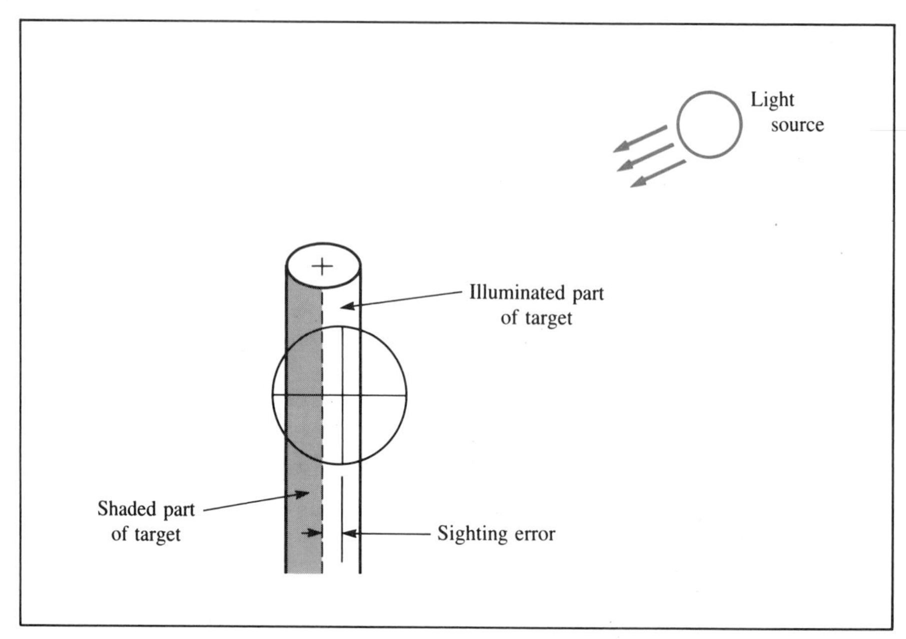

FIGURE 5.40 Uneven lighting of target

are provided. Wood posts or heavy stakes (2 in. × 4 in.) driven to a sound bearing will prove very useful. Differential temperatures in the instrument giving rise to unequal expansions may be controlled by shading the instrument if very intense sunlight prevails.

5. *Reading the Vernier or Micrometer Scale.* Errors in reading a transit's vernier can be minimized by reading both the *A* and *B* vernier, and employing a magnifying glass for reading. The best method of minimizing reading errors in angle measurement is making repeated measurements of the same angle. The number of repetitions required to reach a certain desired level of precision can be estimated using the statistical methods discussed in Chapter 2.

6. *Heat Waves.* Heat waves generated by hot objects, such as buildings and pavement, can significantly degrade the accuracy of angle measurements. During hot weather conditions, observations along lines of sight that pass close to such hot objects should be avoided.

7. *Index Error in Vertical Angles.* For instruments that do not have automatic indexing of the vertical circle, the index error should be read and recorded with each vertical angle reading.

PROBLEMS

5.1 A circular arc with a radius of 375.00 ft has a central angle of 39°14′55″. Find the following:

a. radian value of the angle

b. length of arc

5.2 Convert the following angles from the centesimal system to sexagesimal units in degrees, minutes, and seconds:

a. 47.6025^g

b. 189.3216^g

c. 381.5493^g

d. 263.1124^g

5.3 Convert the following angles from the sexagesimal units to radians:

a. 43°26′41″

b. 89°10′26.3″

c. 293°04′54.1″

d. 0°43′17.0″

5.4 Convert the following angles to sexagesimal units in degrees, minutes, and seconds:

a. 1.432187319 radians

b. 2.419632115 radians

c. 4.173654291 radians

d. 3.733361488 radians

5.5 A circular arc with a radius of 265.45 m has a central angle of 38.1460^g. Find the arc length.

5.6 Find the arc sine in grads of 0.954620.

5.7 Convert 0.00001 radian to sexagesimal units.

5.8 For each of the lines *AB*, *AC*, *AD*, and *AE*, shown in Figure P5.1 on the next page, determine the following:

a. true bearing

b. true azimuth

c. true back bearing

d. true back azimuth

e. magnetic bearing

f. magnetic azimuth

5.9 Given below are the true and magnetic bearings along several boundary lines. Determine the magnetic declination (magnitude and direction) at each of these locations:

a. Line *AB*, true bearing = S 89°22′ W
magnetic bearing = S 89°42′ W

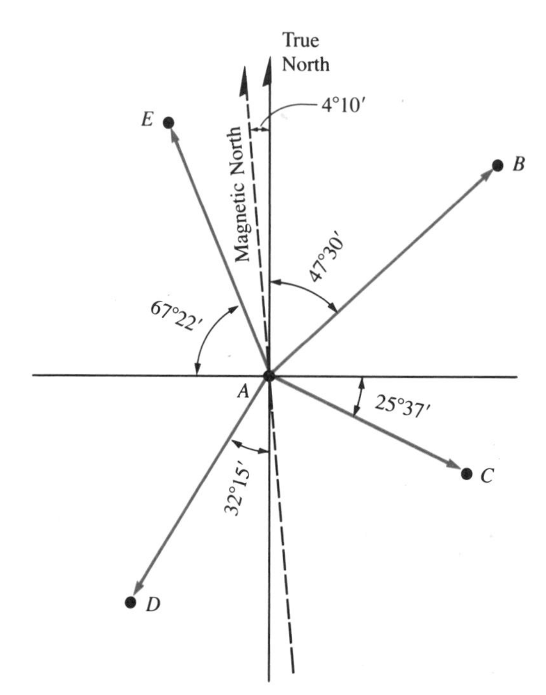

FIGURE P5.1 Bearings

b. Line *DE*, true bearing = N 1°30′ W
magnetic bearing = N 3°19′ E

c. Line *GH*, true bearing = S 47°35′ E
magnetic bearing = S 48°32′ E

d. Line *IJ*, true bearing = N 79°46′ W
magnetic bearing = N 77°22′ W

5.10 Determine the present magnetic bearing for each of the following boundary lines:

Magnetic Bearing in 1870	Magnetic Declination in 1870	Magnetic Declination at Present
a. N 88°15′ E	2°15′ E	0°30′ W
b. N 15°30′ W	1°43′ W	0°20′ E
c. S 32°14′ E	5°03′ E	1°22′ E
d. S 78°43′ W	3°40′ E	2°33′ W
e. N 1°21′ E	0°53′ E	3°40′ W

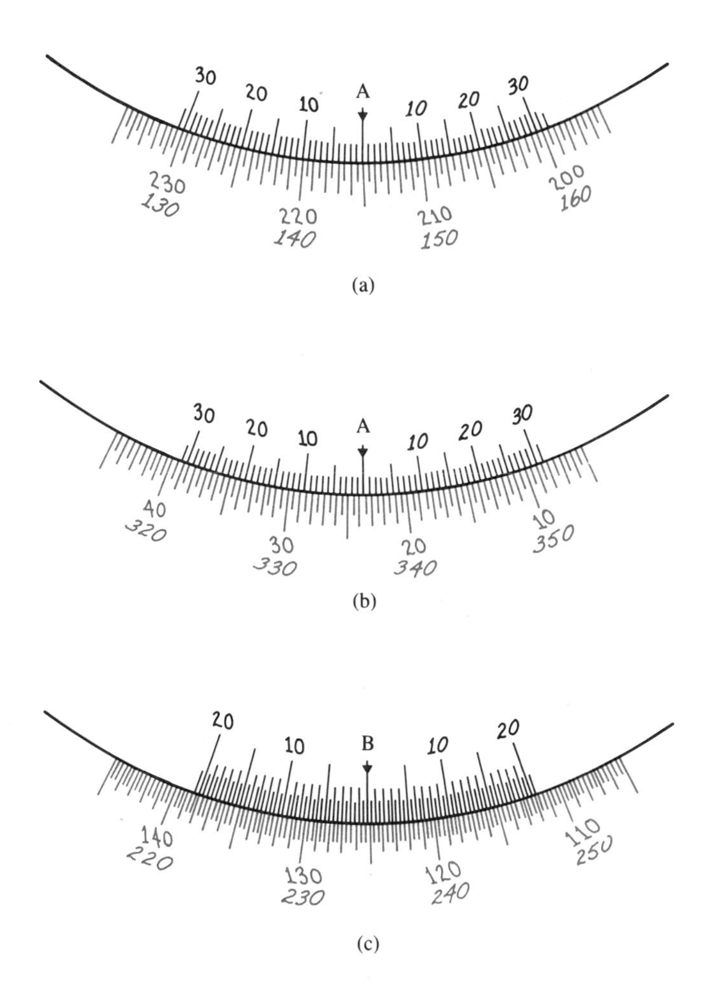

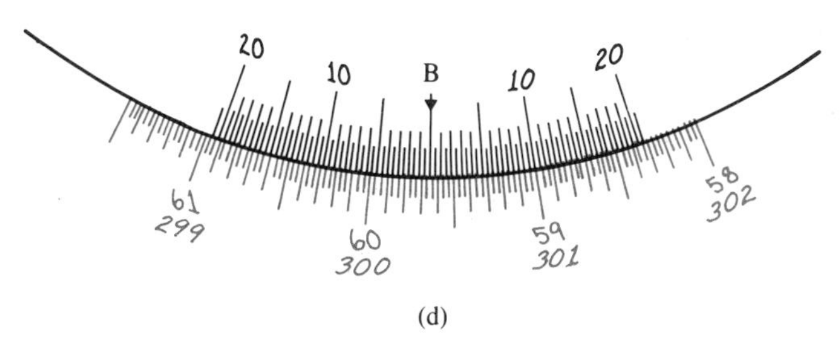

FIGURE P5.2 Vernier settings (courtesy of Keuffel & Esser Company)

5.11 Read the horizontal angles from the vernier settings shown in Figure P5.2.

5.12 Compute the values of the measured angles from the field notes given on the top of the next page. The angles were measured by the repetition method.

Station	Rep.	Tel.	Horizontal Angle	Vernier A	Vernier B
a. Transit at Station *A*					
G	0	*D*	0°00′	00″	00″
	1	*D*	260°43′	00″	
F	4	*R*	322°50′	30″	00″
b. Transit at Station *B*					
D	0	*D*	0°00′	30″	30″
	1	*D*	194°47′	00″	
F	2	*R*	29°35′	00″	30″
c. Transit at Station *C*					
H	0	*D*	0°00′	30″	00″
	1	*D*	71°12′	00″	
K	2	*R*	142°16′	00″	00″

5.13 Determine from the following set of field records:

a. the horizontal angles $A\hat{B}C$, $E\hat{F}G$, and $H\hat{I}J$;

b. the vertical angle along lines *BA*, *BC*, *FE*, *FG*, *IH*, and *IJ*.

Station	Rep.	Tel.	Horizontal Angle	Vernier A	Vernier B	Vertical Angle D	Vertical Angle R	Index Reading
a. ⊼ at Station *B*								
A	0	*D*	0°00′	00″	30″	+33°14′		
	1	*D*	56°14′	30″				
C	6	*R*	337°26′	00″	30″	−15°37′		+1′30″
b. ⊼ at Station *F*								
E	0	*D*	0°00′	20″	40″	+10°20′		
	1	*D*	128°32′	00″				
G	6	*R*	51°03′	40″	00″	+5°13′		−1′20″
c. ⊼ at Station *I*								
H	0	*D*	60°00′	00″	30″	−3°47′		
	1	*D*	273°41′	30″				
J	6	*R*	202°04′	30″	30″	−15°53′		+1′

5.14 Compute the mean, standard deviation, and estimated standard error of the mean for angle $A\hat{O}B$ from the following set of field notes.

Station	Tel.	Direction
⊼ at station 0		
A	*D*	0°00′00″
	R	180°00′02″
B	*D*	131°43′27″
	R	311°43′25″
A	*D*	90°00′00″
	R	269°59′56″
B	*D*	221°43′24″
	R	41°43′25″
A	*D*	180°00′00″
	R	359°59′56″
B	*D*	311°43′28″
	R	131°43′25″
A	*D*	270°00′00″
	R	90°00′03″
B	*D*	41°43′27″
	R	221°43′24″

5.15 Compute the mean, standard deviation, and estimated standard error of the mean for angle $E\hat{O}F$ from the field notes given below.

Station	Tel.	Direction	Station	Tel.	Direction
⊼ at station 0			⊼ at station 0		
E	*D*	0°00′03″	*E*	*D*	180°00′00″
	R	359°59′58″		*R*	359°59′57″
F	*D*	41°53′37″	*F*	*D*	221°53′33″
	R	221°53′33″		*R*	41°53′37″
E	*D*	60°00′00″	*E*	*D*	240°00′00″
	R	240°00′01″		*R*	60°00′02″
F	*D*	101°53′36″	*F*	*D*	281°53′38″
	R	281°53′35″		*R*	101°53′37″
E	*D*	120°00′02″	*E*	*D*	300°00′05″
	R	300°00′00″		*R*	120°00′03″
F	*D*	161°53′38″	*F*	*D*	341°53′37″
	R	341°53′35″		*R*	161°53′39″

5.16 The observed vertical angle to the top of a water tower was 12°14′30″. The telescope was then brought to the horizontal and its bubble carefully centered.

If the vertical circle reading was +0°01′30″, what is the corrected value of the vertical angle?

5.17 The transit was used to wiggle-in on line between two intervisible points as depicted in Figure 5.35. The distances AB' and $B'C'$ are approximately 1,660 ft and 2,210 ft respectively. Following careful double-centering procedures, C' is found to lie 0.19 ft to the left of C. Calculate the distance BB' that the transit must be moved to get on line.

5.18 A random line was run out for a distance of 2,642.50 ft in a direction of N 72°15′ W through timbered country and failed to meet an established marker by a measured offset of 11.25 ft. If the random line fell to the right of the marker, find the bearing of the true line.

5.19 Because of high brush obstructing the sight line, transit pointings were made on the top of a range pole, which was subsequently found to be 0.12 ft out of plumb perpendicular to the line of sight. What directional error does this displacement represent for distances of 950, 450, and 150 ft?

5.20 A check of the optical plummet of a transit revealed that it was out of adjustment by 6 mm at the ground surface. What is the maximum directional error due to faulty centering of the instrument for sight distances of 350, 70, and 5 m?

5.21 In making the layout survey of a large mill building, a 90° angle was established with a single measurement of an optical transit. The preliminary angle was then measured by repetition and found to be 90°00′13″. What offset (nearest 0.001 m) should be made at a distance of 215 m from the instrument to establish the true line?

5.22 Observations were made on a tall radio mast to check its verticality. With the transit telescope direct, the top of the mast was sighted, the telescope rotated about the horizontal axis, and a point was set on the ground 0.37 ft to the left of the base of the mast. The same procedure was followed with the telescope in the inverted position and a point was set 0.68 ft to the right. How much is the top of the mast out of plumb and in which direction, left or right?

ELECTRONIC DISTANCE MEASUREMENT

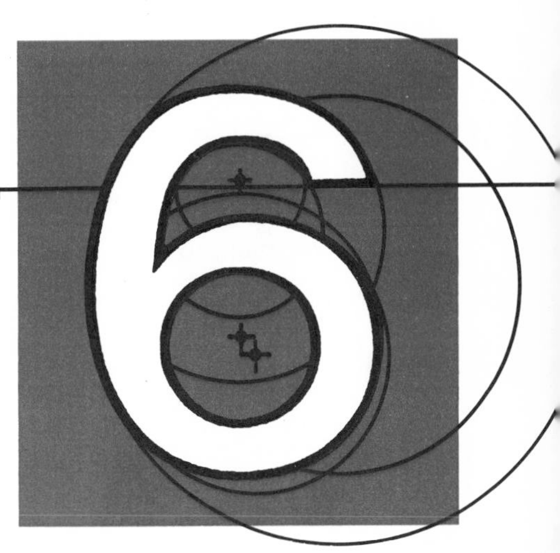

6.1 INTRODUCTION

Until the mid-1950s, one of the most challenging and difficult tasks in surveying was the measurement of long distances with an accuracy better than 1/20,000. There was no alternative to the laborious and costly procedure of invar tape measurements, as described in Section 3.16. In hilly and mountainous terrain, the accurate measurement of distances of more than one or two thousand feet was practicably impossible. Thus the development of electronic distance-measuring (EDM) instruments during the 1960s and 1970s has had just as great an impact on surveying as that of the electronic computer.

There are two basic types of EDM instruments: (1) electro-optical instruments, which use light waves as a measurement medium; and (2) microwave instruments, which use radio waves as the medium. Here the word light is used rather loosely. Light is more exactly defined as that portion of the electromagnetic spectrum to which the human eye is sensitive. This so-called "visible" light has a wavelength of 0.4 to 0.7 μm (micrometers or simply microns). Infrared "light," which is not visible and which is used as the carrier wave by most of the newest short-distance EDM instruments, has a wavelength of 0.7 to 1.2 μm. On the other hand, the microwave used in distance measurements usually has a wavelength between 10 μm and 100 μm.

6.2 ELECTRO-OPTICAL INSTRUMENTS

The first electro-optical distance-measuring device was developed by the Swedish geodesist Erik Bergstrand in 1947 as the outgrowth of an experiment to determine the velocity of light. The instrument was called the *Geodimeter,* which was derived from the words GEOdetic DIstance METER. The basic measurement principle is illustrated in Figure 6.1.

A frequency-modulated light beam is generated from the instrument and projected to a reflector located at the other end of the line. The reflector sends the light beam back to the instrument, which subsequently measures the time (t) required for the beam to travel from its source to the reflector and back. The time measurement is then converted to the distance (s) between the instrument and the reflector by the following simple relationship:

$$s = \frac{1}{2} Vt \tag{6.1}$$

where V is the velocity of light in the atmosphere.

The initial model of geodimeter became commercially available in 1952. The early models used tungsten or mercury lamps as light source. These early instruments were very bulky and expensive. Measurements often had to be made at night because of the weak light signal, and the measurement accuracy was highly questionable for the distances—usually under 1,500 ft—most commonly involved in surveying. For these reasons early application of the geodimeter was limited to special surveys, such as national geodetic survey programs. It was not until the 1960s that the development of the Gallium Arsenide (GaAs) diode, a solid-state device for gener-

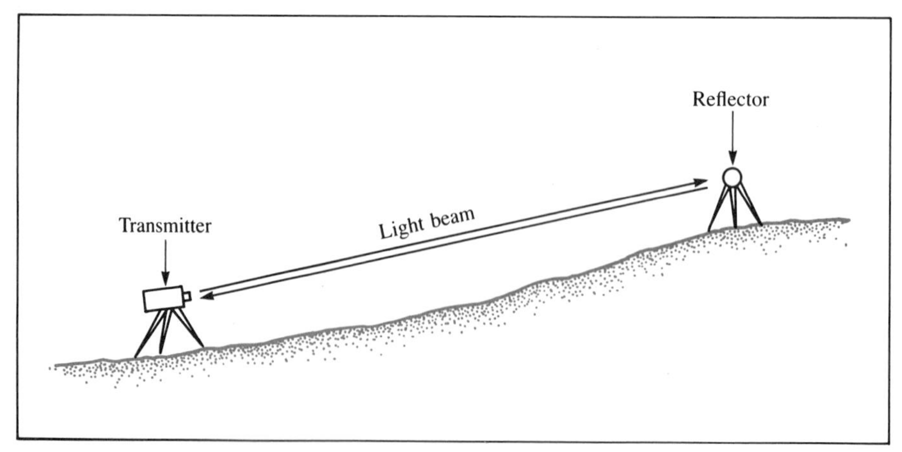

FIGURE 6.1 Measurement principle of an electro-optical distance-measuring instrument

FIGURE 6.2 Hewlett Packard 3805A distance meter (courtesy of Hewlett-Packard, Inc.)

ating infrared radiation, as a light source made low cost electro-optical distance-measuring instruments commonly available.

There is a wide variety of electro-optical distance-measuring instruments available in the market today. Figure 6.2 shows a Hewlett-Packard 3805A Distance Meter, which has a maximum range of 1.6 km and a measurement accuracy, expressed as root-mean-square (rms) error, of ±(7 mm + 7 mm/km); that is, for a measured distance of 1 km, the rms error of the measurement is expected to amount to ±(7 mm + 7 mm), or ±14 mm. The optimum operating temperature is between −10°C and 40°C.

Figure 6.3 illustrates a Geodimeter 112, which is mounted on the telescope of a Wild T2 theodolite. The Geodimeter 112 has a maximum range of 5.5 km and a minimum measurable distance of 0.2 m. It has a measuring accuracy, expressed as rms error, of ±(5 mm + 5 ppm), where ppm is the abbreviation for parts per million of the distance measured. Thus, for a measured distance of 1 km, its rms error is expected to be ±(5 mm + 5/1,000,000 × 1,000 m), or ±10 mm. It has an operating temperature range between −20°C and +50°C.

Figure 6.4 shows the operating panel of the K & E Rangemaster® III, which has a maximum range of 60 km. It uses a helium-neon laser, with a wavelength of 0.6328 μm, as light source. Its measurement accuracy, expressed as root-mean-square error, is ±(5 mm + 1 ppm) for the optimum operating temperature range of −6°C to +43°C.

Three other models of electro-optical distance-measuring instruments are shown in Figures 6.5 to 6.7. The range and accuracy limitations of

FIGURE 6.3 Geodimeter 112 (courtesy of AGA Geodimeter, Inc.)

these models and the ones discussed above are summarized in Table 6.1. These six models represent only a small sample of the many models commercially available. Almost every manufacturer has a whole line of models covering a wide range of accuracy, range limitations, and costs. Moreover, the field of EDM is still in a stage of active development. Beginning in the late 1970s, advances in electronics and computer technology have led to the development of an instrument that can be used to measure the vertical angle, horizontal angle, and distance at a single pointing of the instrument. Such an instrument is called a *total station*.

Figures 6.8, 6.9, and 6.10 show three models of total stations, and Table 6.2 summarizes the technical data on several models. Typically, a total station can display the horizontal angle, vertical angle, slope distance, horizontal distance, or vertical distance at the command of the operator. Most

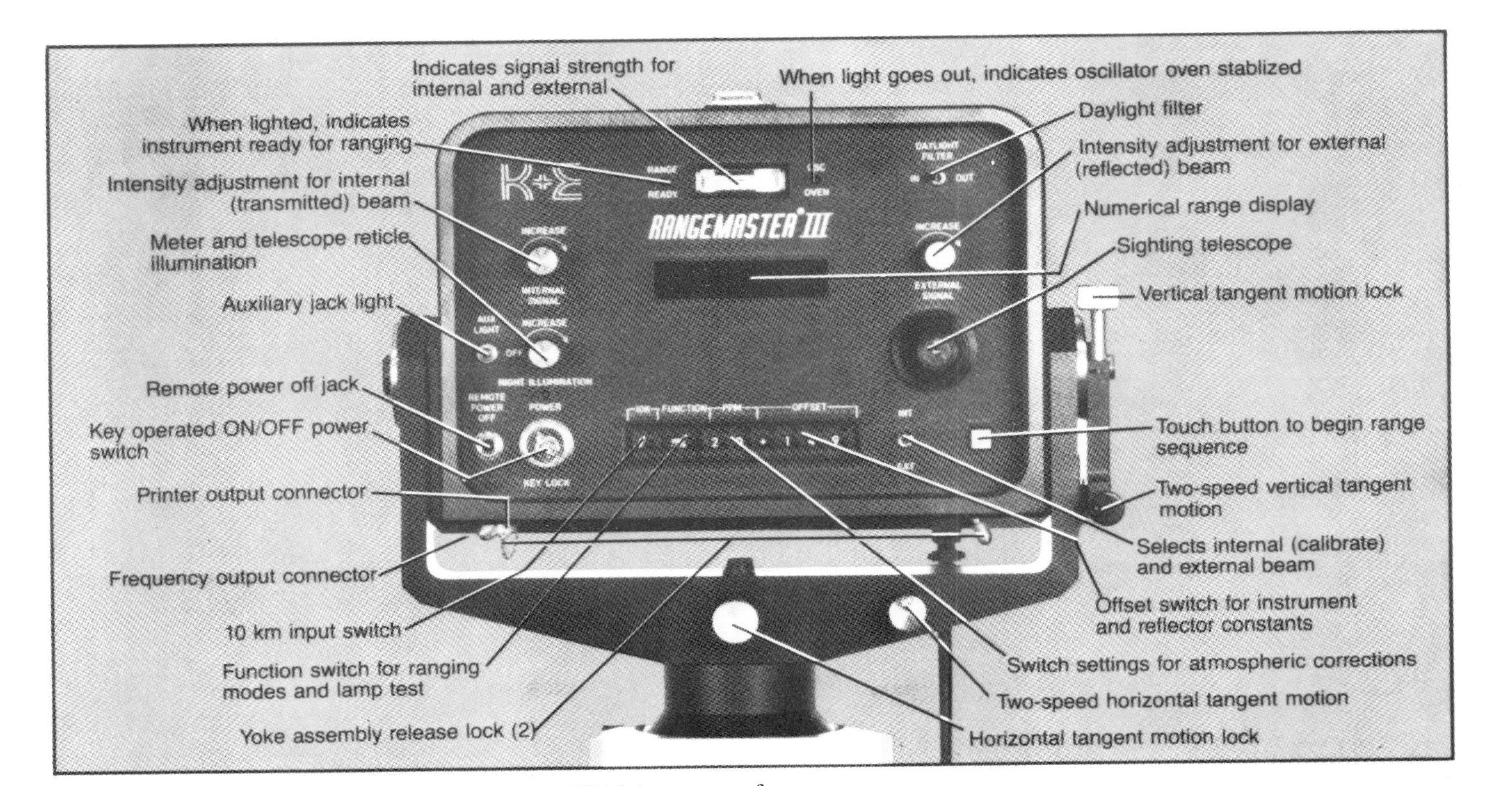

FIGURE 6.4 RANGEMASTER® III EDM (courtesy of Keuffel & Esser Company)

FIGURE 6.5 Wild Distomat DI4 (courtesy of Wild Heerbrugg Instruments, Inc.)

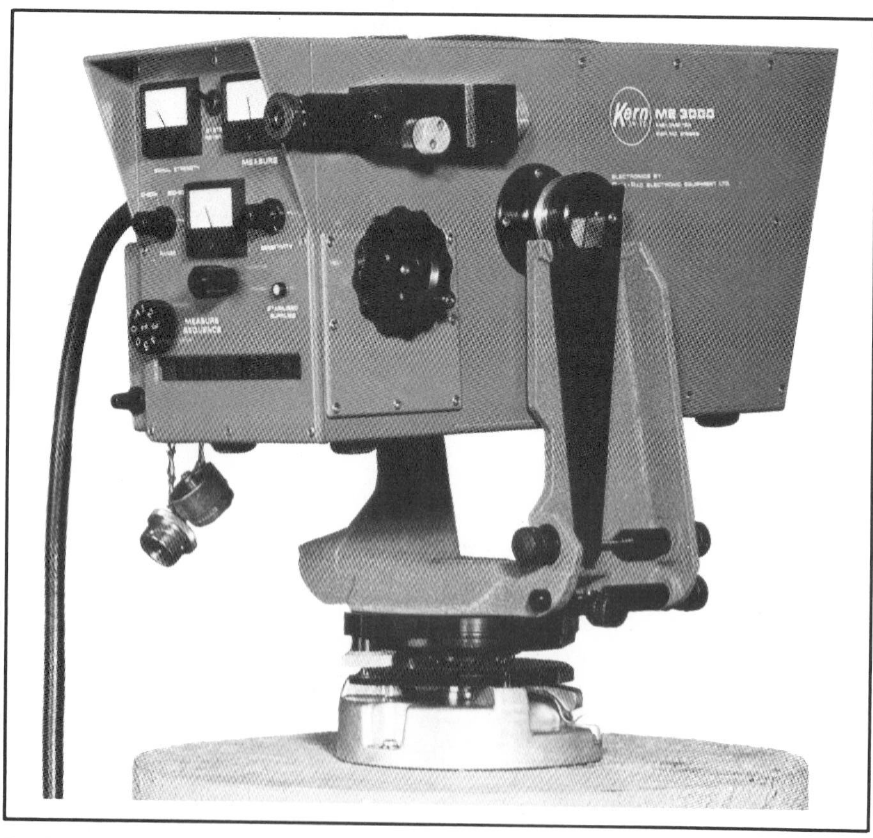

FIGURE 6.6 Kern ME 3000 Mekometer (courtesy of Kern Instruments, Inc.)

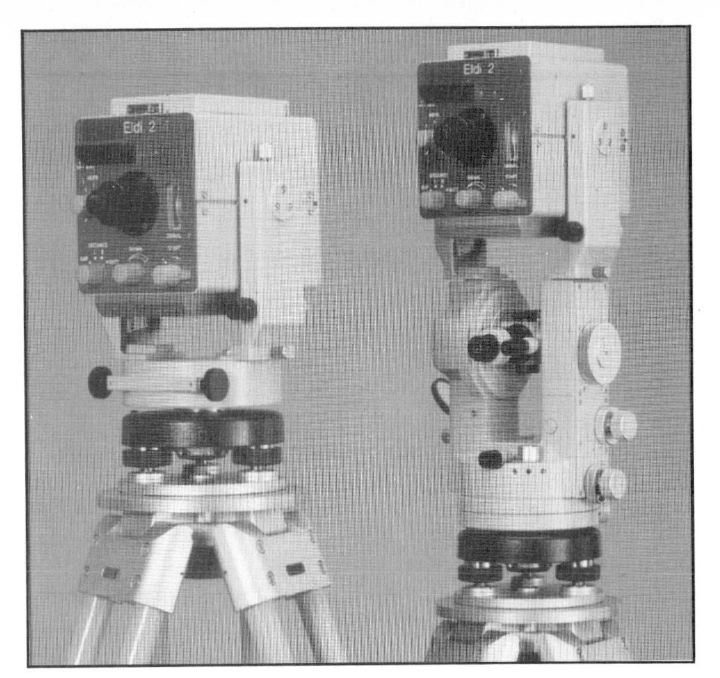

FIGURE 6.7 Zeiss Eldi 2 (courtesy of Carl Zeiss, Inc., Thornwood, N.Y.)

FIGURE 6.8 HP3820A electronic total station with HP 3851A data collector (courtesy of Hewlett Packard, Inc.)

TABLE 6.1 Characteristics of Some Electro-Optical EDM Instruments

Model (Manufacturer)	Range	Accuracy Within Optimum Temperature Range (Root-mean-square error)	Optimum Operating Temperature Range	Weight	Dimensions
Geodimeter 112 (AGA Geodimeter, Inc.)	5.5 km	±(5 mm + 5 ppm)	−20°C to +50°C	2.6 kg	220 × 180 × 90 mm
HP 3805A Distance Meter (Hewlett-Packard, Inc.)	1.6 km	±(7 mm + 7 mm/km)	−10°C to +40°C	7.7 kg	332 × 318 × 275 mm
Rangemaster III (Keuffel & Esser Company)	60 km	±(5 mm + 1 ppm)	−6°C to +43°C	25 kg	317 × 228 × 434 mm
ME 3000 Mekometer (Kern Instruments, Inc.)	2.5 km	±(0.2 mm + 1 ppm)	−9°C to +40°C	14.5 kg	460 × 160 × 220 mm
Wild DI4 Distomat (Wild Heerbrugg Instruments, Inc.)	2.5 km	±(5 mm + 5 mm/km)	−25°C to +50°C	1.9 kg	190 × 60 × 60 mm
Zeiss Eldi 2 (Carl Zeiss, Inc.)	5 km	±(10 mm + 2 ppm)	−20°C to +60°C	4.2 kg	135 × 120 × 155 mm

TABLE 6.2 Characteristics of Some Total Stations

Model (Manufacturer)	Maximum Range	Range Accuracy Within Optimum Temperature Range (RMS error)	Optimum Operating Temperature Range	Least Count in		Data Collector Available
				Horizontal Angle	Vertical Angle	
Geodimeter 140 (AGA Geodimeter, Inc.)	5.5 km	±(5 mm + 5 ppm)	−20°C to +50°C	2″	2″	Yes
HP 3820A (Hewlett Packard, Inc.)	5 km	±(5 mm + 5 ppm)	−10°C to 40°C	1″	1″	Yes
Kern El Electronic Theodolite with DM502 Distance Meter (Kern Instruments, Inc.)	6.5 km	±(3 mm + 5 ppm)	−20°C to +50°C	1″	1″	Yes
Sokkisha SDM3ER (Distributed in the U.S. by The Lietz Company)	2.5 km	±(5 mm + 5 ppm)	−20°C to +50°C	10″	10″	Yes
Wild TC1 Total Station (Wild Heerbrugg Instruments, Inc.)	2.0 km	±(5 mm + 5 ppm)	−20°C to +50°C	2″	3″	Yes
Zeiss Elta 2 (Carl Zeiss, Inc.)	5 km	±(5 mm + 2 ppm)	−20°C to +60°C	0.6″	0.6″	Yes

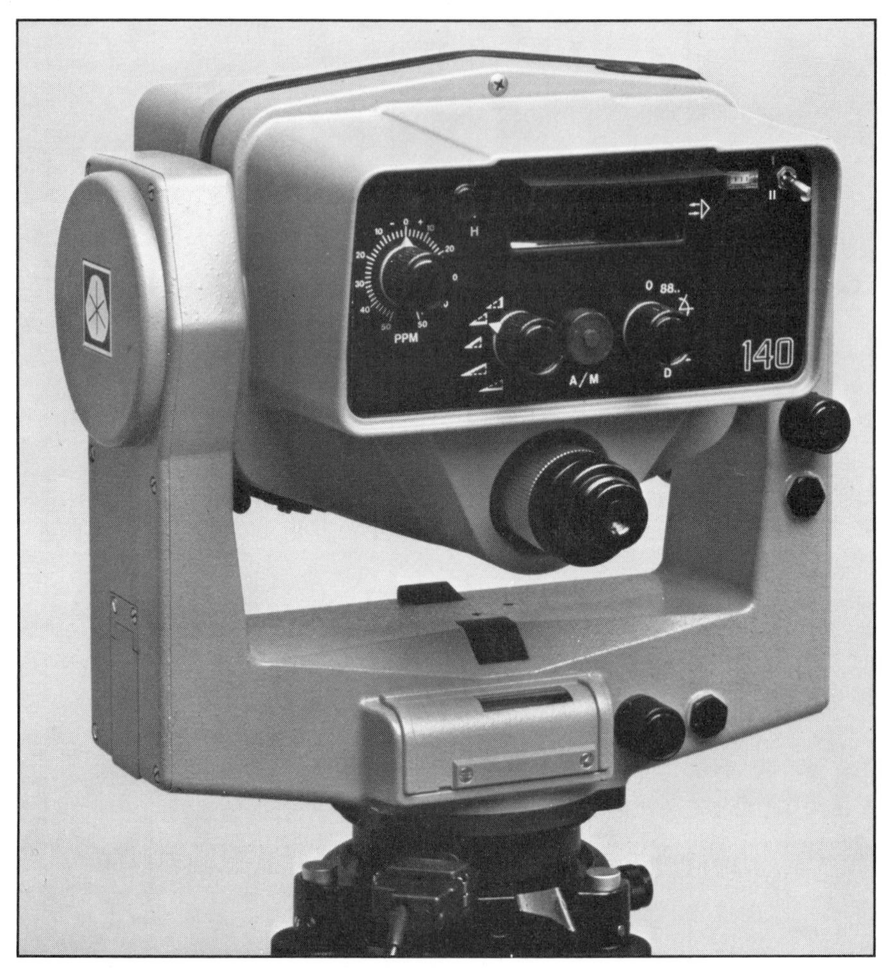

FIGURE 6.9 Geodimeter 140 total station (courtesy of AGA Geodimeter, Inc.)

total stations can be operated with an optional data collector, which automatically records the distance and angle measurements. The data collector can in turn transfer the measurements either directly, or through a telephone line, into a computer for processing. The field book is used primarily to record backup information including sketches, station locations, and weather conditions.

6.3 MICROWAVE INSTRUMENTS

The first successful application of radio waves to the problem of distance measurement took place about 1935. Termed *radar,* it was the

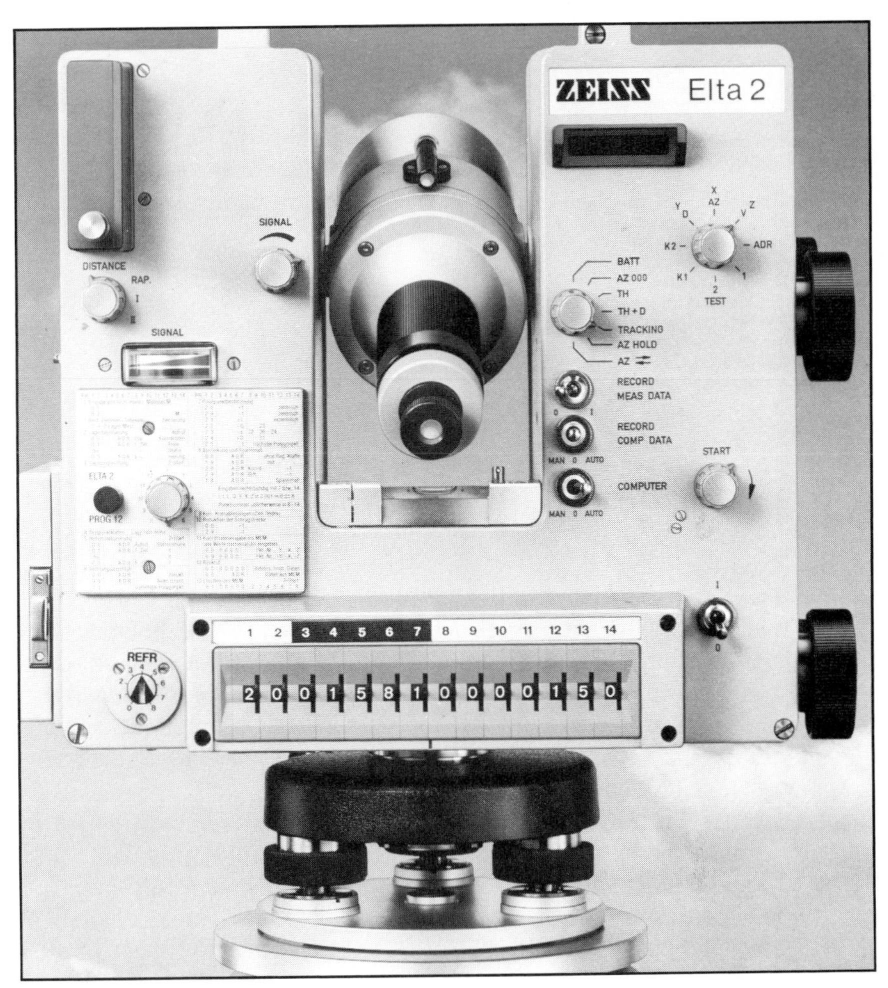

FIGURE 6.10 Zeiss Elta 2 electronic tacheometer (courtesy of Carl Zeiss, Inc., Thornwood, N.Y.)

precursor of most military and commercial radar systems. However, the equipment was too heavy and bulky, and the accuracy was too low to make it adaptable to the needs of surveying and mapping. The first portable distance-measuring instrument using microwave was developed by Dr. T. L. Wadley of the National Telecommunications Research Laboratory, South Africa, and became available in the American market in 1957. The instrument was called the *Tellurometer*. A year later, Cubic Corporation of San Diego introduced a similar instrument and called it the *Electrotape*. Subsequent models of both the Tellurometer and the Electrotape were used extensively for the measurement of long distances.

Figure 6.11 illustrates the measurement principle used by the Tellurometer and the Electrotape. The measurement system consists of two

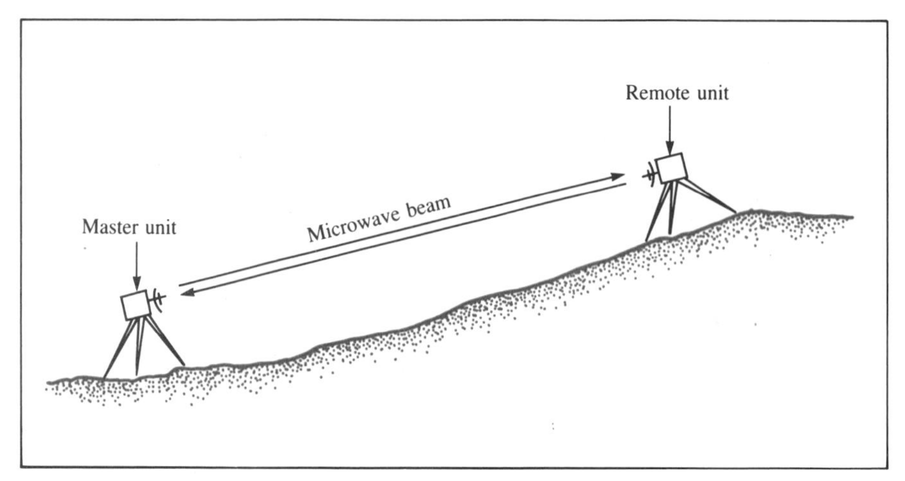

FIGURE 6.11 Measurement principle of a microwave distance-measuring instrument

identical electronic units. One unit is set up at each end of the line to be measured. One unit serves as the master unit, while the other serves as the remote unit. From the master unit's antenna, a frequency-modulated radio wave is sent towards the remote unit. The remote unit receives the signal and retransmits it back to the master unit, which then measures the double transit time of the radio signal and converts the time to distance using the relationship expressed by Eq. (6.1).

Shown in Figure 6.12 is a MRA5 Tellurometer, which has a minimum measureable range of 300 m and a maximum range of 50 km. It provides a fully automatic digital display of distance in meters and has an accuracy, expressed in probable error, of ±(5 cm + 10 ppm) within the temperature range of −32°C to 44°C. The accuracy can be considerably upgraded by making measurements on several different frequencies. The resolution (least count) is 1 cm. The unit has a weight of 12 kg.

Microwave instruments have several major disadvantages when compared with the electro-optical instruments:

1. Since a microwave system consists of two identical electronic units, two operators are required to measure a line. In the case of electro-optical instruments, one person can operate the instrument while another person can have the responsibility of setting up reflectors. The second person does not have to be at the reflector during the measurement.

2. There are many applications (such as measuring structural deformation) where it is either impossible or inconvenient to set up a measuring unit at both ends of the line.

3. The accuracy of microwave instruments is more severely affected by changes in atmospheric conditions.

4. The accuracy of microwave instruments can also be degraded by multipath reflections, resulting in a measurement error called ground swing. The beam transmitted by microwave instruments is in the shape of a cone and measures about $1\frac{1}{2}°$ wide. Any surface struck by part of the beam will cause reflection. The reflections

FIGURE 6.12 MRA 5 tellurometer (courtesy of Telluro-meter-U.S.A.)

that reach the instrument will degrade the readings, since the microwave receiver cannot distinguish between a direct and a reflected wave. The problem is particularly serious in measuring long distances over large bodies of water or flat plains.

5. The measurement accuracy of microwave instruments is generally lower than that of electro-optical instruments of similar costs.

Because of the preceding reasons, electro-optical instruments are gaining increasing popularity over the microwave type. The remaining parts of this chapter will deal exclusively with electro-optical instruments. However, many of the basic principles to be discussed can be applied equally well to microwave instruments.

6.4 PRINCIPLES OF ELECTROMAGNETIC MEASUREMENTS

Associated with the oscillation of energy of an alternating current flowing in an open circuit, such as an antenna, will be the formation of electric fields that radiate electromagnetic energy into space as electromagnetic waves. Both radio waves and light waves are electromagnetic waves having identical velocities in a vacuum.

In 1957 the International Union of Geodesy and Geophysics (IUGG) adopted as standard the value of 299,792.5 ± 0.4 kilometers per second for the propagation velocity in vacuum of visible light and radio microwaves.

An important relationship in wave motion is that expressed by the equation

$$\lambda = \frac{V}{f} \tag{6.2}$$

where λ is the wavelength or the distance traveled during the period of one cycle, V is the velocity of propagation, and f is the frequency or number of cycles per unit of time. The unit of frequency is the Hertz (Hz), which is one cycle per second. A megahertz (MHz) is one million cycles per second. A frequency of 10 MHz indicates that the wave completes 10 million cycles per second. The waveform is sinusoidal.

It was stated in previous sections that EDM instruments determine a distance by measuring the time it takes for a radio wave or light beam to travel from one end of the line to the other and back. In reality, the travelling time is never directly measured by an EDM instrument. To determine a distance accurately, the time would have to be measured with extremely high accuracy. For example, if the distance is to be calculated from the transit time with an accuracy of 0.5 ft, the interval of time would have to be correct to within one billionth of a second or one millimicrosecond. So small an interval would be extremely difficult to measure directly and the problem would become still more acute if the tolerance in distance were realistically smaller, such as 0.01 ft. All EDM instruments measure the time indirectly by measuring the phase difference between outgoing and returning waves.

Phase refers to a portion of a complete cycle of a wave (see Figure 6.13). One complete cycle is represented by 360°, and one-quarter cycle would be 90°. Phase difference is the time in electrical degrees by which one wave leads or lags another. Phase comparison procedures measure the phase of one wave and that of another at the same moment of time.

For example, suppose that an electromagnetic wave (called the *carrier wave*) is caused to pulsate at a frequency of 14.989625 MHz. This carrier wave is said to be frequency modulated with a pattern frequency of 14.989625 MHz. Suppose that at a given instant in time, the wave leaving an EDM instrument is measured to have a phase angle of 30° while the returning wave is measured to have a phase angle of 287°, as shown in Figure 6.14. The phase difference is therefore 257°.

It can be seen from Figure 6.14 that the distance (s) between the transmitter and the reflector can be computed as follows:

$$s = \frac{1}{2}\left(n\lambda + \frac{\Delta}{360^\circ}\lambda\right)$$

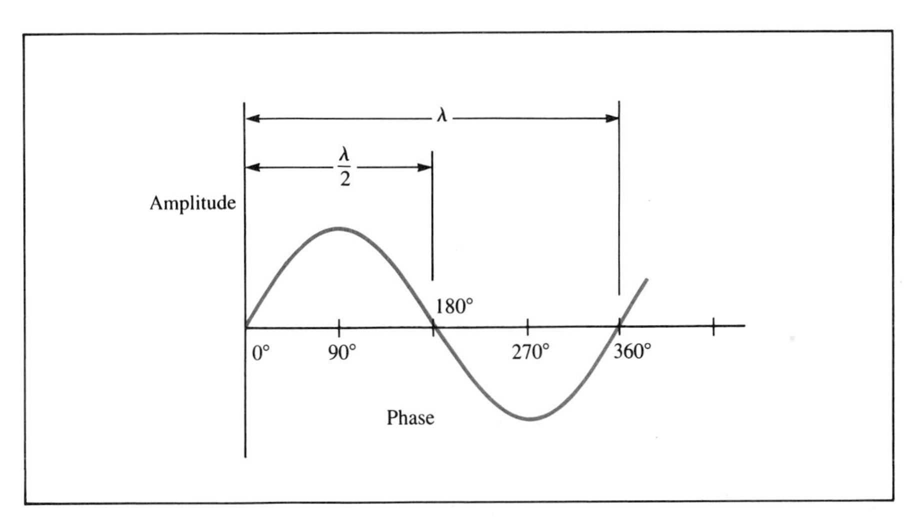

FIGURE 6.13 Phase angle of a sinusoidal wave

where λ is the wavelength, n is the total number of full wavelengths, and Δ is the measured phase difference. The term (Δ/360°)λ thus represents the fractional wavelength.

For the preceding example, the wavelength for a frequency of 14.989625 MHz can be computed from Eq. (6.2) as follows:

$$\lambda = \frac{V}{f} = \frac{299{,}792{,}500 \text{ m/sec}}{14{,}989{,}625 \text{ cycles/sec}} = 20 \text{ m}$$

Therefore, a measured phase difference of 257° means that the distance s is as follows:

$$s = \left[\frac{1}{2}(20) \cdot n + \frac{1}{2} \cdot \frac{257°}{360°} \cdot 20\right] \text{ meters}$$

$$s = [10n + 7.139] \text{ meters}$$

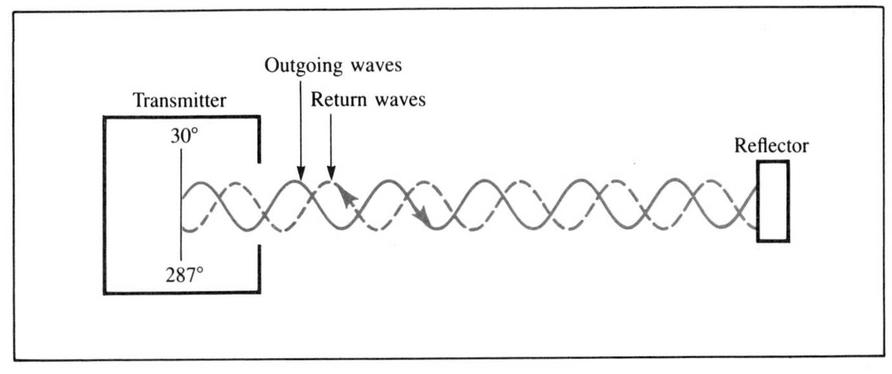

FIGURE 6.14 Measurement of phase difference

Of course, the number of whole wavelengths (n) cannot be determined from phase measurement using a single pattern frequency.

An EDM instrument measures a distance by using several pattern frequencies, which are a multiple of 10 of each other. For example, suppose that an EDM instrument measures the phase difference along a line using the following four frequencies: $F_1 = 14.989625$ MHz, $F_2 = 1.4989625$ MHz, $F_3 = 149.89625$ KHz, and $F_4 = 14.989625$ KHz. Suppose that the measured phase differences are: $\Delta_1 = 257°$, $\Delta_2 = 62°$, $\Delta_3 = 150°$, and $\Delta_4 = 123°$. The distance s can then be computed as follows:

Frequency	Measured Phase Difference	λ	$\frac{1}{2}\frac{\Delta}{360°}\lambda$
$F_1 = 14.989625$ MHz	$\Delta_1 = 257°$	20 m	7.139 m
$F_2 = 1.4989625$ MHz	$\Delta_2 = 62°$	200 m	17 m
$F_3 = 149.89625$ KHz	$\Delta_3 = 150°$	2,000 m	416 m
$F_4 = 14.989625$ KHz	$\Delta_4 = 123°$	20,000 m	3416 m
$s = 3{,}417.139$ meters			

Most modern EDM instruments perform the phase measurements automatically, switching from one frequency to the other without operator assistance, and display the measured distance directly in feet or meters.

6.5 INDEX OF REFRACTION

The ratio between the velocity of propagation of electromagnetic wave in vacuum (V_o) and the velocity in the atmosphere (V) is called the *index of refraction* (N_a); that is,

$$N_a = \frac{V_o}{V} \tag{6.3}$$

The value of N_a is dependent on the wavelength, atmospheric pressure, temperature, and relative humidity.

Microwaves behave differently in the atmosphere than visible or near visible light waves. The refractive index for light waves, to be denoted by N_g, in dry atmosphere at standard temperature (0°C) and at sea level (760 mm Hg) is given by the following expression:

$$N_g = 1 + (287.604 + 4.8864\,\lambda_c^{-2} + 0.068\,\lambda_c^{-4})\,10^{-6} \tag{6.4}$$

where λ_c is the wavelength of light in μm, which is between 0.90 and 0.93 μm for near-infrared light generated by a gallium arsenide diode and equal to 0.6328 μm for light generated by a helium-neon laser.

At other atmospheric conditions, the refractive index for light waves can be computed to an accuracy of 1 ppm by using the following expression:

$$N_a = 1 + \frac{0.359474(N_g - 1)P}{273.2 + t} \tag{6.5}$$

where P is the atmospheric pressure in millimeters of mercury (mmHg) and t is air temperature in degrees Celsius (°C). The effect of relative humidity is generally negligible for computing N_a for light waves.

Thus, for $\lambda_c = 0.90\ \mu m$, $N_a = 1.000294$ at sea level (760 mmHg) and air temperature of 20°C. In EDM applications, the index of refraction is often expressed as follows:

$$N_a = 1 + 10^{-6} N \tag{6.6}$$

If $N_a = 1.000294$, then $N = 294$. The unit of N is often labelled as ppm (parts per million).

EDM instruments are designed to display the measured distance for a specific value of N, and provision is usually provided for the operator to input manually any correction that is needed for the specified value of N. This correction is called *atmospheric correction.* For example, if an EDM instrument uses a design value of $N = 294$ ppm and the actual N during the measurement is 298 ppm, then the atmospheric correction to be dialed in before the measurement is +4 ppm.

6.6 TYPES OF MOUNTS

An electro-optical distance-measuring instrument can be mounted either on a tripod to be operated as a unit standing alone or on top of a theodolite. The different types of mounts are described in the following paragraphs.

1. *As a standing-alone unit, as shown in Figures 6.2, 6.4, and 6.6.*

In this mode the instrument is usually mounted on a frame that permits the instrument to be rotated 360° about a vertical axis and be tilted upward or downward along a horizontal axis. The frame can usually be fitted directly onto a standard tribrach, which consists of an optical plummet and a bubble level (see Section 5.19). By removing the EDM instrument from the tribrach, it can be replaced with a reflector or a theodolite over the same station. Most long-range (about 20 to 60 km) EDM instruments are used in this mode.

2. *Mounted on the standards of a theodolite, as shown in Figure 6.7.*

In this manner both angles and distances can be measured from the same set-up. However, the theodolite and the EDM instruments are operated as separate units. Although the two instruments are theoretically centered over the same survey station, the line of sight of the theodolite does not coincide with the optical axis of the EDM instrument. There is usually sufficient space under the EDM instrument so that the telescope of the theodolite can be reversed or plunged. This method of mounting requires that the two sides of the standard of the theodolite be of the same height.

3. *Mounted on top of the telescope of the theodolite, as shown in Figure 6.3.*

This method of mounting has the advantage that the optical axis of the EDM instrument is always parallel to that of the theodolite, thus reducing the time required to point both instruments to the same target. But it has the disadvantage that the telescope of the theodolite cannot be reversed when the EDM instrument is mounted in place.

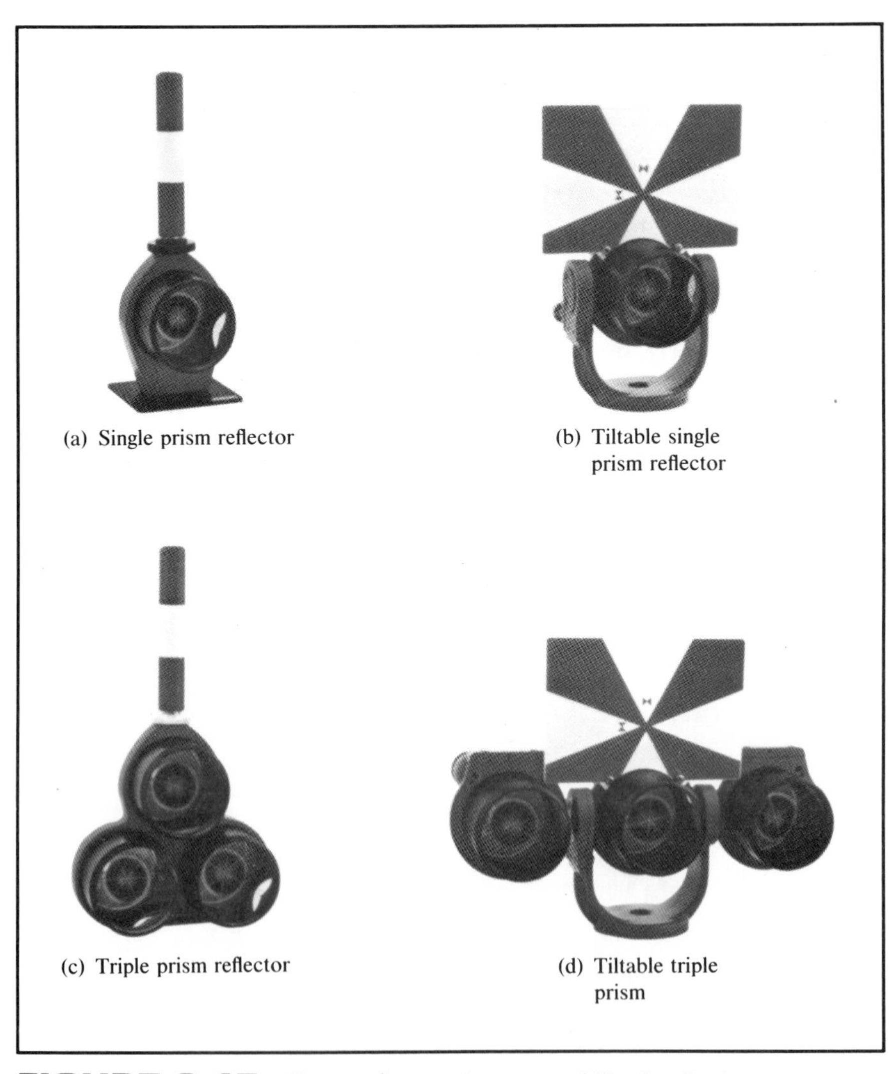

(a) Single prism reflector

(b) Tiltable single prism reflector

(c) Triple prism reflector

(d) Tiltable triple prism

FIGURE 6.15 Retro-reflectors (courtesy of Hewlett-Packard, Inc.)

4. *Mounted around the telescope of the theodolite, as shown in Figure 6.5.* These are usually light-weight EDM instruments designed so that they can be fitted directly onto the telescopes of some specific theodolites. This method of mounting permits the telescope and the EDM instrument to be plunged about the horizontal axis as a single unit. As in the case of modes 2 and 3 discussed previously, the optical axis of the EDM instrument does not coincide with that of the theodolite.

5. *Designed as an integral part of an instrument called total station (see Figures 6.8, 6.9, and 6.10).*

6.7 RETRO-REFLECTORS

The reflectors used for measuring distances with an electro-optical instrument are called *retro-reflectors,* because they are designed to reflect the light ray back along the same direction from which it came. Precisely ground trihedral prisms are used for this purpose. Figure 6.15a shows a single prism reflector mounted in a fixed housing on top of which is fitted a sight pole for angle measurements. Figure 6.15b shows a tilting single prism

FIGURE 6.16 A 15-Prism retro-reflector for long range measurement (courtesy of Keuffel & Esser Company)

retro-reflector assembly. In this arrangement the reflector can be tilted about a horizontal axis for measurement of distances along a slope. Figures 6.15c and d show two different arrangements of a triple prism retro-reflector assembly.

The maximum distance that an electro-optical instrument can measure depends on the design of the instrument, the quality and number of prisms used in the reflector assembly, and the atmospheric conditions. Generally, for a given instrument, the longer the distance to be measured, the larger is the number of prisms required. Figure 6.16 shows a 15-prism assembly for use in long-range measurement.

Figure 6.17 shows two reflector arrangements; each is mounted on a tribrach, can be tilted or rotated, and has a target below the prisms for angular measurements.

Figure 6.18 shows a single prism mounted on a metal rod that also has a circular bubble level for plumbing and a target for angular measurement. This arrangement is used for construction stake-out. In one variation of this arrangement, the prism is mounted near the bottom of the prism rod to reduce distance error due to incorrect plumbing.

FIGURE 6.17 Reflectors with sighting target mounted below the prisms (courtesy of Wild Heerbrugg Instruments, Inc.)

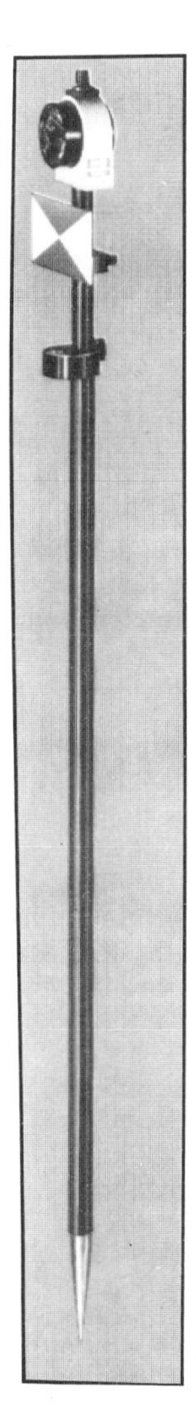

FIGURE 6.18 Prism rod for staking out (courtesy of Keuffel & Esser Company)

6.8 OPERATING PROCEDURE

Modern electro-optical EDM instruments are fully automated to such an extent that the operating procedure can be easily learned in a very short period of time. Usually less than one hour of training time is required. In general, the measurement of a distance requires the following steps:

1. Set up the instrument over the survey station, and record the height of the instrument above the survey station.

2. Set up the reflector at the other end of the line and record its height above the survey station.
3. Point the instrument towards the reflector by means of an alignment telescope.
4. Turn power on.
5. Adjust the pointing using tangent screws to maximize the strength of the returning signal.
6. Balance the strength of the returning signal and an internal reference signal for the purpose of phase-delay measurement.
7. Read and record the field temperature and atmospheric pressure.
8. Dial in the atmospheric correction.
9. Press measure button.
10. Record the distance displayed in numerical digits, either in feet or meters at the option of the operator.

Many instruments can be operated in a tracking mode, which is particularly useful in stake-out work. In this mode the reflector is mounted on a staff. As the reflector is moved forward or backward, the EDM instrument provides an updated measurement of the distance on a regular time interval, which could range from 0.4 second to a few seconds depending on the instrument.

Some instruments have an automatic ranging mode. In this mode the instrument automatically repeats measurement to a stationary reflector and displays the updated arithmetic mean as well as standard deviation of the repeated measurements.

Figure 6.3 shows the operation panel for the Geodimeter 112, whereas the panel for the K & E Rangemaster® III is illustrated in Figure 6.4.

6.9 SOURCES OF MEASUREMENT ERRORS

The sources of measurement errors in using electro-optical devices are the following:

1. Eccentric error due to inexact centering of the instrument and/or reflector over the survey station. Since an optical plummet is usually used to perform this operation, the magnitude of the error caused by this source should be relatively small as long as careful field procedure is followed.

2. Inexactness of the instrument in performing the phase measurements.

3. The zero point of the light ray used in phase measurement does not coincide exactly with the theoretical center of the instrument.

4. The actual center of the reflector does not coincide with the theoretical center.

5. The actual modulating frequencies differ from the theoretical values of these frequencies.

6. Inability to determine the index of refraction of the atmosphere at every point along the measured line during the moment of measurement. In common surveying applications, it is sufficient to measure the air temperature and atmospheric pressure near the measuring instrument. By measuring the temperature to within 1°C and atmospheric pressure to within 2 mmHg, the refractive index can be determined to within one part per million, which is sufficiently accurate for most applications. In high precision work, it may be desirable to determine the refractive index at both ends of the line and use the average value. In special applications, such as in measuring fault movement in earthquake zones, an aircraft is sometimes used to fly along the measured line to determine the average refractive index.

The errors caused by sources 1 and 2 are random in nature, whereas those from sources 3 and 4 are constant errors. The magnitudes of the errors from these first four sources are independent of the length of the distance measured. For example, if an electro-optical instrument is specified to have a measurement accuracy, expressed as root-mean-square error, of ±(5 mm + 3 ppm), then the combined rms error from all of these four sources is expected to be ±5 mm regardless of the length of the distance measured.

The errors resulting from sources 5 and 6 are directly proportional to the length of the distance measured and are equivalent to a scale error. For example, if the measurement accuracy (rms error) of an instrument is said to be ±(5 mm + 3 ppm), then the combined effects of the errors from these two sources is expected to amount to ±3 parts per million (that is, 3/1,000,000) of the distance measured.

By making repeated measurements from the same set-up and taking the arithmetic mean, the effects of the errors resulted from phase measurement and atmospheric conditions can be reduced. By making repeated measurements using different set-ups of the instrument and the reflectors, the effect of errors caused by eccentricity can also be reduced. However, the errors caused by sources 3, 4, and 5 above cannot be reduced by repeated measurements.

6.10 CALIBRATION PROCEDURES

Unlike modern theodolites and automatic levels, which usually remain in excellent calibration for many years under normal use conditions, EDM instruments must be calibrated frequently even under normal careful use. Natural aging of the electronic components can result in changes in the magnitude of the measurement errors, the most serious of which is a scale error caused by changes in the modulating frequencies.

EDM instruments can be accurately calibrated with or without accurately measured base lines. Recognizing the needs for frequent calibration of EDM instruments, the National Geodetic Survey started in 1974 to establish calibration base lines throughout the country. These calibration base lines are usually 1,400 m long with monuments located at 0 m, 150 m, 430 m, and 1,400 m. Latest information concerning the availability of these base lines in any part of the country, their exact locations, their accuracy, and recommended calibration procedures can be obtained from the National Geodetic Survey (see Reference 6.1).

Calibration procedures for determining the systematic errors due to zero centering, pattern frequency, and phase measurements will be described in the following paragraphs.

DETERMINATION OF CORRECTIONS FOR ZERO CENTERING

The combined effect of errors in zero centering at the EDM instrument and at the reflector is a constant error that appears in all the distances measured by that combination of instrument and reflector. The correction needed to compensate for this constant error can be easily determined. Some EDM instruments have a provision for this constant correction to be dialed into the instrument before the measurements so that all displayed distances are automatically corrected for zero centering errors.

The correction needed for any combination of instrument and reflector can be determined by laying out a distance AB of several hundred meters in length (see Figure 6.19). Along the line AB, a third point C is marked so that AB is much longer than AC. The distances AC, CB, and AB are then measured using the EDM instrument and reflector that are being calibrated. Let these measured distances, after slope correction, be denoted as $\overline{AC}$, $\overline{CB}$, and $\overline{AB}$. Furthermore, let c be the unknown correction for zero centering. Then,

$$AC = \overline{AC} + c$$
$$CB = \overline{CB} + c$$
$$AB = \overline{AB} + c$$

However, $AB = AC + CB$. Therefore,

$$\overline{AB} + c = \overline{AC} + \overline{CB} + 2c$$

FIGURE 6.19 Layout for determination of correction for zero centering errors

and

$$c = \overline{AB} - \overline{AC} - \overline{CB} \tag{6.7}$$

Methods for computing slope corrections for EDM measurements will be discussed in Sections 6.11 and 6.16.

EXAMPLE 6.1

To determine the correction for zero centering for a HP 3805A Distance Meter and a triple-prism reflector, the distances *AB*, *AC*, and *CB* were measured. After slope correction, the corresponding horizontal distances were computed as follows:

$$\overline{AB} = 425.704 \text{ m}$$
$$\overline{AC} = 123.277 \text{ m}$$
$$\overline{CB} = 302.425 \text{ m}$$

Points *A*, *B*, and *C* were all located along a straight line, as shown in Figure 6.19.

Determine the correction for zero centering for this combination of instrument and reflector.

SOLUTION

$$\begin{aligned}\text{Correction for zero centering} &= \overline{AB} - \overline{AC} - \overline{CB} \\ &= (425.704 - 123.277 - 302.425) \text{ m} \\ &= +0.002 \text{ m}\end{aligned}$$

EXAMPLE 6.2

The same HP3805A Distance Meter and triple-prism reflector were later used to measure a distance, which was recorded as 1,369.248 m. Compute the corrected distance.

SOLUTION

$$\begin{aligned}\text{Corrected distance} &= 1{,}369.248 + 0.002 \text{ m} \\ &= 1{,}369.250 \text{ m}\end{aligned}$$

DETERMINATION OF CORRECTION FOR ERROR IN MODULATING FREQUENCY

The modulating frequency can be determined in the laboratory with a frequency counter to almost any desired degree of accuracy. If the theoretical frequency is f and the actual frequency is measured to be f', then the measured distances must be multiplied by a scale factor f'/f.

If a frequency counter is not available, the scale factor f'/f can also be determined by using the instrument to measure two or more known distances along a calibrated base line. Suppose that a calibrated base line consists of three stations (A, B, and C), which are located along a straight line, as shown in Figure 6.19. The distances AB and AC have been determined either by using an EDM instrument or by precise taping using invar tapes. The two known distances are measured with the instrument to be calibrated. Let the measured distances, after slope correction, be denoted as $\overline{AB}$ and $\overline{AC}$. Let g be the scale factor f'/f, and let c be the zero centering constant as described previously. Then,

$$AB = g \cdot \overline{AB} + c \tag{6.8}$$

and

$$AC = g \cdot \overline{AC} + c \tag{6.9}$$

These two equations include two unknowns: g and c, which can then be easily computed. If the base line has more than two known distances, a least-squares solution can be used for a more reliable determination of g and c. A detailed discussion of the computation procedure for a least-squares solution of g and c can be found in Reference 6.1.

EXAMPLE 6.3

The horizontal distances AB and AC along a calibration base line are known to be as follows:

$$AB = 1{,}499.8635 \text{ m}$$
$$AC = 149.9921 \text{ m}$$

The same distances were measured with an EDM instrument. After slope correction the measured horizontal distances were as follows:

$$\overline{AB} = 1{,}499.9000 \text{ m}$$
$$\overline{AC} = 149.9935 \text{ m}$$

Determine the scale factor g and constant correction c.

SOLUTION

$$1{,}499.8635 = 1{,}499.9000\ g + c$$

and

$$149.9921 = 149.9935\ g + c$$

Solving for g and c yields:

$$g = 0.999974$$
$$c = +0.0025$$

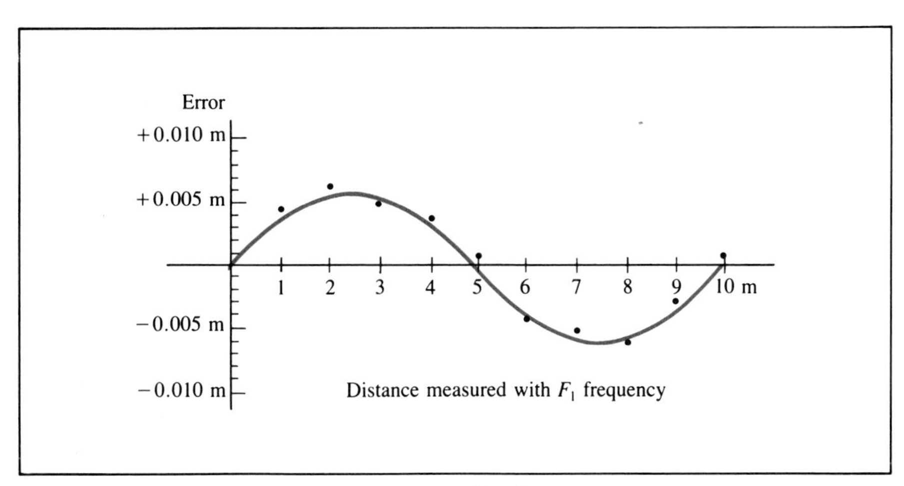

FIGURE 6.20 Systematic error in phase measurement

DETERMINATION OF SYSTEMATIC ERRORS IN PHASE MEASUREMENT

The systematic error in phase measurement is cyclic in nature with a period equal to one-half of the wave length of the F_1 frequency. For an F_1 frequency of 14.989625 MHz that has a wavelength of 20 m, the period of the phase measurement error is 10 m. Figure 6.20 shows a typical error curve for this type of systematic error. Once such an error curve has been determined for an EDM instrument, it can be used to correct any measured distances for systematic errors in phase measurement. This type of measurement error is usually very small and can be ignored in most common applications of EDM instruments. However, by determining the error curve and making corrections on the measured distances, the accuracy potential of an EDM instrument can be fully exploited.

EXAMPLE 6.4

Suppose that the curve in Figure 6.20 represents the systematic errors in phase measurement for an EDM instrument. Correct the following distances, which were measured by the same instrument:

a. 493.325 m

b. 1,725.641 m

c. 847.132 m

SOLUTION

Only the part of the distance that was measured using the F_1 frequency would be used for finding the measurement errors from the curve.

Measured Distance	Distance Measured with F_1 Frequency	Measurement Error from Error Curve	Corrected Distance
a. 493.325 m	3.325 m	+0.005 m	493.320 m
b. 1,725.641 m	5.641 m	−0.003 m	1,725.644 m
c. 847.132 m	7.132 m	−0.006 m	847.138 m

The error curve for an EDM instrument can be determined using a base line as shown in Figure 6.21. The distance *AB* should be about 100 m long, but need not be precisely known. The distance between point *B* and point 10 should be approximately equal to one-half of the wavelength (λ) of the F_1 frequency. The points 1 to 10 should be spaced approximately at a distance of $(1/20)\lambda$. The distance from point *B* to each of the points 1 to 10 should be measured with a tape with a maximum error (3σ) not to exceed ±0.001 m. Let $B1$, $B2$, $B3$, . . . , and $B10$ represent these calibrated distances.

The EDM instrument to be calibrated is then set up at point *A*. The distances *AB*, *A*1, *A*2, . . . , to *A*10 are then successively measured using the same reflector. The distances from point *B* to points 1 to 10 can then be calculated from these measurements. Let these distances be represented as $\overline{B1}$, $\overline{B2}$, . . . , and $\overline{B10}$. Then, the phase measurement error at each of the ten points can be computed as follows:

$$
\begin{aligned}
\varepsilon_1 &= \overline{B1} - B1 \\
\varepsilon_2 &= \overline{B2} - B2 \\
\varepsilon_3 &= \overline{B3} - B3 \\
&\vdots \\
\varepsilon_{10} &= \overline{B10} - B10
\end{aligned}
$$

where ε_i is the phase measurement error at point *i*. The errors ε_1, ε_2, . . . , and ε_{10} are then plotted against the distances $B1$, $B2$, . . . , and $B10$ to give the error curve.

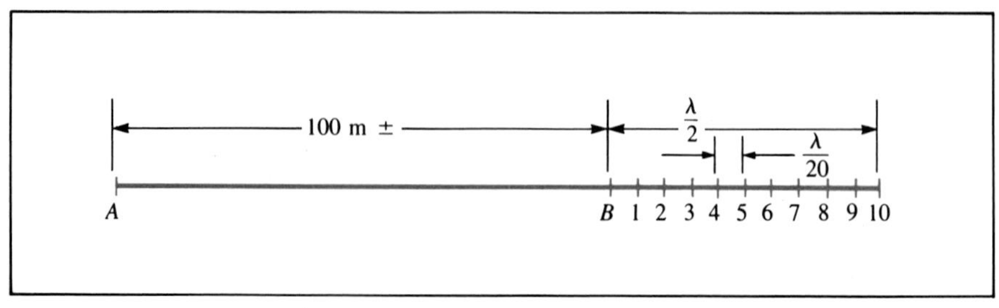

FIGURE 6.21 Base line for calibration of systematic error in phase measurement

6.11 REDUCTION TO HORIZONTAL DISTANCE USING HEIGHT DIFFERENCE

Electromagnetic waves actually travel through the atmosphere along a curved path. The curvature of the path depends on the wavelength as well as the atmospheric conditions. For wavelengths commonly used in microwave distance-measuring instruments, the radius of curvature of the actual path is approximately four times the average radius of curvature of the earth, which is about 6,372 km. For the light waves used in electro-optical instruments, the radius of curvature of the ray path is about 5.5 times that of the earth. By ignoring this curvature, the error introduced in the measured distance is less than 1/4,000,000 of the distance measured for a distance shorter than 60 km. Therefore, for all practical purposes, both radio waves and light waves can be assumed to travel in a straight line in electronic distance measurement.

Figure 6.22 shows a typical situation in which a slope distance is measured between stations A and B. Let HI represent the height of the instrument above station A, and HT represent the height of the reflector target at B. The elevation of stations A and B above a vertical reference datum are represented by h_A and h_B respectively. It can be seen from Figure 6.22 that the height difference (Δh) between the center of the measuring instrument and that of the target can be obtained by the following expression:

$$\Delta h = (h_B + HT) - (h_A + HI) \tag{6.10}$$

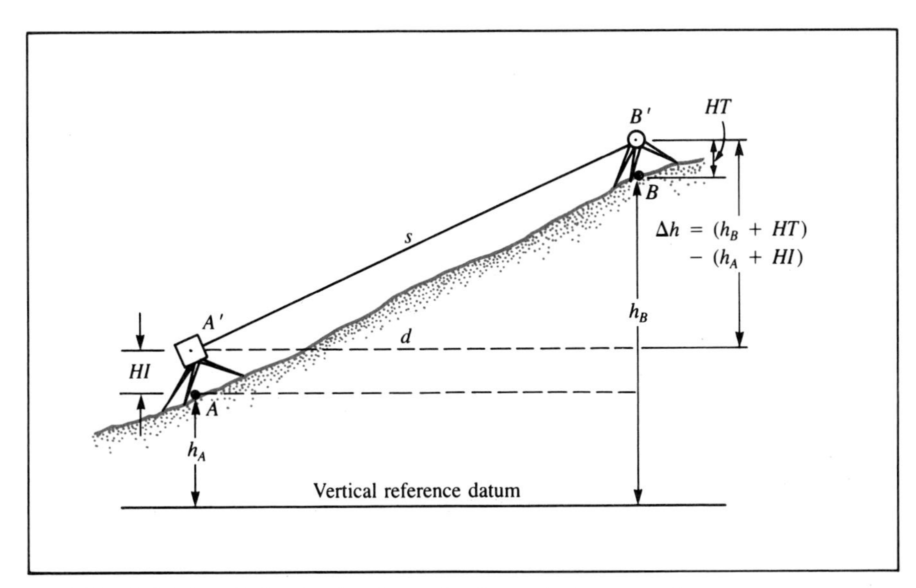

FIGURE 6.22 Reduction to horizontal distance using height difference

The horizontal distance d can then be computed from the measured slope distance s by Eq. (3.1); that is,

$$d = \sqrt{s^2 - \Delta h^2} \tag{6.11}$$

As in the case of reducing slope distances measured by tapes, the accuracy of the computed horizontal distance d is a function of the accuracy with which the parameters s and Δh are measured. The relationship is expressed by Eq. (3.3), which states

$$\hat{\sigma}_d^2 = \left(\frac{s^2}{s^2 - \Delta h^2}\right) \hat{\sigma}_s^2 + \left(\frac{\Delta h^2}{s^2 - \Delta h^2}\right) \hat{\sigma}_{\Delta h}^2$$

It can also be derived from Eq. (6.10) by error propagation that

$$\hat{\sigma}_{\Delta_h}^2 = \hat{\sigma}_{h_B}^2 + \hat{\sigma}_{HT}^2 + \hat{\sigma}_{h_A}^2 + \hat{\sigma}_{HI}^2 \tag{6.12}$$

EXAMPLE 6.5

The slope distance shown in Figure 6.22 has been measured using a Geodimeter 112, which has a measurement accuracy (root-mean-square error) of ±(5 mm + 5 ppm). The measured slope distance is 3,781.298 m. Other known parameters and their rms errors are given below:

$$\begin{aligned} h_A &= 432.564 \pm 0.10 \text{ m} \\ h_B &= 487.273 \pm 0.05 \text{ m} \\ HI &= 1.624 \pm 0.005 \text{ m} \\ HT &= 1.431 \pm 0.005 \text{ m} \end{aligned}$$

Compute the horizontal distance d and its rms error.

SOLUTION

From Eq. (6.10),

$$\begin{aligned} \Delta h &= (h_B + HT) - (h_A + HI) \\ &= (487.273 + 1.431) - (432.564 + 1.624) \\ &= 54.516 \text{ m} \\ d &= \sqrt{s^2 - \Delta h^2} = \sqrt{(3{,}781.298)^2 - (54.516)^2} = \underline{\underline{3{,}780.905 \text{ m}}} \end{aligned}$$

From Eq. (6.12),

$$\begin{aligned} \hat{\sigma}_{\Delta_h}^2 &= \hat{\sigma}_{h_B}^2 + \hat{\sigma}_{HT}^2 + \hat{\sigma}_{h_A}^2 + \hat{\sigma}_{HI}^2 \\ &= (0.05)^2 + (0.005)^2 + (0.10)^2 + (0.005)^2 \\ &= 0.01255 \\ \hat{\sigma}_s &= \pm\left(0.005 \text{ m} + \frac{5}{1{,}000{,}000} \times 3{,}781.298 \text{ m}\right) \\ &= \pm 0.024 \text{ m} \end{aligned}$$

Using Eq. (3.3),

$$\hat{\sigma}_d^2 = \left(\frac{s^2}{s^2 - \Delta h^2}\right) \hat{\sigma}_s^2 + \left(\frac{\Delta h^2}{s^2 - \Delta h^2}\right) \hat{\sigma}_{\Delta h}^2$$

$$= \frac{(3{,}781.298)^2}{(3{,}781.298)^2 - (54.516)^2} (0.024)^2 + \frac{(54.516)^2}{(3{,}781.298)^2 - (54.516)^2} (0.01255)$$

$$= 0.00057612 + 0.00000261$$

$$= 0.00057873$$

$$\hat{\sigma}_d = \pm 0.024 \text{ m}$$

Therefore, $d = 3{,}780.91 \pm 0.02$ m

The heights of the instrument and the target can usually be measured with an rms error better than ±0.005 m, in which case the effects of these errors on the computed horizontal distance are negligible. The following simplified formula can then be used to estimate the required accuracy for measuring the height difference in order to achieve a certain specified accuracy for the computed horizontal distance.

$$\hat{\sigma}_{\Delta h} = \pm \sqrt{\left(\frac{s^2}{\Delta h^2} - 1\right) \hat{\sigma}_d^2 - \frac{s^2}{\Delta h^2} \hat{\sigma}_s^2} \tag{6.13}$$

EXAMPLE 6.6

For the data given in Example 6.5, determine the accuracy with which the height difference must be measured if the computed horizontal distance is to have an rms error of ±0.05 m.

SOLUTION

From Eq. (6.13),

$$\hat{\sigma}_{\Delta_h} = \pm \sqrt{\left[\frac{(3{,}781.298)^2}{(54.516)^2} - 1\right] (0.05)^2 - \frac{(3{,}781.298)^2}{(54.516)^2} (0.024)^2}$$

$$= \underline{\pm 3 \text{ m}}$$

6.12 EARTH CURVATURE EFFECT

When conducting surveys of high accuracy in hilly or mountainous terrains, the earth curvature must be considered in computing horizontal distances from measured slope distances. Figure 6.23, though grossly distorted in scale, illustrates the effect of earth curvature. Let s be the measured slope distance between stations A and B, with the height of the instrument at station A denoted by HI, and the height of the reflector at station B

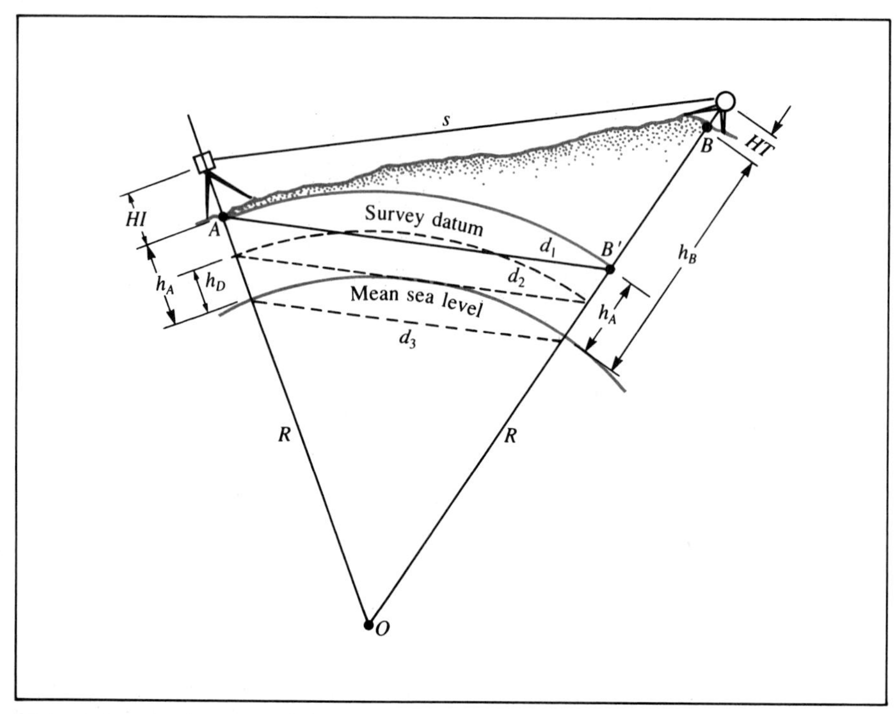

FIGURE 6.23 Effect of earth curvature on slope reduction

denoted by HT. The cord distance d_1 can be computed by the following expression:

$$d_1 = \sqrt{\frac{s^2 - (h_B - h_A + HT - HI)^2}{\left(1 + \frac{HI}{R + h_A}\right)\left(1 + \frac{h_B - h_A + HT}{R + h_A}\right)}} \tag{6.14}$$

where R is the mean radius of the earth and can be approximated to be 6,372,200 m, and h_A and h_B are the elevations of stations A and B, respectively, above mean sea level. Equation (6.14) can be derived by applying the cosine law to the triangles subtended at O. Since HI and HT are usually much smaller than $(R + h_A)$, Eq. (6.14) can be simplified to the following form without any degradation in accuracy:

$$d_1 = \sqrt{\frac{s^2 - (h_B - h_A + HT - HI)^2}{\left(1 + \frac{h_B - h_A}{R + h_A}\right)}} \tag{6.15}$$

By comparing Eq. (6.15) with Eq. (6.11), it can be seen that the earth curvature causes a small change in scale on the horizontal distance. The magnitude of the scale change is directly proportional to the height difference (h_B

$- h_A$) between the two end stations and is independent of the length of the distance s. The scale errors introduced by ignoring the earth curvature for several height differences are given below:

$(h_B - h_A)$ Height Difference	Scale Error by Ignoring Earth Curvature
50 m	1/120,000
100 m	1/60,000
500 m	1/12,000

Normally, the arc distance AB' shown in Figure 6.23 is needed in geodetic calculation. However, the arc distance AB' differs from the chord distance d_1 by only 0.0015 m for a line 10,000 m in length. Therefore, for all practical purposes the arc length AB' can be considered equal to d_1.

EXAMPLE 6.7

For the measured slope distance given in Example 6.5, compute the correct horizontal distance by including the effect of earth curvature.

SOLUTION

From Eq. (6.15),

$$d_1 = \sqrt{\frac{(3{,}781.298)^2 - (487.273 - 432.564 + 1.431 - 1.624)^2}{\left(1 + \dfrac{487.273 - 432.564}{6{,}372{,}200 + 432.564}\right)}}$$

$$d_1 = 3{,}780.889 \text{ m}$$

By ignoring earth curvature, it was computed in Example 6.5 that $d = 3{,}780.905$ m. Thus, error introduced into the computed distance by ignoring curvature is +0.016 m, which is slightly less than the computed rms error ($\hat{\sigma}_s = \pm 0.02$ m) due to the measurements.

6.13 REDUCTION TO DATUM

Referring to Figure 6.23, suppose that the survey datum has been chosen to be at an elevation h_D above the mean sea level. The chord distance d_2 at the datum surface can be computed from the chord distance d_1 at an elevation h_A by proportion:

$$d_2 = \frac{R + h_D}{R + h_A} d_1 \qquad (6.16)$$

where R can be approximated to be 6,372,200 m. Similarly, the chord distance (d_3) at sea level can be computed from chord distance d_1 by the following expression:

$$d_3 = \frac{R}{R + h_A} d_1 \tag{6.17}$$

6.14 TRIGONOMETRIC LEVELING—SHORT LINE

Trigonometric leveling is the process of determining height differences by the measurement of distances and vertical or zenith angles. Figure 6.24 illustrates the geometry of trigonometric leveling when earth's curvature and refraction are ignored. The following relationship exists:

$$h_B - h_A = s \cos z + i - t \tag{6.18}$$

where h_A and h_B are the elevations at points A and B respectively, s is the slope distance along the line of sight, z is the zenith angle, i is the height of the theodolite, and t is the height of the sight target.

Usually, the measured slope distance does not coincide with the line of sight in the angular measurement. The geometry of this general case is illustrated in Figure 6.25. The following relationship exists:

$$s \times \Delta z''(0.4848 \times 10^{-5}) = (HI - i + t - HT)\sin z \tag{6.19}$$

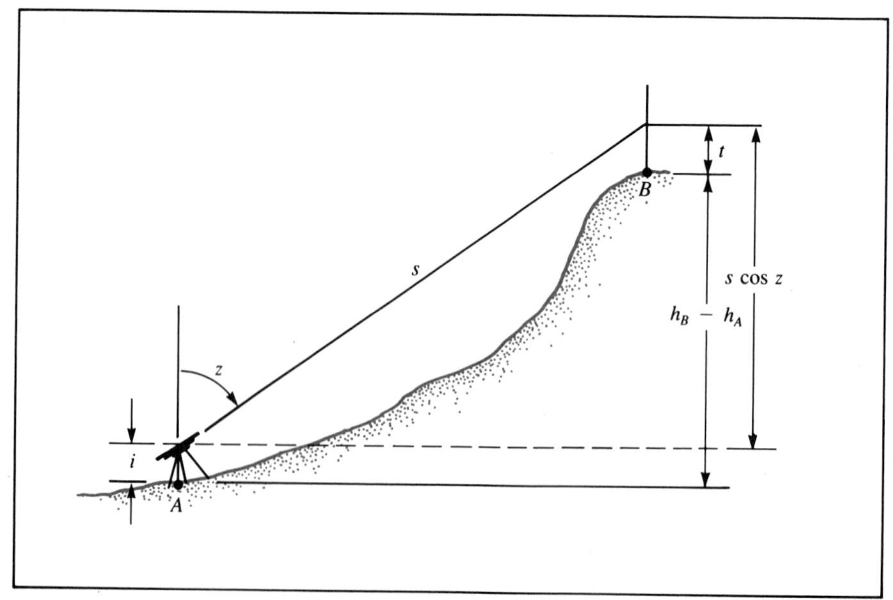

FIGURE 6.24 Trigonometric leveling

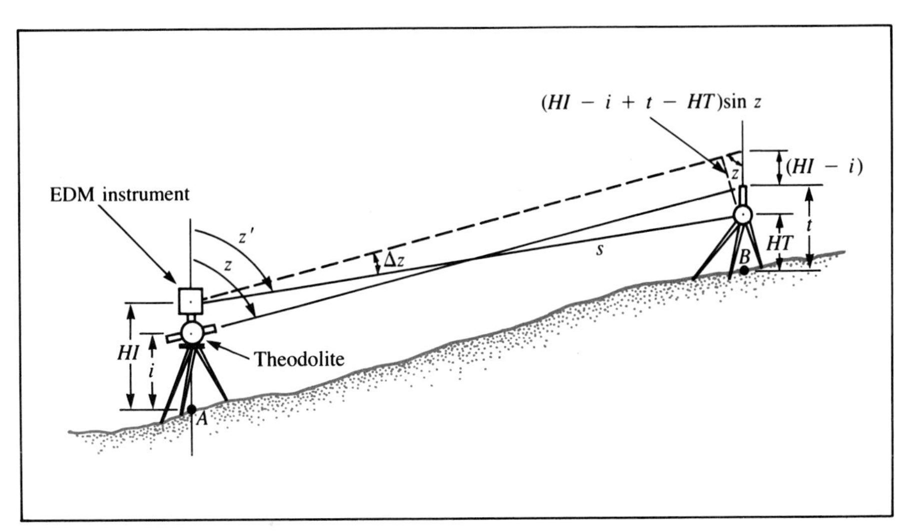

FIGURE 6.25 Trigonometric leveling with EDM instruments

where s is the measured slope distance, $\Delta z''$ is the angular difference in seconds of arc between the measured slope distance and the line of sight of the angular measurement, HI and HT are the respective heights of instrument and reflector used in distance measurement, i and t are the respective heights of instrument and sight target used in angular measurement, and z is the measured zenith angle. Rearranging the terms in Eq. (6.19) yields the following expression:

$$\Delta z'' = \frac{(HI - i + t - HT)\sin z}{0.4848 \times 10^{-5} \cdot s} \tag{6.20}$$

The corrected zenith angle (z') is then computed as follows:

$$z' = z + \Delta z'' \tag{6.21}$$

The term $\Delta z''$ should carry the algebraic sign obtained from Eq. (6.20) and can be either positive or negative. The height difference between the end stations A and B can then be computed as follows:

$$h_B - h_A = s \cdot \cos z' + HI - HT \tag{6.22}$$

By applying the law of propagation of random errors, the following expression can be derived to compute an estimated standard error ($\hat{\sigma}_{\Delta h}$) for the height difference:

$$\hat{\sigma}^2_{\Delta h} = \cos^2 z \cdot \hat{\sigma}^2_s + (23.503 \times 10^{-12})(s \cdot \sin z)^2 \hat{\sigma}^2_z + \hat{\sigma}^2_{HI} + \hat{\sigma}^2_{HT} \tag{6.23}$$

where $\hat{\sigma}_s$ is the estimated standard error of the measured slope distance, $\hat{\sigma}_z$ is the estimated standard error of the measured zenith angle in seconds of arc, and $\hat{\sigma}_{HI}$ and $\hat{\sigma}_{HT}$ are the estimated standard errors for measuring the *HI* and *HT* respectively. The measured angle z is used in Eq. (6.23) instead of the adjusted angle z' without introducing any appreciable error in the calculation.

EXAMPLE 6.8

Referring to Figure 6.25, the slope distance between stations *A* and *B* was measured with a Zeiss Eldi 2 EDM instrument, which has a measurement accuracy (estimated standard error) of ±(10 mm + 2 ppm). The measured slope distance, after correction for atmospheric refraction, was 478.256 m. The other relevant data are given below:

$$HI = 1.365 \pm 0.005 \text{ m}$$
$$HT = 1.482 \pm 0.005 \text{ m}$$
$$h_A = 369.27 \pm 0.00 \text{ m (assumed error free)}$$

The zenith angle measurement was recorded as follows:

$$z = 86°20' \pm 1'$$
$$i = 1.330 \pm 0.005 \text{ m}$$
$$t = 1.436 \pm 0.005 \text{ m}$$

Compute the height difference between stations *A* and *B* and its estimated standard error ($\hat{\sigma}_{\Delta h}$).

SOLUTION

From Eq. (6.20),

$$\begin{aligned}\Delta z'' &= \frac{(HI - i + t - HT)\sin z}{0.4848 \times 10^{-5} \cdot s} \\ &= \frac{(1.365 - 1.330 + 1.436 - 1.482)\sin 86°20'}{0.4848 \times 10^{-5} \times 478.256} \\ &= -5''\end{aligned}$$

Then,

$$\begin{aligned}z' &= z + \Delta z'' \\ &= 86°20'00'' - 5'' \\ &= 86°19'55''\end{aligned}$$

From Eq. (6.22),

$$\begin{aligned}\Delta h = h_B - h_A &= s \cdot \cos z' + HI - HT \\ &= 478.256 \cos 86°19'55'' + 1.365 - 1.482 \\ &= 30.480 \text{ m}\end{aligned}$$

$$\hat{\sigma}_s = \pm\left(10 \text{ mm} + \frac{2}{1{,}000{,}000} \cdot s\right)$$
$$= \pm\left(0.010 + \frac{2}{1{,}000{,}000} \times 478.256\right)\text{m}$$
$$= \pm 0.011 \text{ m}$$

From Eq. (6.23),

$$\hat{\sigma}^2_{\Delta h} = \cos^2 86°20'(0.011)^2$$
$$+ (23.503 \times 10^{-12})(478.256 \cdot \sin 86°20')^2(60)^2$$
$$+ (0.005)^2 + (0.005)^2$$
$$= 0.00000049 + 0.01927378 + 0.000025 + 0.000025$$

$$\therefore \qquad \hat{\sigma}_{\Delta h} = \pm 0.139 \text{ m}$$

Therefore, $\Delta h = 30.5 \pm 0.1$ m

Note that $\hat{\sigma}_z$ in Eq. (6.23) is expressed in seconds. In Example 6.8 above, the large estimated standard error ($\hat{\sigma}_{\Delta h}$) for the computed height difference is primarily due to a large estimated standard error ($\pm 1'$) on the measured zenith angle. The magnitude of $\hat{\sigma}_{\Delta h}$ can be decreased by making more accurate measurement of the zenith angle. As a general rule, the zenith angle should be measured with an estimated standard error smaller than $\pm 1'$ if the full accuracy potential of the EDM instrument is to be used in trigonometric leveling. One of the most common mistakes in EDM applications is overlooking the importance of making accurate angular measurements for computing height differences or horizontal distances.

By neglecting the small errors introduced in measuring the slope distance s and the *HI* and *HT*, the following simplified formula can be used to estimate the required accuracy for measuring zenith angle to achieve a specified accuracy of height difference.

$$\hat{\sigma}_z(\text{in seconds of arc}) = \frac{0.206 \times 10^6 \hat{\sigma}_{\Delta h}}{|s \cdot \sin z|} \tag{6.24}$$

EXAMPLE 6.9

For the data given in Example 6.8, suppose that the height difference must be determined with an estimated standard error of ±0.01 m. What should be the maximum permissible standard error for the measured zenith angle?

SOLUTION

From Eq. (6.24),

$$\hat{\sigma}_z = \pm \frac{0.206 \times 10^6(0.01)}{|478.256 \sin 86°20'|} \text{ sec}$$
$$= \pm 4 \text{ sec}$$

The magnitude of error introduced by ignoring the effects of earth's curvature and refraction can be estimated by the use of Eq. (4.7) or (4.8). For a horizontal distance of 1,000 ft, it amounts to 0.021 ft. For a horizontal distance of 5,000 ft, it is increased to 0.5 ft. Thus, for long lines, the effects of earth's curvature and refraction must be considered.

6.15 TRIGONOMETRIC LEVELING—LONG LINE

Atmospheric refraction is the primary factor that limits the accuracy of trigonometric leveling for long lines. Its effect on the zenith angle measurement can be minimized by making simultaneous reciprocal zenith angle measurements from both ends of the line. Special sighting targets can be mounted on the standards of the two theodolites so that the heights of the targets coincide with the horizontal axes of the theodolites. In this arrangement, the theodolite at one end of the line can sight on the theodolite located on the other end while zenith angle measurements are being made from both ends.

Figure 6.26 illustrates the geometry involved. An EDM instrument is set up over point A, with the height of the instrument being represented by HI. Point C is the center of the EDM instrument over point A. Point O is the center of the earth, and line CO represents the vertical line (or direction of gravity) at point C. Line CF is the horizontal line at C, and is perpendicular to CO. The zenith angle z_A is measured at a height HI above point A. If the actual height of the theodolite is not equal to HI, then the measured zenith angle must be corrected according to Eqs. (6.20) and (6.21). Angle $E\hat{C}D$ is the small angular displacement caused by atmospheric refraction and is equal to $m\gamma$; where m is the coefficient of refraction at point A, and γ is the angle subtended at the center of the earth by the line of sight.

A second theodolite is assumed to be set up over point B, with the height of the theodolite being equal to HT. If the actual height differs from HT, then the measured zenith angle must also be corrected according to Eqs. (6.20) and (6.21). Point D is at a height of HT above point B, and the zenith angle at point D is represented by z_B.

By assuming that the coefficient of refraction at B is the same as at A, angle $E\hat{D}C = m \cdot \gamma$. Since $OC = OB'$, triangle OCB' is an isosceles triangle and line OF is perpendicular to CB' at point G. The radius R should be the radius of the earth at the average latitude along the line joining stations A and B. In most cases it is sufficiently accurate to use the mean radius of the earth; that is, $R = 20{,}906{,}000$ ft or 6,372,200 m.

In triangle CDB', the sum of the three interior angles must add up to 180°; that is,

$$\left(90° - z_A - m\gamma + \frac{\gamma}{2}\right) + (180° - m\gamma - z_B) + \left(90° + \frac{\gamma}{2}\right) = 180° \tag{6.25}$$

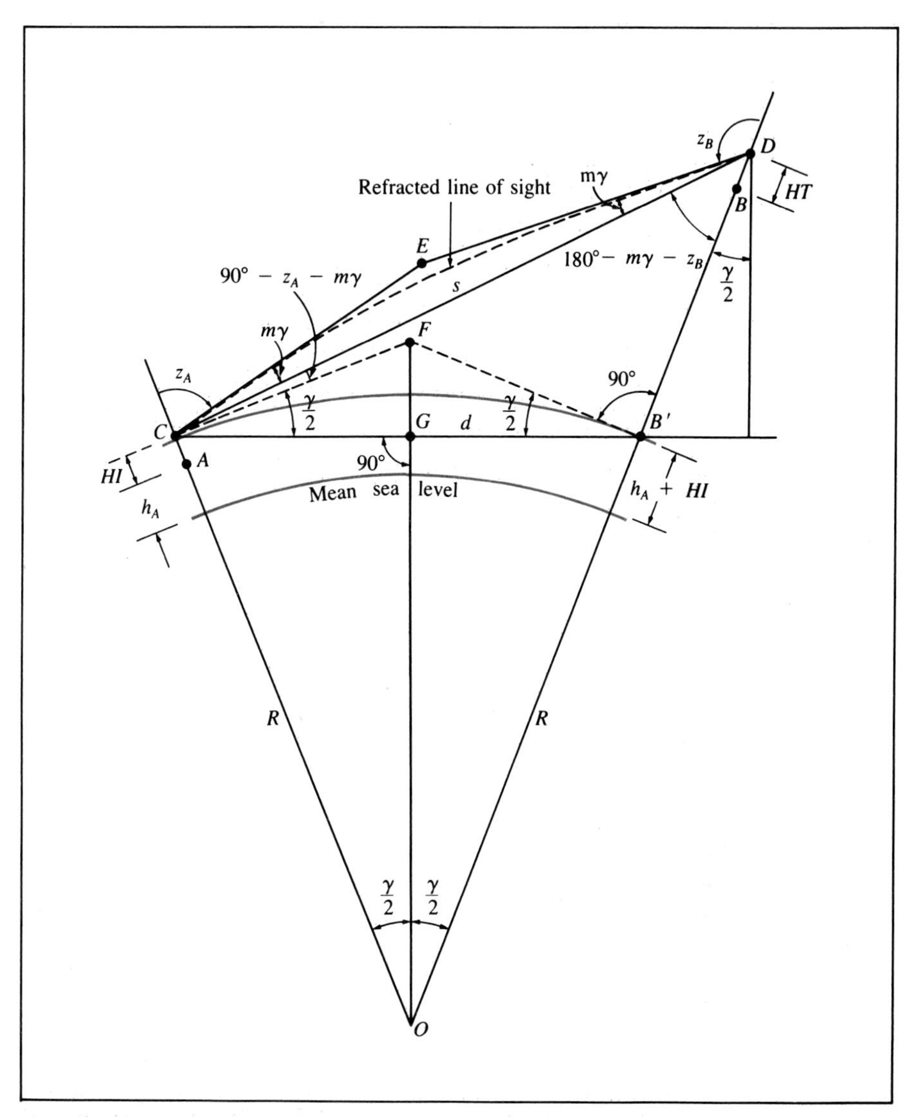

FIGURE 6.26 Trigonometric leveling by simultaneous reciprocal zenith angles

By manipulating the terms in Eq. (6.25), the following expression can be derived:

$$m = \frac{1}{2}\left[1 - \frac{z_A + z_B - 180^\circ}{\gamma}\right] \tag{6.26}$$

Furthermore,

$$D\hat{C}B' = 180^\circ - C\hat{D}B' - D\hat{B}'C \tag{6.27}$$

Therefore, from Figure 6.26,

$$D\hat{C}B' = 180° - (180° - m\gamma - z_B) - \left(90° + \frac{\gamma}{2}\right) \tag{6.28}$$

By substituting Eq. (6.26) into Eq. (6.28), the following expression can be derived:

$$D\hat{C}B' = \frac{1}{2}(z_B - z_A) \tag{6.29}$$

Then, by applying the sine law to triangle CDB', the following expression is derived:

$$DB' = \frac{s \cdot \sin \frac{1}{2}(z_B - z_A)}{\sin\left(90° + \frac{\gamma}{2}\right)} \tag{6.30}$$

Since $DB' = (h_B + HT) - (h_A + HI)$,

$$(h_B + HT) - (h_A + HI) = \frac{s \cdot \sin \frac{1}{2}(z_B - z_A)}{\sin\left(90° + \frac{\gamma}{2}\right)} \tag{6.31}$$

An expression can be derived for the angle γ by using triangle CGO:

$$\sin \frac{\gamma}{2} = \frac{d}{2(R + h_A + HI)} \tag{6.32}$$

The term HI is much smaller than the mean radius of the earth (R) and can be omitted. The distance d can be approximated by the following expression without introducing appreciable error in γ:

$$d \approx s \cdot \sin z_A \tag{6.33}$$

Then, from Eqs. (6.32) and (6.33),

$$\gamma = 2 \sin^{-1}\left[\frac{s \cdot \sin z_A}{2(R + h_A)}\right] \tag{6.34}$$

Equations (6.31) and (6.34) can be used to compute the height difference ($h_B - h_A$) from the measured slope distance s and the zenith angles z_A and z_B. The coefficient of refraction (m) is not involved in any one of these two equations. Thus, by making reciprocal zenith angle measurements from both ends of the line, the need for knowing the value of m is eliminated. However, since the derivation assumed that the refraction is the same from both ends of the line, error due to refraction is minimized by measuring the zenith angles from both ends of the line at the same time.

If zenith angle is measured only from one end of the line, say from point C in Figure 6.26, the following expression can be derived by applying the sine law to triangle CDB':

$$(h_B + HT) - (h_A + HI) = \frac{s \cdot \sin[90° - z_A + (0.5 - m)\gamma]}{\sin(90 \; + 0.5\,\gamma)} \tag{6.35}$$

Then, Eq. (6.35) is used with (6.34) to compute the difference in height $(h_B - h_A)$. However, in this case a value must be assumed for the coefficient of refraction m. The coefficient of refraction at a particular location at a given time can be determined by making simultaneous reciprocal zenith angle measurements between two stations. The value of m can then be computed using Eqs. (6.26) and (6.34).

Remember that the coefficient of refraction depends on atmospheric conditions. It varies from location to location and with time. Thus the inability to precisely determine the coefficient of refraction along a specific line at a specific time is the ultimate limiting factor on the accuracy of trigonometric leveling.

EXAMPLE 6.10

The elevation difference between stations A and B was to be determined by simultaneous reciprocal zenith angle measurements. The following field data were obtained:

$$z_A = 88°01'33'' \qquad i_A = 5.34 \text{ ft}$$
$$z_B = 92°7'11'' \qquad i_B = 5.26 \text{ ft}$$

The slope distance between stations A and B was also measured, with the EDM instrument located at A and the target at B. The following field data were recorded:

$$s = 63{,}285.42 \text{ ft}$$
$$HI \text{ at } A = 5.11 \text{ ft}$$
$$HT \text{ at } B = 4.78 \text{ ft}$$

The elevation (h_A) of station A is known to be 1,423.46 ft. Determine the elevation of station B. Use R = 20,906,000 ft.

SOLUTION

Step 1. Compute correction for z_A. Using Eq. (6.20),

$$\Delta z''_A = \frac{(5.11 - 5.34 + 5.26 - 4.78)\sin 88°01'33''}{0.4848 \times 10^{-5} \times 63{,}285.42}$$
$$= +0.8''$$
$$z'_A = 88°01'33'' + 0.8'' = 88°01'33.8''$$

Step 2. Compute correction for z_B. Using Eq. (6.20),

$$\Delta z''_B = \frac{(4.78 - 5.26 + 5.34 - 5.11)\sin 92°7'11''}{0.4848 \times 10^{-5} \times 63{,}285.42}$$

$$= -0.8''$$

$$z'_B = 92°7'11'' - 0.8'' = 92°7'10.2''$$

Step 3. Compute angle γ. Using Eq. (6.34),

$$\gamma = 2 \sin^{-1} \frac{63{,}285.42 \sin 88°01'33.8''}{2(20{,}906{,}000 + 1{,}423.46)}$$

$$= 0.17332777°$$

Step 4. Compute elevation difference. Using Eq. (6.31),

$$(h_B + HT) - (h_A + HI) = \frac{63{,}285.42 \sin \frac{1}{2}(92°7'10.2'' - 88°01'33.8'')}{\sin[90° + \frac{1}{2}(0.17332777)]}$$

$$= 2{,}260.21$$

Step 5. Compute elevation of station B.

$$h_B = 2{,}260.21 + h_A - HT + HI$$

$$= 2{,}260.21 + 1{,}423.46 - 4.78 + 5.11$$

$$= 3{,}684.00 \text{ ft}$$

EXAMPLE 6.11

The elevation difference between stations G and P is to be determined by trigonometric leveling. A theodolite is set up at station G and a zenith angle of 91°23′46″ is measured to station P. The height of the theodolite (i_G) at station G is recorded to be 1.83 m, and the height of the sighting target (t) at P is 1.47 m.

The slope distance between the two stations is also measured with an EDM instrument. It is recorded as 8,429.216 m. The height of the EDM instrument (HI) at station G is recorded as 1.61 m, and the height of the reflector at station P is recorded as 1.53 m.

The elevation (h_G) of station G is known to be 487.24 m above a mean sea level datum. Determine the elevation of station B. Use $R = 6{,}372{,}200$ m and a value of 0.071 as the coefficient of refraction.

SOLUTION

Step 1. Compute correction for z_G. Using Eq. (6.20),

$$z'' = \frac{1.61 - 1.83 + 1.47 - 1.53}{0.4848 \times 10^{-5} \times 8{,}429.216}$$

$$= -7''$$

$$z'_G = z_G + \Delta z'' = 90°23'46'' - 7''$$
$$= 91°23'39''$$

Step 2. Compute angle γ using Eq. (6.34). From Eq. (6.34),

$$\gamma = 2 \sin^{-1}\left[\frac{8{,}429.216 \sin 91°23'39''}{2(6{,}372{,}200 + 487.24)}\right]$$

$$\gamma = 0.07576325°$$

Step 3. Compute elevation difference. From Eq. (6.35),

$$(h_p + HT) - (h_G + HI) = \frac{8{,}429.216 \sin[90° - 91°23'39'' + (0.5 - 0.071)(0.07576325°)]}{\sin(90° + 0.5 \times 0.07576325°)}$$
$$= -200.31 \text{ m}$$

Step 4. Compute elevation of station P.

$$h_p = -200.31 + h_G - HT + HI$$
$$= -200.31 + 487.24 - 1.53 + 1.61$$
$$= 287.01 \text{ m}$$

6.16 REDUCTION OF SLOPE DISTANCES USING ZENITH ANGLE

Either the zenith angle (z) or the vertical angle $(90° - z)$ can be measured to compute the horizontal distance from a measured slope distance. The difference in height between the two end stations can first be computed from the angular measurement by using the formulas developed for trigonometric leveling (Sections 6.14 and 6.15). The computed height difference can then be used to compute the horizontal distance using Eq. (6.11) or (6.15).

By neglecting the errors on the heights of instruments and targets, the following simplified formula can be used to estimate the required accuracy for measuring the zenith angle in order to achieve a certain specified accuracy for the computed horizontal distance:

$$\hat{\sigma}_z \text{ (in seconds)} = \pm 0.206 \times 10^6 \frac{\sqrt{\hat{\sigma}_d^2 - \sin^2 z \cdot \hat{\sigma}_s^2}}{|s \cos z|} \quad (6.36)$$

EXAMPLE 6.12

For the data given in Example 6.8, suppose that the horizontal distance must be determined with an rms error of ±0.011 m. What should be the maximum permissible rms error for the zenith angle?

SOLUTION

From Eq. (6.36),

$$\hat{\sigma}_z \text{ (in seconds)} = \pm 0.206 \times 10^6 \frac{\sqrt{(0.011)^2 - (\sin 86°20')^2\,(0.011)^2}}{|478.256 \cos 86°20'|}$$

$$= \pm 5''$$

6.17 SUMMARY

Modern EDM instruments can measure distances with great ease and accuracy. However, natural aging of the electronic components under normal use often results in changes in the modulating frequencies causing a scale error on the distance measurements. Therefore, all EDM instruments must be frequently calibrated. Moreover, EDM instruments directly measure the slope distance between two stations. To obtain the same measurement accuracy on the computed horizontal distance, the zenith (or vertical) angle of the line must be measured with an rms error of no more than a few seconds. If height differences are used for slope reduction, the height differences must also be determined with appropriate accuracy. In measuring long lines, the distortion effect of earth curvature and atmospheric refraction on the zenith angle must be taken into consideration. Earth curvature must also be considered in slope reduction when height differences exceed 100 m and the required survey accuracy exceeds 1/50,000.

Continuing developments of EDM instruments and total stations can be expected to result in further miniaturization as well as lower costs.

PROBLEMS

6.1 Compute the wavelength in meters for the following frequencies:

a. 30 KHz

b. 499.5103 MHz

c. 150 MHz

6.2 Compute the frequency for the following wavelengths:

a. 0.910 μm

b. 0.6328 μm

c. 0.865 μm

d. 20 ft

e. 60 cm

6.3 An EDM instrument uses the following modulating frequencies:

F_1 = 14.989625 MHz, F_2 = 1.4989625 MHz, F_3 = 149.89625 KHz, and F_4 = 14.989625 KHz. Compute the distances in meters for the following sets of phase differences:

a. $\Delta_1 = 285.95°$ $\Delta_2 = 244.1°$ $\Delta_3 = 168.1°$ $\Delta_4 = 0°$

b. $\Delta_1 = 348.41°$ $\Delta_2 = 142.9°$ $\Delta_3 = 86.2°$ $\Delta_4 = 44.6°$

c. $\Delta_1 = 151.56°$ $\Delta_2 = 33.9°$ $\Delta_3 = 0°$ $\Delta_4 = 0°$

6.4 An EDM instrument used an F_1 frequency of 14.989625 MHz and a reflective index of N = 294 ppm in computing the distances. Compute the atmospheric refraction correction needed for the following set of field conditions:

a. t = 15°C P = 700 mm Hg

b. t = 10°C P = 720 mm Hg

c. t = −5°C P = 680 mm Hg

d. t = 25°C P = 750 mm Hg

6.5 The following distances were measured with an EDM instrument and a single prism reflector: $\overline{AB}$ = 467.326 m, $\overline{AC}$ = 80.428 m, and $\overline{CB}$ = 386.891 m. Points A, B, and C are located along a straight line. Determine the constant correction for zero centering.

6.6 The following distances were measured with an EDM instrument and a triple-prism reflector: $\overline{AB}$ = 2,489.730 ft, $\overline{AC}$ = 150.427 ft, $\overline{CB}$ = 2,339.312 ft. Points A, B, and C are located along a straight line. Determine the constant correction (in ft) for zero centering.

6.7 The stations A, B, and C are located along a straight line. The calibrated lengths of lines AB and BC were AB = 1,480.306 m and AC = 150.046 m. These two known distances were measured using an EDM instrument and a triple prism reflector and were found to be as follows: $\overline{AB}$ = 1,480.253 m and $\overline{AC}$ = 150.038 m.

a. Determine the scale factor (g) and correction constant (c) for zero centering for the EDM instrument-reflector combination.

b. The same EDM instrument-reflector combination was later used to measure a line as 2,467.134 m. Compute the correct length for this line.

6.8 An EDM base line consists of three stations A, B, and C, which are all located along the line. The lines AB and BC were known to have the following lengths: AB = 2,160.377 ft and AC = 89.218 ft. An EDM instrument and a reflector were used to measure these lines and were found to be as follows: $\overline{AB}$ = 2,160.341 ft and $\overline{AC}$ = 89.219 ft.

a. Compute the scale factor and correction constant for the instrument-reflector combination.

b. Compute the corrected length of a line that was measured using this instrument-reflector combination to be 1,812.369 ft.

6.9 An EDM instrument and reflector combination has been found through calibration to have a scale factor (g) of 1.000027 and a correction constant (c) of +0.002 m. The systematic error due to phase measurement for this instrument-reflector combination was found to be represented by the error curve shown in Figure 6.20. Compute the corrected length for the following distances measured by this instrument-reflector combination:

a. 486.216 m

b. 321.459 m

c. 873.143 ft

d. 1,858.791 ft

6.10 An EDM instrument was set up over station A with $HI = 1.36$ m and a reflector was set up over station B with $HT = 1.24$ m. The slope distance was measured to be 1,365.25 m. The elevations of stations A and B were known to be as follows: $h_A = 151.32$ m and $h_B = 187.25$ m. Compute the horizontal distance between the two stations. Ignore the earth curvature.

6.11 An EDM instrument was set up over station D with $HI = 4.67$ ft, and a reflector wa set up over station E with $HT = 3.61$ ft. The slope distance was measured to be 3,784.215 ft. The elevations of the two stations were known to be as follows: $h_D = 467.21$ ft and $h_E = 433.46$ ft. Compute the horizontal distance between the two stations. Ignore the earth curvature.

6.12 An EDM instrument is known to have an accuracy (rms error) of $\pm$(5 mm + 5 ppm). It was used to measure the slope distance between stations A and B. For each of the following sets of data, compute the horizontal distance and its rms error. Neglect earth curvature in all cases.

a. $s = 1{,}347.215$ m

$HI = 1.56 \pm 0.01$ m

$HT = 1.24 \pm 0.01$ m

$h_A = 132.45 \pm 0.02$ m

$h_B = 163.21 \pm 0.03$ m

b. $s = 2{,}543.214$ ft

$HI = 4.75 \pm 0.01$ ft

$HT = 3.21 \pm 0.01$ ft

$h_A = 761.25 \pm 0.01$ ft

$h_B = 743.78 \pm 0.02$ ft

c. $s = 451.467$ m

$HI = 1.56 \pm 0.01$ m

$HT = 1.48 \pm 0.01$ m

$$h_A = 341.26 \pm 0.02 \text{ m}$$
$$h_B = 315.73 \pm 0.01 \text{ m}$$

6.13 An EDM instrument is to be used to measure distances in surveying projects. The EDM instrument has an accuracy (rms error) of $\pm(10 \text{ mm} + 3 \text{ ppm})$. All the horizontal distances must be determined with an accuracy (rms error) of $\pm 1/25{,}000$. Given below are the approximate slope distance (s) and height difference for four of the lines:

a. $s \approx 341$ m $\quad \Delta h \approx 30$ m

b. $s \approx 1{,}670$ m $\quad \Delta h \approx 45$ m

c. $s \approx 570$ ft $\quad \Delta h \approx 25$ ft

d. $s \approx 1{,}780$ ft $\quad \Delta h \approx 50$ ft

Determine for each case the accuracy (rms error) with which the height difference must be determined. Assume that errors due to measuring the heights of instruments and targets are negligible.

6.14 The following data were obtained from a surveying project in which an EDM instrument was used to measure distances and a theodolite was used to measure the zenith angles:

Line	Measured Slope Distance(s)	*HI*	*HT*	Measured Zenith Angle(s)	Height of	
					theodolite	target
AB	1,352.561 m	1.35 m	1.21 m	86°31′23″	1.24 m	1.30 m
BC	896.287 m	1.46 m	1.35 m	94°26′41″	1.36 m	1.20 m
CD	731.486 ft	5.21 ft	4.65 ft	93°30′26″	5.00 ft	4.13 ft
DE	2,146.293 ft	5.46 ft	5.00 ft	87°15′46″	5.17 ft	4.80 ft

Compute the horizontal distances for the lines. Neglect earth curvature and refraction.

6.15 The distances in Problem 6.14 were measured with an EDM instrument that had an accuracy of $\pm(0.01' + 0.01'$ per $1{,}000'$ of measured distance). The zenith angles were measured with an rms error of $\pm 10''$. The measurement errors on the heights of instruments and targets were negligible. Determine the rms errors for the horizontal distances computed in Problem 6.14.

6.16 The EDM instrument to be used in a surveying project has an accuracy of $\pm(5 \text{ mm} + 5 \text{ ppm})$. All the horizontal distances in the project must be determined with an rms error of $\pm 1/50{,}000$ or smaller. Given below are the approximate slope distance(s) and zenith angles (z) for four of the lines:

a. $s \approx 560$ m $\quad z \approx 91°$

b. $s \approx 1{,}400$ m $\quad z \approx 86°$

c. $s \approx 840$ ft $\quad z \approx 85°$

d. $s \approx 1{,}400$ ft $\quad z \approx 93°$

Determine the accuracy (rms error) with which the zenith angles must be measured. Ignore earth curvature and refraction.

6.17 The following data were obtained from an EDM survey:

Station		Measured Slope	Elevation of			
from	to	Distance(s)	from station	to station	HI	HT
A	B	8,461.325 m	1,431.65 m	1,563.48 m	1.46 m	1.32 m
B	C	3,625.187 m	1,141.24 m	1,035.27 m	1.53 m	1.46 m
C	D	15,489.321 ft	5,421.26 ft	5,341.65 ft	5.25 ft	4.65 ft
D	E	24,138.594 ft	781.23 ft	925.71 ft	5.15 ft	5.32 ft

By considering the earth curvature effect,

a. Compute the horizontal distances at the elevation of the instrument stations (that is, the from stations).

b. Compute the horizontal distances at mean sea level.

6.18 For each of the following sets of data from trigonometric leveling, calculate the difference in elevation between the end stations and its estimated standard error. Ignore earth curvature and refraction.

a. *From station* A *to station* B:
Zenith angle measured at station A

$$z_A = 91°23'44'' \pm 10''$$
$$i = 5.44' \pm 0.01'$$
$$t = 4.61' \pm 0.01'$$

Slope distance measured with EDM at station A

$$s = 461.24' \pm 0.02'$$

HI at station A = $5.41' \pm 0.01'$

HT at station B = $4.89' \pm 0.01'$

b. *From station* C *to station* D:
Zenith angle measured at station C

$z_C = 85°44'35'' \pm 20''$

$i = 1.76$ m

$t = 1.43$ m

Slope distance measured with EDM at station C

$s = 241.66 \text{ m} \pm 0.01 \text{ m}$

HI at station $C = 1.885 \text{ m} \pm 0.005 \text{ m}$

HT at station $D = 1.431 \text{ m} \pm 0.005 \text{ m}$

6.19 The elevation differences must be determined with an estimated standard error of ± 0.05 ft or better in Problem 6.18a and ± 0.02 m or better in Problem 6.18b. Determine the maximum permissible standard error for the measured zenith angle in each case.

6.20 Compute the elevation of station B for each of the following sets of data from trigonometric leveling by simultaneous reciprocal zenith angle measurements:

a. $z_A = 88°57'25''$ $\quad i_A = 1.55$ m

$z_B = 91°02'51''$ $\quad i_B = 1.63$ m

$s = 35{,}711.26$ m

HI at station $A = 1.47$ m

HT at station $B = 1.32$ m

Elevation of station $A = 863.21$ m above mean sea level
Use $R = 6{,}372{,}200$ m

b. $z_A = 77°03'22''$ $\quad i_A = 5.67'$

$z_B = 91°00'28''$ $\quad i_B = 5.83'$

$s = 12{,}431.87'$

HI at station $A = 4.72'$

HT at station $B = 3.14'$

Elevation of station $A = 367.24'$ above mean sea level
Use $R = 20{,}906{,}000$ ft

6.21 Compute the horizontal distance between the end stations in Problem 6.20a and b at mean sea level.

6.22 The following sets of data are obtained from trigonometric leveling by measuring zenith angles at one end of the lines only. Compute the elevation of station B in each case.

a. $z_A = 93°25'42''$ $\quad i_A = 5.24'$

$s = 41{,}279.63'$

HI at station $A = 4.97'$

HT at station $B = 3.21'$

Elevation of station A = 142.61′ above mean sea level
Use m = 0.071 and R = 20,906,000′

b. $z_A = 87°41'23''$ $i_A = 5.21'$

s = 21,439.61 m

HI at station A = 1.47 m

HT at station B = 1.58 m

Elevation at station A = 1,673.29 m above mean sea level
Use m = 0.071 and R = 6,372,200 m

6.23 Compute the horizontal distance of the lines in Problem 6.22a and b at mean sea level.

REFERENCES

6.1 Fronczek, Charles J. "Use of Calibration Base Lines," *NOAA Technical Memorandum NOS NGS-10* (1977) NOAA, National Geodetic Survey, Rockville, MD, Revised 1980.
6.2 Greene, John R. "Accuracy Evaluation in Electro-Optic Distance-Measuring Instruments," *Surveying and Mapping,* Vol. XXXVII, No. 3 (September 1977) 247–256, American Congress on Surveying and Mapping, Washington, D.C.
6.3 Laurila, Simo. *Electronic Surveying and Navigation,* Wiley-Interscience, New York, 1976.
6.4 Moffitt, Francis H. "Calibration of EDM's for Precise Measurement," *Surveying and Mapping,* Vol. XXXV, No. 2 (June 1975) 147–154, American Congress on Surveying and Mapping, Washington, D.C.

COORDINATE GEOMETRY

7

7.1 INTRODUCTION

With the increasing use of computers in all facets of engineering planning and design, the use of coordinates to define geographic positions of survey points can no longer be considered as a convenience, but as a necessity. Computerized land-titles systems now use rectangular coordinates to define the locations of property corners. Highways, dams, mass transit systems, and urban renewal projects are being planned and designed using computerized data files that include such information as topography, land use, drainage features, population distributions, and soil types. The U.S. Bureau of the Census is putting all the census information into a geographic reference framework so that the information can be readily used in all types of planning activities.

There are now computer programs available for performing many of the basic surveying calculations by the use of coordinates. One of the most commonly used program packages is called COGO (**CO**ordinate **G**e**O**metry), which was developed by Professor C. L. Miller of the Massachusetts Institute of Technology in the early 1960s. Some of the more sophisticated total stations also have built-in computers programmed for performing coordinate calculations. In fact, advanced surveying instruments in the near future will directly measure the coordinates of new survey stations to high accuracy (see Chapter 11).

Some of the fundamental principles of coordinate geometry will be discussed in this chapter. Additional methods of coordinate computation will be presented in later chapters wherever it is appropriate. The computation of elevation and elevation differences from vertical angles has already been discussed in Sections 6.14 and 6.15. This chapter will be limited to discussions on the computation of horizontal coordinates.

7.2 LATITUDE, LONGITUDE, AND ELEVATION

One of the most commonly used coordinate systems for defining geographic position on a global basis is the geographic coordinate system. In this system, the position of a point is defined by its latitude, longitude, and elevation. This system of coordinate is illustrated in Figure 7.1, in which the earth is represented by an ellipsoid of revolution about the north-south axis.

The latitude of a point, P, is defined as the angle between the direction of the plumb line at P and the earth's equatorial plane. The latitude is commonly represented by the Greek letter, ϕ, and is measured either to the north or south of the equator. Thus, in Figure 7.1, the latitude of point P is 43°N.

The longitude of a point, P, is defined as the angular distance, measured in the equatorial plane, between the Greenwich meridian and the meridian passing through P. The longitude is measured either east or west of the Greenwich meridian and has a value between 0° and 180°. Longitude is commonly represented by the Greek letter, λ. In Figure 7.1, the longitude of point P is 61°W.

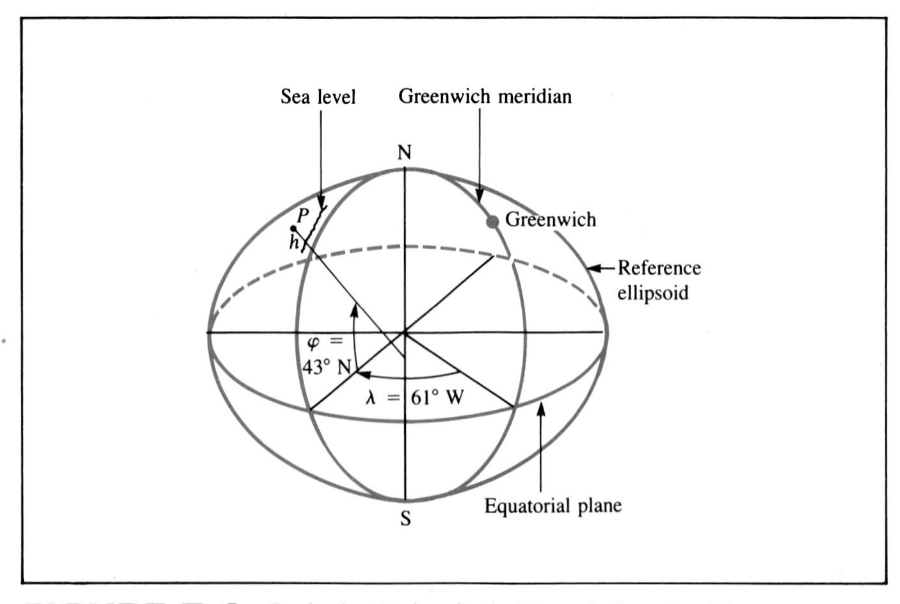

FIGURE 7.1 Latitude (ϕ), longitude (λ), and elevation (h)

The elevation of a point, P, is defined as its vertical distance above a reference datum, such as mean sea level.

The geographic coordinate system is universally used in navigation. It is also commonly used for defining the positions of survey stations on a continental or global basis. Calculations of geographic coordinates necessitate the use of spherical geometry as well as a reference figure (such as an ellipsoid) to approximate the shape of the earth. Its use, therefore, is generally impractical for surveying projects that are of local or regional scope.

7.3 RECTANGULAR COORDINATES

In a rectangular coordinate system, the horizontal position of a point is defined by its X- and Y- coordinates with respect to a set of X- and Y- axes; and its vertical position is defined by its elevation with respect to a reference datum, such as mean sea level. In Figure 7.2, the position of point P is defined by its coordinates X_P, Y_P, and h_P.

A rectangular coordinate system can be arbitrarily defined for a project site by specifying the following:

1. the X- and Y- coordinates of one point;
2. the direction (bearing or azimuth) of one line, or the X-coordinate of a second point, or the Y-coordinate of the second point; and
3. the elevation of one point.

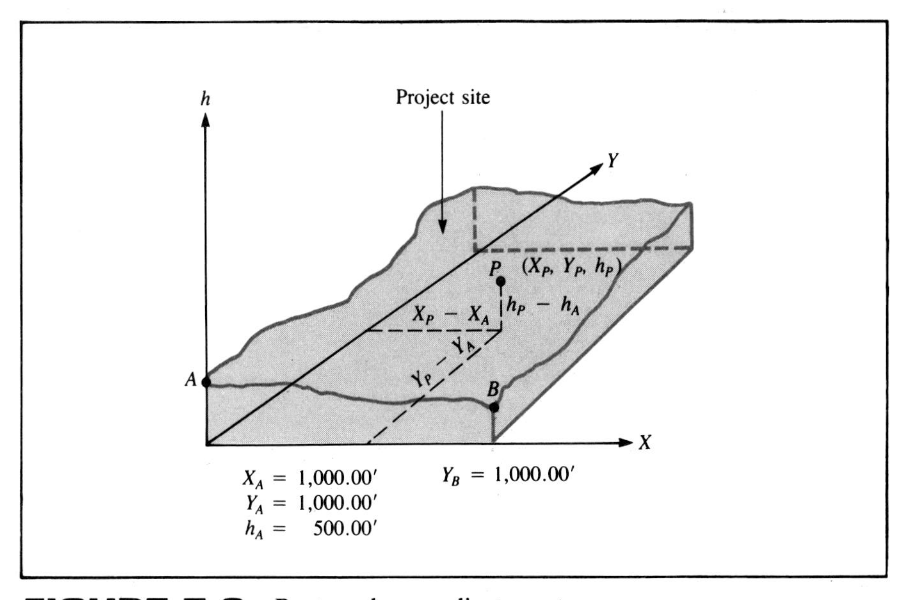

FIGURE 7.2 Rectangular coordinate system

Such an arbitrary local rectangular coordinate system is usually defined so that all the survey points lie in the North-East quadrant. The need for coordinates with negative sign is thus eliminated. For example, the rectangular coordinate system shown in Figure 7.2 is defined by specifying the following parameters:

1. Coordinates of point A:

$$X_A = 1{,}000.00 \text{ ft} \qquad Y_A = 1{,}000.00 \text{ ft}$$

2. Y-Coordinate of point B:

$$Y_B = 1{,}000.00 \text{ ft}$$

3. Elevation of point A:

$$h_A = 500.00 \text{ ft}$$

Although an arbitrarily defined coordinate system usually satisfies the immediate needs of a local engineering project, it is advisable to always use an existing coordinate system that is commonly accepted within the region. One such coordinate system is the State Plane Coordinate System, which will be discussed in detail in Chapter 15.

The principles of coordinate geometry to be discussed in the remaining parts of this chapter apply equally well to all rectangular coordinate systems.

7.4 THE INVERSE PROBLEM

If the X- and Y- coordinates of two points are known, the horizontal distance and azimuth of the line joining the two points can be computed. Let X_i and Y_i be the horizontal coordinates of point i, and X_j and Y_j be the coordinates of point j (see Figure 7.3). Then, the horizontal distance, d_{ij}, between the two points can be computed from the following formula:

$$d_{ij} = \sqrt{(X_j - X_i)^2 + (Y_j - Y_i)^2} \tag{7.1}$$

The azimuth, α_{ij}, of the line going from i to j can be computed from the following expression:

$$\alpha_{ij} = \tan^{-1} \frac{(X_j - X_i)}{(Y_j - Y_i)} \tag{7.2}$$

Traditionally, the difference in X-coordinates $(X_j - X_i)$ is called the *departure* of the line; the difference in Y-coordinates $(Y_j - Y_i)$ is called the *latitude*

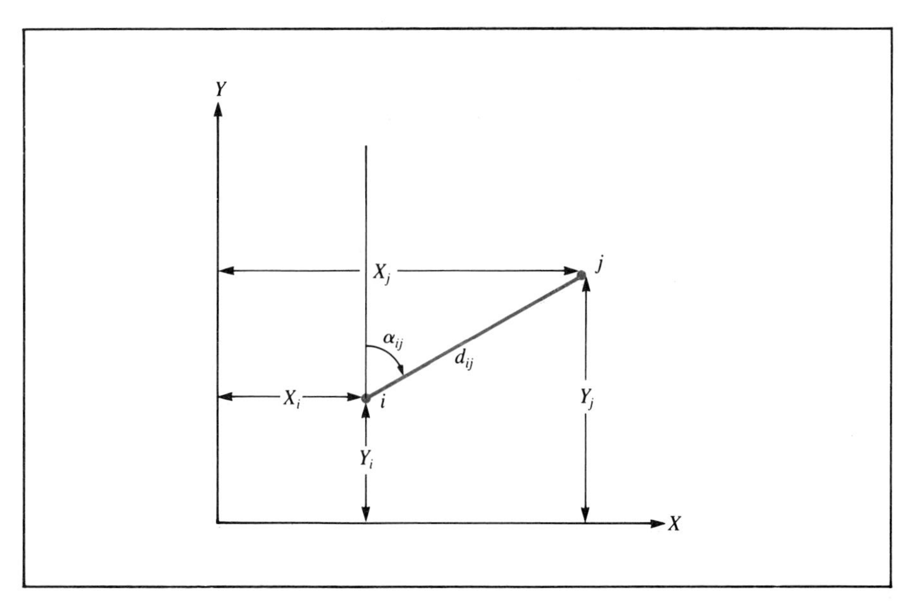

FIGURE 7.3 The inverse problem

of the line. Since the Y-axis may or may not coincide with true north, depending on how it is defined, the azimuth α_{ij} is often referred to as the *grid azimuth*.

When using Eq. (7.2) to compute α_{ij}, care should be exercised to locate the proper quadrant for the angle α_{ij} (see Example 7.1 below). A simple sketch showing the relative positions of the two points can help to identify the proper quadrant.

Survey points of known coordinates are used during construction to determine the proper field locations of the various parts of an engineering structure. Hence such survey points are often referred to as *control points*.

EXAMPLE 7.1

Given the following horizontal coordinates for points i and j:

$$X_i = 1{,}437.21 \text{ ft} \qquad Y_i = 2{,}681.46 \text{ ft}$$
$$X_j = 1{,}169.72 \text{ ft} \qquad Y_j = 2{,}004.53 \text{ ft}$$

Compute the distance (d_{ij}) and azimuth (α_{ij}).

SOLUTION

$$X_j - X_i = 1{,}169.72 - 1{,}437.21 = -267.49 \text{ ft}$$
$$Y_j - Y_i = 2{,}004.53 - 2{,}681.46 = -676.93 \text{ ft}$$

From Eq. (7.1),

$$d_{ij} = \sqrt{(-267.49)^2 + (-676.93)^2} = \underline{727.86 \text{ ft}}$$

$$\alpha_{ij} = \tan^{-1} \frac{-267.49}{-676.93} = 21°33'42'' + 180°$$

$$= \underline{201°33'42''}$$

7.5 LOCATION BY ANGLE AND DISTANCE

Referring to Figure 7.4, let i and j be two points of known coordinates. The horizontal coordinates of a new point, k, can be determined by measuring the horizontal angle β and the distance d_{ik}.

The azimuth, α_{ij}, of the line joining points i and j can be first determined using Eq. (7.2). The azimuth, α_{ik}, is then computed from the measured angle β:

$$\alpha_{ik} = \alpha_{ij} + \beta \tag{7.3}$$

The coordinates (X_k, Y_k) of point k can be computed from the following expressions:

$$X_k = X_i + d_{ik} \sin \alpha_{ik} \tag{7.4}$$

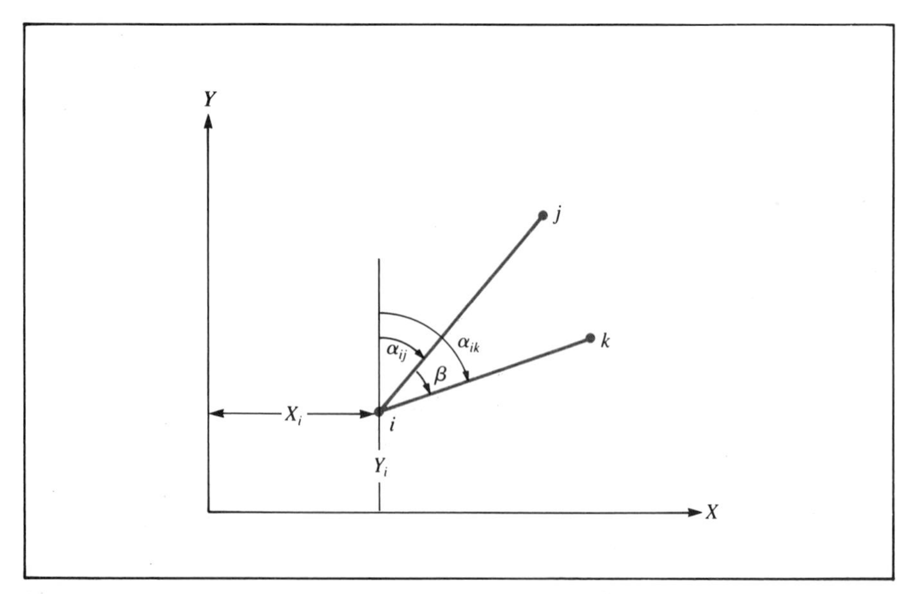

FIGURE 7.4 Location by angle and distance

and

$$Y_k = Y_i + d_{ik} \cos \alpha_{ik} \tag{7.5}$$

EXAMPLE 7.2

Given the following coordinates for points i and j:

$$X_i = 3{,}632.11 \text{ m} \qquad Y_i = 1{,}469.27 \text{ m}$$
$$X_j = 4{,}987.22 \text{ m} \qquad Y_j = 2{,}073.91 \text{ m}$$

The horizontal angle $j\hat{i}k$, measured clockwise and denoted by β, is measured to be 141°27′33″. The horizontal distance d_{ik} is measured to be 1,432.55 m. Compute the horizontal coordinates of point k.

SOLUTION

From Eq. (7.2),

$$\begin{aligned} \alpha_{ij} &= \tan^{-1} \frac{4{,}978.22 - 3{,}632.11}{2{,}073.91 - 1{,}469.27} \\ &= 65°57'14'' \end{aligned}$$

Then

$$\begin{aligned} \alpha_{ik} &= \alpha_{ij} + \beta \\ &= 65°57'14'' + 141°27'33'' \\ &= 207°24'47'' \end{aligned}$$

From Eq. (7.4),

$$\begin{aligned} X_k &= 3{,}632.11 + 1{,}432.55 \sin 207°24'47'' \\ &= 3{,}632.11 - 659.55 \\ &= \underline{2{,}972.56 \text{ m}} \end{aligned}$$

From Eq. (7.5),

$$\begin{aligned} Y_k &= 1{,}469.27 + 1{,}432.55 \cos 207°24'47'' \\ &= 1{,}469.27 - 1{,}271.69 \\ &= \underline{197.58 \text{ m}} \end{aligned}$$

7.6 INTERSECTION BY ANGLES

The horizontal coordinates of a new point can be determined by measuring angles from two points of known coordinates. In Figure 7.5, suppose that the coordinates (X_i, Y_i) for point i and (X_j, Y_j) for point j are

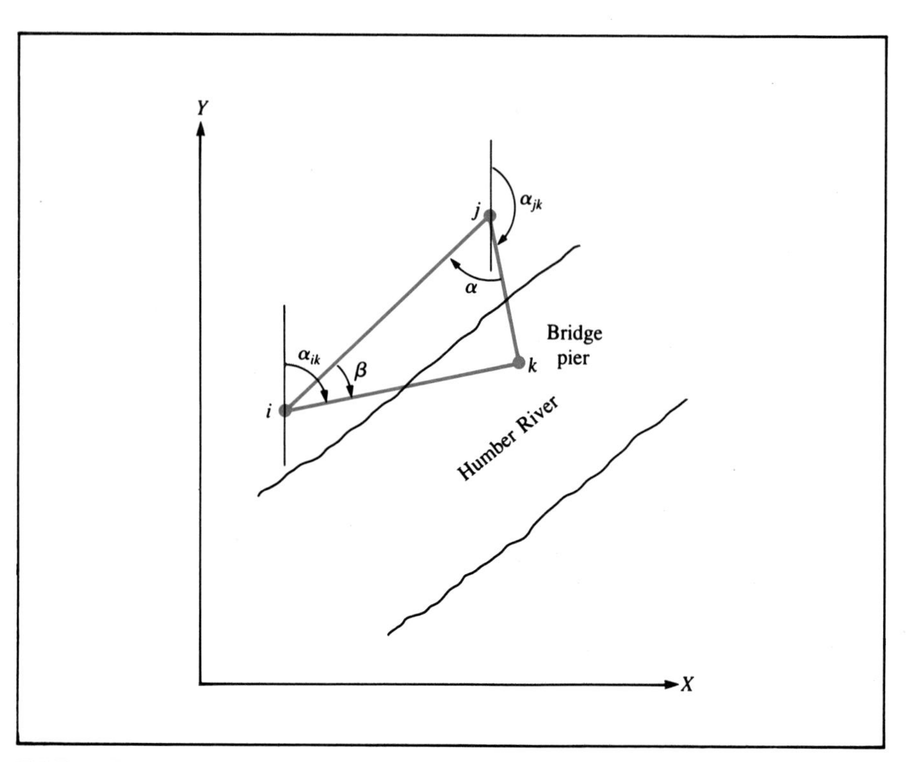

FIGURE 7.5 Intersection by angles

known. The coordinates (X_k, Y_k) for point k can be determined by measuring the angles β and γ.

The azimuth α_{ij} and distance d_{ij} of the line ij can first be computed from the coordinates of points i and j. The azimuths α_{ik} and α_{jk} can then be computed from the azimuth α_{ij} and the measured angles β and γ. Let d_{ik} and d_{jk} represent the lengths of the lines ik and jk, respectively. By sine law:

$$\frac{d_{ik}}{\sin \gamma} = \frac{d_{ij}}{\sin(180° - \beta - \gamma)} \tag{7.6}$$

That is,

$$d_{ik} = \frac{d_{ij} \sin \gamma}{\sin(180° - \beta - \gamma)} \tag{7.7}$$

Then,

$$X_k = X_i + d_{ik} \sin \alpha_{ik} \tag{7.8}$$

$$Y_k = Y_i + d_{ik} \cos \alpha_{ik} \tag{7.9}$$

Similarly, the following expressions can be derived:

$$d_{jk} = \frac{d_{ij} \sin \beta}{\sin(180° - \beta - \gamma)} \tag{7.10}$$

$$X_k = X_j + d_{jk} \sin \alpha_{jk} \tag{7.11}$$

and

$$Y_k = Y_j + d_{jk} \cos \alpha_{jk} \tag{7.12}$$

Either Eqs. (7.7)–(7.9) or Eqs. (7.10)–(7.12) may be used to compute the coordinates (X_k, Y_k) for point k.

EXAMPLE 7.3

Referring to Figure 7.5, let

$$X_i = 5{,}329.41 \text{ ft} \qquad Y_i = 4{,}672.66 \text{ ft}$$
$$X_j = 6{,}321.75 \text{ ft} \qquad Y_j = 5{,}188.24 \text{ ft}$$
$$\beta = 31°26'30''$$
$$\gamma = 42°33'41''$$

Compute the coordinates X_k and Y_k.

SOLUTION

$$X_j - X_i = 6{,}321.75 - 5{,}329.41 = 992.34 \text{ ft}$$
$$Y_j - Y_i = 5{,}188.24 - 4{,}672.66 = 515.58 \text{ ft}$$
$$d_{ij} = \sqrt{(992.34)^2 + (515.58)^2} = 1{,}118.29 \text{ ft}$$
$$\alpha_{ij} = \tan^{-1} \frac{992.34}{515.58} = 62°32'44''$$
$$\alpha_{ik} = \alpha_{ij} + \beta = 62°32'44'' + 31°26'30''$$
$$= 93°59'14''$$
$$180° - \beta - \gamma = 180° - 31°26'30'' - 42°33'41''$$
$$= 105°59'49''$$
$$d_{ik} = \frac{1{,}118.29 \sin 42°33'41''}{\sin 105°59'49''}$$
$$= 786.86 \text{ ft}$$

Then,

$$X_k = 5{,}329.41 + 786.86 \sin 93°59'14''$$
$$= 5{,}329.41 + 784.96$$
$$= \underline{6{,}114.37 \text{ ft}}$$

$$Y_k = 4{,}672{,}66 + 786.86 \cos 93°59'14''$$
$$= 4{,}672.66 - 54.71$$
$$= \underline{4{,}617.95 \text{ ft}}$$

As a check in the calculation, X_k and Y_k can also be computed from point j.

$$\alpha_{ji} = \alpha_{ij} + 180° = 242°32'44''$$
$$\alpha_{jk} = \alpha_{ji} - \gamma = 242°32'44'' - 42°33'41''$$
$$= 199°59'03''$$
$$d_{jk} = \frac{1{,}118.29 \sin 31°26'30''}{\sin 105°59'49''}$$
$$= 606.83 \text{ ft}$$

$$X_k = 6{,}321.75 + 606.83 \sin 199°59'03''$$
$$= 6{,}321.75 - 207.39$$
$$= \underline{6{,}114.36 \text{ ft}} \qquad \text{Check}$$

$$Y_k = 5{,}188.24 + 606.83 \cos 199°59'03''$$
$$= 5{,}188.24 - 570.29$$
$$= \underline{4{,}617.95 \text{ ft}} \qquad \text{Check}$$

7.7 INTERSECTION BY DISTANCES

The coordinates of a new point can also be determined by measuring distances from (or to) two points of known coordinates. In Figure 7.5, the coordinates (X_k, Y_k) of new point k can be determined by measuring the distances d_{ik} and d_{jk} instead of angles β and γ.

The solution method is similar to that of intersection by angles. The angle β can be computed from the measured distances d_{ik} and d_{jk} by use of the cosine law:

$$d_{jk}^2 = d_{ij}^2 + d_{ik}^2 - 2d_{ij}d_{ik} \cos \beta \tag{7.13}$$

Therefore,

$$\beta = \cos^{-1}\left[\frac{d_{ij}^2 + d_{ik}^2 - d_{jk}^2}{2d_{ij}d_{ik}}\right] \tag{7.14}$$

Then, the azimuth α_{ik} can be computed from the azimuth α_{ij} and angle β. The coordinates X_k and Y_k can then be computed from azimuth α_{ik} and distance d_{ik}.

Alternatively, the angle γ can be computed from the following expression:

$$\gamma = \cos^{-1}\left[\frac{d_{ij}^2 + d_{jk}^2 - d_{ik}^2}{2d_{ij}d_{jk}}\right] \tag{7.15}$$

Then, the coordinates (X_k, Y_k) of point k can be computed from the azimuth α_{jk} and distance d_{jk}.

7.8 RESECTION

The horizontal position of a new point can also be determined by measuring angles from the new point to three points of known coordinates. This method is called *resection*. In Figure 7.6, suppose that the horizontal coordinates of points A, B, and C are known and that the coordinates of new point P are to be determined by measuring the angles M and N.

From the known coordinates of points A, B, and C, the distances c and b and the angle R can be computed. Referring to Figure 7.6, let

$$J = \beta + \gamma \tag{7.16}$$

then

$$J = 360° - (M + N + R) \tag{7.17}$$

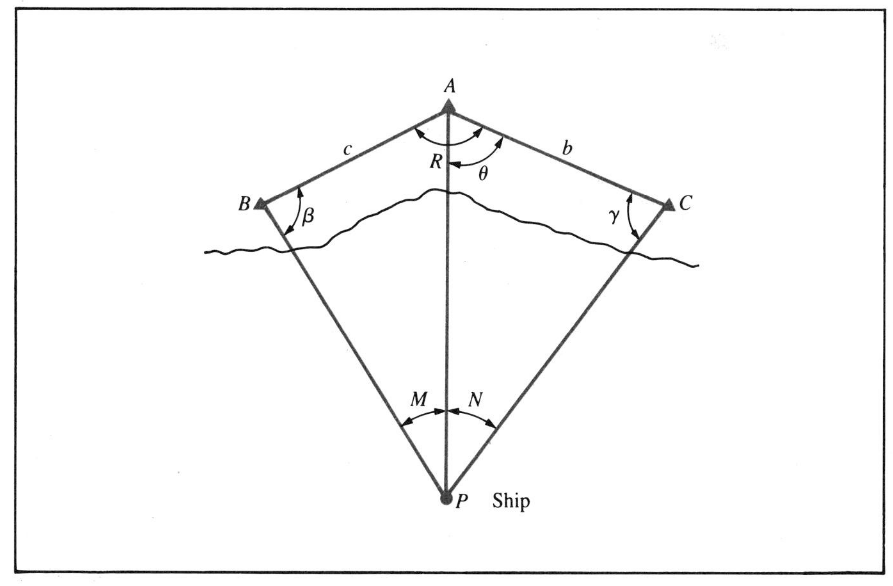

FIGURE 7.6 Resection

By the sine law,

$$\frac{\sin \beta}{AP} = \frac{\sin M}{c}$$

Hence

$$AP = \frac{c \sin \beta}{\sin M}$$

Similarly,

$$AP = \frac{b \sin \gamma}{\sin N}$$

Therefore,

$$\frac{c \sin \beta}{\sin M} = \frac{b \sin \gamma}{\sin N} \tag{7.18}$$

Let

$$H = \frac{\sin \beta}{\sin \gamma} \tag{7.19}$$

Then from Eq. (7.18),

$$H = \frac{b \sin M}{c \sin N} \tag{7.20}$$

From Eq. (7.19),

$$H \sin \gamma = \sin \beta = \sin (J - \gamma)$$

or

$$H \sin \gamma = \sin J \cos \gamma - \cos J \sin \gamma$$

Dividing throughout by $\cos \gamma$ and rearranging terms yield the following expression:

$$\tan \gamma = \frac{\sin J}{H + \cos J} \tag{7.21}$$

Also,

$$\theta = 180° - N - \gamma \tag{7.22}$$

The following procedure can be used to compute the coordinates (X_P, Y_P) of point P:

1. Compute b, c, R, and azimuth α_{AC} from the known coordinates of points A, B, and C.
2. Compute J using Eq. (7.17).
3. Compute H using Eq. (7.20).
4. Compute the angle γ from Eq. (7.21).
5. Compute the angle θ from Eq. (7.22).
6. Compute the azimuth α_{AP} of the line AP.
7. Compute X_P and Y_P:

$$X_P = X_A + AP \sin \alpha_{AP}$$

and

$$Y_P = Y_A + AP \cos \alpha_{AP}$$

Note that, should the new point P lie on a circle that passes through the three known points A, B, and C, the solution will be indeterminate.

EXAMPLE 7.4

Given the following coordinates for points A, B, and C in Figure 7.6:

point A: $X_A = 46{,}732.41'$ $\quad Y_A = 38{,}111.26'$
point B: $X_B = 42{,}139.65'$ $\quad Y_B = 36{,}781.33'$
point C: $X_C = 49{,}822.47'$ $\quad Y_C = 37{,}266.32'$

The measured angles are

$$M = 37°21'33''$$
$$N = 41°03'56''$$

Compute the coordinates (X_P, Y_P) of point P.

SOLUTION

$$c = \sqrt{(42{,}139.65 - 46{,}732.41)^2 + (36{,}781.33 - 38{,}111.26)^2}$$
$$= 4{,}781.44'$$

$$b = \sqrt{(49{,}822.47 - 46{,}732.41)^2 + (37{,}266.32 - 38{,}111.26)^2}$$
$$= 3{,}203.50$$

$$\alpha_{AB} = \tan^{-1} \frac{42{,}139.65 - 46{,}732.41}{36{,}781.33 - 38{,}111.26}$$
$$= 73°51'02'' + 180° = 253°51'02''$$

$$\alpha_{AC} = \tan^{-1} \frac{49{,}822.47 - 46{,}732.41}{37{,}266.32 - 38{,}111.26}$$
$$= 180° - 74°42'25'' = 105°17'35''$$

$$R = \alpha_{AB} - \alpha_{AC} = 148°33'27''$$

$$J = 360° - (M + N + R)$$
$$= 133°01'04''$$

$$H = \frac{b \sin M}{c \sin N} = \frac{3{,}203.50 \sin 37°21'33''}{4{,}781.44 \sin 41°03'56''}$$
$$= 0.61887727$$

$$\tan \gamma = \frac{\sin J}{H + \cos J} = \frac{\sin 133°01'04''}{0.61887727 + \cos 133°01'04''}$$
$$= -11.5416786$$

$$\gamma = 94°57'07''$$

$$\theta = 180° - N - \gamma = 43°58'57''$$

$$AP = \frac{b \sin \gamma}{\sin N} = \frac{3{,}203.50 \sin 94°57'07''}{\sin 41°03'56''}$$
$$= 4{,}858.33'$$

$$\alpha_{AP} = \alpha_{AC} + \theta$$
$$= 105°17'35'' + 43°58'57''$$
$$= 149°16'32''$$

$$X_P = X_A + AP \sin \alpha_{AP}$$
$$= 46{,}732.41 + 4{,}858.33 \sin 149°16'32''$$
$$= 46{,}732.41 + 2{,}482.17$$
$$= \underline{49{,}214.58'}$$

$$Y_P = Y_A + AP \cos \alpha_{AP}$$
$$= 38{,}111.26 + 4{,}858.33 \cos 149°16'32''$$
$$= 38{,}111.26 - 4{,}176.39$$
$$= \underline{33{,}934.87'}$$

PROBLEMS

7.1 Figure P7.1 shows the locations of the beginning point (1) and end point (2) of a new tunnel with respect to four survey control points: A, B, C, and D. The coordinates of the four control points and the design coordinates of points 1 and 2 are given below:

Point	X-coordinate	Y-coordinate
A	41,331.27 ft	36,488.27 ft
B	47,027.95 ft	37,113.55 ft
C	42,816.22 ft	32,136.46 ft
D	49,163.41 ft	32,415.81 ft
1	44,376.34 ft	34,369.29 ft
2	47,979.76 ft	35,557.23 ft

Compute the following:

a. the length and azimuth of the centerline 1-2 of the tunnel

b. the azimuth and distance of the line from control point A to point 1

c. the azimuth and distance of the line from control point C to point 1

d. the azimuth and distance of the line from control point B to point 2

e. the azimuth and distance of the line from control point D to point 2

7.2 Let point M be the mid-point of the tunnel in Problem 7.1. It is located on the centerline halfway between points 1 and 2. Compute the horizontal coordinates (X_M, Y_M) of point M.

7.3 The beginning point (1) of the tunnel in Problem 7.1 is to be located in the field by angle and distance measurements from control point A. A total station will be set

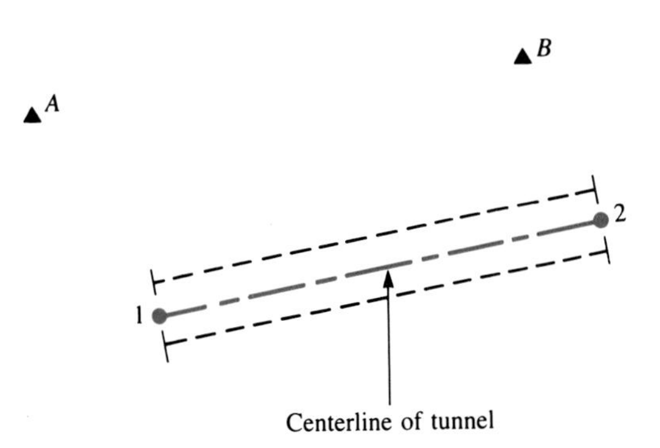

FIGURE P7.1 A tunnel survey

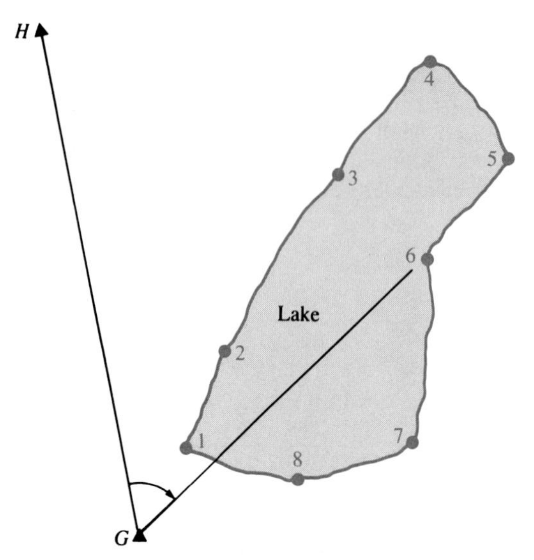

FIGURE P7.2 Shoreline mapping

up at point A and backsight to point B. Compute the clockwise horizontal angle $B\hat{A}1$.

7.4 The end point (2) of the tunnel in Problem 7.1 is to be located in the field by angle and distance measurements from control point B. A total station will be set up at point B and backsight to point C. Compute the clockwise angle $C\hat{B}2$.

7.5 In a mapping project, the shoreline of a small lake is being measured with a total station set up at point G, as shown in Figure P7.2. Backsight is made to point H, and all horizontal angles are measured clockwise. The total station measures the slope distances from point G and automatically reduces them to horizontal distances. The coordinates of points G and H are known to be as follows:

$$X_G = 14{,}163.27 \text{ m} \qquad Y_G = 27{,}249.88 \text{ m}$$
$$X_H = 13{,}877.12 \text{ m} \qquad Y_H = 27{,}878.35 \text{ m}$$

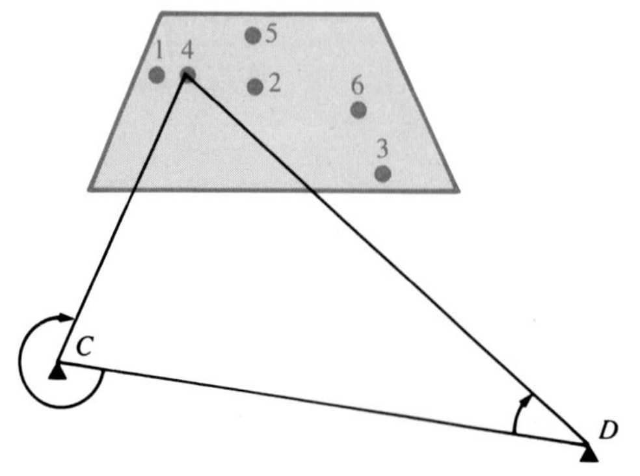

FIGURE P7.3 Dam deformation survey

The measured angles and distances are given below:

Point	Measured Horizontal Angle from Line GH	Measured Horizontal Distance from Point G
1	41°27′36″	141.346 m
2	37°33′50″	272.477 m
3	39°14′00″	523.881 m
4	43°54′42″	749.650 m
5	56°03′18″	711.265 m
6	57°54′24″	544.245 m
7	83°21′45″	387.261 m
8	81°40′20″	223.411 m

Compute the coordinates of the shoreline points.

7.6 In a dam deformation survey, as shown in Figure P7.3, the positions of survey points 1, 2, 3, 4, 5, and 6 located on the concrete face of the dam are to be determined by angle measurements from control points C and D. The horizontal coordinates of points C and D are:

$$X_C = 6{,}311.26 \text{ ft} \qquad Y_C = 4{,}892.33 \text{ ft}$$

$$X_D = 10{,}877.97 \text{ ft} \qquad Y_D = 4{,}631.66 \text{ ft}$$

With the theodolite at point C and backsight to point D, the following clockwise angles are measured:

To Point	Horizontal Angle
1	281°33′16″
2	295°10′51″
3	312°43′32″
4	285°07′08″
5	291°44′25″
6	311°58′01″

The following clockwise angles are measured at point D with backsight to point C:

To Point	Horizontal Angle
1	31°54′33″
2	37°21′00″
3	41°03′14″
4	33°01′08″
5	39°47′09″
6	45°07′53″

Compute the X- and Y- coordinates of points 1, 2, 3, 4, 5, and 6.

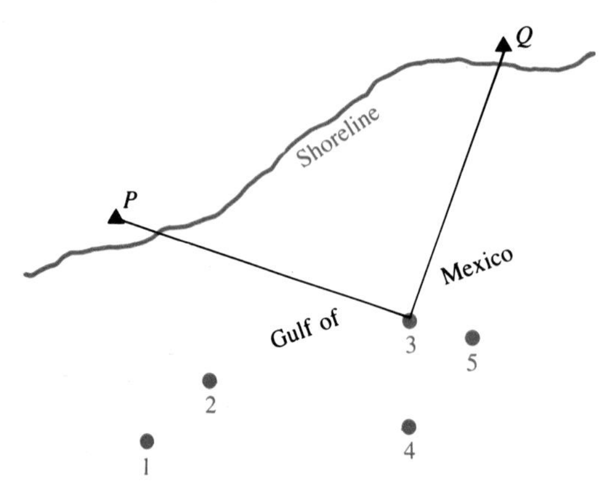

FIGURE P7.4 Offshore survey

7.7 In an offshore construction project, the locations of five offshore platforms are to be determined by measuring distances to two known survey points, P and Q (see Figure P7.4). The coordinates of P and Q are

$$X_P = 73{,}241.69 \text{ m} \qquad Y_P = 47{,}989.21 \text{ m}$$
$$X_Q = 75{,}666.35 \text{ m} \qquad Y_Q = 49{,}273.55 \text{ m}$$

A reflector is set up at both points P and Q, and an EDM instrument is used to measure the distances successively from the five platforms. The EDM instrument performs slope reduction automatically and provides readout of the horizontal distances. The measured distances are given below:

From Platform No.	To Control Point	Measured Horizontal Distance
1	*P*	1,341.26 m
1	*Q*	3,167.94 m
2	*P*	1,152.78 m
2	*Q*	2,637.44 m
3	*P*	1,891.28 m
3	*Q*	1,659.71 m
4	*P*	2,143.88 m
4	*Q*	2,269.76 m
5	*P*	2,487.11 m
5	*Q*	1,648.75 m

Compute the X- and Y- coordinates of each offshore platform.

7.8 In a bridge construction project, the position of the pile-driving barge is determined by the method of resection. A theodolite is set up on the barge, and angles

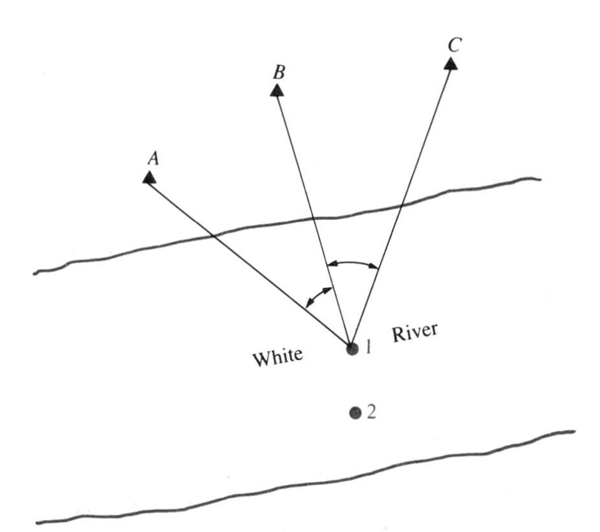

FIGURE P7.5 Location survey

are measured to three control points: A, B, and C (see Figure P7.5). The coordinates of the control points are

Control Point	X-Coordinate	Y-Coordinate
A	143,721.45 ft	89,332.15 ft
B	146,298.36 ft	92,175.27 ft
C	150,036.79 ft	91,628.21 ft

The following angles are measured at barge positions 1 and 2:

$$A\hat{1}B = 33°47'36''$$
$$B\hat{1}C = 37°23'08''$$
$$A\hat{2}B = 29°06'55''$$
$$B\hat{2}C = 31°17'27''$$

Compute the coordinates of the barge at positions 1 and 2.

REFERENCES

7.1 American Society of Photogrammetry. *Automated Photogrammetry and Cartography of Highways and Transport Systems,* Proceeding of workshop held in Portland, Oregon, April 28–May 1, 1981. Published by American Society of Photogrammetry, 105 N. Virginia Ave., Falls Church, VA 22046.

7.2 American Society of Photogrammetry. *Proceedings of the Digital Terrain Models (DTM) Symposium,* held in St. Louis, Missouri, May 9–11, 1978. Published by American Society of Photogrammetry, 105 N. Virginia Ave., Falls Church, VA 22046.

7.3 McEwen, Robert B. "USGS Digital Cartographic Applications Program," *Journal of the Surveying and Mapping Division,* American Society of Civil Engineers, Vol. 106, No. SU 1, Proc. Paper 15846, November, 1980, 13–22.

7.4 Roos, D., and Miller, C. L. *COGO-90: Engineering User's Manual,* Massachusetts Institute of Technology, Department of Civil Engineering, Research Report R64-12, April 1964.

TRAVERSE

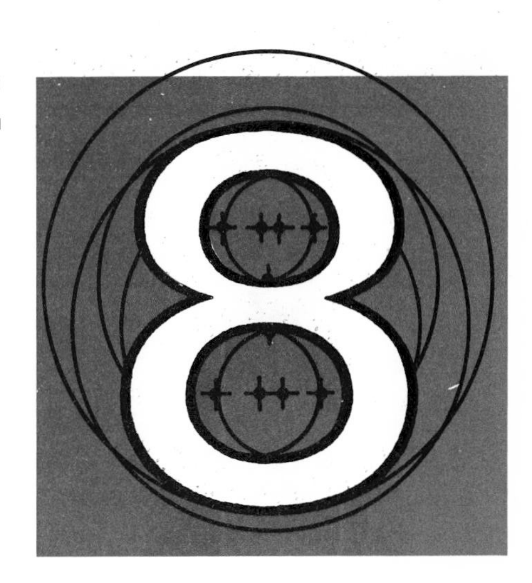

8.1 INTRODUCTION

Traverse survey is one of the most commonly used methods for determining the relative positions of a number of survey points. Basically, it consists of repeated applications of the method of locationing by angle and distance, which was discussed in Section 7.5. By starting from a point of known position and a line of known direction, the location of a new point is determined by measuring the distance and angle from the known point. Then the location of another new point is determined by angle and distance measurement from the newly located point. This procedure is repeated from point to point. The resulting geometric figure is called a *traverse*. Frequently, the traverse is also closed on a point of known position and a line of known direction to provide closure checks on the survey.

A traverse that starts and closes on a point of known position is called a *closed traverse*. A traverse that does not close on a point of known position is called an *open traverse*.

8.2 CONNECTING TRAVERSE

A closed traverse that starts from a known point and closes on a second known point is called a *connecting traverse*. It is often used to establish survey control points for route construction projects such as high-

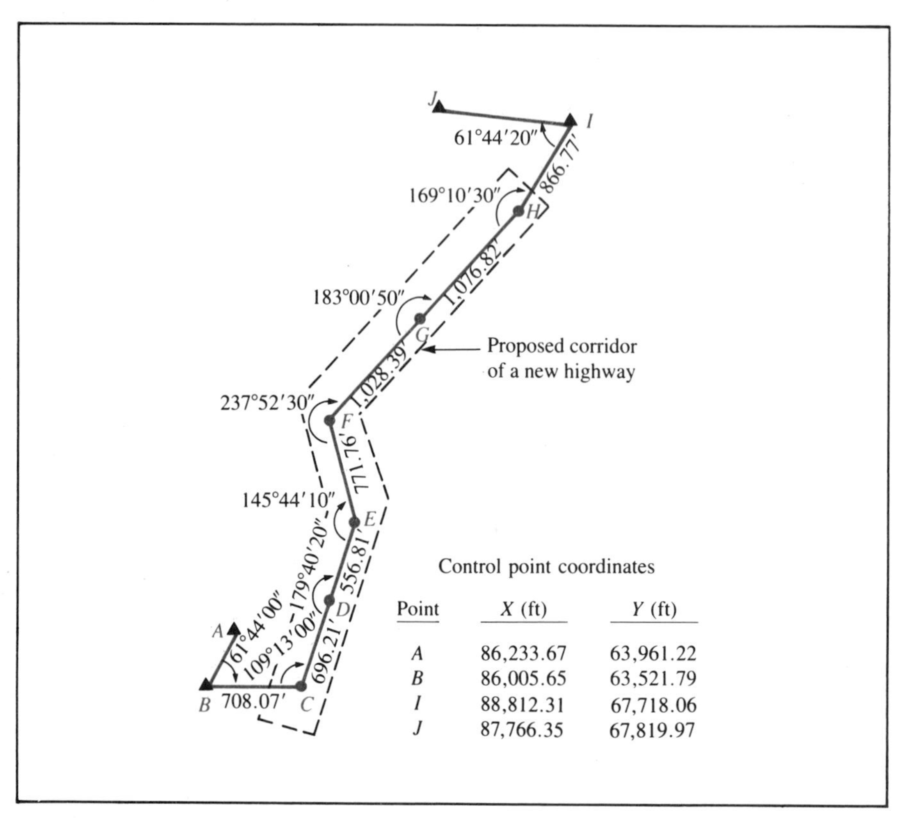

FIGURE 8.1 A connecting traverse

ways, railroads, pipelines, and transmission lines. Figure 8.1 shows a connecting traverse used to determine the locations of six survey points along the proposed corridor of a new highway. Points *A, B, I,* and *J* are survey control points whose horizontal coordinates have already been established from previous surveys. They could also be survey points that have been established by the National Geodetic Survey as part of a national control network. The traverse starts from point *B* and closes on Point *I*. At each point along the traverse, the clockwise (angle-to-the-right) angle is measured. The horizontal length of each line (often referred to as a leg or a course) of the traverse is also measured. The line *BA* is used to provide a known starting azimuth, and the line *IJ* is used to provide a known azimuth for checking the field angle measurements. If the coordinates of station *A* are not known and only the direction of the line *BA* is known, then station *A* is called an *azimuth mark*.

The measured angles and distances along the traverse are used to compute the coordinates of the new points along the proposed corridor. These survey points are then used in topographic mapping of the corridor. The topographic maps are subsequently used by the highway engineers to develop detailed design of the new highway. After the design has been completed, the new highway is then located on the ground using these same survey points as control points. Thus these six survey points play an important role throughout the project.

8.3 LOOP TRAVERSE

A closed traverse that starts and ends on the same points is called a *loop traverse*. It is commonly used to survey the boundaries of a tract of land. It is also used to establish survey controls for the mapping of a project site.

Figure 8.2 shows a loop traverse used to determine the bearings and lengths of the boundary lines for a tract of land. The traverse starts from and closes on point 1. Each of the traverse stations is a boundary corner. At each station the interior angle is measured. The horizontal length of each course of the traverse (that is, boundary line) is also measured. In order to determine the bearings of the boundary lines, a short traverse is run to connect the loop traverse with a line (AB) of known bearing.

The coordinates of points A and B are known. The bearing of the line AB can thus be computed from the coordinates. The angles $B\hat{A}1$ and $A\hat{1}6$ are

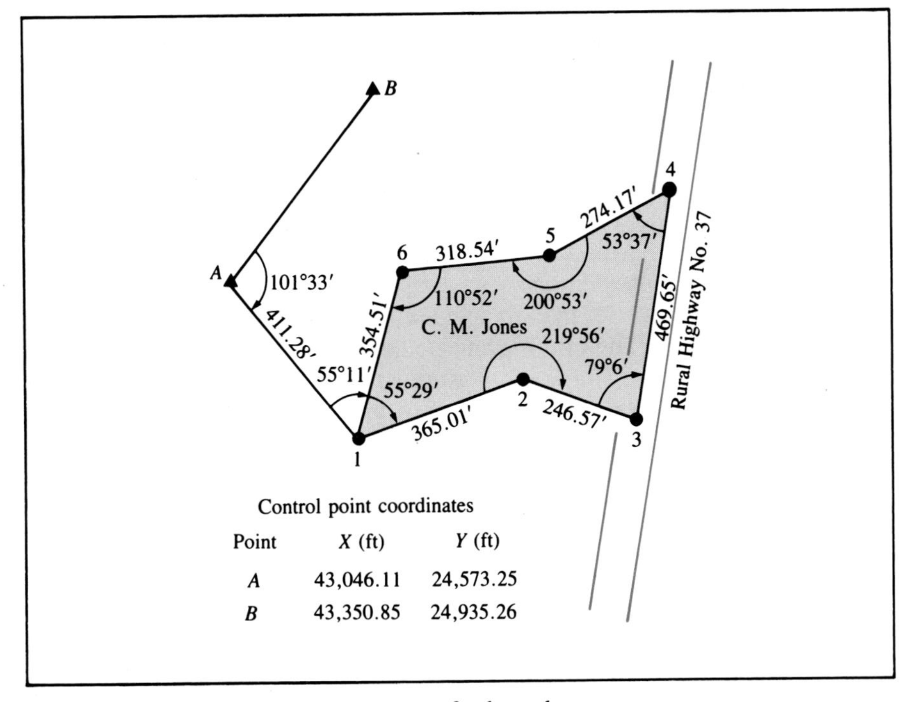

FIGURE 8.2 A loop traverse for boundary survey

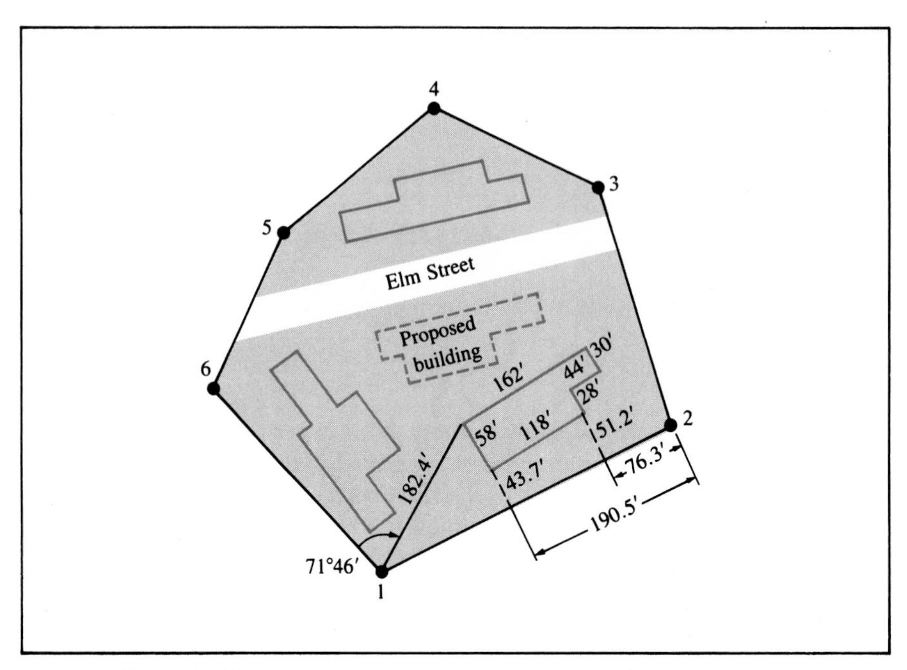

FIGURE 8.3 A loop traverse for survey control

measured. By using these two measured angles and the calculated bearing of line *AB*, the bearing of line 1-6 can be computed. The bearings of the remaining lines included in the traverse can then be computed from the measured traverse angles. By also measuring the length of the line *A*1, the coordinates of the property corners can also be computed.

If a line of known bearing cannot be found nearby, the bearing of one of the boundary lines can be initially determined by astronomical methods, which will be discussed in Chapter 16.

Figure 8.3 shows a loop traverse used to establish a network of control points for mapping an industrial site at which a new building is being considered. The interior angles and the lengths of the sides of the traverse are measured. Relevant features such as building corners, fire hydrants, telephone poles, trees, and roads are then located with respect to the traverse by measuring distances and angles. All these measurements can then be used together to draw a map for the project site. Methods of topographic mapping will be discussed in greater detail in Chapter 9.

8.4 OPEN TRAVERSE

An open traverse starts from a point and ends on another point, the position of which is not known. Thus an open traverse provides no closure check on the accuracy of the field measurements. The short traverse $BA1$ in Figure 8.2 is an example of an open traverse. Open traverse is also often used for locating the centerline of a tunnel during construction. Survey errors accumulated along the traverse can seriously degrade the positioning accuracy of an open traverse. The total length of an open traverse should, therefore, be kept as short as possible.

8.5 TRAVERSE COMPUTATIONS

A closed traverse provides two independent checks on the accuracy of the field measurements. These are:

1. *Azimuth closure:* The azimuth of the closing line computed from the azimuth of the beginning line and the traverse angles must be equal to the known value.
2. *Position closure:* The coordinates of the closing station computed from the coordinates of the starting station and the measured angles and distances must also be equal to the known values.

Traverse computations usually include the following steps:

1. Correct the angle measurements for closure error in azimuth.
2. Compute the preliminary coordinates of the traverse stations.
3. Compute the position closure error at the last point of the traverse.
4. Apply corrections to the preliminary coordinates to compensate for the position closure error.
5. Compute the azimuth (or bearing) and length of each course of the traverse using the final coordinates.

The procedure for performing these calculations will be discussed in the following sections.

8.6 AZIMUTH CLOSURE

By using the known beginning azimuth, the azimuth of each successive course of the traverse is first computed from the measured angles. Let α be the known azimuth of the reference line at the end of the traverse, and α' be the azimuth of the same reference line computed from the traverse angles. Then the closure error in azimuth, to be denoted as ε_α, is computed as follows:

$$\varepsilon_\alpha = \alpha' - \alpha \tag{8.1}$$

The closure error should be distributed equally to all the measured traverse angles. Let there be n measured angles in the traverse; then the correction (C_α) that should be applied to each traverse angle is computed as follows:

$$C_\alpha = -\frac{\varepsilon_\alpha}{n} \tag{8.2}$$

Instead of applying the angular correction directly to the traverse angles, it is more convenient to apply the correction directly to the computed azimuths of the traverse legs. Let α_i' be the initially computed azimuth of the i-th leg in the traverse, then the corrected azimuth α_i'' of the same leg can be computed as follows:

$$\alpha_i'' = \alpha_i' - i \cdot \left(\frac{\varepsilon_\alpha}{n}\right) \tag{8.3}$$

EXAMPLE 8.1

For the connecting traverse in Figure 8.1, compute the azimuth closure and the corrected azimuths of all courses.

SOLUTION

The azimuths of lines BA and IJ are first computed from the known coordinates:

$$\alpha_{BA} = \tan^{-1}\frac{X_A - X_B}{Y_A - Y_B} = \tan^{-1}\frac{86{,}233.67 - 86{,}005.65}{63{,}961.22 - 63{,}521.79}$$

$$\therefore \alpha_{BA} = 27°25'29''$$

$$\alpha_{IJ} = \tan^{-1}\frac{87{,}766.35 - 88{,}812.31}{67{,}819.97 - 67{,}718.06}$$

$$\therefore \alpha_{IJ} = 275°33'54''$$

The calculation can be tabulated as follows:

Line		Azimuth	Correction	Corrected Azimuth
BA		27°25′29″		
	$+ A\hat{B}C$	61°44′00″		
BC		89° 9′29″	− 9″	89°9′20″
		+180°00′00″		
CB		269° 9′29″		
	$+ B\hat{C}D$	109°13′00″		
		378°22′29″		
		−360°		
CD		18°22′29″	−19″	18°22′10″
		+180°		

(continued)

Line		Azimuth	Correction	Corrected Azimuth
DC		198°22′29″		
	$+ C\hat{D}E$	179°40′20″		
		378° 2′49″		
		−360°		
DE		18° 2′49″	−28″	18°2′21″
		+180°		
ED		198° 2′49″		
	$+ D\hat{E}F$	145°44′10″		
EF		343°46′59″	−38″	343°46′21″
		−180°		
FE		163°46′59″		
	$+ E\hat{F}G$	237°52′30″		
		401°39′29″		
		−360°		
FG		41°39′29″	−47″	41°38′42″
		+180°		
GF		221°39′29″		
	$+ F\hat{G}H$	183°00′50″		
		404°40′19″		
		−360°		
GH		44°40′19″	−56″	44°39′23″
		+180°		
HG		224°40′19″		
	$+ G\hat{H}I$	169°10′30″		
		393°50′49″		
		−360°		
HI		33°50′49″	−1′6″	33°49′43″
		+180°		
IH		213°50′49″		
	$+ H\hat{I}J$	61°44′20″		
IJ		275°35′ 9″	−1′15″	275°33′54″
	$-\alpha_{IJ}$	275°33′54″		
Azimuth closure error =		+ 1′15″		

$$\text{Correction/angle} = -\frac{75''}{8} = -9.4''$$

In the preceding example, the traverse angles are measured to nearest 10 seconds. However, the azimuths and azimuth corrections are computed to the nearest second, whereas the correction per angle is computed to 0.1 second. With the computational power of modern calculators, it is a good practice to carry more significant digits than absolutely necessary during intermediate calculations. Only the final results of the computation (in this case, the coordinates for the traverse stations) need to be rounded off to the proper number of significant digits. When an electronic computer is used for survey calculations, the computer automatically carries the same number of significant digits in all steps of the calculation process.

In Example 8.2 below, all the traverse angles are measured to the nearest minute. The azimuths and azimuth corrections for all courses in the traverse are rounded off to the nearest 0.01 minute.

EXAMPLE 8.2

For the loop traverse in Figure 8.2, compute the azimuth closure error and the corrected azimuths of the courses.

SOLUTION

The azimuth of line *AB* is first computed from the known coordinates:

$$\alpha_{AB} = \tan^{-1} \frac{43{,}350.85 - 43{,}046.11}{24{,}935.26 - 24{,}573.25} = 40°5'$$

The azimuth of line 1-6 of the traverse is then computed as follows:

Line		Azimuth
AB		40° 5′
	+ $B\hat{A}1$	101°33′
*A*1		141°38′
		+180°
1*A*		321°38′
	+ $A\hat{1}6$	55°11′
		376°49′
		−360°
1-6		16°49′

The azimuths of the traverse courses can be calculated and corrected as follows:

Line		Azimuth	Correction	Corrected Azimuth
1-6		16°49′		
	+ $6\hat{1}2$	55°29′		
1-2		72°18′	+ 1.17′	72°19.17′
		+180°		
2-1		252°18′		
	+ $1\hat{2}3$	219°56′		

(*continued*)

Line	Azimuth	Correction	Corrected Azimuth
	472°14′		
	−360°		
2-3	112°14′	+ 2.34′	112°16.34′
	+180°		
3-2	292°14′		
$+\ 2\hat{3}4$	79°06′		
	371°20′		
	−360°		
3-4	11°20′	+ 3.51′	11°23.51′
	+180°		
4-3	191°20′		
$+\ 3\hat{4}5$	53°37′		
4-5	244°57′	+ 4.68′	245° 1.68′
	−180°		
5-4	64°57′		
$+\ 4\hat{5}6$	200°53′		
5-6	265°50′	+ 5.85′	265°55.85′
	−180°		
6-5	85°50′		
$+\ 5\hat{6}1$	110°52′		
6-1	196°42′	+ 7.0′	196°49.0′
	−180°		
1-6	16°42′		
	−16°49′		
Azimuth closure error =	− 7′		

$$\text{Correction/angle} = +\frac{7'}{6} = +1.17'$$

In the case of a loop traverse, the interior angles of the traverse can also be adjusted by using the condition that the sum of the interior angles of a loop traverse of n courses must be equal to $(n - 2) \times 180°$. The loop traverse in Example 8.2 has six courses (that is, $n = 6$). The sum of the six interior angles should then be equal to $(6 - 2) \times 180° = 720°$. The angular closure error (ε_a) can be computed by the following expression:

$$\varepsilon_a = (n - 2) \times 180° - \Sigma \text{ interior angles}$$

The correction to be applied to each angle is then computed as follows:

$$\text{correction/angle} = \frac{-\varepsilon_a}{n}$$

The correction is applied to each angle, and then the azimuths of the courses are computed with the corrected angles.

EXAMPLE 8.3

For the loop traverse in Figure 8.2, compute the angular closure error by summing the interior angles. Also compute the corrected angles.

SOLUTION

Station	Measured Angle	Correction	Corrected Angles
1	55°29′	+1.17′	55°30.17′
2	219°56′	+1.17′	219°57.17′
3	79°06′	+1.17′	79° 7.17′
4	53°37′	+1.17′	53°38.17′
5	200°53′	+1.17′	200°54.17′
6	110°52′	+1.17′	110°53.17′
Sum =	−719°53′	Sum =	720°00′
	+720°		
Closure error =	− 7′		

$$\text{Correction/angle} = +\frac{7'}{6} = +1.17'$$

In Example 8.3, the angular correction per angle and the corrected angles are rounded to 0.01′. This was done to avoid any inconsistency in the calculation due to round-off errors. If the angular correction had been rounded off to +1′ per angle, the sum of the corrections to the six angles would amount to +6′, and the sum of the corrected angles would amount to only 719°59′. The exact closure condition can then be achieved only by applying an additional correction of +1′ to one of the angles. This angle can be chosen either randomly or selectively according to one's perception of the relative accuracy of angles. However, in this age of computer processing, it is advisable to adopt a procedure (such as that used in Example 8.3) that can be used whether the calculation is to be performed by hand or by an electronic computer.

In Example 8.3, the adjusted angles can be used to compute the azimuths of the traverse courses as in Example 8.2. The computed azimuths can then be rounded off, if desired, to 1 minute. The azimuth closure error should be equal to zero, and no further adjustment of the azimuths is needed.

8.7 POSITION CLOSURE

After the angular corrections have been completed, the corrected azimuths are used with the measured distances to compute the coordinates of the traverse stations. If the distances are measured by taping, the distances should be corrected for systematic errors due to tension, temperature, slope, and sag. If the distances are measured by EDM equipment, corrections should be applied for atmospheric conditions, scale error, and zero constants. Furthermore, if State Plane Coordinates are used to define the positions of the traverse stations, scale corrections should be applied to the field-measured distances (see Chapter 15).

Starting from the beginning known station, the coordinates of each successive traverse station are computed using Eqs. (7.4) and (7.5). At the last station, the differences between the computed and known coordinates provide a measure on the magnitude of positional closure error. Let X and Y be the known coordinates of the last station, and X' and Y' be the corresponding computed coordinates. The closure error can be computed as follows:

$$\text{closure error in } X\text{-coordinate } (\varepsilon_X) = X' - X \tag{8.4}$$

$$\text{closure error in } Y\text{-coordinate } (\varepsilon_Y) = Y' - Y \tag{8.5}$$

The linear error of closure (ε_d) is computed as follows:

$$\varepsilon_d = \sqrt{\varepsilon_X^2 + \varepsilon_Y^2} \tag{8.6}$$

The linear error of closure (ε_d) is the straight-line distance between the known and the computed positions of the last traverse point. It provides a measure on the quality of the traverse measurements. Obviously, a traverse 5.0 miles long having a linear closure of 4.60 ft will be of greater accuracy than a traverse that is only 2.5 miles long and has the same magnitude of linear closure. It is common, therefore, to measure the accuracy of the traverse by calculating the relative error of closure, which is defined as follows:

$$\text{relative error of closure} = \frac{1}{\dfrac{D}{\varepsilon_d}} \tag{8.7}$$

where D is the sum of the distances along the traverse and ε_d is the linear error of closure. The relative error of closure is expressed in the form of a ratio with unity as the numerator, and the denominator (that is, D/ε_d) is rounded off to either one or two significant digit(s).

Table 8.1 shows the calculation performed for the connecting traverse in Figure 8.1. The linear error of closure is 0.30 ft, and the relative error of closure amounts to 1/18,000.

Table 8.2 shows the calculation performed for the loop traverse in Figure 8.2. This loop traverse has a linear closure error of 0.18 ft over a total traverse distance of 2,028.45 ft. Thus, the relative error of closure is 1/11,000. The traverse starts and ends on station 1. The side $A1$ is not included in the

TABLE 8.1 Computing Coordinate Closure Errors for a Connecting Traverse

Station	Corrected Azimuth α_{ij}	Distance d_{ij} (ft)	Departure $\Delta X_{ij} = d_{ij} \sin \alpha_{ij}$ (ft)	Latitude $\Delta Y_{ij} = d_{ij} \cos \alpha_{ij}$ (ft)	Preliminary Coordinates X (ft)	Preliminary Coordinates Y (ft)
B					86,005.65	63,521.79
	89° 9′20″	708.07	707.99	10.44		
C					86,713.64	63,532.23
	18°22′10″	696.21	219.41	660.73		
D					86,933.05	64,192.96
	18° 2′21″	556.81	172.43	529.44		
E					87,105.48	64,722.40
	343°46′21″	771.76	−215.67	741.01		
F					86,889.81	65,463.41
	41°38′42″	1,028.39	683.38	768.49		
G					87,573.19	66,231.90
	44°39′23″	1,076.82	756.85	765.98		
H					88,330.04	66,997.88
	33°49′43″	866.77	482.54	720.03		
I					88,812.58	67,717.91
				(known)	−88,812.31	−67,718.06
		$\Sigma d_{ij} = 5,704.83$			closure error $\varepsilon_x = +0.27$	$\varepsilon_y = -0.15$

$$\text{linear error of closure} = \sqrt{(0.27)^2 + (-0.15)^2} = 0.31 \text{ ft}$$

$$\text{relative error of closure} = \frac{1}{\frac{5,704.83}{0.31}} = \frac{1}{18,000}$$

TABLE 8.2 Computing Coordinate Closure Errors for a Closed Traverse

Station	Corrected Azimuth α_{ij}	Distance d_{ij} (ft)	Departure $\Delta X_{ij} = d_{ij} \sin \alpha_{ij}$ (ft)	Latitude $\Delta Y_{ij} = d_{ij} \cos \alpha_{ij}$ (ft)	Preliminary Coordinates	
					X (ft)	Y (ft)
A					43,046.11	24,573.25
	141°38′	411.28	255.28	−322.47		
1					43,301.39	24,250.78
	72°19.17′	365.01	347.77	110.86		
2					43,649.16	24,361.64
	112°16.34′	246.57	228.17	− 93.45		
3					43,877.33	24,268.19
	11°23.51′	469.65	92.76	460.40		
4					43,970.09	24,728.59
	245° 1.68′	274.17	−248.54	−115.75		
5					43,721.55	24,612.84
	265°55.85′	318.54	−317.74	− 22.60		
6					43,403.81	24,590.24
	196°49.0′	354.51	−102.56	−339.35		
1					43,301.25	24,250.89
		$\Sigma d_{ij} = 2{,}028.45$*		(known)	−43,301.39	−24,250.78
				closure error	$\varepsilon_x = -0.14$	$\varepsilon_y = +0.11$

$$\text{linear error of closure} = \sqrt{(-0.14)^2 + (0.11)^2} = 0.18 \text{ ft}$$

$$\text{relative error of closure} = \frac{1}{\dfrac{2{,}028.45}{0.18}} = \frac{1}{11{,}000}$$

* The distance 411.28 ft for side $A1$ is not included in the sum.

loop traverse and is therefore not included in computing the sum of the traverse distances.

8.8 COMPASS RULE

After the closure errors in position have been computed, the preliminary coordinates of the traverse stations must be corrected so that the closure errors are reduced to zero. This procedure is often referred to as *balancing the traverse*. Its purpose is to make the traverse a mathematically correct figure. One of the most commonly used methods of balancing a traverse is the *compass rule*. It is sometimes called the *Bowditch Rule* after the eminent American navigator Nathaniel Bowditch (1773–1838), who is generally credited with its formulation. It assumes that the accuracy of angular and distance measurements are approximately the same. The compass rule may be mathematically expressed as follows:

$$\begin{array}{c}\text{correction to departure}\\ \text{of side } ij\end{array} = -\frac{\text{length of side } ij}{\begin{array}{c}\text{total length of}\\ \text{traverse}\end{array}} \times \begin{array}{c}\text{closure error}\\ \text{in departure}\end{array} \tag{8.8}$$

$$\begin{array}{c}\text{correction to latitude}\\ \text{of side } ij\end{array} = -\frac{\text{length of side } ij}{\begin{array}{c}\text{total length of}\\ \text{traverse}\end{array}} \times \begin{array}{c}\text{closure error}\\ \text{in latitude}\end{array} \tag{8.9}$$

When coordinates are used in traverse computation, as shown in Tables 8.1 and 8.2, the compass rule can be more conveniently used to compute corrections to the station coordinates directly by the following formulas:

$$CX_i = -\frac{L_i}{D} \times \varepsilon_X \tag{8.10}$$

and

$$CY_i = -\frac{L_i}{D} \times \varepsilon_Y \tag{8.11}$$

where CX_i and CY_i are the corrections to be applied to the preliminary coordinates X_i and Y_i of station i, L_i is the cumulative traverse distance up to station i, D is the total traverse distance, and ε_X and ε_Y are the closure errors in X and Y coordinates respectively. The corrected coordinates (X_i, Y_i) of station i are then computed as follows:

$$X_i = X_i' + CX_i \tag{8.12}$$

$$Y_i = Y_i' + CY_i \tag{8.13}$$

where X_i' and Y_i' are the preliminary coordinates of station i.

Table 8.3 presents the calculation involved in balancing the connecting traverse of Figure 8.1 by compass rule, and Table 8.4 presents the calculation for the loop traverse in Figure 8.2.

8.9 TRANSIT RULE

Another commonly recognized, but seldom used, method for balancing a traverse is the *transit rule,* which is based on the assumption that the angular measurements in the traverse are made with higher accuracy than the distance measurements. The transit rule may be mathematically expressed as follows:

$$\begin{array}{c}\text{correction for departure}\\ \text{of side } ij\end{array} = -\frac{\begin{array}{c}\text{departure of}\\ \text{side } ij\end{array}}{\begin{array}{c}\text{arithmetic sum of}\\ \text{the departures of}\\ \text{all sides}\end{array}} \times \begin{array}{c}\text{closure error}\\ \text{in departure}\end{array} \qquad (8.14)$$

$$\begin{array}{c}\text{correction for latitude}\\ \text{of side } ij\end{array} = -\frac{\begin{array}{c}\text{latitude of}\\ \text{side } ij\end{array}}{\begin{array}{c}\text{arithmetic sum of}\\ \text{the latitudes of}\\ \text{all sides}\end{array}} \times \begin{array}{c}\text{closure error}\\ \text{in latitude}\end{array} \qquad (8.15)$$

For coordinate computations the transit rule can be more conveniently expressed as follows:

$$CX_i = -\frac{\begin{array}{c}\text{cumulative arithmetic sum of}\\ \text{departures up to station } i\end{array}}{\begin{array}{c}\text{arithmetic sum of the}\\ \text{departures of all sides}\end{array}} \times \begin{array}{c}\text{closure error}\\ \text{in departure}\end{array} \qquad (8.16)$$

$$CY_i = -\frac{\begin{array}{c}\text{cumulative arithmetic sum of}\\ \text{latitudes up to station } i\end{array}}{\begin{array}{c}\text{arithmetic sum of the}\\ \text{latitudes of all sides}\end{array}} \times \begin{array}{c}\text{closure error}\\ \text{in latitude}\end{array} \qquad (8.17)$$

where CX_i and CY_i are the respective corrections for the preliminary coordinates X_i' and Y_i' of station i. In Eqs. (8.14) to (8.17), the arithmetic sum of departures (or latitudes) is computed as the sum of the absolute values of the departures (or latitudes).

Application of the transit rule is illustrated in Tables 8.5 and 8.6 for the two traverses previously discussed in this chapter. The corrected station coordinates do not differ significantly from those obtained by compass rule.

TABLE 8.3 Balancing a Connecting Traverse by Compass Rule

Station	Cumulative Distance	*X*-Coordinate			*Y*-Coordinate		
		Preliminary	Correction	Final	Preliminary	Correction	Final
B	0	86,005.65	0	86,005.65	63,521.79	0	63,521.79
C	708.07	86,713.64	−0.03	86,713.61	63,532.23	+0.02	63,532.25
D	1,404.28	86,933.05	−0.07	86,932.98	64,192.96	+0.04	64,193.00
E	1,961.09	87,105.48	−0.09	87,105.39	64,722.40	+0.05	64,722.45
F	2,732.85	86,889.81	−0.13	86,889.68	65,463.41	+0.07	65,463.48
G	3,761.24	87,573.19	−0.18	87,573.01	66,231.90	+0.10	66,232.00
H	4,838.06	88,330.04	−0.23	88,329.81	66,997.88	+0.13	66,998.01
I	5,704.83	88,812.58	−0.27	88,812.31	67,717.91	+0.15	67,718.06

TABLE 8.4 Balancing a Closed Traverse by Compass Rule

Station	Cumulative Distance	*X*-Coordinate			*Y*-Coordinate		
		Preliminary	Correction	Final	Preliminary	Correction	Final
1	0	43,301.39	0.0	43,301.39	24,250.78	0.0	24,250.78
2	365.01	43,649.16	+0.03	43,649.19	24,361.64	−0.02	24,361.62
3	611.58	43,877.33	+0.04	43,877.37	24,268.19	−0.03	24,268.16
4	1,081.23	43,970.09	+0.07	43,970.16	24,728.59	−0.06	24,728.53
5	1,355.40	43,721.55	+0.09	43,721.64	24,612.84	−0.07	24,612.77
6	1,673.94	43,403.81	+0.12	43,403.93	24,590.24	−0.09	24,590.15
1	2,028.45	43,301.25	+0.14	43,301.39	24,250.89	−0.11	24,250.78

TABLE 8.5 Balancing a Connecting Traverse by Transit Rule

Station	Cumulative Departure	Cumulative Latitude	X-Coordinate			Y-Coordinate		
			Preliminary	Correction	Final	Preliminary	Correction	Final
B	0	0	86,005.65	0	86,005.65	63,521.79	0	63,521.79
C	707.99	10.44	86,713.64	−0.06	86,713.58	63,532.23	0.00	63,532.23
D	927.40	671.17	86,933.05	−0.08	86,932.97	64,192.96	+0.02	64,192.98
E	1,099.83	1,200.61	87,105.48	−0.09	87,105.39	64,722.40	+0.04	64,722.44
F	1,315.50	1,941.62	86,889.81	−0.11	86,889.70	65,463.41	+0.07	65,463.48
G	1,998.88	2,710.11	87,573.19	−0.17	87,573.02	66,231.90	+0.10	66,232.00
H	2,755.73	3,476.09	88,330.04	−0.23	88,329.81	66,997.88	+0.12	66,998.00
I	3,238.27	4,196.12	88,812.58	−0.27	88,812.31	67,717.91	+0.15	67,718.06

TABLE 8.6 Balancing a Closed Traverse by Transit Rule

Station	Cumulative Departure	Cumulative Latitude	X-Coordinate			Y-Coordinate		
			Preliminary	Correction	Final	Preliminary	Correction	Final
1	0	0	43,301.39	0.0	43,301.39	24,250.78	0.0	24,250.78
2	347.77	110.86	43,649.16	+0.04	43,649.20	24,361.64	−0.01	24,361.63
3	575.94	204.31	43,877.33	+0.06	43,877.39	24,268.19	−0.02	24,268.17
4	668.70	664.71	43,970.09	+0.07	43,970.16	24,728.59	−0.06	24,728.53
5	917.24	780.46	43,721.55	+0.10	43,721.65	24,612.84	−0.08	24,612.76
6	1,234.98	803.06	43,403.81	+0.13	43,403.94	24,590.24	−0.08	24,590.16
1	1,337.54	1,142.41	43,301.25	+0.14	43,301.39	24,250.89	−0.11	24,250.78

8.10 LENGTHS AND AZIMUTHS OF TRAVERSE COURSES

After the traverse is balanced and the corrected coordinates are computed for the traverse stations, the lengths, azimuths, and/or bearings of the courses in the traverse can be computed using the final coordinates. In the case of boundary survey, these computed lengths and bearings are then used to describe the location of the boundary lines. These computed lengths and bearings should always be rounded off to represent the accuracy levels of the field measurements.

8.11 WEAKNESS OF THE TRAVERSE FIGURE

The azimuth and position closure errors in a closed traverse should not be considered as foolproof indicators of the accuracy of the traverse survey. The geometric figure of a traverse is inherently weak. Large undetected measurement errors can mutually compensate each other and result in misleadingly small closure errors in both azimuth and position, as shown in Figure 8.4. For this reason measurement errors must be kept at an accept-

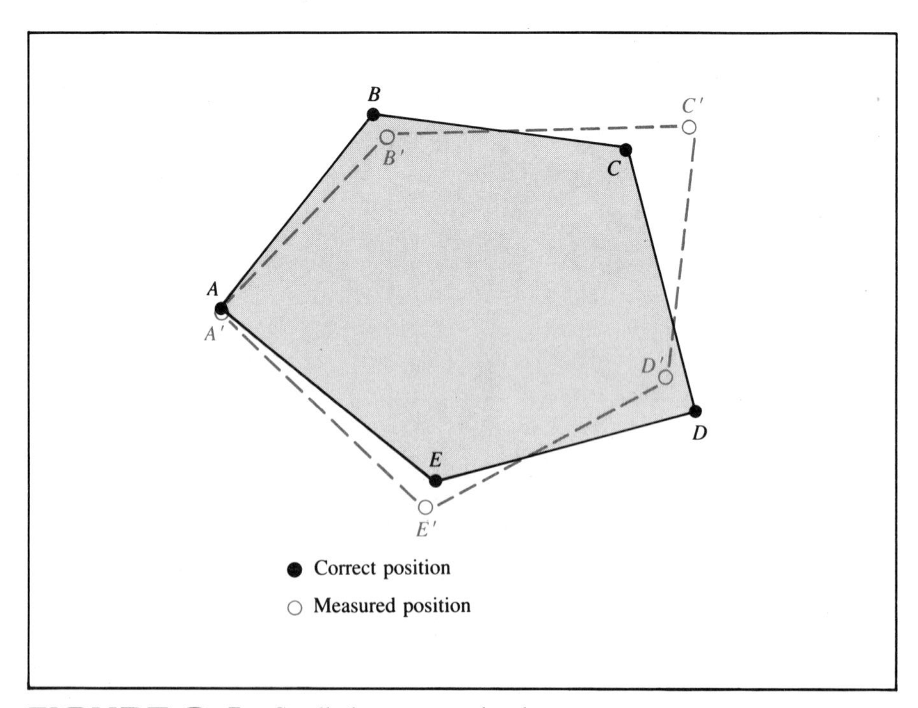

FIGURE 8.4 Small closure error but low accuracy

	First-order	Second-order		Third-order	
		CLASS I	CLASS II	CLASS I	CLASS II
Number of courses between azimuth checks, not to exceed	5–6	10–12	15–20	20–25	30–40
Horizontal directions or angles					
Least count	0.2″	1.0″	1.0″	1.0″	1.0″
No. of observations	16	12	8	4	2
Length measurements (standard error)	$\frac{1}{600,000}$	$\frac{1}{300,000}$	$\frac{1}{120,000}$	$\frac{1}{60,000}$	$\frac{1}{30,000}$
Azimuth closure at azimuth check-points, not to exceed	1.0″/station or $2'' \sqrt{N}$	1.5″/sta. or $3'' \sqrt{N}$	2.0″/sta. or $6'' \sqrt{N}$	3.0″/sta. or $10'' \sqrt{N}$	8″/sta. or $30'' \sqrt{N}$
After azimuth adjustment, position closure not to exceed	$0.04\text{m} \sqrt{K}$ or 1 : 100,000	$0.08\text{m} \sqrt{K}$ or 1 : 50,000	$0.2\text{m} \sqrt{K}$ or 1 : 20,000	$0.4\text{m} \sqrt{K}$ or 1 : 10,000	$0.8\text{m} \sqrt{K}$ or 1 : 5,000

N is the number of stations for carrying azimuth.
K is the distance in Kilometers.

FIGURE 8.5 Accuracy standards for traverse (Federal Geodetic Control Committee)

able level by the use of proper field procedures, properly adjusted instruments, and repeated measurements.

8.12 GRADES OF TRAVERSE

Figure 8.5 on the previous page summarizes the traverse accuracy standards specified by the Federal Geodetic Control Committee (FGCC) for use by the federal surveying and mapping agencies. *Third-order* accuracy is most commonly required for small engineering projects, topographic mapping of small project sites, and boundary surveys. *Second-order* accuracy is usually required for large engineering projects such as highways, urban renewal and development projects, interchanges, short tunnels, and small reservoirs and for the measurement of ground and structural movement. *First-order* accuracy is usually used in large engineering projects such as dams, tunnels, and high-speed rail systems.

In Figure 8.5, the tolerances for closure errors in azimuth and position are given in two forms. The expression containing the square root is designed for long lines, where higher proportional accuracy is required. The formula that gives the smaller permissible tolerance should govern. More specific details on the accuracy standards can be found in the two FGCC publications listed as References 4.2 and 4.3 in Chapter 4.

8.13 COMPATIBILITY OF DISTANCE AND ANGLE MEASUREMENTS

In traversing, the accuracy of distance and angle measurements must be mutually consistent. Since horizontal positions of survey points are computed from both angle and distance measurements, it would be inconsistent to make one type of measurement significantly more accurate than the other.

The effects of errors from angle and distance measurements on the calculated position of a new point is illustrated in Figure 8.6. Point A is fixed in position, and the position of point B is to be determined by measuring both the angle θ from a reference line and the horizontal distance d. Suppose that the angle θ is measured with a standard error of $\pm\,\sigma_\theta$ expressed in radians. The corresponding position error is represented by the linear distance $BB'' = d \cdot \sigma_\theta$. Let σ_d denote the standard error of the measured distance d and be represented by the linear position error BB'. Equating BB' with BB'' yields the following general relationship:

$$\sigma_\theta \text{ (in radians)} = \frac{\sigma_d}{d} \qquad (8.18)$$

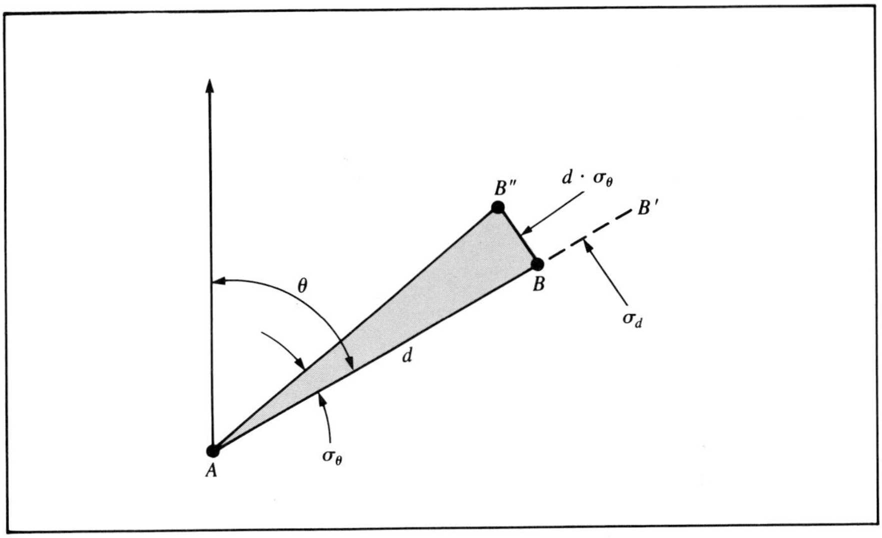

FIGURE 8.6 Angular and linear errors

EXAMPLE 8.4

The angles in a traverse project are to be measured with a standard error of ±5 seconds. If the distance and angle measurements are to be of a consistent level of accuracy, what should be the relative accuracy of the distance measurements at 1 standard error?

SOLUTION

$$\frac{\sigma_d}{d} = \sigma_\theta = 5'' \times 0.000004848 \text{ radian/sec}$$

$$= 0.00002424$$

$$= \frac{1}{41{,}000}$$

Hence, the distances should be measured with a relative accuracy of 1/41,000 at 1σ.

EXAMPLE 8.5

The distances in a traverse project are to be measured with a relative accuracy of 1/100,000 at 1σ. What should be the standard error of the angle measurements in seconds?

SOLUTION

$$\sigma_\theta = \frac{\sigma_d}{d} = \pm\frac{1}{100{,}000} \text{ radian}$$

$$= \pm 2 \text{ sec}$$

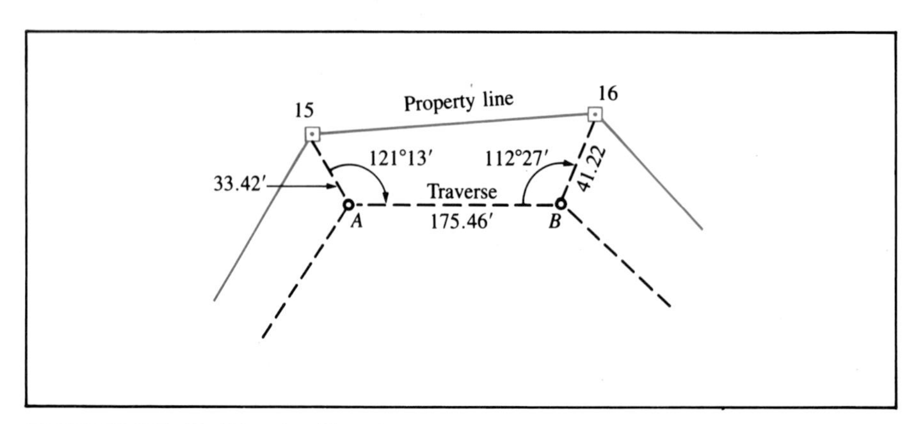

FIGURE 8.7 Auxiliary traverse

8.14 AUXILIARY TRAVERSE

Sometimes in executing a land survey it is very difficult to measure directly along the property lines because of obstructions. The lengths and bearings of these lines can be calculated from an auxiliary traverse whose stations are situated in the clear and conveniently close to the property corners. This situation is depicted in Figure 8.7. Angle and distance ties are carefully made from each transit hub to the nearby corner. The calculation of the traverse will yield coordinates for the hubs and from such values the coordinates of the corners are determined. The inverse problem is then solved to obtain the lengths and bearings of the property lines.

8.15 DEFLECTION ANGLE TRAVERSE

A deflection angle is measured either to the left or right of the extension of the preceding line. Figure 8.8 shows a traverse in which deflection angles are measured. At station 1, the deflection angle of 20°43′10″ is measured to the right, whereas at station 2 the deflection angle is measured to the left of the extension of the preceding line. To measure a deflection angle, say at station 1, the theodolite is pointed to the backsight station (in this case, station *B*) with the telescope in the inverted position and the horizontal graduation circle set to read 0°00′. The telescope is then plunged and pointed to the foresight station. The direction of deflection (left or right) must be recorded with the numerical value of each deflection angle.

Deflection angles provide no advantage over the conventional method of measuring angles to the right or angles to the left. In fact, the measurement procedure for deflection angles is quite awkward when a number of repetitions are desired on each angle.

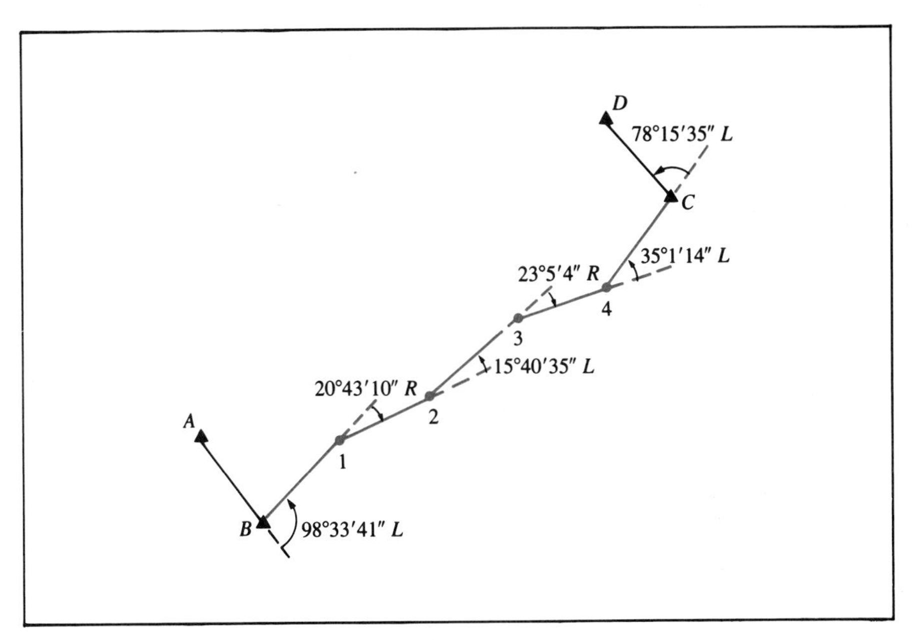

FIGURE 8.8 Deflection angle traverse

Deflection angles are used primarily for the layout of the straight lines along a proposed route for highway, railroad, or pipelines.

8.16 CONVERGENCE OF TRUE MERIDIANS

When long traverses are oriented primarily in the east-west direction, the convergence of meridians must be taken into consideration in calculating closure error in azimuth. Shown in Figure 8.9 is such a traverse. It starts at station A, and the azimuth (α_{A1}) of the line $A1$ with respect to the true north is known. The traverse extends eastward and closes on point B. The azimuth (α_{BC}) of the line BC with respect to the true north at station B is also known. Let α'_{BC} be the azimuth of the line BC computed from α_{A1} and the traverse angles. Even if the traverse angles can be measured so that they are free of errors, the computed azimuth α'_{BC} would differ from the correct azimuth α_{BC} by the amount θ. The quantity θ is the angular convergence of the true meridians.

The magnitude of θ depends on the east-west distance (that is, departure) between the two end stations A and B and the general latitude in which the traverse is located.

In Figure 8.10, PCA and PDB represent two meridians that intercept the arc AB on the equator and the arc CD at the latitude ϕ at which the angular convergence is desired. CT and DT are lines tangent to the meridians

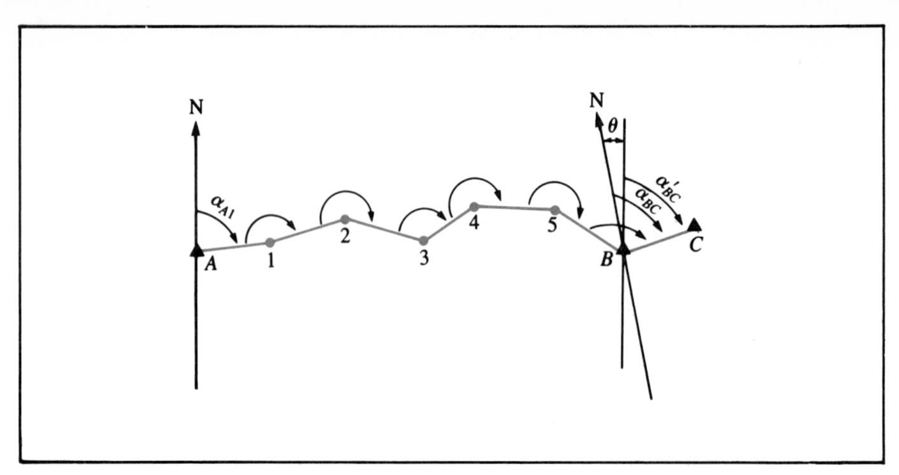

FIGURE 8.9 Convergence of true meridians

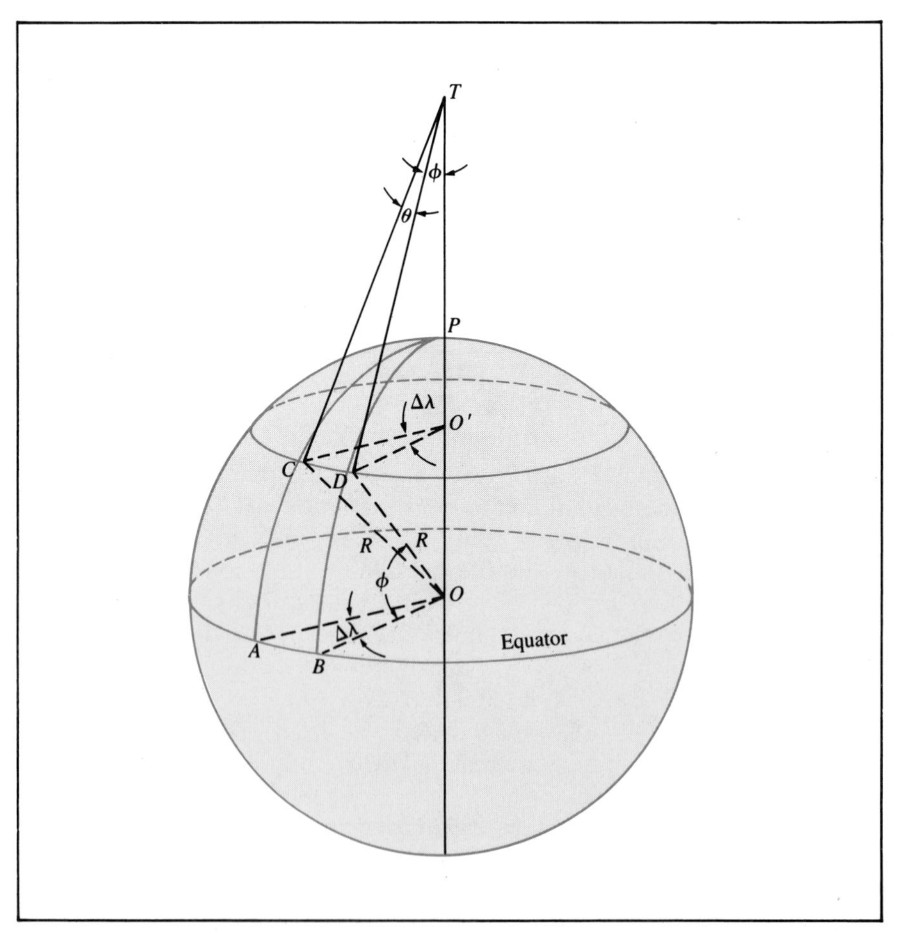

FIGURE 8.10 Longitude, latitude, and convergence of meridians

at C and D, which, if prolonged, meet the earth's polar axis at T. The angle $CO'D$ is the difference in longitude $\Delta\lambda$ of points C and D. The angle DTO equals the angle DOB or the latitude ϕ.

The difference in longitude between C and D is

$$\Delta\lambda = \frac{CD}{DO'} \text{ or } CD = DO' \cdot \Delta\lambda \tag{8.19}$$

Also,

$$\sin \phi = \frac{DO'}{DT} \text{ or } DT = \frac{DO'}{\sin \phi} \tag{8.20}$$

The angle of convergence is

$$\theta = \frac{CD}{DT} \tag{8.21}$$

Substituting Eqs. (8.19) and (8.20) into Eq. (8.21) yields

$$\theta = \Delta\lambda \cdot \sin \phi \tag{8.22}$$

where both θ and $\Delta\lambda$ are in radians. Let θ'' and $\Delta\lambda''$ denote the corresponding angles θ and $\Delta\lambda$ expressed in seconds. Then,

$$\theta'' = \Delta\lambda'' \cdot \sin \phi \tag{8.23}$$

Equation (8.23) is the basic expression for calculating convergence. Since points C and D for a typical convergence problem will not be in the same latitude, the term $\sin \phi$ is taken as the sine of the mean latitude of the two points.

Consider further that

$$\Delta\lambda = \frac{CD}{DO'} = \frac{L}{R \cos \phi} \tag{8.24}$$

where L is the departure between the two end stations expressed in miles and R is the mean radius of the earth expressed in miles.

Substituting Eq. (8.24) into Eq. (8.22) yields

$$\theta = \frac{L \sin \phi}{R \cos \phi} = \frac{L \tan \phi}{R}$$

where θ is in radians.

Letting R = 20,890,000 ft and changing radians to seconds, yields

$$\theta'' = 52.13\ L \tan \phi \tag{8.25}$$

The term tan ϕ is the tangent of the mean latitude of the two end points.

EXAMPLE 8.6

Calculate the angular convergence of the meridians between two points whose geographic coordinates are

$$\phi = 45°15'15''\ \text{N} \qquad \lambda = 75°13'28''\ \text{W}$$

$$\phi = 45°10'45''\ \text{N} \qquad \lambda = 75°10'12''\ \text{W}$$

SOLUTION

The mean latitude is 45°13′N; the difference in longitude is $\Delta\lambda = 3'16'' = 196''$.

$$\theta'' = \Delta\lambda'' \sin \phi$$

$$\theta'' = (196)(0.70978) = 139'' = 2'19''$$

EXAMPLE 8.7

In Figure 8.9, traverse station B is 8.35 miles east of station A. The azimuth (α'_{BC}) of line BC as computed from the traverse angles is 44°10′20″. The azimuth (α_{BC}) is known to be N 44°16′50″ E. The mean latitude of stations A and B is 36°43½′ N. Compute the azimuth closure error due to the angle measurements.

SOLUTION

From Figure 8.9,

$$\text{azimuth closure error} = \alpha'_{BC} + \theta - \alpha_{BC}$$

$$\theta'' = 52.13\ L \tan \phi$$

$$= 52.13 \times 8.35 \times 0.74606$$

$$= 325 \text{ sec}$$

$$\therefore\ \theta = 0°05'25''$$

$$\text{azimuth closure error} = 44°10'20'' + 0°05'25'' - 44°16'50''$$

$$= -01'05''$$

8.17 MONUMENTATIONS

The intended use and expected life of the traverse stations will govern whether they are to be permanently monumented or only temporarily marked. In small engineering projects and in the topographic mapping of small areas, the traverse stations are usually intended to be used for a relatively short time, such as a few weeks or months. Then temporary markers such as wood stakes may be used. They are usually 1 in. by 2 in. and at least 18 in. long, with one end sharpened. They are driven flush with the surface of the ground and have a tack on top to mark the exact position of the point. The term *hub* is commonly applied to such a stake. Semipermanent markers may be iron pipe or steel rod and heavy wood hubs that have been treated with a preservative to inhibit rot. Points on concrete surfaces may be defined by chisel marks, and those on bituminous surfaces by nails driven through small metal disks called shiners or pieces of colored flagging.

When a traverse station is expected to be used as position reference for a long period of time and/or its stability must be maintained to assure high positioning accuracy on its computed coordinates, then an appropriately constructed permanent station marker must be used. A 4-in. circular bronze tablet, appropriately inscribed and having a $3\frac{1}{2}$-in. shank, makes a very satisfactory marker when anchored in a concrete post, a masonry structure, or

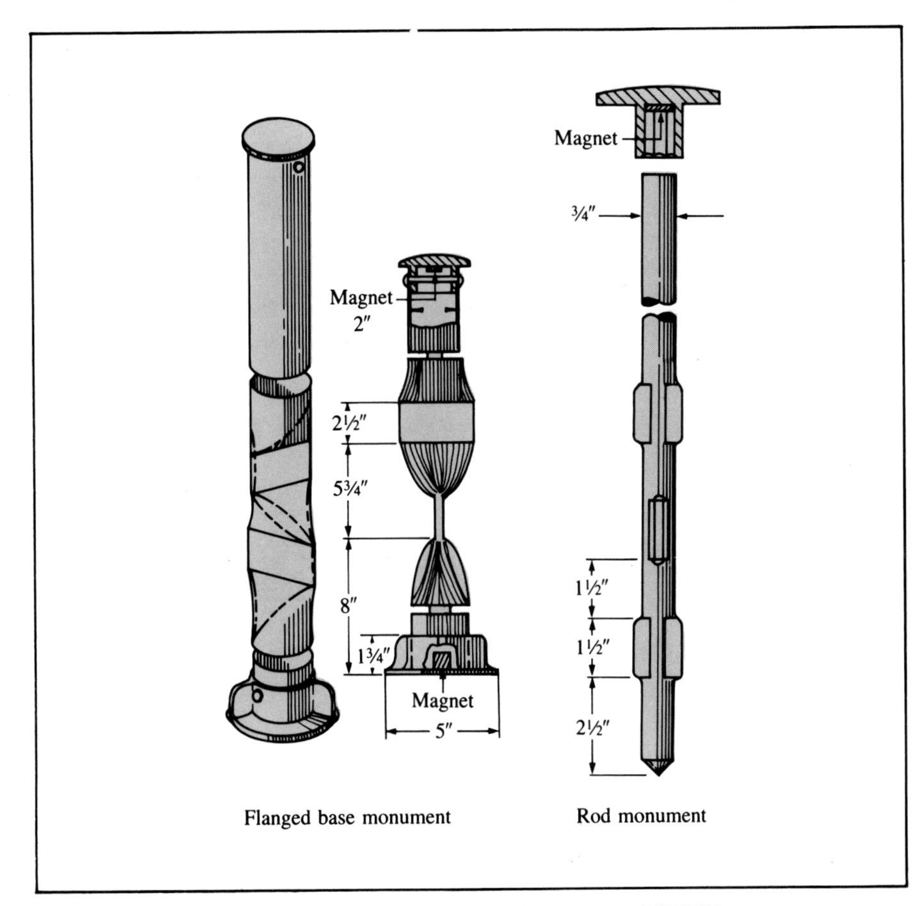

FIGURE 8.11 Survey monuments (courtesy of SURV-KAP/Allied Manufacturing Company)

solid rock in place. Survey monuments that are designed for easy installation and to meet a variety of requirements with respect to permanency and stability are commercially available. Figure 8.11 on page 317 shows the designs for two aluminum monuments. More detailed discussions on the construction of permanent survey markers will be presented in Section 11.10.

8.18 FIELD OPERATIONS

A traverse party may consist of as few as two persons and as many as eight to ten, depending on the types of instruments being used and on the field conditions. When transits and engineer's tapes are used, the field party is commonly divided into two groups. One is assigned to measure the angles while the other is assigned to measure the distances. When an EDM instrument is available for distance measurement, one person can be assigned to set up reflector targets, while another person operates both the EDM instrument and the theodolite and a third person serves as notekeeper. The accessibility of timber and the amount of timber cutting along a route are other factors influencing the size of the traverse party.

To promote speed and increase accuracy on traverse operations, special targets like those in Figure 5.39 are used. The target assembly includes a tribrach with three-screw leveling head, circular bubble, and optical plummet. This arrangement permits the targets and theodolite heads to be interchangeably mounted on the tripod. Thus only a single setup of the tripod is required at all stations. Provision is made for artificial illumination of the target.

8.19 FIELD COMMUNICATIONS

In all surveying operations the various members of the party must be able to communicate with each other. This is particularly necessary in transit traverse and other surveying activities that widely separate the personnel. Survey party members, particularly the instrument operator and the rod and tape holders, generally communicate by hand signals over distances beyond conversational limits or where noise seriously interferes with conversation. Prolonged shouting should be avoided, whether on campus or on the job. It identifies the beginner.

A few of the more common hand signals are as follows:

Give a foresight. When the instrument operator desires a foresight on a given point or a rod reading on a benchmark or turning point, a signal as in Figure 8.12a may be used.

Plumb pole or rod. The instrument operator will raise an arm straight above head and move it slowly in the desired direction.

FIGURE 8.12 Hand signals

Move right or left. When lining in a range pole, the operator may move one arm outward horizontally in the direction that change is desired. Slow movement means a large distance and quick movement a small distance.

All right. Both arms are extended sideward and waved up and down. This signal may be given by any member of the survey party.

Pickup, come in, or come ahead. The arms are extended sideward and downward and then raised quickly as in Figure 8.12b.

Where distances are too great for good visual identification or when the noise of construction activities and road traffic make direct voice call-outs ineffective, portable radio equipment may be used. It is a standard accessory on surveys involving the use of electronic distance-measuring devices. Two-way radio communication systems have been found very helpful in coordinating and accelerating surveying operations on extensive route engineering projects. They also provide the means for a central office to keep in constant touch with several field parties and to expedite the dispatch of personnel to a new assignment.

8.20 FIELD SAFETY

Attention to safety is an important consideration in the conduct of all surveying operations. At every level of responsibility from project engineer to rodholder there must be a clear recognition of the hazards of surveying, whether on a busy highway, in a transformer yard, on top of a triangulation tower, or in a remote area having dangerous terrain features. It is the primary responsibility of the chief of party to recognize the obscure as well as the more apparent hazards on a job site and to train the subordinates to be alert at all times, to avoid dangerous practices and situations filled with peril, and to know what to do in case of trouble. Each survey party should have a first-aid kit and some personnel trained in first-aid measures.

Safety is the concern of every member of a survey party. Every person should observe sensible safeguards in handling equipment, dealing with a climatically hostile environment, and working under hazardous conditions. This discussion cannot deal with such special situations as survival in the Arctic, helicopter transport to isolated mountain peaks, traveling over muskeg, or snakebite in the desert (see Figure 8.13). Only the more common hazards, especially those associated with engineering construction and along highways, will be mentioned.

Because of noise, dust, and movement of all kinds of equipment one must be especially vigilant around construction activities. Make it obvious to operators of cranes (see Figure 8.14), trucks, dozers, and other mobile equipment that you are working in the vicinity.

Gloves should be worn when working in poison oak and ivy or brushing line through briars. Everyone should know the symptoms of heat exhaustion and sunstroke, and some members of the party should be trained in the proper first-aid treatment.

Insects such as bees, wasps, and yellowjackets probably constitute a greater hazard than snakes. Some persons are highly allergic to their stings

FIGURE 8.13 Use of leggings in snake-infested areas (courtesy of Tennessee Valley Authority)

FIGURE 8.14 Construction survey (courtesy of Dravo Corporation)

and bites. If a member of a survey party is known to be sensitive to the attack of a particular insect, that individual should be advised to carry an antidote.

The ocular hazards associated with the use of some laser EDM equipments make it advisable that operating personnel be properly trained and that all recommended safeguards be observed.

All power lines should be regarded with great respect. Direct measurement of the height of a power line wire with an inverted rod, even one made from fiberglass, is extremely dangerous.

When working near others, carry range poles and rods vertically against the body so that another's head or eyes will not be struck if you suddenly turn. Avoid guiding a stake with your hand while another person is driving it with a sledge hammer. Permitting a tape to slide rapidly through the hands may cut them. Sheath cutting tools like the machete when not in use or put them in a safe place. When cutting brush (Figure 8.15), be at least 10 ft away from the nearest person. Use metal tree climbers to ascend trees and poles.

Protective headgear (see Figure 8.16) should be worn whenever there is danger of head injury from falling objects or impact with stationary features. When walking up steep and rocky slopes or out of excavations, do not climb directly behind another person. A spill or a rock knocked loose by the leading person could cause trouble for those following.

FIGURE 8.15 Line clearing (courtesy of California Division of Highways)

FIGURE 8.16 Surveying in sewer tunnel (courtesy of Chicago Metropolitan Sanitary District)

Do not enter manholes unless there is assurance that dangerous gases are not present. Provide a suitable barrier around the manhole if the cover has been removed. Do not throw any kind of tape across electric wires.

When working along a highway, use signs and flags to notify motorists of your presence, but do not depend on such markers to slow traffic. All party personnel must be continually alert and ready to leap off the road. The wearing of jackets or vests of distinctive colors, such as fluorescent red or orange, will increase visibility to motorists. Painting of tripods in contrasting color bands is another safety measure. In general, an effort should be made to select a time for the survey when the traffic is slack, and the party chief should minimize the time spent on the traveled roadway by working along offset lines on the shoulders or sidewalks. The safest solution in especially hazardous situations is the protection of a law enforcement officer.

PROBLEMS

8.1 A loop traverse originated and ended at point *A*. The bearing of line *AB* is fixed at S 42°16½′ W. Points *C*, *D*, and *E* lie westward from line *AB*. The following interior angles were measured with a 30″ engineer's transit:

Sta.	Angle
A	99°15′00″
B	111°42′30″
C	97°36′00″
D	102°14′00″
E	129°15′00″

Distribute the angular error of closure and calculate both bearings and azimuths of the remaining traverse sides.

8.2 The tabulated deflection angles in a loop traverse were measured with an optical transit. The azimuth of line *AB* is known to be 315°16′10″. Distribute the angular error of closure and calculate the azimuths of the remaining traverse sides.

Sta.	Angle
A	115°02′25″ R
B	92°52′15″ R
C	47°18′30″ L
D	121°47′05″ R
E	77°36′20″ R

8.3 The angles-to-right of an open traverse are as follows:

Sta. Occupied	from	to	Angle
A	*B*	1	60°32′
1	*A*	2	220°14′
2	1	3	178°36′
3	2	4	220°01′
4	3	5	155°10′

If the azimuth from station *A* to station *B* is 310°11′, calculate the azimuths of the remaining traverse courses.

8.4 The angles-to-right of a connecting traverse beginning at station *B* and closing at traverse station *W* were as follows:

Sta. Occupied	from	to	Angle
B	*A*	1	139°16′21″
1	*B*	2	182°11′12″
2	1	3	176°37′29″
3	2	4	162°48′39″
4	3	*W*	192°01′56″
W	4	*M*	92°14′03″

If the bearing from station *B* to station *A* is N 59°14′16″ W, calculate the azimuths of the traverse courses.

8.5 Perform the following calculations for the traverses shown in Figures P8.1 to P8.5:

a. Distribute the angular error of closure and compute the corrected azimuths of the courses.

b. Calculate the linear and relative errors of closure.

c. Balance the traverse by compass rule and compute the coordinates of the traverse stations.

d. Compute the distance, azimuth, and bearing of each course using the final coordinates.

8.6 An open traverse *ABCD* was executed in the cultivated area skirting a heavy forest, through which the line *AD* passes. There is need to mark out on the ground the line *AD* whose ends are not intervisible.

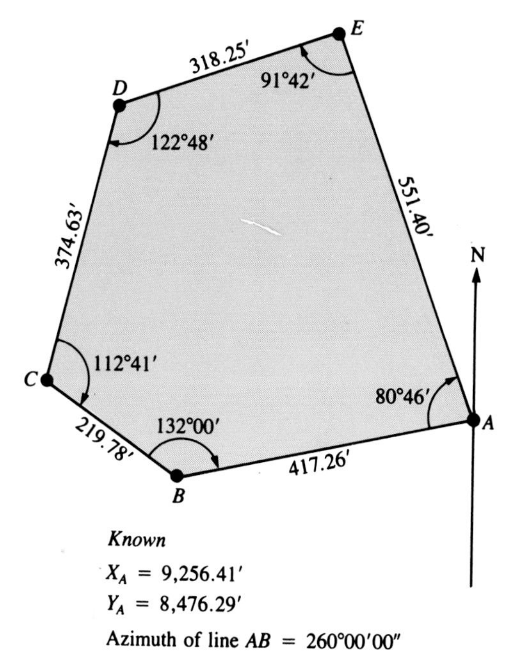

FIGURE P8.1 Loop traverse 1

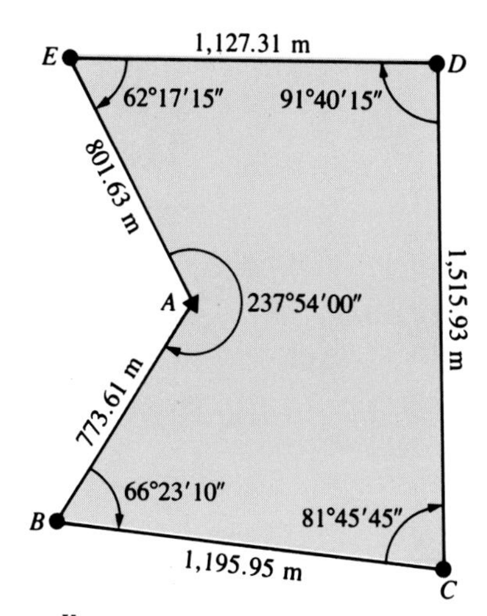

FIGURE P8.2 Loop traverse 2

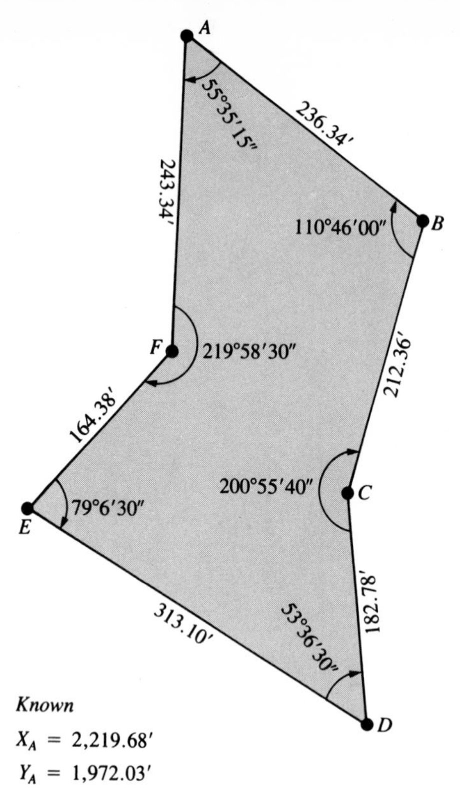

Known

X_A = 2,219.68′

Y_A = 1,972.03′

Azimuth of line AB = 126°30′10″

FIGURE P8.3 Loop traverse 3

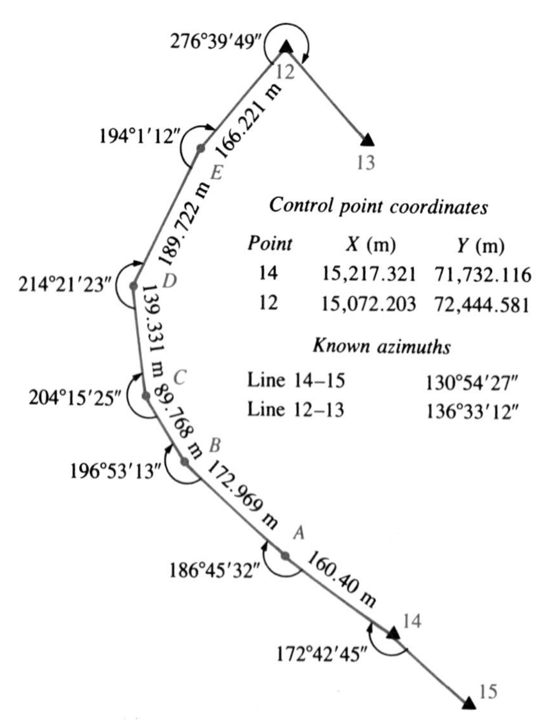

Control point coordinates

Point	*X* (m)	*Y* (m)
14	15,217.321	71,732.116
12	15,072.203	72,444.581

Known azimuths

Line 14–15	130°54′27″
Line 12–13	136°33′12″

FIGURE P8.4 Connecting traverse 1

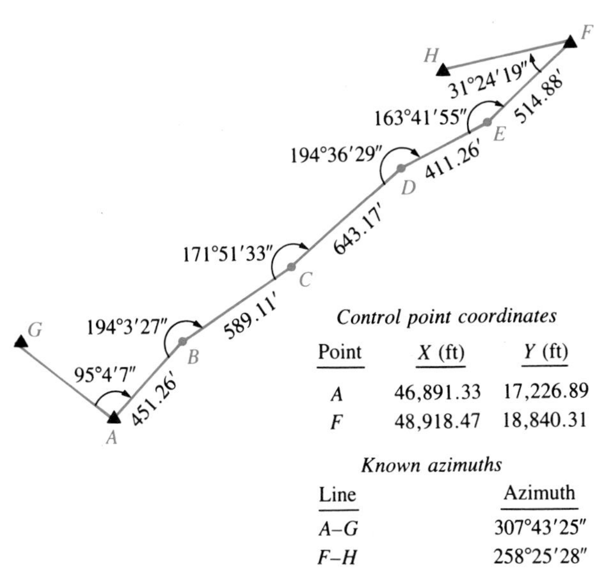

Control point coordinates

Point	X (ft)	Y (ft)
A	46,891.33	17,226.89
F	48,918.47	18,840.31

Known azimuths

Line	Azimuth
A–G	307°43′25″
F–H	258°25′28″

FIGURE P8.5 Connecting traverse 2

The results of the traverse are as follows:

Point	Distance (ft)	Deflection Angle
A		
	372.35	
B		60°20′ R
	781.15	
C		93°45′ R
	628.85	
D		

Calculate (a) the angle, to nearest 01′, at *A* needed to define the direction of *AD* with respect to *AB* and (b) the length of *AD* to the nearest 0.01 ft.

8.7 The four corners of *EFGH* of a small tract of land have been recovered and a traverse of the perimeter is desired. However, an old stone fence and several large trees make measurements along the boundaries very difficult. Therefore, an auxiliary traverse *ABCD* is conducted within the borders of the parcel, and distance and direction ties are obtained from the traverse stations to the property corners (see Figure P8.6).

a. Distribute the angular closure error and balance the traverse by compass rule. Compute the coordinates of the traverse stations *A*, *B*, *C*, and *D*. Point *D* is assumed to have the following coordinates:

$$X_D = 450.00 \text{ ft}$$
$$Y_D = 350.00 \text{ ft}$$

The azimuth of line *DA* is known to be 34°23′.

b. Compute the coordinates of the property corners *E*, *F*, *G*, and *H*.

c. Compute the length and bearing of each side of the parcel.

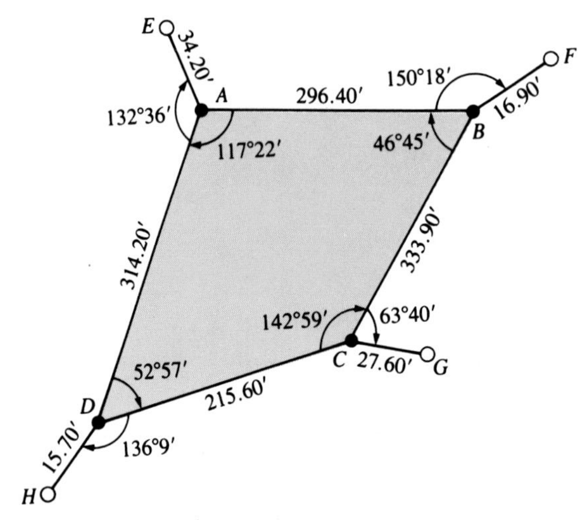

FIGURE P8.6 Boundary survey

8.8 The sides of a traverse are measured with an EDM instrument that can measure with an accuracy of ±(4 mm + 3 ppm) at 1 standard error. Determine the standard error with which the angles should be measured if the average length of the sides of the traverse is as follows:

a. 200 ft

b. 500 ft

c. 2,000 ft

8.9 Determine the required relative accuracy of distance measurement if the traverse angles are measured with a standard error of:

a. ±1 min

b. ±30 sec

c. ±15 sec

d. ±1 sec

8.10 Calculate the angular convergence of the true meridians between survey points *A* and *B* whose latitudes and longitudes are as follows:

a. $\phi_A = 39°11'15''$ N $\quad \lambda_A = 88°14'17''$ W
$\phi_B = 39°18'25''$ N $\quad \lambda_B = 88°19'35''$ W

b. $\phi_A = 31°15'30''$ N $\quad \lambda_A = 76°43'20''$ W
$\phi_B = 31°10'16''$ N $\quad \lambda_B = 76°51'31''$ W

c. $\phi_A = 45°23'16''$ N $\quad \lambda_A = 120°33'10''$ W
$\phi_B = 45°24'49''$ N $\quad \lambda_B = 120°21'43''$ W

8.11 Calculate the angular convergence of the true meridians passing through two points whose average latitude is 40° N if the departure (east-west) distance between the points is as follows:

a. 6.0 miles

b. 1.65 miles

c. 2,000 ft

d. 250 ft

8.12 In Problem 8.4, the true bearing from station *W* to station *M* is known to be N 13°55′12″ W. Find the closure in azimuth due to angle measurement error if:

a. Station *W* is 12.622 miles due east of station *B* in a latitude of 39°50′ N.

b. Station *W* is 5 miles due east of station *B* in a latitude of 45°23′ N.

c. Station *W* is 2 miles due east of station *B* in a latitude of 31°45′ N.

9 STADIA, PLANE TABLE, AND TOPOGRAPHIC MAPPING

9.1 INTRODUCTION

Cartography may be defined as the science and art of expressing graphically, by means of maps and charts, knowledge of the earth's surface and its varied features. *Cartographic surveying* is, therefore, that branch of surveying that includes all the operations leading to the production of topographic maps and nautical charts.

The predominant characteristic of a *topographic map* is that it portrays by some means the shape and elevation of the terrain. The most common way of representing land forms is by contour lines. In addition to depicting relief, topographic maps show drainage features, natural vegetation, transportation facilities, and cities, and towns, as well as any other features that may be of interest to the map users. A variety of symbols and specific colors are utilized to promote clarity of cartographic expression.

A *nautical chart* delineates the submarine topography or *bathymetry* of a given water area by means of depth curves and portrays other significant marine features, as well as those of the shore area. The term *nautical chart* is usually employed when large and navigable bodies of water are involved. Charts of rivers, harbors, reservoirs, and relatively small lakes are frequently known as *hydrographic maps*.

Before maps or charts can be drawn, the information to be shown on them must be obtained from a survey with such detail and accuracy as is appropriate to the scale on which they are to be published. These surveys are termed *topographic* and *hydrographic surveys*.

The topographic mapping of large areas such as counties, cities, dam sites, and proposed corridors for highways and rapid transit systems, as well as entire countries, is now most frequently accomplished by the use of aerial photographs. Yet for the mapping of small areas such as building sites, recreational parks, and industrial plants, the traditional methods of stadia and plane table are still commonly used. These methods require only simple instrumentation and can be used in practically any time of the year and in a wide variety of natural environments. Vegetation and trees present no major obstacle to these methods. In contrast, aerial survey methods require that the photography be obtained under cloudless sky and when the ground is free from snow. Vegetation and trees can seriously degrade the accuracy of mapping and in many instances render aerial mapping impossible. Thus, in spite of major advances in aerial mapping, the methods of stadia and plane table can be expected to continue to play an important role in surveying and mapping.

The methods of stadia and plane table will be presented in this chapter. In addition, the chapter also includes discussions on the use of contour lines for representing topographic relief, map scales, map drafting, and the national mapping programs in the United States. The principles of mapping from photographs, called photogrammetry, will be discussed in Chapter 14. Methods and problems of hydrographic surveys will be presented in Chapter 17.

9.2 STADIA METHOD

The *stadia* method is a simple and rapid method for determining the location of a new point with respect to a fixed station and a fixed reference line. Figure 9.1 illustrates this basic principle. Suppose that the location of point A and the direction of the line AC are fixed, and the position of point B is to be determined. A transit or a theodolite is set up over point A. This angle-measuring instrument must be equipped with a reticle that has three horizontal and one vertical cross-wires, as shown in Figure 9.2. The upper and lower horizontal cross-wires are called *stadia wires*.

A graduated rod is held over point B. The rod divisions are usually in units of feet and tenths of feet and arranged in various patterns, one of which is shown in Figure 9.3. Of the many patterns of rods that have been designed, probably none is better for general use than that shown here. For short sights, up to about 500 ft, the ordinary leveling rod may be used; beyond that distance the fine markings of a level rod become indistinct, and a pattern of larger divisions is necessary.

After backsighting to point C, the transit is turned and sighted on the rod held at point B. The horizontal angle, θ, and vertical angle, α, are read

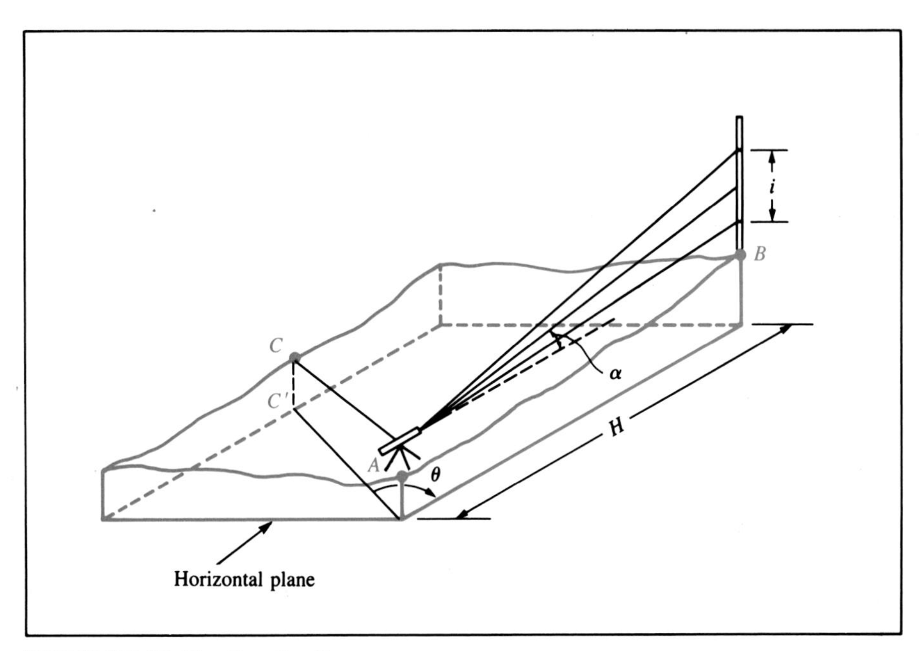

FIGURE 9.1 Stadia method

and recorded. The graduated rod is also read at all three horizontal cross-hairs. The difference between the upper and lower cross-wire readings is called the *interval,* or *intercept*. Its magnitude is a function of the distance the rod is from the transit. Thus the horizontal distance, H, can be computed from this interval. Under favorable conditions the error in the computed distance will not exceed 1/500. In addition, from the middle cross-wire reading, the difference in elevation between points B and A can be computed by the equations developed for trigonometric leveling. Thus the three-dimen-

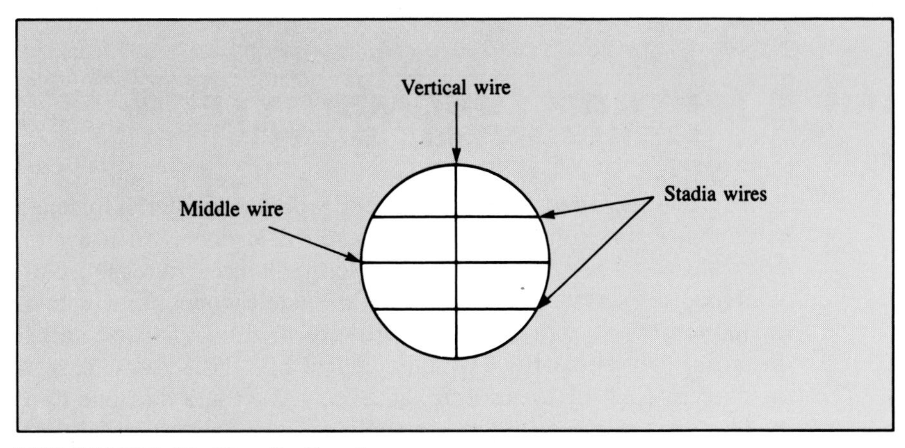

FIGURE 9.2 Stadia wires

FIGURE 9.3 Stadia rod (courtesy of Keuffel & Esser Company)

sional position of point B is determined. In practice, stadia calculations are simplified through the use of proper field procedures and computational aids such as stadia tables and slide rules.

9.3 STADIA GEOMETRY FOR HORIZONTAL SIGHT

The theory of the stadia method will be considered for the case of a horizontal sight with an *external-focusing* telescope. Such a telescope is easily identified because the objective lens will move in response to rotation of its focusing screw. Figure 9.4 illustrates a telescope, the plumb line, and a rod intercept r. Two rays of vision are shown, namely those emanating from the cross-wires and parallel with the optical axis. These rays are refracted by the objective lens and pass through a focal point at a distance F in front of the lens, and intersect the rod as shown.

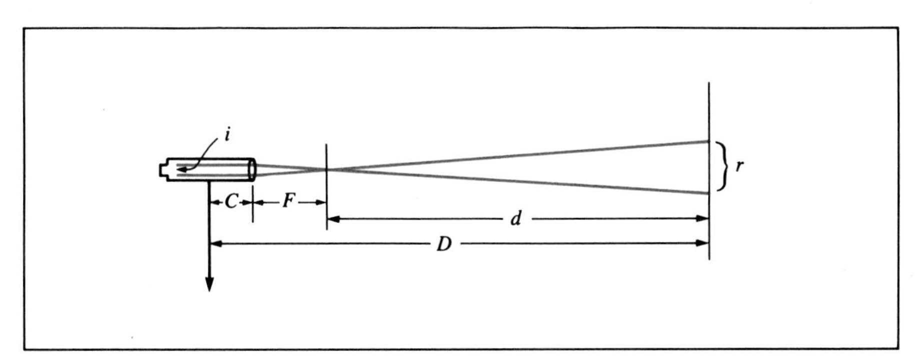

FIGURE 9.4 Stadia geometry

If the distance between the stadia cross-wires is represented by i, then in the two similar triangles we have the relation

$$\frac{d}{r} = \frac{F}{i}$$

from which $d = (F/i)r$, and from the figure, the total distance from the rod to the plumb line is

$$D = \left(\frac{F}{i}\right) r + (F + c) \tag{9.1}$$

The distance F is the focal length of the lens and is a constant. Also, the quantity $(F + c)$ is practically a constant, since the distance c varies only a small and negligible amount when the telescope is focused on different objects. The value of this constant varies somewhat for different instruments, but the range is between about 0.8 ft and 1.2 ft, or practically 1.0 ft. Since the uncertainties in determining distance by stadia, even for sights of moderate length, are at least 1 ft, inclusion of an allowance for $(F + c)$ is not critical. For most surveys this quantity can be safely neglected.

Most instruments of recent design have *interior focusing* telescopes, and for such telescopes the quantity $(F + c)$ is a constant and can be disregarded under all conditions.

Since the principal focal length F and the distance i between cross-wires are constants for any telescope, the ratio (F/i) also is a constant and is represented by k. Then, neglecting $(F + c)$, Eq. (9.1) may be simplified to the following form:

$$D = kr \tag{9.2}$$

The constant k is sometimes called the *stadia coefficient* or the *stadia multiplying factor* of the instrument. For convenience, most instruments are designed so that $k = 100$. It is of no importance to know the value of either F or i separately.

9.4 STADIA GEOMETRY FOR INCLINED SIGHT

On sloping ground the difference in elevation and the horizontal distance between two points can be found by the stadia method if, in addition to the rod interval, the angle of inclination of the line of sight is read on the vertical circle.

Refer to Figure 9.5 and suppose that the transit is in position at station A and that the rod is held vertically at station B, that the rod intercept is $mn = r$, and that the inclination of the line of sight as measured on the vertical arc is α; also that $m'n' = r'$ is the intercept that would be read on the rod if it were held perpendicular to the line of sight.

Since the imaginary line $m'n'$ is perpendicular to the line of sight IO, angle $Om'm = On'n$ is not exactly a right angle. However, in the following demonstration it will be so considered, and the error thus introduced is negligible. With this approximation understood, the following equations may be written:

$$m'n' = mn \cos \alpha, \text{ or } r' = r \cos \alpha \tag{9.3}$$

$$D = kr' + (F + c) \tag{9.4}$$

Substituting Eq. (9.3) into Eq. (9.4) yields

$$D = kr \cos \alpha + (F + c) \tag{9.5}$$

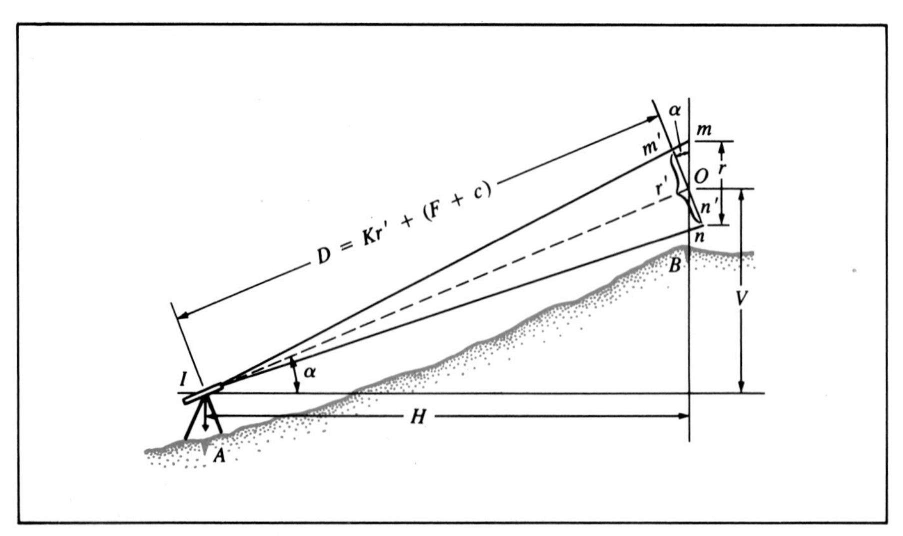

FIGURE 9.5 Stadia principle for inclined sight

The difference in elevation between I and O is given by the vertical projection V of the slope distance D. Hence,

$$V = D \sin \alpha \text{ or } V = kr \sin \alpha \cos \alpha + (F + c) \sin \alpha$$

from which

$$V = kr \tfrac{1}{2} \sin 2\alpha + (F + c) \sin \alpha \tag{9.6}$$

Also, the horizontal distance H is equal to the horizontal projection of the slope distance D, and

$$\begin{aligned} H &= D \cos \alpha \\ &= kr \cos^2 \alpha + (F + c) \cos \alpha \end{aligned} \tag{9.7}$$

From the figure it is evident that the difference in elevation (Δh) between A and B is:

$$\Delta h = V + IA - OB$$

If distance OB is taken equal to IA, then $\Delta h = V$. This is easily accomplished by always setting the middle cross-hair on the rod graduation equal to the height of the transit (HI). Thus, Eqs. (9.6) and (9.7) permit the determination of the difference in elevation and the horizontal distance between the two end points.

For reasons stated above and under usual conditions, sufficiently satisfactory results will be secured if the constant $(F + c)$ is disregarded. Hence, Eqs. (9.6) and (9.7) become

$$V = \tfrac{1}{2}kr \sin 2\alpha \tag{9.8}$$

and

$$H = kr \cos^2 \alpha \tag{9.9}$$

9.5 STADIA REDUCTION TABLES

In Eqs. (9.8) and (9.9), let

$$V_t = \tfrac{1}{2}k \sin 2\alpha \tag{9.10}$$

and

$$H_t = k \cos^2 \alpha \tag{9.11}$$

then

$$V = r \cdot V_t \tag{9.12}$$

and

$$H = r \cdot H_t \tag{9.13}$$

Stadia calculations can be simplified by using stadia reduction tables, which list the values of V_t and H_t for various values of α and k. Table VII in Appendix B lists the values of these parameters for $k = 100$.

EXAMPLE 9.1

Given: $r = 3.17$, $k = 100$, and $\alpha = 4°20'$; find H and V.

SOLUTION

From Table VII,

$$H_t = 99.43 \qquad V_t = 7.53$$

Therefore,

$$H = 3.17 \times 99.43 = 315 \text{ ft}$$
$$V = 3.17 \times 7.53 = 23.9 \text{ ft}$$

EXAMPLE 9.2

Given: $r = 6.37$, $k = 101.2$, and $\alpha = 3°40'$; find H and V.

SOLUTION

From Table VII, for $k = 100$,

$$H_t = 99.59 \qquad V_t = 6.38$$

Hence,

$$H = 6.37 \times 99.59 \times \frac{101.2}{100} = 642 \text{ ft}$$

$$V = 6.37 \times 6.38 \times \frac{101.2}{100} = 41.1 \text{ ft}$$

The uncertainties in stadia work make it both illogical and misleading to express horizontal distances and vertical heights to closer than the nearest full foot and one-tenth foot, respectively.

To avoid the use of tables and further simplify the reductions of stadia notes, special slide rules have been designed. These need not be described in detail here, but they are most convenient and should be available if many transit-stadia notes are to be reduced. This remark applies in particular to plane table surveys where the reductions of notes are made in the field.

9.6 ERRORS IN STADIA

Some of the errors associated with stadia surveying are common to various basic operations with the transit, such as measuring vertical angles. The stadia determination of horizontal distances and vertical heights is, however, affected by other errors such as the following:

READING THE ROD

The inability of the instrument operator to estimate exactly the rod intercept is a source of an accidental error. Its magnitude varies with the weather conditions and the length of sight; for average conditions the average error will be, perhaps, for distances up to 300 ft, ±0.01 ft, and ±0.01 to ±0.03 ft for distances between 300 and 800 ft. Beyond that distance the error is subject to large variations. This error is minimized by limiting the length of sight.

ROD NOT PLUMB

If the rod is not plumb, the intercept is too large and the corresponding distance is too great. This is a systematic error.

THE CONSTANT *k*

Since all rod intercepts are multiplied by the constant $F/i = k$, any error in the determination of this constant is a systematic error. The resulting error in measured distances is directly proportional to the error in the value of k. Thus an error of 1% in k will introduce an error of 1% in all measured distances.

The value of k is usually provided by the manufacturer for each transit or theodolite. It can also be easily determined in the field by setting stakes at distances of about 100, 300, 500, and 700 ft from the instrument. The ground should be nearly level so that horizontal line of sight can be used. With the stadia rod held over each stake successively, the stadia intercepts are read to 0.01 ft. From these readings four values of k can be determined from Eq. (9.2). The mean and estimated standard error of the mean are then computed. If the transit has an external focusing telescope, Eq. (9.1) can be used to calculate k with $(F + c) = 1$.

LENGTH OF ROD

It is obvious that any error in the length of a stadia rod is multiplied by k, or about 100, in the horizontal distance. Thus, an error of 0.1 ft in a 12-ft rod will introduce an error of $0.1 \times 100 = 10$ ft in a distance of 1,200 ft. It is therefore important that the lengths of all stadia rods in use should be checked occasionally with a steel tape.

PARALLAX

Obviously, parallax, if present, is an important source of error in stadia work. It is prevented by keeping the eyepiece accurately focused on the cross-wires at all times.

NATURAL ERRORS

Such natural errors as wind, differential refraction, moisture, and temperature changes affect stadia measurements more or less depending on weather conditions. The two latter effects are usually not important. The wind is frequently troublesome, and good results cannot be expected when a high wind is blowing.

Differential refraction is that effect whereby the line of sight as fixed by one cross-wire is refracted more than the line of sight fixed by the other. This condition is caused by the difference in temperature and density of the air strata just above the ground. It is evidenced occasionally by so-called heat waves, but is present in lesser amounts at all times. It may be minimized by keeping the line of sight fixed by the lower cross-wire well above the ground; by reducing the lengths of sights when the heat waves are apparent; or, if practicable, by avoiding stadia work when this condition is serious.

9.7 TOPOGRAPHIC MAPPING BY STADIA

Topographic mapping by the stadia method requires initially the establishment of a network of strategically positioned points over the area to be mapped. These points serve as horizontal and vertical control points and constitute the framework to which the location of all cultural, drainage, and topographic details is tied. Without such control points, the topographic survey would lack coordination and its separate parts would not fit together.

Horizontal control is usually provided by traversing, as shown in Figure 8.3. Distances along the traverse are usually measured with steel tapes or EDM instruments. Elevations of the traverse stations are determined by differential leveling. For low accuracy mapping, the method of stadia may be used to measure the distances and elevations along the traverse.

For mapping a small area, a single base line consisting of two control points sometimes provides the needed control framework. Stadia measurements are made from the two end stations of the fixed base line. The length of the base line can be measured by an engineer's tape, and the direction of the base line can either be arbitrarily fixed or approximately determined by the use of a magnetic compass.

The field procedure for locating topographic details is as follows:

1. The transit or theodolite is set up over a control point. A backsight is made to an adjacent station with the horizontal circle set to read 0°00′ or the known azimuth of the line. The height of the instrument (*HI*) above the control point is measured with a tape.

2. The rodholder selects a topographic point to be located and holds the stadia rod plumb at that point.

3. The transit is next sighted on the rod using the upper motion. The bottom cross-wire is set on a full footmark, and the stadia interval is read by noting where the upper cross-wire cuts the rod. If possible, the instrument operator then sets the middle cross-wire at a point on the rod that has the same height above the ground as the *HI* of the instrument. The operator then motions the rodholder to find another point.

4. While the rodholder is finding another point, the instrument operator reads the vertical and horizontal angle of the point observed.

5. The recorder keeps the notes and, if time permits, computes the elevations of and horizontal distances to all points. Sketches are often better than written descriptions of the topographic points, and should be used freely.

A typical set of field notes is shown in Figure 9.6. The instrument is assumed to be in position at station C, whose position has been located by traverse and whose elevation is 633.3. The transit is oriented by a backsight on station B, the azimuth being 5°35′. The *HI* is found to be 4.4 ft, and $k = 100.00$.

A reading is then taken on object 1, which is a point on the property line fence. The bottom cross-wire is set on a full footmark and the intercept is read at the upper cross-wire. The middle cross-wire is then set to read 4.4 on the rod, and the instrument operator then motions the rodholder forward. Since the azimuths of side details are subsequently plotted with protractors commonly subdivided into 30′ spaces, such side azimuths are read to the nearest 5′ only and without the assistance of the vernier.

At point 2 a horizontal sight is taken. The foresight reading of 3.9 is enclosed in parentheses to indicate this. Since the *HI* is 4.4, the difference in elevation is +0.5 ft.

At point 5, it was impossible, for some reason, to sight the rod at 4.4, and a reading was taken at 8.0 ft. Accordingly, a correction of 3.6 ft is applied to the calculated difference in elevation.

Before leaving the station a check reading is taken on B and found to be 5°37′. This assures the transit operator that the instrument has not been disturbed while observations were being taken.

Locating details Lake Como Reservoir Inst. at C. Elev. 633.3 HI = 4.4						Gurley Transit No. 7 July 12, 1984	G.B. Righter, ⊼ 43 H.R. Linder, Notes Dan Cohen, Rod
Object	Az	Rod	Vert. angle	Hor. dist.	Diff. in elev.	Elev.	
B	5°35'						
1	208°40'	2.22	-2°21'	222	-9.1	624.2	
2	258°30'	1.10	(3.9)	110	+0.5	633.8	
3	333°40'	2.43	-1°14'	243	-5.2	628.1	
4	347°15'	4.98	-2°23'	497	-20.7	612.6	
5	8°25'	2.32	(on 8.0) -3°05'	431	-26.8	606.5	
6	14°15'	3.63	3°01'	362	-19.1	614.2	
			etc.				
B	5°37'	check					

FIGURE 9.6 Field notes for stadia surveying

The reductions of the notes, shown in the last three columns, are usually made in the office or drafting room.

9.8 ELECTRONIC TOTAL STATIONS

The basic procedure of stadia surveying can also be adopted for use with electronic total stations (see Section 6.2). Instead of a stadia rod, a reflector prism is used as sight target. Once the telescope of the total station is centered on the reflector, a direct readout of the horizontal distance, height difference, and azimuth angle can be obtained. In some instruments these measured data can be automatically recorded by a data collector and later transferred to a computer for processing. The field book is used primarily to record backup information including sketches, station locations, and weather conditions.

9.9 PLANE TABLE

Plane table surveys are operations combining the stadia method and field drafting on a portable drawing board mounted on a tripod. The plane table instruments (see Figure 9.7) include an alidade, a board, and a tripod.

FIGURE 9.7 Plane table instrument (courtesy of U.S. Geological Survey)

The alidade consists of a telescope mounted on a column that is fixed to a steel straightedge about 2½ in. × 15 in. The telescope is similar in all respects to the transit telescope and is supported by a horizontal axis resting on short standards above the supporting column. By this arrangement the telescope may be rotated in a plane perpendicular to the horizontal axis through a given arc. A vertical arc, vernier, clamp, and tangent screw are provided, as well as a striding, or an attached level, which permits the instrument to be used for direct leveling. A compass needle is usually provided, housed in a small metal box on the straightedge blade. Placed on the blade, a small bubble tube having a spherical surface is used to level the table. A bubble tube mounted on the vernier frame facilitates the measurement of vertical angles.

The drawing board is a plane board available in different sizes for different field conditions. The most common size is 24 in. × 30 in. A flat metal disc is let into the underside of the board by which it is screwed onto a threaded bolt of the tripod, thus securing the board to the tripod.

The tripod is provided with a special device by which the board is first leveled and then rotated, thus orienting the instrument for field use.

Figure 9.8 shows an optical-reading, self-indexing plane table alidade. A pendulum device automatically corrects for slight tilts of the plane table board.

SETTING UP AND ORIENTING

In plane table surveys, the locations of lines and points are plotted directly upon the drawing paper and, accordingly, a pencil dot on the paper

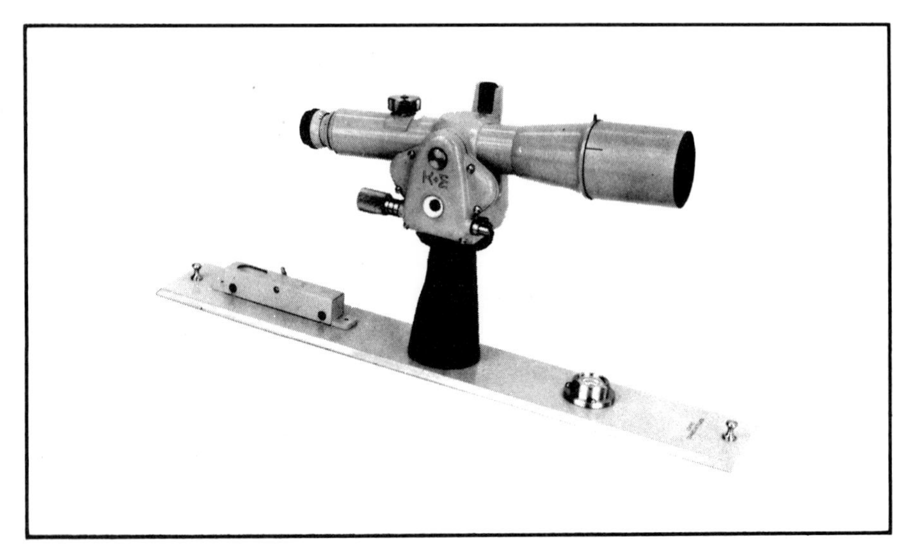

FIGURE 9.8 Optical-reading plane table alidade (courtesy of Keuffel & Esser Company)

will represent a small or a large dimension on the ground according to the scale to which the map is drawn. For example, a fine pencil-point dot may be 0.01 in. in diameter. If the scale of the drawing is 1 in. = 100 ft, evidently such a dot represents an area 1 ft in diameter on the ground. Thus it is evident that in centering the drawing board over a ground station, it is not necessary to use the care that is required in setting up a transit. Usually the table is centered over the ground point by estimation only. However, on careful work and if the scale of the map is large (that is, 1 in. = 20 ft), a plumb bob may be suspended from the underside of the table to center a plotted point over the corresponding ground point.

In setting up the table, it is first turned into its proper relation to the field being surveyed and placed over the ground station as indicated above. The tripod legs are firmly planted in the ground and spread at such an angle as to bring the drawing board to a comfortable height. This height should be such that the observer can work easily on all parts of the drawing without leaning against the board. The table is then leveled by whatever devices are provided for that purpose.

As the transit must be oriented by means of a backsight, so the plane table must be properly oriented before any mapping is done. In small-scale mapping and rough surveys, this may be done by use of the compass needle, but on most work a backsight is used, as is the case with the transit.

If the compass needle is used, a line representing the magnetic meridian is drawn on the plane table sheet at the initial station. The alidade straightedge is then placed along this reference meridian and the table is turned in azimuth until the needle indicates north. The table is then clamped in this position and mapping proceeds. At the next instrument station, after the table is set up and leveled, the board is again oriented as explained above.

If a backsight is used to orient the table, the procedure is as follows: Suppose the instrument is at station *B* and is to be oriented by a backsight to station *A* (Figure 9.9). After the table is set up and leveled, the alidade is placed along line *ba* and the table is turned in azimuth until the observer, looking through the telescope, sights station *A*. The board is then clamped and thus properly oriented.

TRAVERSING

It has been shown that in transit operations the location of instrument stations is effected by the use of traversing or triangulation; likewise, in plane table surveys the instrument stations may be located by traversing or by *graphical triangulation*. Also, the plane table makes convenient use of two other principles of locating points—namely, *intersection* and *resection*.

The procedure of traversing with the plane table is simple and is illustrated by Figure 9.9. Having chosen a suitable scale for the map, a point (*a*) is chosen arbitrarily to represent the initial ground point *A* of the traverse. The table is then set up and oriented so that the sheet has a proper relation to the field. The straightedge of the alidade is then pivoted about

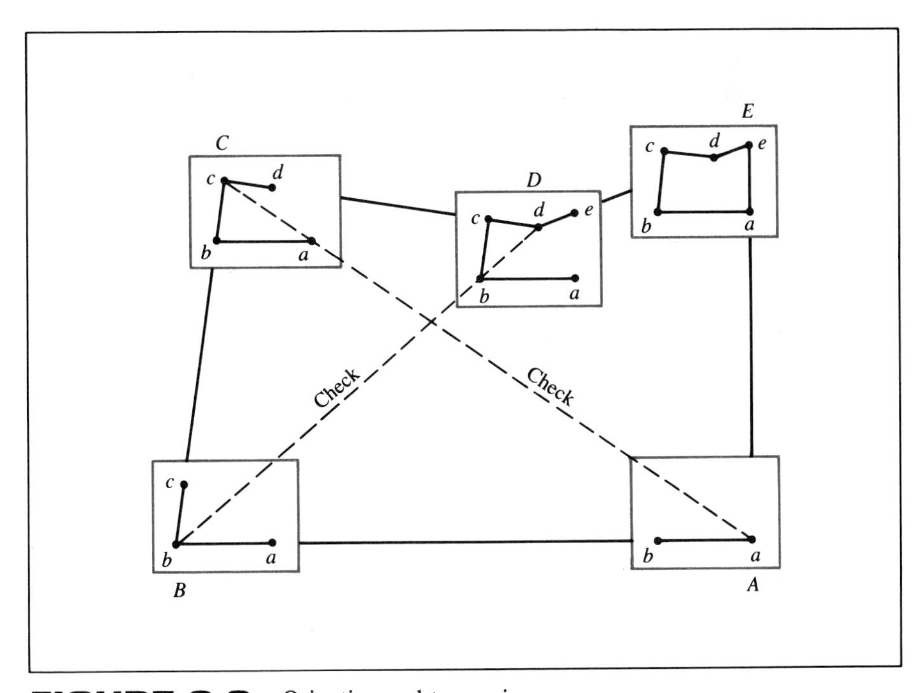

FIGURE 9.9 Orienting and traversing

point *a* on the map and a sight is taken toward the forward station *B*, and a line is drawn along the straightedge in that direction. The distance *AB* is measured by stadia and the corresponding map distance is plotted to locate point *b*.

The table is then moved to station *B*, set up, and oriented by a backsight on *A*. A foresight is then taken on *C* by pivoting the alidade about *b*, sighting *C*, and drawing a line on the map in that direction. The distance *BC* is then measured, and point *c* plotted on the map. In this manner the traverse proceeds, and if the field is a closed one, the final position of *a* should fall on the initial position. The error of closure is, therefore, at once apparent as soon as the last observation is made.

INTERSECTION

The principle of intersection is a simple application of the condition that the position of a point is fixed by lines of direction drawn toward it from any two other known points. The method is especially convenient in plane table work. Thus, in Figure 9.10 let *A, B, C* represent instrument stations, located by the traverse *ABC*. At station *A* direction lines (*rays*) are drawn from *a* to two distant objects *E* (a fence corner) and *D* (a tree). At station *B*

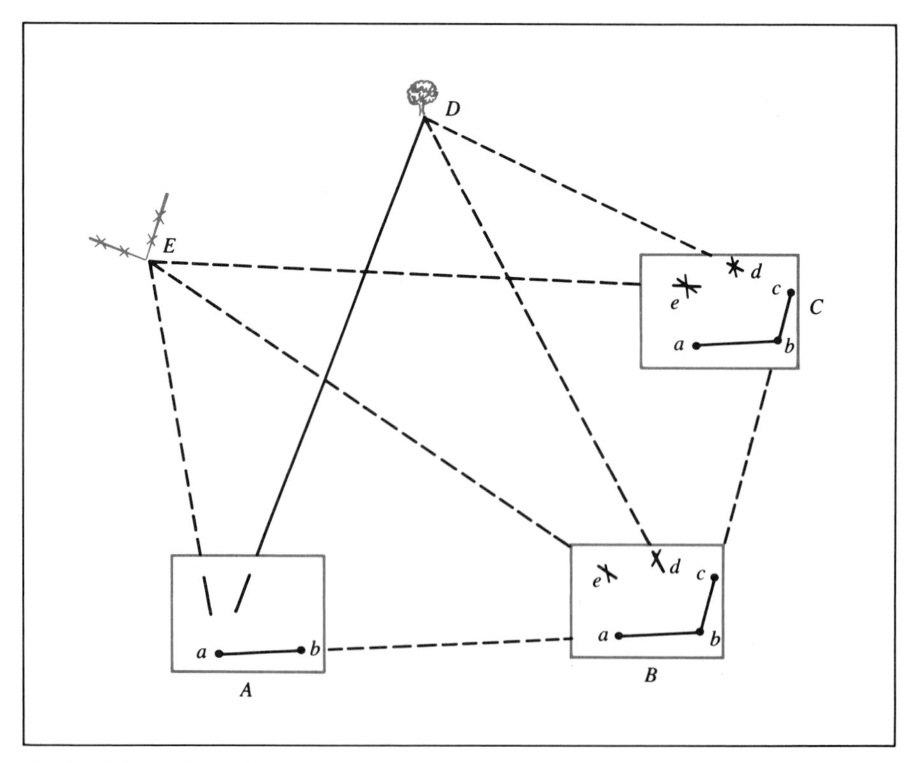

FIGURE 9.10 Intersection

rays are again drawn from b toward E and D. These rays intersect those previously drawn from station A, and these intersections fix the positions e and d, being the plotted positions of E and D. A third station C is occupied and rays again drawn from c toward E and D; if these rays pass through the points previously located, the plotted positions e and d are proved or checked.

It will be noticed that these points are located without the necessity of any distance measurements or of a rod being held there. Obviously, any number of details like E or D, which are visible from two or more stations on the traverse ABC, can be located in this manner.

RESECTION

The principle of intersection as described in the preceding article is useful in locating a distant point from known plane table stations. The principle of resection is opposite to that of intersection; that is, it is used to locate the plane table station from known distant points.

In mapping details it is frequently desirable that the instrument should occupy an advantageous point that has not previously been located and, therefore, is not plotted on the map. The plane table offers a number of solutions for this problem, but it will be sufficient here to describe a simple one only.

Referring to Figure 9.11, suppose that it is desired to locate the position of C from sights on two points that have previously been located or plotted on the map. Also, suppose that A and B are two adjacent stations or objects whose positions have been located and plotted at a and b, respectively.

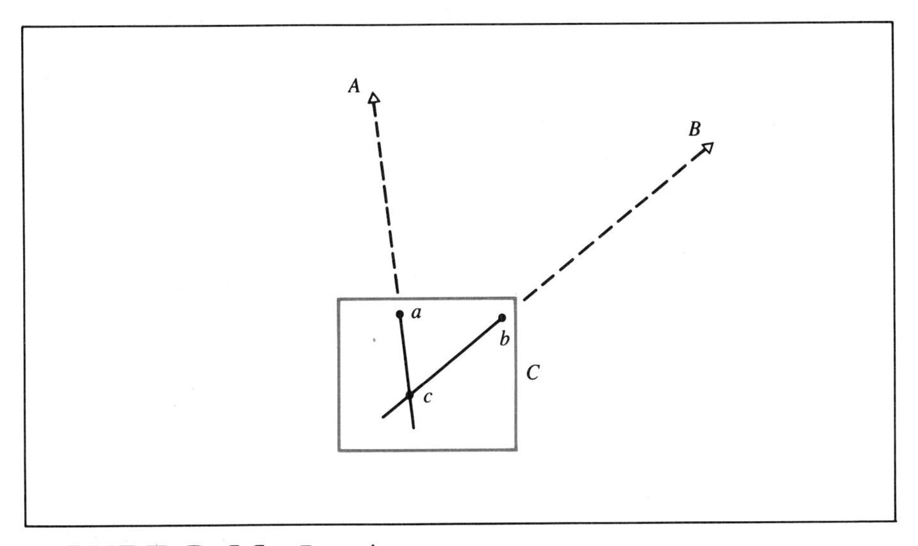

FIGURE 9.11 Resection

The procedure of locating point *c* on the map and orienting the board is as follows: The plane table is set up at station *B* and a ray is drawn toward station *C*, but no distance is measured. Next, the instrument is set up at *C* and oriented by a backsight along ray *cb*, toward station *B*. Then the alidade is pivoted about point *a* on the map until object *A* is sighted in the field, then ray *ac* is drawn to resect ray *bc* at point *c*. This fixes the position of *c* on the map as being the located position of station *C* over which the instrument is set up. This procedure is called location by *resection*.

LOCATING DETAILS

From the previous description of the plane table instrument and methods, it is a simple matter to describe its use in locating details. The instrument is set up at a given station and oriented as has been explained. The stadia method is universally used in this work, although on maps to a large scale, certain definite and important objects such as lot corners, water hydrants, and sewer manholes may be located by taped measurements.

With the rod held vertical at a given object to be mapped, the instrument operator pivots the alidade about the plotted position of the station occupied, sights on the rod, and draws a short portion of a line toward it (see Figure 9.12). Then the alidade is moved to a convenient position on the table and sighted on the rod again. The stadia intercept, the middle wire reading, and the vertical angle are then read. The horizontal distance to and the elevation of the point are then computed. The correct scaled distance is plotted to locate the point on the map, and its elevation is written beside it.

FIGURE 9.12 Mapping with plane table (courtesy of U.S. Geological Survey)

Meanwhile, the rodholder chooses another point, and the work thus proceeds.

As the detail points are plotted, the operator draws the map, sketches the contours, and by suitable symbols represents the features of the area to be mapped. The map is drawn, in all its essential details, in the field while the terrain is in the view of the observer.

9.10 REPRESENTATION OF RELIEF

CONTOURS

The most distinctive characteristic of a topographic map that differentiates it from other maps is that it portrays the configuration of the land surface. Various devices have been employed to express topography, but the most important one is the contour.

A *contour* is a line connecting points having the same elevation. It may be considered as the trace of the intersection of a level surface with the ground. The shoreline of a body of still water is an excellent illustration of a contour.

Contour interval is the vertical distance, or difference in elevation, between two adjacent contours. The contour interval in Figure 9.13 is 10 ft.

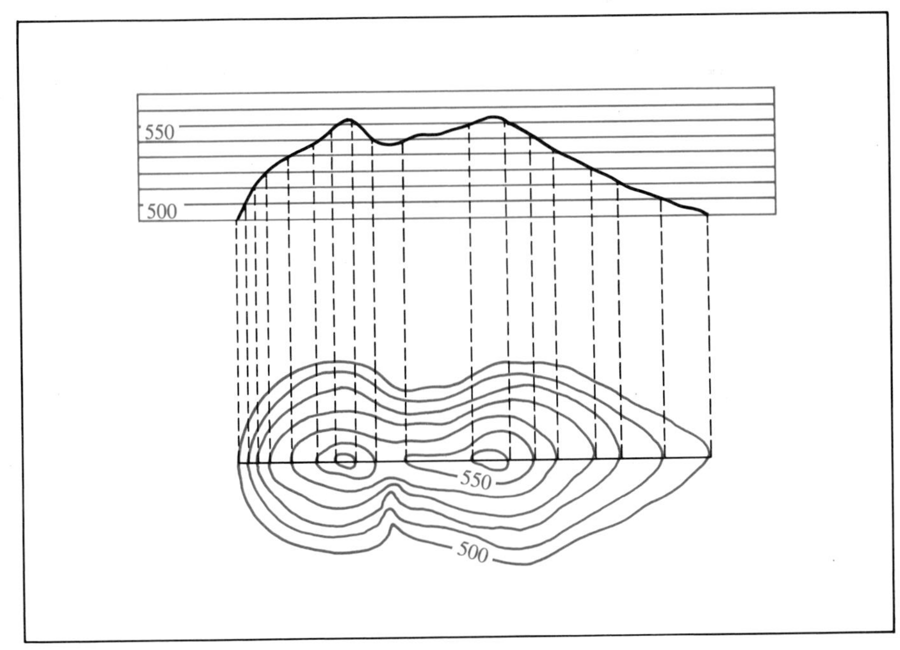

FIGURE 9.13 Contours and ground profile

The contour was probably first introduced in connection with sea soundings by the Dutch surveyor Cruquius in 1729. Its use as applied to terrain representation was initially suggested by Laplace in 1816, and it is considered today to be superior to all other topographic symbols for engineering purposes.

The basic datum for the vertical location by contours or depth curves of ground or marine features on maps and charts, respectively, is furnished by tidal planes of reference. For topographic maps, the most satisfactory plane is mean sea level. Hence, contours express height above (with rare exceptions) that datum.

For the soundings and depth curves of a nautical chart, a reference surface must be employed that incorporates a measure of safety to the principal user, the mariner. Consequently, a low water plane is more satisfactory because the chart then will show the least depth to be expected regardless of cyclic variations in the water surface elevation. Hence, fathom or bathymetric curves express depth of water below the adopted datum.

The principal characteristics are as follows:

1. Contours spaced closely together represent a steep slope. When spaced far apart, they indicate a flat slope (see Figure 9.13).

2. If the terrain is rough and uneven, the contours will be irregular. If the ground surface is even, as on earthwork slopes, the contours will be uniformly spaced and parallel (see Figure 10.10).

3. Contours that portray summits or depressions are closed lines. Usually the identification of adjacent contours or the presence of a pond or lake will indicate whether the feature is a summit or a depression. To dispel confusion a *depression contour* may be used. This is a closed contour line with short ticks inside it. It can be concluded that all contours are closed lines, either within or outside the borders of a map.

4. Contours do not cross each other nor do they merge.

5. Contours are perpendicular to the direction of the maximum slope.

6. Contours cross a watershed or ridge line at right angles. The concave side of the curve is toward the higher ground.

7. In valleys and ravines the contours run up the valley on one side, cross the stream at right angles, and run downward on the other side. The curving portion of the contour as it crosses the valley is convex toward the higher ground.

Figure 9.14 shows both a perspective view and the corresponding topographic map of the same terrain. The principal features include a river lying between two hills and emptying into a bay of the sea. Nearly all the characteristics of contour lines listed above are shown in this map, which is drawn to a relatively small scale with a contour interval of 20 ft.

OTHER RELIEF SYMBOLS

Several other devices are used to indicate relief. They include hachures, shading, color tints, and form lines. The most dramatic expression of topography is provided by the terrain model shown in Figure 9.15.

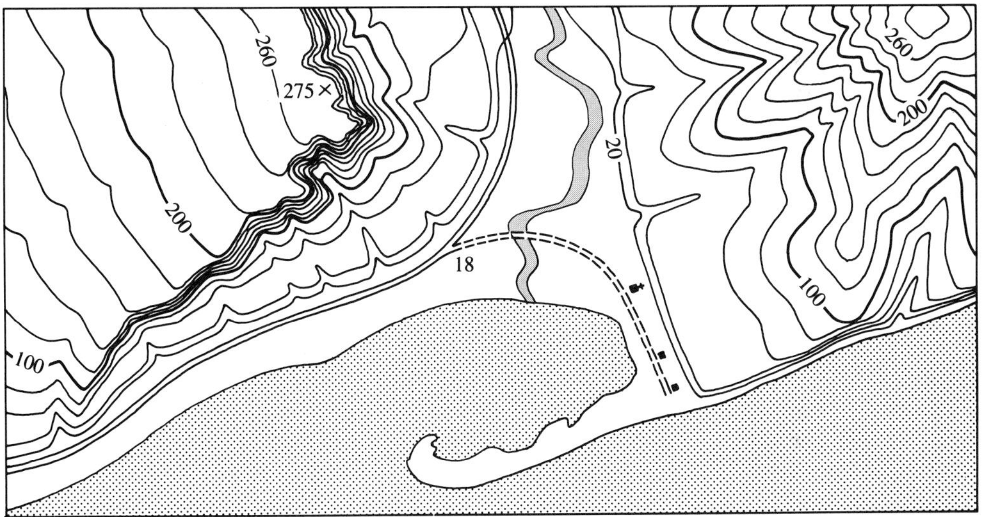

FIGURE 9.14 Perspective view and topographic map (courtesy of U.S. Geological Survey)

Hachures are short lines drawn in the direction of the ground slope. They are not often used on modern topographic maps except in those cases where the scale of the map is too small to permit successful contouring of such features as gravel or borrow pits, mine dumps, and highway or railroad cuts and fills. If drawn properly, hachures convey a good conception of the terrain but their value is largely pictorial. They are of little use if elevations are to be scaled from a map.

Shading is accomplished by the proper placement on the map of different shades of gray tints. The map can be regarded as a picture of a relief model illuminated by a light source directly overhead or from the northwest. If vertical illumination is assumed, less light will fall upon the slopes than on level land. Therefore, the effect is similar to hachuring because the steeper slopes will appear darker on the map. If oblique illumination is assumed, the

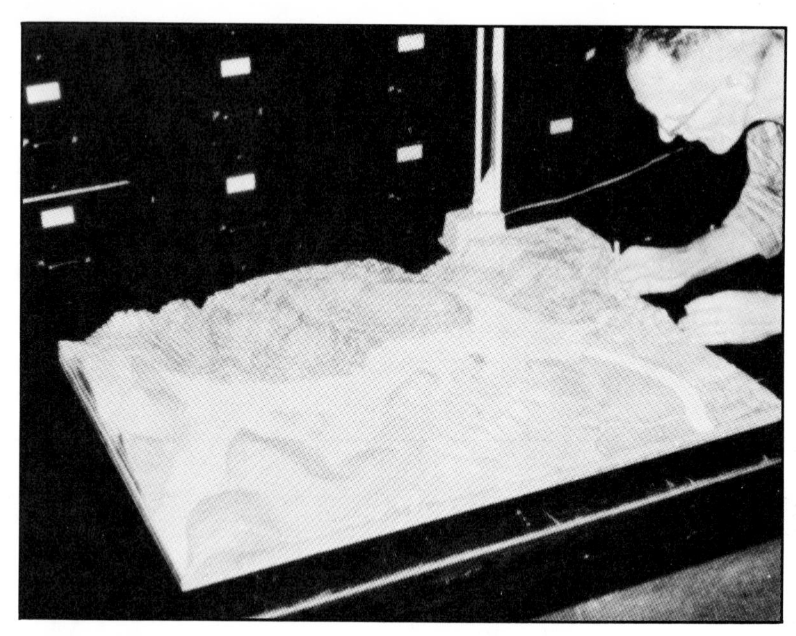

FIGURE 9.15 Terrain model (courtesy of Tennessee Valley Authority)

illusion of solid, three-dimensional topography is especially striking. This is particularly so in mountainous country. Relief shading can be provided as an overprint on a conventional contour map in order to assist the layperson with the interpretation of contours.

A *color tint system* is in common use for aerial navigation charts and on small-scale maps of the world. A scale of graded color tints or a system of different colors is chosen to show different zones of elevation. Each zone is bounded by contours that are usually shown on the map. Color tints, when used in conjunction with contour lines, give pictorial effect by accentuating the areas of different elevation.

When the land surface is too irregular or intricate to contour, as in the case of sand dunes, open strip mines, and lava beds, various *symbol patterns* are used. These are made up of dots, hachures, and form lines in a manner that is expressive of the typical appearance of the particular area. *Form lines* are similar to contour lines but are less accurate and have no definite interval.

9.11 SYSTEMS OF CONTOUR POINTS

A contour can be drawn on a topographic map that is being prepared if the horizontal position and elevation of properly selected ground points are known. The manner of obtaining the required data provides a basis for defining four systems of contour points. They are as follows:

SYSTEM A—GRID POINTS

System A consists of an established grid of squares, usually marked by stakes in the ground (see Figure 9.16; a-1, a-2). The elevations of the corners are then determined, thus forming a system of coordinate points from which the contour lines may be drawn.

In field surveys, this system is used when the available personnel are relatively unacquainted with mapping procedures and the tract is open and of limited extent. The survey of a building site would be a suitable example. The elevations of the coordinate points may be found with the engineer's level or transit. Sometimes the hand level is employed. In irregular terrain the accuracy of the map will be increased if the contours are drawn in the field either on a sketch board or a plane table.

This system is also commonly used by a modern automatic mapping instrument to generate a dense network of elevation points from aerial photographs. The resulting set of elevation data is called a *digital elevation model (DEM),* or *digital terrain model (DTM),* of the terrain. The use of aerial photographs for topographic mapping will be discussed in Chapter 14.

SYSTEM B—TRACING OUT CONTOURS

If a series of points having the same elevation are located on the ground and plotted on a map, the line joining these points will be a contour line (see Figure 9.16; b-1, b-2). Thus, if the series of points having the elevation 914 ft are plotted, the 914 contour line is found by drawing a smooth line through these points. The delineation of relief by tracing out each individual contour is accurate but very time consuming. The method is used when the map scale is large and the project area is rather small and open.

A transit is set up on a control point and oriented along the line to another traverse station. The elevation of the line of sight is determined by a backsight on a benchmark. The proper foresight to locate a point on a given contour is found by subtracting the elevation of that contour from the *HI*. Suppose, for example, the *HI* (elevation of line of sight) is 437.6 ft and the 430 contour is to be traced. This means that a foresight reading of 7.6 must be obtained if the rod is on that contour. As the rod holder proceeds to move along the contour, regular determinations of the rod position are made by reading the stadia intercept and the horizontal circle of the transit.

The use of a plane table is much more efficient because the rod positions can be plotted directly on the map in the field.

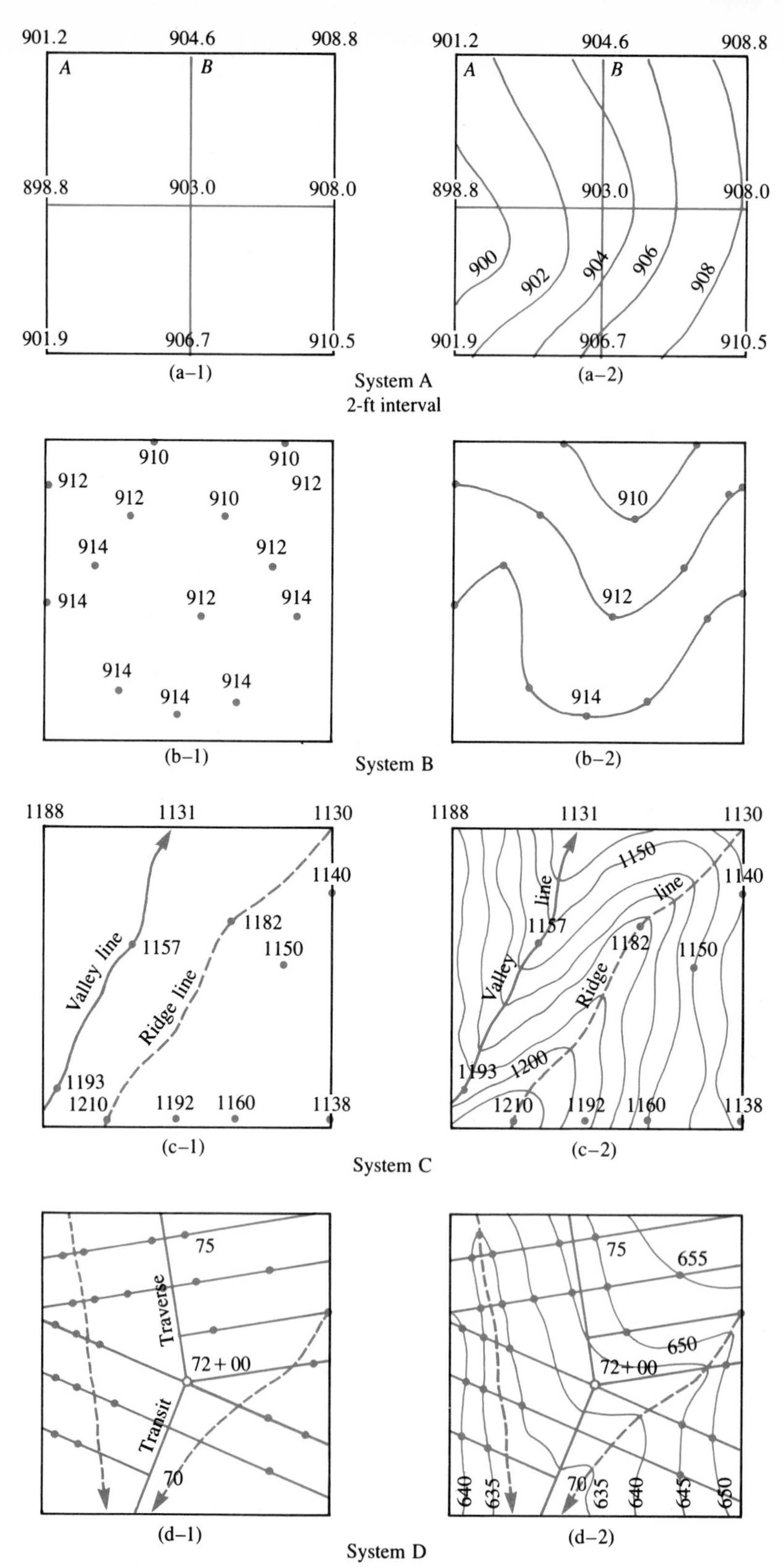

FIGURE 9.16 Systems of contour points

SYSTEM *C*—CONTROLLING POINTS

Although System B provides an accurate contour map, it requires the location of many points (see Figure 9.16; c-1, c-2). Where such great accuracy is not necessary, a more expeditious method is that in which a few *controlling points* only are located and the contour lines are interpolated and sketched in, to represent the ground surface. Such points are summits, depressions, changes in slope, and especially points along ridge and valley lines.

This is the most commonly used system in stadia survey. The field procedure is the same as that described in Section 9.7.

SYSTEM *D*—CROSS-SECTION PROFILING

To establish System D, (see Figure 9.16; d-1, d-2), a transit traverse is first run, with stakes set at 100-ft intervals, over which profile levels are taken. At these points cross-sections are taken to locate contour points, valley lines, and so on. From this system of points, the contour lines may be drawn.

The method used for locating points according to System D makes use of the hand level instrument. The field party usually includes the topographer and two tapeholders. The equipment includes a hand level, a 100-ft steel tape, a rod (10 to 15 ft long) graduated in one-foot divisions, and a 5-ft staff. The record is kept in the regular transit notebook, or in a special topography book made up of cross-ruled pages.

This method is used principally in connection with route surveys and it is therefore assumed that a located line has been established by means of a transit traverse, with stakes set at 100-ft intervals, and that profile levels have been run over the line.

In the field cross-sections are taken at the 100-ft stations or at irregular intervals, depending on the existing conditions. The procedure may be described by reference to Figure 9.17. At station 40 the elevation (521.5) of the ground at the stake is known from the profile levels. Assuming that the contour points to the right of the centerline are to be located, the front tapeholder takes the hand level and moves out on a line perpendicular to the centerline. The rear tapeholder takes the rod and remains at the center stake. The ground slopes uphill, and by sighting on the rod held at the center stake, the front tapeholder is directed to a point that is 0.5 ft higher than station 40, and thus locates a point on the 522 contour. The distance from the center stake is 27 ft, and this point is plotted by the topographer in the notebook. The rear tapeholder then takes the position at this point and the front tapeholder goes forward to find a point on the next higher (524-ft) contour. The distance between the two contours is 58 ft, or the latter point is 85 ft from the center stake. The method of taping may be that of measuring from one contour point to the next, or the topographer may serve as rear

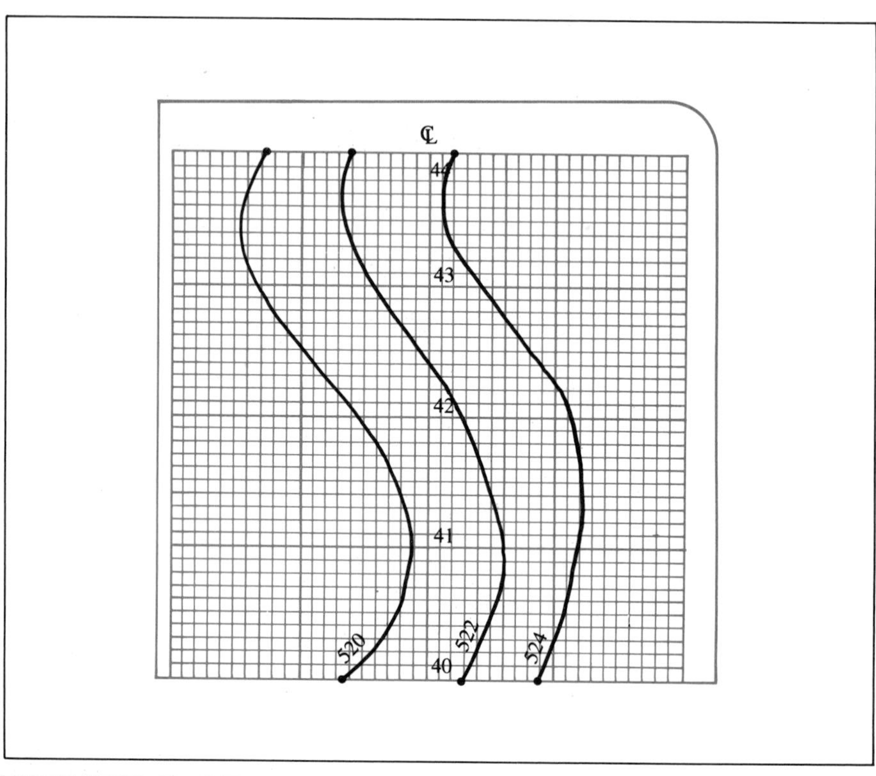

FIGURE 9.17 Locating contours with a hand level

tapeholder, thus permitting the rear tapeholder to become a rodholder, and all distances measured continuously from the center stake.

Locating points to left of the centerline, the slope is downhill, and either the tapeholders reverse positions, or the front tapeholder carries the rod and the rear tapeholder the hand level.

Distinguishing marks are usually placed on the rod at each contour interval above and below the *HI* of the hand level. The *HI* of the hand level may be the natural height of the topographer's eyes above the ground, or a 5-ft staff may be used to support the hand level. The latter practice is better, because it permits the tapeholder to reverse positions without confusion.

9.12 CONTOUR INTERPOLATION

Given a set of elevation points as shown in (a-1) and (c-1) in Figure 9.16, contour lines can be drawn by interpolation. The interpolation can be performed by estimation, by calculation, or by graphical means.

1. Estimation. The method of estimation is used where the highest accuracy is not desired, where the ground forms are quite regular, and where the scale of the map is intermediate or small. Thus the contours of Figure 9.16 (c-2) have been interpolated by this method. For example, on the valley line between the two points whose elevations are 1,157 and 1,193, it is at once noticed that the next higher contour above 1,157 is 1,160; also, that the additional contours, 1,170, 1,180, and 1,190, fall on the stream line below 1,193. Accordingly, the four points where these contours cross the stream are spaced by estimation and marked on the map. Possibly the first trial is seen to be erroneous, and a second spacing will be desirable. Similarly, the contour lines are spaced between the other controlling points, after which the contour lines may be sketched in to complete the map.

An important principle relating to the correct portrayal of ground forms by means of contour lines is that the salient or controlling features of the topography should be sketched in before the contour lines are drawn. Thus in Figure 9.18 the ridge and valley lines were located and the contour crossings determined along these lines before any contours were drawn.

2. Calculation. The method of calculation is used where high accuracy is desired, and where the scale of the map is intermediate or large. For example, suppose that it is desired to space the contours between points A and B along the top line of Figure 9.16 (a-1). It is noticed that the difference in elevation is 3.4 ft; also that the difference in elevation between point A and the next higher contour (902) is 0.8 ft. The distance (d) from A to the 902-ft contour line is given by the relation $d = 0.8/3.4 \times AB = 0.24\,AB$. If it is assumed that the distance $AB = 100$ ft, then $d = 24$ ft and the 902 contour line is plotted at that distance from A. Likewise, the 904 contour line is plotted at the distance $2.8/3.4 \times 100 = 82$ ft from A.

3. Graphical Means. If many interpolations are to be made, and if a relatively high accuracy is desired, it will be more rapid and convenient to provide a proportional scale by which the contour points may be interpolated. Such a scale is drawn on tracing cloth or tracing paper, showing parallel lines (to any convenient scale) to represent the desired contour interval. Figure 9.19 illustrates such a scale drawn for plotting 2-ft interval contour lines.

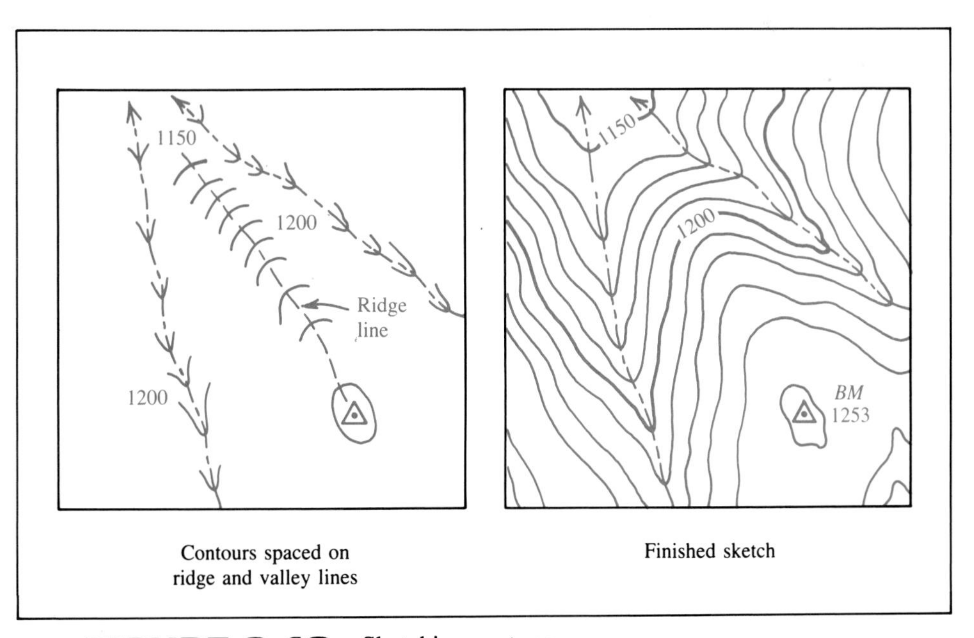

FIGURE 9.18 Sketching contours

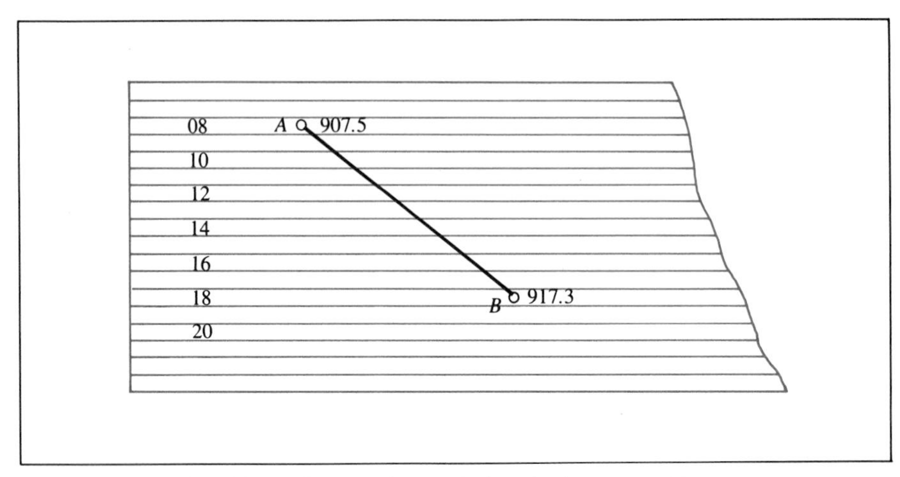

FIGURE 9.19 Contour interpolation device

The scale is used as follows: Suppose *A* and *B* are two points, 907.5 and 917.3 ft respectively, and it is desired to interpolate the 2-ft contour lines between these points. The tracing-paper scale is shifted over the points *A* and *B* until the point *A* shows at elevation 07.5 on the scale, and point *B* shows at elevation 17.3. In this position, the 908, 910, 912 (and so on) contour lines are found at the intersections of the 08, 10, 12-ft (and so on) lines with the plotted line *AB*. These points are fixed by pricking through the tracing paper into the drawing.

Obviously it is feasible to plot only a few points on any given contour line and the line must be drawn freehand between such points. Contours are fairly regular curved lines except for the rough surfaces of outcropping rock strata; contour lines, therefore, are smooth curves that conform to one another more or less closely, depending on the regularity of the ground forms. Thus the lines joining the contour points of Figure 9.16 (b-1) and (b-2) are not straight but are smooth curves.

It will usually aid in giving the proper character to the contour lines if every *fifth* contour lines is carefully sketched in. These lines will then serve as guides when the intermediate lines are being drawn.

The portrayal of relief on a topographic map by means of contour lines is both an art and a science. It is a science in the sense that various vertical and horizontal field dimensions are measured, which when plotted on a map base control the placement of the contour lines. It is an art with respect to the range of discretion and judgment that the topographer may exercise in determining the configuration of the contour lines between points of known horizontal position and elevation. Good topographic expression results when the contour lines convey to the map reader the typical characteristics of the ground surface.

9.13 MAP SCALE AND CONTOUR INTERVAL

Many considerations may govern the scale of a map, but, in general, a proper scale is just sufficiently large to permit all desired features to be shown clearly and all dimensions to be scaled with the desired accuracy. Engineers, accustomed to the large-scale drawings of structures, are apt to choose a much larger scale for a map than is necessary. Possibly little harm results from such practice except that oftentimes dimensions are scaled from such maps with a precision that the accuracy of the map does not warrant.

The proper choice of the contour interval for a topographic survey depends on the slopes of the terrain to be represented, the scale of the map, and the purpose of the survey. In hilly regions, if the contour interval is too small in relation to the scale of the map, the contour lines become so crowded as to become illegible and to obscure other features. Also, the accuracy and number of field measurements should bear a consistent relation to the contour interval. Consequently, the cost of a map to a small interval is much greater, for a given area, than one to a large interval.

No definite rule can be given that will meet all conditions, but for most purposes and field conditions, the contour intervals for large-scale maps (1 in. = 10 ft to 100 ft) will be taken as $\frac{1}{2}$, 1, 2, or 5 ft for slopes which range from flat to hilly. For intermediate-scale maps (1 in. = 10 ft to 1,000 ft) and for corresponding slopes the contour interval may be taken as 1, 2, 5, or 10 ft.

As metric conversion proceeds in the United States it seems likely that the following scales, or representative fractions, and contour intervals are likely to be employed in the preparation of special-purpose topographic maps of large and intermediate scale:

Present Scale	New Scale (*RF*)
1 in. = 10 ft	$\frac{1}{100}$
1 in. = 20 ft	$\frac{1}{250}$
1 in. = 100 ft	$\frac{1}{1,000}$
1 in. = 400 ft	$\frac{1}{5,000}$
Present Contour Interval	**Metric Contour Interval**
1 ft	1/2 m
5 ft	1 m
10 ft	5 m

9.14 PLATS AND MAPS

The distinction should be made between a plat and a map. A plat is a dimension drawing showing the data that pertain to a land survey or a subdivision of land. A map is a graphical representation, to scale, of the

FIGURE 9.20 Drawing a plat (courtesy of Bureau of Land Management)

relative positions and character of the features of a given area. A plat is primarily a legal instrument being made a matter of public record and used in the description and conveyance of real estate, having all important dimensions recorded. Although a plat (see Figure 9.20) is usually drawn to scale, little if any use is made of this condition. A map, however, shows no recorded dimensions, but a large part of its usefulness depends on the accuracy with which distances, areas, and elevations can be derived from it.

Excessive ornateness is not desired on maps that are to serve practical purposes; nevertheless the cartographer should strive for such harmonious effects that the finished map will have a pleasing appearance. This is attained by the use of good taste in the matters of form, proportion, and color in the symbols and lettering used.

9.15 PLOTTING

Plotting refers to the transfer of survey data to the map or plat. In the preparation of a topographic map, the first plotting task is to define accurately the positions of the horizontal control points on the map. This is best accomplished by preparing a rectangular grid and plotting each traverse point by means of its computed coordinates.

The plotting of side details, including contour points, can be effected most easily with the use of a circular protractor. It is properly oriented at a control station and fastened to the map sheet. A measuring scale can be utilized in conjunction with the protractor to locate the points.

9.16 TITLES

A title for a map usually provides the following items of information: (1) the organization or company for whom the map is made, (2) the name of the tract or feature that has been mapped, (3) the name of the engineer in charge, (4) the name of the person doing the drafting, (5) the place or office where the map was drawn, (6) the date, and (7) the scale, both numerical and graphical. Two examples are shown in Figure 9.21.

In executing a title, proper emphasis should be given to the different items listed above by varying the weight and the size of the letters used. A principal purpose a title serves is to identify a given map in a file with other similar maps. Hence the most important item in the title is the name of the tract or feature that the map represents, and this item should be given the most emphasis. For example, in Figure 9.21b the feature shown on this map is that portion of the location between stations 425 and 563 on Route 47. Accordingly, this item is given the most prominence in the title.

A map title is placed anywhere on the drawing where it will balance the map as a whole. It is not boxed in at the lower right-hand corner as are titles on mechanical drawings. The size of the letters should be consistent with the size of the drawing. A general tendency with beginners is to make the title too bold, with letters too large or the weight of the lines too heavy. Another common fault is too much space between the lines, which gives the title a loose and disjointed appearance.

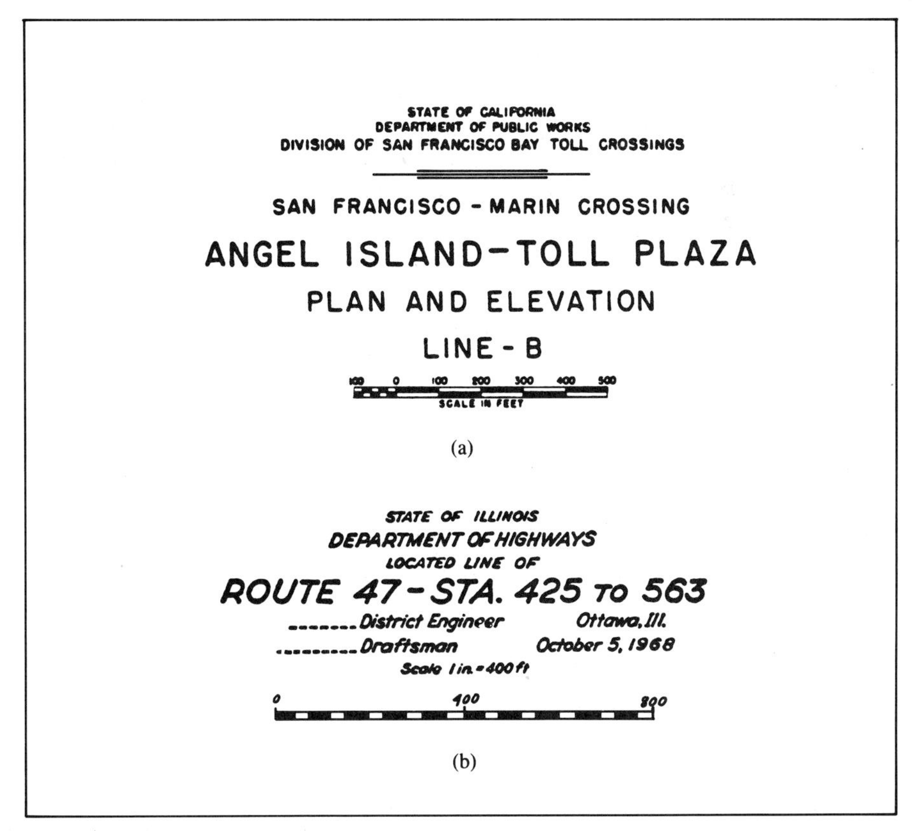

FIGURE 9.21 Map titles

FIGURE 9.22 Meridian arrow

Mechanical lettering is now used very widely for map drafting. This practice has important advantages when more than one person performs drafting tasks in the preparation of the same map.

A meridian should appear on every map. It is indicated by an arrow, somewhat as shown in Figure 9.22, which, unless otherwise specified, indicates true north.

9.17 SYMBOLS

Symbols are used to portray various features on maps. Different draftsmen may use different symbols to represent the same feature; it is desirable, therefore, that the most common features be represented by symbols that are widely accepted and understood. In this matter it is natural to follow the practice of those governmental organizations engaged in mapping work, such as the U.S. Geological Survey, the National Ocean Survey, and others. The symbols shown in Figure 9.23 are those commonly used.

Since map scales vary so widely, the character of the symbols will vary somewhat to suit the scale of the map. Care must be taken in executing the symbols so that they do not obscure and render illegible other features on the map.

On topographic maps it is quite necessary to draw the various symbols in different colors. The standard practice is as follows: *black* for letter-

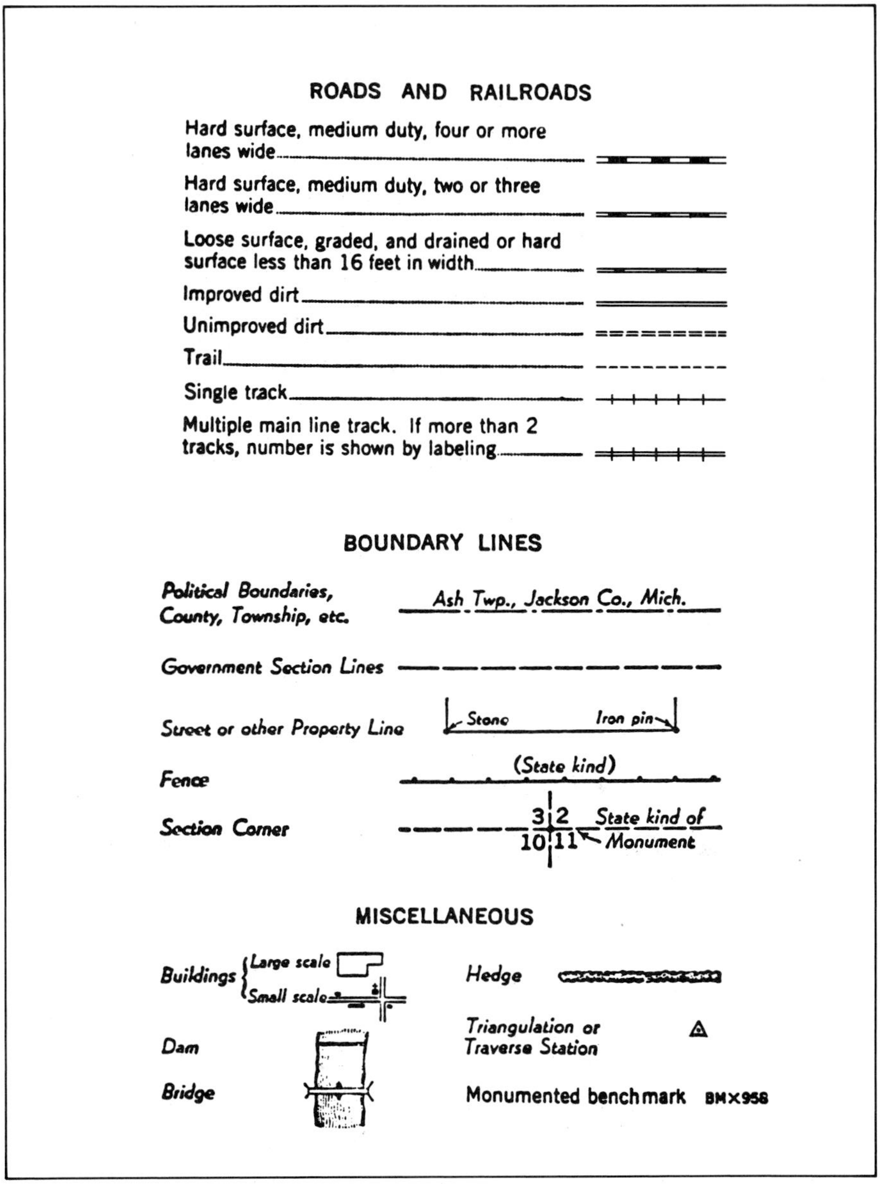

ROADS AND RAILROADS

Hard surface, medium duty, four or more lanes wide

Hard surface, medium duty, two or three lanes wide

Loose surface, graded, and drained or hard surface less than 16 feet in width

Improved dirt

Unimproved dirt

Trail

Single track

Multiple main line track. If more than 2 tracks, number is shown by labeling

BOUNDARY LINES

Political Boundaries, County, Township, etc. — *Ash Twp., Jackson Co., Mich.*

Government Section Lines

Street or other Property Line — *Stone* — *Iron pin*

Fence — *(State kind)*

Section Corner — 3 2 10 11 — *State kind of Monument*

MISCELLANEOUS

Buildings *Large scale* *Small scale*

Hedge

Triangulation or Traverse Station

Dam

Bridge

Monumented benchmark BMX958

FIGURE 9.23 Map symbols

ing and cultural features, such as roads, railways, houses, and other structures; *brown* for all land forms, such as contours and hachures; *blue* for water features, such as streams, lakes, marsh, and ponds; and *green* for woodland cover.

To secure greater uniformity of appearance of the finished map and to save time and money through the elimination of much hand lettering,

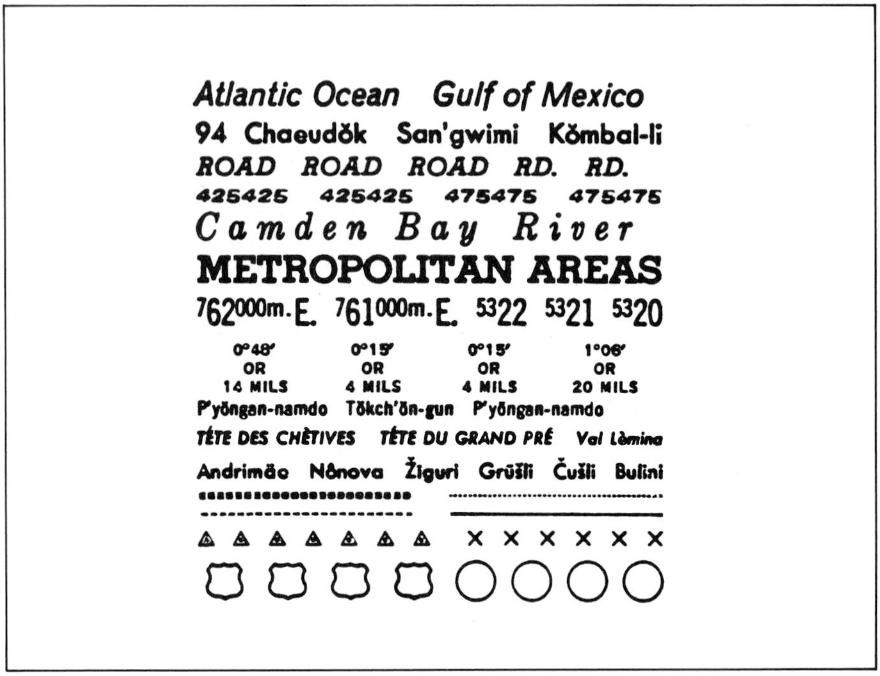

FIGURE 9.24 Adhesive map type

increasing use is being made of printed names, symbols, titles, meridian arrows, and so on. These are prepared on a tough, transparent plastic sheet (see Figure 9.24). It is a simple matter to cut out the required name or symbol from this sheet, put it in the proper position on the map, and press it down.

9.18 U.S. NATIONAL MAPPING PROGRAM

Probably one of the first attempts to provide national topographic coverage was that initiated by Robert Erskine, geographer-surveyor to the Continental Army, who not only established a Military Cartographic Headquarters at Ringwood, New Jersey, in 1777, but later urged the Congress to provide for a permanent staff of surveyors and geographers.

Following the exploratory surveys in the early part of the nineteenth century such as those by Lewis and Clark, Pike, Bonneville, and Nicollet, there was a lull until after the Civil War. Four major territorial surveys were then authorized by Congress. Although primarily geological surveys, consideration was given to acquiring topographic and other information as well. These surveys—namely, Geological Survey of the Fortieth Parallel (King Survey), 1867–79; Geological and Geographic Survey of the Territories (Hayden Survey), 1867–79; Geographical and Geological Survey of the

Rocky Mountain Region (Powell Survey), 1867–79; and the Geographical Survey West of the 100th Meridian (Wheeler Survey), 1872–79—were the precursors of later unified topographic surveys by a single organization, the U.S. Geological Survey. This agency was established in the Interior Department in 1879 with Clarence King as its first director. After one year Major John Wesley Powell (1834–1902), a vigorous proponent of national topographic mapping, became the director and laid the foundation for what has become the primary federal map-making organization. The Topographic Division of the U.S. Geological Survey is directly responsible for the mapping mission.

Topographic maps can be classified into two broad categories: special-purpose maps, and general-purpose maps. Special-purpose maps are prepared for specific projects and are compiled and published on various scales. Examples of special-purpose topographic maps are those prepared of reservoir sites and of metropolitan areas. General-purpose maps are designed to satisfy a wide range in public needs. The selection of the publication scale is based upon considerations of the economic character of the area, its cultural development, and the amount of map detail that is required for engineering, scientific, military, industrial, and commercial purposes.

Most of the topographic maps published by the U.S. Geological Survey are general-purpose maps. Collectively these maps constitute the national topographic map series. The two primary map series are:

1. 7½-Minute Quad Maps. These maps cover areas of great public importance, such as metropolitan and industrial areas and other regions where detailed map information is needed. The publication scale is 1 : 24,000 (1 in. = 2,000 ft) or 1 : 31,680 (1 in. = ½ mile), although the latter scale is used only infrequently for new mapping. Topographic maps published on these scales have dimensions of 7½′ in both latitude and longitude. As of 1983, published maps in this group were available for 68.6% of the area of the fifty states.

2. 15-Minute Quad Maps. These maps cover areas of average public importance. The publication scale is 1 : 62,500 (1 in. = nearly 1 mile). Geological Survey maps published on this scale measure 15′ in both latitude and longitude. This scale once was considered quite adequate for all uses where detailed studies are not contemplated. Much of the past national mapping has been to this scale. Coverage in 1983 was 25.2%. Published maps of Alaska to a scale of 1 : 63,360 are available for 83% of that state. The objectives of the *National Mapping Program* are to complete nationwide coverage (except Alaska) at 1 : 24,000 scale and to keep this series up-to-date by periodic revision. In time it is expected that the 1 : 62,500-scale maps will be phased out.

9.19 U.S. NATIONAL MAP ACCURACY STANDARDS

A knowledge of the accuracy specifications governing national topographic mapping is particularly important to engineers who may use the standard topographic maps as source data for the purpose of compiling special-purpose maps. If such specifications were better known, engineers would be much more cautious in the use of standard topographic maps, which have sometimes been enlarged to several times their intended usable scale in order to provide a base for detailed planning studies.

The accuracy specifications for special-purpose maps may cover a wide range in keeping with the requirements of the engineering firm that contracts to have such work done or performs it with its own forces. The specifications may be far above, equal to, or far below what are termed standard map accuracies.

The important features of accuracy specifications for federal topographic mapping in domestic United States areas are as follows:

HORIZONTAL ACCURACY

The horizontal position tolerance for 90% of all well-defined planimetric features is 40 ft of ground distance for both the 1 : 62,500 and 1 : 24,000 scale maps. This is equivalent to an error of 1/50 in. on the map at the publication scale of 1 : 24,000, and somewhat less than 1/100 in. at the 1 : 62,500 scale.

VERTICAL ACCURACY

Ninety percent of all elevations interpolated from the map shall be correct within one-half the contour interval. However, in checking elevations taken from the map, the apparent vertical error may be decreased by assuming a horizontal displacement within the permissible horizontal error for a map of that scale. Thus any contour may be shifted (imaginarily) either uphill or downhill by the map distance equivalent to a ground distance of 40 ft.

9.20 NATIONAL CARTOGRAPHIC INFORMATION CENTER

In July 1974 the *National Cartographic Information Center (NCIC)* was established to provide information about the availability of U.S. maps and charts, aerial photographs, and related cartographic data. Engineers and others seeking map information not readily available from local sources are encouraged to direct their inquiries to this facility of the U.S. Geological Survey, whose address is National Center, Reston, Virginia 22092.

PROBLEMS

9.1 Given the following data for two stadia observations with a transit having an external-focusing telescope ($k = 100$, $F + c$ negligible):

a. $r = 6.12$ ft $\qquad \alpha = 3°10'$

b. $r = 8.35$ ft $\qquad \alpha = 8°13'$

Compute the horizontal (H) and Vertical (V) distances by (1) formulas and (2) stadia table.

9.2 Calculate the distance from a transit having an internal-focusing telescope with a stadia multiplying factor of 100.3 to the rod. The sight is horizontal and the half-stadia interval is 7.45 ft.

9.3 Intervening brush obscuring the lower portion of the stadia rod make it necessary to read the vertical angle of $-11°12'$ to the 10.0-ft mark instead of the HI (height of the telescope above the survey station), which was 4.5 ft. If the half-intercept is 4.65 ft and the stadia constant is 100, find the difference in elevation.

9.4 A field determination was made of the stadia multiplying factor k of the internal-focusing telescope of an engineer's transit ($F + c = 0$). The stadia intercepts and the corresponding taped distances from the transit station were as follows:

taped distance (ft)	116.3	305.1	528.0	714.8	834.1
intercept (ft)	1.16	3.04	5.27	7.13	8.33

Compute the mean, standard deviation, and estimated standard error of the mean for the five determinations of k.

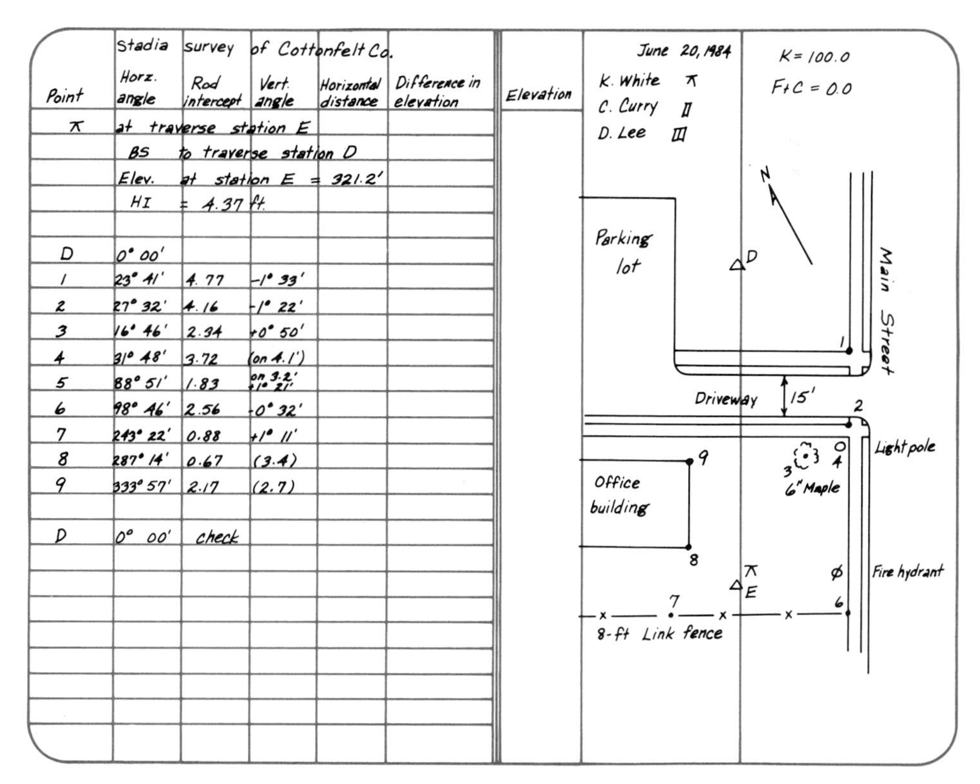

Stadia survey of Cottonfelt Co.

Point	Horz. angle	Rod intercept	Vert. angle	Horizontal distance	Difference in elevation	Elevation
⊼	at traverse station E					
	BS to traverse station D					
	Elev. at station E = 321.2'					
	HI = 4.37 ft.					
D	0° 00'					
1	23° 41'	4.77	−1° 33'			
2	27° 32'	4.16	−1° 22'			
3	16° 46'	2.34	+0° 50'			
4	31° 48'	3.72	(on 4.1')			
5	88° 51'	1.83	on 3.2' +1° 21'			
6	98° 46'	2.56	−0° 32'			
7	243° 22'	0.88	+1° 11'			
8	287° 14'	0.67	(3.4)			
9	333° 57'	2.17	(2.7)			
D	0° 00'	check				

FIGURE P9.1 Stadia survey 1

Point	Horizontal angle	Rod intercept	Vert. angle	Horz. distance	Diff. in elevation	Elevation
	Stadia	survey	of Riverside	Park		
	⊼ at	traverse station B				
	BS	to traverse station A				
	Elevation at station B = 714.3′					
		HI =	4.15′			
A	0° 00′					
1	25° 14′	4.16	−1° 47′			
2	81° 26′	3.27	−1° 35′			
3	123° 31′	1.87	(2.1)			
4	301° 56′	0.83	+3° 24′ (5.6)			
5	342° 27′	4.91	2° 26′			
A	0° 00′	check				

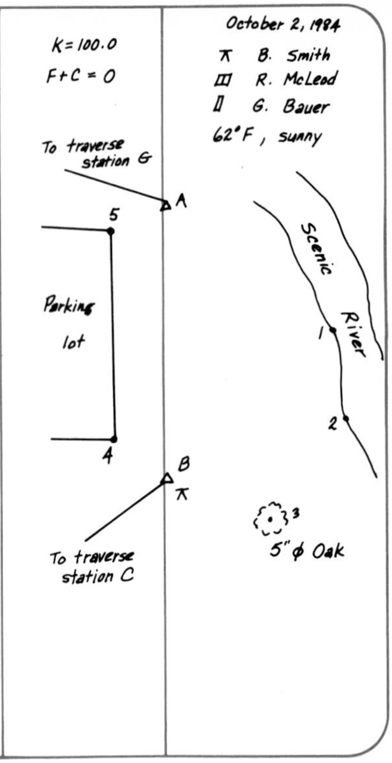

K = 100.0
F + C = 0

October 2, 1984
⊼ B. Smith
⌶ R. McLeod
▯ G. Bauer
62°F, sunny

To traverse station G
A
5
Parking lot
4
B
⊼
To traverse station C
Scenic River
1
2
3
5″ ϕ Oak

FIGURE P9.2 Stadia survey 2

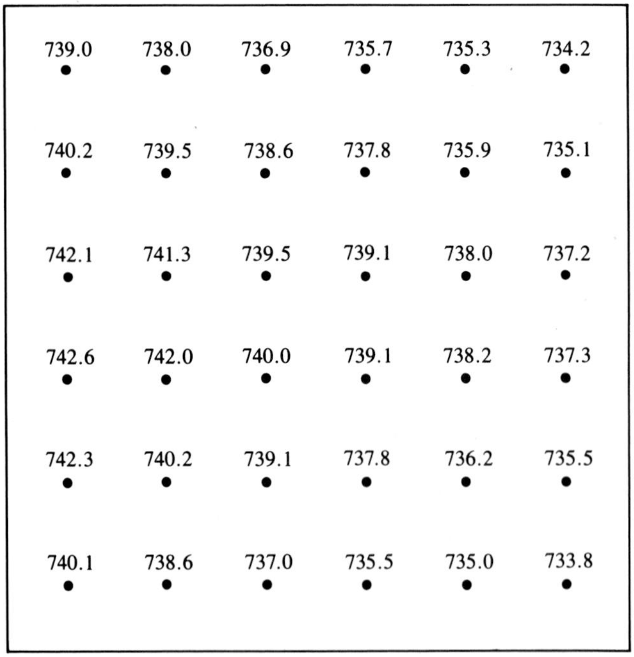

739.0	738.0	736.9	735.7	735.3	734.2
740.2	739.5	738.6	737.8	735.9	735.1
742.1	741.3	739.5	739.1	738.0	737.2
742.6	742.0	740.0	739.1	738.2	737.3
742.3	740.2	739.1	737.8	736.2	735.5
740.1	738.6	737.0	735.5	735.0	733.8

Scale: 1″ = 50′ Contour interval = 2 ft

FIGURE P9.3 Elevation points for an area

9.5 Complete the field note for stadia surveys given in Figure P9.1 and P9.2. Compute the horizontal distances to and elevations of the stadia points.

9.6 Draw the contour lines for the areas in Figures P9.3, P9.4, and P9.5.

9.7 Plot the elevation profiles along the cross-sections *A-A*, *B-B*, *C-C*, and *D-D* in Figure P9.6. Use a horizontal scale of 1″ = 2,000 ft and a vertical scale of 1″ = 50 ft.

9.8 Express the following map scales as representative fractions:

a. 1″ = 25′

b. 1″ = 100′

c. 1″ = 2,000′

d. 1″ = 5,280′

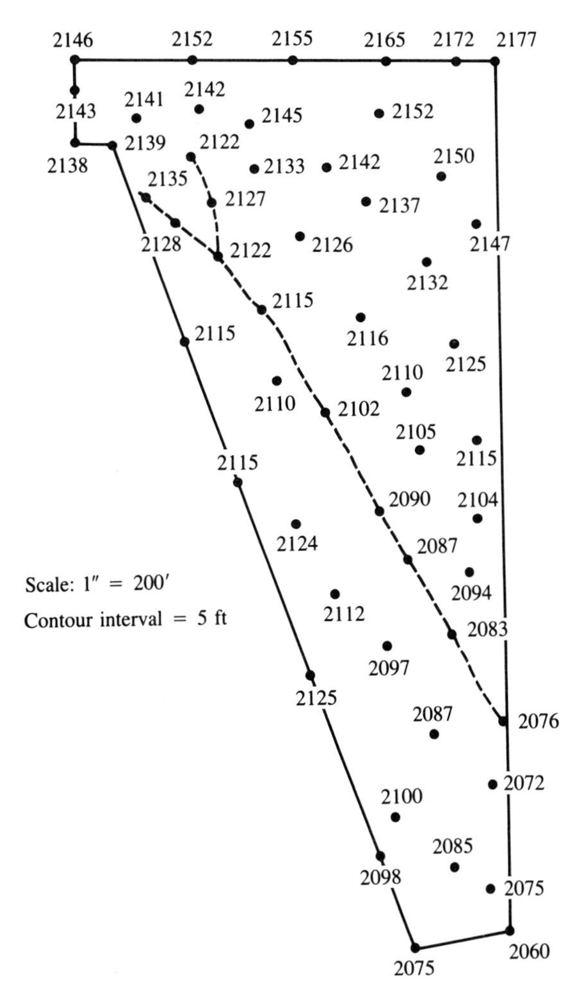

FIGURE P9.4 Elevation points for a residential site

9.9 Express the following map scales as so many feet to an inch:

a. 1/100

b. 1/5,000

c. 1/10,000

d. 1/1,000,000

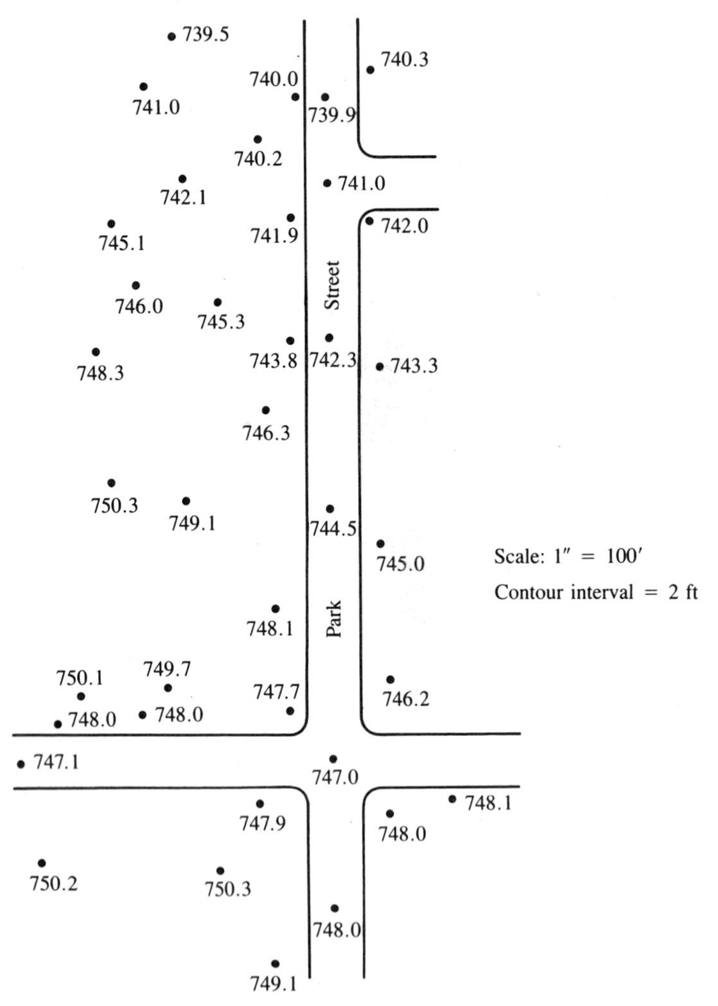

FIGURE P9.5 Elevation points for a street

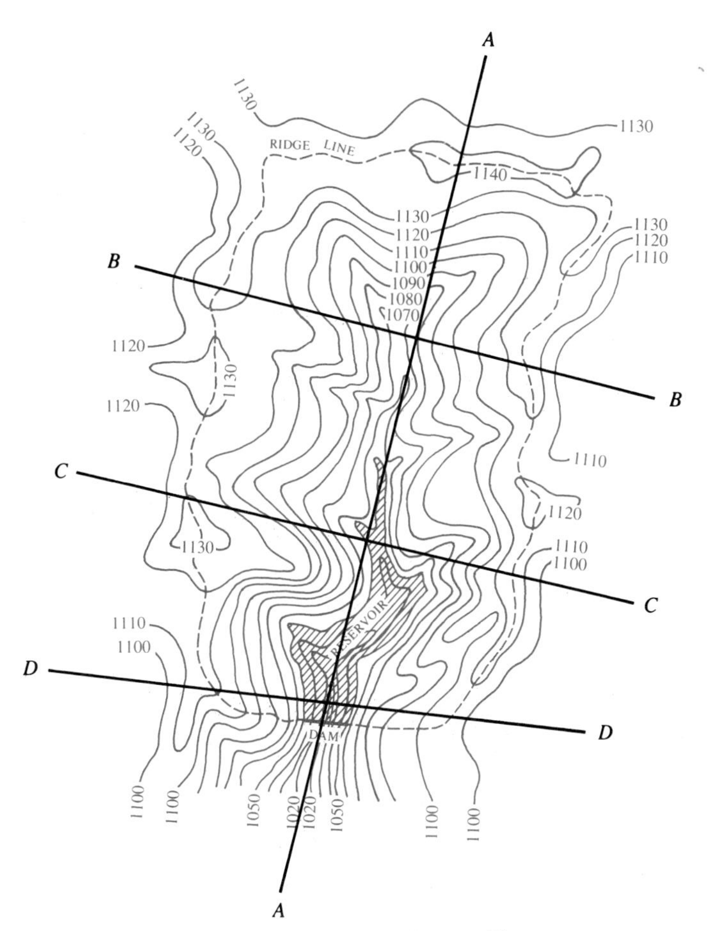

FIGURE P9.6 Cross-section profiles

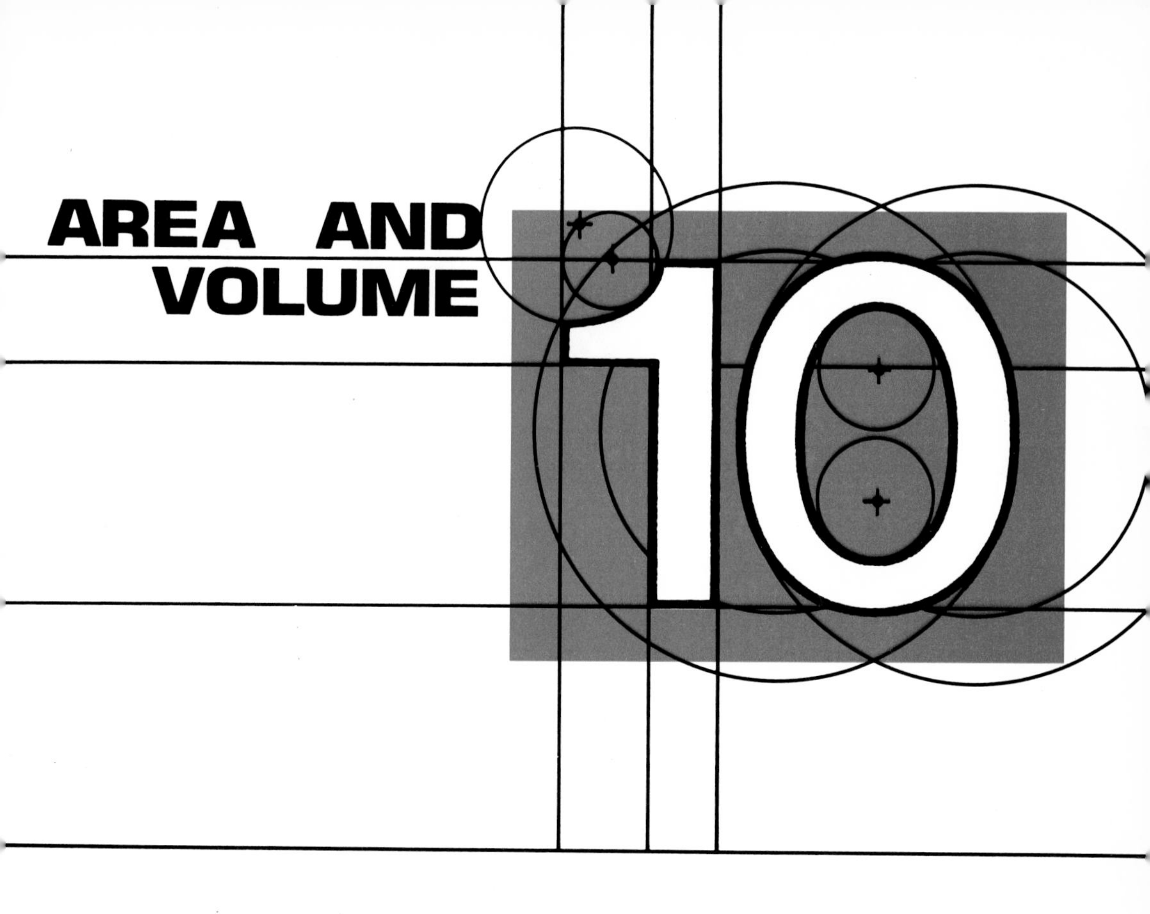

10.1 INTRODUCTION

This chapter will deal with the most commonly used methods for determining land areas and volume of materials.

10.2 AREA ENCLOSED IN A TRAVERSE

One of the principal purposes of conducting a boundary survey is to acquire the data needed to determine the area of a tract of land. Usually, a traverse is run along the perimeter of the tract, and the traverse is balanced according to the procedures described in Chapter 8. Since the coordinates of the corner points are thus already available, it is particularly convenient to use these traverse coordinates to compute the area enclosed in the traverse.

Consider the loop traverse shown in Figure 10.1. The traverse stations are numbered sequentially in a clockwise direction. The reason for this numbering scheme will become obvious shortly. The lines *a*1, *b*2, *c*6, *d*3, *e*4, and *f*5 are parallel to the *X*-axis, and are drawn from the traverse stations to the *Y*-axis. Suppose that the X- and Y-coordinates of the traverse stations have already been calculated from the traverse measurements. It can be seen from Figure 10.1 that the following relationship is true:

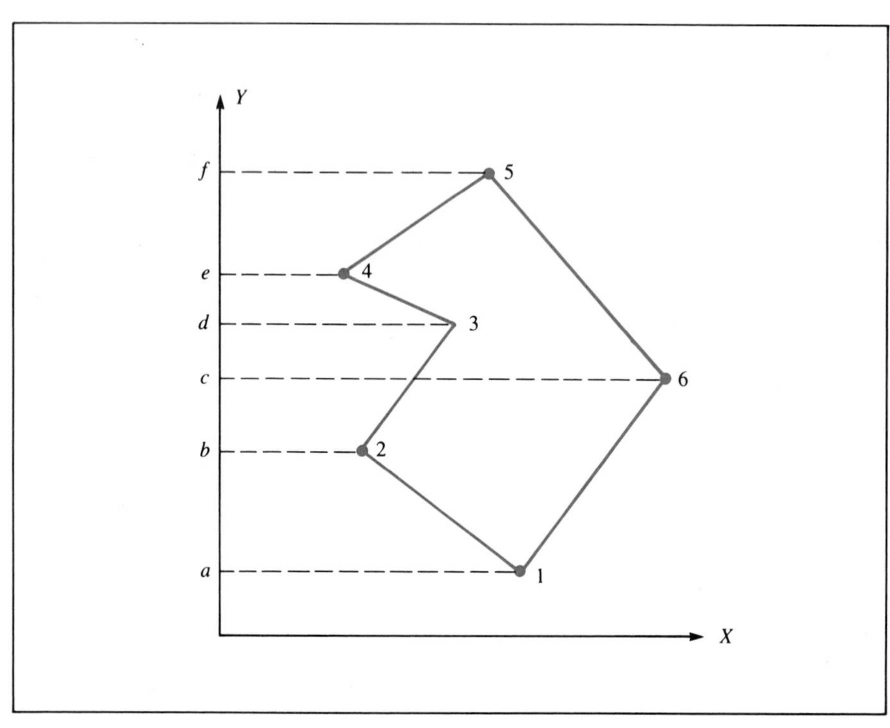

FIGURE 10.1 Area enclosed in a traverse

$$\text{area enclosed in the traverse} = \text{sum of the areas of trapezoids } a16c \text{ and } c65f - \text{sum of areas of the trapezoids } a12b, b23d, d34e, \text{ and } e45f \qquad (10.1)$$

The area of any of the trapezoids can be expressed in terms of the known station coordinates. For example,

$$\text{area of trapezoid } a16c = \tfrac{1}{2}(Y_6 - Y_1)(X_6 + X_1) \qquad (10.2)$$

Therefore,

$$\begin{aligned}\text{area enclosed in traverse} = {} & \tfrac{1}{2}[(Y_6 - Y_1)(X_6 + X_1) \\ & + (Y_5 - Y_6)(X_5 + X_6)] \\ & - \tfrac{1}{2}[(Y_2 - Y_1)(X_2 + X_1) \\ & + (Y_3 - Y_2)(X_3 + X_2) \\ & + (Y_4 - Y_3)(X_4 + X_3) \\ & + (Y_5 - Y_4)(X_5 + X_4)] \end{aligned} \qquad (10.3)$$

From the preceding expression, the following simplified equation can be derived:

$$\text{area} = \tfrac{1}{2}[Y_1X_2 + Y_2X_3 + Y_3X_4 + Y_4X_5 + Y_5X_6 + Y_6X_1 - X_1Y_2 - X_2Y_3 - X_3Y_4 - X_4Y_5 - X_5Y_6 - X_6Y_1] \quad (10.4)$$

The computation can be tabulated as follows:

Point	X	Y	Positive Product	Negative Product
1	X_1	Y_1		
2	X_2	Y_2	Y_1X_2	X_1Y_2
3	X_3	Y_3	Y_2X_3	X_2Y_3
4	X_4	Y_4	Y_3X_4	X_3Y_4
5	X_5	Y_5	Y_4X_5	X_4Y_5
6	X_6	Y_6	Y_5X_6	X_5Y_6
1	X_1	Y_1	Y_6X_1	X_6Y_1
			sum 1	sum 2
area = $\tfrac{1}{2}$[sum 1 − sum 2]				

In the preceding tabulation, the solid line joining Y_1 and X_2 indicates that the product Y_1X_2 is included in the positive product column. Similarly, the broken line joining X_1 and Y_2 indicates that the product X_1Y_2 is negative. Note also that the traverse stations are listed sequentially and that the first station (station 1) is listed both at the beginning and at the end.

Thus, by numbering the traverse stations sequentially in a clockwise direction, a convenient formula is derived for computing the area. In practice, it is not necessary to *number* the traverse stations in this manner. Any number or name can be assigned to any of the stations. It is only necessary to list the stations in either a clockwise or counterclockwise order for performing the preceding tabulation.

Sometimes, because of the shape of the traverse, the area computed by the preceding method can be a negative quantity. The negative sign has no significance and can be ignored in reporting the area.

The unit in which the area is initially calculated is related to the unit of length. Hence, if lengths are in meters, the area will be in square meters (m^2). There is no officially sanctioned unit of area in the SI system although the *hectare* (10,000 m^2) has some recognition. One hectare equals 2.47104 acres, and 1 acre = 43,560 sq ft.

When state plane grid coordinates define the positions of points in a traverse along the boundary of a tract of land, it may be desirable to effect a correction, usually small, to the calculated grid area in order to obtain the actual ground area. This is done by dividing the grid area by the square of the grid factor (see Section 15.19). For example, suppose the grid area is

2,655,478 ft² and the grid factor is 1.0000875. The area at the average elevation of the property is

$$\frac{2,655,478}{(1.0000875)^2} = 2,655,013 \text{ ft}^2$$

EXAMPLE 10.1

Find the area enclosed in the loop traverse of Figure 8.2 by using the coordinates computed in Table 8.4.

SOLUTION

Note that in Table 8.4 the X-coordinates of all the traverse stations start with the two identical numerical digits 43. Thus the calculation can be simplified by subtracting 43,000.00 from each of the X-coordinates. Similarly, the number 24,000 is subtracted from each Y-coordinate.

Point	X-43,000.00	Y-24,000.00	Positive Product	Negative Product
	(ft)	(ft)	(ft²)	(ft²)
1	301.39	250.78		
6	403.93	590.15	101,298	177,865
5	721.64	612.77	425,876	247,516
4	970.16	728.53	594,485	525,736
3	877.37	268.16	639,190	260,158
2	649.19	361.62	174,087	317,275
1	301.39	250.78	108,989	162,804
			2,043,925	1,691,354

$$\begin{aligned}\text{area} &= \tfrac{1}{2}[2,043,925 - 1,691,354] \text{ sq ft} \\ &= 176,286 \text{ sq ft} \\ &= 4.0470 \text{ acres}\end{aligned}$$

By rearranging the terms in Eq. (10.4), the following variation of the area equation can be derived:

$$\begin{aligned}\text{area} = \tfrac{1}{2}[&Y_1(X_2 - X_6) + Y_2(X_3 - X_1) + Y_3(X_4 - X_2) \\ &+ Y_4(X_5 - X_3) + Y_5(X_6 - X_4) + Y_6(X_1 - X_5)]\end{aligned} \tag{10.5}$$

Eq. (10.5) is also commonly used.

The accuracy of the area computed from traverse coordinates can be evaluated by applying the law of propagation of random error (see Section 2.9) to Eq. (10.4). To simplify the accuracy analysis, it can be assumed that the standard error of the computed X- and Y-coordinates is approximately equal and is denoted as $\hat{\sigma}_{X,Y}$. Then, the following approximate equation can be derived:

$$\hat{\sigma}_A \cong \pm \frac{\hat{\sigma}_{X,Y}}{2} [(X_1 - X_3)^2 + (X_2 - X_4)^2 + (X_3 - X_5)^2$$
$$+ (X_4 - X_6)^2 + (X_5 - X_1)^2 + (X_6 - X_2)^2$$
$$+ (Y_1 - Y_3)^2 + (Y_2 - Y_4)^2 + (Y_3 - Y_5)^2$$
$$+ (Y_4 - Y_6)^2 + (Y_5 - Y_1)^2 + (Y_6 - Y_2)^2]^{1/2} \qquad (10.6)$$

where $\hat{\sigma}_A$ is the estimated standard error of the computed area. Note that the subscripts for the *X*- and *Y*-coordinates in Eq. (10.6) follow a regular sequence. The equation can be easily applied to any loop traverse.

For a closed traverse, the average standard error of the computed station coordinates can be approximated by the following expression:

$$\hat{\sigma}_{X,Y} \cong \pm\tfrac{1}{4} D \cdot R \qquad (10.7)$$

where *D* is the sum of the distances measured along the traverse, and *R* is the relative precision (at 1σ) of the distance measurement.

For example, suppose that the distances in the traverse discussed in Example 10.1 were measured with a relative precision of 1/10,000 at 1σ. From Table 8.2, the total traverse distance (*D*) was 2,028.45 ft. Thus,

$$\hat{\sigma}_{X,Y} = \pm\tfrac{1}{4} \times 2{,}028 \times \frac{1}{10{,}000} = \pm 0.05 \text{ ft}$$

Then, for the area computed in Example 10.1,

$$\hat{\sigma}_A \cong \pm \frac{0.05}{2} [(301.39 - 877.37)^2 + (649.19 - 970.16)^2$$
$$+ (877.37 - 721.64)^2 + (970.16 - 403.93)^2$$
$$+ (721.64 - 301.39)^2 + (403.93 - 649.19)^2$$
$$+ (250.78 - 268.16)^2 + (361.62 - 728.53)^2$$
$$+ (268.16 - 612.77)^2 + (728.53 - 590.15)^2$$
$$+ (612.77 - 250.78)^2 + (590.15 - 361.62)^2]^{1/2}$$
$$\hat{\sigma}_A \cong \pm 30 \text{ sq ft or } \pm 0.0007 \text{ acre}$$

10.3 AREA OF IRREGULAR TRACTS

The area of a tract with an irregularly shaped boundary can be found by measuring perpendicular offsets from a base line. For example, in Figure 10.2, Brown's property is bounded by lines *EA*, *AB*, *BF*, and the section of the shoreline represented by *EF*. To determine the land area included in Brown's property, points *D* and *C* are located along the boundary lines *EA*

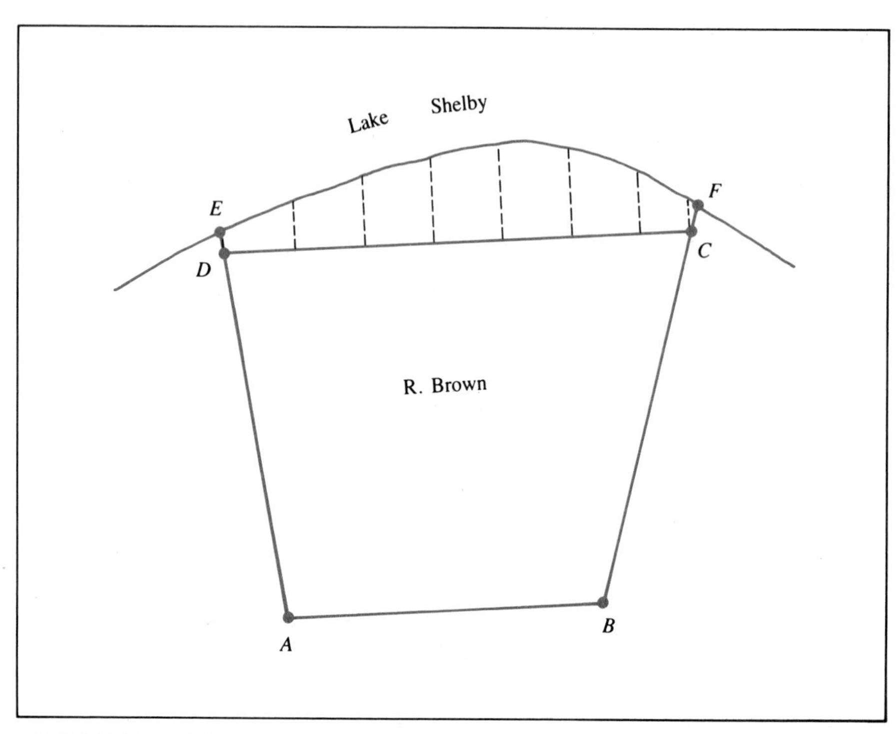

FIGURE 10.2 Area of irregular tract

and *FB* respectively. A traverse is run along the four-sided figure *ABCD*, and the area enclosed in the traverse can be determined using the procedure described in Section 10.2. The remaining portion of the tract, represented by figure *DEFC*, can be determined by measuring offset distances from the base line *DC* to the shoreline.

TRAPEZOIDAL RULE

Figure 10.3 shows the offset measurements made to determine the area of the partial tract bounded by the shoreline. To facilitate the area calculations, the offsets are spaced at regular intervals of 20 ft, except the last offset (*HC*), which is located 10.2 ft away from the adjacent offset *MN*. By approximating the shoreline with a series of straight lines joining adjacent offsets, the figure formed by any two adjacent offsets is a trapezoid. The area of the section *GDNM* can thus be computed as the sum of the trapezoids between offsets *GD* and *MN*. This area can be computed by the *trapezoidal rule:*

$$A = b\left[\frac{h_1 + h_n}{2} + h_2 + h_3 + h_4 + \cdots + h_{n-1}\right] \tag{10.8}$$

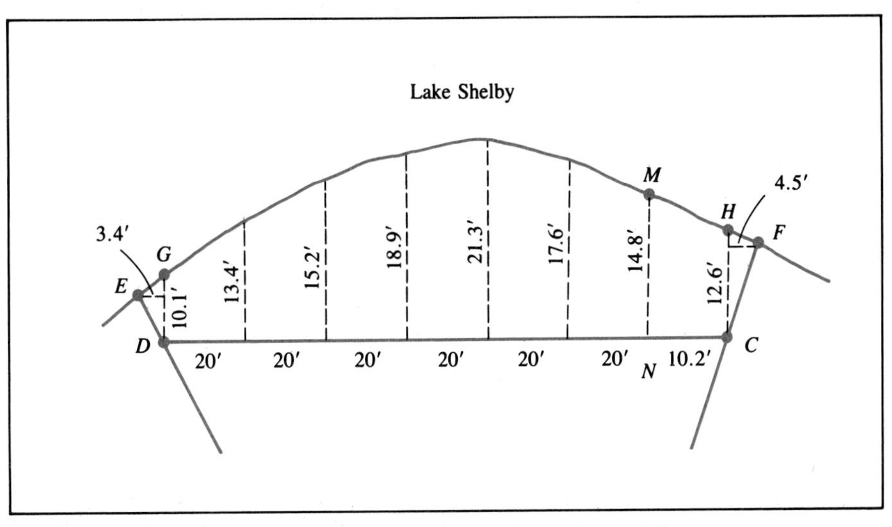

FIGURE 10.3 Offset measurements for an irregular tract

where A is the area, b is the common distance between offsets, h_i are the offsets, and h_n is the last offset. Thus,

$$\text{area of } GDNM = 20\left[\frac{10.1 + 14.8}{2} + 13.4 + 15.2 + 18.9 + 21.3 + 17.6\right] \text{ sq ft}$$
$$= 1{,}977 \text{ sq ft}$$

And,

$$\text{area of } EDCF = \text{area } EDG + GDNM + MNHC + HCF$$
$$= \tfrac{1}{2}(10.1 \times 3.4) + 1{,}977 + \frac{14.8 + 12.6}{2} \times 10.2 + \tfrac{1}{2}(12.6 \times 4.5)$$
$$= 2{,}162 \text{ sq ft}$$

SIMPSON'S ONE-THIRD RULE

Area of the section $GDNM$ can also be computed by *Simpson's one-third rule:*

$$A = \frac{b}{3}\left[h_1 + h_n + 2(h_3 + h_5 + \cdots + h_{n-2}) + 4(h_2 + h_4 + \cdots + h_{n-1})\right] \tag{10.9}$$

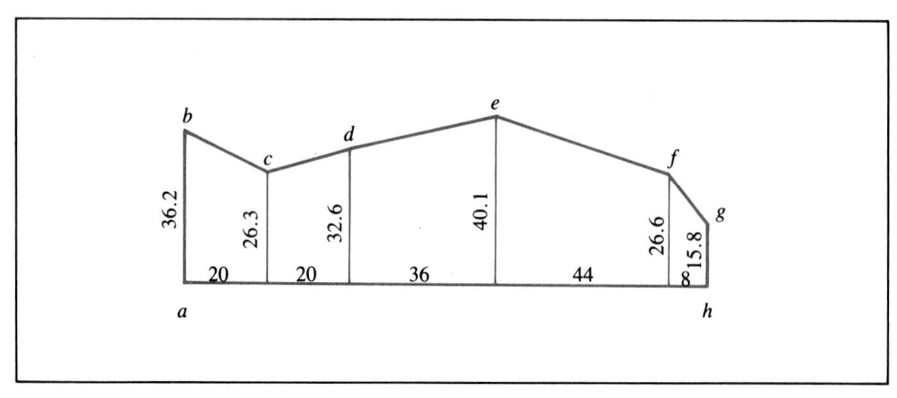

FIGURE 10.4 Offsets at irregular intervals

In Eq. (10.9), n must be an *odd* number. This rule is based on the assumption that the curve passing through the ends of the first three offsets is a parabola; the same is true for the curve through the ends of offsets 3, 4, and 5, and through the ends of offsets 5, 6, and 7, and so on. It is supposed that this series of parabolic curves will approximate the boundary line more closely than do straight lines and so yield a more accurate value for the area.

Using Simpson's one-third rule:

$$\begin{aligned}\text{area of } GDNM &= \frac{20}{3}[10.1 + 14.8 + 2(15.2 + 21.3) \\ &\quad + 4(13.4 + 18.9 + 17.6)] \\ &= 1{,}983 \text{ sq ft}\end{aligned}$$

The relative merits of the two area formulas are as follows:

1. The trapezoidal method is simplest and is sufficiently accurate for most areas, provided proper care is taken in the measurements. Note that the calculated area will be slightly too large for trapezoids where the boundary line is concave upward and too small for trapezoids where the line is concave downward, as is the case in Figure 10.3. Thus this method is more accurate where the boundary line consists of segments of contrary flexure.

2. Simpson's rule is more laborious to apply than the trapezoidal rule, but it is more accurate for all conditions. It is especially applicable to figures like the one shown in Figure 10.3.

If the boundary is very irregular, then the offsets should be measured wherever there is a change in the direction of the boundary line, which results in irregular intervals between offsets (see Figure 10.4). The area may then be calculated as a series of separate trapezoids. For example, the area of the tract shown in Figure 10.4 can be computed as follows:

$$
\begin{aligned}
\text{area} &= \left(\frac{36.2 + 26.3}{2}\right) \times 20 + \left(\frac{26.3 + 32.6}{2}\right) \times 20 \\
&\quad + \left(\frac{32.6 + 40.1}{2}\right) \times 36 + \left(\frac{40.1 + 26.6}{2}\right) \times 44 \\
&\quad + \left(\frac{26.6 + 15.8}{2}\right) \times 8 \\
&= 4{,}160 \text{ sq ft}
\end{aligned}
$$

The area can also be computed using the coordinate method described in Section 10.2. The lower left corner (point a in Figure 10.4) of the tract is arbitrarily assigned the coordinates $x = 0.0$ and $y = 0.0$. The coordinates of the remaining corners are then computed from the offset measurements yielding the following results:

Point	x	y	Positive Product	Negative Product
a	0.0	0.0		
b	0.0	36.2	0	0
c	20.0	26.3	724	0
d	40.0	32.6	1,052	652
e	76.0	40.1	2,478	1,604
f	120.0	26.6	4,812	2,022
g	128.0	15.8	3,405	1,896
h	128.0	0.0	2,022	0
a	0.0	0.0	0	0
			14,493	6,174

$$
\begin{aligned}
\text{area} &= \tfrac{1}{2}[14{,}493 - 6{,}174] \text{ sq ft} \\
&= 4{,}160 \text{ sq ft}
\end{aligned}
$$

10.4 AREA BY POLAR PLANIMETER

The planimeter is an instrument by means of which the area of a plotted, closed figure may be determined directly by tracing the perimeter and reading the result from the scale.

This instrument is illustrated in Figure 10.5. Its essential features are an anchor point or pole P, a tracing point T, and a roller R, which has a graduated scale on a drum.

The two arms connecting these points are movable about the connecting pivot. In the older type, the tracing arm is of fixed length and capable of reading areas in one unit only, usually square inches. In the type shown in Figure 10.5, the tracing arm is adjustable in a sleeve, to read areas in differ-

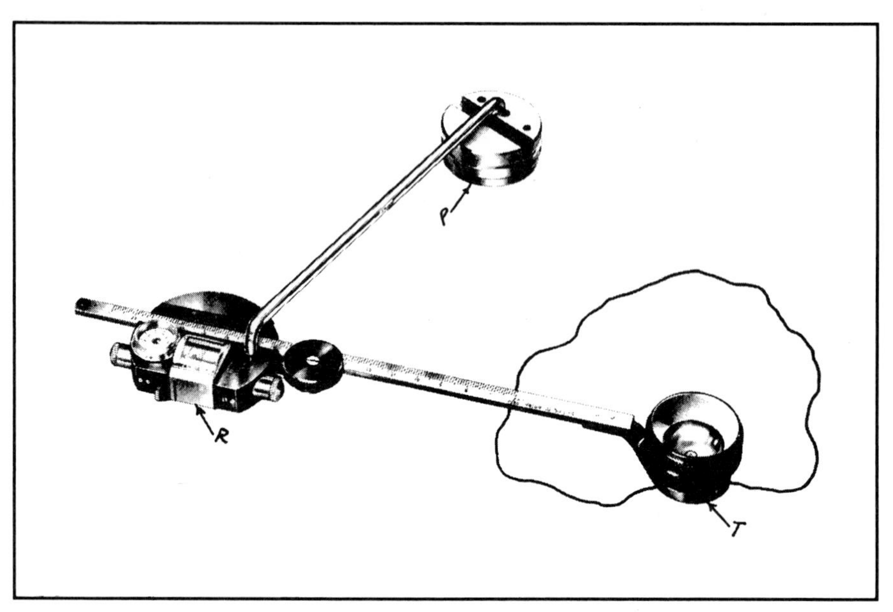

FIGURE 10.5 Optical polar planimeter (courtesy of National Surveying Instruments, Inc.)

ent units, corresponding to the scale of the map. The arms are usually adjusted so that one revolution of the roller measures an area of 10 sq in.; the scale is divided into 100 parts, and the vernier reads to 1/10 of a scale division, or 1/1,000 of 10 sq in. = 0.01 sq in.

With some instruments there is provided a proving bar, which is a flat metal bar with a needle point at one end and a hole drilled through the bar at a distance from the needle point equal to the radius of a circle of 10 sq in. area. With the needle point pressed into the drawing paper and the tracing point inserted in the hole, an area of 10 sq in. can be quickly and accurately traced, and, if necessary, the tracing arm can be precisely adjusted. If the instrument is of the fixed-arm type, the tracing arm cannot be adjusted, but a constant can be determined by which all results are to be multiplied to determine correct values.

If a proving bar is not provided, a square of known area can be carefully drawn and the perimeter traced a few times, either to adjust the tracing arm or to find the instrument's constant.

In use the pole is placed in any convenient position *outside* the area and the tracing point is placed at some initial point on the perimeter to be traced. The scale is then read by means of the index and vernier provided on the roller frame. This reading of the scale is recorded as the initial reading. The tracing point is then moved carefully around the perimeter, during which process the roller will both turn and slide, until the initial point is again reached. The scale is again read, and recorded as the final reading. The difference between the two readings is a measure of the area within the perimeter traced. If the direction of movement of the tracing point is clockwise, the second reading will be greater than the first; if counterclockwise, it will be less. Two or more determinations of each area should be made to provide a check and to secure a more accurate result.

Because of backlash in the instrument, no attempt should be made to set the fixed-arm planimeter at zero. The optical planimeter, however, is quickly and accurately set to zero by depressing a button.

EXAMPLE 10.2

A fixed-arm polar planimeter is used to determine the area of a parcel of land shown on a map whose scale is 1 in. = 400 ft. The initial reading is 5,678 and the final reading is 7,215. If one unit in the fourth place equals 0.01 sq in., find the land area in acres.

SOLUTION

$$\text{map area} = (7{,}215 - 5{,}678).01 = 15.37 \text{ sq in.}$$

$$\text{land area} = \frac{(15.37)(160{,}000)}{43{,}560} = 56.5 \text{ acres}$$

If the area is too large to be included by the tracing arm for one position of the pole, the area can be divided into the requisite number of subdivisions.

In highway applications the cross-section of a road is usually drawn on cross-section or profile paper using different scales for the vertical and horizontal dimensions (see Figure 10.6). Suppose that the cross-sections are drawn with a horizontal scale of 1 in. = 10 ft and a vertical scale of 1 in. = 5 ft. Then, the measured area would have a scale of 1 sq in. = 10 × 5 = 50 sq ft.

EXAMPLE 10.3

The areas of plotted highway cross-sections are determined with an optical planimeter whose adjustable arm is set so that one unit in the fourth place equals 0.015 sq in. The horizontal and vertical scales of the cross-sections are 1 in. = 10 ft and 1 in. = 5 ft, respectively. The mean difference of readings of several runs around the figure is 1,280. What is the end area?

SOLUTION

$$\text{area} = (1{,}280)(0.015)(10 \times 5) = 960 \text{ sq ft}$$

The precision of an area measured by a planimeter can be determined by measuring the same area several times, and then computing the mean, standard deviation, and estimated standard error of the mean according to the procedures described in Chapter 2.

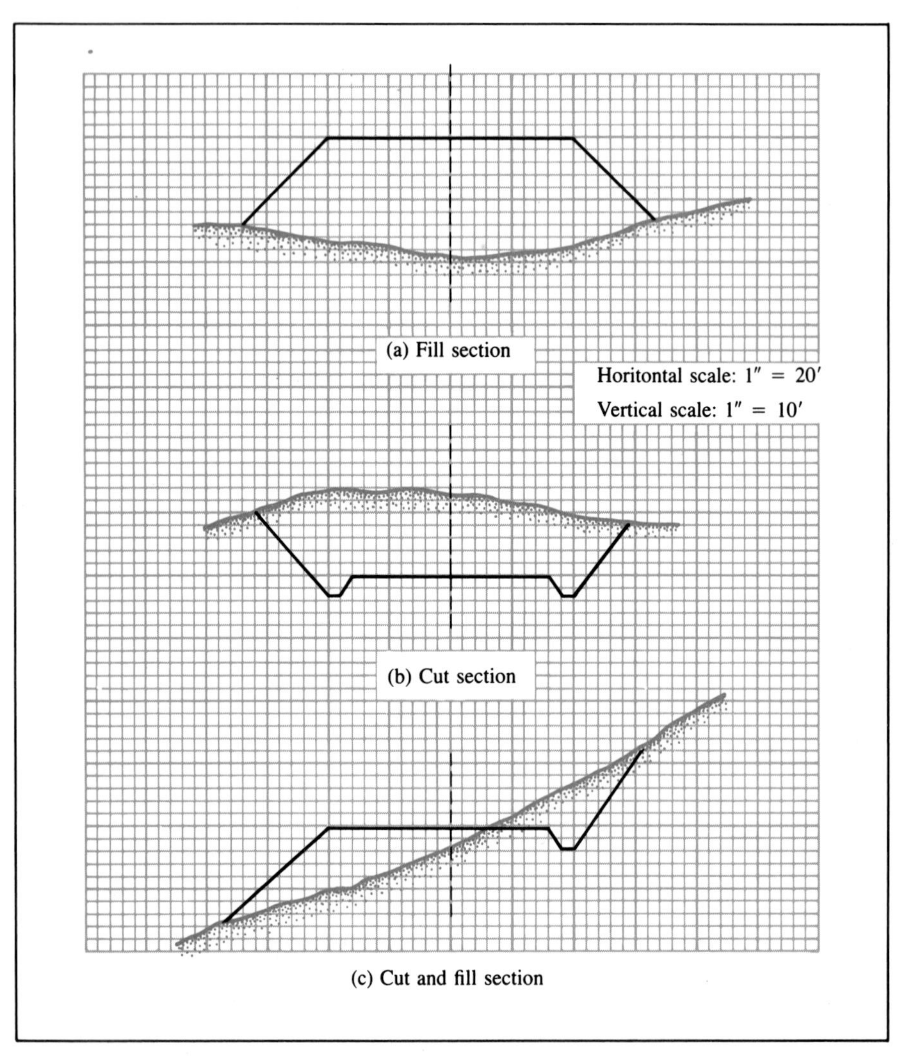

FIGURE 10.6 Highway cross-sections

When carefully manipulated, planimeter results are surprisingly accurate. The percentage of error decreases as the size of the area increases; one reason is, of course, that the area increases with the square of the perimeter. The mean of two determinations should be correct within 1% for small areas and may easily be within 0.1% for areas of 20 sq in. or more. This precision is sufficient for such purposes as determining drainage areas, cross-section and contour areas for earthwork, and reservoir areas and volumes. Its precision and facility in operation make this a most useful instrument for determining plotted areas of any shape whatsoever.

Electronic planimeters with digital displays can further increase both the speed and accuracy of area measurement. Figure 10.7 shows the

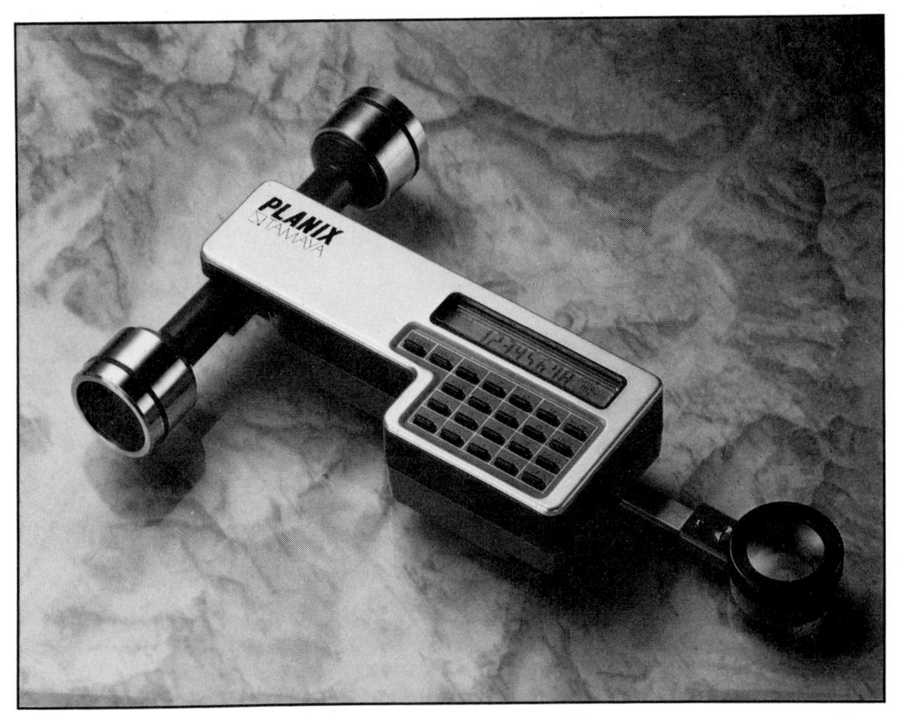

FIGURE 10.7 The TAMAYA Planix digital planimeter (courtesy of Tamaya Technics, Inc. and the Lietz Company)

TAMAYA Planix 7 digital planimeter. It can be operated either with rechargeable battery or on a standard AC adaptor. Among some of the convenient features are digital display of the area measurements, measurement in either metric or English units, automatic scale conversion, and automatic averaging of multiple measurements.

10.5 VOLUME BY AVERAGE-END-AREA METHOD

All highway and railroad construction projects involve a considerable amount of earthwork usually indicated by the terms *cut* (excavation) and *fill* (embankment). The quantity of earthwork on a particular job reflects some of the project design features and, most importantly, is a significant element of total cost.

To ascertain before construction the preliminary cut and fill amounts, it is necessary to fix the longitudinal gradeline, select a roadway design cross-section, and perform cross-sectioning as described in Section

4.29. The various cross-sections are then plotted and the outline (see Figure 10.6) of the proposed roadway, as traced from a template, superimposed on them. The areas (in sq ft) of cut and fill are then planimetered and the volumes computed by the *average-end-area method* or by the *prismoidal formula*. The former method is generally used by the engineering and construction community. The *A.E.A.* formula is

$$V = \left(\frac{A_1 + A_2}{2}\right) \frac{L}{27} \tag{10.10}$$

in which V is the volume in cubic yards, A_1 and A_2 are the areas of the end sections in square feet, and L is the distance between the sections in feet.

If the ground is uneven or changes slope abruptly, intermediate sections are taken such that the errors in resulting volumes will not be serious. Theoretically, this method is not exact unless the two end areas are equal, but except as indicated in the following article, the resulting errors are insignificant.

In the case of side-hill sections like that in Figure 10.6c, it is necessary to calculate the cut and fill separately because payment for earthwork is based on excavated quantities only. Fill quantities are needed for studies of balancing excavation and embankment.

EXAMPLE 10.4

The planimetered areas (ft^2) at two side-hill cross-sections of a proposed highway are as follows:

Sta. 42 + 00	*C*82	*F*112
Sta. 42 + 76	*C*214	*F*78

C denotes cut, and *F* denotes fill. Calculate the quantities of earthwork.

SOLUTION

$$\text{cut volume} = \left[\frac{82 + 214}{2}\right] \frac{76}{27} = 417 \text{ cy}$$

$$\text{fill volume} = \left[\frac{112 + 78}{2}\right] \frac{76}{27} = 267 \text{ cy}$$

When a roadway section changes from cut to fill or vice versa, the stationing of the *grade point* must be determined. This is the point where the grade line intersects the original ground. Then, the earthwork quantities are calculated for the fractional part of a full station length up to and, also, beyond the grade point.

Figure 10.8 shows a transition from a fill section at station 17 + 10 to a cut section at station 17 + 93, with a side-hill section at station 17 + 52,

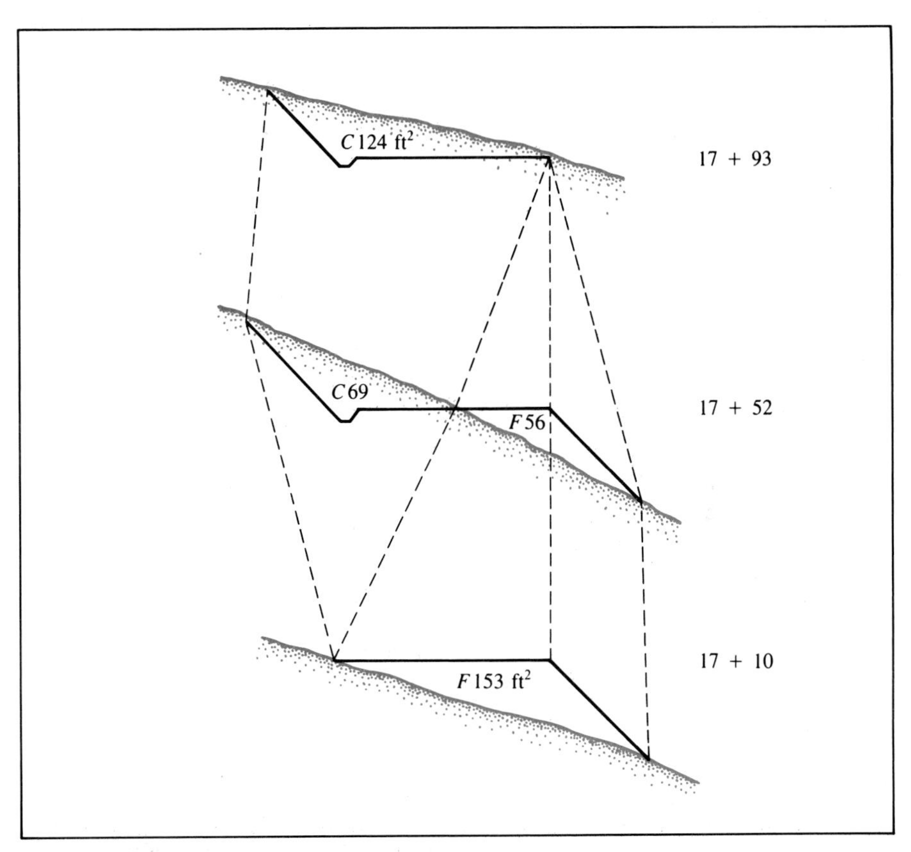

FIGURE 10.8 Earthwork volumes

which is a grade point. The volume of fill between stations 17 + 10 and 17 + 52 can be computed by the average-end-area formula:

$$\text{fill volume } (17 + 10 \text{ to } 17 + 52) = \left(\frac{153 + 56}{2}\right) \times \frac{42}{27}$$

$$= 163 \text{ cy}$$

The cut volume between stations 17 + 10 and 17 + 52 is a pyramid with a triangular base. Its volume should not be computed by the average-end-area formula. Instead, the following prismoidal formula for a triangular pyramid should be used.

$$V = \frac{1}{3 \times 27} A \times L \tag{10.11}$$

where A is the area of the triangular base in sq ft, L is the distance between the two sections in ft, and V is the volume in cubic yards. Thus,

$$\text{cut volume } (17 + 10 \text{ to } 17 + 52) = \frac{1}{3 \times 27} \times 69 \times 42 \text{ cy}$$
$$= 36 \text{ cy}$$

Similarly,

$$\text{fill volume } (17 + 52 \text{ to } 17 + 93) = \frac{1}{3 \times 27} \times 56 \times 41 \text{ cy}$$
$$= 28 \text{ cy}$$
$$\text{cut volume } (17 + 52 \text{ to } 17 + 93) = \left(\frac{124 + 69}{2}\right) \times \frac{41}{27}$$
$$= 147 \text{ cy}$$

Therefore, the total volumes of cut and fill between stations 17 + 10 and 17 + 93 are

$$\text{total cut volume} = 36 + 147 \text{ cy} = 183 \text{ cy}$$
$$\text{total fill volume} = 163 + 28 \text{ cy} = 191 \text{ cy}$$

10.6 PRISMOIDAL VOLUME

When the ground surface is such that the two end areas are widely different in area, or when high precision is desired, as in the cases of rock quantities in excavation and of volumes of concrete structures, the *prismoidal formula* may be used to compute the volume.

A prismoid may be defined as a solid having parallel, plane bases, and sides that are plane surfaces. Thus, by assuming straight lines joining the corners of two successive cross-sections, the following prismoidal formula may be derived:

$$V_P = \frac{L}{27}\left(\frac{A_1 + 4A_m + A_2}{6}\right) \tag{10.12}$$

in which V_P is the prismoidal volume in cubic yards, L is the perpendicular distance between end sections in feet, A_1 and A_2 are the areas of the end sections in square feet, and A_m is the area of a midsection in square feet. The midsectional area, A_m, will *not* ordinarily be the same as the mean of the two end areas. The dimensions of this midsection are computed as the average of the corresponding dimensions of the end sections. Figure 10.9 shows the calculation of the dimensions of the midsection between sections 17 + 10 and 17 + 52 in Figure 10.9. Using the computed dimensions, the fill area for

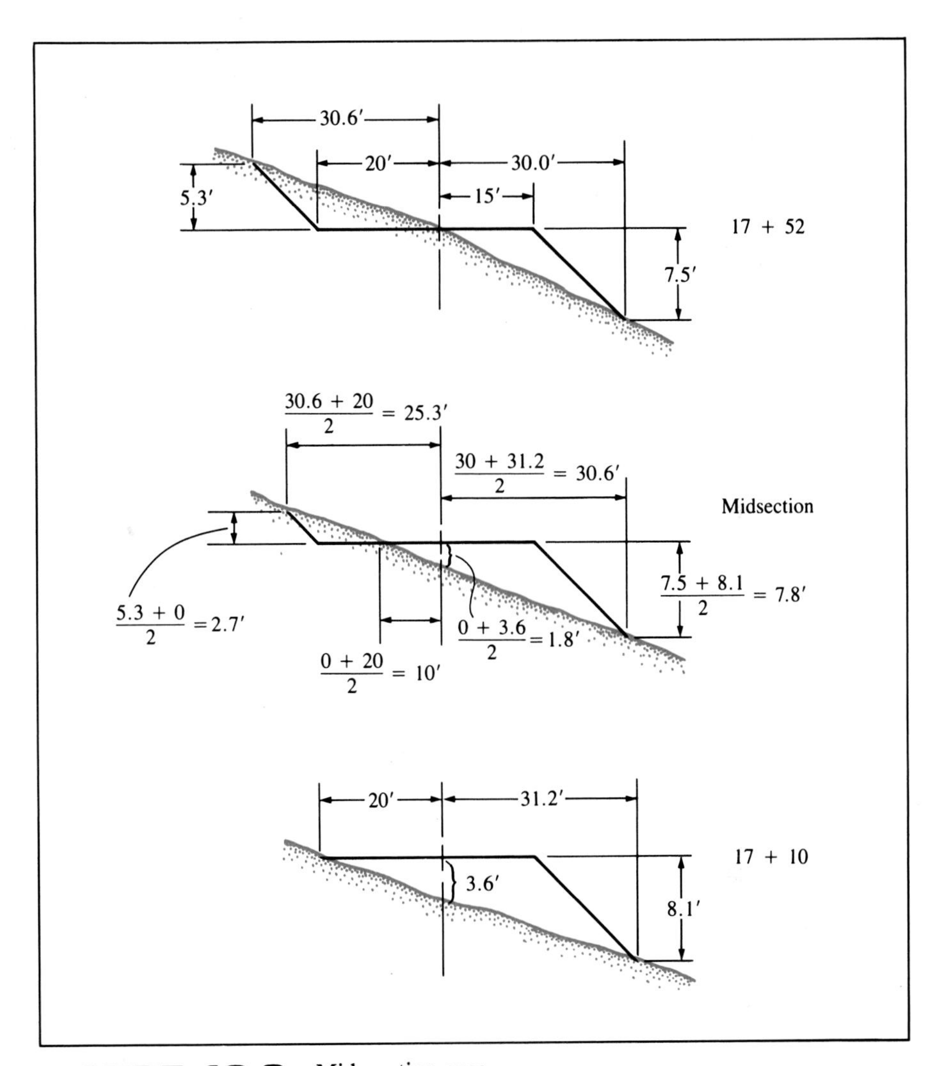

FIGURE 10.9 Mid-section area

the midsection is determined to be 95 sq ft. Thus the prismoidal volume of the fill section between station 17 + 10 and 17 + 52 is computed as follows:

$$V_P = \frac{42}{27}\left(\frac{153 + 4 \times 95 + 56}{6}\right) = 153 \text{ cy}$$

which is slightly smaller than the value of 163 cy previously computed by the average-end-area formula. The average-end-area volume is usually larger than the prismoidal volume, although the reverse can sometimes be true.

Note that the cut volume between stations 17 + 10 and 17 + 52 has already been computed previously using a prismoidal formula for a triangular pyramid—that is, Eq. (10.11).

10.7 AREA AND VOLUME FROM CONTOUR MAPS

Since a contour map is a representation of the earth's surface in its three dimensions, it is commonly used in engineering to determine volume of earthwork and water storage capacity of proposed reservoir sites.

Figure 10.10 illustrates the use of a contour map to estimate the amount of cut and fill necessary to construct a parking lot on a hillside. Scale of the map is 1″ = 500′. The outline of the parking lot is represented by the heavy line rectangle. The elevation of the lot's surface is to be 908 ft and the side slopes are to be 3 : 1—that is, 1 ft drop (or rise) for every 3 ft of horizontal distance. From these specifications, the contour lines representing the completed parking lot are drawn as shown. The contour lines above the 908-ft elevation represent cut (shown cross-hatched) and those below that elevation represent fill.

The volume of cut is found by determining with a planimeter the area of the closed 908-ft contour line in cut and that of the 910-ft contour line. The volume of earthwork between these two contours is then found as the average of the two areas multiplied by the contour interval, 2 ft.

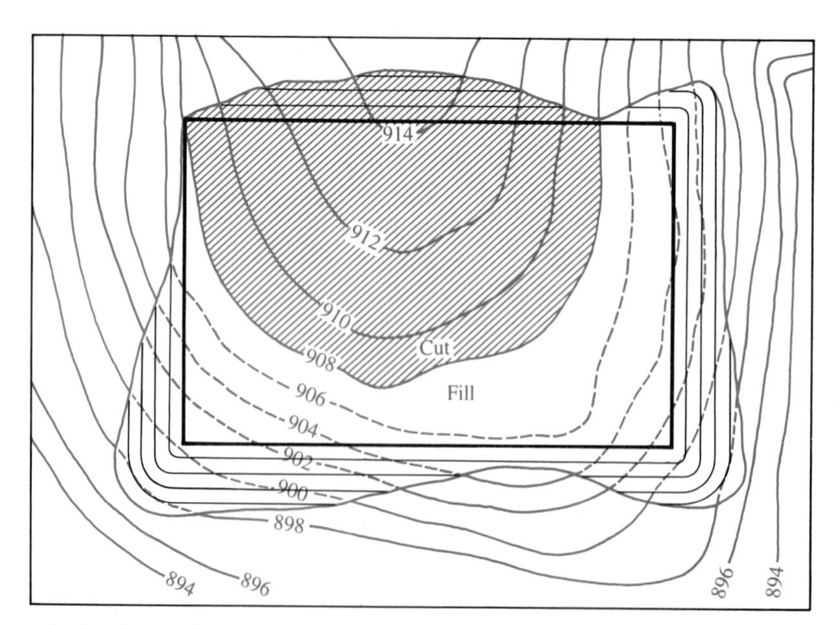

FIGURE 10.10 Earthwork volume from contours

Similarly, the volume between the 910- and 912-ft contours is found, and so on.

The volume of fill is found in a manner similar to that for the cut. It may be noted that, because of the ridge line through the parking space, the area of the 904-ft contour line is divided into two parts—one on the left and one on the right of the ridge. The same is true for the contour lines below 904 ft. The results may be recorded as follows:

Contour Line	AREA		Volume cu ft
	sq in.	sq ft	
CUT			
908	10.15	25,400	
			42,300
910	6.78	16,900	
			25,200
912	3.32	8,300	
			9,600
914	0.52	1,300	
			77,100 = 2,850 cu yd
FILL			
908	7.70	19,200	
			32,700
906	5.40	13,500	
			23,100
904	3.94	9,600	
			16,300
902	2.68	6,700	
			9,800
900	1.24	3,100	
			3,600
898	0.22	500	
			85,500 = 3,170 cu yd

The surface area of the parking lot is equal to the area enclosed in the 908-ft contour—that is, 25,400 sq ft.

In the design of reservoirs for water supply, power, or irrigation projects, the studies are made on contour maps to locate the dam, to determine the volume of water to be impounded, to locate the boundary of the area to be inundated, and to find the drainage area. The necessary maps will be drawn to different scales suitable to the different studies that are made. Thus a large-scale map will be used to fix the dam site, an intermediate-scale map will be used to determine the area and volume of the reservoir, and a small-scale map will be used to delineate the drainage area.

The method of finding the area and volume of the water to be stored is illustrated in Figure 10.11. As the water is impounded and rises by 10-ft stages to the elevations of 1,030, 1,040, and 1,050, it will have as its shoreline the corresponding contour lines shown. It may be noted that each contour

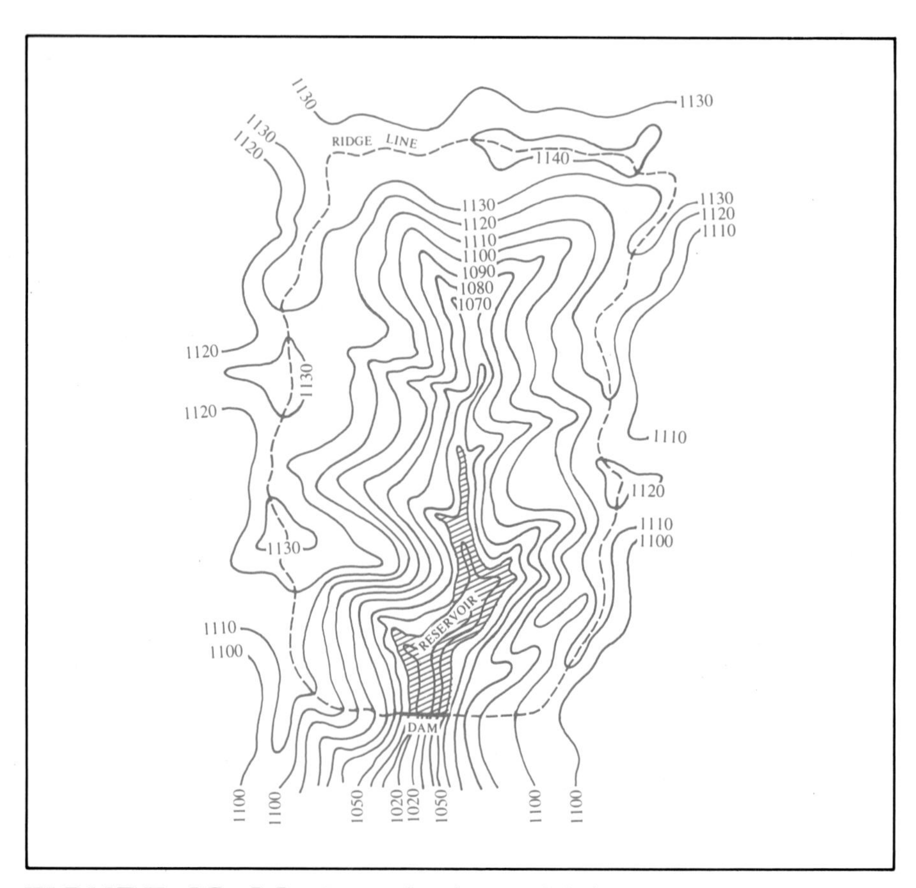

FIGURE 10.11 Reservoir volume and drainage area

line is a closed line within the reservoir area. If the areas within the 1,030- and the 1,040-ft contour lines are determined with a planimeter, averaged, and multiplied by the vertical distance between them (10 ft), the result is the volume of water included between these contours.

The total reservoir capacity, usually expressed in *acre-feet,* is the sum of the volumes between successive contours.

Obviously, the maximum flood line of the reservoir will be given by that contour having the elevation of the crest of the dam, increased by whatever head of water may exist above it.

The drainage area of the proposed reservoir can be found by locating the ridge line around the watershed, as shown in Figure 10.11. This line may not always be evident at all places on the map, and some field measurements are sometimes necessary. The drainage area is found by planimeter measurement around the ridge line. The area determined in this manner is not the land surface area, but is the area of the projection of the land surface onto a horizontal plane.

Point	BS	HI	FS	Elev	Old elev.	Cut (c)	No. of squares (n)	c x n
BM 4129	4.1	745.3		741.2	745.3			
A1			3.1	742.2	746.4	4.2	1	4.2
A2			4.7	740.6	744.2	3.6	2	7.2
A3			5.6	739.7	742.2	2.5	1	2.5
B1			4.2	741.1	745.7	4.6	2	9.2
B2			6.3	739.0	745.8	6.8	4	27.2
B3			8.1	737.2	740.2	3.0	2	6.0
C1			4.7	740.6	746.4	5.8	2	11.6
C2			7.8	737.5	744.7	7.2	4	28.8
C3			9.1	736.2	738.8	2.6	3	7.8
C4			8.6	736.7	739.1	2.4	1	2.4
D1			6.3	739.0	745.4	6.4	1	6.4
D2			5.1	740.2	746.1	5.9	2	11.8
D3			4.7	740.6	745.9	5.3	2	10.6
D4			4.2	740.1	741.7	1.6	1	1.6
BM 4129			4.1	741.2	Check			
							Sum =	137.3

June 12, 1983
ⲡ D. Baines
Ⅱ C. Smith
Ⅲ K. Mac Innes

1 2 3 4
A B C D
50' 50' 50' 50' 50' 50'

U.S. Route 15

$$\text{Volume} = \frac{50 \times 50}{4 \times 27} \times 137.3$$
$$= 3{,}178 \text{ cy}$$

FIGURE 10.12 Borrow pit survey

10.8 BORROW PITS

In engineering construction, it is frequently necessary to excavate material from areas adjacent to the project to form the embankments. Such excavations are called *borrow pits*. The volume of excavation may be determined by taking conventional cross-sections before and after removal of the material. However, the computation can be simplified by first dividing the area into squares of suitable size and then taking level readings at the corners of the squares both before and after the excavation. Figure 10.12 shows the field measurements and calculations involved in a borrow pit survey.

The volume removed from any square is computed as the average of the heights of the four corners times the area. Some corners are common to more than one square; thus *A*3, in Figure 10.12, is common to one, *B*3 is common to two, *C*3 is common to three, and *B*2 is common to four squares. By summing up the volumes within all the squares, the following equation results:

$$V = \frac{A}{4 \times 27} (\Sigma h_1 + \Sigma 2h_2 + \Sigma 3h_3 + \Sigma 4h_4)$$

in which V is the volume in cubic yards; A is the area of one square in square feet; h_1, h_2, h_3, and h_4 are the corner heights common to one, two, three, and four squares respectively.

PROBLEMS

10.1 Find the area to the nearest 0.001 acre enclosed in the following traverses:

a. Traverse *ABCDEA*

Station	X (ft)	Y (ft)
A	67,129.31	84,277.64
B	67,831.29	84,342.91
C	67,900.65	84,597.85
D	67,478.63	84,623.47
E	67,311.54	84,462.14

b. Traverse 1-2-3-4-5-6-1

Station	X (ft)	Y (ft)
1	17,431.22	26,189.27
2	17,000.56	26,344.51
3	16,439.44	26,516.34
4	16,663.17	26,729.41
5	16,600.25	26,598.63
6	16,812.83	26,314.78

c. Traverse *FGHIJF*

Station	X (ft)	Y (ft)
F	−57.41	−231.66
G	−311.26	−79.49
H	−31.66	123.48
I	62.35	309.11
J	172.76	−19.44

10.2 To determine the area of a tract of land bordered by the high-water line of a sinuous stream, offsets from a transit line were measured at regular intervals of 30 ft as follows: 20.8, 16.7, 21.5, 29.3, 31.0, 25.1, 15.7, 18.0, and 23.2. Find the area by: (a) the trapezoidal rule, (b) Simpson's one-third rule.

10.3 The area of an irregular parcel of land is to be determined from the following field record of offsets from a transit line *AB*:

Distance from Point A (ft)	Offset (ft)
0	14.0
15	22.5
35	12.7
70	24.2
105	27.3
120	12.0
155	5.2

Find the area by the method of coordinates.

10.4 A fixed-arm polar planimeter is used to measure the area of a highway cross-section that has been plotted with a horizontal scale of 1 in. = 20 ft and a vertical scale of 1 in. = 5 ft. The mean difference for several runs between the initial and final readings was 1,578 with one unit in the fourth place being equal to 0.01 sq in.

Find the end area to the nearest square foot.

10.5 A planimeter was used to determine the area of a tract of timber shown on an aerial photograph whose scale is 1 in. = 400 ft. If the initial reading was 3,781 and the final reading (going clockwise) was 6,359 and the unit's place has a value of 0.01 sq in., find the area to the nearest acre.

10.6 The shoreline of a reservoir depicted on a contour map was traced with a polar planimeter in order to determine the area of the water surface. The mean difference (final minus initial, going clockwise) in the readings was 2,462 and the smallest unit had a value of 0.015 sq in. The scale of the map was 1 cm = 500 m. Calculate the area and express it to the nearest 1,000 m^2.

10.7 The end areas along a proposed drainage ditch are as follows:

Station	End Area (sq ft)
16 + 22	85
17 + 00	174
17 + 81	260

Find the total volume (nearest cubic yard) of excavation between 16 + 22 and 17 + 81.

10.8 The end areas along a proposed levee are as follows:

Station	End Area (sq ft)
37 + 00	596
38 + 00	1,372
38 + 80	978

Find the total volume (nearest cubic yard) of embankment between 37 + 00 and 38 + 80.

10.9 The planimetered areas, in square feet, of cross-sections along a short portion of a proposed highway improvement project are as follows:

Station	Cut	Fill
117	92	0
118	60	0
118 + 30	11	24
119	0	36
120	0	202

Calculate the total volumes (nearest cubic yard) of cut and fill between 117 + 00 and 120 + 00.

10.10 The cross-section areas, in square feet, for four consecutive sections are as follows:

Station	Cut	Fill
512 + 00	137	122
513 + 00	102	68
513 + 75	94	0
514 + 00	151	0

Calculate the total volumes (nearest cubic yard) of cut and fill between 512 + 00 and 514 + 00.

10.11 Figure P10.1 shows the field notes from a borrow pit survey. Determine the volume of earth removed from the pit between the two surveys.

Point	BS	HI	FS	Elev.	Old elev.	Cut (c)	No. of squares (n)	c × n	
BM 420	4.6			531.4	531.4				
A1			4.6		530.2				
A2			5.3		531.0				
A3			7.2		529.8				
A4			6.7		525.3				
B1			5.1		531.2				
B2			7.9		633.6				
B3			8.4		531.5				
B4			5.2		630.6				
C1			4.4		529.6				
C2			6.1		528.2				
C3			7.2		530.3				
C4			5.8		529.5				
D1			3.1		527.6				
D2			5.2		528.1				
D3			7.6		530.4				
E1			4.5		529.5				
E2			6.7		531.0				
E3			7.5		528.7				
BM 420			4.6		Check				

June 12, 1983
⊼ D. Baines
Ⅱ C. Smith
Ⅲ K. MacInnes

FIGURE P10.1 Field notes from a borrow pit survey

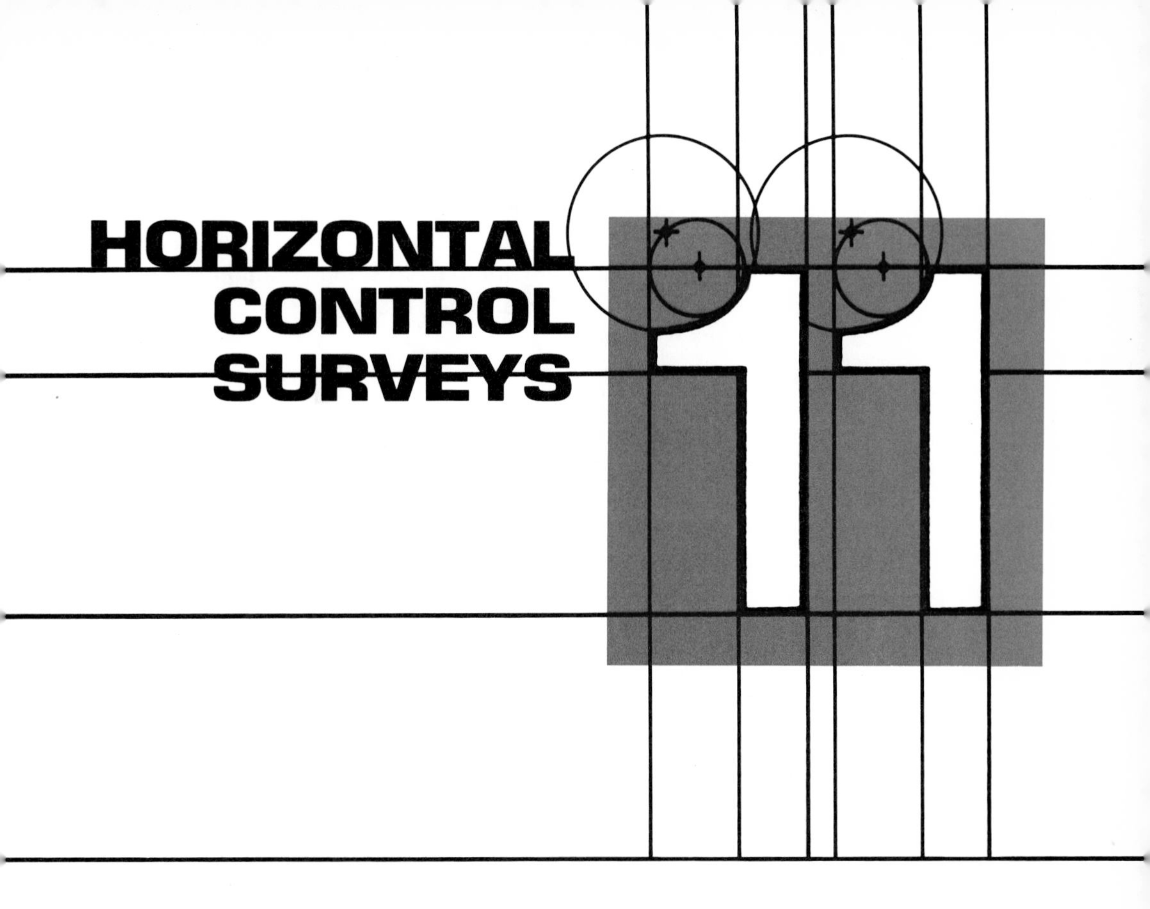

11.1 INTRODUCTION

A horizontal control survey is the surveying process required to establish the horizontal positions of a number of points located on or near the surface of the earth. Such points, after their positions have been determined, are referred to as *horizontal control points*.

A horizontal control survey is usually one of the first steps involved in an engineering project. For example, in a dam construction project, a control survey is conducted to establish a number of control points within the project area. These control points are then used in conjunction with aerial photographs (see Chapter 14) to prepare topographic maps of the area. After feasibility studies have been completed and the final design drawings prepared, the same control points are then used to locate the dam and other engineering facilities on the ground. Thus these control points provide the important link between the physical surface of the earth and the engineering designs. Even after the completion of the project, these control points are often used to monitor the performance of the engineering facilities. It is not unusual for such control points to have a useful life of tens of years.

A horizontal control survey usually involves the measurement of angles and distances. For a small project area, the control survey may simply involve one or more interconnecting traverses. For large project sites, however, a horizontal control network consisting of triangles and quadrilaterals is usually required to achieve the required positioning accuracy.

Before the development of EDM instruments, it was extremely difficult to measure long distances with high accuracy. Consequently, most horizontal control surveys conducted in the past relied on the measurement of angles and on a method called *triangulation*. With the common availability of EDM instruments, most horizontal control surveys now involve both distance and angle measurements.

Recent developments in satellite positioning systems are on the threshold of completely revolutionizing the horizontal control survey process. Two miniature portable antennas set up over two points to continuously track a group of satellites for a period of about two hours can determine the relative positions of the two points with an accuracy equal to, if not exceeding, that of conventional ground survey methods. It is now highly likely that a "black box" surveying system will be available in the near future. By positioning the "black box" over a survey point for a period of less than one hour, it will likely be possible to obtain the geographic position of the point with an error of less than 1 cm in all three coordinates.

Nevertheless, traditional methods of control survey can be expected to continue to play important roles in many engineering projects for years to

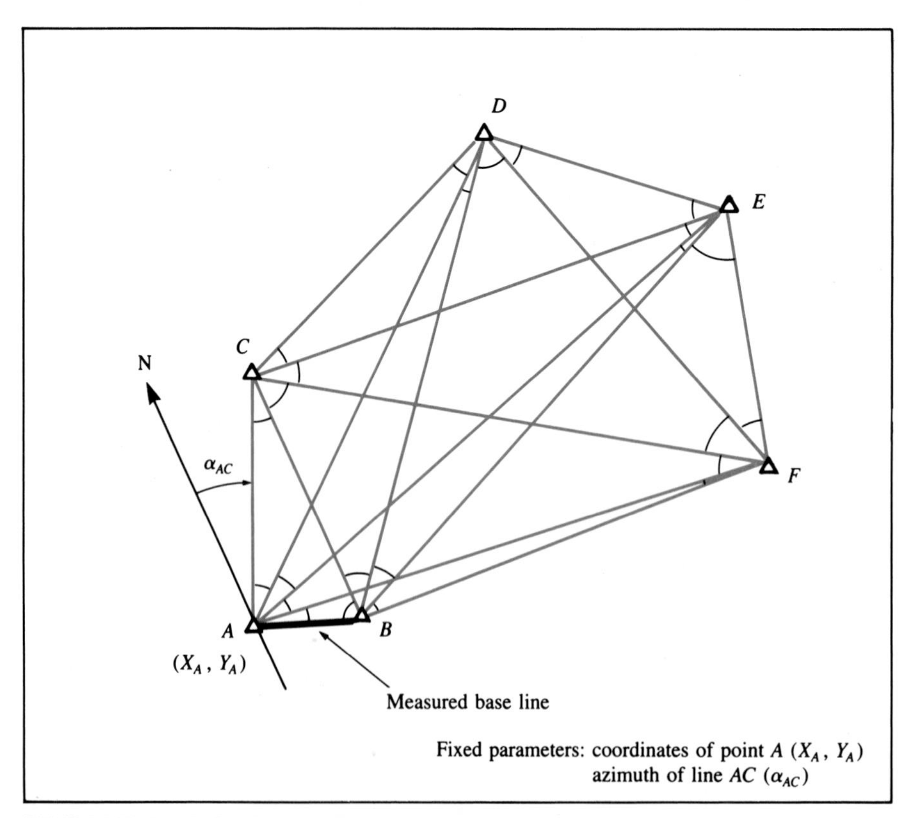

FIGURE 11.1 A triangulation network

come. It is beyond the scope of this book to deal in depth with the methods of control surveys. This chapter is intended primarily to present the basic principles of the traditional methods as well as the potential capability of the evolving satellite positioning systems.

11.2 TRIANGULATION

Figure 11.1 shows a triangulation network consisting of six survey stations: A, B, C, D, E, and F. The length of the line AB is measured. Since it provides the scale for the network, it is called the *baseline*. All the horizontal angles subtended at each station in the network are also measured. These angles, together with the baseline measurement, are then used to compute the horizontal positions of the survey stations.

Before the common availability of EDM instruments, baselines were usually measured by using invar tapes. Such baselines were usually kept very short (few hundred feet) and had to be located in relatively flat terrain to achieve accuracy in the order of 1/50,000. In large triangulation networks, usually more than one baseline was measured.

To define the survey datum for the triangulation network, the position of at least one point and the direction of at least one line must be either known or arbitrarily defined. Most commonly, one or more of the stations in a triangulation network are so-called NGS stations, which are part of the national geodetic network and whose positions have already been determined by the National Geodetic Survey (NGS).

11.3 TRILATERATION

With the availability of EDM instruments, it is now considerably more convenient to measure distances than angles. One can determine the positions of new control points by measuring distances only. For example, instead of measuring the horizontal angles in the network in Figure 11.1, the lengths of all the lines can be measured. A control network in which only distances are measured is called a trilateration network, and the survey process is called *trilateration*.

11.4 HORIZONTAL CONTROL NETWORK

The present practice in horizontal control survey is to measure as many angles and distances in a network as possible to achieve maximum survey accuracy. All the survey measurements are then used in a least-squares adjustment to determine the positions of the survey stations.

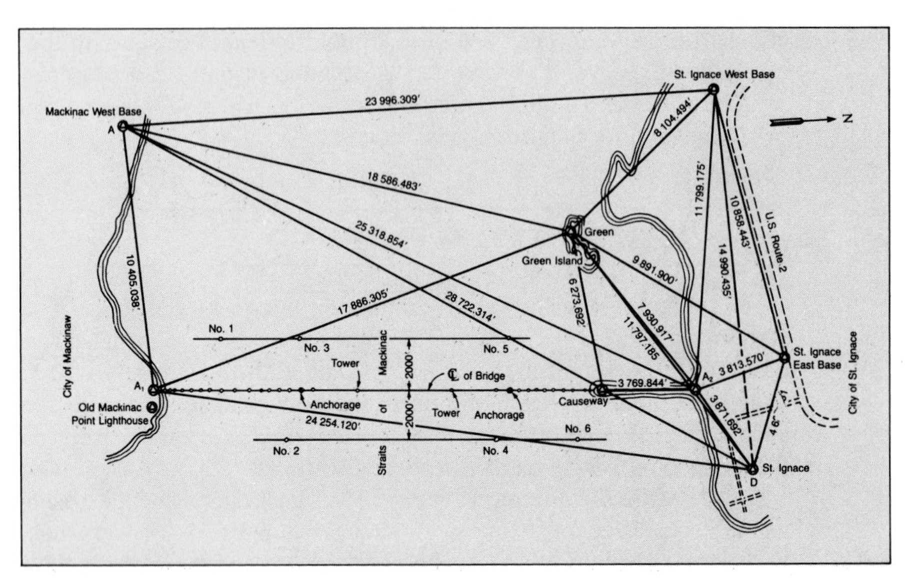

FIGURE 11.2 Control network for Mackinac Straits bridge (*Civil Engineering*)

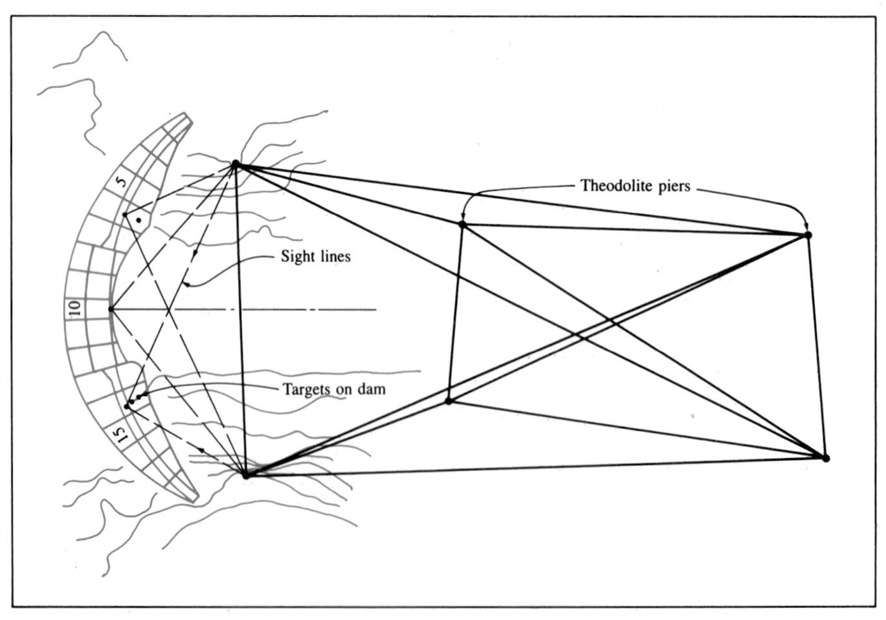

FIGURE 11.3 A control network for dam deformation survey (courtesy of Bureau of Reclamation)

Figure 11.2 shows a control network used to determine the positions of eight survey stations. The locations of the survey stations in a control network are carefully selected with due consideration to topographic constraints, engineering requirements, and proper geometric shape of the network.

Figure 11.3 shows a control survey network used to establish the positions of six control points for monitoring dam deformation. After the positions of these six control points have been determined, the position of a point on the dam structure is determined by intersection from two or more of these control points.

11.5 PRELIMINARY CHECKS OF FIELD MEASUREMENTS

In control surveys several repetitions are usually made of each angle and distance that are to be measured. The number of repetitions needed depends on the precision of the surveying instruments and on the accuracy requirements of the project (see Chapter 2). To minimize the effects of atmospheric conditions on the survey measurements, an angle or distance is sometimes measured on two or more different days and during different times of the day. The mean and standard deviation are computed in the field for each group of repeated measurements to assure that the expected precision is achieved.

After all the angles and distances have been measured, the averaged values of these measurements are then used in a least-squares adjustment to determine the most probable coordinates of the new stations in the control network. There is usually a time period of several days to several months that exists between the completion of the field measurements and the final computation of the station coordinates. Should blunders or large errors be found in the field measurements during the computation phase, expensive return trips to the project site might be needed to obtain additional measurements. To avoid such expenses, one should always perform preliminary checks on the field measurements while the measuring program is in progress and the survey parties are still on the project site. The accuracy of the field measurements can be checked by using the following two geometric conditions:

1. The sum of the three included angles in a spherical triangle must be equal to $180° + \varepsilon$, where ε is called the *spherical excess* and is due to the curvature of the earth. For a triangle having an area of 75 square miles, ε amounts to 1″. Thus, in most control surveys associated with engineering projects, the spherical excess can be ignored.

2. Given any two angles in a triangle and the length of the side included by those two angles, the lengths of the two remaining sides can be computed by the sine law.

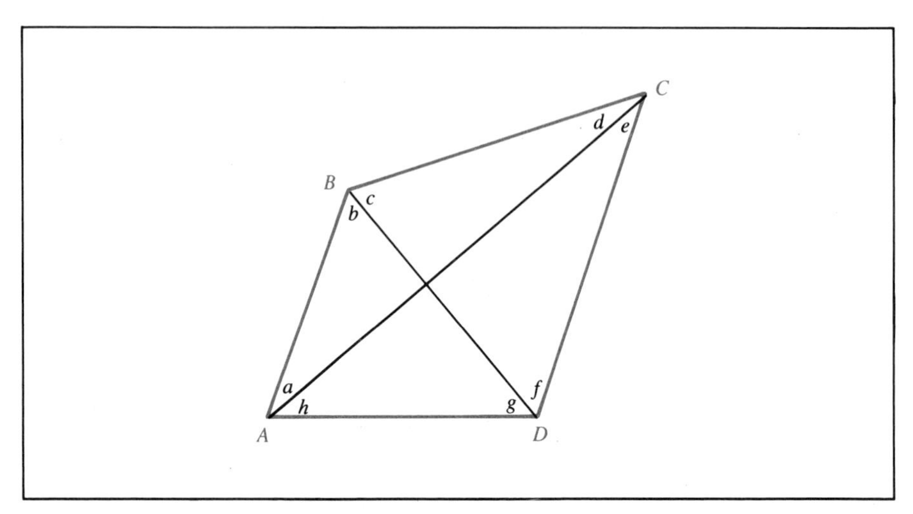

FIGURE 11.4 A quadrilateral

For example, consider the quadrilateral shown in Figure 11.4. The averaged values of the measured angles and distances are as follows:

Angles	°	′	″	Distances (ft)	
$a =$	29	25	34	$AD =$	984.684
$b =$	58	41	20	$AB =$	890.540
$c =$	69	36	05	$DB =$	1,087.883
$d =$	22	17	02	$AC =$	1,843.258
$e =$	30	29	25	$BC =$	1,153.857
$f =$	57	37	33	$DC =$	1,280.632
$g =$	50	35	44		
$h =$	41	17	21		

The following closure checks can be performed on the angles:

$$(a + b + c + d) - 180° = +0°00'01''$$
$$(e + f + g + h) - 180° = +0°00'03''$$
$$(a + b + g + h) - 180° = -0°00'01''$$
$$(c + d + e + f) - 180° = +0°00'05''$$

Suppose that all the angles in the quadrilateral are measured with the same accuracy and are each expected to have a standard error of σ. Then the sum of the four measured angles in each of the triangles in the above example would have a standard error of $\sqrt{4}\,\sigma$. The maximum error in the sum of the angles in a triangle should be $3\sqrt{4}\,\sigma$, or 6σ. Thus the maximum closure error in a triangle should not exceed 6σ. In this example suppose that the standard error of each angle is expected to ±1″. Then the maximum allowable closure error in a triangle is ±6″.

To check the measured distances in the quadrilateral, one of the distances is assumed to be correct and the lengths of the remaining sides are computed by the sine law. Assuming that the measured value of distance *AD* is correct, the length of side *AB* can be computed as follows:

$$AB = \frac{AD \sin g}{\sin b} = 890.552 \text{ ft}$$

Thus the difference (ΔAB) between the computed and measured values of $AB = 890.540 - 890.552 = -0.012$ ft. The measured values of the remaining sides in the quadrilateral can be similarly checked as follows:

$$\Delta DB = 1{,}087.883 - \frac{AD \sin (a + h)}{\sin b} = +0.011 \text{ ft}$$

$$\Delta AC = 1{,}843.258 - \frac{890.552 \sin (b + c)}{\sin d} = -0.057 \text{ ft}$$

$$\Delta BC = 1{,}153.857 - \frac{890.552 \sin a}{\sin d} = +0.024 \text{ ft}$$

$$\Delta DC = 1{,}280.632 - \frac{AD \sin h}{\sin e} = +0.058 \text{ ft}$$

The differences between the computed and measured distances are attributed to errors in both the measured distances and the angles. Generally, these computed differences should be within two times the expected maximum errors in the distance measurements. For example, if the distances were all measured with a relative precision of 1/100,000 at 1σ, then the maximum expected error (3σ) in distances *AB*, *DB*, *AC*, *BC*, and *DC* are ±0.03, 0.03, 0.06, 0.03, and 0.04 ft respectively. The acceptable differences between the computed and measured values of *AB*, *DB*, *AC*, *BC*, and *DC* should then be ±0.06, 0.06, 0.12, 0.06, and 0.08 respectively.

11.6 LEAST-SQUARES ADJUSTMENT

In a horizontal control network, there are usually more angles and distances measured than the absolute minimum number required. The method of least squares is used almost universally to determine the most probable solution of the station coordinates from the angle and distance measurements. Basically, in a least-square adjustment, an equation is written for each measured angle or distance in the network. All the equations are then used to compute the station coordinates, which will result in minimizing the sum of squares of the residual errors in the measurements—that is, $\Sigma P_i V_i^2$, where V_i is the residual error in the i-th measurement and P_i is the weight.

The method of least-squares adjustment and its application in horizontal control survey are well documented in the literature (References 11.9 and 11.1). Computer programs for performing least-squares adjustment of horizontal control networks are also available (References 11.1, 11.6, 11.7, and 11.12). Such a computer program usually provides a general solution to the problems of horizontal control surveys; that is, the solution is independent of the geometric shape of the survey network, which can be a traverse, a quadrilateral, or a network of triangular figures. The only requirement is that sufficient angle and distance measurements exist to determine the coordinates of the new survey stations. The input to such a computer solution usually includes the angle and distance measurements, estimated standard errors of these measurements, measured azimuths, and known coordinates of the existing stations. The output usually includes the most probable coordinates of the new survey stations as well as the estimated standard errors of these computed coordinates.

11.7 ERROR ELLIPSE

Some computer programs for least-squares adjustment of horizontal control networks also provide in the outputs the parameters for the error

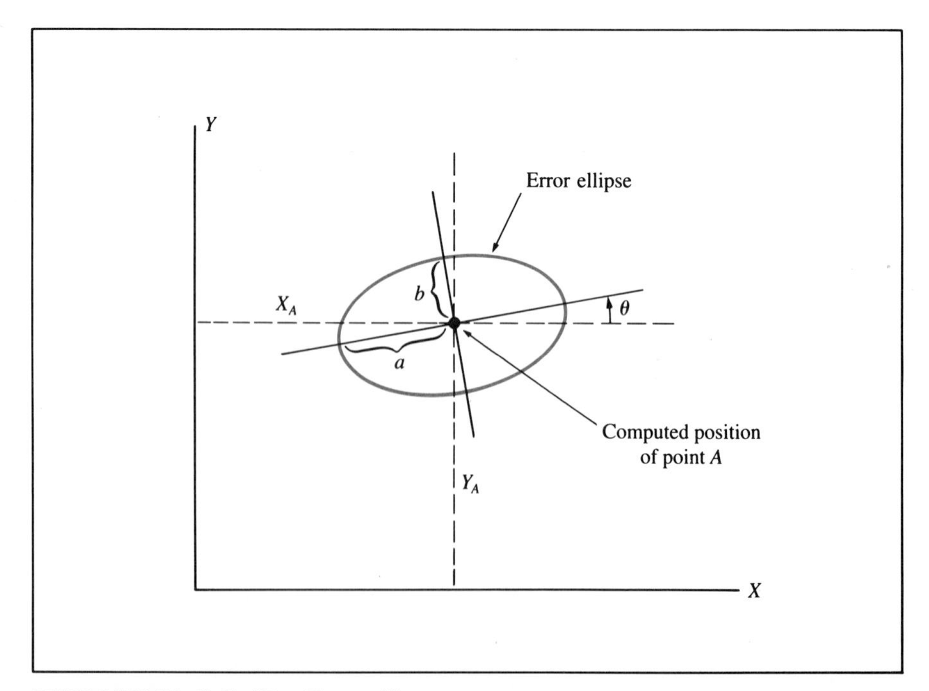

FIGURE 11.5 Error ellipse

ellipse of each new survey station. An error ellipse is defined by its semi-major diameter (a), semi-minor diameter (b), and the direction of the semi-major diameter, as shown in Figure 11.5. The error ellipse for a new survey station in the control network can be computed at any specified confidence level. For example, an error ellipse of the 95% confidence level defines a boundary around the computed position of the new station within which there is 95% chance that the survey station is actually located. The sizes and shapes of the error ellipses depend on the accuracy of the field measurements and shape of the geometric network.

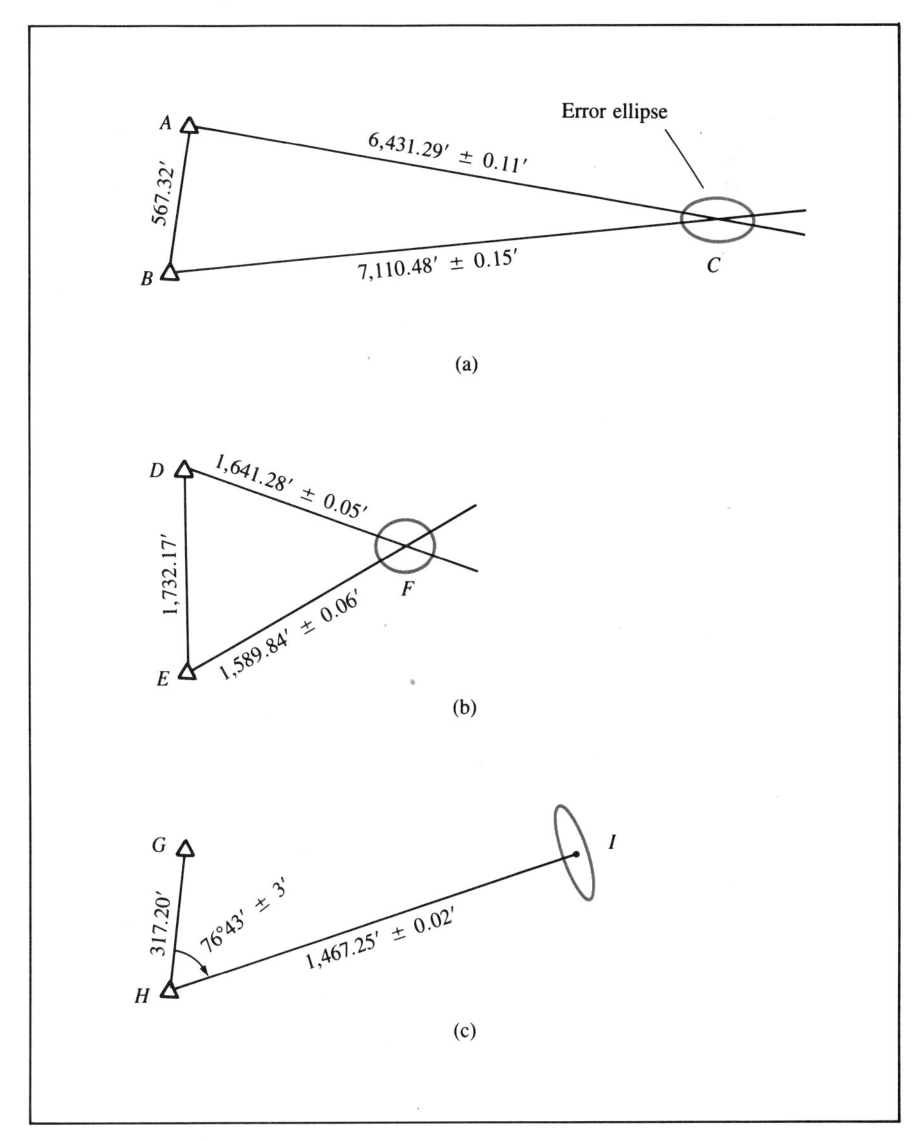

FIGURE 11.6 Effects of geometry and measurement accuracy on error ellipse

Figure 11.6 on the previous page illustrates the effects of network geometry and measurement accuracy on the size, shape, and orientation of the error ellipse. In Figure 11.6a, the location of station *C* is to be determined by distance intersection from a short baseline *AB*. Because of the poor intersection geometry along the direction perpendicular to the base line *AB*, the computed position of point *C* has a much higher uncertainty (low accuracy) along that direction. Thus the error ellipse of point *C* will have the long dimension oriented along that direction.

In Figure 11.6b, the location of station *F* is also determined by distance intersection from base line *DE*. However, since the three sides *DE*, *EF*, and *DF* are approximately equal (thus providing good intersection geometry), the error ellipse almost resembles a circle.

In Figure 11.6c, the location of station *I* is to be determined by angle and distance measurements from station *H*. It is assumed that the distance *HI* is to be measured much more accurately than the angle $G\hat{H}I$. Thus the error ellipse for the computed position of station *I* will have its large dimension perpendicular to the line *HI*.

11.8 ORDERS OF ACCURACY

Table 11.1 summarizes the accuracy standards recommended by the Federal Geodetic Control Committee (see Section 4.23) for horizontal control surveys. The following orders of accuracy were established:

1. First-Order
2. Second-Order, Class I
3. Second-Order, Class II
4. Third-Order, Class I
5. Third-Order, Class II

Complete details on the FGCC standards and specifications can be found in References 11.4 and 11.5.

11.9 NETWORK DESIGN AND PLANNING

During the design and planning of a horizontal control survey, the following questions must be raised and answered:

1. What is the required positional accuracy of the new survey stations?
2. How many existing control points are available and what is their accuracy?

TABLE 11.1 Accuracy Standards for Horizontal Control Surveys (According to Federal Geodetic Control Committee)

	First-Order	Second-Order		Third-Order	
		Class I	Class II	Class I	Class II
Relative positioning accuracy between directly connected adjacent points at the 2σ level	$\frac{1}{100{,}000}$	$\frac{1}{50{,}000}$	$\frac{1}{20{,}000}$	$\frac{1}{10{,}000}$	$\frac{1}{5{,}000}$
Direction Measurements					
1. Least count of instrument	0.2″	0.2″	1″	1″	1″
2. No. of sets (one direct and one reverse per set)	16	16	12	4	2
3. Maximum triangle closure seldom to exceed	3″	3″	5″	5″	10″
Recommended Uses	Primary national control network, metropolitan control networks; crustal movement studies; large engineering projects	Secondary national control network; metropolitan control networks; large engineering projects	Controls along coastlines and inland waterways; highway construction; land subdivision	Local engineering projects; topographic and hydrographic surveys	

3. Which coordinate system is to be used?
4. Where are the new survey stations to be located?
5. Which angles and distances are to be measured?
6. Which instruments are to be used for field measurements?
7. How many repetitions are to be made for each angle and distance in the control network?
8. How are the field measurements to be processed to determine the coordinates of the new survey stations?

The proper design of control survey networks requires a good understanding of the practical limitations of field conditions and surveying personnel, as well as the technical fundamentals discussed throughout this book.

Some computer programs for the least-squares adjustment of horizontal control networks can also be used for simulation studies during the design process. A topographic map is usually used in conjunction with field reconnaissance to determine the general locations of the new survey stations. The approximate coordinates of these stations can be scaled from the maps. By using these coordinates, the exact values of the angles and distances in the network can be computed. These angles and distances are then treated as "simulated" field measurements and are assumed to have a certain desired level of accuracy. These simulated data are then used as input to a least-squares adjustment. The output from the solution then provides the estimated error ellipse parameters for all the new stations based on the given set of simulated data. In essence, the computer program is used to conduct an error propagation study. By means of this approach, several different network designs can be tested to arrive at one that best satisfies all the project requirements.

11.10 MONUMENTATION

A horizontal control point must be so constructed that its anticipated movement is well within the accuracy with which it is surveyed. It must also remain intact and be protected from damage throughout the expected duration of the engineering project. For these reasons wooden grade stakes and rod-type monuments, such as those used as leveling bench marks, are seldom used as horizontal control points except in small and short-duration engineering projects.

Figure 11.7 shows a basic design for a concrete horizontal control monument. It consists of a $2' \times 2' \times 6''$ thick base buried below the frost line and an 8-inch diameter column. A brass survey marker (see Figure 11.8) is embedded in the concrete at the top of the column. An indentation mark on the brass marker indicates the exact location of the survey point. To mini-

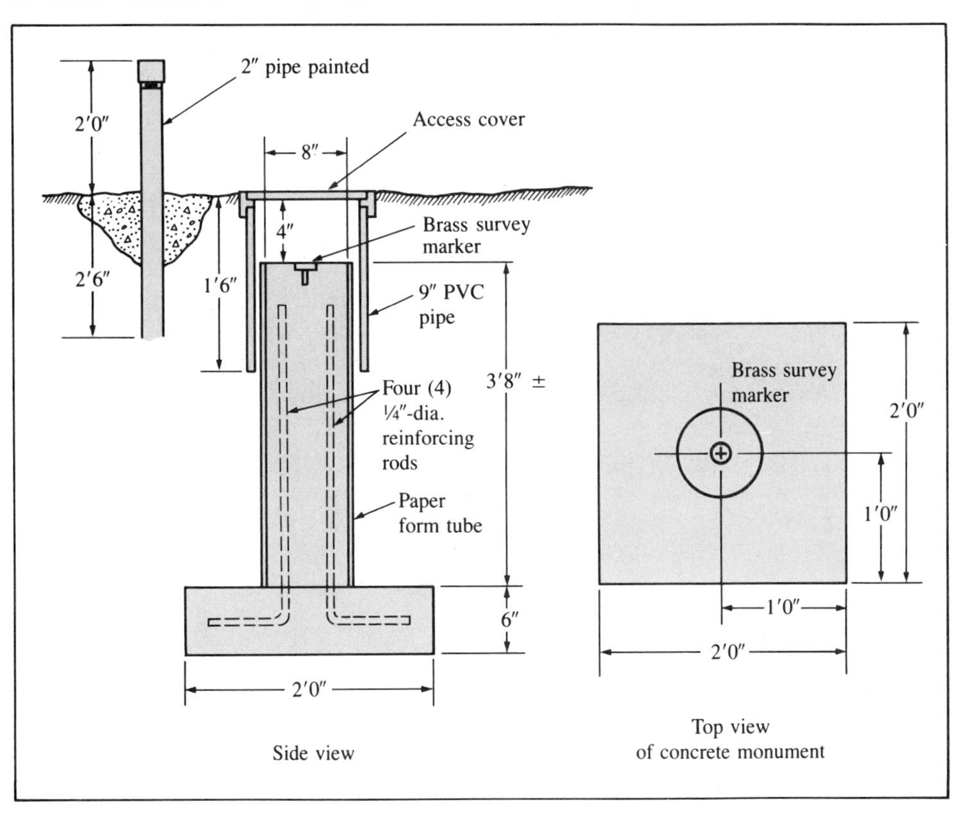

FIGURE 11.7 Concrete horizontal control monument

FIGURE 11.8 Horizontal control mark (courtesy of National Geodetic Survey)

mize the possibility of damage by surface vehicles, the top surface of the concrete column is positioned about 4 in. below the ground surface and is protected by a 9-inch diameter outer pipe and an access cover. In rural areas a witness post can be installed within two or three feet of the concrete monument to increase the ease of its location by survey crews. Unfortunately, such a witness post also draws the undesirable attention of vandals. Regardless of whether a witness post is installed, a good sketch should always be drawn to illustrate the exact location of the survey monuments. The sketch should provide sufficient information for a survey crew to positively locate the monument (see Figure 11.9).

Figure 11.10 shows a pillar concrete monument used in a dam deformation survey. The concrete column is extended to a height of about 3 ft above ground elevation. On top of the column is mounted a stainless steel plate and a threaded bolt for accepting a theodolite or a target. In this manner the use of a tripod is eliminated and the error from centering the

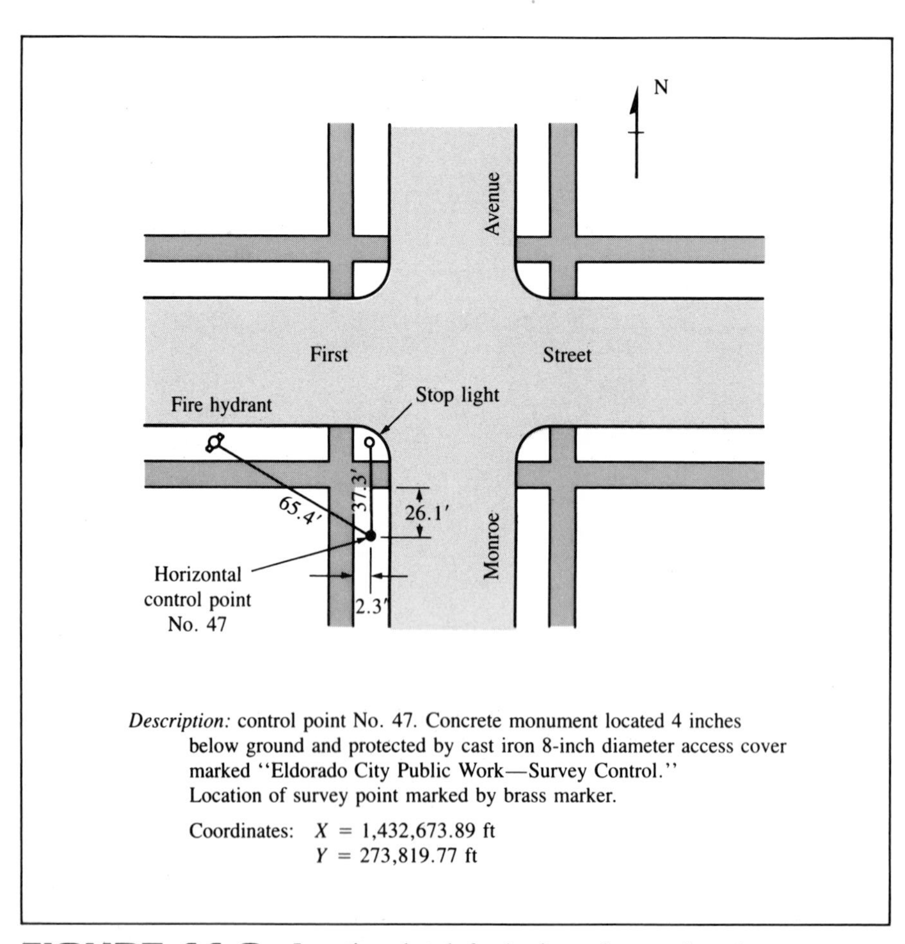

FIGURE 11.9 Location sketch for horizontal control mark

FIGURE 11.10 Pillar monument for dam deformation observations (courtesy of Bureau of Reclamation)

theodolite over the survey station is practically reduced to zero. The concrete pillar also provides a more stable support for the theodolite than a regular tripod.

11.11 CONTROL DENSIFICATION

After a basic network of control points has been established over a project area by the methods described in the previous sections, additional control points can be established by traversing, intersection, or resection. Such control points are usually needed for topographic mapping as well as layout purposes. They can be intended for either temporary or long-term uses. In all instances the basic network of control points is used to determine the location of the additional points. This survey process is usually called *control densification.*

11.12 BLACK BOX POSITIONING SYSTEMS

Traditional methods of horizontal control survey are both time consuming and labor intensive. Even for moderate size engineering projects, horizontal control surveys may require 3 to 6 months for completion. The surveying process can be particularly difficult under adverse weather and/or terrain conditions. Thus the development of black box positioning systems, which are capable of determining the three-dimensional positions of survey stations without measuring either angles or distances, represents a major breakthrough in surveying and mapping. To the surveyor such a positioning system can be considered a black box. It usually consists of a hardware component, which is used to collect field data, and a software component, which consists of computer programs used to compute the positions of the survey stations. Such systems have been in operational use since the early 1970s. However, it was not until the early 1980s that a system was developed that could achieve the accuracy level of traditional methods at compatible costs and at considerable savings in time.

The measurement principles, capabilities, and limitations of three major types of black box positioning systems will be described in the following sections. They include:

1. satellite doppler positioning systems
2. radio interferometric systems
3. inertial positioning systems

11.13 SATELLITE DOPPLER POSITIONING SYSTEMS

Figure 11.11 illustrates the operational principle of a satellite doppler positioning system. A highly portable antenna is mounted over a survey station whose position is to be determined. The antenna receives radio signals emitted by an earth-orbiting satellite. The radio signal is transmitted in a stable frequency, say f_T. However, as the satellite rises above the horizon and passes over the survey station, the frequency of the radio signal received by the antenna (to be denoted as f_D) changes continuously due to the motion of the satellite. For example, at time t_1 and t_2 in Figure 11.11 the satellite is traveling toward the antenna, and the frequency of the received signal is higher than that of the initially transmitted signal. At time t_4 and t_5 the satellite is moving away from the antenna; therefore f_D is smaller than f_T. The receiving and recording unit generates its own reference signal at a stable frequency, say f_R. The received signal is superimposed with the reference signal, resulting in a signal which has a frequency of $(f_D - f_R)$. The recording unit actually measures the number of full cycles (to be denoted as N) of the resulting signal for a given time interval.

If the receiver tracks the satellite continuously between time t_1 and t_2, then the following simple relationship exists:

$$\begin{aligned}&\sqrt{(X_D - X_1)^2 + (Y_D - Y_1)^2 + (Z_D - Z_1)^2}\\&\quad - \sqrt{(X_D - X_2)^2 + (Y_D - Y_2)^2 + (Z_D - Z_2)^2}\\&\quad = \frac{v}{f_R}[N - (f_R - f_T)(t_2 - t_1)]\end{aligned} \tag{11.1}$$

where X_D, Y_D, and Z_D are the coordinates of the survey station; X_1, Y_1, and Z_1 are the coordinates of the satellite position at time t_1; X_2, Y_2, and Z_2 are the coordinates of the satellite position at time t_2; v is the velocity of propagation of radio waves in the atmosphere; N is the number of full cycles of the resultant signal recorded between time t_1 and t_2; and f_R and f_T are the reference and transmitted frequencies as defined above. The satellite coordinates (X_1, Y_1, Z_1, X_2, Y_2, and Z_2) can be determined from orbital data. Thus the only unknowns in the above equation are the coordinates (X_D, Y_D, and Z_D) of the survey station. By tracking the satellite over three different time intervals, three equations of the form of Eq. (11.1) can be computed.

Typically, a receiver is programmed to automatically turn itself on whenever a satellite comes into view and to receive its signal until it passes over the horizon. Once it is set up over the survey station, such a system can operate unattended for a period lasting from a few hours to several days. The longer the system operates, the more accurate will be the derived coordi-

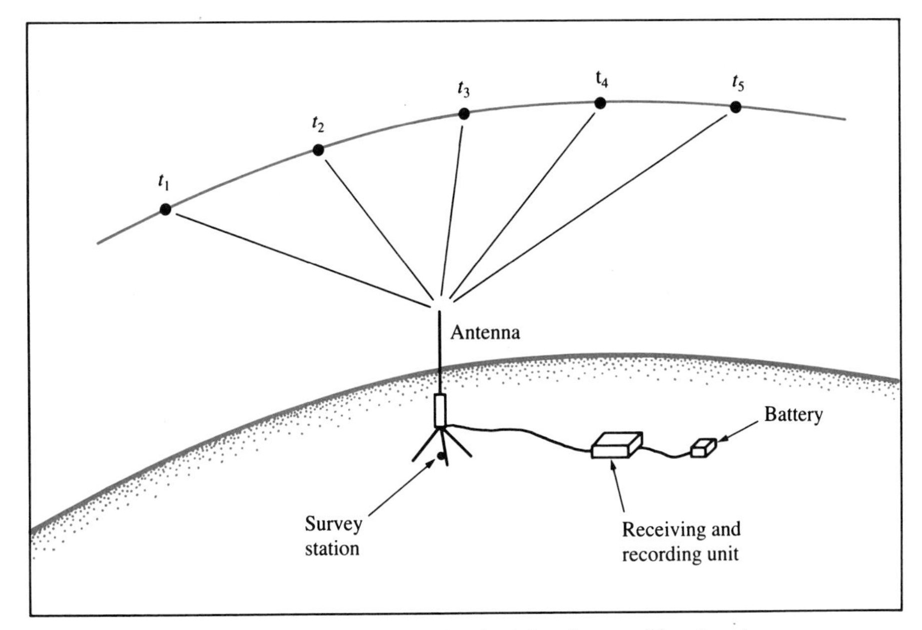

FIGURE 11.11 Operational principle of a satellite doppler positioning system

nates of the survey station. The receiver can usually provide a real-time readout of the coordinates of the survey station by using the predicted orbits of the satellites. However, for the highest accuracy in positioning, the recorded data are processed later with updated orbital data provided by the U.S. Department of Defense and with more sophisticated computer programs. Shown in Figure 11.12 is a satellite doppler positioning system manufactured by Magnavox, called the Magnavox MX1502 Geoceiver Satellite Surveyor.

There are, in early 1984, two systems of satellites that may be used for geodetic positioning by the doppler method. These are the Transit System and the Navstar Global Positioning System. Both systems are operated

FIGURE 11.12 Magnavox MX1502 Geoceiver satellite surveyor (courtesy of Magnavox)

by the U.S. Department of Defense. The Transit System was initially developed as an all-weather navigation system by the Applied Physics Laboratory in Johns Hopkins University for the U.S. Navy. It has been operational since 1964 and available for civilian use since 1967. It consists of four to five satellites in near polar orbits at an altitude of about 660 miles and an orbital period of about 110 minutes. The success of the Transit System led to the development of the Navstar Global Positioning System (GPS). The first GPS satellite was launched in 1978. The system is expected to be completed in 1990 with 18 satellites in 12-hour circular orbits at an altitude of about 12,000 miles. It will allow a receiver located anywhere on the earth's surface to track the signals from at least four satellites during any part of the day or night.

Doppler positioning systems can be operated in either the point positioning or translocation mode. In the point positioning mode, a single receiver is used to determine the position of a single point. The best point positioning accuracy is about 1.5 to 3.3 ft at one standard error for all three coordinates. In the translocation mode, two or more receivers are used to track the same satellites simultaneously and the relative positions of the survey stations are determined. The best accuracy for relative positioning is about 0.5 to 1.0 foot at one standard error. Thus, for two survey stations located at a distance of 50 miles apart, their relative positions can be determined with an accuracy better than 1/250,000 (standard error) in all three coordinates! But for a distance of 1 mile, the relative positioning accuracy is degraded to only 1/5,000. Thus the method has been used primarily for the establishment of widely spaced horizontal control stations. It has been found to be particularly useful for establishing geodetic controls in remote areas and for the positioning of off-shore platforms.

The accuracy of the doppler method is limited primarily by the following factors:

1. accuracy of the orbital data of the satellites
2. atmospheric effects on the radio signals
3. accuracy of the doppler signal measurements
4. orbital geometry of the satellites that are available for tracking

As research and development continue to result in improved doppler systems, the positioning accuracy cited in the above paragraphs can be expected to improve in the future.

11.14 RADIO INTERFEROMETRIC SYSTEMS

The principle of radio interferometry has been used since 1969 to measure very long baselines. The method is commonly referred to as *very*

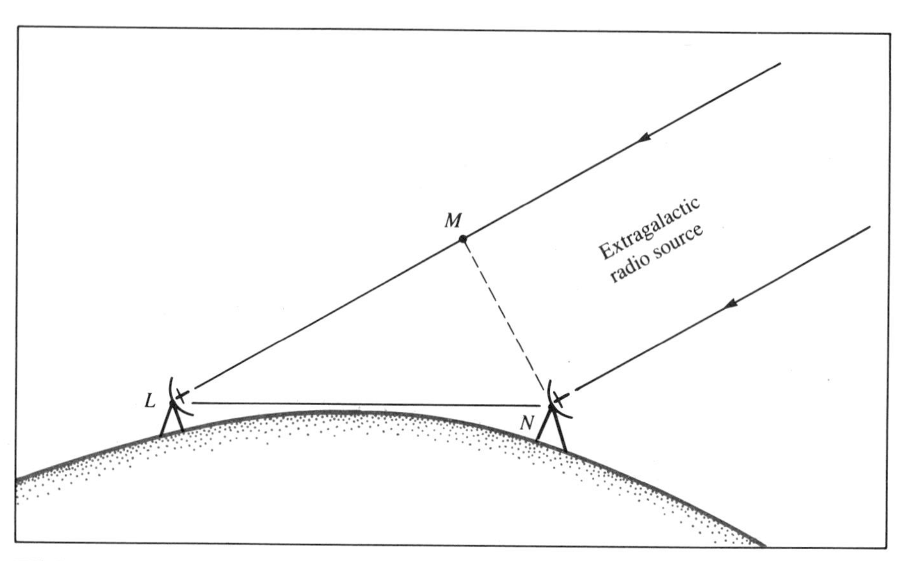

FIGURE 11.13 Very long baseline interferometry

long baseline interferometry (VLBI). The operational principle is illustrated in Figure 11.13. A large-diameter radio telescope is located at each end of the baseline and both are directed to receive signals from the same extragalactic radio source, typically a quasar. The emission source is located at such a great distance (billions of light years) from earth that its position in the sky can be assumed to be fixed. The signal is actually a complicated random noise. As the signal is received at each station, the arrival time is precisely recorded. In Figure 11.13, the signal arriving at station L will have to travel the extra distance LM as compared to that arriving at station N. The time delay in arrival of this signal is measured by comparison of the signals received at the two stations. Then the distance LM may be computed as the product of the time delay and the velocity of propagation of radio waves in the atmosphere. Since the direction to the star (signal source) is known, the direction and length of the baseline (LN) can be calculated. Baselines longer than 600 miles can usually be measured with a standard deviation better than 1 part in 10^8. This method is being used to measure tectonic plate motion, regional land subsidence, and polar motion of the earth. Although radio telescope systems that can be transported in mobilized trailers have been developed, the instruments are still so bulky and large that this method will find its applications limited to special scientific studies.

However, the principle of radio interferometry can also be applied to radio signals emitted from earth-orbiting satellites, such as the Transit and GPS satellites discussed in Section 11.13. Because of the much shorter distance between the radio source and the receiver on earth, small portable antennas and receivers can be used to track the radio signals. The orbits of the satellites, however, must be accurately known.

Shown in Figure 11.14 is a radio interferometric surveying system called the *Macrometer*™. The field unit consists of an antenna that can be mounted on a tripod over the survey point and a receiver and a power source, which are transported on a vehicle. To determine the location of a new survey station, one such field unit occupies a survey point whose position is known. A second field unit occupies the new station. The two units are programmed to receive radio signals from the GPS satellites simultaneously at some predetermined time. The field data can be sent on pocket-sized tape cartridge or by telephone for processing by an office data processor, which computes the longitude, latitude, and height above a reference ellipsoid of the new station. The processing time for each new station is less than an hour. Within the continental United States, all three coordinates of the new station can be measured with an accuracy better than $\pm(5 \text{ mm} + 5 \text{ ppm})$ in three hours of field observation.

The system can operate in any kind of weather conditions and does not require line-of-sight intervisibility between stations. It does require a clear view of the sky, although objects such as power lines and utility poles present no major problem. The survey cost is only about 20 to 30% that of the traditional field survey method. With the possible significant saving in time and costs, such survey systems can be expected to replace the tradi-

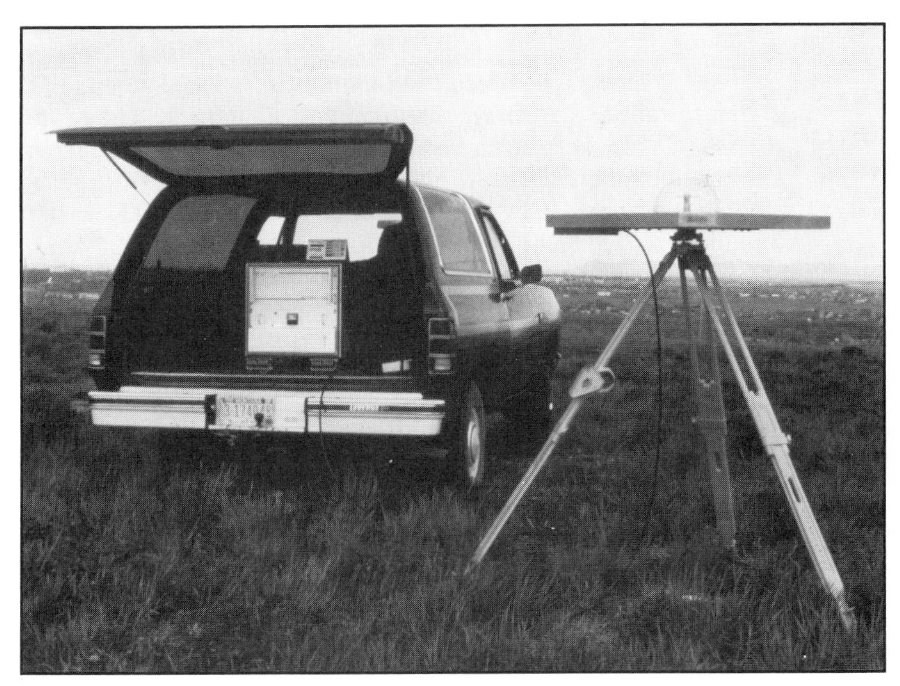

FIGURE 11.14 The MACROMETER™ interferometric surveying system (courtesy of Aero Service Division, Western Geophysical Company of America)

tional methods of triangulation and trilateration in all but some special applications.

11.15 INERTIAL POSITIONING SYSTEMS

For several decades inertial positioning systems have been used widely as navigational aids on ships and aircraft. It was not until 1972 that a system was developed to provide sufficient accuracy for surveying purposes. Since 1975 inertial positioning systems have been widely used to locate boundary corners in remote areas, for control surveys in urban areas, and for pipeline and transmission line surveys.

An inertial positioning system consists of three accelerometers mounted on three mutually perpendicular axes, a gyro system for finding the north direction, and an on-board computer for real-time computation of position coordinates. The system is usually mounted on a land vehicle or a helicopter. A survey must start from a point of known position. At this initial point, the gyro system is used to orient the instrument to the north. Then, as the vehicle moves away from the known point, the accelerometers measure the amounts of acceleration along the three orthogonal axes. The on-board computer then integrates the acceleration measurement with time to compute the change in position along the three axes. To determine the position of a new station, the vehicle simply travels from the initial known station to the new station, and the on-board computer provides a real-time readout on the longitude, latitude, and height of the new station. The vehicle must be stopped every four to five minutes for a so-called zero-velocity-updating, to provide a zero-velocity reference for the accelerometers.

An inertial survey is usually conducted in the traverse mode. It starts from a known station, then traverses from station to station, and finally closes again on a known station. The closure error at the known station can be used by the on-board computer to apply corrections to the computed coordinates of the previous stations along the traverse. The field data can be recorded on magnetic tape and a more rigorous adjustment of the measurements can be performed by a larger computer to give more refined coordinates for the traverse stations. Positioning accuracy ranging from 0.2 to 1.0 foot (root-mean-square error) for each coordinate can be achieved. Figure 11.15 shows the Litton Auto-Surveyor Dash II inertial surveying system mounted on a vehicle.

One key advantage of inertial positioning systems is the speed of operation. They are particularly suited for establishing a large number of control points within a network of controls established by either the doppler or radio interferometric method. They can also be used in heavily forested areas where other systems depending on satellite signals cannot easily operate.

FIGURE 11.15 The Litton Auto-Surveyor Dash II inertial surveying system (courtesy of International Technology Ltd.)

11.16 NATIONAL CONTROL NETWORK

A vast network of triangulation of various orders of accuracy and different densities of distribution covers the entire United States. Although the major portion of the higher-order work has been executed by the U.S. Coast and Geodetic Survey, other federal agencies have assisted with the extension of the triangulation, particularly with third-order work. Such organizations include the U.S. Geological Survey, the U.S. Bureau of Reclamation, the Mississippi River Commission, the Bureau of Land Management, the U.S. Corps of Engineers, and the U.S. Forest Service.

Figure 11.16 depicts a small part of the national system of first- and second-order triangulation. Currently (1984) the horizontal control framework of the United States has approximately 250,000 stations. This includes both triangulation and traverse points.

The beginnings of geodetic surveys in this country stem from the early efforts of a Swiss scientist, Ferdinand R. Hassler, first superintendent of the federal department that in later years became known as the Coast and Geodetic Survey. Hassler initiated triangulation operations in 1816 in the vicinity of New York City. Under the direction of his successors, progress continued in the extension of the national network.

By the end of the nineteenth century two notable accomplishments in triangulation were effected. One was the completion of the Eastern Oblique Arc, which stretched for a distance of more than 1,600 miles from

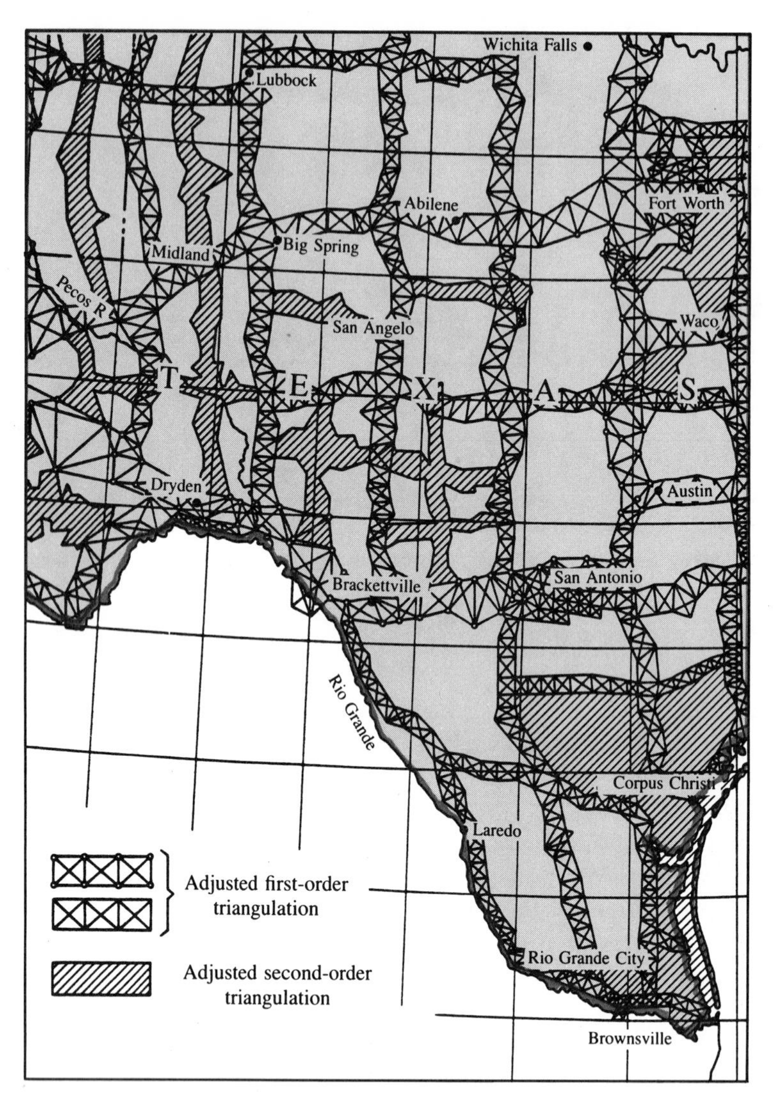

FIGURE 11.16 Network of triangulation (courtesy of National Geodetic Survey)

Calais, Maine, southwestwardly along the Appalachian Mountains to the vicinity of New Orleans. A second milestone was the execution of the transcontinental arc of triangulation from Cape May, New Jersey, to Point Arena Lighthouse, California. This network extended over a distance of 2,750 miles along the 39th parallel of latitude and is sometimes called the 39th Parallel Arc. Many of the triangle sides in the region of the Sierras have lengths in excess of 100 miles. The longest, between Mt. Shasta and Mt. Helena, is $192\frac{1}{2}$ miles.

The early part of the twentieth century witnessed the first systematic efforts to extend arcs of first-order triangulation north to south and east to west at approximately 100-mile intervals and to subdivide the intervening areas with second-order arcs. A significant divergence from the triangulation effort to provide high-order horizontal control was the execution of several thousand miles of first-order traverse in the period from 1917 to 1926, when the cost of lumber to build the towers needed in flat and timbered areas for triangulation rose to prohibitive levels.

The development of the Bilby portable steel tower (see Figure 11.17) in 1926 effected a reversal from traverse to triangulation procedures and materially increased the productivity of field personnel. The geodetic mission of the Coast and Geodetic Survey continues to be performed, although the name of the organization is now the National Geodetic Survey of the National Ocean Survey. However, the traditional methods of triangulation and traversing have now been largely replaced by space-age positioning methods such as satellite doppler positioning, radio interferometry, and inertial positioning.

FIGURE 11.17 Bilby steel tower used for triangulation (courtesy of National Geodetic Survey)

11.17 NORTH AMERICAN DATUM OF 1927

The geographic positions of control points in the U.S. national geodetic network were computed on a reference datum called the North American Datum of 1927 (NAD27). It is defined by an ellipsoid of revolution called the Clarke Ellipsoid of 1866, which has a semi-major diameter (a) of 6,378,206.4 m, and a semi-minor diameter (b) of 6,356,583.8 m. The equatorial plane is a circular plane with a radius of 6,378,206.4 m (see Figure 11.18). A triangulation station located in central Kansas called *Meade's Ranch* was assigned a latitude of 39°13′26.686″ North and a longitude of 98°32′30.506″ West. The azimuth, measured from the south, from Meade's Ranch to a nearby triangulation station called *Waldo* was fixed at 75°28′14.52″.

All measured angles and distances were reduced to corresponding angles and distances on the surface of this reference ellipsoid. The longitude and latitude of all geodetic stations were then computed on the defined datum.

The NAD27 has also been used by Canada, Mexico, and the Central and South American countries.

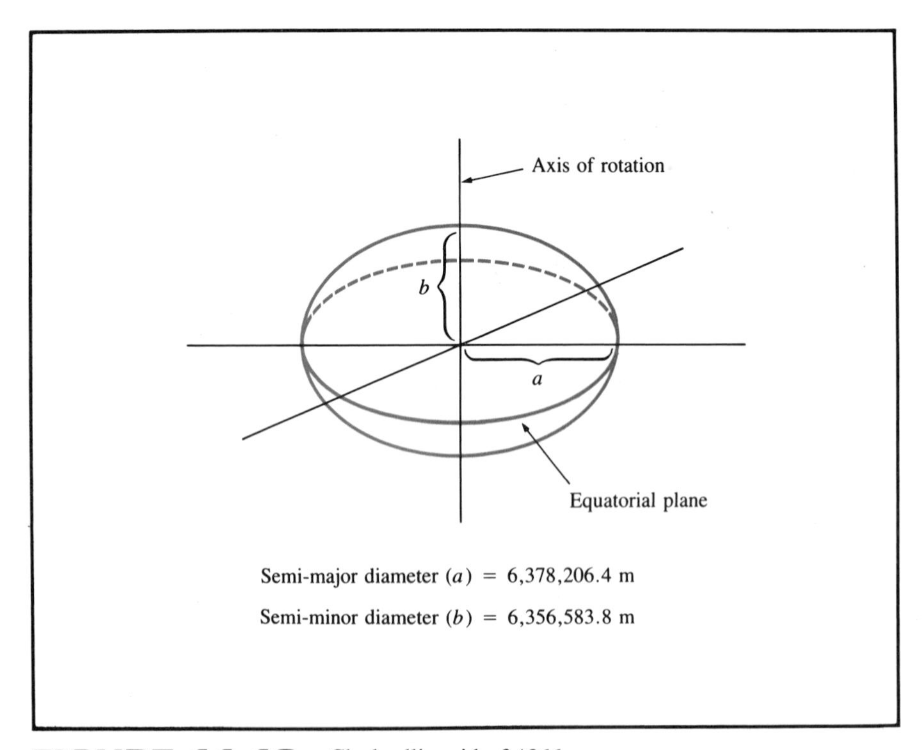

FIGURE 11.18 Clark ellipsoid of 1866

11.18 NORTH AMERICAN DATUM OF 1983

On July 1, 1974, the National Geodetic Survey started a program to perform a new adjustment of the North American Datum. The project combines both old and new geodetic measurements from the United States, Canada, Greenland, Mexico, and Central and South America into a single adjustment. About 2,750,000 geodetic measurements involving 250,000 geodetic stations will be included. The results will be revised geographic coordinates for the control stations and a new reference datum. The project was originally scheduled to be completed in 1983. Hence the new datum has been officially designated as the North American Datum of 1983 (NAD83). However, because of delays, the project is not expected to be completed until late 1984.

The new datum will use a best-fitting ellipsoid with the center located at the center of mass of the earth and the axis of rotation coinciding with the mean axis of rotation of the earth. The Greenwich meridian will be used as zero longitude.

The revised geographic positions of the geodetic stations will be published in three systems:

1. longitude and latitude
2. plane coordinates (X, Y) in the Universal Transverse Mercator (UTM) System
3. plane coordinates (X, Y) in the State Plane Coordinates System

11.19 SOURCE OF GEODETIC INFORMATION

Information on control stations included in the national network can be obtained from the National Geodetic Survey, Rockville, Maryland 20852. Geodetic control diagrams showing the general locations of national control points are published for various regions of the country. Users should first consult a geodetic control diagram covering the area of interest and identify the specific control points that can be used for their project. Detailed information concerning those control points can then be requested from NGS. The following specific information can be provided by NGS for each control point:

1. horizontal position in the three coordinates systems stated in Section 11.18
2. azimuths to some nearby visible stations
3. conditions of the survey monument
4. detailed description on the exact location of the station

PROBLEMS

11.1 What are the differences between triangulation and trilateration?

11.2 What minimum parameters are needed to define the datum for a horizontal control network?

11.3 Given the following measured values for the angles in the quadrilateral shown in Figure 11.4 on page 404:

	°	′	″
a =	44	59	07
b =	69	36	44
c =	43	33	32
d =	21	50	25
e =	39	29	06
f =	75	06	58
g =	34	14	58
h =	31	09	02

Perform a closure check on these angles.

11.4 In a triangulation project, the measured angles were estimated to have a standard error of ±0.5″. What should be the maximum allowable closure error for the triangles in a quadrilateral?

11.5 If maximum allowable closure error for the triangles in a quadrilateral is ±3″, what should be the standard error of the measured angles?

11.6 The following are the measured angles and distances for a quadrilateral as shown in Figure 11.4:

	°	′	″		
a =	52	11	06	AD =	3,803.20 ft
b =	32	17	04	AB =	6,442.31 ft
c =	23	47	32	AC =	5,629.27 ft
d =	71	44	13	DB =	7,066.18 ft
e =	39	28	25	DC =	3,057.91 ft
f =	44	59	47	BC =	5,359.36 ft
g =	64	47	29		
h =	30	44	21		

a. Perform closure checks on the angles.

b. Perform checks on the distances by assuming the distance AD to be free of error.

c. What should be the maximum acceptable differences between the computed and measured distances if the distances were measured with an EDM instrument that has an accuracy of ±(0.01 ft + 5 ppm) at 1σ?

11.7 **a.** What is an error ellipse?

b. What factors affect the dimensions and orientation of an error ellipse?

11.8 **a.** What is the accuracy of the primary control network in the United States?

b. Approximately how many control points are included in the U.S. national control network?

c. How and where can information be obtained concerning a control point within the U.S. national network?

d. What kind of information can usually be obtained on each control point?

11.9 What factors should be considered in constructing monuments for control stations?

11.10 Briefly describe the operating principle and obtainable accuracy of the following methods of control survey:

a. satellite doppler method

b. radio interferometry method

c. inertial method

REFERENCES

11.1 Anderson, Walter L. *Weighted Triangulation Adjustment,* Geological Survey Computer Contribution No. 1, Program No. W8250, U.S. Geological Survey, Computer Center Division, Washington, D.C.: 1969.

11.2 Bossler, John D. "New Adjustment of North American Datum," *Journal of the Surveying and Mapping Division,* American Society of Civil Engineers, Vol. 108, No. SU2 (August 1982) 47–52.

11.3 Counselman, Charles C., et al. "Centimeter-Level Relative Positioning with GPS," *Journal of Surveying Engineering,* American Society of Civil Engineers, Vol. 109, No. 2 (August 1983) 81–89.

11.4 Federal Geodetic Control Committee. *Classification, Standards of Accuracy, and General Specifications of Geodetic Control Surveys,* National Oceanic and Atmospheric Administration, 1974. For sale by Superintendent of Documents, U.S. Government Printing Office, Washington, D.C. 20402.

11.5 Federal Geodetic Control Committee. *Specifications to Support Classification, Standards of Accuracy, and General Specifications of Geodetic Control Surveys,* National Oceanic and Atmospheric Administration, 1975, Revised 1980. For sale by Superintendent of Documents, U.S. Government Printing Office, Washington, D.C. 20402.

11.6 Holdahl, Jeanne H., and Dubester, Dorothy E. *A Computer Program to Adjust a State Plane Coordinate Traverse by the Method of Least Squares.* Available from National Geodetic Information Center, NOAA, 6001 Executive Boulevard, Rockville, MD 20852; 232 pp., 1972.

11.7 McLellan, C. D., Peterson, A. E., and Katinas, G. E. *GALS, A Computer Program to Adjust Horizontal Control Surveys,* a report of Geodetic Survey of

Canada, Department of Energy, Mines and Resources, Ottawa, Canada, January, 1970.

11.8 Mezera, David F. "Geodetic Surveying: The Next Decade," *Journal of the Surveying and Mapping Division,* American Society of Civil Engineers, Vol. 105, SU1 (November 1979) 93–108.

11.9 Mikhail, Edward M. *Observations and Least Squares,* Harper & Row, New York: 1976.

11.10 Root, Edward F. "Inertial Survey Systems," *Journal of Surveying Engineering,* American Society of Civil Engineers, Vol. 109, No. 2 (August 1983) 116–135.

11.11 Saxena, N. K., and Vonderohe, A. P. "Three-Dimensional Densification for Local Control," *Journal of the Surveying and Mapping Division,* American Society of Civil Engineers, Vol. 104, No. SU1 (November 1978) 63–77.

11.12 Vincenty, T. *HOACOS: A Program for Adjusting Horizontal Networks in Three Dimensions,* National Oceanic and Atmospheric Administration, Technical Memorandum NOS NGS-19, 13 pp., 1979.

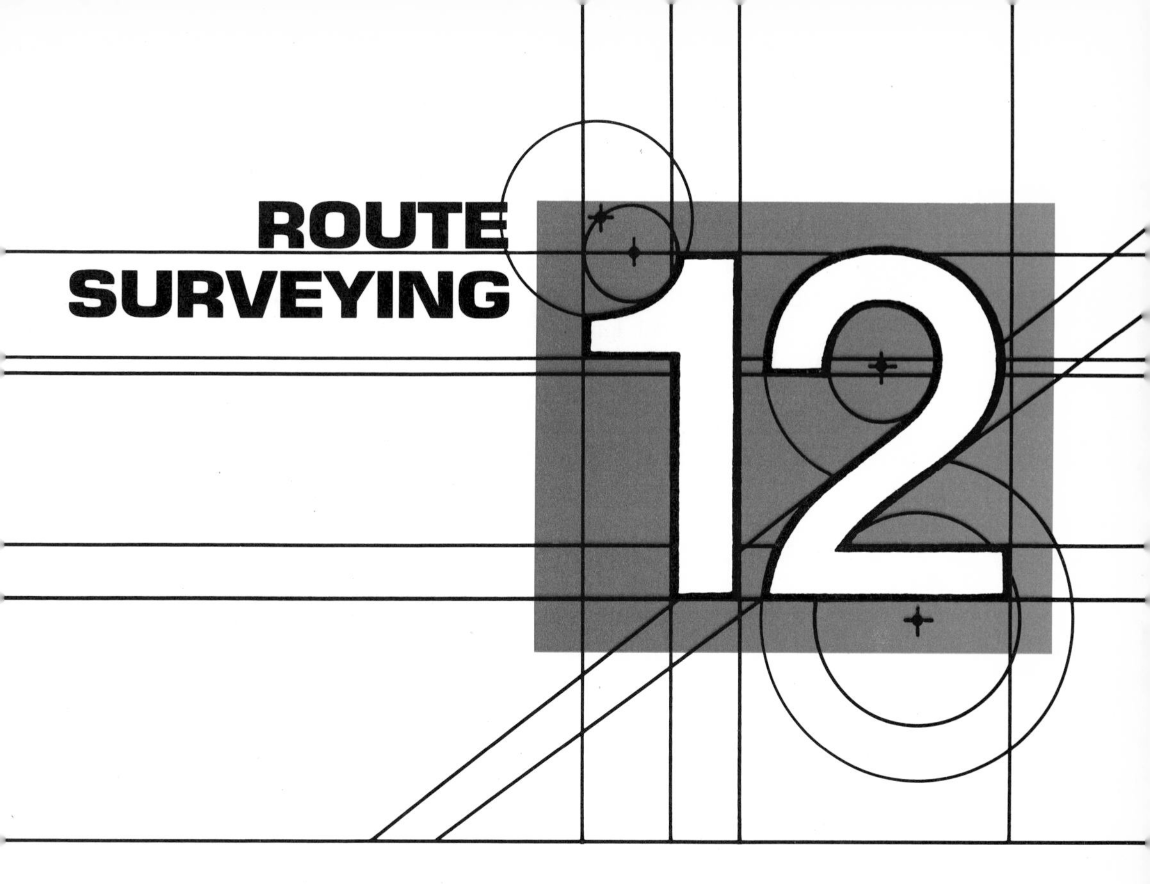

12.1 INTRODUCTION

Route surveying includes all the surveying and mapping activities that are performed for the planning, design, and construction of any route of transportation. Routes, such as those for railroads and highways, are designed to satisfy specific geometric criteria with respect to horizontal and vertical alignment.

The horizontal alignment of such routes consists of straight lines termed *tangents,* connected by curves. The curves are usually arcs of circles. Easement curves in the form of spirals are frequently used to provide gradual transitions between tangents and the circular curves along high-speed routes.

The vertical alignment consists of straight sections of grade line connected by *vertical curves*. These curves are always parabolic in form because certain characteristics of the parabola facilitate the calculation and layout of the curve. The transverse section of highway pavement is likewise built to a parabolic form.

Although the treatment of route surveying presented here is directed primarily to highways and railroads, the principles are also generally applicable to such routes as pipelines, waterways, and electric transmission lines.

12.2 GEOMETRY OF CIRCULAR CURVES

The geometry along the centerline of a circular curve is illustrated in Figure 12.1. Point *A* is the beginning of the route and is designated as station 0+00. A point located at a distance of 100 ft from the beginning point is said to be station 1+00. Thus one station is equivalent to 100 ft.

The back tangent (*AB*) intersects the forward tangent (*BC*) at point *B*, which is referred to as the point of intersection (*PI*). Since the back tangent is 1,423.54 ft long measured from station 0+00, the *PI* is said to be station 14+23.54.

The *Intersection Angle, I,* is the angle formed by the intersection of the two tangents at the *PI*. The central angle of the curve is the angle subtended by the curve at the center of the circle (Point *O*). The central angle is numerically equal to the intersection angle *I*.

Point of Curve, PC, is the point at which the curve departs from the back tangent as one proceeds around the curve in the direction in which the stationing increases.

Point of Tangent, PT, marks the end of the curve and the beginning of the forward tangent.

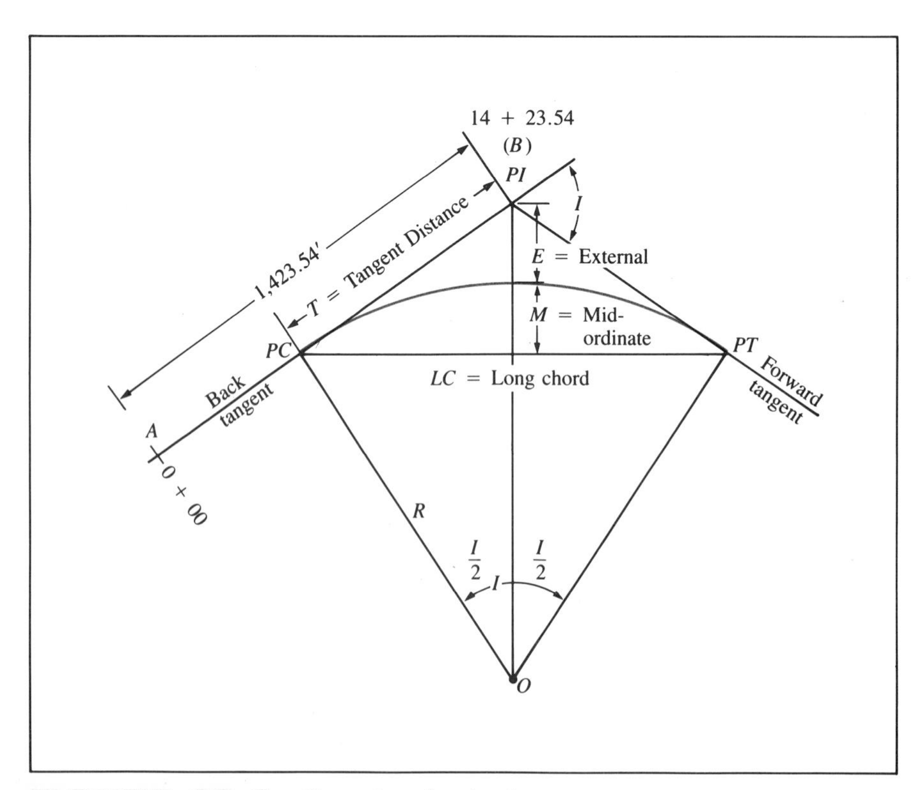

FIGURE 12.1 Geometry of a circular curve

The *Tangent Distance, T,* is the distance along the tangent from the *PC* or *PT* to the *PI*. From Figure 12.1 it is evident that

$$T = R \tan \frac{I}{2} \tag{12.1}$$

The External, E, is the distance from the mid-point of the curve to the *PI*. Evidently,

$$\frac{R}{E + R} = \cos \frac{I}{2}$$

or

$$E = R\left(\frac{1}{\cos \frac{I}{2}} - 1\right) \tag{12.2}$$

The Mid-Ordinate, M, is the perpendicular distance from the mid-point of the curve to the long chord. Then

$$\frac{R - M}{R} = \cos \frac{I}{2} \qquad \text{or} \qquad M = R - R \cos \frac{I}{2}$$

Hence

$$M = R\left(1 - \cos \frac{I}{2}\right) \tag{12.3}$$

The Long Chord, LC, is the chord joining the *PC* and *PT*

$$\frac{LC}{2} = R \sin \frac{I}{2}$$

or

$$LC = 2R \sin \frac{I}{2} \tag{12.4}$$

12.3 DEGREE OF CURVATURE—ARC DEFINITION

The degree of curvature of a curve may be defined as the central angle subtended by an arc distance of 100 ft along the curve (see Figure

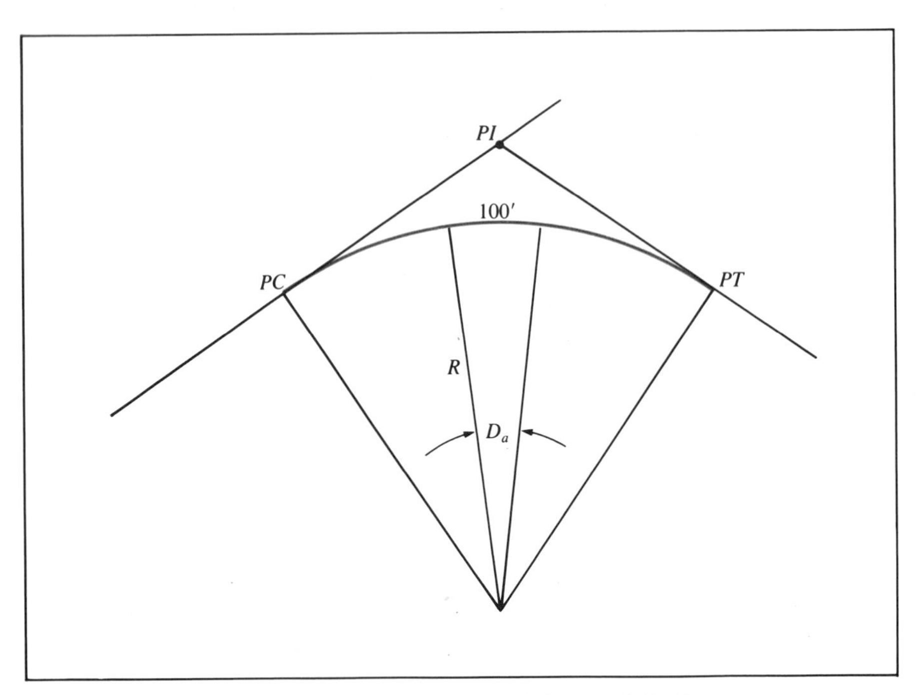

FIGURE 12.2 Degree of curvature (D_a)—arc definition

12.2). This is referred to as the *arc definition of degree of curvature*. In this book the degree of curvature expressed in the arc definition will be denoted as D_a. When $D_a = 1°$, the curve is said to be a *one-degree curve;* when $D_a = 2°$, it is called a *two-degree curve,* and so on. The arc definition is most commonly used in highway applications. The degree of curvature on modern high-speed highways is usually less than 4°.

The following relationship exists between the degree of curvature, D_a, and the radius, R, of the curve:

$$100 \text{ ft} = R \cdot \frac{D_a}{57.2958}$$

Hence

$$R = \frac{5{,}729.58}{D_a} \tag{12.5}$$

Eq. (12.5) expresses the exact inverse relationship between R and D_a. It is to be noted that a sharp (short radius) curve will have a greater value of D_a than a flat (long radius) curve. Also, if D_a is 1°, the radius is 5,729.58 ft.

According to the definition of D_a, the length of curve, L, can be computed from the following expression:

$$L = \frac{I}{D_a} \times 100 \tag{12.6}$$

where L is the arc length of the curve.

12.4 DEGREE OF CURVATURE—CHORD DEFINITION

In railroad applications the degree of curvature is usually defined as the central angle subtended by a *chord* distance of 100 ft, as shown in Figure 12.3. It is called the *chord definition of the degree of curvature* and will be denoted as D_c in this book.

The following relationship exists between the radius, R, and the degree of curvature, D_c:

$$100 = 2R \sin \frac{D_c}{2}$$

Therefore

$$R = \frac{50}{\sin \frac{D_c}{2}} \tag{12.7}$$

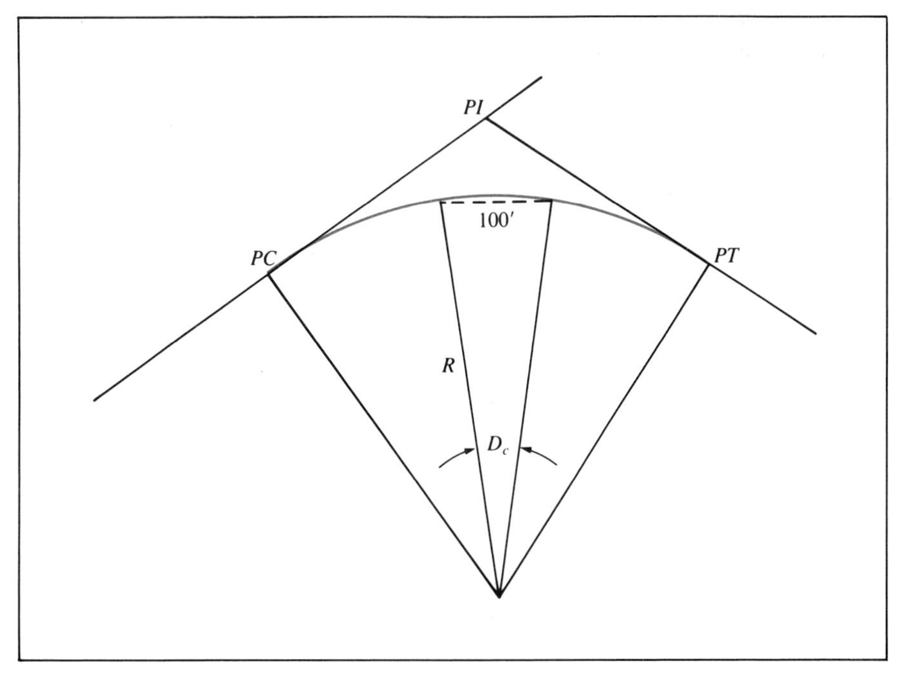

FIGURE 12.3 Degree of curvature (D_c)—chord definition

The length of curve, L, is computed by the following formula:

$$L = \frac{I}{D_c} \times 100 \tag{12.8}$$

However, L here represents the sum of the chord lengths along the curve rather than the actual arc length as in Eq. (12.6).

12.5 PRINCIPLE OF DEFLECTION ANGLES

Stations along a circular curve are most frequently located in the field by measuring deflection angles from the *PC*. The method is based on the geometric principle that the angle between a tangent and a chord at a point on a circle is equal to one-half the angle subtended by the chord. Figure 12.4 depicts the essential relationship between a deflection angle ($B\hat{A}P$) at the *PC* and the corresponding central angle ($A\hat{O}P$).

Let angle $A\hat{O}P$ be equal to d, and let point Q be the intersection of the chord AP and the perpendicular from point O. Then angle $A\hat{O}Q = P\hat{O}Q = d/2$. Since $B\hat{A}P + P\hat{A}O = 90°$, and $P\hat{A}O + A\hat{O}Q = 90°$; then $B\hat{A}P = A\hat{O}Q = d/2$.

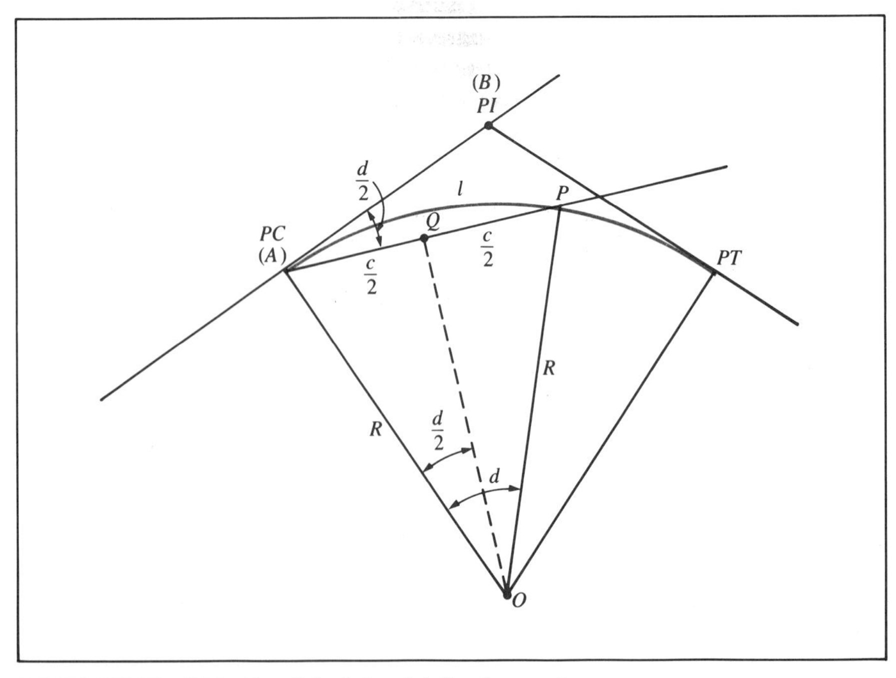

FIGURE 12.4 Principle of deflection angles

If the degree of curvature is defined by arc definition, then the central angle d can be calculated by direct proportion:

$$d = \frac{l}{100} \times D_a \tag{12.9}$$

where l is the arc length subtending the central angle d.

Let c denote the chord length AP. Then it can be seen from Figure 12.4 that

$$c = 2R \sin\left(\frac{d}{2}\right) \tag{12.10}$$

EXAMPLE 12.1

Given the following parameters for a circular curve (see Figure 12.5):

$I = 21°17'30''\ R$

$D_a = 2°10'$

Station $PI = 47 + 10.82$

Compute the curve parameters and the deflection angles for locating all full stations along the curve from PC.

Station	Point	Chord	Deflection angle		
51 + 96.43	PT	96.44	10° 39'	$= \frac{I}{2}$	check
51		100	9° 36'		
50		100	8° 31'		
49		100	7° 26'		
48		100	6° 21'		
47		100	5° 16'		
46		100	4° 11'		
45		100	3° 06'		
44		100	2° 01'		
43		86.26	0° 56'		
42 + 13.74	PC		0° 00'		

Curve parameters

$I = 21°\ 17'\ 30''R$

$D_a = 2°\ 10'$

$R = 2{,}644.42'$

$T = 497.08'$

$E = 46.31'$

$M = 45.52'$

$LC = 977.05'$

$L = 982.69'$

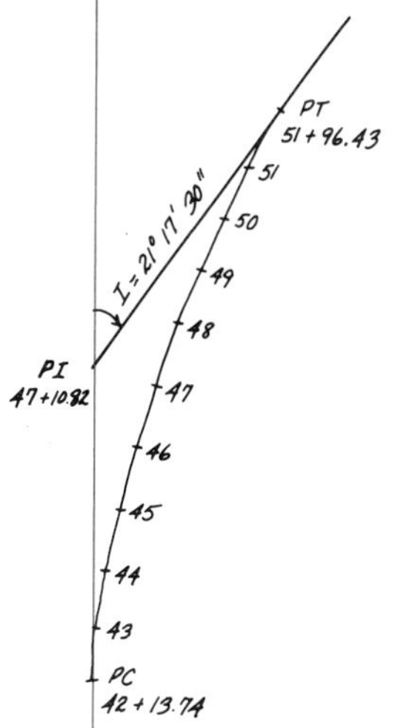

PT 51 + 96.43

51

50

49

48

47

46

45

44

43

PC 42 + 13.74

PI 47+10.82

$I = 21°17'30''$

FIGURE 12.5 Field notes for layout of a circular curve

SOLUTION

1. Calculate curve parameters:

$$R = \frac{5{,}729.58}{D_a} = 2{,}644.42'$$

$$T = R \tan \frac{I}{2} = 497.08'$$

$$E = R\left(\frac{1}{\cos \frac{I}{2}} - 1\right) = 46.31'$$

$$M = R\left(1 - \cos \frac{I}{2}\right) = 45.52'$$

$$LC = 2R \sin \frac{I}{2} = 977.05'$$

$$L = \frac{I}{D_a} \times 100 = 982.69'$$

2. Calculate stations for PC and PT:

$$\begin{aligned} \text{station } PI &= 47 + 10.82 \\ -T \quad & \underline{4 + 97.08} \\ \text{station } PC &= 42 + 13.74 \\ +L \quad & \underline{9 + 82.69} \\ \text{station } PT &= 51 + 96.43 \end{aligned}$$

Note that the stationing of PT is obtained by adding the length (L) of the curve to the station of PC. It is incorrect to compute the station of PT by adding the tangent distance (T) to the station of PI. It should also be noted that the length of the curve is denoted by L, and not LC.

3. Calculate Deflection Angles

The first full station located on the curve is station 43+00, which is at an arc distance of 4,300.00 − 4,213.74 = 86.26′ from PC. Thus the deflection angle to station 43+00 can be computed as follows:

$$\frac{d_1}{2} = \left(\frac{1}{2}\right)\frac{86.26'}{100'}(2.16667°) = 0°56'$$

The next full station on the curve is station 44+00. Since the central angle subtended by the chord from PC to station 44+00 is $(d_1 + D_a)$, the deflection angle from PC to station 44 is then equal to $d_1/2 + D_a/2$. Thus the deflection angle is increased by an amount equal to $D_a/2$ for each successive station.

The last full station on the curve is station 51+00, which is located at an arc distance of 96.43 ft from the PT. The corresponding central angle (d_2) subtended by this arc length of 96.43 ft is calculated by proportion:

$$d_2 = \frac{96.43}{100}(2.16667°) = 2°5.4'$$

Thus

$$\frac{d_2}{2} = 1°3'$$

The deflection angle to station 51+00 is 9°36′. Then the deflection angle to station *PT* is equal to 9°36′ + 1°3′ = 10°39′. The deflection angle from *PC* to *PT* should be equal to $I/2$ plus or minus round-off error, since the long chord subtends the central angle I. This computational check should always be performed to provide a check on the calculation.

Figure 12.5 shows how the deflection angles are tabulated for use in the field. On the right-hand page of the field note, a sketch shows the orientation of the curve and the computed curve parameters. The sketch is customarily drawn so that the stations increase from the bottom to the top of the page. In this manner the sketch depicts a proper view of the curve to the instrument operator standing over station *PC*. The deflection angles are also listed in that order on the left-hand page so that the listing corresponds to the sketch.

The chord lengths tabulated in Figure 12.5 represent the chord distance between adjacent stations on the curve and are calculated by Eq. (12.10). Since the curve is very flat, there is negligible difference between the computed chord lengths and the corresponding arc lengths.

The deflection angles in Figure 12.5 are rounded off to 01′ for use with a 1-minute transit. Should the instrument to be used in the field have a least count smaller than 1 minute, the deflection angles should be rounded off to the least count value.

12.6 CURVE LAYOUT BY DEFLECTION ANGLES

The first step in the curve layout is the placement of hubs marking the *PC* and *PT*. The *PC* can be advantageously located by measuring from the nearest flanking stakes along the back tangent. Hence it will be positioned 13.75 ft ahead of station 42+00. Of course, the *PC* can be also located by measuring the tangent distance from the *PI*. The *PT* is fixed by measuring T along the forward tangent.

Then, the *PC* is occupied with the transit and an initial pointing is taken toward the *PI* with the vernier set at 0°00′. The first stake on the curve, station 43+00, is located with a deflection angle of 0°56′ and a chord distance of 86.26 ft. The remaining stations are located with the simultaneous use of directions corresponding to the successive deflection angles and a common chord length of 99.99 ft measured from the preceding stake. The last deflection angle of 10°39′ and the employment of a final subchord of 96.42 ft should locate a point exactly over the *PT* hub whose position was fixed earlier by direct measurement from the *PI*. Any discrepancy in reaching the *PT* reflects the relative quality of the curve layout and no general or standard criteria can be applied to determine whether the field work is satisfactory. This difference in specified stakeout accuracy can be illustrated by the stringent requirement governing high-speed railway alignments as compared with curves on low-traffic, rural roads that are to be surfaced with inexpensive local materials like sand-gravel mixtures, which will continuously shift over the subgrade.

Of course, the curve could be laid out from the *PT* just as well as from the *PC*. The notes would be arranged so as to accumulate the deflection angles in the reverse direction.

12.7 INTERMEDIATE SETUP ON CURVE

Occasionally, because of obstructions or the great length of the curve, it may be necessary to make intermediate setups. The procedures for making two intermediate setups on the curve of Section 12.5 are described as follows:

1. *At station 44+00 with backsight on PC*. With telescope inverted and *A* vernier set at 0°00′, backsight on *PC* using lower motion. Release upper motion clamp, bring telescope to direct position, and set off 3°06′ in order to locate station 45+00.

The theory supporting the procedure is explained with the use of Figure 12.6. The angle at 44+00 between the auxiliary tangent to the curve at that point and the chord to the *PC* is the same as the angle, 2°01′, that was used to set 44+00.

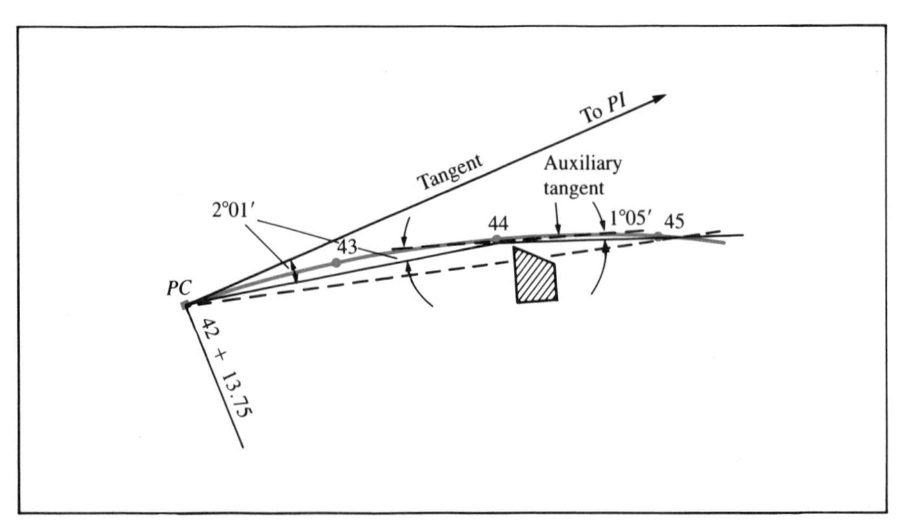

FIGURE 12.6 Intermediate setup

Following the backsight on the *PC*, the telescope can be considered to be rotated in azimuth until it reaches the auxiliary tangent. It is then brought to the normal position and rotated in azimuth by the amount $D_a/2$ so that the reading used to set station 45+00 is 3°06′ or the same angular quantity opposite that point in the original notes.

2. *At station 45+00 with backsight on 44+00.* With telescope inverted and *A* vernier set at 2°01′ (with zero to inside of curve), backsight on 44+00 using lower motion. Release upper motion clamp, bring telescope to direct position, and set off 4°11′ in order to locate station 46+00.

The procedure described above can be summarized by the following general rule: *With the transit in position at any station on a curve, the backsight is taken with the telescope inverted and the* A *vernier set to read the deflection angle of the point sighted. The telescope is then brought to normal position and the following stations on the curve are located by using the deflection angles previously computed and recorded in the notebook.*

12.8 CURVE LAYOUT USING TOTAL STATIONS

The procedure for curve layout can be significantly simplified by the use of total stations, which can measure both distance and angle with a single pointing of the instrument (see Section 6.2). A reflector prism mounted on one end of a plumbing rod, such as that shown in Figure 6.18, may be used as sight target. The procedure is illustrated in Figure 12.7. The total station is

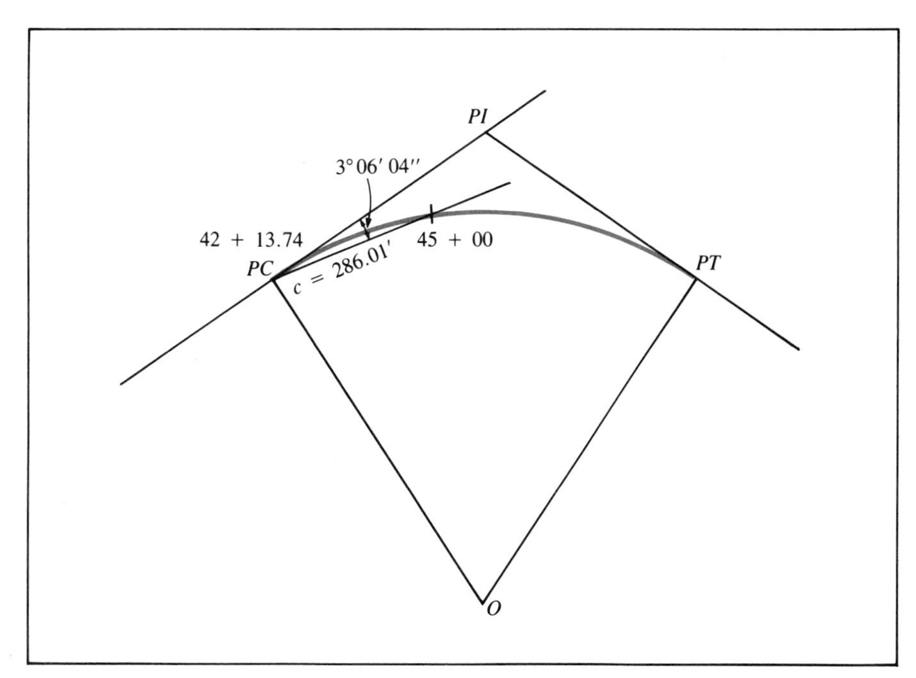

FIGURE 12.7 Curve layout with a total station

Station	Point	Chord $c = 2R \sin\frac{d}{2}$	Deflection angle $(\frac{d}{2})$		
51 + 96.43	PT	977.05	10° 38′	45″	= $\frac{I}{2}$ check
51		882.12	9° 36′	04″	
50		783.37	8° 31′	04″	
49		684.34	7° 26′	04″	
48		585.06	6° 21′	04″	
47		485.58	5° 16′	04″	
46		385.92	4° 11′	04″	
45		286.12	3° 06′	04″	
44		186.22	2° 01′	04″	
43		86.22	0° 56′	04″	
42 + 13.74	PC		0° 00′	00″	

Curve Parameters

$I = 21° 17' 30''R$

$D_a = 2° 10'$

$R = 2,644.42'$

$T = 497.08'$

$E = 46.31'$

$M = 45.52'$

$LC = 977.05'$

$L = 982.69'$

PT 51 + 96.43; PI 47 + 10.82; PC 42 + 13.74; $I = 21° 17' 30''$

FIGURE 12.8 Field notes for curve layout with a total station

set up over *PC* and is backsighted to *PI* with the horizontal angle set to read 0°00′. To locate station 45+00, for example, a deflection angle of 3°06′04″ is turned. The reflector prism is then positioned along the line-of-sight of the instrument. It is moved either away or towards the *PC* until the required chord distance *c* is displayed on the total station. The chord *c* can be computed from Eq. (12.10) as follows for station 45+00:

$$c = 2R \sin\left(\frac{d}{2}\right)$$

$$= 2 \times 2,644.42 \sin(3°06'04'')$$

$$= 286.12'$$

Figure 12.8 shows sample field notes for laying out an entire curve with a total station. Since most total stations have a least count of a few seconds or less, the deflection angles are rounded off to 1 sec.

12.9 EVEN-RADIUS CURVE

Given the intersection angle *I*, a circular curve is defined by specifying the value of one of the following parameters: *R*, *T*, *E*, *M*, *L*, *LC*, D_a, or

D_c. In highway and railroad applications, circular curves are most often defined by specifying some even value for D_a or D_c. In some applications, such as streets and pipelines, the curves are often defined by choosing some even values for the radii. Such curves are called *even-radius curves*.

12.10 STATIONING

The procedure for computing the stations along the centerline of a route can be illustrated by the example shown in Figure 12.9. The route consists of four straight tangents and three *PI*s. The lengths of the tangents and the magnitude of the intersection angles are shown in Figure 12.9a. The values of D_a, L, R, and T for each of the three curves are shown in Figure 12.9b. The stations for *PC*s, *PI*s, and *PT*s are computed as follows:

beginning station =	0 + 00	
	+24 + 21.69	
station *PI* 1 =	24 + 21.69	
$-T_1$	− 5 + 11.75	
station *PC* 1 =	19 + 09.94	
$+L_1$	+10 + 00.00	
station *PT* 1 =	29 + 09.44	
	+18 + 61.65	station *PT* 1 + distance
	47 + 71.09	from *PT* 1 to *PI* 2
$-T_1$	− 5 + 11.75	
station *PI* 2 =	42 + 59.34	
$-T_2$	− 5 + 05.14	
station *PC* 2 =	37 + 54.20	
$+L_2$	+10 + 00.00	
station *PT* 2 =	47 + 54.20	
	+21 + 48.32	station *PT* 2 + distance
	69 + 02.52	from *PT* 2 to *PI* 3
$-T_2$	− 5 + 05.14	
station *PI* 3 =	63 + 97.38	
$-T_3$	−10 + 42.70	
station *PC* 3 =	53 + 54.68	
$+L_3$	+20 + 00.00	
station *PT* 3 =	73 + 54.68	
	+22 + 72.66	station *PT* 3 + distance
	96 + 27.34	from *PT* 3 to ending station
$-T_3$	10 + 42.70	
ending station =	85 + 84.64	

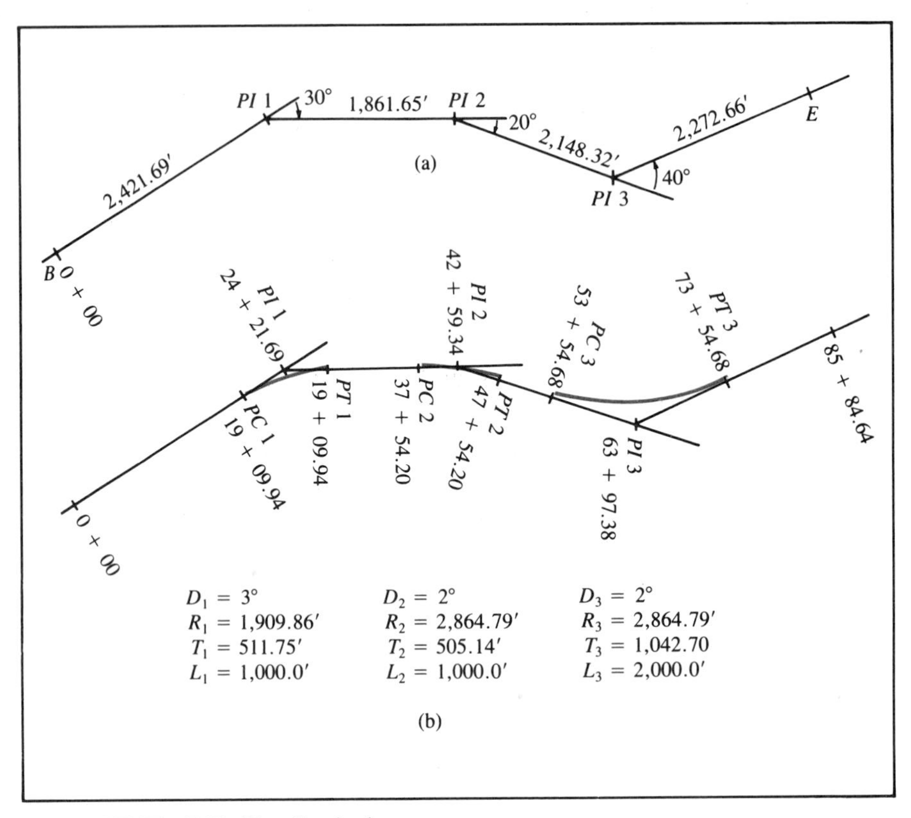

FIGURE 12.9 Stationing

In this manner the stations along the center line of the route are numbered through the curves and not through the *PI*s. If the arc definition of degree of curvature is used, then the stationing provides the actual center line distance from the beginning point of the project.

12.11 EASEMENT CURVES

For a simple circular curve, an abrupt change in curvature occurs at the *PC* and *PT*. For example, at the *PC* of a 5-degree curve, the curvature is abruptly changed from 0° to 5°. For high-speed routes, an easement curve is often used to provide a gradual transition from 0° to the degree of curvature of the curve. An easement curve also provides a gradual transition from a regular crowned cross-section to a superelevated cross-section along the curve, which is sometimes needed to counterbalance the effect of centrifugal force acting on a vehicle. One of the most commonly used easement curves is the spiral. A spiral has the desirable characteristic that its curvature changes at a constant rate.

12.12 EQUAL-TANGENT SPIRALED CIRCULAR CURVE

The geometry of an equal-tangent spiraled circular curve is illustrated in Figure 12.10. It consists of an approach spiral of length L_s, a circular curve, and a leaving spiral also of length L_s. The point where the spiral departs from the tangent is called *TS* (tangent to spiral). At *TS*, the curvature of the spiral is 0°. The curvature gradually increases along the spiral until it reaches the degree of curvature of the circular curve at point *SC* (spiral to curve). The point at which the circular curve runs into the spiral is denoted as *CS* (curve to spiral); and the leaving spiral meets the forward tangent at point *ST* (spiral to tangent).

Point *O* in Figure 12.10 is the center of the circular curve. The line *OB* is perpendicular to the back tangent and is parallel to the line joining *O′* and *TS*. When the circular curve is extended beyond *SC*, it meets line *BO* at point *G*. The distance *BG* is called the *throw* of the spiral and is usually denoted by the letter *o*. The distance along the back tangent between points *B* and *TS* is denoted as X_o. The position of point *SC* with respect to the point *TS* is defined by the distances *X* along the tangent and the offset distance *Y*. The angle, Δ, subtended by the line *BO* and the radius at *SC* is called the spiral angle.

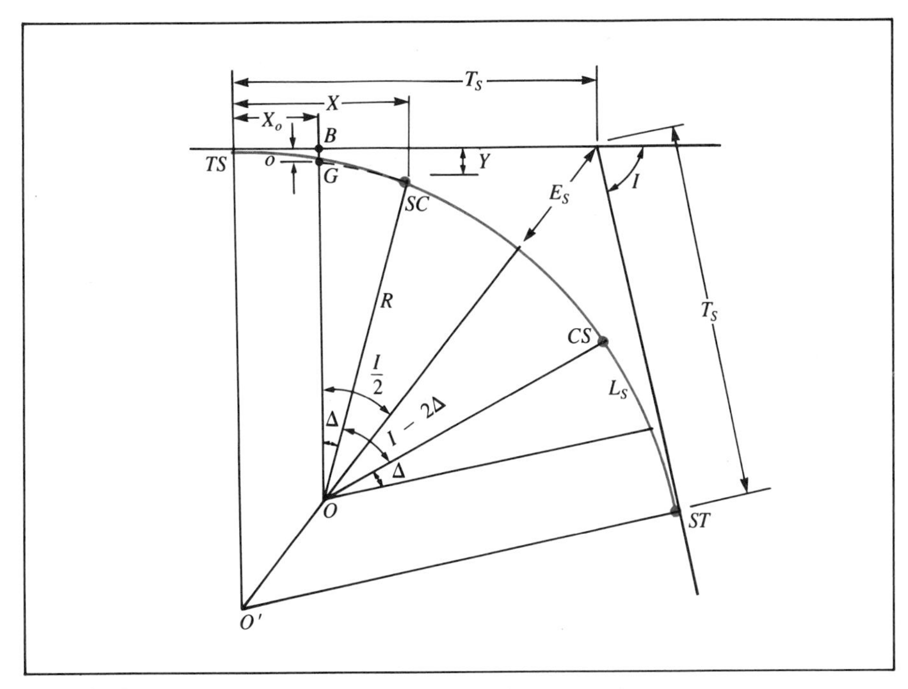

FIGURE 12.10 Equal-tangent spiraled circular curve

Let T_s denote the tangent distance between *PI* and *TS*, E_s denote the external distance, *I* denote the intersection angle, and *R* denote the radius of the circular curve. Then the following relationships can be derived from Figure 12.10:

$$X_o = X - R \sin \Delta \tag{12.11}$$

$$o = Y - R(1 - \cos \Delta) \tag{12.12}$$

$$T_s = (R + o)\tan \frac{I}{2} + X_o \tag{12.13}$$

$$E_s = (R + o)\left(\frac{1}{\cos \frac{I}{2}} - 1\right) + o \tag{12.14}$$

A spiral is usually defined by its length L_s. The spiral angle, Δ, for a spiral of length L_s and connecting to a circular curve of D_a degrees can be computed from the following expression:

$$\Delta \text{ (in degrees)} = \frac{L_s D_a}{200} \tag{12.15}$$

The values of *X* and *Y* in Eqs. (12.11) and (12.12) can be computed using Eqs. (12.17) and (12.18) given in the next section.

12.13 EQUATIONS OF A HIGHWAY SPIRAL

Let *x* and *y* denote the tangent and offset distances respectively of a point *P* located on the spiral (see Figure 12.11). Suppose that *P* is located at a distance *l*, measured along the spiral, from *TS*. Furthermore, let δ be the angle, measured in radians, subtended by the distance *l*. Then the following relationships exist:

$$x = l\left[1 - \frac{\delta^2}{5(2!)} + \frac{\delta^4}{9(4!)} - \frac{\delta^6}{13(6!)} + \cdots\right] \tag{12.16}$$

and

$$y = l\left[\frac{\delta}{3} - \frac{\delta^3}{7(3!)} + \frac{\delta^5}{11(5!)} - \frac{\delta^7}{15(7!)} + \cdots\right] \tag{12.17}$$

where

$$\delta = \left(\frac{l}{L_s}\right)^2 \Delta \tag{12.18}$$

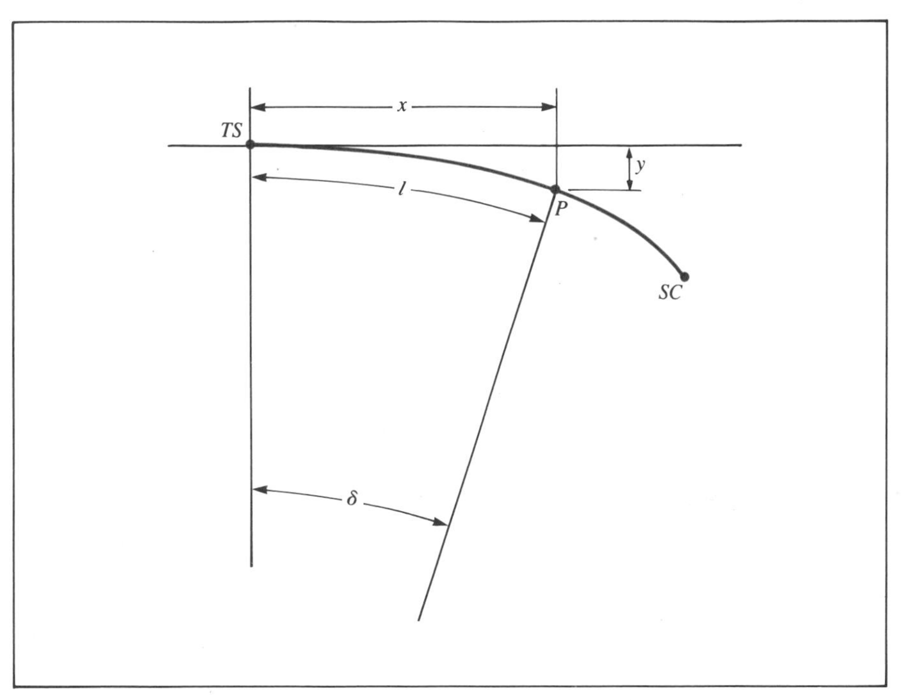

FIGURE 12.11 Coordinates of a point on a spiral

The preceding equations describe a spiral that is used universally in highway applications. A slightly different spiral is used in railroads.

Of particular interest is the tangent and offset distances (X and Y) of point SC. Since $l = L_s$ and $\delta = \Delta$ at SC,

$$X = L_s \left[1 - \frac{\Delta^2}{5(2!)} + \frac{\Delta^4}{9(4!)} - \frac{\Delta^6}{13(6!)} + \cdots \right] \tag{12.19}$$

$$Y = L_s \left[\frac{\Delta}{3} - \frac{\Delta^3}{7(3!)} + \frac{\Delta^5}{11(5!)} - \frac{\Delta^7}{15(7!)} + \cdots \right] \tag{12.20}$$

Remember that δ and Δ are expressed in radians in Eqs. (12.16) to (12.20).

For the derivations of Eqs. (12.15) to (12.20), readers are referred to References 12.1 and 12.2.

12.14 LAYOUT OF A SPIRAL

A spiral is usually located in the field using a number of equal sections. Figure 12.12 shows a spiral located in five equal sections. Point 1 on the spiral is located using the deflection angle a_1 and a chord distance equal to $L_s/5$. Point 2 is then located using deflection angle a_2 and a chord distance

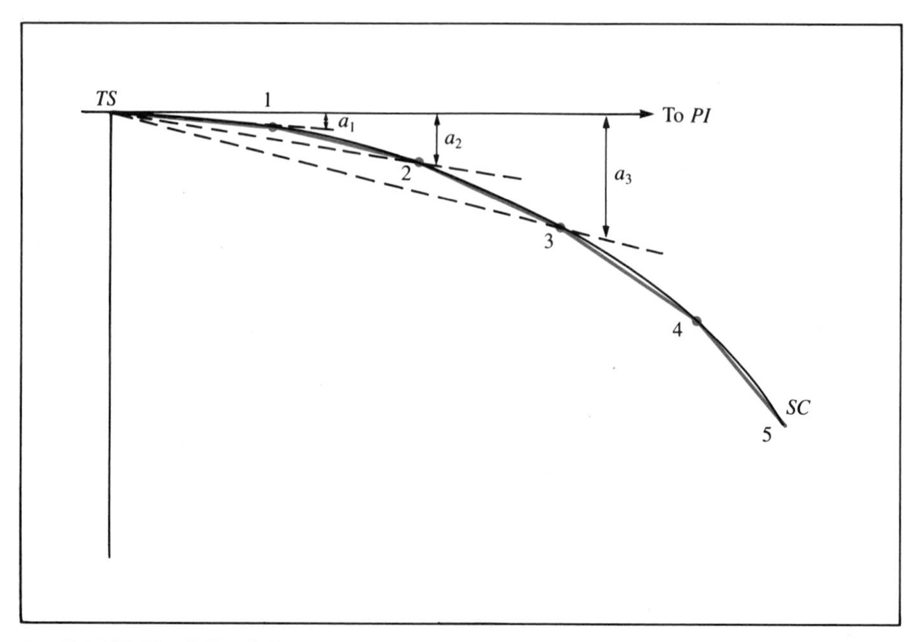

FIGURE 12.12 Layout of a spiral

$L_s/5$ measured from point 1. Each successive point is similarly located until the *SC* is reached. The exact deflection angle, a, to any point on the spiral can be computed as follows:

$$a = \tan^{-1} \frac{y}{x} \tag{12.21}$$

where x and y are the tangent and offset distances to the point and can be computed using Eqs. (12.16), (12.17), and (12.18).

Alternatively, the deflection angles can be computed by the following approximate formulas:

$$\begin{aligned}
a_1 &\cong \frac{20\Delta^\circ}{n^2} \\
a_2 &\cong (2)^2 a_1 \\
a_3 &\cong (3)^2 a_1 \\
a_4 &\cong (4)^2 a_1 \\
&\vdots \\
a_n &\cong (n)^2 a_1
\end{aligned} \tag{12.22}$$

where Δ° is the spiral angle expressed in degrees; n is the number of equal sections along the spiral; and a_i is the deflection angle to point i expressed in minutes. These formulas are sufficiently accurate for most highway applications when $\Delta° \leq 5°$. For derivation of these equations, see References 12.1 and 12.2.

Should a total station be used for layout work, the distance (s) from TS to a point on the spiral can be computed as follows:

$$s = \sqrt{x^2 + y^2} \tag{12.23}$$

where x and y are tangent and offset distances as computed from Eqs. (12.16) to (12.18).

12.15 AN EXAMPLE

EXAMPLE 12.2

Given the following parameters for an equal-tangent spiraled circular curve:

$$I = 25°00'$$

$$\text{station } PI = 142 + 63.15$$

$$D_a = 3°$$

$$L_s = 300 \text{ ft}$$

a. Compute the parameters X, Y, o, X_o, T_s, and E_s.

b. Compute the stationing for TS, SC, CS, and ST.

c. Compute the deflection angles for locating the spiral in five equal sections.

d. Prepare field notes for layout of the entire curve from TS to ST.

SOLUTION

a. From Eq. (12.15),

$$\Delta = \frac{L_s D_a}{200} = \frac{300 \times 3}{200} = 4.5°$$

$$= 0.078540 \text{ radian}$$

From Eqs. (12.19) and (12.20),

$$X = 300[1 - 0.000617 + 0.000000 - 0.000000] = 299.81 \text{ ft}$$

$$Y = 300[0.026180 - 0.000012 + 0.000000 - 0.000000] = 7.85 \text{ ft}$$

From Eq. (12.5),

$$R = \frac{5{,}729.58}{3} = 1{,}909.86 \text{ ft}$$

From Eqs. (12.11) to (12.14),

$$X_o = 299.81 - 1{,}909.86 \sin 4.5° = 149.96 \text{ ft}$$

$$o = 7.85 - 1{,}909.86(1 - \cos 4.5°) = 1.96 \text{ ft}$$

$$T_s = (1{,}909.86 + 1.96) \tan \frac{25°}{2} + 149.96 = 573.80 \text{ ft}$$

$$E_s = (1{,}909.86 + 1.96)\left(\frac{1}{\cos \frac{25°}{2}} - 1\right) + 1.96 = 48.38 \text{ ft}$$

b. Central angle subtended by circular curve $= I - 2\Delta$
$= 25° - 2 \times 4.5°$
$= 16°$

$$\text{Length of circular curve } (L) = \frac{I - 2\Delta}{D_a} \times 100 = \frac{16°}{3°} \times 100 = 533.33 \text{ ft}$$

station PI	=	142 + 63.15
$-T_s$		−5 + 73.80
station TS	=	136 + 89.35
$+L_s$		3 + 00.00
station SC	=	139 + 89.35
$+L$		+5 + 33.33
station CS	=	145 + 22.68
$+L_s$		+3 + 00.00
station ST	=	148 + 22.68

c. The deflection angles can be computed by the approximate formulas given in Eq. (12.22):

$$n = 5$$

$$a_1 \cong \frac{20 \times 4.5°}{5 \times 5} = 3.60 \text{ minutes}$$

$$a_2 \cong (2)^2 \times 3.60 = 14.4 \text{ minutes}$$

$$a_3 \cong (3)^2 \times 3.60 = 32.4 \text{ minutes}$$

$$a_4 \cong (4)^2 \times 3.60 = 57.6 \text{ minutes}$$

$$a_5 \cong (5)^2 \times 3.60 = 90 \text{ minutes} = 1°30'$$

To check the accuracy of the preceding deflection angles, the exact value of the deflection angle (a_5) to point SC can be computed using Eq. (12.21):

$$a_5 = \tan^{-1} \frac{Y}{X} = \tan^{-1} \frac{7.85}{299.81} = 1°29'59.4''$$

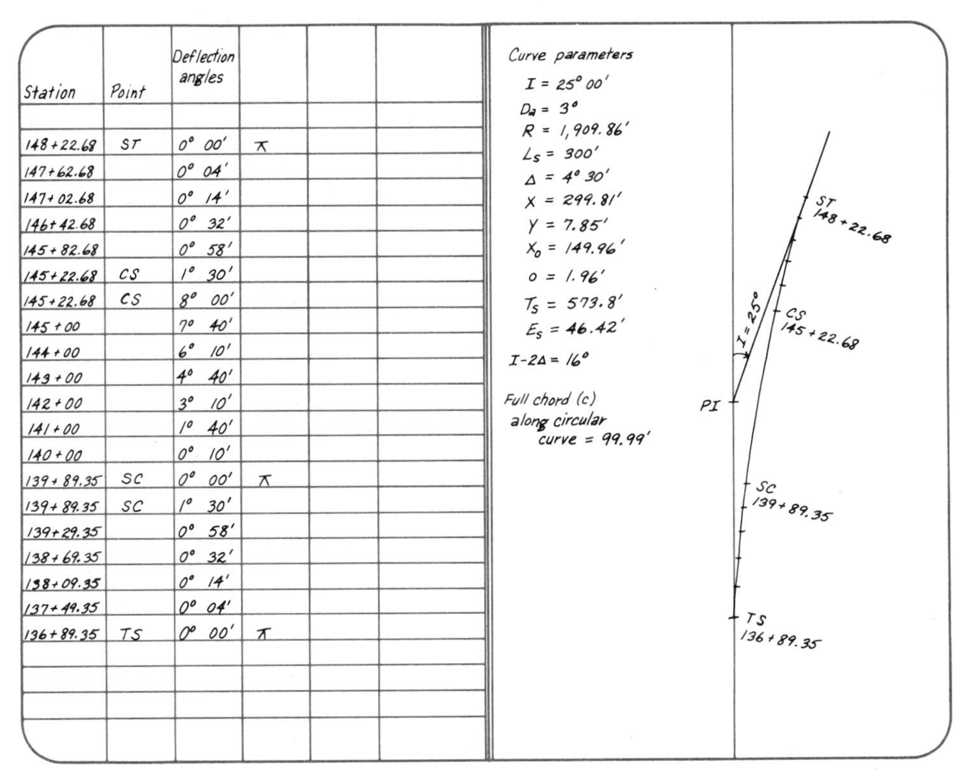

Station	Point	Deflection angles			
148 + 22.68	ST	0° 00′	⊼		
147 + 62.68		0° 04′			
147 + 02.68		0° 14′			
146 + 42.68		0° 32′			
145 + 82.68		0° 58′			
145 + 22.68	CS	1° 30′			
145 + 22.68	CS	8° 00′			
145 + 00		7° 40′			
144 + 00		6° 10′			
143 + 00		4° 40′			
142 + 00		3° 10′			
141 + 00		1° 40′			
140 + 00		0° 10′			
139 + 89.35	SC	0° 00′	⊼		
139 + 89.35	SC	1° 30′			
139 + 29.35		0° 58′			
138 + 69.35		0° 32′			
138 + 09.35		0° 14′			
137 + 49.35		0° 04′			
136 + 89.35	TS	0° 00′	⊼		

FIGURE 12.13 Field notes for an equal-tangent spiraled circular curve

Thus the error introduced by using the approximate formula to compute a_5 amounts to only 0.6 sec. The corresponding errors in a_1, a_2, a_3, and a_4 should even be less than 0.6 sec.

Figure 12.13 shows field notes prepared for the layout of the preceding spiraled circular curve. The deflection angles for the circular curve are calculated in the normal manner and are to be measured at *SC* with reference to the tangent to the curve at *SC*. The deflection angles for the leaving spiral are identical to those for the approach spiral and are measured from *ST*. The field layout procedure is described in the next section.

12.16 LAYOUT OF A SPIRALED CIRCULAR CURVE

The following procedure is usually used for the layout of a spiraled circular curve:

1. Locate stations *TS* and *ST* along the tangents.

2. With the instrument set up at *TS*, locate points along the approach spiral including *SC* with the procedure described in Section 12.14.

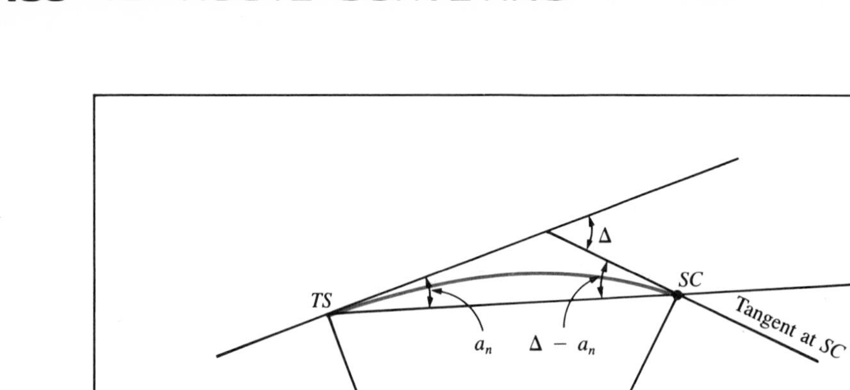

FIGURE 12.14 Local tangent at *SC*

3. Set the instrument up at *SC* and align the telescope so that its line-of-sight is along the local tangent and its horizontal angle reads 0°00′. The geometry is illustrated in Figure 12.14. Let the deflection angle from *TS* to *SC* be denoted by a_n. Then, the deflection angle at *SC* to *TS* must be equal to $(\Delta - a_n)$. The following procedure may be used to align the telescope:

Backsight to *TS* with the telescope plunged and the horizontal angle set to read $(\Delta - a_n)$ on the negative side of the circle (that is, to the left of the zero index). Plunge the telescope and rotate it with the upper motion. When the horizontal angle reads 0°00′, the line-of-sight is aligned along the local tangent.

For the example in Section 12.15 above, $(\Delta - a_n) = 4°30' - 1°30' = 3°00'$.

4. Locate all points along the horizontal curve (including *CS*) from *SC* using the computed deflection angles.

5. Move the instrument to *ST*.

6. Backsight to the *PI* with the horizontal angle set to read 0°00′.

7. Locate points along the leaving spiral using the same procedure as in Step 2. However, in this case the deflection angles must now be measured to the left (that is, counter-clockwise). The closure at *CS* provides a check on the accuracy of the layout.

12.17 VERTICAL CURVES

Following the plotting of the ground profile (see Section 4.28) along the selected route of a highway or railroad in new location or along an existing line of transportation that is to be improved, a study is made of various gradelines. These are the longitudinal lines representing usually the vertical position of the graded soil surface (subgrade) upon which the roadway surfacing material rests. Vertical curves are used to connect the gradelines. These curves are always parabolic and generally have tangents of equal length.

The usual information needed in the calculation of the vertical curve is as follows:

1. The gradients, in percent, of the gradelines.
2. The elevation and stationing of the *PI*. Sometimes, this point is called the *PVI* as in Figure 12.15, which shows plan and profile for an airport taxiway.

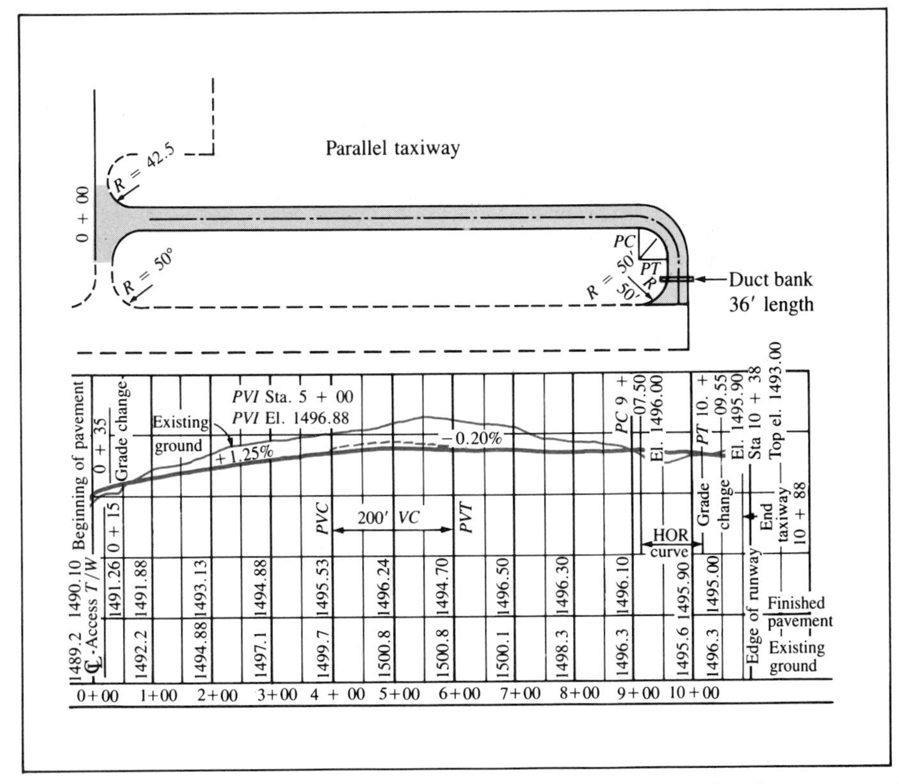

FIGURE 12.15 Taxiway plan and profile (courtesy of Federal Aviation Administration)

3. A selected length of curve. This is the horizontal distance from *PC* to *PT*, and it is usually a whole number of stations or an integral number of feet such as six stations or 840 ft, respectively.

The objective of the calculation is the determination of the elevations, usually to the nearest 0.01 ft, of selected points on the curve.

Referring to Figure 12.16, the elevation of any point *P* along a vertical curve can be calculated by the following equation:

$$y = \frac{50(g_2 - g_1)}{L} x^2 + g_1 x + h_A \tag{12.24}$$

where y is the elevation of point *P* in feet; x is the horizontal distance of point *P* from the beginning point (*A*) of the vertical curve; g_1 and g_2 are the gradients (in percent) of the back and forward tangents respectively; L is the horizontal distance in feet between the two ends of the vertical curve; and h_A is the elevation of point *A*. The values of g_1 and g_2 must carry algebraic signs into Eq. 12.24. A +2.0% grade means a 2 ft rise per 100 ft (or per station) of horizontal distance. A −2.0% grade means a drop of 2.0 ft per station.

The derivation for Eq. 12.24 is given below.

DERIVATION

The general equation for a parabola is as follows:

$$y = ax^2 + bx + c \tag{12.25}$$

At $x = 0$, $y = c = h_A$

$$\therefore \quad c = h_A$$

Differentiating Eq. (12.25) with respect to x yields the following:

$$\frac{dy}{dx} = 2ax + b \tag{12.26}$$

At $x = 0$, $dy/dx = b = g_1$

$$\therefore \quad b = g_1$$

From Eq. 12.26,

$$\frac{d^2y}{dx^2} = 2a \tag{12.27}$$

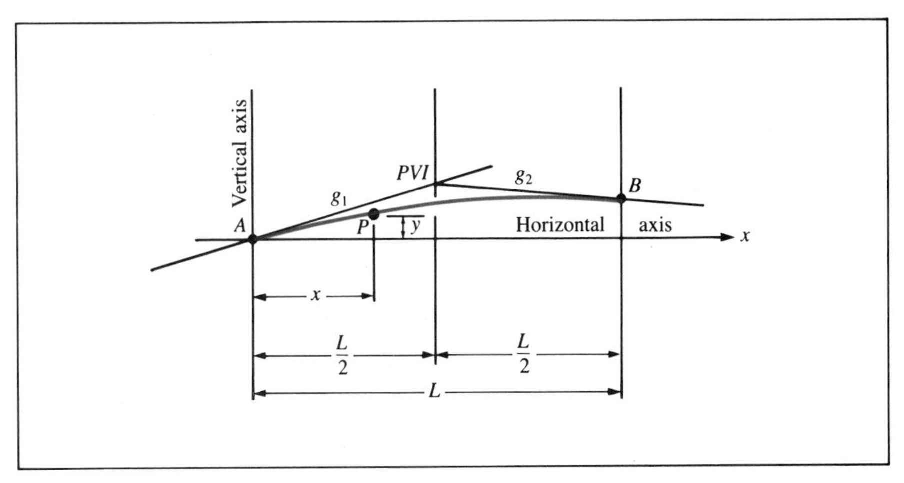

FIGURE 12.16 Vertical curve

The term $d^2y/(dx)^2$ represents the rate of change in gradient and Eq. 12.27 shows that this rate is constant. In fact, it is this characteristic that makes the parabola ideal for vertical curves. Since the gradient along the curve must change from g_1 to g_2 over a distance of $L/100$ stations,

$$\frac{d^2y}{dx^2} = \frac{100(g_2 - g_1)}{L} = 2a$$

Therefore,

$$a = \frac{50(g_2 - g_1)}{L}$$

Hence

$$y = \frac{50(g_2 - g_1)}{L} x^2 + g_1 x + h_A$$

which is Eq. 12.24.

EXAMPLE 12.3

Given the following parameters for a vertical curve (see Figure 12.17):

$g_1 = -1.00\%$ $\quad g_2 = +2.20\%$

station $PVI = 52 + 50.00$

elevation of station $PVI = 432.34$ ft

$L = 800$ ft

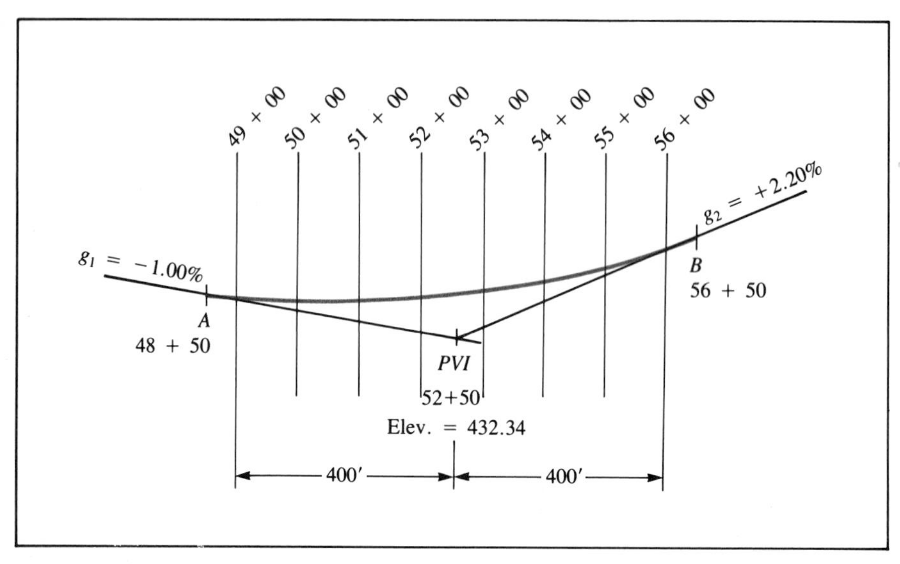

FIGURE 12.17 Vertical curve in Example 12.2

a. Compute the elevation at the beginning (A) and end (B) points of the vertical curve.
b. Compute the elevation at all full stations along the curve.

SOLUTION

a. Elevation at beginning point (A) = 432.34 + (1.0 × 4) = 436.34 ft

Elevation at end point (B) = 432.34 + (2.20 × 4) = 441.14 ft

$$\begin{aligned}\text{station } PVI &= 52 + 50.00\\ -L/2 &\quad \underline{-\ 4 + 00.00}\\ \text{station } A &= 48 + 50.00\\ +L &\quad \underline{\ 8 + 00.00}\\ \text{station } B &= 56 + 50.00\end{aligned}$$

b. From Eq. 12.24,

$$y = \frac{50(2.20 + 1.00)}{800} x^2 - x + 436.34$$

or

$$y = 0.2000\, x^2 - x + 436.34$$

To retain sufficient accuracy in the calculation, the coefficient of x^2 should have at least four decimal digits. The elevation calculation can be tabulated as follows:

Station	x	y	First Difference $\left(\frac{dy}{dx}\right)$	Second Difference $\left(\frac{d^2y}{dx^2}\right)$
48 + 50.00	0.0	436.34		
49	0.5	435.89	—	
50	1.5	435.29	−0.60	0.40
51	2.5	435.09	−0.20	0.40
52	3.5	435.29	0.20	0.40
53	4.5	435.89	0.60	0.40
54	5.5	436.89	1.00	0.40
55	6.5	438.29	1.40	0.40
56	7.5	440.09	1.80	
56 + 50.00	8.0	441.14	—	check

When manual computation is performed, the second difference between successive full stations should always be calculated to provide a check on the calculation. According to Eq. 12.27, the second differences (that is, $d^2y/(dx)^2$) should be all equal to two times the coefficient of x^2.

The parabola is also used in the design of crowned pavements. Knowing the total rise or *crown* at the center of a pavement and the width, it is a simple matter to find the elevation of any intermediate point along a cross-section of the roadway.

Thus in Figure 12.18 the crown is represented by O_2, the rise or height of the center of the pavement above the gutter elevation. The offset O_1, at any other distance d_1, is given by the relation

$$\frac{O_1}{O_2} = \frac{d_1^2}{\left(\frac{W}{2}\right)^2}$$

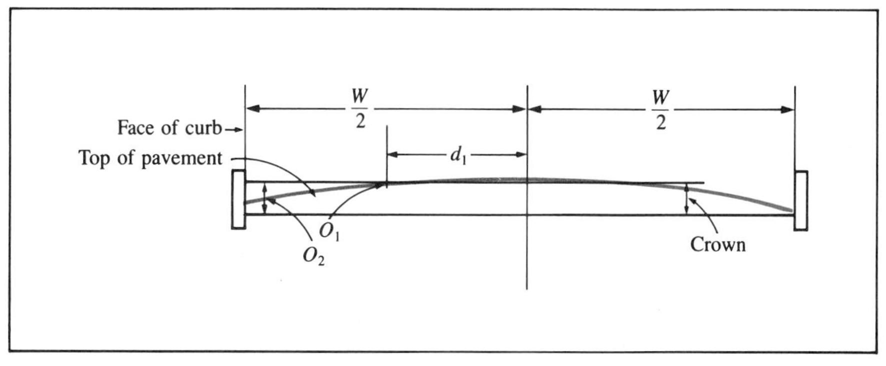

FIGURE 12.18 Pavement crown

For example, a pavement has the following dimensions: $W = 28$ ft, $d_1 = 7$ ft, and $O_2 = 4$ in. Then $O_1 = \frac{1}{4} \times O_2 = 1$ in.

12.18 LAYOUT OF A VERTICAL CURVE

The layout of a vertical curve involves (1) the locating of a stake at the beginning and end points of the vertical curve and (2) the marking of the amount of cut or fill required at a station to reach the designed subgrade elevation. For example, suppose that the design elevation at station 8 + 00 along the vertical curve shown in Figure 12.14 is 1,496.30 and that the existing ground elevation is 1,498.30. Then, a cut of 2.0 ft is needed to reach the design elevation for the subgrade.

12.19 ROUTE SURVEYING OPERATIONS

In addition to the layout of the centerline of a route, which is called location survey, there are many other surveying operations involved in route construction. These are briefly described in the following paragraphs.

1. Horizontal Control Survey. Horizontal control survey in route construction is usually accomplished by traversing, as shown in Figure 8.1. The traverse starts from a national horizontal control station at one end of the route and closes on another control station at the other end. The state plane coordinate system is usually used.

2. Vertical Control. Project elevations should be referenced to the National Geodetic Vertical Datum, and suitable ties should be made to governmental benchmarks at both ends of the project and, if feasible, at intermediate points also. It is customary to execute a line of "bench levels" along the route setting permanent benchmarks at intervals not to exceed one mile and in the vicinity of each structure. Supplementary or temporary benchmarks (TBM) are frequently set at intervals of one-fourth to one-third mile. Leveling (see Figure 12.19) should be of third-order quality.

3. Profile Leveling. Profile levels originate from and close upon permanent and temporary benchmarks (see Section 4.28). In addition to determining ground elevations to the nearest 0.1 ft at all full stations, rod readings should be obtained at all significant changes in the surface profile and at intersections of the centerline with railroads and other highways.

4. Cross-Section Leveling. Cross-sections (see Section 4.29) are usually taken at all full stations and at such intermediate points as are required to ensure adequate coverage for the subsequent calculation of earthwork. Cross-sections extend at least to the right-of-way line. Various expedients are used when cross-

FIGURE 12.19 Construction leveling (courtesy of California Division of Highways)

sectioning through brush and rugged terrain. For example, a fiberglass telescoping rod up to 25 ft in length may be useful.

5. Planimetric Detail. Maps must be prepared showing the positions of all dwellings, buildings, fences, streams, roads, large ornamental trees, and any other pertinent drainage and cultural features within the limits of the proposed right-of-way. For small projects this task can be accomplished by the stadia method. However, for large projects, it is more conveniently performed by photogrammetry (see Chapter 14). Profile and cross-section leveling are also now mostly accomplished by photogrammetry.

6. Slope Staking. If earthwork is to conform to a given alignment and side slope, it is necessary to set *slope stakes* to guide the contractor. Thus, at a given cross-section for a roadway, slope stakes are set where the proposed side slopes meet the ground surface.

If, following the execution of cross-section levels, all cross-sections have been plotted and the design cross-section of the proposed roadway superimposed (see Figure 4.57), it is necessary merely to scale the horizontal distance from the centerline to the intersection of the side slopes with the original ground surface and subsequently to lay off this distance in the field with a tape. However, slope stakes can also be located without such plotted cross-sections by a trial-and-error procedure that will now be explained.

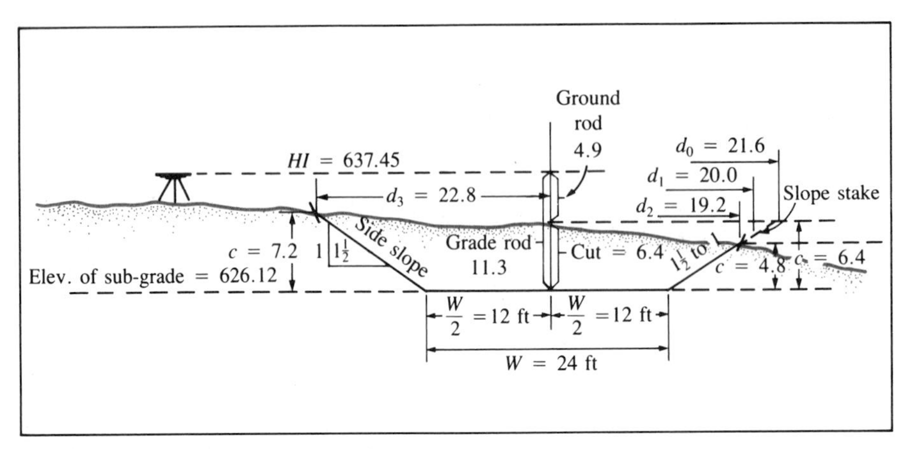

FIGURE 12.20 Slope staking

Figure 12.20 shows the conditions for a roadway in excavation, 24 ft wide, with side slopes to the ratio of 1½ (horizontal) to 1 (vertical). The height of instrument has been found by the usual process of differential leveling. The elevation of *subgrade* is determined by the position of the grade line of the roadway as fixed from the profile study by the engineer.

The field party can consist of four persons: a supervisor, who also keeps the note; a level instrument operator; a rodholder, who also carries a cloth tape; and a fourth person who aids the rodholder in measuring distances and marks and drives stakes as directed by the supervisor.

It is more convenient to find the difference in elevation between the ground surface and the subgrade by means of rod readings than by computed elevations. This procedure makes use of a quantity called the *grade rod,* which is found by subtracting the elevation of subgrade from the *HI,* all rod readings and measured distances being taken to the nearest one-tenth of a foot only. Thus the grade rod for the station shown is $637.4 - 626.1 = 11.3$. In other words, the grade rod is the reading that would be found on a rod if it could be held on the subgrade.

The *ground rod* at the center stake is 4.9, and obviously the cut is 6.4.

Having thus found the amount of cut at the center stake, the party proceeds to locate a slope stake either to the right or left. On the right it is obvious that if the ground were level, as indicated by the dashed line, the distance to the slope stake would be $W/2 + 3/2 \times 6.4 = 12 + 9.6 = 21.6$ ft. Since the ground actually slopes downward, the position of the slope stake will be found at a distance less than 21.6 ft. Hence, as a first trial, the rod can be held at a distance out d_1, say 20.0 ft, for a ground-rod reading. Suppose this reading is 6.5; then the indicated cut (grade rod–ground rod) is 4.8, for which the calculated distance out from the center d_2 is $12 + 3/2 \times 4.8 = 19.2$ ft. However, the rodholder estimated the distance and held the rod at a measured distance of 20.0 ft. There is a discrepancy, therefore, of 0.8 ft between the measured distance and the calculated distance from the centerline. Accordingly, the rod must be moved inward; if the ground is assumed to be level for that small distance, the cut will be, as before, 4.8 ft, and the correct location of the slope stake will be 19.2 ft.

Sometimes two or three trials are necessary before the correct location of the slope stake is found, but it is always fixed at that point where the *measured* distance is equal to the *calculated* distance from the centerline.

In a similar manner the slope stake on the left side is found at a point where the cut (that is, the distance above subgrade) is 7.2 ft and d_3 = 22.8 ft.

The amount of the cut at the center is marked on the back of the center stake. Each slope stake has the cut (or fill) marked on the side facing the centerline and the station number on the back.

The procedure is similar in the case of a fill, except that the ground rod will be greater than the grade rod, and it is this condition that enables the party to determine whether any doubtful point marks a cut or a fill.

To avoid confusion it is always customary to subtract the ground rod from the grade rod. If the result is positive, it represents a cut; if negative, it represents a fill.

7. Land Ties. The location of all section and property corners is determined with respect to the highway centerline. At the intersection with a section line, for example, the stationing is determined, the angle between the section line and the centerline is measured, and the distances along the section line to the nearest flanking land corners are ascertained.

Figure 12.21 shows a portion of a typical highway survey plat prepared for a section of interstate highway.

8. Interchange Sites. It is desirable to prepare a contour map of all interchange sites to facilitate the study of drainage and all design features.

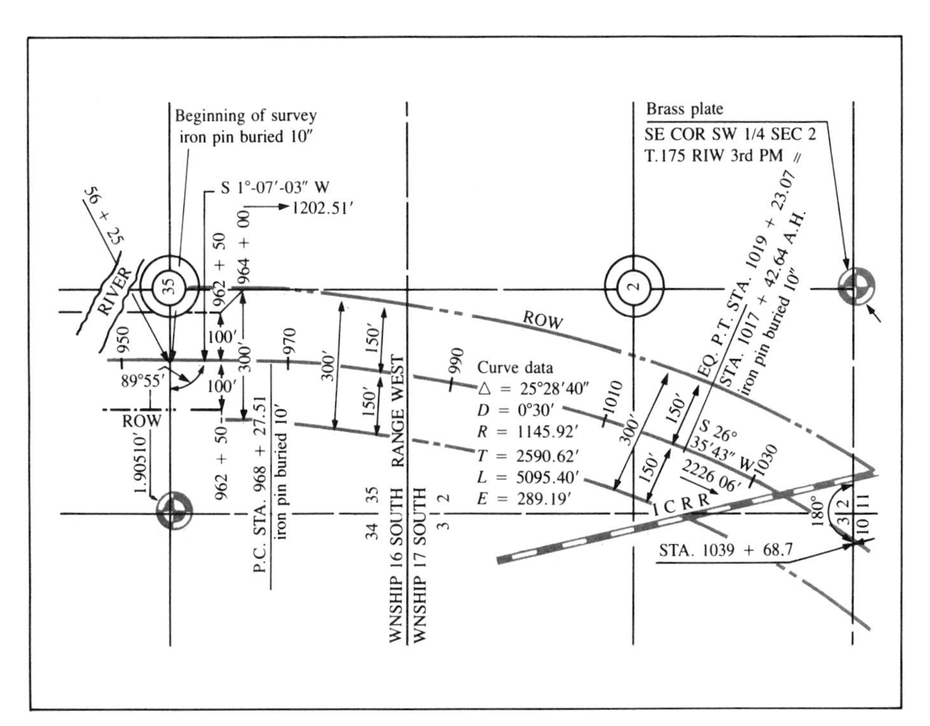

FIGURE 12.21 Plat of highway survey (courtesy of Illinois Department of Highways)

9. Drainage Surveys. Information concerning the stream profile, tributary area, and dimensions of culverts under nearby highways and railroads upstream and downstream from the project centerline is secured.

10. Utility Surveys. Such surveys consist of locating power, telephone and telegraph lines, transmission lines, and sewer, water, and oil pipelines wherever they cross the project centerline or run diagonally or parallel to the centerline within the limits of the right-of-way. Elevations are obtained for the inverts of sewers and for the tops of the manholes.

11. Miscellaneous. To preserve the accepted project centerline during the interval between the initial stakeout and the beginning of construction, all *PI*s must be durably monumented. Adequate reference ties to ensure the accurate recovery of these markers should be secured and documented. On long tangents additional monuments termed points on tangent, *POT*s, are usually set and suitably referenced.

Changes in alignment affecting the stationing generally require the use of a *station equation* in order to avoid changing existing stationing. Hence the station equation at a particular centerline point might be *237+16.42 Back = 237+81.05 Ahead*. This means that station 237+00 is 16.42 ft back of this point and station 238+00 is 18.95 ft ahead.

12.20 GRADE STAKES

In constructing any project to a given grade, it is necessary that *grade stakes* be set to guide the contractor. A grade stake is one driven until its top has the same elevation as the grade of the finished work, or until it has a known relation to that grade. In many cases, also, it fixes the alignment of the project. Examples are grade stakes for street pavements, sidewalks, sewers, railways, and highways.

STREET PAVEMENTS

Grade stakes for street pavements are usually set outside (that is, away from the centerline) of the curb about 2 or 3 ft and are driven to fix the elevation of the top of the curb. The stakes are set at 50-ft intervals when the grade is uniform and at 25-ft intervals on vertical curves, and are carefully set to fix the alignment of the back of the curb.

The level party then sets up the level in a convenient location, its *HI* being determined by differential levels from a nearby benchmark. The difference between the *HI* and the grade elevation of any given grade stake is the rod reading, or grade rod, for that stake. The stake is then driven down until, after repeated trials, the top of the stake has the desired elevation. Finally, the alignment is fixed on the stakes by tacks carefully lined in with a transit.

The tops of grade stakes are usually colored with red or blue keel to distinguish them from other stakes and to assure the contractor that they are at grade. If the location of the stakes is on a high bank such that much excavation would be necessary to set the stakes to grade, they may be set at

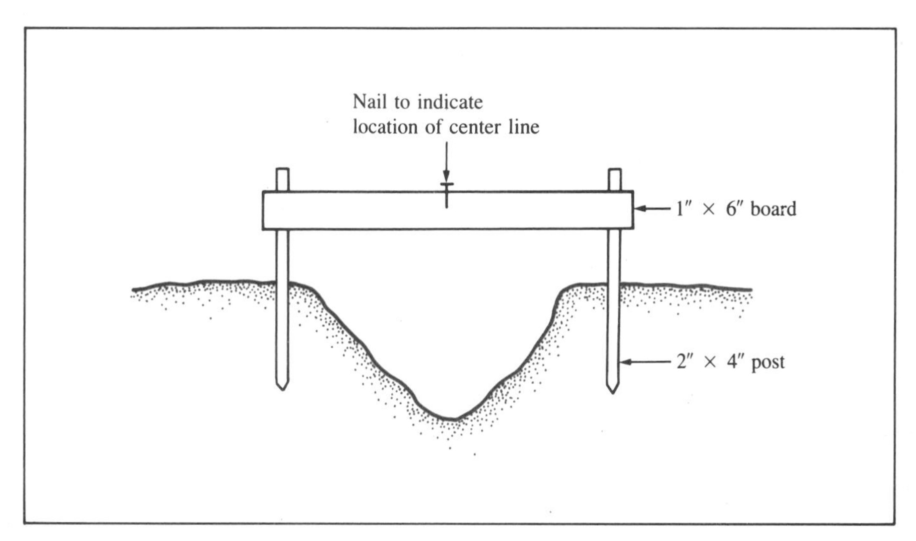

FIGURE 12.22 Batter board

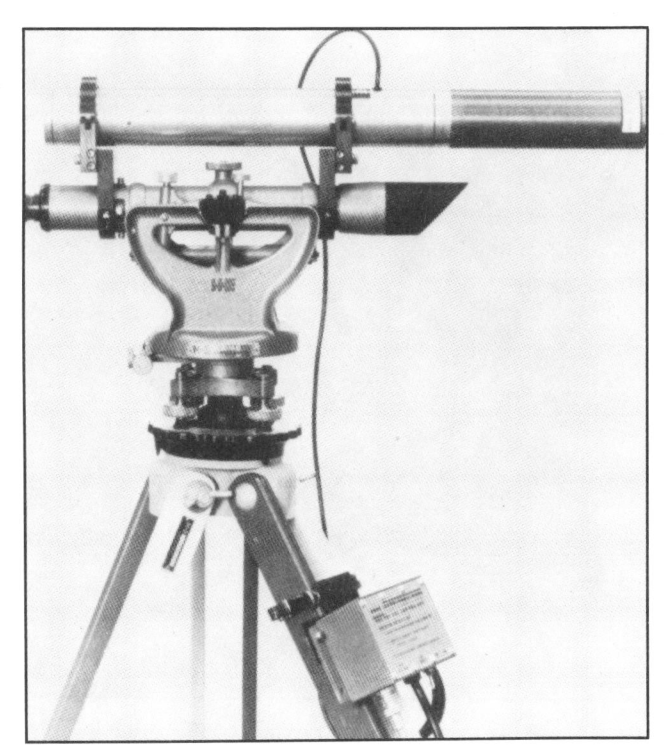

FIGURE 12.23 Builder's transit-level with a laser attachment (courtesy of Keuffel & Esser Company)

a height of, say, 2 ft above grade. The contractor then measures down this amount to fix the elevation of the forms for the pavement.

SEWERS

The grade line for a sewer is commonly established by fixing a line of sight, or by stretching a string a known distance above the grade. At regular intervals of perhaps 50 ft along the centerline of the sewer, two stout stakes are driven, one on either side of the centerline. Having determined the grade rod for a given station as indicated above, the rod is slid up or down along the stakes and a mark is made to establish the desired elevation. Then a crosspiece or ''batter board'' (see Figure 12.22 on page 461) is nailed to the two stakes, its top made level and the elevation indicated by the mark. Thus a series of batter boards are established and, if a string is stretched taut over the tops of these boards, it fixes a grade line at a known distance above the grade of the sewer. The workers then measure this distance down from any point along the string to establish a point on the grade of the sewer.

FIGURE 12.24 Laser beam in tunnel construction (courtesy of Metropolitan Sanitary District of Chicago)

Instead of using a string, the grade line can also be represented by a laser beam. Figure 12.23 on page 461 shows a laser attachment mounted on top of the telescope of a builder's transit-level. By setting the line-of-sight of the telescope at a desired inclination and turning on the laser attachment, a visible reference line of the desired inclination is established. Figure 12.24 shows a laser beam used in tunnel alignment.

RAILWAYS

Railway track is brought to its final grade by means of grade stakes. These are usually driven to the elevation of the top of rail and the track is then raised by tamping ballast underneath until the top of rail is level with the grade stakes.

12.21 STAGES OF A ROUTE CONSTRUCTION PROJECT

The same basic surveying operations, but at different levels of accuracy, are usually required for different stages of a route construction project. The different stages and the corresponding surveying operations are summarized below.

1. Feasibility Study. This phase of activity is concerned with the assessment of the engineering and economic feasibility of a suggested project. At this stage maximum use is made of published maps, such as the 1 : 24,000 scale maps published by the U.S. Geological Survey. Aerial photographs can be obtained to provide up-to-date information on the cultural developments of the area. Generally, only a limited amount of field surveying is conducted during this stage of the project.

2. Preliminary Design. At this stage preliminary designs of the vertical and horizontal alignments are made for several promising alternative routes. Generally, topographic maps having scale of 1″ = 100′ to 1″ = 200′ and a contour interval of 2 ft to 10 ft, are needed. Field surveys may include horizontal and vertical control surveys, profile leveling, cross-section leveling, and drainage surveys.

3. Final Design. Design drawings are prepared for final construction. Topographic maps having a scale of 1 ″ = 20′ to 1″ = 50′ and a contour interval of 1 to 2 ft are usually needed. Field surveys usually include layout of centerline and right-of-way, slope staking, profile leveling, and cross-section leveling.

4. Construction Layout. Surveying activities include setting stakes and reference marks to denote the position of various parts of the project.

5. Postconstruction. Periodic surveys are made during the course of construction to determine the extent of finished work or the partial payment quantities. Upon the completion of all construction, final pay quantities are calculated from the "as-built" survey data. Occasionally, postconstruction surveys are made at regular intervals to measure structural behavior, such as the settlement of a building.

PROBLEMS

12.1 Given: $I = 32°10'R$, $D_a = 1°30'$, $PI = 62 + 05.24$. Prepare a page of notes needed for the layout of this curve from the *PC* with stakes at the full stations.

12.2 Given: $I = 20°22'L$, $D_a = 2°40'$, $PI = 37 + 18.96$. Prepare a page of notes needed for the layout of this curve from the *PC* with stakes at the full and half-stations.

12.3 For the preceding problem, calculate the station number of the mid-point of the curve, the deflection angle needed to locate it, and the value of the external distance.

12.4 Given: $I = 16°23'L$, $D_a = 2°12'$, $PI = 81 + 14.52$. (1) Prepare a page of notes needed for the layout of this curve from the *PC* with stakes at the full stations. (2) Assuming obstructions are encountered, explain completely how to make an intermediate setup at (a) station 79 + 00 with backsight on *PC* and (b) station 82 + 00 with backsight on 80 + 00.

12.5 For the curve in Problem 12.1, compute the deflection angles and chord distances needed to lay out the curve from *PC* using a total station. Tabulate the results in the form of a field note.

12.6 For the curve in Problem 12.2, compute the deflection angles and chord distances needed to lay out the curve from *PC* using a total station. Tabulate the results in the form of field notes.

12.7 Given: $I = 18°22.5'\ L$, $R = 2{,}400$ ft, $PI = 92 + 04.37$. Prepare a page of notes for the layout of this curve from the *PC* with stakes at the full stations. Use arc definition.

12.8 Given $I = 14°32.5'R$, $R = 1{,}300$ ft, $PI = 81 + 12.02$. Prepare a page of notes for the layout of this curve from the *PC* with stakes at all full and half stations. Use arc definition.

12.9 Given: $I = 33°14'15''L$, and the maximum allowable value of E is 82.5 ft. Calculate D_a (to the nearest full minute) of the flattest curve that will satisfy this condition.

12.10 Given the following data along a proposed highway:

Station	Tangent Distances	Intersection Angles (I)	Degrees of Curvature (D_a)
A (0 + 00)			
	6,439.22′		
PI 1		45°30′ *R*	5°30′
	5,008.17′		
PI 2		35°00′ *L*	3°00′
	4,872.65′		
B			

Compute the stationing for all *PI*s, *PC*s, *PT*s, and point *B*. All curves are circular curves.

12.11 Given the following data along a proposed highway route:

Station	Tangent Distances	Intersection Angles (I)	Degrees of Curvature (D_a)
A (0 + 00)			
	4,362.56		
PI 1		35°30′ *R*	3°00′
	10,429.77		
PI 2		28°00′ *L*	2°30′
	8,318.29		
PI 3		30°45′ *R*	4°00′
	3,211.61		
B			

Compute the stationing for all *PI*s, *PC*s, *PT*s, and point *B*. All curves are circular curves.

12.12 Prepare a set of field notes for layout of the following equal tangent-spiraled circular curves. All spirals are to be located in ten equal sections, and all circular curves are to be located at full stations.

a. $I = 40°20'\ R$ $D_a = 4°00'$
$L_s = 200$ ft station $PI = 48 + 97.32$

b. $I = 35°00'\ L$ $D_a = 3°30'$
$L_s = 250$ ft station $PI = 143 + 16.22$

c. $I = 46°15'\ R$ $D_a = 5°30'$
$L_s = 300$ ft station $PI = 266 + 15.37$

12.13 At station 237 + 14.62 along a −2.67% gradeline (having no vertical curves), the elevation of subgrade is 827.67 ft. Find the elevation of subgrade at station 235 + 00.

12.14 A +2.40% grade meets a −1.70% grade at station 32 + 50, elevation 522.14. Calculate the elevations of all full stations on a vertical curve 600 ft long.

12.15 A −1.20% grade meets a +2.50% grade at station 51+00, elevation 127.60. Calculate the elevations of all full and half-stations on a vertical curve 450 ft long.

12.16 A +4.25% grade meets a +1.65% grade at station 60+50, elevation 171.20. Calculate the elevations of all full stations on a vertical curve 950 ft long.

12.17 The elevation of the top of pavement on the centerline at station 20 + 00 is 645.30. The grade is −0.80%, the pavement is 32 ft wide face-to-face of curb, and the crown is 4 in. What will be the grade elevation of a point at station 20 + 40 and 12 ft distance, at right angle, from the centerline?

REFERENCES

12.1 Hickerson, Thomas F. *Route Surveying and Design,* McGraw-Hill Book Company, New York: 1964.
12.2 Meyer, Carl F., and Gibson, David W. *Route Surveying and Design,* 5th ed., Harper & Row, New York: 1980.

13.1 INTRODUCTION

Land surveying deals with land boundaries. If the term *boundary surveying* is employed, its usage is frequently restricted to surveys of boundary lines between political divisions.

A land boundary is a line of demarcation between adjoining tracts of land. It is usually marked on the ground by various kinds of monuments placed specifically for that purpose. A boundary line between privately owned parcels of land is generally called a *property line*. Hence, the designation *property surveying* is usually considered to be synonymous with *land surveying*. A boundary line between contiguous plots of land in a subdivision of a city, town, or village block is usually known as a *lot line*.

A broader and somewhat more classical term embracing all kinds of land surveys is *cadastral surveying*. It is derived from the old Roman "cadastre," which is an official register of the quantity, value, and ownership of real estate. Basically cadastral surveys create, mark, define, and re-establish land boundaries. The term is used primarily to designate surveys of the public lands of the United States as described in later sections of this chapter.

Land surveying comprises the determination of the location of land boundaries and the drawing of plats portraying the subdivision of tracts into smaller parcels. It includes the preparation and interpretation of land descriptions for incorporation in leases, deeds, and other legal instruments. Certain surveys such as those required in the planning, design, and construction of highways, railroads, bridges, and other engineering works are not construed to be land surveys. However, since building and right-of-way lines must be carefully delineated before construction operations commence, the role of land surveying is seen to be an important supporting one.

The practice of land surveying probably began as early as 2500 B.C. Numerous references are made to land surveying problems in Holy Writ (see Reference 3). Although George Washington is known to students of history as a soldier and statesman, his first vocational interests were in the area of surveying. He was active in this profession for only five years (1749–1754), but his knowledge of surveying and mapping proved useful to him throughout his life. Abraham Lincoln executed numerous land surveys in central Illinois before entering politics. Today land surveying is a widely recognized professional activity whose practice is regulated by law in most states.

This chapter will provide a brief introduction to the basic concepts of rural and urban land surveys and describe the U.S. Public Lands Survey System. Because land surveys, their sufficiency, and their quality, are often subject to judicial review, some consideration will be given to certain legal aspects of boundary surveying.

13.2 BOUNDARIES

The function of boundaries is to define areas of jurisdiction. They serve as lines of division ranging from international and state boundaries to simple lot lines in an urban area. Obviously only boundaries that are well marked and described can properly serve the purpose for which they were established.

Property boundaries are needed to define the areal extent of estate rights and its obligations; they are essential in the maintenance of good will and cordial relations in community life. Reliable delineation of such boundaries is a primary function of land surveying. Indefinite property lines can be a source of dispute and controversy, especially to resident landowners.

Many of the difficulties associated with poorly defined boundaries and conflicting descriptions of properties can be traced back to inadequacies in the original surveys. These troubles have been further compounded by frequent subdivision, disappearance of markers, blunders in retracement surveys, and the use of old descriptions that were not modified to reflect later surveys.

13.3 LAND TRANSFERS

Ownership of movable (personal) and immovable property by the individual is usually regarded as one of the most fundamental of human rights. *Real property* or land is one of a person's most precious possessions.

Whether used as a homesite or a means to produce goods, it is traditionally considered as the most important type of property.

The gradual development of individual landownership was not uniform throughout the world. Generically, all real property was owned by the government or sovereign. When the American colonies were established or unappropriated land was settled, title to the land was secured either through grant or purchase from the state. It is to be emphasized that in all cases valid title to land derived from the authority of organized government, whatever its form.

Title to land is the legal basis for ownership. Such titles may be conveyed or transferred in various ways. The most common method of conveying interests in real estate is by a written instrument known as a *deed*. There are various kinds of deeds, but all embody a description of the land whose transfer is involved.

Documentary evidence of land transfers is kept in a public depository, such as a county registry of deeds. It is customary to maintain here an alphabetic index, by years, of the names of the sellers (*grantors*) of property and of the purchasers (*grantees*), together with copies of the deeds. The ready accessibility of such records to the public and particularly to land surveyors is a major factor in the perpetuation of boundaries.

Litigation over property lines has stemmed not only from inadequate descriptions but also from legal entanglements of title. The mere execution of a deed to real estate does not invest title in the grantee unless the grantor owns the property. It is necessary to study the chain of title to the property from the time the United States government made the first conveyance. For a consideration title companies will make an examination of the history of ownership of a lot or tract of land and issue a policy of *title insurance,* which essentially is a contract to indemnify against loss or damage occasioned by defects in the title. Persons buying real property and lenders obtaining a mortgage on real property to secure their loans customarily require such a policy in order to protect their investment.

13.4 KINDS OF LAND SURVEYS

Land surveying deals with the measurement, establishment, and description of the boundaries of real property. Land surveys are made for many specific purposes, such as locating on the ground described boundaries, obtaining data for a deed description, calculating area, and securing information needed for wills, mortgages, leases, tax assessments, and condemnations for public use. In general, all land surveys are classified as being either *original* or *resurveys*. They are described as follows:

1. Original surveys are executed to define the size, shape, and relative location of a tract of land whose general boundaries as evidenced by occupancy and use and as delineated by such landmarks as rivers, fences, walls, and trees have been generally accepted by adjacent owners or adjoiners. It is necessary to mark or monument the corners of the property and to determine the lengths and bearing of the boundaries as the prelude to preparing a description. Original surveys are also executed in order to create new (smaller) parcels of land from a given tract. Such surveys, frequently termed *subdivisional surveys,* are made to subdivide a tract of land according to some plan. The U.S. Public Lands Surveys and private subdivision surveys are examples.

2. Resurveys are executed for the purpose of locating the boundaries of parcels of land already described by existing documents. Such retracement surveys are essential prior to conveyancing by deed from one party to another. It is particularly necessary in the case of urban surveys to ascertain whether driveways are wholly within the boundary lines and whether there are any encroachments upon the property from buildings or structures on the adjoining lots.

Resurveys of satisfactory quality can be both difficult and time-consuming. Frequently there will be no single and unique solution. Honest differences in the evaluation of the same evidence about line location by two equally proficient and licensed land surveyors can lead to dissimilar conclusions. Final judgment is rendered by the courts.

13.5 RURAL AND URBAN LAND SURVEYS

Sometimes land surveys are categorized as being either rural or urban. The legal considerations affecting both types are essentially the same, although there may be differences in the technical aspects occasioned by the size of the tract, its location and topography, survey equipment used, methods employed, accuracy of the results, and other factors. The problems associated with original and resurveys in both rural and urban settings will be detailed in subsequent sections. Only a few broad comments will be made here.

In general, any land survey will utilize the concepts and measurements of a traversing operation because the lengths and directions of the defining perimeter lines will be involved. When the property lines are obstructed, an auxiliary traverse is executed nearby and sufficient tie measurements are made to the corners so that the lengths and bearings of the boundaries can be calculated as already mentioned in Section 8.14. Quite obviously there will be drastic changes in the field procedures between a survey of a large and remotely situated tract (see Figure 13.1) of uncertain economic value and that of a small parcel of land (see Figure 13.2) on which a multimillion dollar building is to be erected in the downtown area of a large city. In the latter situation the search for nearby defining property marks will be most thorough, the measurements, although not extensive, exceedingly precise, and the establishment of lot and building line positions highly accurate.

The terms *municipal surveying* and *city surveying* have connotations that are usually not the same as those of urban land surveying. Both city

FIGURE 13.1 Rural land survey (Bureau of Land Management)

FIGURE 13.2 Lot survey (Houston Lighting & Power Company)

surveying and municipal surveying refer to comprehensive mapping programs culminating in the production of large-scale topographic maps, which are extremely important in city planning. These terms are also employed to indicate various construction or layout surveys, such as those for new streets and public utilities. A cadastral-related aspect of the city surveys is the location of all street lines in view of the fact that such lines represent the boundaries of public property.

13.6 HISTORICAL DEVELOPMENTS IN THE UNITED STATES

In the North American colonies, the manner of acquiring title to the land from established government authorities was far from uniform. The boundary lines of private property were vague and consisted principally of such natural features as tidewater shorelines, streams, highways, fences, trees, and stones. The tracts thus bounded were irregular in shape except for subdivisions within the limits of towns. There was no general system to serve as a control for the positions of these boundaries and, when they became obliterated with the passage of time and the construction of public improvements, it was often difficult to restore them. Furthermore, the deed descriptions of such land parcels were complex and subject to many errors of interpretation and identification. These undesirable conditions had already prevailed for some time over most of the eastern territory of the nation when the *U.S. Public Lands Survey System* was created by an ordinance passed by the Continental Congress on May 20, 1785. The objective of this legislation was to prevent a repetition of the colonial disorder, confusion, and litigation connected with land ownership and to effect a reliable and simple system of describing and identifying land. The ordinance complemented policies of land disposal, which were inaugurated to stimulate settlement and encourage internal improvements. It was the instrument of law providing for the extension of what is loosely termed the *rectangular system* over the entire *public domain*. This is the vacant land held in trust by the federal government for the people.

The extent of federally owned land is enormous. Approximately one-third of all state acreage is still held by the federal government. This includes roughly a third of Colorado, Montana, New Mexico, and Washington; almost one-half of Arizona, California, Oregon, and Wyoming; more than two-thirds of Idaho and Utah; and approximately 90% of Nevada. In Alaska less than 1% of the land is in private ownership.

The U.S. rectangular system was never applied to the original thirteen colonies because of the impracticability of a change in land descriptions where so many boundaries would be affected. However, the idea of the rectangular system undoubtedly stemmed from the plan of settlement adopted in some of the crown colonies. It became apparent to the governments of the colonies that some definite policy of land disposal was essential to aid in the development of the lands. The Colony of Massachusetts took the first action in 1634 to put such a policy in practice. Small grants of land of various sizes, known as "towns," were made to individuals and groups of persons. The first town, containing as much as 36 square miles but with boundaries not along the cardinal directions, was laid out in 1656. In 1749, Bennington, Vermont, then part of New Hampshire, was established as the first town with north-south and east-west boundaries. This tract of land is comparable with the townships in the U.S. rectangular system.

The Bureau of Land Management, an Interior Department agency formed in 1946 through the consolidation of the General Land Office created in 1812 and the Grazing Service, is responsible for surveying the public lands and is charged with the management, leasing, and disposal of the vacant public lands. Most land titles in the United States, except in the colonial states and Kentucky, Tennessee, and Texas, derive from the rectangular system.

13.7 U.S. PUBLIC LANDS SURVEY SYSTEM

Although the regulations for the subdivision of the public domain have been modified from time to time, the general plan has remained substantially the same. A synopsis of the main features of the rectangular system will be provided. A detailed account is available in Reference 13.9.

The basic scheme provided for:

1. The establishment of primary axes designated as *principal meridians* and *baselines* passing through an origin called the *initial point,* as shown in Figure 13.3. Each of the thirty-four principal meridians governing the entire rectangular survey system of the public lands has been named in some distinctive manner, such as the Third Principal Meridian, which governs surveys in much of Illinois. The geographic extent of the surveys originating from a given initial point is depicted in Figure 13.4.

A basic part of the legal description of any tract of land included in the U.S. rectangular system is the name of the principal meridian and the state. Without this information the descriptions of two different tracts may be identical. For example, a parcel of land in California may have the same description as one in Oregon, except that the former is governed by the Mount Diablo and the latter by the Willamette baseline and principal meridian.

2. The establishment of secondary axes known as *guide meridians* and *standard parallels* or *correction lines* at intervals of 24 miles east and west of the principal meridian and at similar intervals north and south of the baseline, respectively.

3. The subdivision of the 24-mile quadrangles (see Figure 13.3) into *townships*. A row of townships extending north and south is termed a *range;* a row extending east and west is called a *tier*. A township is identified by the number of its tier and range and by the designation of its principal meridian.

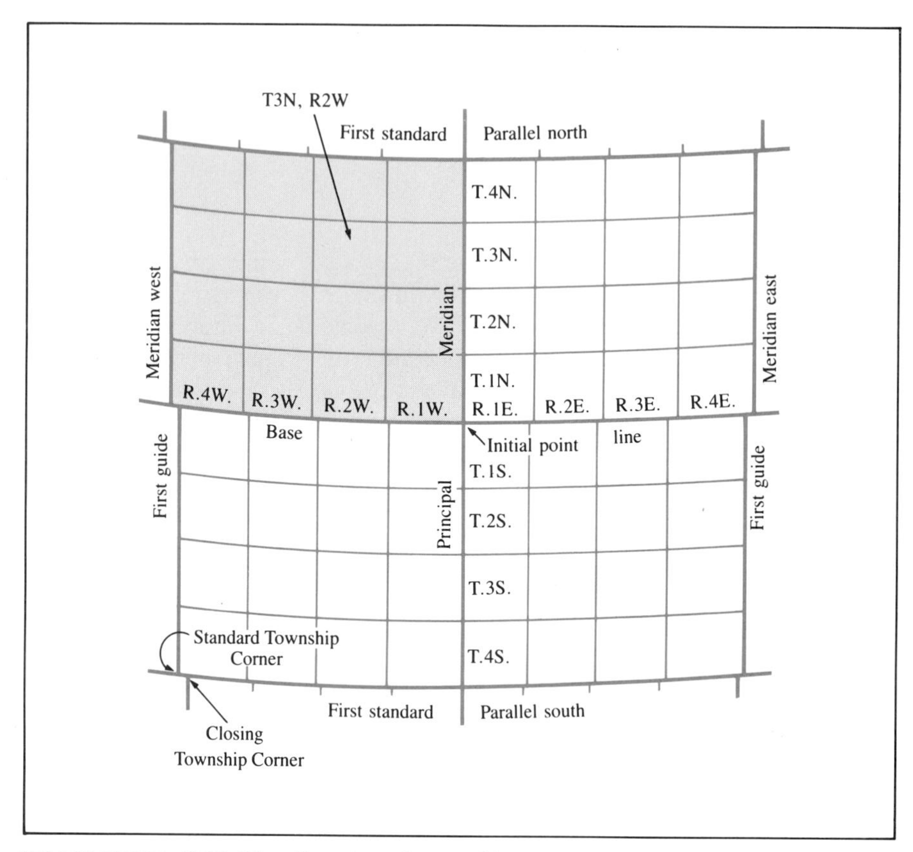

FIGURE 13.3 Creation of townships

The township is the basic unit of the rectangular system. A regular township is approximately 6 miles on a side and, therefore, has a nominal area of 36 square miles. In an effort to minimize the effect of convergency of meridians, *standard township corners* were set. These control the meridional subdivision of the land situated between each standard parallel and the next one to the north. *Closing township corners* were established on the baselines and standard parallels where the converging meridional lines from the south intersected them.

4. The subdivision of the townships into 36 units, known as *sections,* by running parallel lines through the townships from south to north and from east to west at distances of one mile. For a detailed explanation of this procedure, Reference 13.9 should be consulted. In general, the process of subdividing the township was such as to throw the effect of convergency of meridians and of surveying errors into the northern tier and western range of sections, as indicated in Figure 13.5.

The regulations prescribed suitable monumentation of the *section* and *quarter-section corners* along all lines. Quarter corners were nominally in most cases at the mid-point of the section lines.

With this step in the subdivisional process the role of the federal surveyor ended. The subdivision of the section was performed by the private land surveyor.

5. The subdivision of regular sections, containing nominally 640 acres, into quarter sections of 160 acres each by straight lines connecting the quarter-section

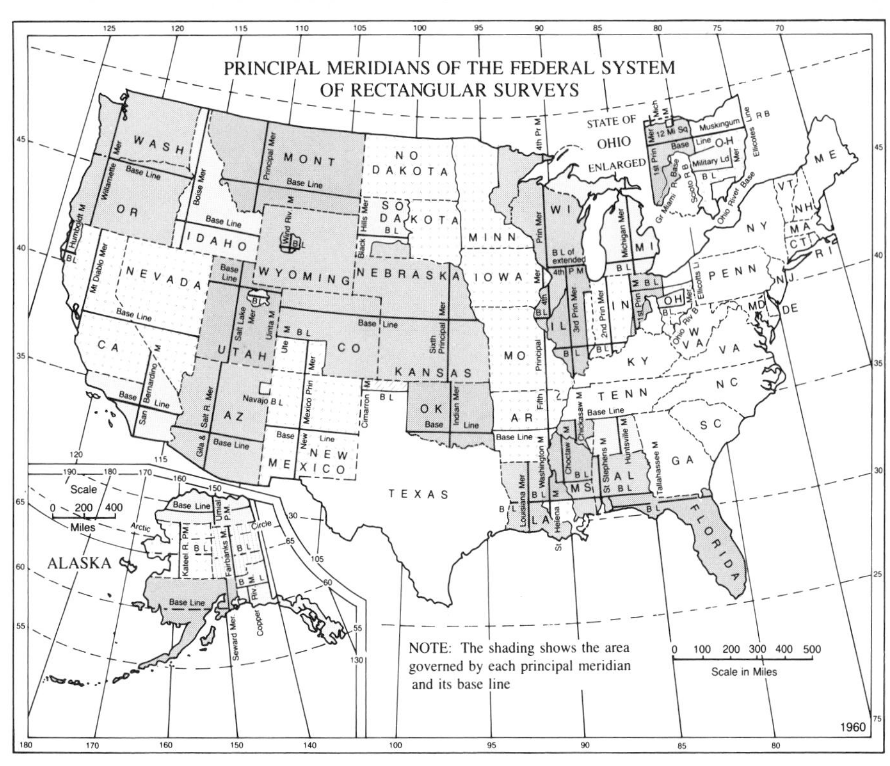

FIGURE 13.4 Principal meridians (Bureau of Land Management)

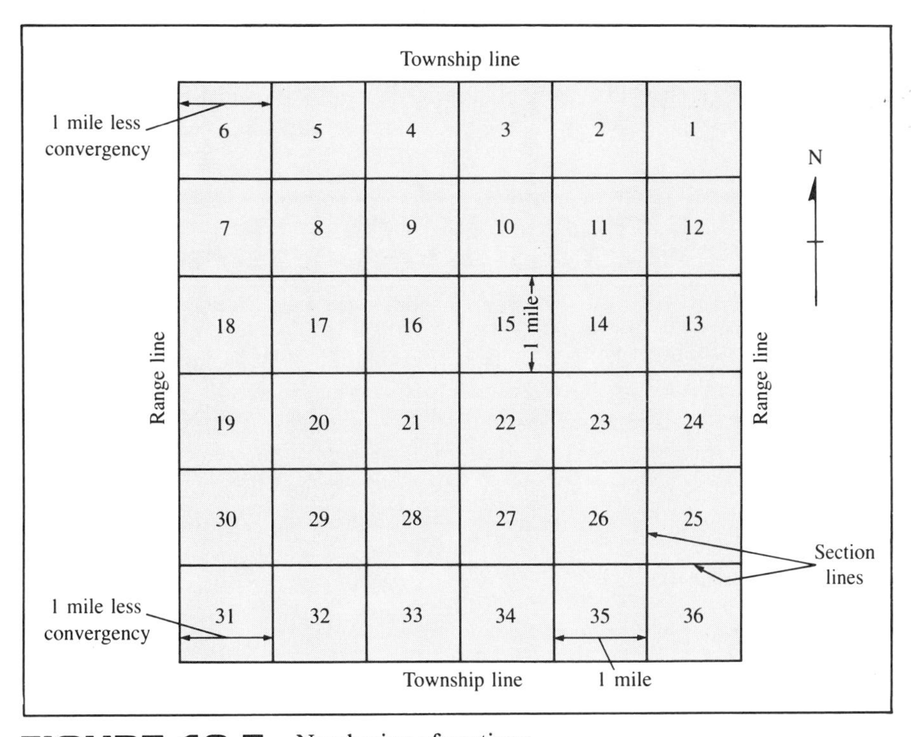

FIGURE 13.5 Numbering of sections

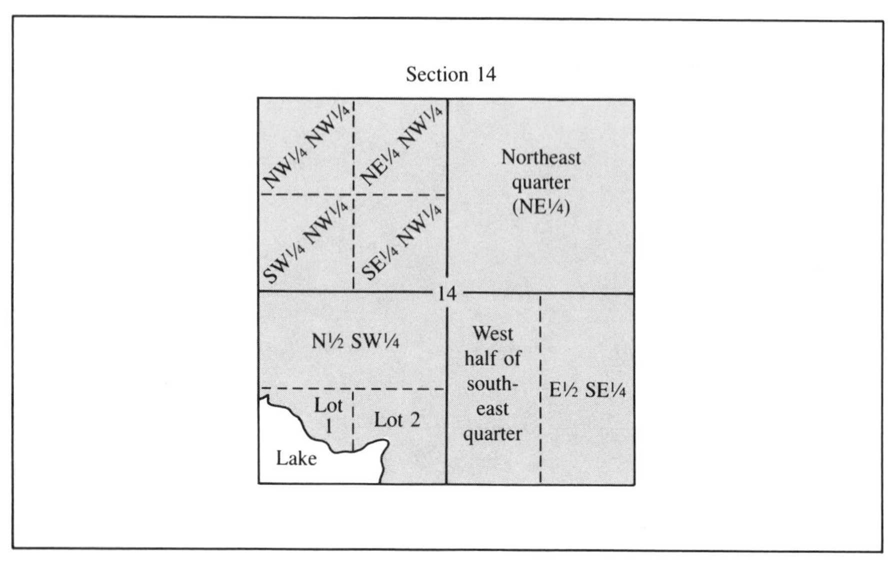

FIGURE 13.6 Aliquot parts of a section (Bureau of Land Management)

corners on opposite boundaries. The corner at the intersection of the quarter lines is known as the *interior quarter corner*. The 40-acre parcel (see Figure 13.6) results from the subdivisions of the quarter sections into quarter-quarter sections by the mid-point procedure. Furthermore, the quarter-quarter sections may be subdivided into 10-acre units by continuing this procedure.

The subdivision of sections that are not regular, such as those adjacent to the north and west boundaries of a normal township, is explained in Section 13.14.

13.8 EXTENT OF THE RECTANGULAR SYSTEM

The territory of the United States that has been or is being surveyed (see Figure 13.7) under the rectangular system includes thirty states, as shown in Figure 13.4. Also depicted are the several principal meridians with their corresponding base lines. The shading indicates the areas governed by each principal meridian and its baseline.

The rectangular system has been modified, to some extent, in each of these states by grants from Congress to individuals, grants to Native American tribes, or claims to mineral lands, and in the southern and western states by early grants from the governments of France and Spain. Many of the Native American treaties include grants of land to individuals, both American and Native; also, many individual property rights had become established, either by purchase or occupation, in the states west of the Mississippi before the date of the Louisiana Purchase. The federal govern-

FIGURE 13.7 Cadastral surveying in Alaska (Bureau of Land Management)

ment early adopted the policy of recognizing many acceptable Spanish, Mexican, French, and Indian titles and the land represented by them was generally excluded from the rectangular system. The United States system has never been applied in the state of Texas, although a modified rectangular system has been used by that state in subdividing tracts as they existed when it was admitted to the Union.

In certain cases departures from the rectangular system of surveys have been made to meet land management problems. The authority for such action vested in the Secretary of the Interior by legislation approved by the Congress on April 29, 1950, is intended for limited use. It has been used, for example, where it has become necessary to identify a particular tract of land by survey, and the extension of the rectangular net to cover such lands would be unduly expensive, in cases where lands adjacent to lakes and rivers were erroneously omitted from the original surveys, and in special situations where the departure from the normal system may be considered essential in promoting the settlement of project areas.

No single description of the U.S. Public Lands Survey System will apply to all the states where it has been used, because the system has undergone many changes since it was instituted in eastern Ohio. From time to time the U.S. General Land Office and its successor, the Bureau of Land Management, have issued instructions under which the work of subdivision has been done, but many changes have been made as the work progressed. These instructions in the early years were in the form of circulars and letters to the deputy surveyors, and it is now impossible to recover all of them, although careful search has supplied many. In later years the U.S. General Land Office and the Bureau of Land Management published "Manuals of Instruction," which prescribed the procedure to be used in the western states and Alaska, where subdivision work still is in progress. But it should be noted that the manner of establishing the land lines in practically all of the middle-western and southern states was very different from that described in any current manual.

13.9 BENEFITS OF THE NATIONAL CADASTRAL SYSTEM

Adequate cadastral surveys are necessary for the administration and wise use of land. The cadastral survey is the first step in any form of land management. It is an essential ingredient in the development of any region and a necessity in the transfer of title to land and in the establishment of land ownership.

The use of the rectangular system ensures that every aliquot part of a section, whether it contains $1\frac{1}{4}$, $2\frac{1}{2}$, 5, 10, 40, or 160 acres, has a definite description that is not duplicated. If the parcel of land is described with reference to the correct section, township, range, principal meridian, and state, it will be the only tract in the entire public domain with that particular description. Also, the description will readily indicate to anyone familiar with the system the general location of the parcel.

The simplicity of describing the individual tracts has made the U.S. rectangular system one of the most efficient and practical means ever devised by any nation for land identification. It has reduced the confusion and litigation over land titles, aided in the maintenance of good land title records, and stimulated the orderly settlement of the public domain.

13.10 INSTRUMENTS AND METHODS

The measurements of directions and distances in the early surveys were very crude. The magnetic compass, Figure 13.8a, was the universal instrument used for establishing the directions of all lines, including the principal meridians and baselines, until William Burt (1792–1858) invented the solar compass in 1836. At about the same time, the transit, in something like its present form, came into use.

Of course, from the beginning the attempt was made to establish important lines as true meridians or parallels by taking astronomical observations and finding the declination of the needle, but these observations were crude and the procedure used to prolong these lines was imperfect so

FIGURE 13.8 (a) Early surveyor's compass, (b) 66-ft chain

that even the principal meridians of the early surveys were not straight lines on the ground.

In running out lines with the compass, peep sights were used and a foresight, determined by a compass bearing, was taken on a range pole or some landmark ahead. The distance was then measured and the compass was moved to that point, where another foresight was taken and the work continued. Thus no backsights were used; and, if a tree was found to be on line, the instrument was carried past it, set up by estimation on the line produced, and a new foresight taken. It is evident that any line run by this procedure would not be a straight line, but would have a small angle introduced at every instrument station.

Hence, by this procedure, no line would be established as a straight line for a distance greater than the segment between two adjacent corners; and since, from an early date, the laws prescribed quarter-section corners to be set on all lines at half-mile intervals, it may be said that for all land lines established before 1894, when the use of the compass was abolished, a section of land is not a square tract having four sides but an irregular tract having eight sides.

There is no definite date when it may be said that the telescope came into general use or when the use of backsights was introduced to establish straight lines. The various surveyors employed different instruments and methods, so that in a given year one surveyor might be using a Burt solar compass, another might be using a transit, and another might be using a peep-sight compass for doing the same kind of work.

The chain used on the early surveys was two "poles," "perches," or "rods," 33 ft long, composed of 50 links. All recorded distances, however, were expressed in units of a four-pole (66 ft) chain (see Figure 13.8b). It was made of wire, and the many links exposed many wearing surfaces so that the length of the chain increased, perhaps as much as a half-foot in a season. A means of adjusting the length of the last link was provided, and comparisons with a "standard" chain were prescribed, but the crude methods used to deal with the errors of measuring are indicated by the inadequate instructions governing the field work.

The Manual of Instructions of 1902 uses the phrase "field chains or steel tapes," which is the first evidence of the use of the steel tape in the work of the government land surveys.

Further insight into the methods and instruments employed by early federal cadastral surveyors can be obtained by consulting References 13.4 and 13.8.

13.11 CONDITIONS AFFECTING EARLY SURVEYS

Permissible limits of error in the field measurements of the public lands were prescribed from a very early date, but at first these were vague and indefinite. Subsequent instructions became more specific, providing limits of error for the lengths and directions of the various lines run. The fact that the measurements actually made were often in error far beyond the limits prescribed arose from many conditions other than limitations of the instruments or procedure in use. They were as follows:

1. *Land Values.* The common allotment of land to early settlers on the public lands was a "homestead" of a quarter section, 160 acres, at a cost of $1.25 per acre, including certain requirements as to improvements to be made within a specified time. Accordingly, it was believed that high accuracy in the field measurements was not warranted by such a low land value.

2. *The Contract System.* From the beginning, haste was imposed principally by the condition that the surveys were made under contracts at a specified sum per mile. Different rates were paid depending on the importance of the lines (that is, whether they were section lines or range lines, and so on) and on the character of the terrain (that is, whether it was open, wooded, flat, hilly, or swampy). At all times, however, the financial return to the contracting surveyor depended on the speed with which the lines could be run.

3. *Supervision.* All contracts were executed under oath as to the completeness and accuracy of the results, but supervision and inspection was at first totally lacking and remained inadequate until about 1880.

It may be added that unsatisfactory work often resulted from pressure for haste exerted by settlers who occupied unsurveyed land and were anxious to have their boundaries fixed; Indians frequently were hostile; and competent surveyors were not available in sufficient number for the vast amount of work to be done.

However, having referred to the many sources of error in the original surveys, it should be said that much of the work was surprisingly well done; in view of the adverse conditions, those surveyors who, over wide areas, achieved such excellent results, have well earned the gratitude and respect of their successors.

13.12 FIELD NOTES AND PLATS

The official record of a cadastral survey ordinarily consists of a drawing or map portraying the lines surveyed and showing their lengths and directions as well as the area of the tract of land.

The importance of keeping an accurate and intelligible record of the field measurements of the public land surveys was recognized from the beginning, and careful and explicit instructions have always governed this work. Of course, the notes returned did not always conform to specifications; either because of adverse field conditions, or shoddy or fraudulent work, the actual present field measurements often vary widely from those shown in the notes or on the accompanying plats. However, the original record as shown in the field notes and on the plats is constantly required by present surveyors as they attempt to retrace the lines established by the original surveys; many serious mistakes are now made by inexperienced surveyors who fail to make proper use of these sources of information. The official government township plats prepared by the federal land surveyor were generally drawn to a scale of 1 in. equal to 40 chains and showed section boundaries, subdivisions of sections, meander lines, and other information, as well as a limited amount of topography.

On the completion of the public land surveys in a state, the original field notes and township plats were transferred to the state. Copies of such notes and plats can usually be obtained either from the state official, usually the Secretary of State, or from the Director of the Bureau of Land Management, Washington, D.C. Much more specific information about the availability of notes and plats as well as other records of the public land surveys is to be found in Reference 13.2.

13.13 MEANDER LINES

The traverse of the margin of a permanent natural body of water is termed a *meander line*. Such lines were run as nearly as possible to conform to the mean high water line. They served the purpose of providing data from which to calculate the areas of tracts of land made fractional by bodies of water. The conditions for and methods of running such lines were originally prescribed by the General Land Office. In general, they follow the margins of lakes whose areas are as large as 25 acres, and of streams whose right-angle widths are 3 chains or more.

Note that, although meander lines are shown on the plats and are used to calculate the areas of land lots, they do not constitute property lines. The ownership of property bordering on bodies of water is discussed under the subject of riparian rights in Section 13.34.

13.14 SUBDIVISION OF SECTIONS

On the plat of all regular sections the boundaries of the quarter sections are shown by straight lines connecting the opposite quarter-section corners. However, the sections bordering the north and west boundaries of the township, except section 6, are subdivided into two regular quarter sections, two regular half-quarter sections, and four fractional quarter-quarter units usually designated as lots (see Figure 13.9). In section 6 the plan of subdivision will show one regular quarter section, two regular half-quarter sections, one regular quarter-quarter section, and seven fractional quarter-quarter units. The purpose of such subdivision is to place the excess or

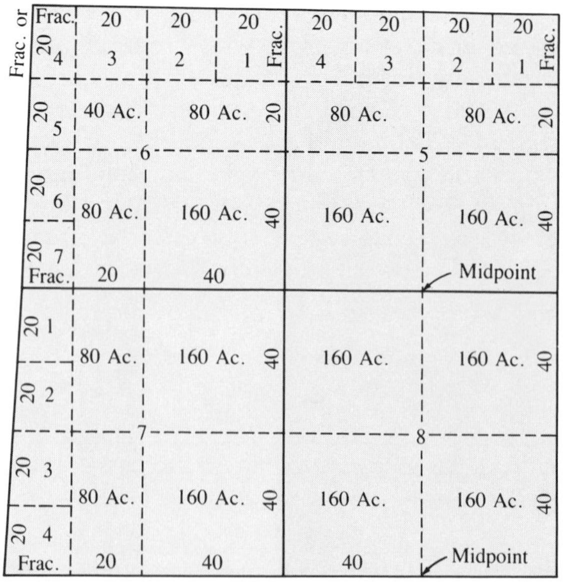

FIGURE 13.9 Subdivision of fractional sections

deficiency in measurements against the north and west boundaries of the township. It is incumbent upon the local surveyor to set the corners defining the subdivisions of a section in the same relative position they would have occupied, as indicated on the plat, had they been set by the federal surveyor.

13.15 KINDS OF CORNERS

The term *corner* has two meanings: (1) it refers to the point fixed on the ground by measurement along a line from another established point, or by the intersection of established lines; and (2) it refers to the physical object or marker, which serves as a more or less permanent monument at the given point. In the following paragraphs, the term has the first meaning.

MARKING CORNER

A marking corner is the point established by the survey measurements as the actual location of one corner of a regular tract of the subdivisional system. Such corners are designated by many different terms, depending on the location of each one in the system, such as a quarter corner, section corner, township corner, or closing corner.

WITNESS CORNER

Where a corner fell in a body of water or other place where it was impracticable to fix the corner itself, a witness corner was set on each one of the survey lines leading to the inaccessible corner and at a known distance from it. Thus the corner itself was fixed by means of the witness corner.

MEANDER CORNERS

Every meander line began at one point, and ended at another, on a line of the regular subdivisional system. At each of these two points a meander corner was established. Also, as the meander line was extended, if it intersected any of the regular subdivisional lines, a meander corner was established at each intersection. Of course, by coincidence, a marking corner or a witness corner might also serve as a meander corner.

13.16 CORNER MARKERS

An essential element of the public land surveys was the establishment of markers on the ground at township, section, and quarter section corners. At first, such markers were the most substantial that local materials afforded. In timbered areas, posts of the most resistant types of wood,

FIGURE 13.10 Land corner

charred to minimize decay, were used. In other areas, cut or natural stones were used. In the plains areas where neither of these were available, small pits were dug and filled with charcoal, which can often be found several decades later. Since 1910, a uniform marker, consisting of an iron pipe (see Figure 13.10) topped with a brass cap on which is stamped the designation of the particular corner, is used. Both the iron and the brass markers are chosen for their ability to withstand corrosion from the elements, and it is believed that such markers will remain and be legible for 100 years or more. A pile of stones is usually erected around the iron post, when stones are available, to assist in its location. In all cases the government surveyor was instructed to record in the field notes the material and method used to mark each corner. It should be said in this connection that the original corner marker is incontestable evidence that fixes the position of a corner. Accordingly, the most satisfactory restoration of the original corners requires the surveyor to know the instructions under which these corners were established and to consult the field notes returned from the original survey.

13.17 CORNER ACCESSORIES

In the attempt to establish the position of a corner as permanently as possible, other objects in the immediate vicinity were used to evidence the location of the corner itself. Such objects are called *corner accessories,* and these may be such natural or artificial objects as were used for the corner materials described above. Of course, they were given specified markings, and careful descriptions were entered in the field notes.

In timbered areas, bearing trees (see Figure 13.11) were universally used. Clearance of the land for lumber or for cultivation, forest fires, and other agencies have destroyed many of these trees, all of them in some regions, but the skillful surveyor may often discover slight bits of evidence which re-establish beyond doubt the position of the original corner.

FIGURE 13.11 Bearing tree (Bureau of Land Management)

13.18 PERPETUATION OF LAND CORNERS

From ancient times the evidence of land ownership has consisted of physical objects on or in the ground that mark the corners or boundaries of the tract owned. That condition still prevails. A legal title deed may confer ownership of the land described in the deed, but that which determines what is really owned is the physical evidence on the ground. This condition makes it absolutely necessary for the purchaser to see the physical objects that mark the boundaries to know exactly what land is being bought.

This necessity of maintaining the identity of land corners is beset with many difficulties resulting from weather, building and construction operations, and other agencies that may disturb or destroy evidence of the original location of a corner or a boundary. Perpetuation of corners is one of the major problems confronting the land surveying profession. With the expansion of the national horizontal control network and the increasing use of state plane coordinates, it is becoming more feasible to fix the position of

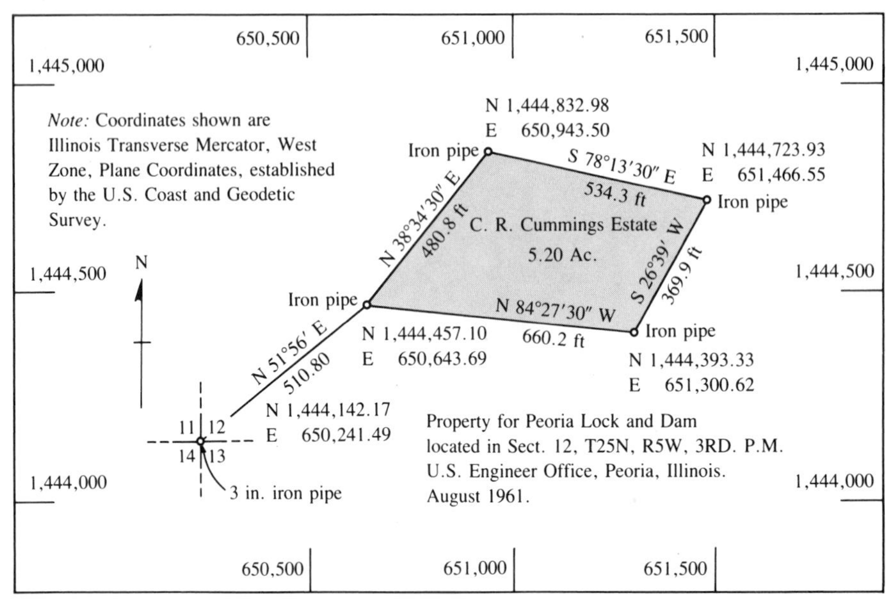

FIGURE 13.12 Property corners fixed by state plane coordinates

land boundary markers by suitable ties to such stations (Figure 13.12). Even though the marker is destroyed, it can be replaced by means of its known coordinates. It should be understood, however, that no horizontal control survey, whether its results are expressed in terms of state plane coordinates or not, can have in itself cadastral qualities. There are certain definite legal procedures, as described in Section 13.27, that govern the restoration of lost or obliterated section and other corners. However, when the corner monument is missing because of destruction or decay, it should be apparent that the accurately known state plane coordinates of the corner may become the best available evidence of the position of the original corner and thus provide the most satisfactory means for its restoration or recovery. This does not mean that the use of state plane coordinates in land conveyancing implies the abandonment of the older methods of describing land, such as by metes and bounds or by parts of a section. State plane coordinates can be used, however, to supplement the other kinds of land descriptions.

Several notable steps have been taken in recent years to perpetuate the positions of corners in the nation's fundamental cadastral network. First of all and most significantly, several governmental and quasi-governmental agencies have begun systematic programs of recovering corner positions and setting new monuments, which are then tied to the state plane coordinate systems. Examples of such comprehensive actions are the programs of the Southeastern Wisconsin Regional Planning Commission in the nine-county area of southeastern Wisconsin; the corner restoration project conducted by

Hennepin County, Minnesota; and the enactment of legislation in Missouri on July 1, 1970, creating a State Land Survey Authority. The duties and responsibilities of the authority are "to restore, maintain, and preserve the land survey monuments, section corners, and quarter corners established by the United States public land survey within Missouri" and to perform certain other related functions.

In several states laws have been passed requiring land surveyors to take various measures to protect and improve the marking of the corners. In Indiana, for example, the General Assembly enacted in March, 1965 the "Perpetual Corner Records Act," which provides that the county surveyor shall check and establish or reestablish and reference at least 5% of all original government corners in the county annually and record the actions in a Corner Record Book.

Another aspect of corner perpetuation activities has been an increased recognition by the private land surveyor of the great importance of preserving evidence of corner positions through the use of more durable monument material. New designs have been effected and different materials have been employed. Although field studies have been conducted to develop a marker that is portable; durable; resistant to frost heaving, corrosion, and rust; capable of being detected magnetically; and relatively inexpensive, a completely satisfactory solution has not yet been found. One of the relatively new developments is a cast-iron monument with a flanged bottom and side vanes, which defies extraction, resists frost action, and can be permanently magnetized. A similar type made of aluminum consists of extendable parts permitting increases in height when necessary without disturbing the original monument setting. It is designed to break off at predetermined points in the event of surface impact, and permanent magnets are incorporated into the unit.

Associated with improvements in monumentation of corners have been advances in the design of metal detectors, which are used by the land surveyor to detect and locate a small mass of ferrous metal. One type is the magnetic detector, such as a dip needle, which is relatively inexpensive but whose effectiveness is limited to indicating objects lying quite close to the ground surface. Another type is the electronic metal detector, which is more expensive but will sense ferrous objects at far greater depths than needles.

13.19 WHAT PRESENT SURVEYS REVEAL

In the preceding pages something has been said of the conditions under which the original surveys were made. Some of these conditions may be listed in summary here: (1) the instruments used (that is, the compass and two-pole chain) rendered accurate measurements impossible; (2) the contract system placed upon the surveyor the incentive for speed rather than accuracy; (3) the lack of training on the part of many surveyors rendered them incapable of interpreting or properly applying the instructions that were intended to govern their work; (4) the lack of supervision or inspection permitted careless and fraudulent surveys to stand.

An important part of the land surveyor's work now is the retracing and restoration of the original lines and corners of the government surveys. In some areas where those surveys have been made in more recent years this work can be done with considerable ease and satisfaction, but in many of the central and southern states where the original surveys are more than a hundred years old, this work is difficult and requires a thorough knowledge of the procedure used in the original work and resourcefulness and good judgment in dealing with whatever evidence can now be found about the location of the original corners.

As some indication of what the surveyor may expect to find, two examples of conditions revealed by present surveys are given below. These are extreme examples, but they will serve to emphasize the fact that the present surveyor must expect to find that (1) few present measurements will agree with those shown on an original plat; (2) undetected mistakes are likely to be present anywhere in the original measurements and in the recorded notes; (3) no line can be regarded as being straight for more than one-half mile; (4) whereas all quarter corners, except those in the north and west sections of a township, are intended to be equidistant between adjacent section corners, such is frequently not the case.

Figure 13.13 shows two plats of the same section, the first as returned by the deputy surveyor in 1884, and the second as returned by a later survey by the General Land Office.

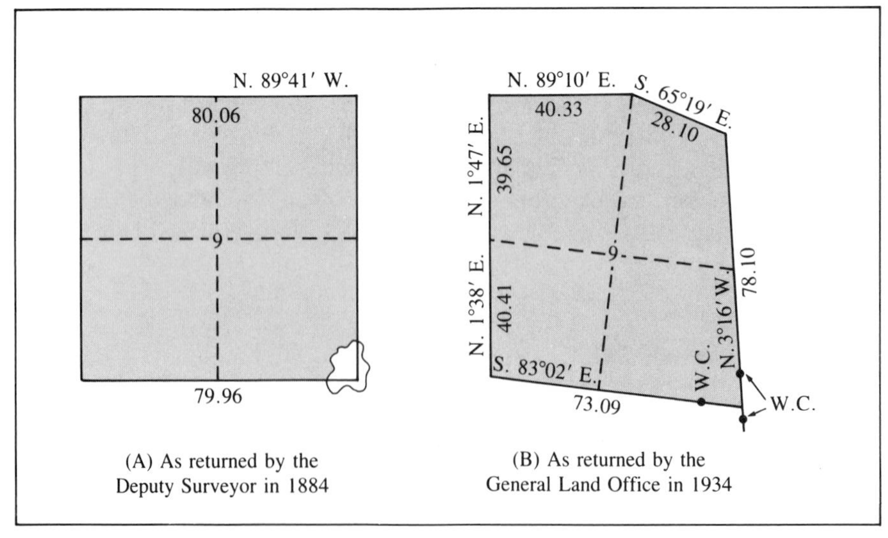

FIGURE 13.13 Two plats of Sec. 9, T.64 N.,R.1W., 4th PM in Minnesota

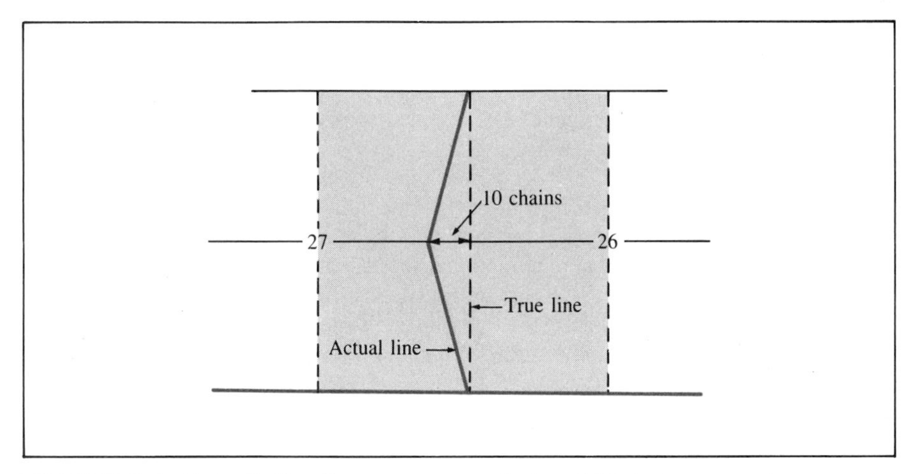

FIGURE 13.14 Quarter corner misplaced 10 chains west of correct location

Figure 13.14 shows a quarter corner in Wisconsin misplaced 10 chains west of its correct position. However, the original quarter corner was found in place with two bearing trees to witness it, so that the owner of section 27 concluded he could not contest the line in court.

13.20 GENERAL RULES GOVERNING SURVEYS OF PUBLIC LANDS

Cadastral surveys, in general, create, reestablish, mark, and define boundaries of tracts of land. Cadastral surveys, unlike informational surveys such as those associated with mapping projects, which may be amended and revised with changing conditions or because they were not executed according to more exacting standards of accuracy, cannot ordinarily be repudiated, altered, or corrected. Furthermore, the boundaries created or reestablished cannot be changed except under very special circumstances.

The foregoing principles are embodied in the following list of general rules contained in the Manual of Instructions for the Survey of the Public Lands of the United States (Reference 13.9):

First. That the boundaries and subdivisions of the public lands as surveyed under approved instructions by the duly appointed engineers, the physical evidence of which survey consists of monuments established upon the ground, and the record evidence of which consists of field notes and plats, duly approved by the authorities constituted by law, are unchangeable after passing of the title by the United States.

Second. That the physical evidence of the original township, quarter section, and other monuments must stand as the true corners of the subdivisions, which they were intended to represent, and will be given controlling preference over the recorded directions and lengths of lines.

Third. The quarter-quarter section corners not established in the process of the original survey shall be on the line connecting the section and the quarter-section corners, and mid-way between them, except on the last half-mile of section lines closing on the north and west boundaries of the township, or on other lines between fractional or irregular sections.

Fourth. That the center lines of a regular section are to be straight, running from the quarter-section corner on one boundary of the section to the corresponding corner on the opposite section line.

Fifth. That in a fractional section where no opposite corresponding quarter-section corner has been or can be established the centerline of such section must run from the proper quarter-section corner as nearly in a cardinal direction to the meander line, reservation, or other boundary of such fractional section, as due parallelism with section lines will permit.

Sixth. That lost or obliterated corners of the approved survey must be restored to their original locations whenever it is possible to do so. Actions or decisions by surveyors, Federal, State, or local, which may involve the possibility of changes in the established boundaries of patented lands, are subject to review by the State courts upon suit advancing that issue.

13.21 LAND TITLE RECORDS

Many documents, particularly those that may affect persons other than the holder, are made a part of public record. For example, title to land, although apparently a private matter, is of public interest. The legal recognition of the public significance of land titles came into being with the recording acts of the several states. Probably the earliest of these acts in the United States was enacted in 1640 by the Massachusetts Bay Colony.

The intent of most land recording acts is to give constructive notice of the contents of the record to the general public. Such notice provides the opportunity for everyone to become knowledgeable concerning the facts contained in the recorded or filed instrument such as a deed, mortgage, or bankruptcy action and is intended to protect an innocent purchaser against secret liens and encumbrances.

Although there are numerous statutes requiring that public records be kept and prescribing the manner of maintaining them, there have been widespread differences in the procedural aspects of indexing such material so it can be readily retrieved and so that data from several sources relating to a single land parcel can be properly coordinated. Hence, for years a comprehensive and unified system of land records has been the subject of study by members of the legal and land surveying professions as well as city planners and others concerned with land management.

Improvements in *land parcel identification* are closely associated with research into the development of a modern system of land records. In its most elementary form, land parcel identification can be illustrated by the street address or house number of a city lot. For various reasons this system is unsatisfactory and lacks legal approval. Another plan possessing much more merit is to use an eleven-digit identification number such as 03-007-05-37-18 for a given parcel. The components of the identifier are as follows: the first two digits indicate the state; the three-digit group indicates the county; the next two digits relate to the ward, district, or township; the next two digits are for the so-called "block"; and, the last two digits identify the parcel as a lot in that block.

Another technique that has been strongly advocated is one involving a coordinate-based system—namely, CULDATA, which is the acronym for *Comprehensive, Unified, Land Data System.* The state plane coordinate systems (SPCS) form the foundation of CULDATA (see Reference 13.5). Probably the most formidable problem in implementing a coordinate system for land parcel identification is that of deciding what point in a parcel will represent that unit of land.

The land surveyor is interested in both land parcel identification and a related topic, *land parcel description.* The latter, frequently termed simply as *land description,* is primarily a written statement recognized by law that definitely fixes or defines the physical location of real property.

13.22 LAND DESCRIPTIONS

The purpose of a deed to real property is to convey ownership of a tract of land from one party to another and to evidence title in the owner. The purpose of the description, which must be a part of the deed, is to furnish the information that is necessary to identify the boundaries of that particular tract on the ground, both at the time the conveyance is made and at any future time. It is evident, therefore, that the proper composition of a deed description requires a knowledge both of the law and of surveying, and something of these requirements is indicated in the following paragraphs.

Land descriptions may be grouped into four general kinds as follows: description by reference to (1) natural objects and adjoiners, without numerical data, (2) metes and bounds, (3) the public lands survey system, and (4) urban subdivisions.

This classification is not a definite one, for many descriptions will include characteristics of two or more of these classes, but it will serve the purpose of indicating the principal features of the various kinds of descriptions.

The *first class* includes those descriptions that refer to such natural objects as a tree, the centerline of a road, the thread of a stream, or a boundary line that may be described by giving no information except the names of the adjoining owners. Such descriptions may contain some numerical data that may aid in fixing the boundaries, but in descriptions so obviously weak the numerical data are likely to be faulty. No reputable surveyor would return a description of this kind now, but in early times, when land values were small, such descriptions were thought sufficient. The following is an example:

Beginning at a stone in the highway leading from Portsmouth to Springfield, and on the east fence line of land of Levi Brown; thence east by the highway to land of James Green; thence north by land of Green to Spring Creek; thence westerly along Spring Creek to the east fence of land of Levi Brown, thence by land of Brown to the point of beginning; containing 22.40 acres, more or less.

The *second class* of descriptions consists of a recital of the lengths and directions of the various courses around the perimeter of a tract of land starting with a careful description of the point of beginning. If one of the corners can be found and the true direction of one of the boundary lines is known, a *metes and bounds* description will permit the relocation of the sides of a parcel of land.

Metes and bounds descriptions define land boundaries in those states where the public land surveys were not executed—namely, the original colonies, Texas, Kentucky, and Tennessee. They are also used, however, to define irregular tracts of land in states that have been sectionized.

The first of the following two examples refers to Figure 13.15, and the second to Figure 13.16 on page 494.

Beginning at a point, marked by an iron pin, 584.10 feet North of the S.E. corner of the N.W. $\frac{1}{4}$ of the S.E. $\frac{1}{4}$ of Sec. 2, T. 19 N., R. 8 E., 3rd P.M.; thence North 607.86 feet to the centerline of Bloomington Road, marked by a cross cut in the pavement; thence in a Northwesterly direction along the centerline of said Bloomington Road, 422.41 feet to another cross cut in the pavement; thence South 817.47 feet to an iron pin; thence East 366.74 feet to the point of beginning; containing 6.00 acres.

All that part of Island No. 239 lying in Fractional Sec. 12, Twp. 27 N., R. 1 W. of the 4th Principal Meridian, Jo Daviess County, Illinois. Area of tract = 30.4 ± acres.

The *third class* deals with aliquot or regular tracts within the areas covered by the U.S. public lands system. Thus, a 40-acre tract may be described as: "N.E.$\frac{1}{4}$, of N.W.$\frac{1}{4}$ of Section 9, Township 64 North, Range 1 West of the Fourth Principal Meridian."

The *fourth class* includes all regular parts of urban subdivisions. Thus, a tract may be described as "Lot No. 9, Block 4, in Hills and Dales Addition to the Town, now city of West Lafayette, Indiana."

In some cases descriptions (see Figure 13.12) relating to irregular parts of a section are supplemented by expressing the positions of the corners in state plane coordinate systems (SPCS), which will be discussed in Chapter 15. Thirty-five state legislatures have enacted laws that provide that the SPCS may be used in deed descriptions. It is emphasized that this legislation is permissive and not mandatory.

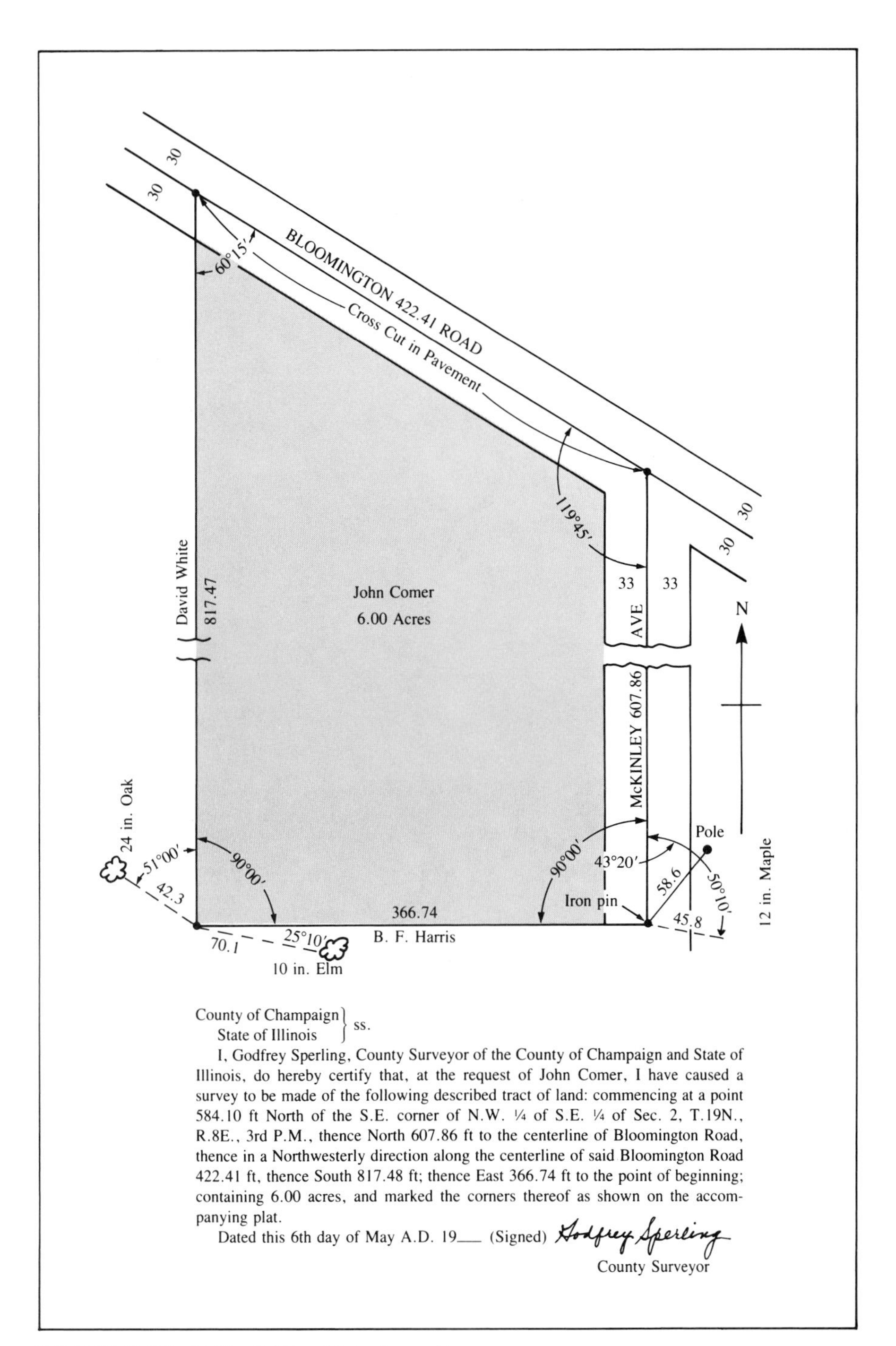

County of Champaign }
State of Illinois } ss.

I, Godfrey Sperling, County Surveyor of the County of Champaign and State of Illinois, do hereby certify that, at the request of John Comer, I have caused a survey to be made of the following described tract of land: commencing at a point 584.10 ft North of the S.E. corner of N.W. ¼ of S.E. ¼ of Sec. 2, T.19N., R.8E., 3rd P.M., thence North 607.86 ft to the centerline of Bloomington Road, thence in a Northwesterly direction along the centerline of said Bloomington Road 422.41 ft, thence South 817.48 ft; thence East 366.74 ft to the point of beginning; containing 6.00 acres, and marked the corners thereof as shown on the accompanying plat.

Dated this 6th day of May A.D. 19___ (Signed) Godfrey Sperling
County Surveyor

FIGURE 13.15 Plat of rural tract

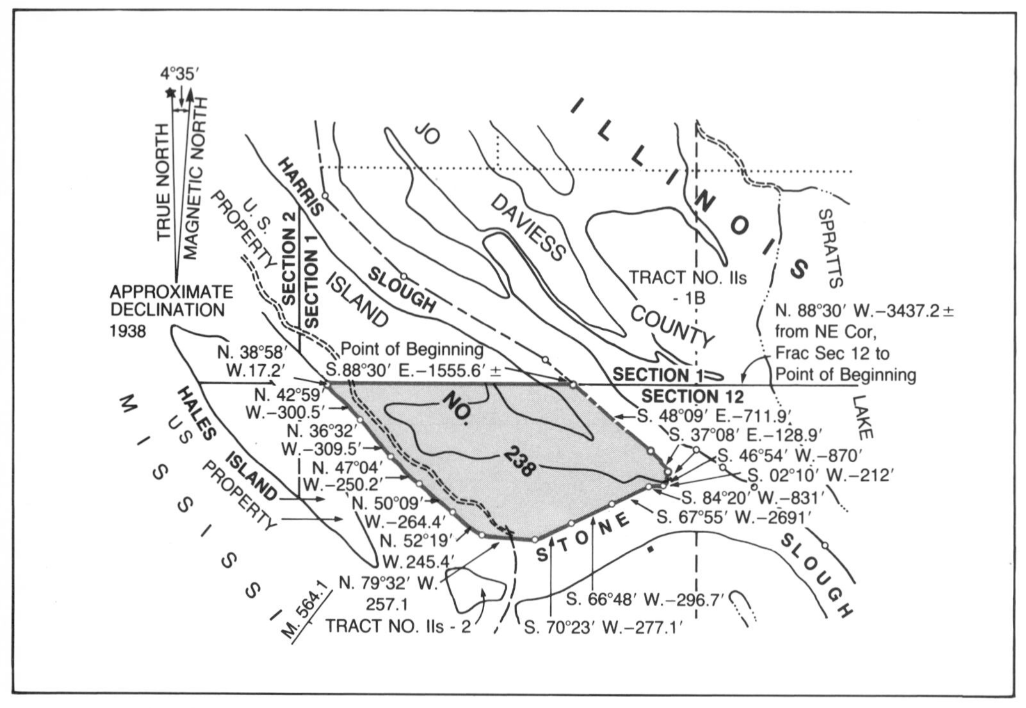

FIGURE 13.16 Plat of island (Corps of Engineers)

In recent years the widespread development of the condominium concept of ownership has literally added a new dimension to the traditional horizontal division of land into lots. In the condominium description it is usually necessary to define the units in three dimensions, thus adding the element of elevation. The unique nature of condominium surveys and descriptions merit special treatment in Section 13.30.

13.23 REQUIREMENTS OF A VALID DESCRIPTION

The essential requirements of a valid land description are that it shall be clear, accurate, and brief.

The great advantage of the United States public lands system is that it makes possible the use of descriptions so excellent in these essentials. The identification on the ground of a given tract may not be a simple matter, but that condition does not detract from the excellence of the system insofar as the description of any regular tract is concerned. The same advantages apply to regular lots of urban subdivisions. The land surveying difficulties that apply to such land parcels, either rural or urban, arise from other sources, which are discussed elsewhere, but the descriptions of these tracts can hardly be improved.

Faulty land descriptions are probably most likely to occur where any irregular tract must be described by "metes and bounds," and some of the considerations that apply to such descriptions will be discussed in the following paragraphs.

CLARITY

A land description should be so clear that it is subject to but one, and that the correct, interpretation; this condition should obtain not only at the time the description is written, but at any future time. Therefore, the writer should maintain the point of view that his description must, if followed explicitly by the surveyor, mark out the correct boundary on the ground either now or at any later time.

So much confusion has arisen from the lack of clarity in deed descriptions that many legal principles have become established as a necessary means of giving the correct interpretation to faulty descriptions. Some of these principles are discussed in the next section.

A first requirement for clarity is the accurate use of words and phrases. For example, in giving directions or bearings, the word "north" should be used to indicate due north only (that is, the direction parallel with the reference meridian of the survey); whereas, "northerly" may indicate any direction in the first or fourth quadrants. It is extremely essential to identify or define the reference meridian such as true (astronomic) north or state plane grid north. Also, the terms, "at right angles to" or "parallel with" should refer only to lines between which the angles are 90° or 0°, respectively. Other examples might be given, but it is evident that only confusion can result from the incorrect or inexact use of words or phrases.

Another source of confusion and ambiguity in land descriptions results from the incorrect use of punctuation marks. Of course, these are often misplaced in copying and recording, but great care should be taken in the sentence structure and phrases used so that the correct meaning will depend as little as possible on the aid of punctuation marks.

A correct plat is always helpful and sometimes necessary for clarity in descriptions. Hence it is desirable that a correct plat should accompany any resurvey, and reference made to it in the description, whereby it becomes a part of the deed. In many cases a plat may be a legal requirement.

To fix the location of a tract, it is frequently necessary to refer to some established monument, road, or street line in the near vicinity. Care must be taken that such monument or line is definite, and identifiable upon the ground, and that the "point of beginning" of the survey shall refer to a corner of the tract surveyed and not to the monument to which the tract as a whole is referred.

ACCURACY

A correct survey should, of course, precede the writing of any metes and bounds description. Any attempt to scale dimensions shown on a map or plat, for the purpose of writing a current description, is sure to result in error and confusion.

For example, assume the following conditions: a city lot is shown on a plat as being 100 ft long and 50 ft wide, its length being in a north and south direction; a previous deed has conveyed out the west 12 ft of this lot; and the actual width of the lot by resurvey is only 49 ft.

The plat indicates the remainder of the lot as being 38 ft wide, but, as a matter of fact, it is only 37 ft wide. It is evident that it would be incorrect for the lawyer to describe this remainder as the "east 38 ft of said lot," because by the previous deed the remainder is only 37 ft wide. Since a deed cannot convey that which is not owned by the grantor, his deed is impotent with respect to the 1-ft strip in question. This remaining tract should properly be described as "the whole of said lot, except the west 12 ft."

The distances and directions of a metes and bounds description should be accurate and complete so that the error of closure of the boundary can be computed. This will ensure that no essential data have been omitted and will indicate the quality of the survey. The description should in no way depend on the data contained in the descriptions of adjoining tracts.

BREVITY

Brevity is essential in land descriptions because brevity enhances clarity, and it reduces expense and mistakes in subsequent records. For this reason it is essential as a part of a description, for it conveys in small space and in legible form information that otherwise would require many words.

In summary, the final test of any description is whether or not the particular tract described can be satisfactorily identified on the ground.

13.24 INTERPRETATION OF DEED DESCRIPTIONS

In making a land survey the surveyor must refer to the deed description, which in many cases is faulty because of omissions of essential data or because of conflicting calls. Court decisions have established a few general rules, which should govern the surveyor in making interpretations.

1. The best interpretation is that which most plainly and completely gives effect to the intentions of the parties to the deed, as revealed by all of the evidence available.

2. As regards conflicts between the calls of a description, the order of precedence is as follows: (a) a natural corner or boundary, such as a tree or a stream, will stand against an artificial boundary, such as a stake, stone, or a fence; (b) an artificial corner or boundary, if it can be identified, will stand against other conflicting calls as to direction, distance, or area; (c) the corner or line of an adjoiner, if it can be definitely identified, will control over calls for direction or distance or area; (d) in case there is a conflict between the boundary dimensions and the calculated area, the former will prevail over the latter, assuming of course that the boundary dimensions are consistent with the evidence as to the corner monuments.

3. If a description is faulty by reason of any obvious errors or omission of essential data, the attempt is made to render it valid rather than void. Thus, if a dimension is incorrect by a full tape length, or if the length or bearing of one side of a field has been omitted, and if otherwise the description and evidence on the ground are satisfactory, the obvious omission or mistake will be corrected and the deed will be held valid.

4. If two interpretations are possible, the one that favors the purchaser will be taken.

13.25 GENERAL SCOPE OF A LAND SURVEY

Land surveys differ in setting, purpose, and complexity, and it would be impossible to describe here all phases of even the most representative types. Certain broad procedural principles, however, can be presented that apply to most property surveys. The experience and judgment of the land surveyor will indicate the manner of application of such principles.

The scope of a land survey comprises three basic steps as follows:

1. *Preliminary study*. This phase involves the collection, study, and interpretation of all available data of record, including old field notes, subdivisional maps, descriptions of adjoining parcels, legal documents, and other information. In the case of private surveying practice, this step should be preceded by a conference with the client to determine the nature and extent of the professional surveying service desired and to make an estimate of its cost.

2. *Field survey*. This operation comprises a search for monuments, the delineation of lines of possession, the location of encroachments, and the execution of a closed traverse with all corners durably monumented.

3. *Preparation of plat and deed description*. The plat, which is a special form of a map, portrays the tract boundaries and depicts other information essential to the description and identification of the property.

13.26 SEARCH FOR CORNERS

At the present time, every survey requires the location, if possible, of existing corners. This may be a simple matter if the work of the previous surveyor was well done and if the lapse of time has not been great. Unfortunately, both of these conditions are frequently adverse, and it requires careful and thoughtful procedure to establish the position of a corner for which much of the original evidence is missing.

The following paragraphs discuss the matter of searching for boundary markers. Although the term *corner* has been applied particularly to monuments set in the course of subdividing the public lands, it has here the broader connotation of being any kind of property marker.

Some fundamental definitions of corners with special reference to the U.S. rectangular system are as follows:

An *obliterated corner* is one at whose position there are no traces of the marker but whose location may be recovered by the acts of competent surveyors, the testimony of witnesses, or the use of some acceptable record evidence.

A *lost corner* is a survey point whose position cannot be determined in any way except by reference to one or more interdependent corners.

The general principles that affect the legal significance of monuments set in the subdivision of the public domain and the manner of restoring lost corners are stated explicitly in the first, second, and sixth rules governing the resurveys of the public lands as provided by the various acts of Congress and reproduced in Section 13.20.

In general, while searching for any kind of property marker the land surveyor is required: (1) to know the procedure used in the original surveys, (2) to have available original survey notes and plats and the records of more recent surveys in the vicinity of the obliterated corner, and (3) to exercise good judgment and discernment in discovering and evaluating all possible evidence pertaining to the obliterated corner.

The kinds of evidence that may be used to restore a corner in rural regions include: (a) the corner itself, (b) accessories, (c) fences, (d) roads, and (e) living witnesses.

Of course the corner itself is the best evidence, and every reasonable effort should be made to find it. If it was a wooden stake, care must be taken that this evidence not be destroyed or overlooked. Having determined by preliminary measurements, as nearly as may be, the probable location of the corner, the search is begun cautiously by slicing off the surface material. If the stake has rotted, its position may be indicated by the rotted wood or discolored soil; or, sometimes in firm soil, the hole formerly occupied by the stake will be found plainly marked. If either of these evidences can be found, the position of the corner is as well recovered as though the stake itself were in place. If it appears that the soil has filled-in over the corner, this condition is usually made apparent by the different quality and color of the filled material. If the corner object is a stone, an excellent tool with which to search is a rod about $\frac{1}{4}$ in. in diameter, 3 or 4 ft long, having an oval bead or ferrule slightly larger than the rod fitted over one end and a suitable handle at the other end. This rod can be pushed down vertically through 3 or 4 ft of nearly any kind of soil, including a packed roadway, to find a buried stone. By systematic procedure, a considerable area can be investigated much more efficiently by this means in a short time than can be done with a spade.

The records of any bearing trees or other reference objects should be found in the notes of the original survey or on the plats of other previous surveys. In this search the descriptions of adjacent property should be investigated, since these might contain indispensable information. If any bearing trees or other accessories can be found, often the corner can be restored as satisfactorily as though the corner mark itself had been found.

Fences offer good evidence about the location of a corner, since it is probable that the fence was built on the line when the corners were apparent. However, fences must be looked on with suspicion, for, in early times, a mutual agreement was frequently made whereby farmer *A* would build a line fence a rod or two over on *B*'s land, and thus secure the use of this strip of land in return for the expense of building the fence. After the lapse of years, the parties who had knowledge of this arrangement may have died or moved away and their successors left in ignorance of the true location of the property line. Or, the fence may not have been a property line at the time it was built and no attempt was made to place it on the line. However, if no such evidence of this kind can be found, it is a reasonable assumption that a fence marks the original land line.

A corner known to have been located in the center of a road will naturally be looked for on the centerline between existing fences. But again, the history of the fence lines must be carefully investigated to determine whether or not they have been moved since the corner was placed. Frequently when a road is widened, the change will be made on one side only, and hence the centerline between fences no longer marks the original property line.

Living witnesses frequently can give valuable information about the location of an obliterated corner. It often happens that individuals living in the vicinity were present when the corner was set, or have seen it on subsequent occasions, and can recall its exact location. There is a wide difference, however, in the ability of persons to remember the position of a marker they have previously seen, and considerable allowance must be made for this. In any case, such evidence must be supported by other evidence, but it may be of great assistance in relocating a corner.

Examination of aerial photographs can be extremely useful in revealing evidences of occupation such as traces of old fence lines and roadways that may not be readily recognizable from ground observation.

The kinds of evidence used to search for a corner within city limits are much the same as those in rural regions, although conditions are quite different. Building operations and street improvements effect many changes that destroy evidence about the location of property lines; for this reason every engineer should be most careful to set and reference any corner so as to render it as permanent as possible. Lot corners are likely to be covered with filled-in material that is usually of a quality or color different from the original subsoil. Fences are seldom on the lot lines and frequently are not parallel therewith; also, pavements and sidewalks are often not parallel with adjacent property lines. All these and many other conditions make it necessary that the work be done by a competent and experienced surveyor if satisfactory results are to be obtained.

13.27 RURAL SURVEYS

Most surveys of rural property are resurveys or retracements. Proficiency in satisfactorily re-establishing boundaries in their original locations is not easily attained. It is necessary that the land surveyor not only have adequate records of former surveys but also be able to interpret them correctly. The land surveyor should be familiar with land surveying practices in the given area, possess the tenacity needed to make thorough field explorations, be judicially minded in assessing conflicting field evidence, and be familiar with the decrees of the local courts in settling boundary disputes.

Some of the difficulties encountered in retracing old boundary lines are due to:

1. Faulty original measurements
2. Indefinite point of beginning
3. Indefinite meridian
4. Obliterated corners
5. Inadequate descriptions
6. Enmity between adjacent property owners
7. Errors in transcribing data from field books
8. Inability to traverse along boundaries

Resurvey procedures will be considered for (a) a parcel described by metes and bounds, and (b) the restoration of a lost corner of the U.S. rectangular system.

PARCEL DESCRIBED BY METES AND BOUNDS

When a tract of land is described by a recital of the bearings and lengths of all boundaries, it is said to be defined by metes and bounds. This method of land description is still widely employed, especially in the nonsectionized eastern states. Also, irregular tracts of land in areas governed by the rectangular survey system are described in the same manner. Figure 13.16 furnishes an example of this. An adequate description of this type starts at a point that should be monumented and referenced to nearby fixed objects. The direction and length of each course is given in a consistent direction back to the point of beginning. The type of bearing used should be stated. The markers placed at each corner must be described to aid in their recovery and identification. In the re-establishment of the boundaries it is assumed that deed descriptions are at hand, both of the property itself and of adjacent properties; also, that one or more corners are in place. Beginning with this evidence, the surveyor retraces the existing boundaries, as nearly as may be, measuring the distances, angles, and bearings. These measurements, when compared with those of the deed descriptions or plats, will not agree exactly but will provide a proportion, which, it is reasonable to suppose, will apply to the remaining boundaries to be restored. Likewise, a comparison between present and old bearings will indicate a variation, which will serve as a guide in establishing the directions of the boundary courses. By the use of these proportionate measurements, the entire boundary is retraced, setting temporary corners as the work proceeds. If the location of these temporary corners seems to be consistent with all evidence at hand, the corners are then established as the permanent corners. However, it is possible that much conflicting evidence will remain after the temporary corners are set, and this is to be harmonized by a further study of all the evidence and by secondary proportionate measurements. In conformance with these adjustments, the final corners are set and carefully referenced. A traverse is then run measuring all the distances and angles, from which the error of closure and the area are calculated. A new description of the boundary is written and a plat prepared to complete the survey.

RESTORATION OF A LOST CORNER IN THE U.S. RECTANGULAR SYSTEM

Resurveys in sectionized areas of the nation may require the restoration of lost corners. Restoration by proportionate measurement as subsequently described should be employed only as a last resort—that is, when all other measures have failed to indicate the position of a desired corner. This procedure requires a thorough understanding of the manner in which the original surveys were executed, and it would be impossible here to describe in detail the many conditions that may be encountered. Because of the complications that may arise, the land surveyor is cautioned about attempting this work without full information regarding the original surveys. However, the procedure for one or two simple cases will be given to indicate the general principles involved.

EXAMPLE 13.1

Single Proportionate Measurement. Suppose that in Figure 13.9 the quarter corner on the south boundary of Section 7 is lost and the westerly fractional distance of the original survey was 38.42 chains; also that the two section corners are in place, and that the present measurement between them is found to be 5,183.6 ft. The original measurement was 5,175.7 ft (78.42 chains), and the lost corner was originally placed at a recorded distance of 2,640.0 ft from the east section corner. Let the west, center, and east corners be represented by letters A, B, and C, respectively. Then the relations to be established are as follows:

$$\frac{BC(\text{new})}{AC(\text{new})} = \frac{BC(\text{old})}{AC(\text{old})}$$

or

$$BC(\text{new}) = \frac{2{,}640.0}{5{,}175.7} \times 5{,}183.6$$

$$= 2{,}644.03 \text{ ft}$$

Accordingly, quarter corner B is to be restored by the measurement of 2,644.03 ft from east corner C. This is called a *single proportionate measurement.*

EXAMPLE 13.2

Double Proportionate Measurement. Let it be supposed that the section corner between Sections 5, 6, 7, and 8, as illustrated by point O in Figure 13.17a, is lost, and that the corners at A, B, C, and D are in place. Then, for the purpose of illustration, it may be supposed that the four quarter corners on the interior section lines are lost. Also, assume that the original measurements are as shown and that resurvey measurements of distances AB and CD are 10,545.1 ft and 10,571.8 ft, respectively. As above, the relations between the new and old measurements are:

$$\frac{AO(\text{new})}{AB(\text{new})} = \frac{AO(\text{old})}{AB(\text{old})}$$

or

$$\begin{aligned} AO(\text{new}) &= \frac{77.60}{157.60} \times 10{,}545.1 \text{ ft} \\ &= 5{,}192.26 \text{ ft} \end{aligned}$$

Let a temporary stake O' be set at this point (see Figure 13.17b). Similarly,

$$\begin{aligned} CO(\text{new}) &= \frac{79.3}{159.3} \times 10{,}571.8 \text{ ft} \\ &= 5{,}262.67 \text{ ft} \end{aligned}$$

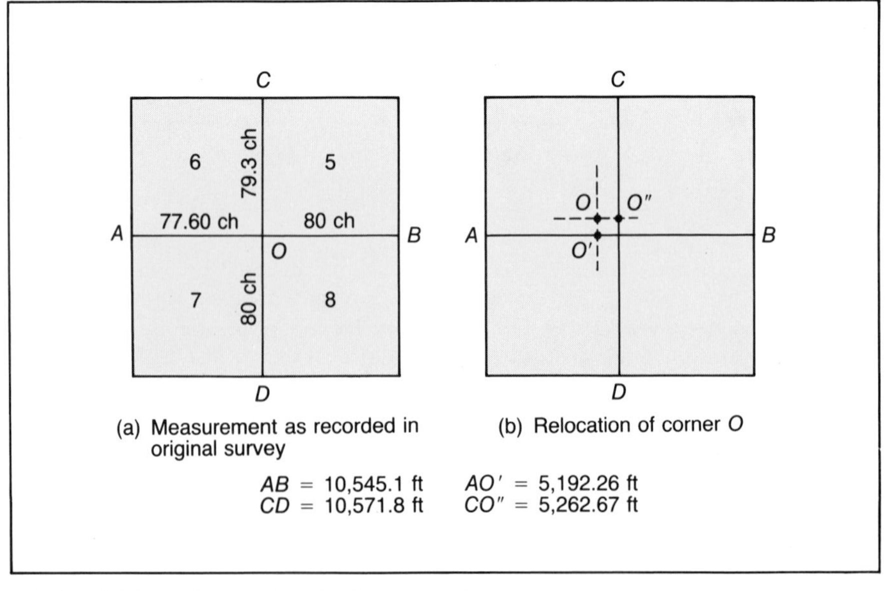

FIGURE 13.17 Double proportionate measurement

Let a temporary stake O'' be set at this point. It is evident that the temporary stake O' marks the longitude of the corner, and O'' marks its latitude. Hence the corner is restored at the intersection of a north-south line passed through O' and an east-west line passed through O''. This procedure is termed *double proportionate measurement.*

In establishing the lines of the original government surveys it was required that the precision of the measurements should be greater on the controlling lines of the larger units than for the lines that subdivided those units. This higher precision not only was specified in the statutes under which the work was executed, but was recognized in the pay that the surveyor received. Thus, a higher price per mile was paid for establishing a base line than for a township line; also, a higher rate was paid for a range line than for a section line. Accordingly, it is required in restoring corners that lines intended to be more precise than others shall have greater consideration and effect.

For example, suppose the southeast corner of Section 12, Figure 13.5, is to be restored. This corner lies on a range line that was intended to be a straight line for 6 miles and was required to be a more accurate line than the connecting section lines. Hence, to restore this corner, single proportionate measurements are taken between the nearest recoverable corners along the range line, and any conflicting evidence contributed by adjacent section lines will have no effect.

Likewise, a corner on a township line, the northwest corner of Section 3, for example, will be replaced by single proportionate measurement along the township line, and any conflicting evidence contributed by adjacent section lines will have no effect.

However, the northeast corner of Section 1 is a corner common to four intersecting lines of equal precision, and, hence, this corner will be restored by double proportionate measurements from the nearest recoverable corners in all four directions. Similarly, an interior section corner, as the northwest corner of Section 15, will be restored by double proportionate measurements from the nearest corners in all four directions.

It is a general principle, therefore, in restoring corners, that if the lines which meet at a given corner are of different degrees of precision, single proportionate measurements are used, but if the lines are all of the same importance, double proportionate measurements are used.

13.28 URBAN SURVEYS

As a city expands, the outlying land is subdivided into lots according to a specific plan. Such subdivisional surveys represent an important function of the land surveyor in urban areas.

In general, the owner or subdivider engages a land surveyor who proceeds to execute a boundary survey of the tract and also a topographic survey in order to delineate the relief and locate the governing features of the terrain. A detailed plan of the subdivision showing all streets, blocks, lots, and their significant dimensions, as well as such other pertinent information as easements and building setback lines, is submitted to the appropriate planning body and municipal authority for approval.

Since the 1960s engineering and land surveying firms have used the computing services offered by various consultants that have the capabilities of making computer-aided subdivision designs in a fraction of the time re-

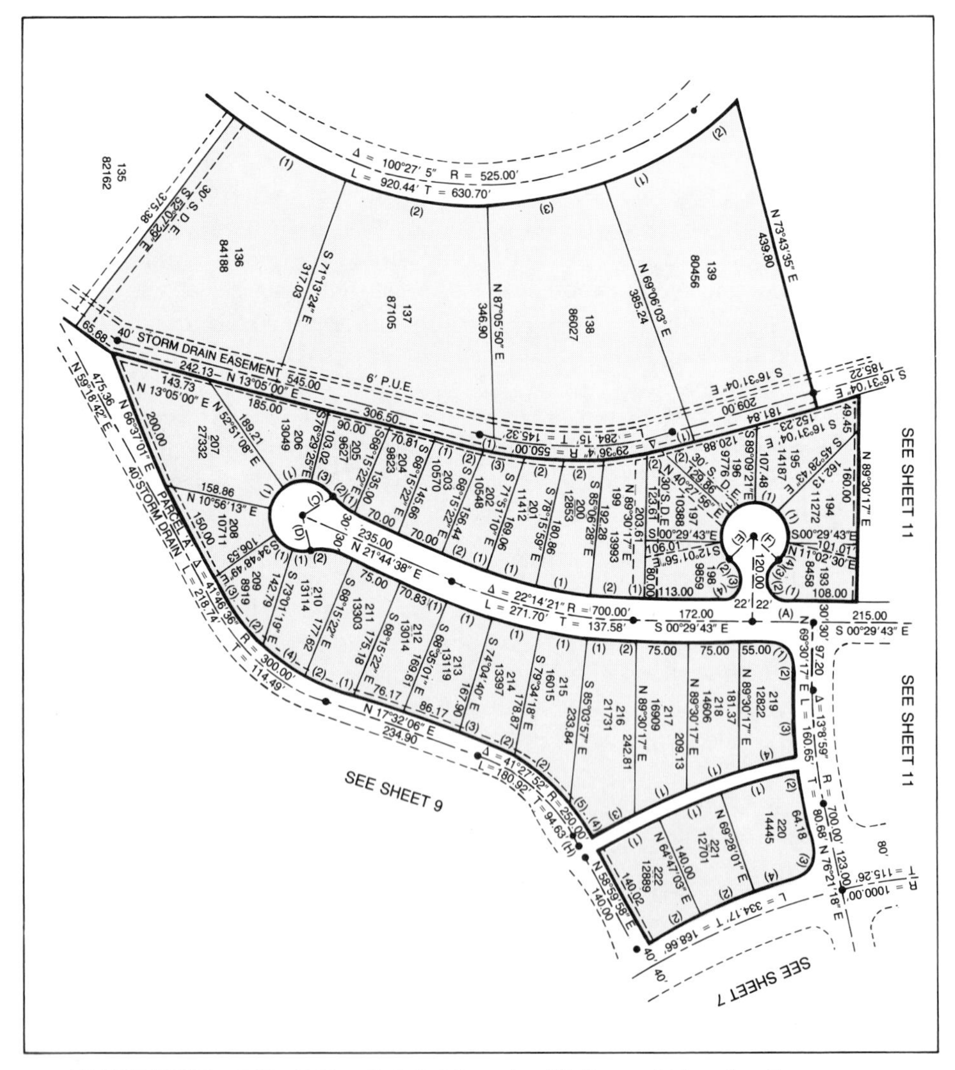

FIGURE 13.18 Subdivision plat (McDonnell Douglas Company)

quired by conventional platting methods. The basic input information needed by such companies consists of a preliminary plan of the proposed subdivision, a mathematically closed boundary survey, lot and plat specifications, the street and easement criteria that must be satisfied, and pertinent guidelines for meeting the platting requirements of the governmental regulatory agencies. On the completion of the design phase, the computing facility furnishes the land developer or subdivision engineer with the completed plat sheets. A portion of such a plat is shown in Figure 13.18. In addition to the plat sheets a computer-generated tabulation of all the points in the subdivision is

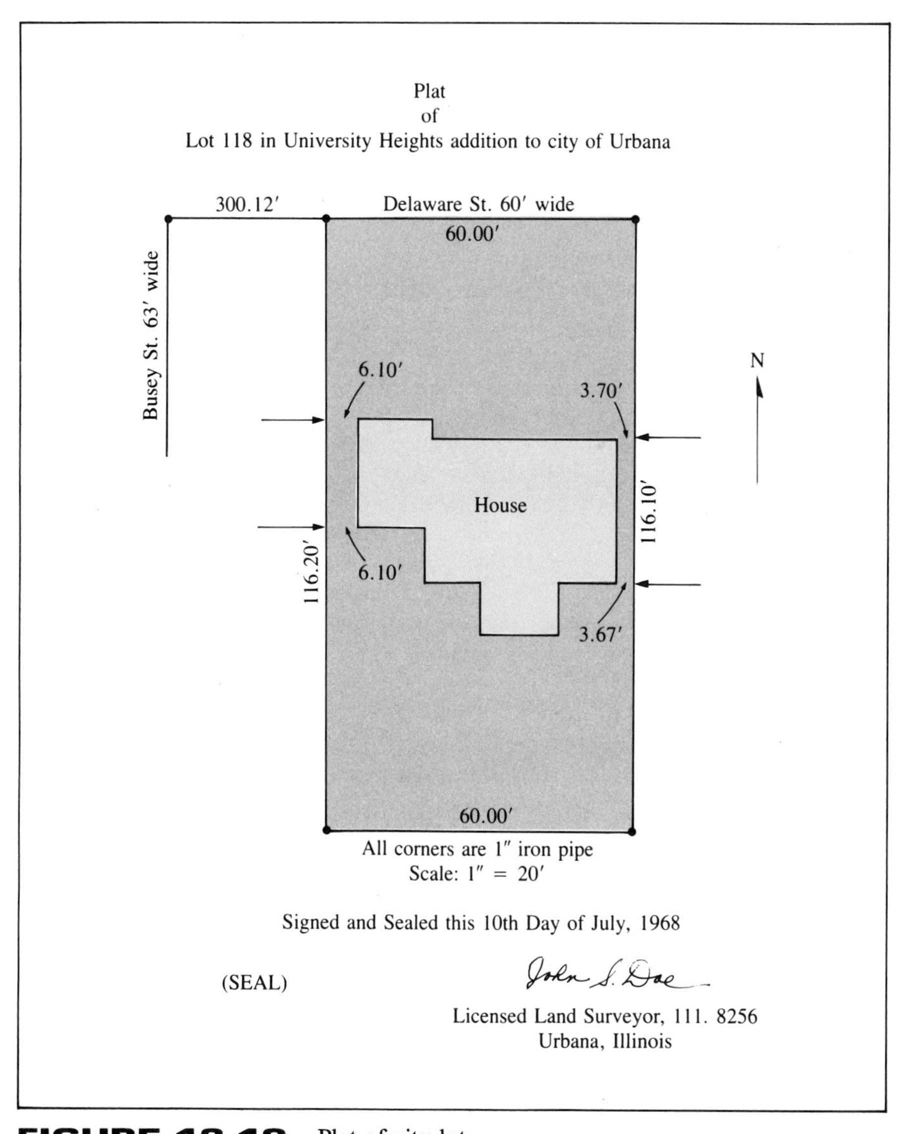

FIGURE 13.19 Plat of city lot

provided. Such an array of geometric data comprising not only point coordinates but also bearings of lines, lengths of curves, and so on facilitates the staking out of the streets and lots. The application of the inverse method in setting lot corners from selected instrumental positions has promoted efficiency in the layout phase.

On the completion of the layout surveys and the adequate monumentation of all property lines, the drawing or plat is recorded by the county recorder and officially becomes part of the public records. For a thorough introduction to platting laws and the characteristics of subdivision plats, the reader is advised to consult Reference 13.3.

Frequently the urban land survey involves the retracement of the boundaries of a single city lot for title insurance and other purposes. Following the completion of the field work, the surveyor may prepare a plat similar to that shown in Figure 13.19 on the previous page.

13.29 SURVEY FOR A DEED

It will be helpful to summarize the general procedure for making a land survey. A parcel of land, either rural or urban, is to be conveyed and a deed description is needed. The procedure is somewhat as follows:

1. The descriptions of the tract to be surveyed and of adjacent tracts are obtained and carefully examined.

2. The corners of the tract to be sold are established by resurvey if necessary, as explained in previous articles.

3. A resurvey of the tract is made in which the following data are obtained: (a) the length of each side of the parcel, (b) the angles at each corner, (c) the kind and position of accessories at each corner, (d) the calculated bearing of each side, referred to true north if possible, (e) the names of adjacent owners.

4. From these data the area is calculated, a plat is drawn, and a new description is written.

The plat generally will show the following: (a) the length of each side, (b) the angle at each corner, (c) the calculated area of the tract, (d) the kind of markers set and the references to the accessories at each corner, (e) the names of adjacent property owners, (f) the positions of buildings, roads, sidewalks, or other permanent objects that would help to perpetuate the property lines, (g) a title and meridian, (h) a description of the property lines, (i) a surveyor's certificate.

13.30 CONDOMINIUM SURVEYS

The term *condominium* refers to a type of ownership of real property whose legal concept was developed centuries ago. That concept is inherently based on a principle of communal ownership that was recognized in modern times only recently and became the cornerstone of enabling state legislation in the early 1960s.

The condominium concept is illustrated by the outright or fee simple ownership of a specific individual airspace unit (apartment) within a multi-unit building together with an undivided interest in the common elements. The latter are those features of the building and its surrounding area that are shared in common by all the unit owners. Examples are the land, bearing walls, stairways, halls, lobbies, and conduits. The term *condominium* is also applied to the entire building and to its individual units. In general, the relief enjoyed by the condominium dweller from the vexing maintenance problems and higher tax burdens associated with ownership of single-family homes has stimulated a strong interest in this mode of housing. Each of the units in a condominium complex is privately owned, can be legally sold or conveyed to another, and is separately taxed and mortgageable.

Historically the land surveyor has been primarily concerned with the division of land horizontally into parcels or lots on which homes have been erected. The condominium concept has broadened the mission of the land surveyor, because now real property, such as a multistory building, must be divided vertically as well and unique descriptions prepared for the three-dimensional space of a specific unit. These developments have challenged the ingenuity of the legal and land surveying professions.

Usually before real property can be admitted to the condominium form of ownership it is necessary to record a *declaration of condominium*. This document is the instrument by which a condominium is created and is sometimes called the *master deed*. The role of the land surveyor in the preparation of the declaration is significant because the document usually includes a legal description of the tract to be submitted to the provisions of the state's condominium property act, the description of each unit as shown on an accompanying plat of survey, and the percentage of ownership interest in the common elements allocated to each unit. This percentage is ordinarily based upon the relative floor area of the unit. The declaration also includes the by-laws, restrictions on use and occupancy, transfer of units, and various other provisions to facilitate the administration and management of the complex and to protect and maintain its desirability and attractiveness.

Generally the map or plat of survey is recorded at the same time as the declaration and is customarily attached as an exhibit to the document. The plat depicts the exterior boundaries of the land parcel in much the same manner as a traditional subdivisional plat and shows the positions of the various buildings, as in a multistructure complex, relative to the boundary lines of the tract. In order to effect the vertical delineation of all the units it is necessary to execute what might be termed a *vertical survey*. Although this is sometimes accomplished by inspection of architectural drawings, it is more accurate to execute an as-built survey. This would involve determining the elevations, with respect to an official datum, of the finished floor and ceiling of each unit. At the same time it would be appropriate to obtain the horizontal dimensions of the finished interior surfaces of the perimeter walls of each unit. A common way of portraying the linear measurements of the units is shown in Figure 13.20. This is a floor plan prepared from an as-built survey of units 401, 402, and 403 on the first floor of one of the buildings in a 10-building condominium complex. Elevations of the finished ceiling and floor are shown in the middle of each apartment. All the buildings are depicted on the cover sheet (not shown) of the plat of condominium survey. This cover sheet is essentially a plot plan showing the location and identification of all the buildings relative to the boundary lines, as well as providing information regarding the benchmarks to which elevations are referred. Condominium surveys are not routine operations. Occasionally, existing structures are converted from conventional apartment rental units to condominium ownership and the necessary surveys are intricate and time-consuming. In the case of some high-rise buildings, the execution of the measurements and the determination of the common elements have been extremely challenging. Last, but not least, the land surveyor must work closely with the attorney to make certain that all requirements of the state condominium recording laws are met.

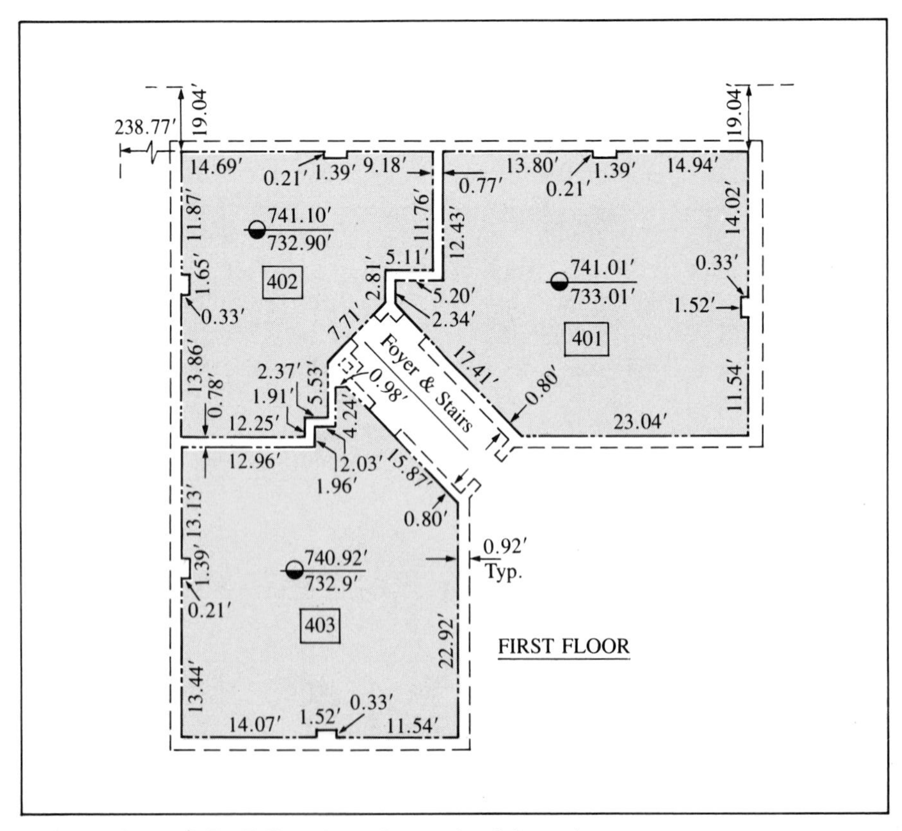

FIGURE 13.20 Part of a condominium plat

13.31 EXCESS AND DEFICIENCY

Every land survey, either urban or rural, requires the retracement and the remeasurement of property lines and, because of the inherent errors in all observations and possible mistakes, the resurvey measurements never agree exactly with the original; therefore, there is always some excess or deficiency to be adjusted. The discrepancy between a recent and a previous measurement may be small and the adjustment a simple one; but frequently the discrepancy is considerable and the adjustment complex, especially where the line has been divided into a number of segments. In such cases the aid of established legal principles is helpful.

The general rule is well stated by the Supreme Court of the State of Nebraska as follows:

> On the line of the same survey, and between remote corners, the whole length of which is found to be variant from the length called for, it is not to be presumed that the variance was caused from a defective survey of any part, but it must be presumed, in the absence of circumstances showing the contrary, that it arose from imperfect measurement of the whole line, and such variance must be distributed between the several subdivisions of the line in proportion to their respective lengths.

This rule has a common application where urban property is subdivided into lots and blocks, with dimensions shown on a plat and where a given parcel is described by its lot number.

The rule applies to all the lots in a given block, even though one at the end may have a frontage dimension different from all the others. It is sometimes supposed that such a lot, being irregular, should be considered a remnant and be assigned all the excess or deficiency found within the block. But, unless the dimension of such a lot is omitted, or other modifying conditions apply, it is held that any discrepancy shall be apportioned among all of the lots within the block.

The rule applies also where a tract described as containing a given acreage is subdivided by stating the number of acres in each subdivided tract. Thus, if an owner wills his farm of 160 acres to his two sons, giving to one the north 60 acres and to the other the south 100 acres, and if a survey shows the whole tract to contain 180 acres, it will be divided between the sons in the proportion of 60 to 100, each receiving his proportionate share of the discrepancy.

The rule also applies commonly to the relocation of lost corners in the U.S. rectangular system, where proportionate measurements are to be made between the nearest adjacent existing corners.

The general rule that any excess or deficiency shall be apportioned throughout the tract in which it occurs does not apply in some cases. Four such cases are stated on the next page.

1. If all of the lots in a block are given dimensions except one. For such a case it is plainly evident that the irregular lot was intended to include whatever remnant was left.

2. If streets intervene between the blocks of a subdivision and if the streets are monumented and have been used for some time, any discrepancy found in one block may not be apportioned among other blocks across street lines, because street lines used by the public soon become fixed.

3. If a tract of land has been partitioned by separate deeds by metes and bounds at different times. In this case, the tract last conveyed will be apportioned any excess or deficiency that may be found.

4. If erection of buildings and other acts of occupation show that ownership has been claimed according to platted dimensions.

13.32 ADVERSE POSSESSION

Adverse possession is a legal term that applies to the condition where the property line between adjacent owners and the titles of the adjacent properties are fixed by occupation and use of the land, as opposed to the descriptions of the properties in the deeds of ownership. Thus, it may be supposed that *A* and *B* are owners of adjacent tracts of land where a common boundary line has the same description in each of the deeds of titles vested in both *A* and *B*. However, when a fence was built, supposedly along this line, it was built over on *B*'s land, say 10 ft. Then if *A* has occupied all the land up to the fence, believing it was his own, for the statutory term of years (20 years in Illinois), and meeting other necessary conditions, he has acquired title to the land thus affected; and the fence, erroneously placed, has become the true boundary. This means of transfer by occupation of ownership and title from *B* to *A* is termed *adverse possession.*

Conditions under which title is acquired by adverse possession are as follows: occupation must be (a) actual, (b) open and notorious, (c) exclusive, (d) hostile, and (e) each of the foregoing conditions must be continuous for the statutory period fixed by the state. Each of these terms needs some further definition.

Actual occupation means that written or oral statements claiming ownership are not valid. There must be actual and visible evidence of occupation on the ground.

Open and notorious occupation can be evidenced in different ways, such as a fence, cultivation of the ground, erection of buildings, and posted boundary markers.

Exclusive possession means that it cannot be shared with anyone. Thus, a driveway used in common with someone else could not be used as evidence for a claim under adverse possession.

Hostile possession does not necessarily imply ill will, but it means there must be no agreement between adjacent owners, or knowledge by the claimant of the true conditions. Accordingly, the claimant will consider as a trespass any undue entry upon the property.

Continuous occupation requires that there shall be no lapses or gaps during the statutory period of possession. Thus, occasional acts of ownership will not be sufficient to establish title, nor can two or more different persons at different times exercise the rights of ownership. A possible exception to the last requirement is that in which the title to property passes through successive owners; that is, if the property is sold to a new owner who meets all the other conditions of adverse possession, the period of occupancy may be added to that of the predecessor. This procedure is called *tacking* and may extend through two or more adverse claimants to meet the statutory period necessary to acquire title. However, there must be no gaps between the successive periods of occupation.

In reviewing the conditions necessary to effect ownership under adverse possession, it is evident that this change of ownership is conditioned entirely on the bona fide intent and belief of the claimant. Thus, in the case of the two owners *A* and *B* mentioned above, it was supposed that each one believed the fence to be in its correct location. It frequently happens, however, that both owners of adjacent properties are not certain about the correct location of the boundary line. In such a case, if there is a mutual understanding between them that the boundary fence in place may not be the true line, but they agree to use it as the boundary, then as long as these conditions exist neither owner can claim ownership to the fence under adverse possession. But if at a later time *A* sells his property to another person *C*, who is not informed of the uncertain location of the fence and who believes it to be the true boundary, then *C* can claim ownership and title to the fence under adverse possession if he occupies his land according to all the necessary conditions stated above. In the latter case the fence has become the true boundary line.

As between the public and individuals, it is a general principle that, although the government, either federal or state, may acquire title by adverse possession, the reverse is not true; that is, the statute does not run against the state. Hence individuals or private corporations cannot acquire public property by means of adverse possession.

13.33 HIGHWAYS AND STREETS

An engineer or surveyor frequently has occasion to deal with public improvements, such as pavements, sidewalks, and sewers, to be constructed in streets or highways, or to establish property lines along a highway or street. An engineer or surveyor should, therefore, be fully informed of the conditions under which title is held for such property.

Sometimes title to property bordering on a highway or a street extends to the center of such highway or street unless there exists an explicit restriction. Such a restriction might be imposed by a metes and bounds description, or by the ordinance or statute under which the street or road was laid out. Some statutes and city ordinances provide for no greater interest in a street by the public than an easement or right to its use as a thoroughfare, and title remains in the owners of the adjacent lots. Also, a street or a highway opened by condemnation proceedings may be governed by statutes that provide for an easement only by the public. Accordingly, when the street or highway is abandoned and the public can show no further need to use it as a thoroughfare, the right to use it reverts to the adjacent property owners.

However, most ordinances and statutes prescribe that title to property used as streets shall be held by the city, and similarly, that title to the right-of-way for improved highways shall be vested in the state.

EXAMPLE 13.3

An 80-ft right-of-way situated in the southwest quarter of Section 15, Township 46 North, Range 12 East, Third Principal Meridian, Lake County, Illinois, is to be acquired by a state highway department. The details of the parcel are shown in Figure 13.21. The designations *PC* (point of curvature) and *PT* (point of tangency) refer to the points of beginning and end of the circular curve as explained in Chapter 12.

It is required (a) to prepare a centerline description of this tract, and (b) to calculate its area.

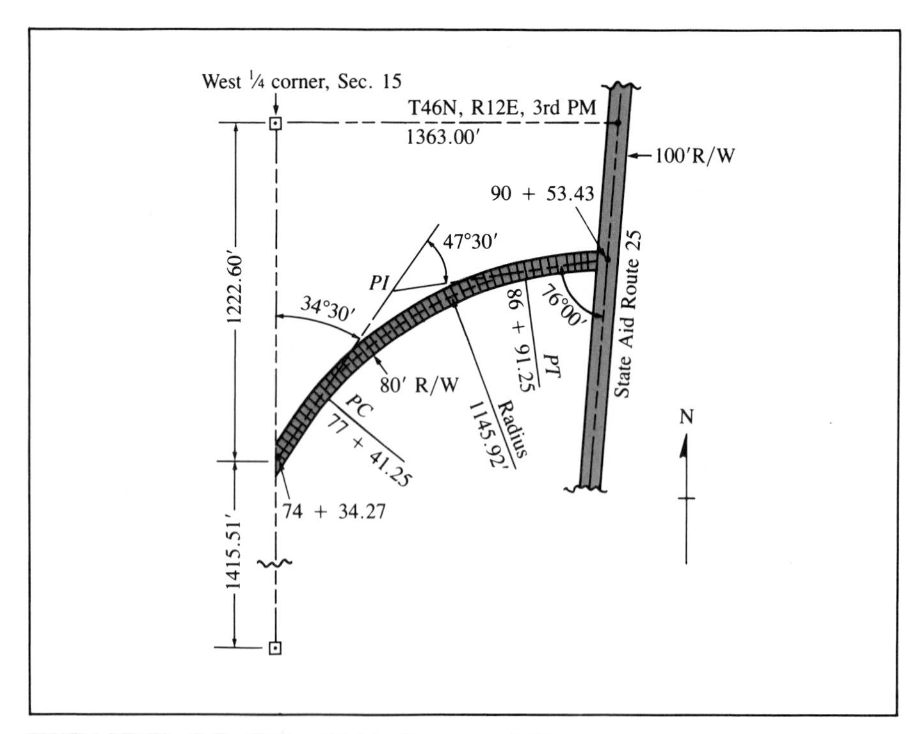

FIGURE 13.21 Right-of-way acquisition

SOLUTION

(a) A strip of land 80 ft wide being 40 ft on both sides of the following described survey of centerline: Beginning at Station 74 + 34.27, which is a point on the west line of the SW $\frac{1}{4}$ of Sec. 15, T46N, R12E, 3rd PM, Lake County, Illinois, a distance of 1,222.60 ft southward of the west quarter corner of said section, thence N.34°30′E, a distance of 306.98 ft to the *PC* (Sta. 77 + 41.25) of a curve of radius 1,145.92 ft having a central angle of 47°30′, thence a distance of 950.00 ft along the circular arc to the *PT* (Sta. 86 + 91.25) of said curve, thence N.82°00′E, a distance of 310.65 ft to a point (Sta. 90 + 01.90) along the west right-of-way line of State Aid Route 25 as recorded in Document No. 2712 as filed with the Lake County Recorder. All bearings refer to the west line of the SW $\frac{1}{4}$ of Sec. 15 as due north.

(b) Inside and outside right-of-way radii are 1,105.92 and 1,185.92 ft, respectively.

Distance along centerline of 2nd tangent from intersection with west right-of-way line of Route 25 to Sta. 90 + 53.43 is

$$\frac{50}{\cos 14°} = 51.53 \text{ ft}$$

length of first tangent centerline = 306.98 ft

length of second tangent centerline = 310.65 ft

area along first tangent = 306.98 × 80 = 24,558 sq ft

$$\text{area along curve} = \pi \times \frac{47.50°}{360°} \times [(1{,}185.92)^2 - (1{,}105.92)^2] = 76{,}000$$

area along second tangent = 310.65 × 80 = 24,852

total = 125,410 sq ft

or area = 2.88 acres

Monumentation of the rights-of-way of streets and highways is being given much more attention because the perpetuation of lines flanking the public roadway deserves the same responsible concern as that for any other property lines. Several governmental bodies including municipalities and counties have embarked on comprehensive programs of marking right-of-way lines. State highway organizations have done this, as a matter of policy, especially under the requirements of the Federal Highway Administration with respect to interstate highways.

The determination of the centerline of a highway right-of-way, however, is not always an easy task nor is the result necessarily agreeable to abutting property owners. In some counties in sectionized states, statutory law permits the establishment of the right-of-way limits 33 feet on both sides of the centerline of the traveled way. This approach is valid if the road has been kept in repair for at least 6 years by the governing body and used continuously as a public thoroughfare. Following public notice to any protesting property owners and the resolution of any controversies over line

FIGURE 13.22 Measuring along a section line (California Division of Highways)

location, the rights-of-way are durably monumented and the markers preferably tied to the state plane coordinate system.

It is to be observed that in Example 13.3 the distance ties, 1,222.60 ft and 1,415.51 ft, were obtained by measuring along the west line of the SW $\frac{1}{4}$ of Sec. 15 from the point of intersection of the proposed highway's centerline with the west line to the nearest recovered flanking corners in the public land survey system. Figure 13.22 depicts such a measurement by EDM along a section line from the centerline of an existing highway.

13.34 RIPARIAN RIGHTS

Riparian rights are rights of ownership of properties that border on the bank or shore of a stream or body of water. Such rights are often of great value, as when they endow owners with many shore privileges, including the right to construct docks and wharves. Because of these values occurring from riparian ownership and because of the irregular and changeable nature of such boundary lines, the surveyor must be well informed concerning the laws and customs that govern the establishment and maintenance of these lines within the state in which the survey is located.

13.35 RIPARIAN BOUNDARIES

NONNAVIGABLE STREAMS

In dealing with streams as boundaries, the procedure depends frequently on whether or not the stream is "navigable." In early instructions of the U.S. Land Office regarding the running of meander lines, it was prescribed that a river was to be considered as "navigable" if the water surface had a width of three chains. Obviously, the width of the water surface for a given stream varies widely with the stage of flow; accordingly, it is quite impossible in many situations to say whether a stream is to be regarded as navigable or nonnavigable. Perhaps there is no better interpretation at present than to regard a stream as navigable if it was meandered in the original survey, and as nonnavigable if it was not meandered.

Where a nonnavigable stream serves as a boundary line, it is common law that ownership extends to the thread, or center, of the stream, but the courts differ somewhat in defining the term "thread of stream." This is sometimes defined as the line midway between the usual shorelines, regardless of the position of the main channel, but the more common interpretation defines it as the center of the main channel.

NAVIGABLE STREAMS

The riparian rights of an owner of property along a navigable river are fixed largely by the statutes of the state in which the property is situated. In regard to the extent of such boundary lines, laws vary from state to state between four limitations—namely: (1) to the high-water mark, or bank, (2) to the low-water mark, (3) to the center of the stream, or (4) to the thalweg.

The *high-water mark,* or *bank,* may be defined as the line where "the presence and action of water is so common and usual, and so long continued in all ordinary years, as to mark upon the soil of the bed a character distinct from that of the banks."

The *low-water mark* may be defined as the line to which a river usually recedes at its lowest stage, unaffected by drought.

Thalweg is the line along the deepest part of a river channel as contrasted from the line midway between the banks.

It should be added that meander lines of the public lands system are not property lines, and, unless they are specifically designated as such in a deed description, they do not limit riparian boundaries.

PONDS AND LAKES

The limits of property bordering on a pond or lake are fixed by the laws of the state in which it is situated. In a few states, ownership is limited by the shoreline, but in most states, if a pond or a lake is nonnavigable, title to property bordering on it extends to the center of it. If the land tract includes all the pond or lake, the bed is included in the title. However, if a metes and bounds description uses the expression "along the east shore of said pond" or "thence by the edge of the lake," ownership extends to the shore only.

If a lake is navigable, title of riparian owners extends to the shore, or to the center, according to the laws of the state in which it is located.

13.36 ALLUVIUM, RELICTION, AND ACCRETION

Alluvium may be defined as the change in the location of a shoreline by the gradual and imperceptible deposition of soil so as to increase the area of the contiguous land.

Reliction refers to the gradual and imperceptible recession of a shoreline due either to the rising of the shore or to the subsidence, or drying up, of the body of water.

Accretion results in either of the above cases, and it is the general rule that accretion so gained is lawful and the boundary of property so affected will change with the movement of the shoreline.

Such accretion, however, has its adverse counterpart. If gradual and imperceptible erosion takes place, the riparian owner will suffer a loss by the change in the shoreline. Likewise, if the shore subsides, or the body of water rises, gradually and imperceptibly, the contiguous property will be inundated and the owner's acreage will thereby be diminished.

13.37 AVULSION

Avulsion refers to the sudden and perceptible change in a shoreline, due usually to a river changing its course during a flood or freshet. It is the general rule that avulsion effects no changes in property lines. Thus, if a river changes its course, or if its banks are suddenly eroded during a flood, property lines will not be disturbed thereby. This rule applies to river boundaries between states as well as to private property lines.

13.38 EXTENSION OF RIPARIAN BOUNDARY LINES

The surveyor is frequently required to extend the land lines of riparian owners in accordance with their rights. This problem may be simple or complex, depending upon conditions, and calls for good judgment and an understanding of the procedures that have been established in similar cases.

RIVERS

Where property rights extend to the center of a river, the general rule is to prolong each boundary line to the high-water mark or bank and then extend it in a direction perpendicular to the centerline of the river. Thus the west boundary of Lot 1, Figure 13.23, is prolonged to *A* and then extended to *B*, direction *AB* being perpendicular to the centerline of the river.

In the public lands system when a bank has been meandered, it is sometimes held that the boundary line should be extended from its intersection with the meander line to the centerline of the river, as from *C* to *D*, but most court decisions are contrary to this view and specify the bank as the proper place at which to change the direction of the boundary line.

ACCRETION

Where boundary lines are to be extended to include an area added by accretion, many complex situations arise, but a general rule is indicated by Figure 13.24, where *A* to *F* represents the old shoreline, and *A′* to *F′* the new line to be apportioned to lots 7 to 11. This is done by dividing the new shoreline so that each new lot is to the whole new shoreline as each old lot line is to the whole old shoreline. Thus, distances *AB*, *BC*, *CD*, *DE*, and *EF* are measured, and the distance *A′* to *F′* is found. Then *A′B′* is to *A′F′* as *AB* is to *AF*; parts *B′C′*, *C′D′*, *D′E′*, and *E′F′* are determined in a similar manner.

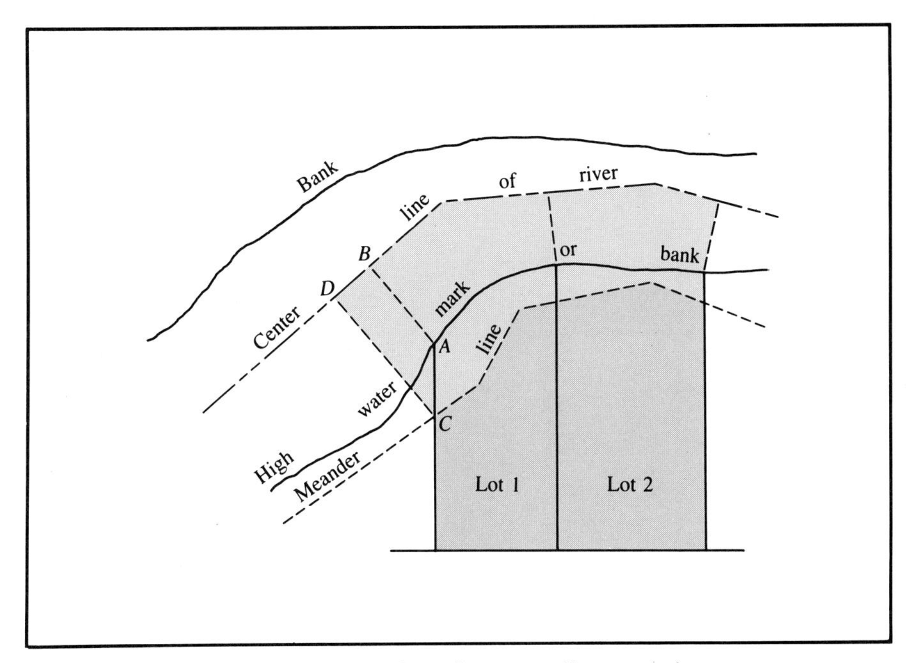

FIGURE 13.23 Extension of property lines at a river

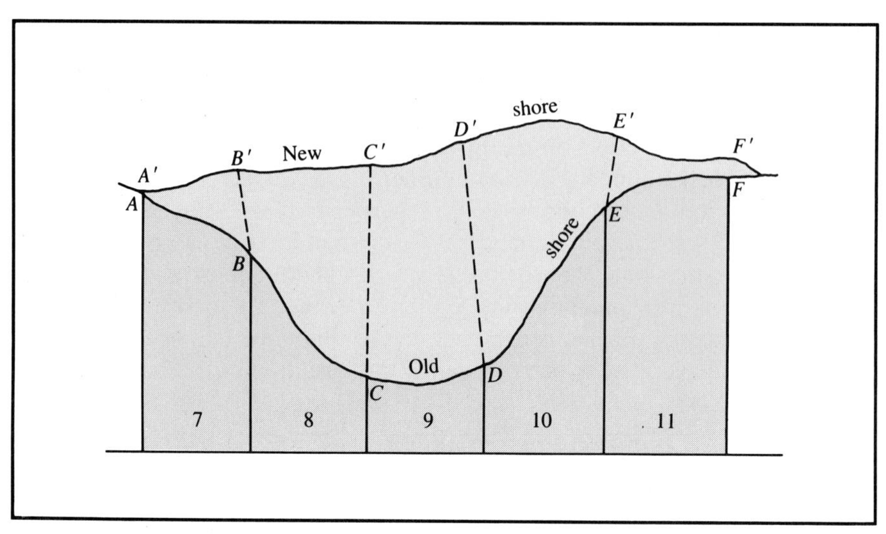

FIGURE 13.24 Lines to partition accretion

LAKES

When property rights include the bed of a lake and where the shape of the lake provides a fairly definite axis, or centerline, the boundary lines are extended from the shore perpendicularly to the centerline (Figure 13.25). But if the lake is circular in shape or has a round area as in the figure, then the boundary lines are extended from the shore to the center of the rounded part. The courts have sometimes called this the "pie-cutting" method of subdivision.

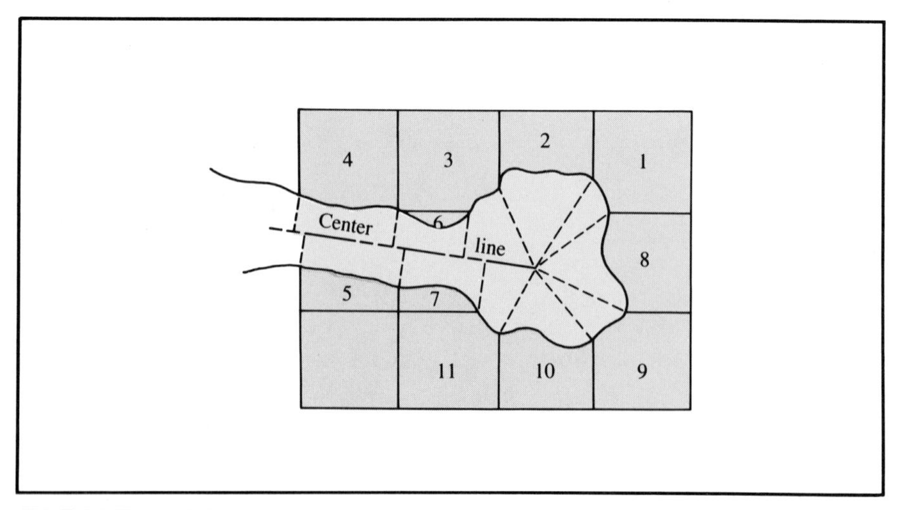

FIGURE 13.25 Extension of riparian lot lines for a lake

13.39 LEGAL AUTHORITY OF THE SURVEYOR

The surveyor has no judicial authority in the event of a dispute as regards the location of a corner of property line. A competent surveyor will gather the evidence and interpret it in such a manner as usually to bring the parties to agreement out of court. However, if no agreement can thus be reached, the surveyor has no authority to impose a settlement. This is strictly the prerogative of the court, and the surveyor serves only as an expert witness in the case.

Also, to avoid unnecessary litigation and to prevent mistakes in present surveys arising from ignorance of the legal principles involved, the surveyor is required to be well informed in the laws relating to land surveys within the state and jurisdiction where the property is located. This fact is evidenced by the preponderant weight in examinations for the land surveyor's license that is given to the legal aspects of surveys as compared with the technical details of field measurement and office calculations.

13.40 LIABILITY

Land surveyors have the responsibility of performing their duties with competence and honesty. If they are negligent in their work, they may be held liable for damages suffered by their clients. Their responsibilities in this respect are similar to those of doctors and lawyers. Thus, a surveyor who is employed to make a survey for a building site may be held liable for damages resulting from an erroneous location of the building. Many practitioners purchase professional liability insurance to protect them against the financial impact of lawsuits involving malpractice.

13.41 REGISTRATION OF LAND SURVEYORS

To safeguard the public interest, the practice of land surveying in many states is limited to persons who have demonstrated their professional competency. The general requirements that must be satisfied are graduation from an accredited engineering curriculum followed by at least four years of approved land surveying experience and the successful passage of a two-day examination. An applicant who has had no formal education beyond high school but has a record of long-established practice in land surveying may be admitted, however, to the examination.

In general, the National Council of Engineering Examiners (NCEE) recommends an upgrading of the educational requirements for licensing as a land surveyor. The Model Law of that organization points to an optimum qualification consisting of graduation from an approved four-year surveying curriculum and a record of an additional four years of combined field and office experience of suitable quality as a prerequisite for admission to the examination.

13.42 LAND PARTITIONING

There are many geometric problems associated with the partitioning of lands. Two of the more common problems are

1. to subdivide a parcel into two parts of specified sizes with a line passing through a given point, and
2. to subdivide a parcel into two parts of specified sizes with a line of given direction.

The general approach to the solution of these problems will be illustrated by two examples below.

EXAMPLE 13.4

Determine the position of the line that will subdivide the area enclosed in the loop traverse of Figure 8.2 into two equal parts. One end of the line is to pass through the mid-point of the line joining corners 1 and 6.

SOLUTION

Step 1. Determine the area of the entire parcel:

from Example 10.1, area of entire parcel = 176,286 sq ft

Step 2. Determine an approximate location of the subdividing line by either graphical or analytical means.

In Figure 13.26, let point M be the mid-point of the line joining corners 1 and 6. Furthermore, let N be the mid-point of the line joining corners 3 and 4. The line MN may be used as an approximate location of the dividing line. From the coordinates computed in Table 8.4, the coordinates of points M and N can be computed as follows:

$$X_M = \tfrac{1}{2}(X_1 + X_6) = 43{,}352.66 \text{ ft}$$
$$Y_M = \tfrac{1}{2}(Y_1 + Y_6) = 24{,}420.47 \text{ ft}$$
$$X_N = \tfrac{1}{2}(X_3 + X_4) = 43{,}923.77 \text{ ft}$$
$$Y_N = \tfrac{1}{2}(Y_3 + Y_4) = 24{,}498.35 \text{ ft}$$

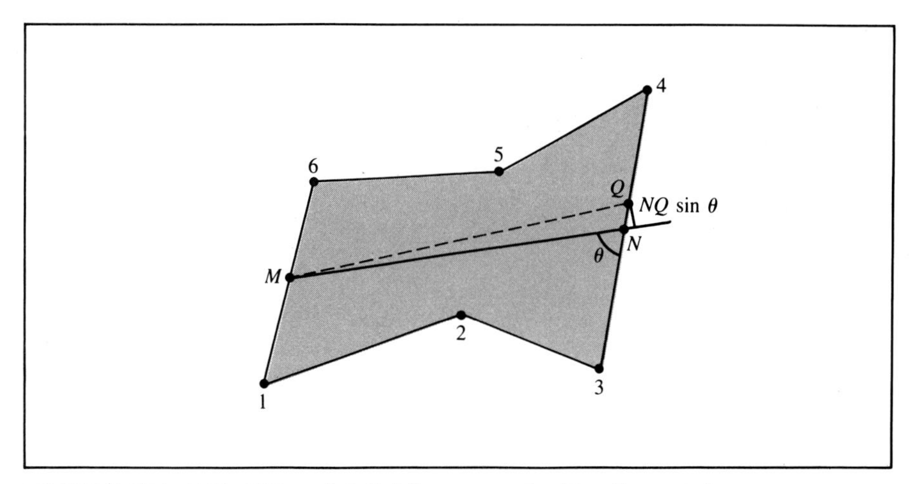

FIGURE 13.26 Subdividing a parcel with a line passing through a given point

The area of the two resulting parcels, $M123N$ and $MN456$, can be computed by the coordinate method discussed in Section 10.2 with the following results:

$$\text{area } M123N = 83{,}450 \text{ sq ft}$$

$$\text{area } MN456 = 92{,}836 \text{ sq ft}$$

$$\tfrac{1}{2} \times \text{area of entire parcel} = \tfrac{1}{2}(176{,}286)$$

$$= 88{,}143 \text{ sq ft}$$

Step 3. Determine the correct position of the subdividing line. Let point Q be the correct position of the unknown end of the dividing line, and θ be the angle between lines NM and $N3$. Then, it can be seen from Figure 13.26 that the following relationship must exist:

$$\tfrac{1}{2}MN \times NQ \sin\theta = (88{,}143 - 83{,}450) \text{ sq ft}$$

But,

$$MN = \sqrt{(X_M - X_N)^2 + (Y_M - Y_N)^2} = 576.40 \text{ ft}$$

$$\text{azimuth } \alpha_{MN} = \tan^{-1}\frac{X_N - X_M}{Y_N - Y_M} = 82.334700°$$

$$\text{azimuth } \alpha_{3N} = \tan^{-1}\frac{X_N - X_3}{Y_N - Y_3} = 11.396545°$$

and

$$\theta = \alpha_{MN} - \alpha_{3N} = 70.838155°$$

Hence

$$NQ = \frac{(88{,}143 - 83{,}450) \times 2}{576.40 \sin \theta}$$
$$= 17.24 \text{ ft}$$

$$\text{Distance between corners 3 and } N = \sqrt{(X_N - X_3)^2 + (Y_N - Y_3)^2}$$
$$= 234.82 \text{ ft}$$

Therefore, point Q should be located at a distance of 234.82 + 17.24 = 252.06 ft from Corner 3.

EXAMPLE 13.5

Determine the position of a line that will divide the area enclosed in the loop traverse of Figure 8.2 into two equal parts. The dividing line is to be parallel to the line joining corners 6 and 5.

SOLUTION

Step 1. Determine the area of the entire parcel.

from Example 10.1, area of entire parcel = 176,286 sq ft

$\therefore\ \frac{1}{2} \times$ area of entire parcel = 88,143 sq ft

Step 2. Determine an approximate location of the subdividing line by either analytical or graphical means. In Figure 13.27, let MN be the approximate location of the dividing line. Point M is located along the line joining corners 1 and 6 and is 150.00 ft

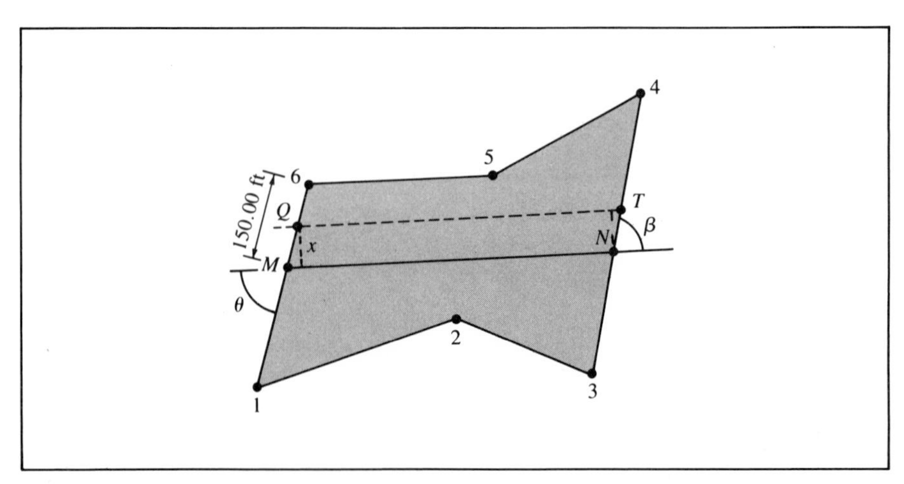

FIGURE 13.27 Subdividing a parcel with a line of given direction

from corner 6. The azimuth of line MN is the same as the azimuth of line joining corners 6 and 5. Thus, by using the coordinates computed in Table 8.4,

$$\alpha_{MN} = \alpha_{65} = \tan^{-1}\frac{X_5 - X_6}{Y_5 - Y_6} = 85.927584°$$

Also

$$\alpha_{6M} = \alpha_{61} = \tan^{-1}\frac{X_1 - X_6}{Y_1 - Y_6} = 196.812092°$$

$$X_M = X_6 + 150.00 \sin \alpha_{6M} = 43{,}360.54 \text{ ft}$$

$$Y_M = Y_6 + 150.00 \cos \alpha_{6M} = 24{,}446.56 \text{ ft}$$

The coordinates of point N can be computed by the method of intersection (see Chapter 7) from point M and corner 4 as follows:

$$\alpha_{4N} = \alpha_{43} = \tan^{-1}\frac{X_3 - X_4}{Y_3 - Y_4} = 191.395590°$$

$$\alpha_{M4} = \tan^{-1}\frac{X_4 - X_M}{Y_4 - Y_M} = 65.177852°$$

$$\text{distance } \overline{M4} = \sqrt{(X_4 - X_M)^2 + (Y_4 - Y_M)^2} = 671.67 \text{ ft}$$

$$4\hat{M}N = \alpha_{MN} - \alpha_{M4} = 20.749732°$$

$$M\hat{4}N = \alpha_{4M} - \alpha_{4N} = 53.782262°$$

$$\overline{MN} = \frac{671.67 \sin 53.782262°}{\sin(180° - 4\hat{M}N - M\hat{4}N)} = 562.25'$$

$$X_N = X_M + 562.25 \sin \alpha_{MN} = 43{,}921.37 \text{ ft}$$

$$Y_N = Y_M + 562.25 \cos \alpha_{MN} = 24{,}486.49 \text{ ft}$$

By using the coordinate method described in Section 10.2, the areas of the two parcels $M123N$ and $MN456$ can be computed to be as follows:

$$\text{area } M123N = 87{,}315 \text{ sq ft}$$

$$\text{area } MN456 = 88{,}970 \text{ sq ft}$$

Thus the actual position of the dividing line should be located slightly north of line MN.

Step 3. Determine the correct position of the subdividing line. In Figure 13.27, let QT be the correct position of the subdividing line. It can be seen from Figure 13.27 that the following relationship exists:

$$\tfrac{1}{2}(QT + MN) \cdot x = (88{,}143 - 87{,}315) \text{ sq ft}$$

where x is the perpendicular distance between lines MN and QT. But,

$$QT = MN - x \cot \theta + x \cot \beta$$

$$\theta = \alpha_{MN} - \alpha_{M6} = 69.115492°$$

and

$$\beta = \alpha_{MN} - \alpha_{N4} = 74.531994°$$

Hence

$$QT = MN - 0.104830x$$

Therefore,

$$\tfrac{1}{2}(MN - 0.104830x + MN) \cdot x = 828$$
$$562.25x - 0.052415x^2 = 828$$

Rearranging terms yields the following:

$$-0.052415x^2 + 562.25x - 828 = 0$$

Solving for x yields:

$$x = 1.47 \text{ ft}$$

From Figure 13.27,

$$MQ = \frac{x}{\sin\theta} = 1.57 \text{ ft}$$

$$NT = \frac{x}{\sin\beta} = 1.53 \text{ ft}$$

Therefore, point Q should be located at a distance of $150.00 - 1.57 = 148.43$ ft from corner 6; and point T should be located $246.91 - 1.53 = 245.38$ ft from corner 4.

PROBLEMS

13.1 By means of a separate sketch, show the location and nominal dimensions of: (a) Township T4N, R3E of the Third Principal Meridian; (b) Section 24 in Township T4N, R3E; and (c) the plot of land described as W½NW¼ of Section 14.

13.2 Use a sketch to show the location of a parcel of land NW¼NW¼ of Section 15, Township T15N, R17W of the Third Principal Meridian.

13.3 What are the *dimensions* and *nominal area* (in acres) of the parcel of land described as E½NE¼ of Section 1, Township T5N, R10W of the Third Principal Meridian?

13.4 What is the nominal area of each of the following parcels of land:
a. W½NE¼SE¼ of Section 5, T3N, R15E
b. SW¼SW¼ of Section 22, T11S, R4W

13.5 What is the nominal distance of that section of an east-west road joining the southeast corner of Section 17, T3N, R2E and the southwest corner of Section 14, T3N, R5E?

13.6 A transmission line is to run along a straight line from the southeast corner of the NW$\frac{1}{4}$ of Section 10, Township T6N, R4E, of the Third Principal Meridian to the northeast corner of the W$\frac{1}{4}$ of Section 22, Township T10N, R7E of the same principal meridian. Determine the nominal length of the transmission line.

13.7 The position of the lost quarter corner common to Sections 5 and 6 is to be restored by the principle of proportionate measurement. The original recorded distance (see Figure 13.9) along the line between them is 81.02 chains and the resurvey distance is 5,359.47 ft. Calculate the distance (in feet) that should be measured northward from the SW corner of Section 5 to fix the position of the lost corner.

13.8 Figure P13.1 shows the distances between boundary corners as recorded in the original field survey notes. In a recent resurvey, it was found that corners *A*, *B*, and *C* have been lost. The adjacent corners *D*, *E*, *F*, and *G* were found in place, and the following distance measurements were made with EDM equipment:

$$DE = 7{,}917.2 \text{ ft}$$
$$GF = 7{,}802.6 \text{ ft}$$

The original survey was conducted using the standard procedure for the U.S. Public Land Survey System.

a. Compute all the necessary distances for relocating the lost corners *A*, *B*, and *C*.
b. Outline the procedure for relocating the lost corners *A*, *B*, and *C*.

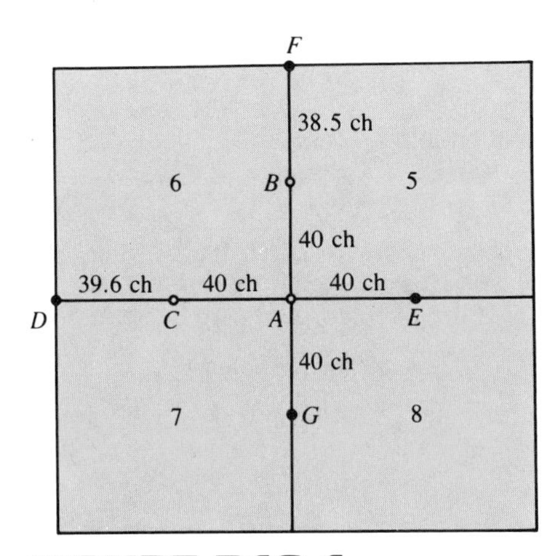

FIGURE P13.1

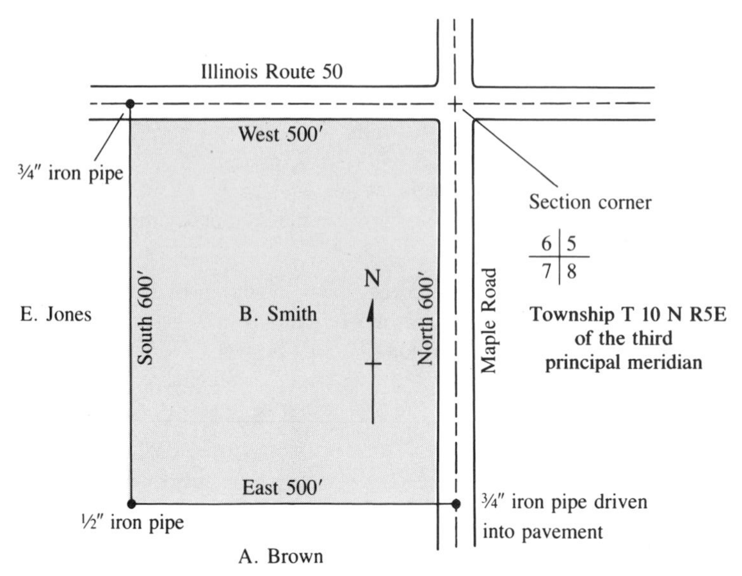

FIGURE P13.2

13.9 Write a metes and bounds description for the parcel of land shown in Figure P13.2.

13.10 Prepare a metes and bounds description of the tract of Example 13.3. Take the point of beginning as the intersection of the west line of the section with the north right-of-way line of the parcel and proceed clockwise around the figure.

13.11 Prepare a sketch of the tract of land having the following legal description: That part of the Northeast Quarter of Section 8, Township 4 North, Range 7 West of the Third Principal Meridian, Macoupin County, Illinois, described as follows:
Commencing at the Southwest Corner of the Northeast Quarter of Section 8, Township 4 North, Range 7 West of the Third Principal Meridian, Macoupin County, Illinois; thence N.1°13′51″W. along the West line of the Northeast Quarter of said Section 8, 1,432.27 ft to the point of beginning; thence N.1°13′51″W. along the West line of the Northeast Quarter of said Section 8, 181.94 ft; thence S.66°18′41″E., 1,908.55 ft to the Westerly right-of-way line of Illinois Route 157; thence S.22°39′54″W. along the Westerly right-of-way line of Illinois Route 157, 165.03 ft; thence N.66°18′41″W., 1,834.84 ft, more or less, to the point of beginning, said tract being situated in Macoupin County, Illinois. All bearings are grid in the Illinois Coordinate System, West Zone.

13.12 Calculate the area of the tract described in Problem 13.11.

13.13 Determine the position of the line *EF* in Figure P13.3 such that *EF* subdivides the panel *ABCD* into two panels having equal area. *EF* is to be parallel to the side *BA*.

13.14 The owner of lot *ABCD* shown in Figure P13.4 wished to have *CD* laid out so that 50,000 sq ft will be enclosed with *CD* parallel to the street. Determine the lengths of sides *BC, CD,* and *DA*.

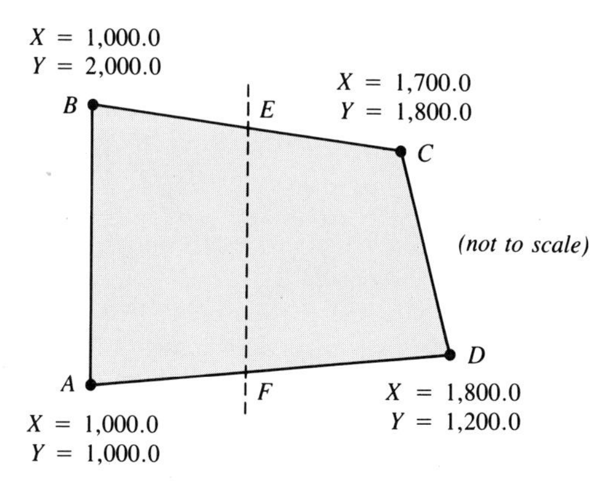

FIGURE P13.3

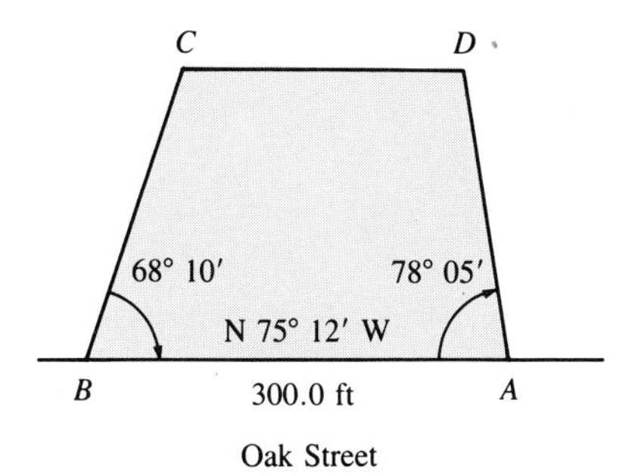

FIGURE P13.4

13.15 For the loop traverse given in Problem 10.1a, determine the location of the line that subdivides the parcel into two equal parts if (a) the line passes through station *A,* (b) the line is parallel to the line joining stations *D* and *C.*

REFERENCES

13.1 Bauer, K. W. "Progress toward Restoration and Revitalization of the U.S. Public Land Survey System in Southeastern Wisconsin," *Proceedings of the 1968 Fall Meeting of ACSM,* 48–61.

13.2 Bouman, Lane J. "The Survey Records of the General Land Office and Where They Can Be Found Today," *Proceedings of the 1976 Annual Meeting of ACSM,* 263–272.

13.3 Brown, C. M., Robillard, W. G., and Wilson, D. A. *Evidence and Procedures for Boundary Location,* 2nd ed., Wiley, New York: 1981.

13.4 Dodds, John S. *Original Instructions Governing the Public Land Surveys in Iowa,* Collegiate Press, Ames, Iowa: 1943.

13.5 Howe, R. T. *Fundamentals of a Modern System of Land Parcel Records.* Department of Civil Engineering, University of Cincinnati, Cincinnati, Ohio: 1968.

13.6 Keith, William V. "Condominium Surveys," *Surveying and Mapping,* ACSM, Vol. 32, No. 4 (Dec. 1972) 443–451.

13.7 McEntyre, John G. *Land Survey Systems,* Wiley, New York: 1978.

13.8 Stewart, L. O. *Public Land Surveys: History, Instructions & Methods,* Collegiate Press, Ames, Iowa: 1935.

13.9 U.S. Bureau of Land Management. *Manual of Instructions for the Survey of the Public Lands of the United States,* Government Printing Office, Washington, D.C.: 1973.

13.10 U.S. Bureau of Land Management. *Restoration of Lost or Obliterated Corners and Subdivision of Sections,* Government Printing Office, Washington, D.C.: 1974.

14.1 INTRODUCTION

Photogrammetry is the art, science, and technology of obtaining information about physical objects and the environment by means of photographic and electromagnetic images.

One of the most common applications of photogrammetry is topographic mapping. Photographs are taken from an aircraft flying over the area to be mapped. The aerial photographs are then used to reconstruct a scaled three-dimensional optical model of the land's surface using an instrument called a stereoplotter. Contour lines, centerline profiles, and cross-section elevations as well as planimetric maps of all cultural and topographic features can be derived directly from the optical model.

Rigorous mathematical methods have also been developed to make precise three-dimensional measurements from photographs. The three-dimensional positions of survey stations located on the ground can be determined from aerial photographs by a process called phototriangulation. Second-order accuracy can be achieved.

Aside from topographic mapping, photogrammetry has found applications in many areas of engineering and scientific studies. It has been used for the geometric measurement of human bodies, artificial human hearts, large radio telescopes, ships, aircrafts, and dams.

Through a process called photo-interpretation, valuable information concerning land use, mineral resources, landforms, soil types, urban and regional developments, and transportation systems can be derived directly from aerial photographs. Photo-interpretation has long been recognized as an indispensable tool in gathering military intelligence.

The imaging device is not limited to conventional photographic cameras. Other devices include synthetic aperture radar and multispectral scanners, which record radiations outside the visible range of the electromagnetic spectrum. The development of non-photographic imaging systems and computerized image processing techniques has led to the development of a new area of specialty called remote sensing.

It is beyond the scope of this chapter to deal in detail with the many methods and applications of photogrammetry. This chapter will present two simple measurement methods that do not require any sophisticated instruments but are extremely useful in engineering planning and reconnaissance survey. The basic principles of topographic mapping by photogrammetry will also be discussed. Readers who wish to go beyond the subject matters included in this chapter are referred to two excellent publications by the American Society of Photogrammetry, References 14.1 and 14.2.

14.2 AERIAL MAPPING CAMERAS

Figure 14.1 shows an aerial camera in use during a photographic mission, and Figure 14.2 is a sketch showing the basic elements of an aerial camera. Typically the lens assembly has a focal length of 3, 6, 8, or 12 inches

FIGURE 14.1 Wild RC 10 Universal Film Camera (courtesy of Wild Heerbrugg Instruments, Inc.)

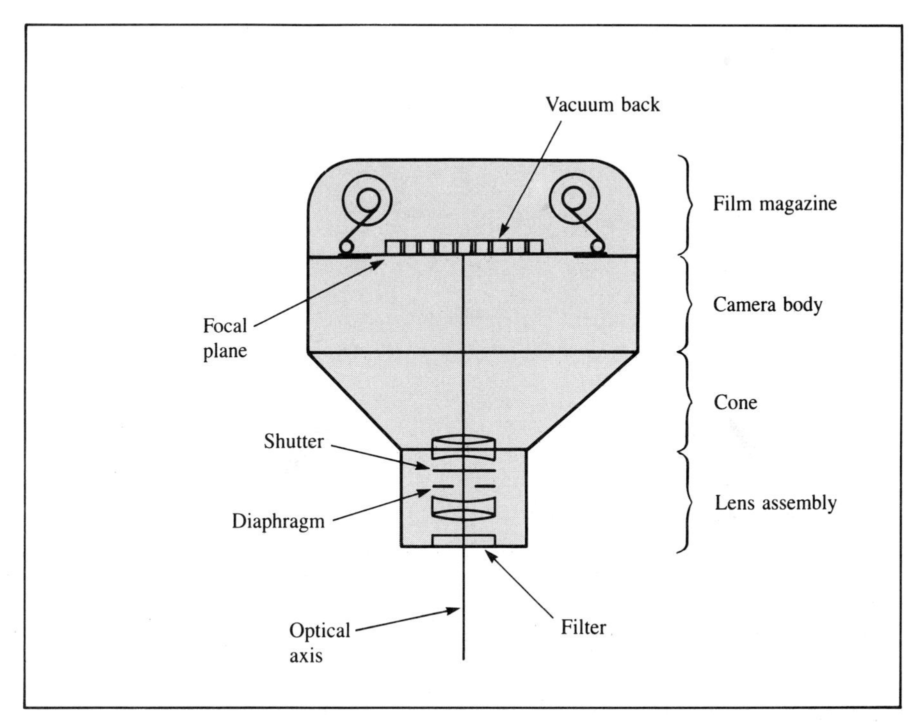

FIGURE 14.2 Elements of an aerial mapping camera

and an aperture ranging from f/4.0 to f/6.3. It has a between-the-lens shutter, which can be operated at speeds ranging from about 1/100 to 1/800 second. The film magazine can carry a film load of about 100 to 400 ft or more. The standard picture format is 9 in. by 9 in. A vacuum system located behind the image plane keeps the film flat against the vacuum platen. Geometric distortion caused by the lens assembly is usually less than 0.030 mm on the image plane.

One index mark is usually mounted on the center of each edge of the image plane. Images of these index marks, which are called *fiducial marks,* appear in every photograph as shown in Figure 14.3. The intersection of the two straight lines joining the opposite pairs of fiducial marks defines the exact location of the principal point; that is, the point at which the optical axis intersects the image plane. Typical mapping projects require that the two intersecting lines be mutually perpendicular to within ± 1 minute and that their intersection defines the true location of the principal point to within ± 0.030 mm. The focal length of the camera must also be known to within ± 0.020 mm. In some cameras the fiducial marks are located at the four corners of the picture. There are also cameras equipped with eight fiducial marks—four at the corners and four at the edges.

Also shown in Figure 14.3 is the relationship between the negative image formed on the film plane and a positive print of the same image. For illustration purposes it is sometimes more convenient to use the positive image in drawings.

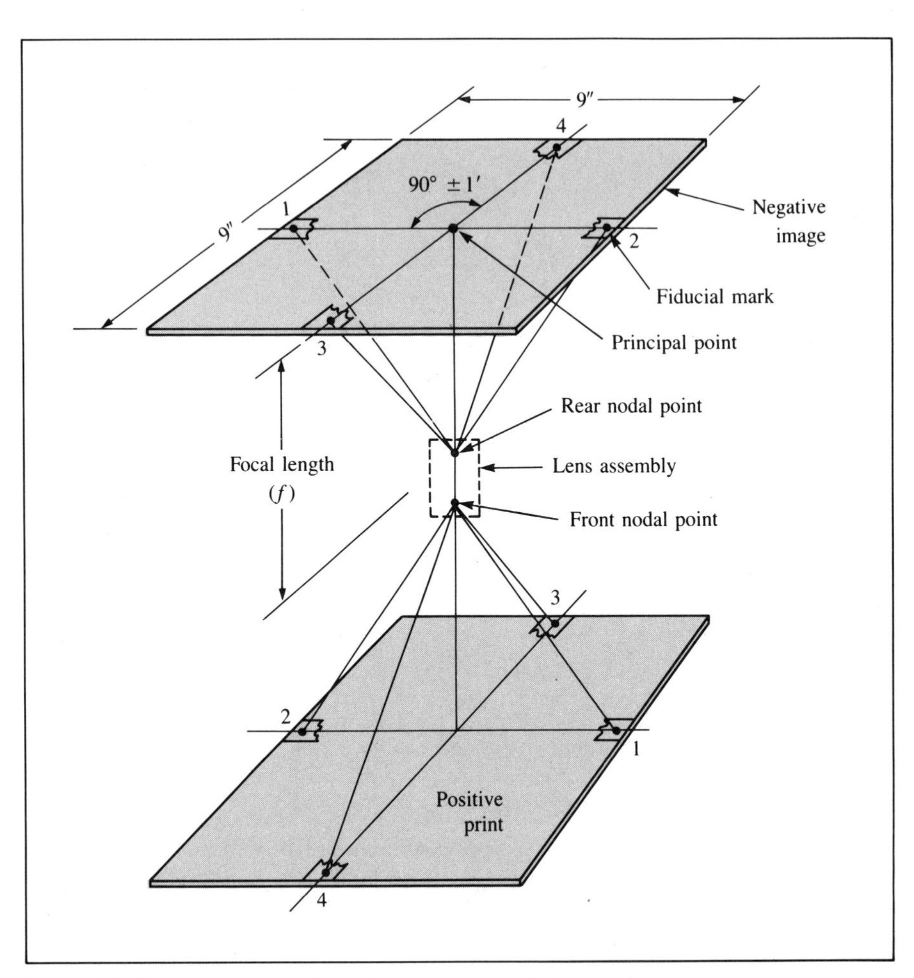

FIGURE 14.3 Fiducial marks and principal point

In topographic mapping it is important that the aerial photographs be taken with the optical axis of the camera pointing as nearly vertical as possible. The maximum tolerance for tilting of the optical axis is usually less than 4 degrees. The camera can be mounted on a gyro-controlled platform that maintains the optical axis in a near vertical direction. Camera accessories include an intervalometer, which regulates the interval between exposures, and a view finder, which assures proper orientation of the camera with respect to the ground.

Figures 14.4 and 14.5 show two typical aerial photographs used in topographic mapping. Both were originally 9″ × 9″ and were printed here at the reduced scale in order to fit into the page format. Marked on the photograph shown in Figure 14.5 are the locations of three control points whose ground coordinates are known.

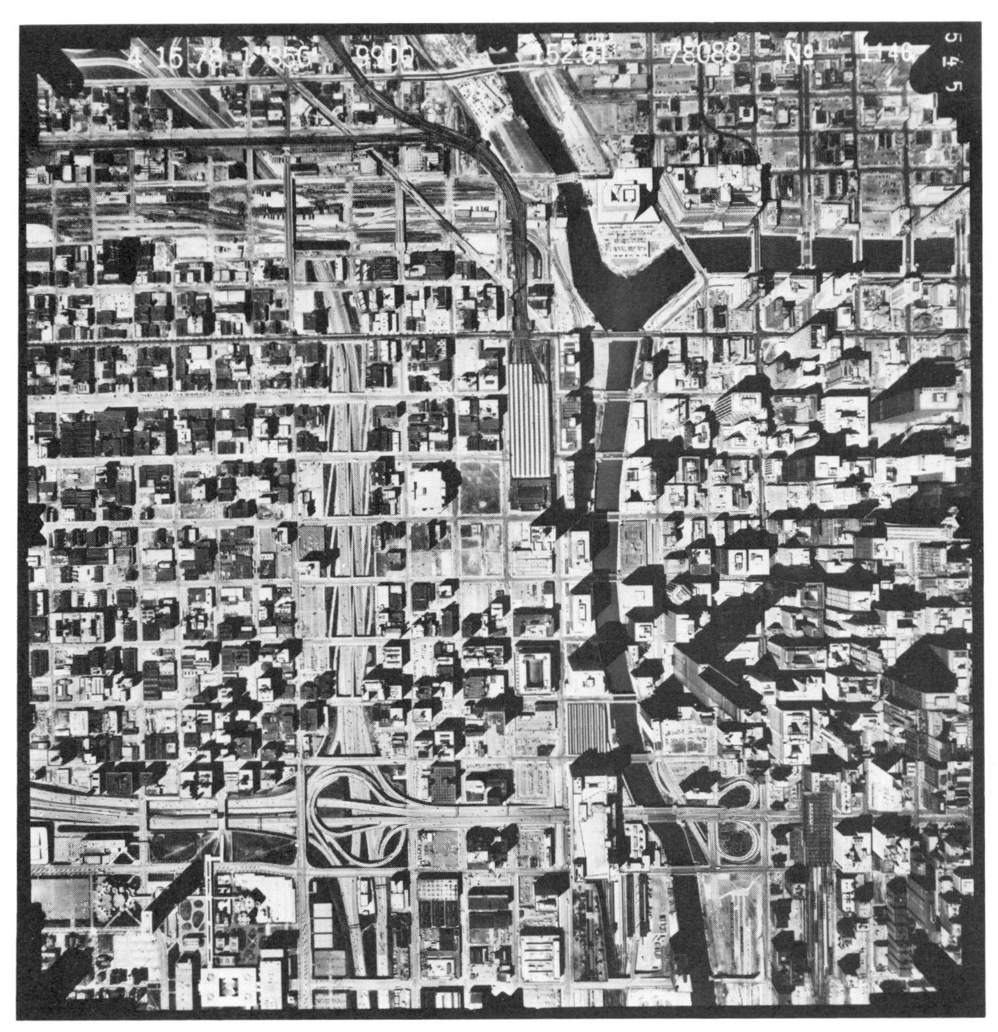

FIGURE 14.4 Vertical aerial photograph (original size 9 in. × 9 in.) of a part of the city of Chicago (courtesy of Chicago Aerial Survey)

14.3 SCALE OF A VERTICAL AERIAL PHOTOGRAPH

The central perspective geometry of a vertical aerial photograph is shown in Figure 14.6. A light ray travels from the ground, passes through the theoretical center of the camera's lens, and exposes an image on the photographic film. Thus the bundle of light rays that forms the photographic image theoretically converges on the lens center.

FIGURE 14.5 Vertical aerial photograph (original 9 in. × 9 in.) used in a highway mapping project (courtesy of Chicago Aerial Survey)

The scale of any image point on the photograph depends on the distance of the corresponding ground point from the camera at the moment the photograph was taken. Unlike a topographic map, which is an orthogonal projection and has a uniform scale everywhere on the map, an aerial photograph therefore does not have a uniform scale. The relationship between photographic scale and the flying height of the aircraft is illustrated in Figure 14.7. Let H represent the flying height of the aircraft above an elevation datum, h_A and h_B represent the ground elevations of points A and B respectively, and f denotes the focal length of the camera lens. By geometric proportion the following relationships can be derived:

$$\text{scale of photo at image point } a = \frac{f}{H - h_A} = \frac{1}{\left(\dfrac{H - h_A}{f}\right)} \tag{14.1}$$

$$\text{scale of photo at image point } b = \frac{1}{\left(\dfrac{H - h_B}{f}\right)} \tag{14.2}$$

The units for H, h_A, h_B, and f must be identical. For example, suppose that $H = 3{,}500$ ft, $h_A = 500$ ft, and $f = 6'' = 0.5$ ft. Then the scale of the photograph at image point a is 1/6,000 or $1'' = 500$ ft.

For convenience, the scale of an aerial photograph is often referred to by its average scale, which is computed by using the average ground elevation (h_{avg}); that is,

$$\text{average photo scale} = \frac{1}{\left(\dfrac{H - h_{avg}}{f}\right)} \tag{14.3}$$

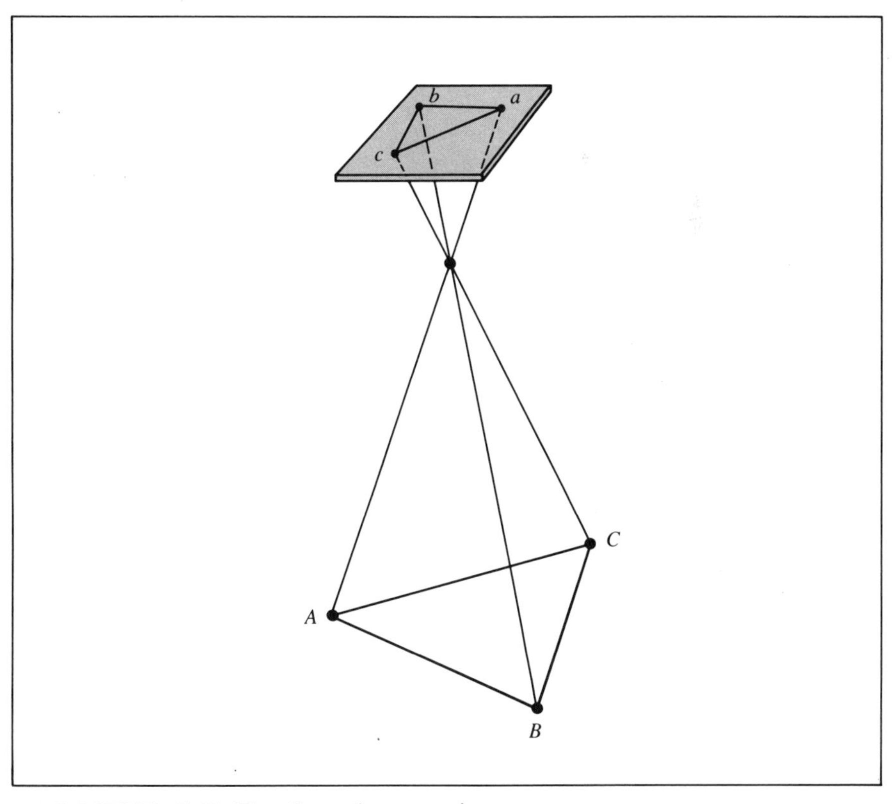

FIGURE 14.6 Central perspective geometry

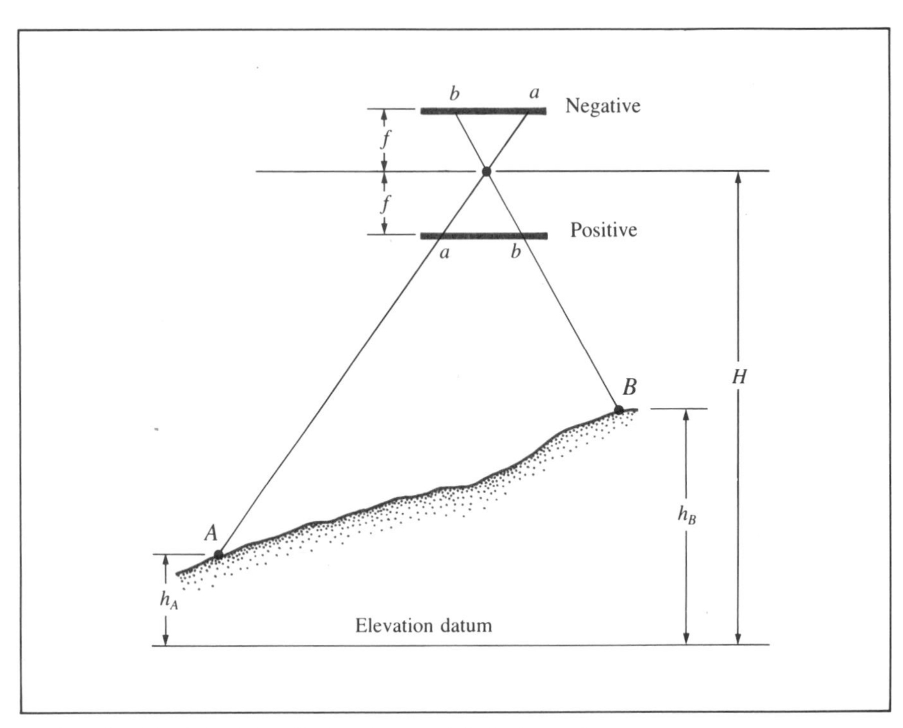

FIGURE 14.7 Scale of an aerial photograph

14.4 HEIGHT DETERMINATION FROM A SINGLE PHOTOGRAPH

For a perfectly vertical aerial photograph, the images of vertical objects such as buildings and utility poles radiate outward from the *principal point* (see Figures 14.4 and 14.8). This geometric condition can be used to measure the height of vertical objects by using a single aerial photograph.

In Figure 14.8, let points t and b represent the top and bottom respectively of a corner edge of a tall building. Let r represent the radial distance measured in inches from the principal point to image point t; Δr represents the radial distance in inches from image point b to image point t. Furthermore, let H denote the flying height in feet of the aircraft above an elevation datum, and let h_B be the elevation in feet of the bottom corner of the building above the same datum. Then it can be shown that the following relationship exists:

$$\Delta h = \frac{\Delta r}{r}(H - h_B) \tag{14.4}$$

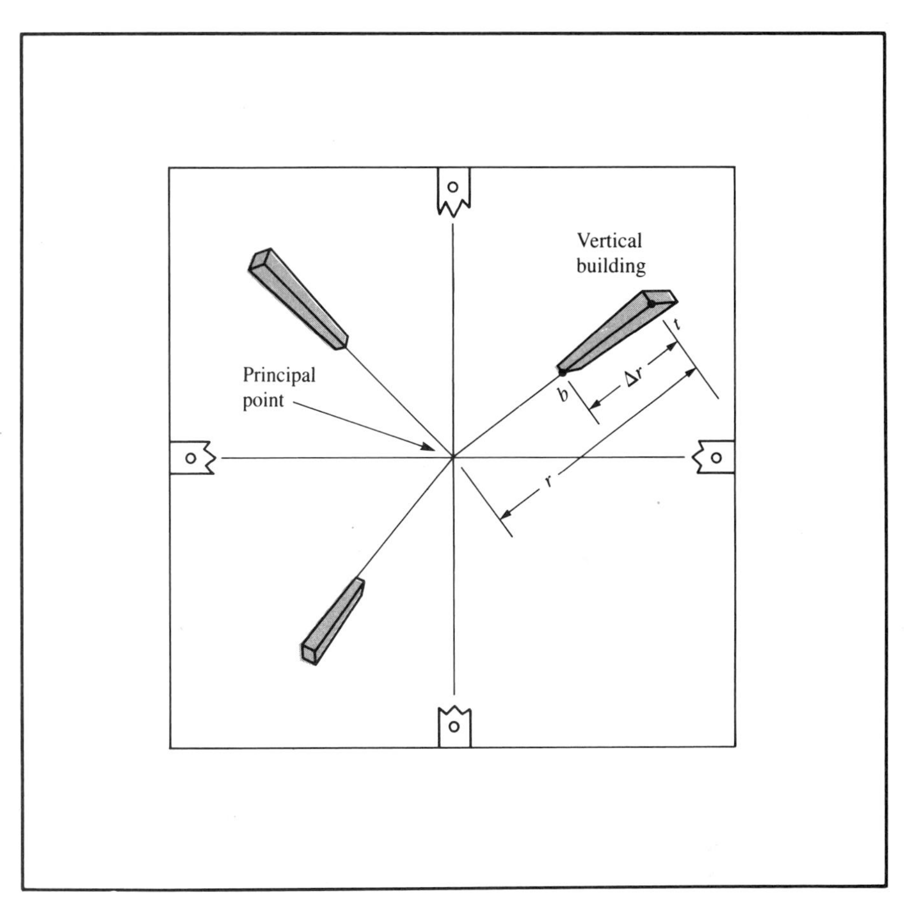

FIGURE 14.8 Height displacements radial outwards from principal point

where Δh is the height of the tall building in feet. For example, let $\Delta r = 0.45''$, $r = 4.75''$, $h_B = 700$ ft, and $H = 2{,}000$ ft. Then,

$$\Delta h = \frac{0.45''}{4.75''}(2{,}000 - 700) = 123 \text{ ft}$$

The accuracy of the computed value of Δh depends largely on the accuracy of the measured values of Δr and r and on the accuracy with which the term $(H - h_B)$ is known. Note that $(H - h_B)$ represents the flying height of the aircraft above the bottom of the building. In this type of measurement, the average ground elevation (h_{avg}) is often used in place of h_B in Eq. (14.4). The formula assumes a perfectly vertical photograph. Therefore the actual tilt of the photograph also contributes to error in the computed height value.

The derivation of Eq. (14.4) is given on the next page.

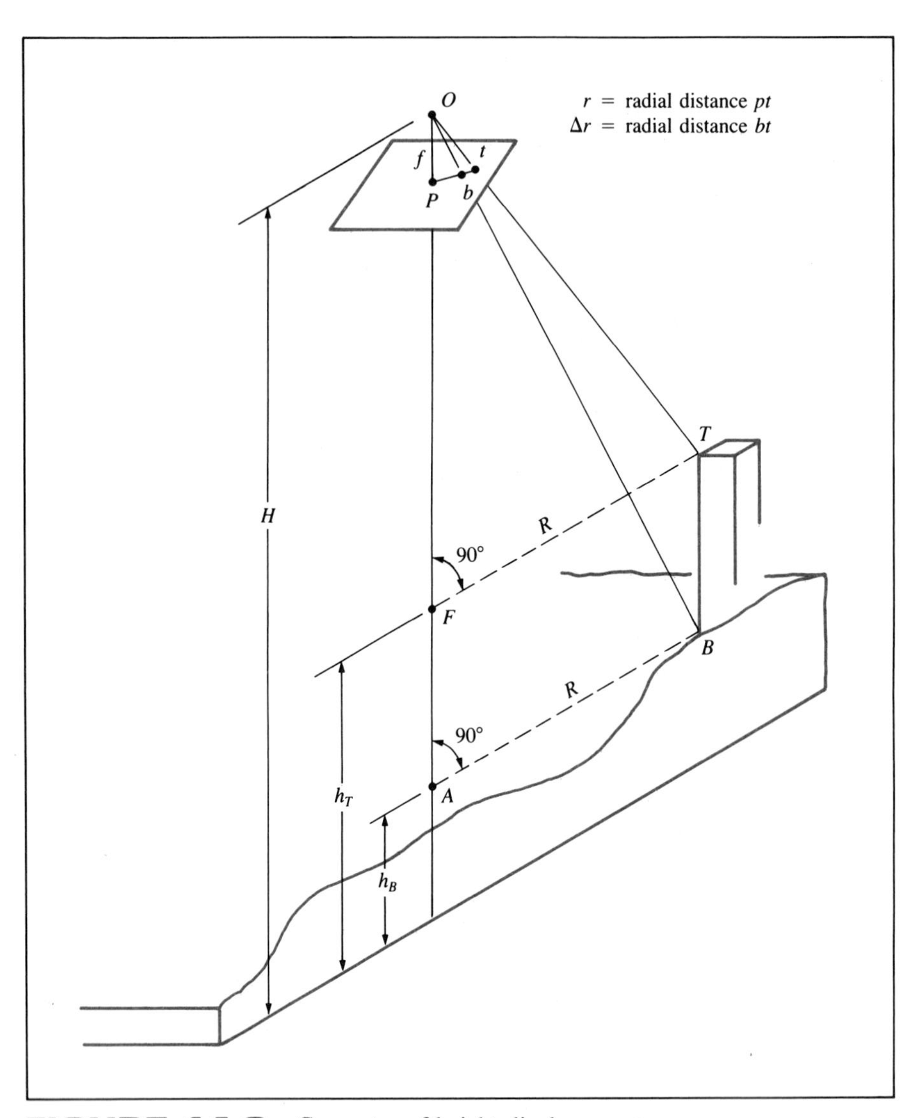

FIGURE 14.9 Geometry of height displacement

DERIVATION

In Figure 14.9, ΔOpb and ΔOAB are similar triangles. Therefore

$$\frac{R}{r - \Delta r} = \frac{H - h_B}{f}$$

or

$$H - h_B = \frac{f}{r - \Delta r} R \tag{14.5}$$

Similarly, from ΔOpt and ΔOFT,

$$\frac{R}{r} = \frac{H - h_T}{f}$$

or

$$H - h_T = \frac{f}{r} R \tag{14.6}$$

From Eqs. 14.5 and 14.6,

$$(H - h_B) - (H - h_T) = \frac{f}{r - \Delta r} R - \frac{f}{r} R$$

Collecting and rearranging terms yield the following expression:

$$h_T - h_B = \frac{\Delta r f R}{r(r - \Delta r)} \tag{14.7}$$

From Eq. 14.5,

$$R = \frac{r - \Delta r}{f} (H - h_B) \tag{14.8}$$

Substituting Eq. 14.8 into 14.7 yields the following:

$$h_T - h_B = \frac{\Delta r}{r} (H - h_B)$$

Since $h_T - h_B = \Delta h$,

$$\Delta h = \frac{\Delta r}{r} (H - h_B)$$

which is Eq. (14.4).

14.5 STEREOSCOPIC VISION

Depth perception in humans is due partly to the relative apparent sizes of near and far objects and to the effects of light and shade. But the most important condition is that a given object is viewed simultaneously with two eyes that are separated in space; hence the two rays of vision converge at an angle upon the object viewed. The angle of convergence of the two rays of vision is called the *parallactic angle* and its magnitude has an important effect upon the accuracy with which the observer can judge the true distance of a given object.

Figure 14.10 illustrates the effects of parallactic angles on depth perception. Points I_1 and I_2 denote the positions of the two eyes, separated by an eye base, b. Points A and B are located in the field of vision. The two eyes subtend the parallactic angles ϕ_1 and ϕ_2 at points A and B respectively. Since the angle ϕ_2 is larger than ϕ_1, point B appears closer to the eyes than point A. The angular difference $\delta\phi = \phi_2 - \phi_1$ is called the *differential parallax*. It provides a direct measure of the difference in distance of the two objects from the two eyes—that is, the distance AB. The human brain recognizes this angular difference immediately and translates it to a difference in distance of the two objects.

It is evident that as $\delta\phi$ becomes small, there is a limiting value below which the sense of stereoscopic vision is nil and the observer is unable to judge which is the nearer of two objects. This limiting value of $\delta\phi$ for most observers is about 20″. If the distance between the observer's eyes is 2-1/2 in., then for an angle of 20″, the rays, I_1P_1 and I_2P_2 meet at a distance of about 2,100 ft; at that distance or beyond, the sense of stereoscopic vision becomes inoperative and the relative distances to objects must be judged by their apparent sizes or by other factors.

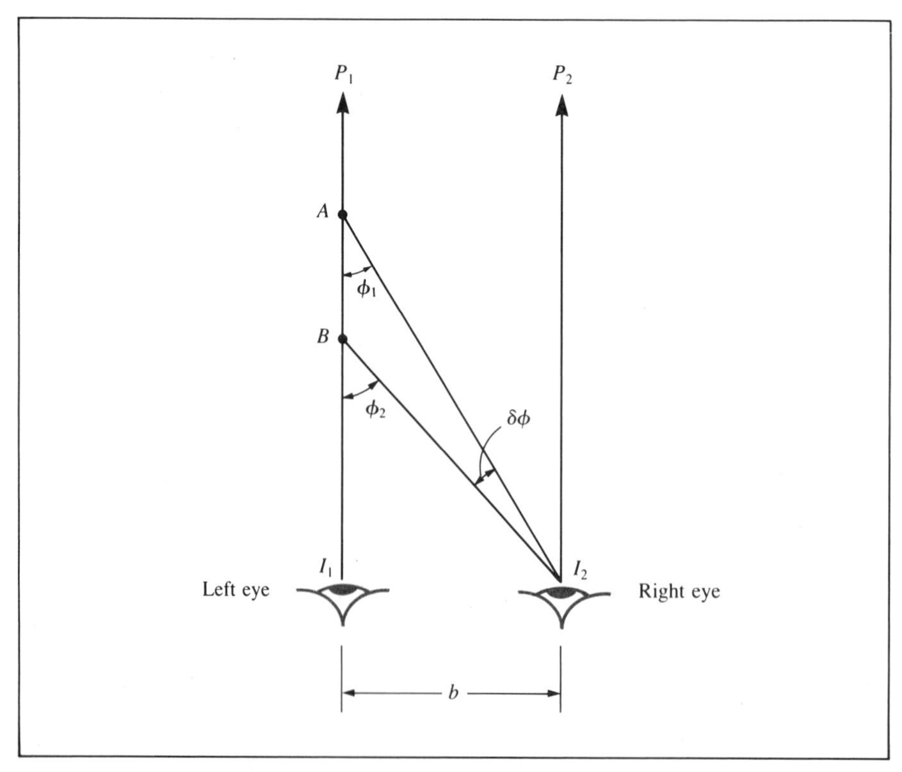

FIGURE 14.10 Parallactic angle

However, the range and intensity of stereoscopic perception can be increased in two ways—either by apparently increasing the base between viewpoints or by magnifying the field of view by the use of lenses. Some binoculars use both of these principles, having prisms that apparently spread the base b of vision and lenses that magnify the field. If the base is thus apparently increased two times and if the lenses magnify the field three times, then the effect of stereoscopic perception is increased six times.

In aerial photogrammetry an area to be mapped is photographed from two different positions, as shown in Figure 14.11. With the help of an optical system, the left photograph can be presented to the left eye while the

FIGURE 14.11 Stereoscopic coverage of aerial photographs (courtesy of Wild Heerbrugg Instruments, Inc.)

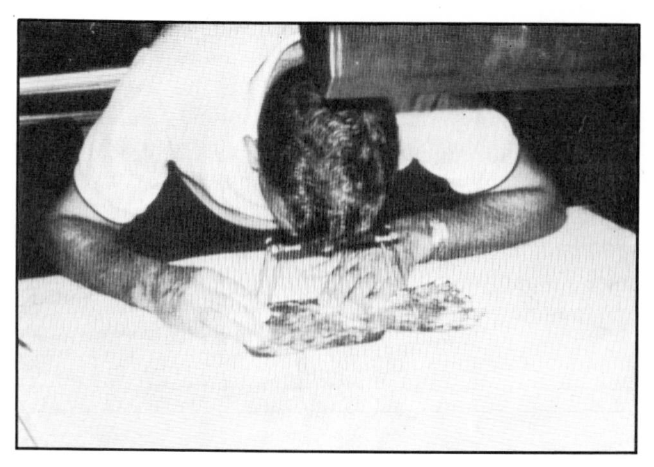

FIGURE 14.12 Pocket stereoscope (courtesy of Tennessee Valley Authority)

right photograph is presented separately to the right eye. The human brain then merges the two images into a three-dimensional vision of the ground. This is one of the fundamental principles of photogrammetry.

One of the simplest optical devices for viewing aerial photographs in stereo is the *pocket stereoscope,* shown in Figure 14.12. It consists of two simple lenses supported in a foldable frame. The device simply allows each eye to see a separate picture with a small magnification.

To illustrate the stereoscopic viewing process, a simple experiment can be conducted with Figure 14.13. By looking at the two triangles in the normal manner with both eyes, no stereoscopic vision is possible since both eyes see both triangles. Now, hold a piece of cardboard perpendicular to the

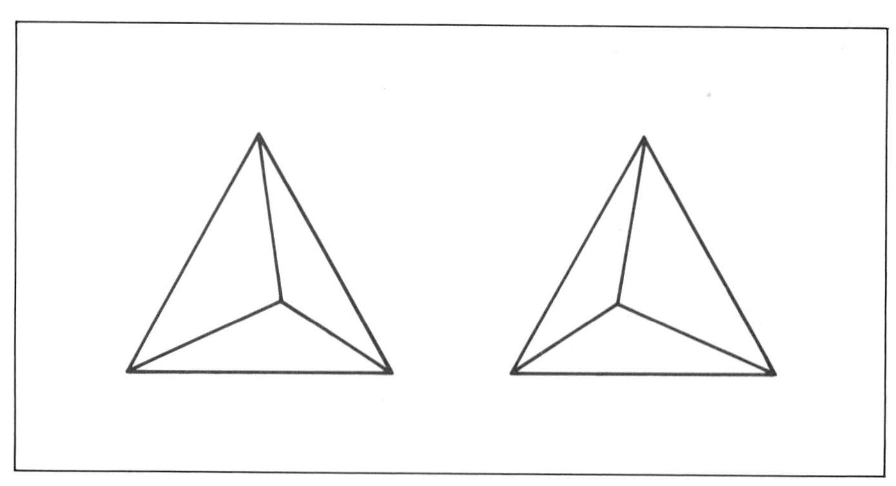

FIGURE 14.13 Stereoscopic vision

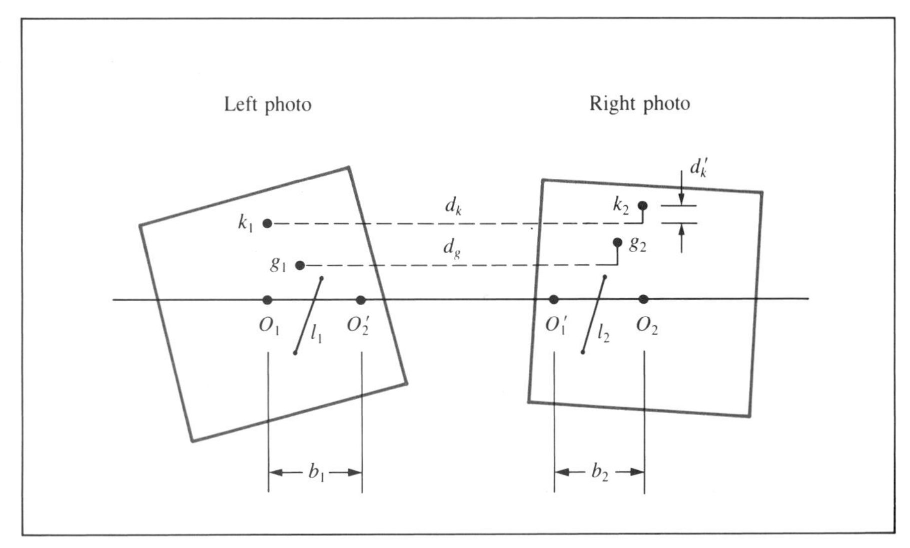

FIGURE 14.14 Measurement of parallax differences

page and between the two triangles, so that the right triangle is seen only by the right eye and the left triangle is seen only by the left eye. Focus both eyes on the respective triangle. After a short time, a stereoscopic view should appear showing a solid pyramid with the apex standing above the base.

14.6 ELEVATION FROM PARALLAX DIFFERENCE

A very simple procedure can be used to determine the approximate elevations of ground points that are imaged on two aerial photographs. The only requirement is that at least one ground point of known elevation and one known distance be identifiable on the two photographs. It does not require any sophisticated equipment. A simple measuring scale and a straight edge are all that is needed.

Figure 14.14 illustrates the setup. The measurement procedure is as follows:

1. Locate the principal point on the left photograph by drawing lines connecting the opposite pairs of fiducial marks. Denote this point as O_1 on the left photograph. By examining the photographic detail, identify the location of point O_1 on the right photo. Denote this point as O_1'.

2. By following the procedure in step 1 above, determine the location of the principal point (O_2) on the right photo and its corresponding point (O_2') on the left photo.

3. Lay the two photographs flat on a tabletop. Rotate the two photos so that the four points O_1, O_2', O_1', and O_2 all lie on a straight line. A straight edge may be used to perform this alignment. Move the two photographs sideways, without disturbing the above alignment, until the distance between O_1 and O_2 is about 10 inches.

4. Use masking tape to tape both pictures to the table.

5. Measure the lengths of a known ground distance on both photos. These are marked l_1 and l_2 in Figure 14.14. Some convenient ground distances that can be used include distances between street intersections, adjacent utility poles, bridge abutments, and corners of a large building. Preferably, the known ground distance should be located near average ground elevation. Compute the average distance measured on the photos:

$$l = \frac{l_1 + l_2}{2}$$

Using the known and measured distances, compute the average scale of the photographs. Then, using the known focal length of the camera, determine the average flying height $(H - h_{avg})$ above ground.

6. Measure the base distances b_1 on the left photo and b_2 on the right photo. Compute the average distance b as follows:

$$b = \frac{b_1 + b_2}{2}$$

The distance b is the distance, measured at photo scale, between the two camera stations at the moments the pictures were taken. It is commonly referred to as the air base.

7. Identify the image of the ground point, whose elevation is known, on each photograph. These image points are marked as k_1 and k_2 in Figure 14.14. Use a scale to measure the distance d_k, which should be parallel to the line defined by points O_1, O_2', O_1', and O_2. The distance d_k is called the *x-parallax*. The small distance d_k', measured perpendicular to the base line, is called the *y-parallax* and serves no useful purpose in this measurement program.

8. Identify the image of a ground point, whose elevation is to be determined, on each photo. These image points are marked as g_1 and g_2 in Figure 14.14. Measure the distance d_g as described in step 6.

9. Repeat step 8 for as many points as desired.

10. The elevation (h_g) of an unknown point, g, can be computed by the following formula:

$$h_g = h_k + \frac{(H - h_{avg})}{b}(d_k - d_g) \tag{14.9}$$

The term $(d_k - d_g)$ is called the *x-parallax difference*, since the x-direction is defined as being parallel to the air base between the two pictures. It is this difference in x-parallax that accounts for the difference in elevation. It corresponds to the angular difference, $\delta\phi$, shown in Figure 14.10. When $d_k - d_g = 0$, the two points have the same elevation.

The accuracy of the parallax measurement can be improved by using a *parallax bar* and a *mirror stereoscope* as shown in Figure 14.15. The mirror stereoscope uses two mirrors to enlarge the eye base for stereo viewing. The

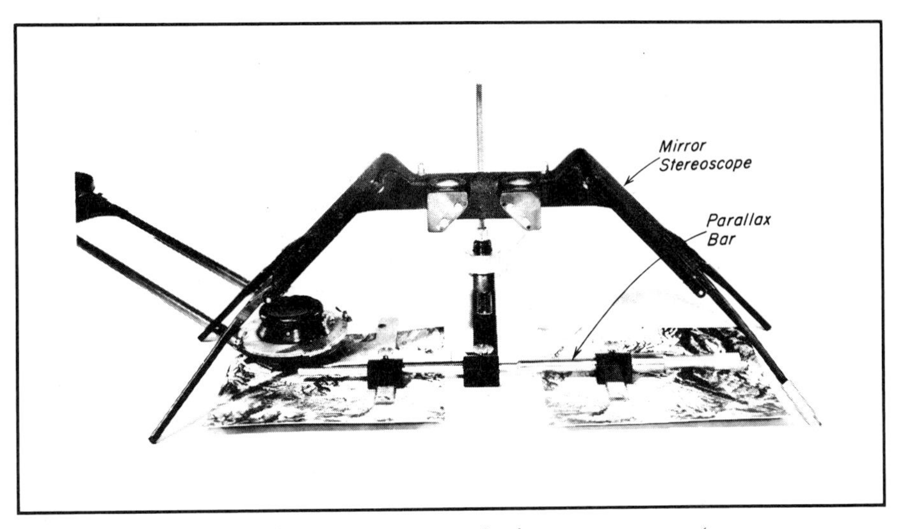

FIGURE 14.15 Parallax bar and mirror stereoscope (courtesy of Fairchild Camera and Instrument Corporation)

parallax bar consists of two glass discs mounted on a bar. On each glass disc is a black measuring mark. When these two dots are viewed properly under a stereoscope, they fuse into a single dot called the *floating mark*. As the right-hand dot is moved toward the left one, the floating mark appears to move vertically upward. Also, as the right-hand dot is moved to the right, the floating mark will appear to move vertically downward. By placing the floating mark exactly on ground level, the two measuring marks would be exactly over the same image point on both photos. The parallax bar is equipped with a micrometer for measuring the change in distance between the two measuring marks. Thus the distances d_k and d_g are measured separately with the parallax bar, and the difference $d_k - d_g$ is used in Eq. 14.9 for computing the unknown elevation.

Even with the help of a parallax bar, the preceding method is of relatively low accuracy. Since the two photographs are laid down flat on the table, the photographs are assumed to be perfectly vertical, which is seldom the case. However, the method does illustrate well the basic principle of measuring elevation differences from parallax differences. When carefully executed, the method can provide spot elevations sufficiently accurate for many preliminary engineering studies.

EXAMPLE 14.1

Referring to Figure 14.14, let the measured parameters be as follows:

$b_1 = 4.15''$ $\quad$ $b_2 = 4.17''$

known ground distance $(L) = 743.2$ ft

corresponding distance measured on the two photos:

$l_1 = 3.45''$ $\quad l_2 = 3.51''$

ground elevation of point k = 743.1 ft

average ground elevation = 700.0 ft

focal length of camera (f) = 6 in.

measured x-parallax for points k, g, and c:

$d_k = 11.46''$

$d_g = 11.10''$

$d_c = 11.63''$

Compute the ground elevation of points g and c.

SOLUTION

1. Compute air base b:

$$b = \frac{4.15'' + 4.17''}{2} = 4.16''$$

2. Compute average photo scale:

$$\text{average distance measured on the photographs} = \frac{l_1 + l_2}{2} = \frac{3.45'' + 3.51''}{2}$$

$$= 3.48'' = 0.29'$$

$$\text{average photo scale} = \frac{0.29'}{743.2'} = \frac{1}{2{,}563}$$

From Eq. 14.3,

$$\frac{1}{\left(\dfrac{H - 700}{0.5}\right)} = \frac{1}{2{,}563}$$

$$H = 1{,}982 \text{ ft}$$

3. Compute elevations of unknown points. From Eq. (14.9),

$$h_g = 743.1' + \frac{1{,}982' - 700.0'}{4.16''}(11.46'' - 11.10'')$$

$$= \underline{854.0 \text{ ft}}$$

$$h_c = 743.1' + \frac{1{,}982' - 700.0'}{4.16''}(11.46'' - 11.63'')$$

$$= \underline{690.7 \text{ ft}}$$

The derivation of Eq. 14.9 is given on the next page.

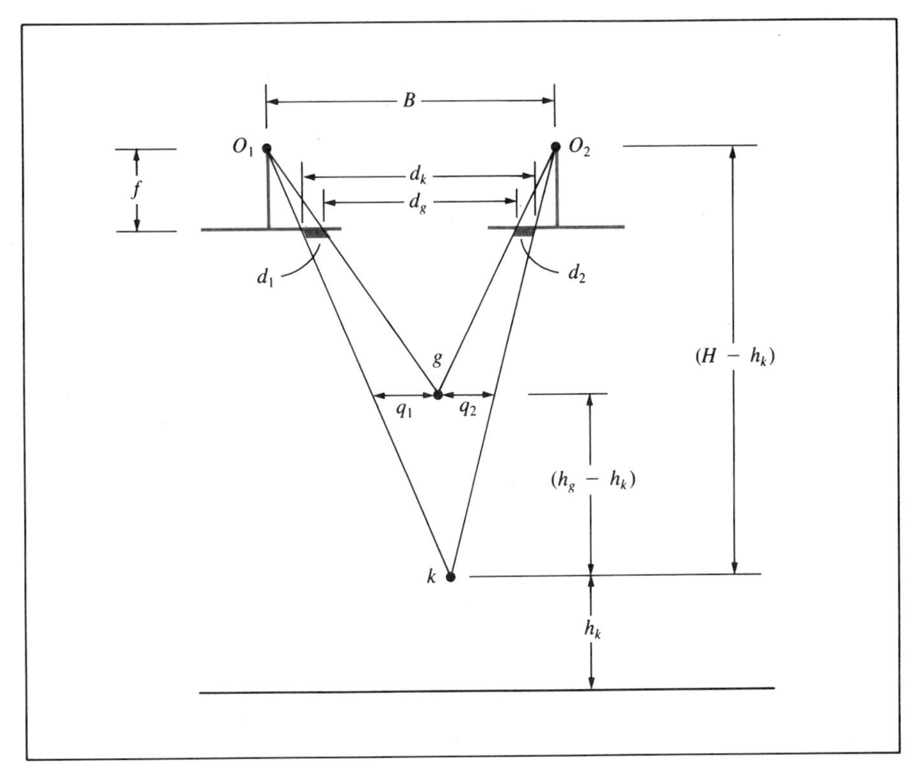

FIGURE 14.16 Elevation from parallax difference

DERIVATION

Figure 14.16 shows a cross-sectional view of two aerial photographs. Points 0_1 and 0_2 are the projection centers of the lens for the left and right photographs respectively; h_k and h_g are the ground elevations of points k and g respectively; H is the flying height above the elevation datum; d_k and d_g are the x-parallaxes measured for points k and g respectively; and B is the distance between 0_1 and 0_2.

By similar triangles:

$$\frac{d_1}{f} = \frac{q_1}{(H - h_k) - (h_g - h_k)} \tag{14.10}$$

and

$$\frac{d_2}{f} = \frac{q_2}{(H - h_k) - (h_g - h_k)} \tag{14.11}$$

Adding corresponding terms on Eqs. (14.10) and (14.11) and collecting terms yield:

$$d_1 + d_2 = \frac{f(q_1 + q_2)}{(H - h_k) - (h_g - h_k)}$$

But

$$d_1 + d_2 = d_k - d_g$$

Therefore,

$$d_k - d_g = \frac{f(q_1 + q_2)}{(H - h_k) - (h_g - h_k)} \tag{14.12}$$

Again, by similar triangles,

$$\frac{q_1 + q_2}{(h_g - h_k)} = \frac{B}{(H - h_k)}$$

$$q_1 + q_2 = \frac{(h_g - h_k)}{(H - h_k)} \cdot B \tag{14.13}$$

Substituting Eq. 14.13 into Eq. 14.12 and collecting terms yield the following expression:

$$(h_g - h_k) = \frac{(H - h_k)(H - h_g)}{f \cdot B} (d_k - d_g)$$

But

$$(H - h_k) \cong (H - h_g)$$

and

$$B = \frac{(H - h_k)}{f} \cdot b \tag{14.14}$$

where b is the air base measured on the photographs. Therefore,

$$h_g = h_k + \frac{(H - h_k)}{b} (d_k - d_g) \tag{14.15}$$

The term $(H - h_k)$ can be approximated by $(H - h_{avg})$. Then Eq. (14.15) can be expressed as follows:

$$h_g = h_k + \frac{(H - h_{avg})}{b} (d_k - d_g)$$

which is Eq. (14.9). Either Eq. (14.15) or (14.9) may be used to compute the elevation of point g.

14.7 OPERATIONAL PRINCIPLE OF A STEREOPLOTTER

A *stereoplotter* is an instrument used to draw topographic maps using overlapping aerial photographs. In the simplest case, a stereoplotter has two projectors mounted over a drawing table. A positive print of each aerial photograph is made on either glass plate or film. The positive prints

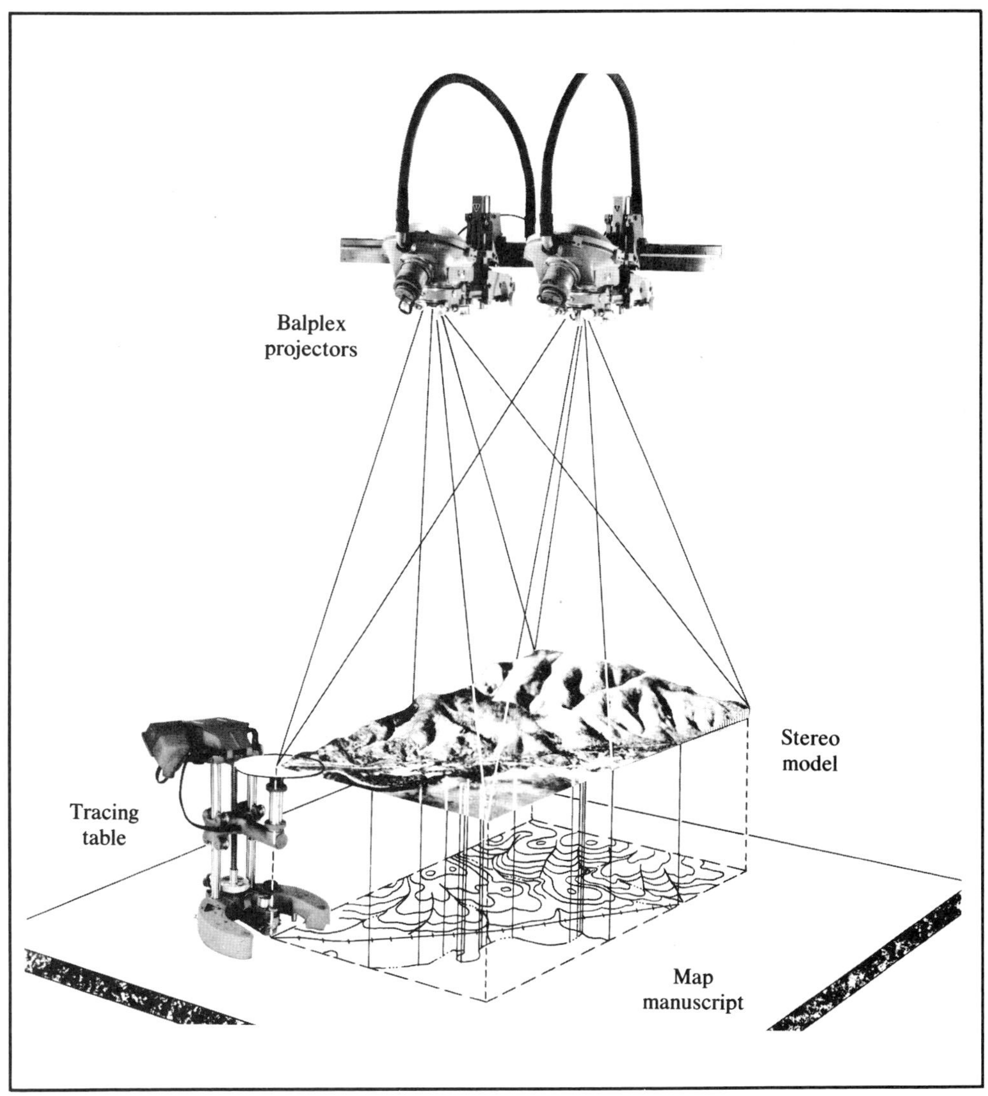

FIGURE 14.17 Operational principle of a Bausch & Lomb Balplex stereoplotter (courtesy of Bausch & Lomb)

are each inserted into a projector in which a light source projects the image down to the drawing table. By orienting the two projectors so that they assume the same relative orientation as the camera during the moments of exposure, a three-dimensional optical model of the terrain is constructed over the drawing table (see Figure 14.17 on the previous page). The optical model can be scaled and leveled with respect to the drawing table, and then features on the terrain projected point by point down to the map sheet by a tracing table. This operation is assisted by a measuring mark, called floating mark, located at the center of the tracing table and a drawing pencil located directly below the measuring mark.

Figure 14.18 is a schematic diagram showing the basic elements of a simple stereoplotter. Each projector mount has six degrees of freedom: three rotations (roll, pitch, and yaw) and three translations (*X*-, *Y*-, and *Z*-motions). Then the whole frame supporting the two projectors can be tilted to achieve proper orientation of the optical model with respect to the map sheet.

The principle of using the floating mark to map features from a three-dimensional optical model is shown in Figure 14.19. In a, the tracing table is set too high and the floating mark appears above the terrain in the optical model. The tracing table can be lowered so that the floating mark is at the level of the terrain as shown in b. A height counter on the tracing table

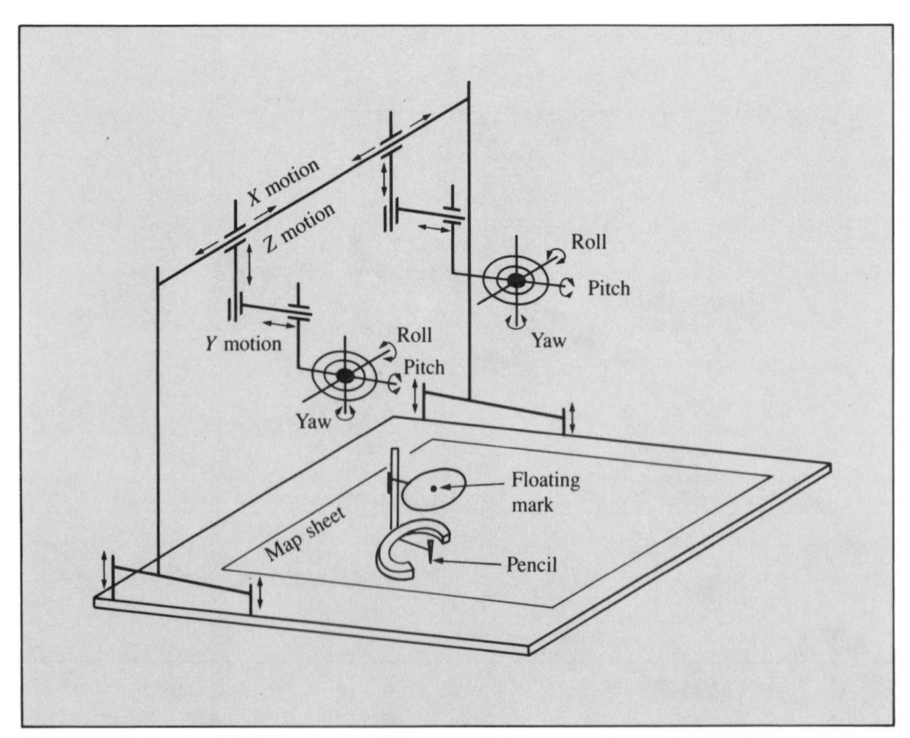

FIGURE 14.18 Schematic diagram of a simple stereoplotter

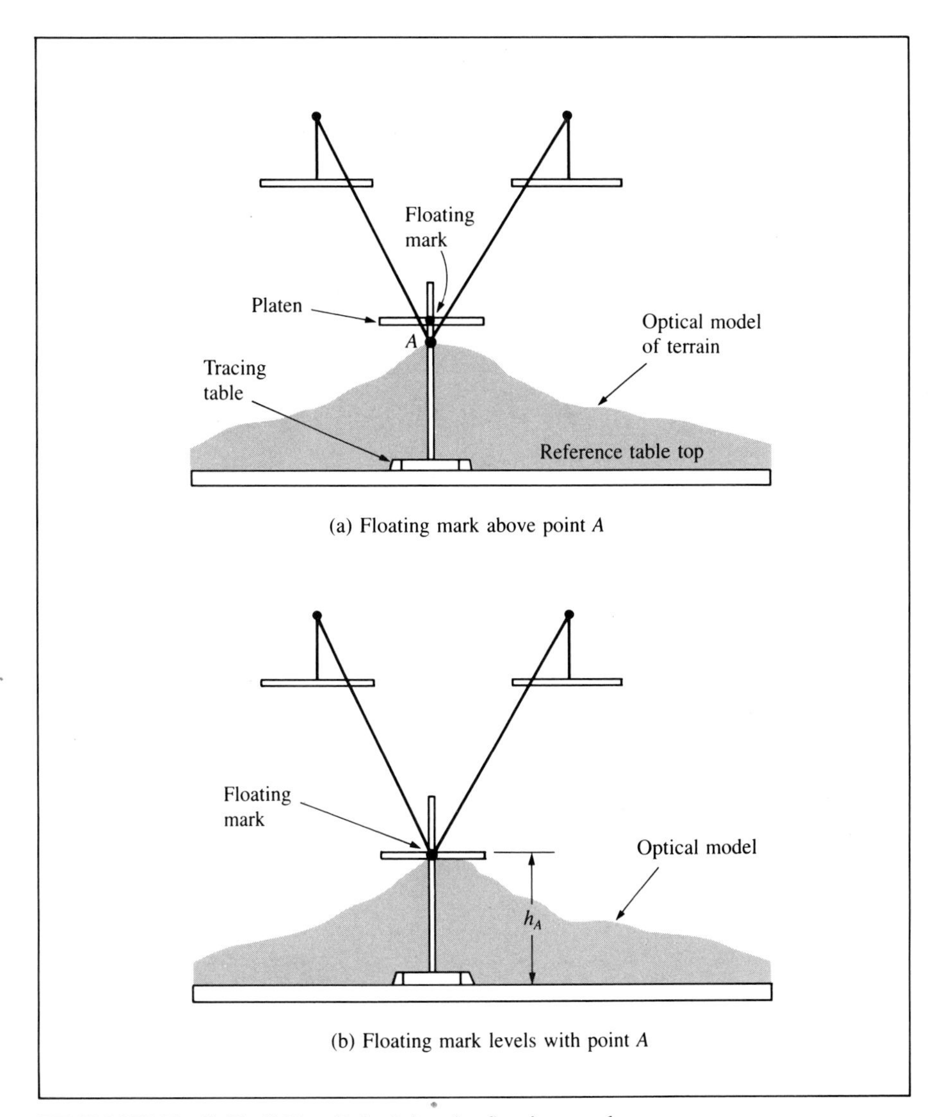

FIGURE 14.19 Principle of a floating mark

provides a direct readout of the height of the floating mark above the drawing table. The height counter can be properly scaled so that the readout directly gives the elevation of the terrain point above an elevation datum.

The operational procedure on a stereoplotter may be summarized as follows:

1. Mount the positive prints of the aerial photographs in the projectors.

2. Perform *relative orientation* of the two projectors. Initially, the two bundles of rays from the projectors are not properly oriented, as shown in Figure 14.20a. The projectors are then translated and rotated so that corresponding pairs of

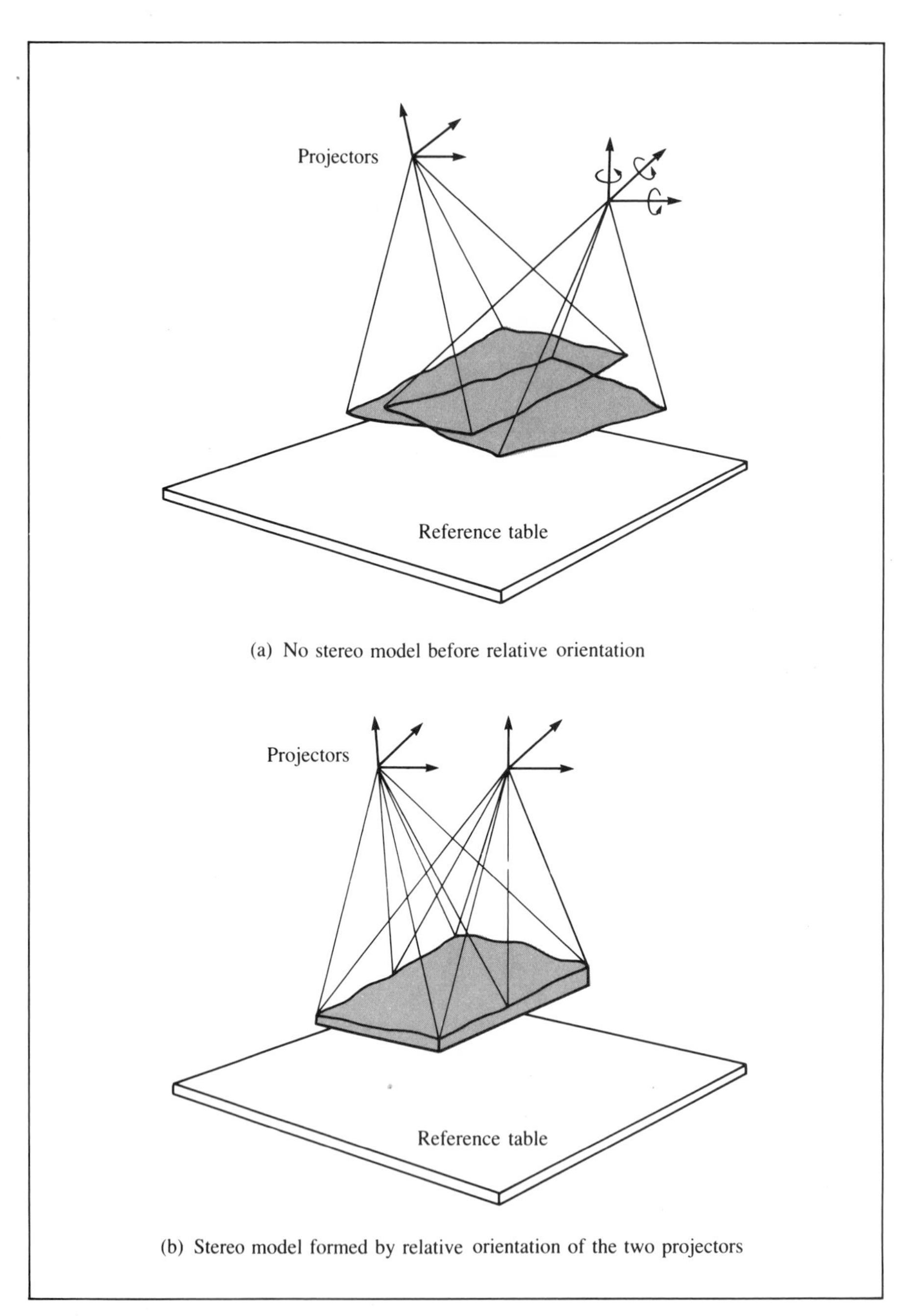

FIGURE 14.20 Relative orientation

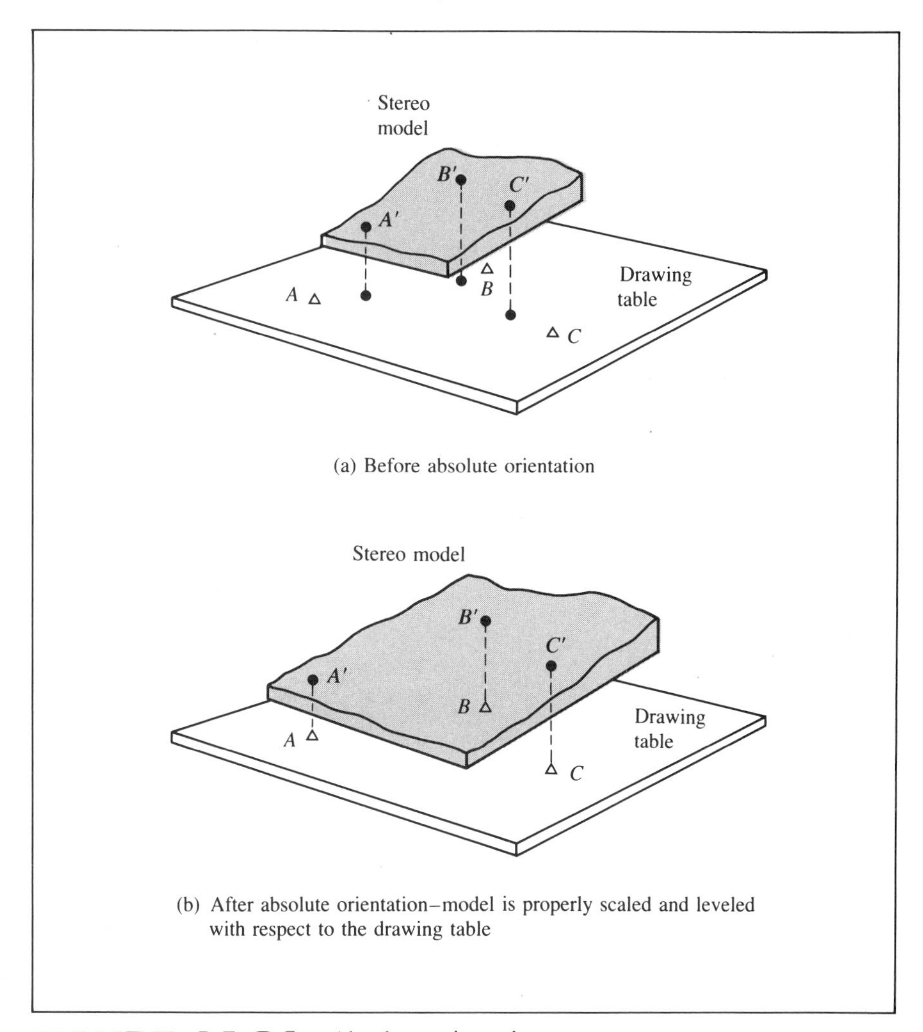

FIGURE 14.21 Absolute orientation

rays intersect in space, resulting in a three-dimensional optical model as shown in Figure 14.20b. This is accomplished by making five pairs of rays intersect properly. This procedure is called relative orientation.

3. Perform *absolute orientation* of the model (see Figure 14.21). After relative orientation the model is not properly scaled and is not properly leveled with respect to the drawing table. To scale the model, the length of at least one ground distance must be known. This is usually provided by two or more horizontal control points of known *X*- and *Y*-coordinates. To level the model properly with respect to the drawing table, which in effect is the reference elevation datum, at least three ground points of known elevations must be identified in the model area.

4. Planimetric details are then drawn using the following procedure:

a. Set the floating mark at "ground" level on one edge of a road.

b. Move the floating mark along the edge of the road. It is kept at "ground" level by moving the tracing table up or down with respect to change in elevation.
c. The pencil, which is mounted vertically below the floating mark, will then draw the edge of the road on the map sheet.
d. The same approach is used to draw outlines of buildings, streams, lakes, and so on.

5. Contour lines are drawn using the following procedure:
a. Put the floating mark at the desired elevation by moving the tracing table up or down.
b. By keeping the height of the floating mark fixed at the desired elevation, it is then moved along a direction such that it stays touching the "ground" at all times. It thus traces out a line of equal elevation on the ground. This line is drawn by the pencil on the map sheet.

14.8 TYPES OF STEREOPLOTTERS

There is a wide variety of stereoplotters, ranging from simple instruments that have only the basic components shown in Figures 14.17 and 14.18 to fully automated ones.

Shown in Figure 14.22 is the Kern MAPS 200 Microcomputer Assisted Plotting System. It includes a basic stereoplotter (in the middle), a microcomputer with dual floppy disk (left), and an automatic drafting table (right). The automatic drafting table has a built-in microcomputer so that it

FIGURE 14.22 The Kern MAPS 200 Microcomputer Assisted Plotting System (courtesy of Kern Instruments, Inc.)

can plot line and character map symbols, draw straight or best-fitting lines between points, and draw parallel lines. The microcomputer can be used to assist in map sheet preparation and in performing relative and absolute orientation of the model. It enables the line drawing data to be digitally recorded for later editing and replotting at different scales. Figure 14.23 shows a profiling system that may be attached to the Kern stereoplotter for collecting cross-section profiles.

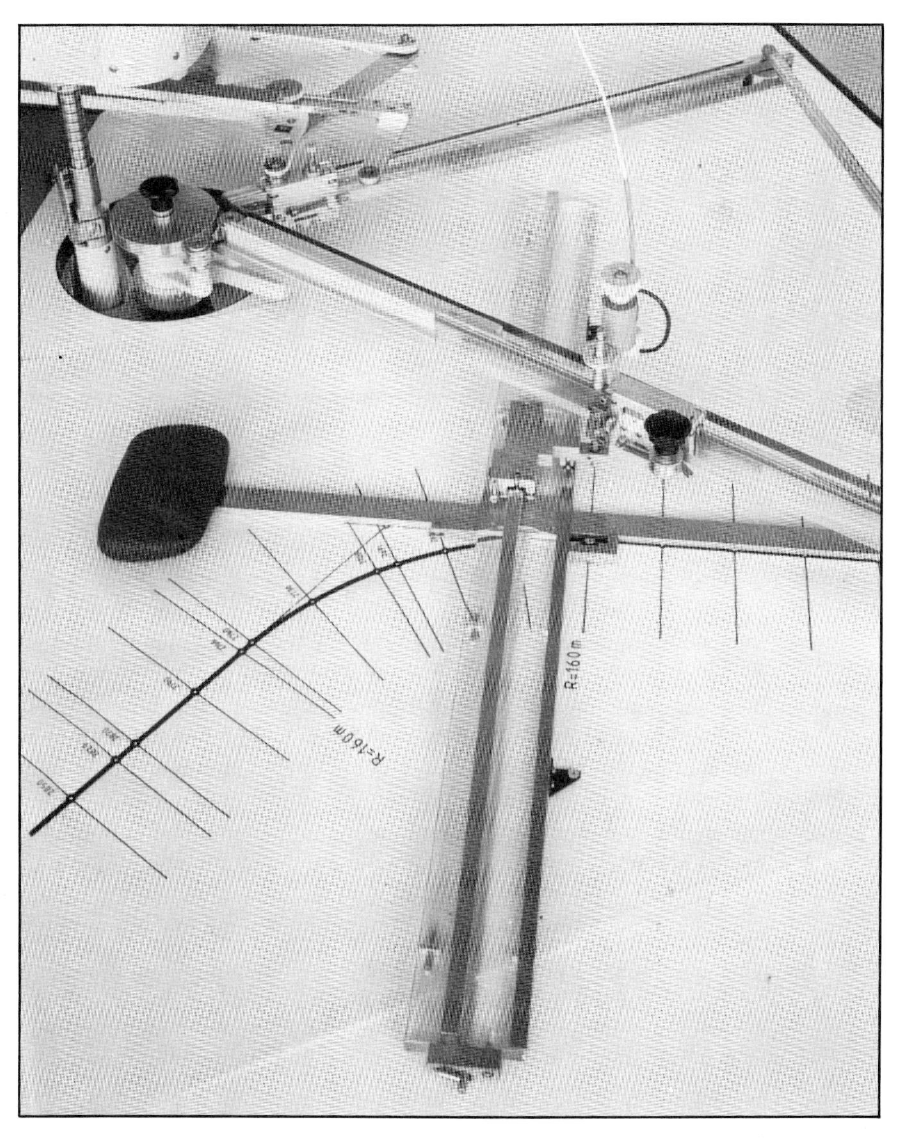

FIGURE 14.23 The Kern PS Profiling System (courtesy of Kern Instruments, Inc.)

FIGURE 14.24 The Zeiss C130 Planicomp Analytical Plotter (courtesy of Carl Zeiss, Inc., Thornwood, N.Y.)

Figure 14.24 shows a Zeiss C130 Planicomp Analytical Plotter. It represents a new generation of stereoplotters, which deviate significantly from the basic design shown in Figure 14.18. It has no optical projectors that can be rotated and translated to reverse the photographic function of the aerial camera. Instead it has an optical system that allows the operator to view the aerial photographs individually with each eye and a microcomputer that mathematically computes the point of intersection (that is, ground coordinates) of each corresponding pair of image points. Because of this, the analytical plotter can accept photographs taken with any camera type and any degree of tilt angle.

14.9 PRECISION OF STEREOPLOTTERS

The precision of stereoplotters is commonly expressed by an empirical factor called *C-factor,* which is defined as follows:

$$C\text{-factor} = \frac{\text{maximum allowable flying height above average ground elevation}}{\text{least contour interval}} \qquad (14.16)$$

The C-factor can range from 600 for a low precision plotter to 3,000 or higher for a high precision plotter. For example, suppose that topographic maps with 1-ft contour interval are to be made of an area. If the stereoplotter to be used for the project has a C-factor of 600, then the maximum allowable flying height above average ground elevation for the aerial photography is $600 \times 1 = 600$ ft. On the other hand, if the stereoplotter has a C-factor of 3,000, then the airplane can fly as much as $3{,}000 \times 1 = 3{,}000$ ft above average ground elevation. Thus, by using a higher precision stereoplotter, the aerial photography can be done at a higher altitude. Each photograph then covers a larger area, and fewer photographs are needed to cover the entire area. The end result is a reduction in time needed for performing relative and absolute orientation of the stereo models.

14.10 PROCUREMENT OF AERIAL PHOTOGRAPHS

Aerial photographs to be used for topographic mapping purposes are usually procured through commercial firms specializing in aerial photography. Proper camera equipment, as discussed in Section 14.2, must be used. Aerial photographs are usually taken in early spring or late autumn to minimize visual obstructions by trees and snow. Only clear, cloudless days are acceptable for aerial photography. To avoid long shadows cast by tall objects such as trees, towers, and buildings, photography must be done during the part of the day when the sun is at least 30° above the horizon. In addition, the aerial photographs must be taken according to a specific set of geometric requirements in order to achieve the desired stereoscopic coverage and to avoid gaps in the coverage.

Figure 14.25 shows the normal geometric configuration for aerial photography in topographic mapping. The aerial photographs are taken along straight lines to form *strips* of photographs that overlay each other. The flight lines are usually oriented along the longest dimension of the area in order to minimize the number of times the aircraft is required to turn around.

Along each flight line, adjacent photographs overlap each other usually by about 60 to 65%, which is called the *forward overlap*. The overlap between adjacent strips of photographs is called the *side overlap,* which usually ranges from 20 to 30%. The amount of overlap is measured with respect to the dimension of the photograph. Thus, for a 60% forward overlap, adjacent photographs along a strip would overlap each other by $9'' \times 0.6 = 5.4''$ for standard $9'' \times 9''$ photographs.

The geometry of forward and side overlaps is illustrated in Figure 14.26. Let O_f and O_s represent the forward and side overlaps expressed as fractional parts of the dimension of a photo, respectively. For example, if the forward and side overlaps are 60% and 20% respectively, then $O_f = 0.6$ and $O_s = 0.2$. Furthermore, let W represent the ground distance covered by the width of a photograph. Then,

$$D_e = (1 - O_f)W \tag{14.17}$$

and

$$D_s = (1 - O_s)W \tag{14.18}$$

where D_e is the ground distance between adjacent exposures along a flight line and D_s is the ground distance between adjacent flight lines.

The following expressions can also be derived from the geometry illustrated in Figures 14.25 and 14.26:

$$N_s = \frac{L_1 - W}{(1 - O_s)W} + 1 \tag{14.19}$$

$$N_p = \frac{L_2 - 2(O_f - 0.5)W}{(1 - O_f)W} + 1 \tag{14.20}$$

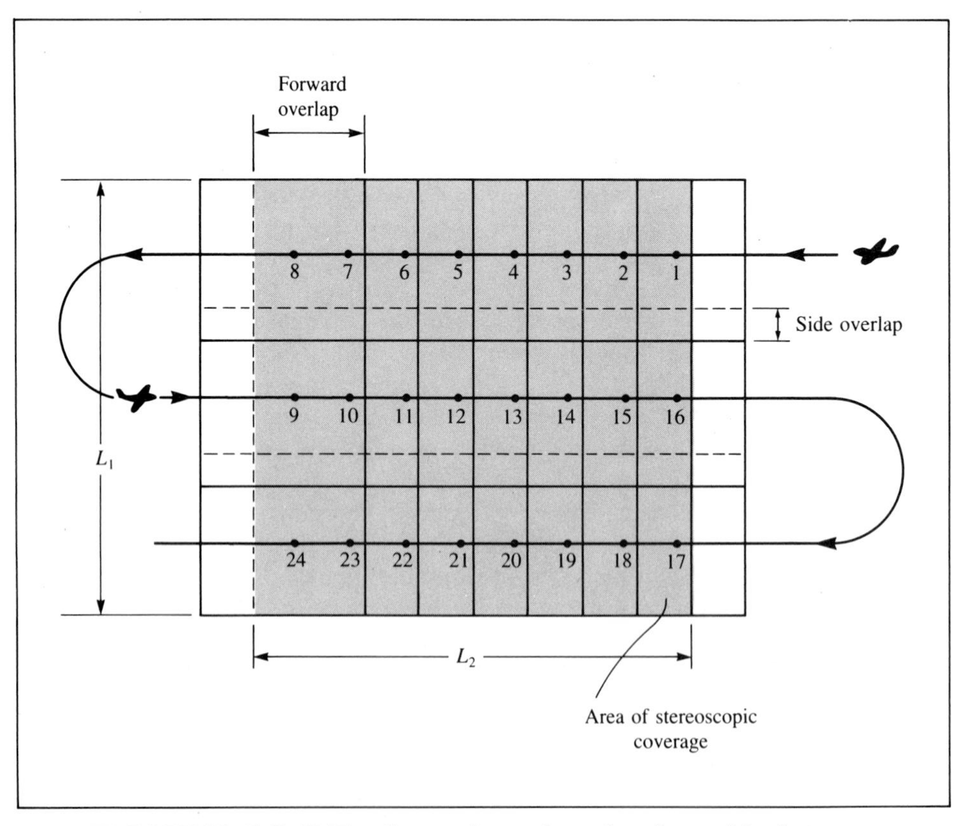

FIGURE 14.25 Geometric configuration for aerial photography

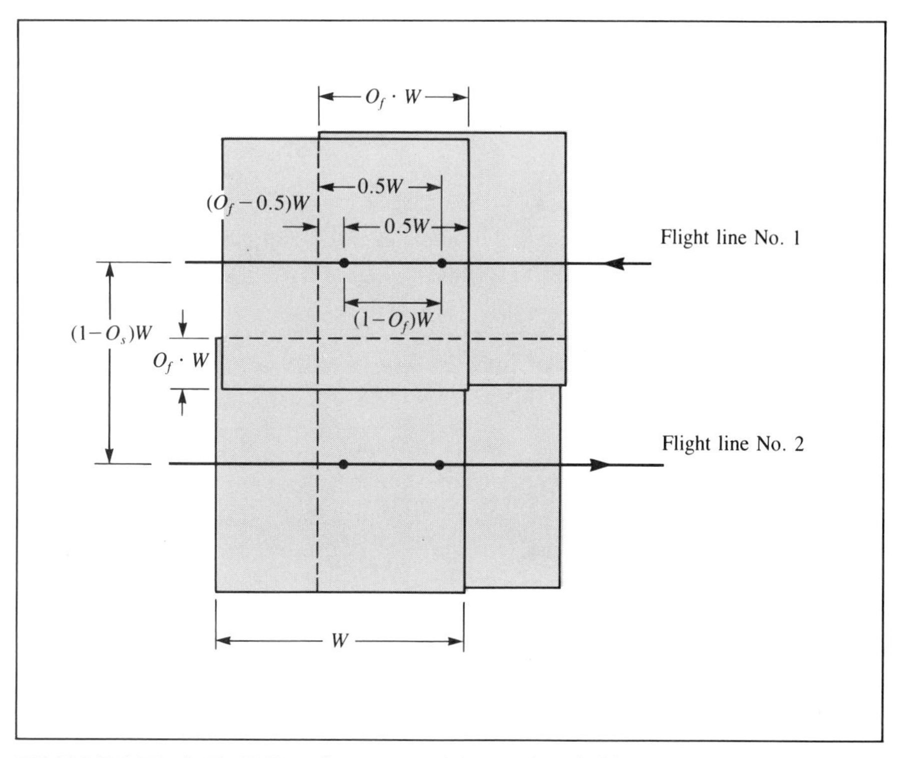

FIGURE 14.26 Geometry of forward and side overlaps

where N_s denotes the number of flight lines, N_p denotes the number of photographs per flight line, and L_1 and L_2 are the outside dimensions of the stereoscopic coverage area.

EXAMPLE 14.2

Aerial photographs are required to provide full stereoscopic coverage of a rectangular area that measures 3.7 miles in the east-west direction and 2.75 miles in the north-south direction. The desired average scale of the vertical photograph is 1 : 9,600. The terrain is relatively flat and has an average elevation of 750 ft above sea level. A 6-inch focal length camera with a 9″ × 9″ picture format is to be used. The flight lines are to be along the east-west direction. Forward and side overlaps are to be 60% and 20% respectively.

Determine the following:

a. the flight altitude above sea level;

b. the ground area, in acres, covered by each photograph;

c. the number of flight lines;

d. the number of photographs per flight line; and

e. the time interval between exposures if the ground speed of the aircraft is to be 100 mph.

SOLUTION

a. From Eq. (14.3),

$$\frac{1}{\left(\frac{H - h_{avg}}{f}\right)} = \frac{1}{9{,}600}$$

$$\text{flying height above sea level } (H) = 9{,}600 \times 0.5 \text{ ft} + 750 \text{ ft} = 5{,}550 \text{ ft}$$

b. Average scale of photograph = 1/9,600, which is equivalent to 1″ = 800′
ground dimension of a photo (W) = 9 × 800 ft = 7,200 ft
ground area covered by 1 photo = 7,200 ft × 7,200 ft
= 1,190 acres

c. From Eq. (14.20),

$$\text{number of flight lines } (N_s) = \frac{2.75 \times 5{,}280 - 7{,}200}{(1 - 0.2) \times 7{,}200} + 1 = 2.27 = 3 \text{ lines}$$

$$\text{ground distance between flight lines} = (1 - 0.2) \times 7{,}200 = 5{,}760 \text{ ft}$$

With three strips of photographs and a ground distance of 5,760 ft between flight lines, the width of the area of stereoscopic coverage would be 7,200 + 0.8(3 − 1) × 7,200 = 18,720 ft = 3.55 miles.

d. From Eq. (4.20),

$$\text{number of photos per strip } (N_p) = \frac{3.7 \times 5{,}280 - 2(0.6 - 0.5)7{,}200}{(1 - 0.6)7{,}200} + 1 = 7.3 \text{ photos} = 8 \text{ photos}$$

Thus there must be at least three strips of eight photographs each with a total of 24 photographs, as is the case shown in Figure 14.25.

e. $\text{aircraft speed} = \frac{100 \times 5{,}280}{3{,}600} = 147 \text{ ft/sec}$

ground distance between exposures = (1 − 0.6) × 7,200 = 2,880 ft

$\text{time interval between exposure} = \frac{2{,}880}{147} = 19.6 \text{ sec}$

In planning for aerial photography, it is important that the area of stereoscopic coverage be larger than the area to be mapped to allow for a safe margin of error in flying the photography. It is a common practice to take one or two extra photographs at each end of a flight line.

This section is intended primarily to illustrate the basic geometric requirements for stereoscopic coverage. For more detailed discussion of this topic, see Reference 14.1. Details on standard specifications for procuring aerial photography can be found in Reference 14.3.

14.11 SOURCES OF AERIAL PHOTOGRAPHS

Aerial photographs that are used for topographic mapping purposes are invariably flown specifically to satisfy the requirements of each individual project. Aerial photographs two or more years old usually have little value if the latest topographic and cultural details are to be accurately mapped. However, there are many engineering and scientific studies that actually require the use of dated (or historical) aerial photographs. The National Cartographic Information Center (NCIC) of the U.S. Geological Survey maintains a summary record of existing, in-progress, and planned aerial photography in the United States. Included in the record are aerial photographs acquired or to be acquired by agencies of the federal and state governments as well as some commercial firms. The record provides information on the locations of the photographic coverage, scales, types of camera, dates of photography, and addresses for making purchase inquiry. The summary record is continuously updated by NCIC and is an excellent source for information concerning the availability of existing aerial photography for a given location.

14.12 ORTHOPHOTOGRAPHS

Only in very rare cases would an ordinary perspective vertical aerial photograph show the images of objects in their true orthographic position as in a map. This would occur only when the ground is perfectly level and the optical axis of the camera is truly vertical at the moment of exposure.

An orthophotograph is essentially a copy of a conventional perspective aerial photograph that has been processed to show the photographic details in their proper orthographic positions. Although the displacements of vertical objects (such as walls of buildings) cannot be removed, an orthophotograph has uniform scale and is geometrically equivalent to a regular map. In addition, it contains all the details of an aerial photograph.

The production of an orthophotograph basically involves the division of an aerial photograph into rectangular patches measuring as small as a few millimeters in each dimension, and then each patch is individually corrected for tilts of the camera and slope of the terrain. The production time is considerably less than that required to produce a regular line map from the same photographs.

Orthophotographs are particularly useful in landuse planning, urban

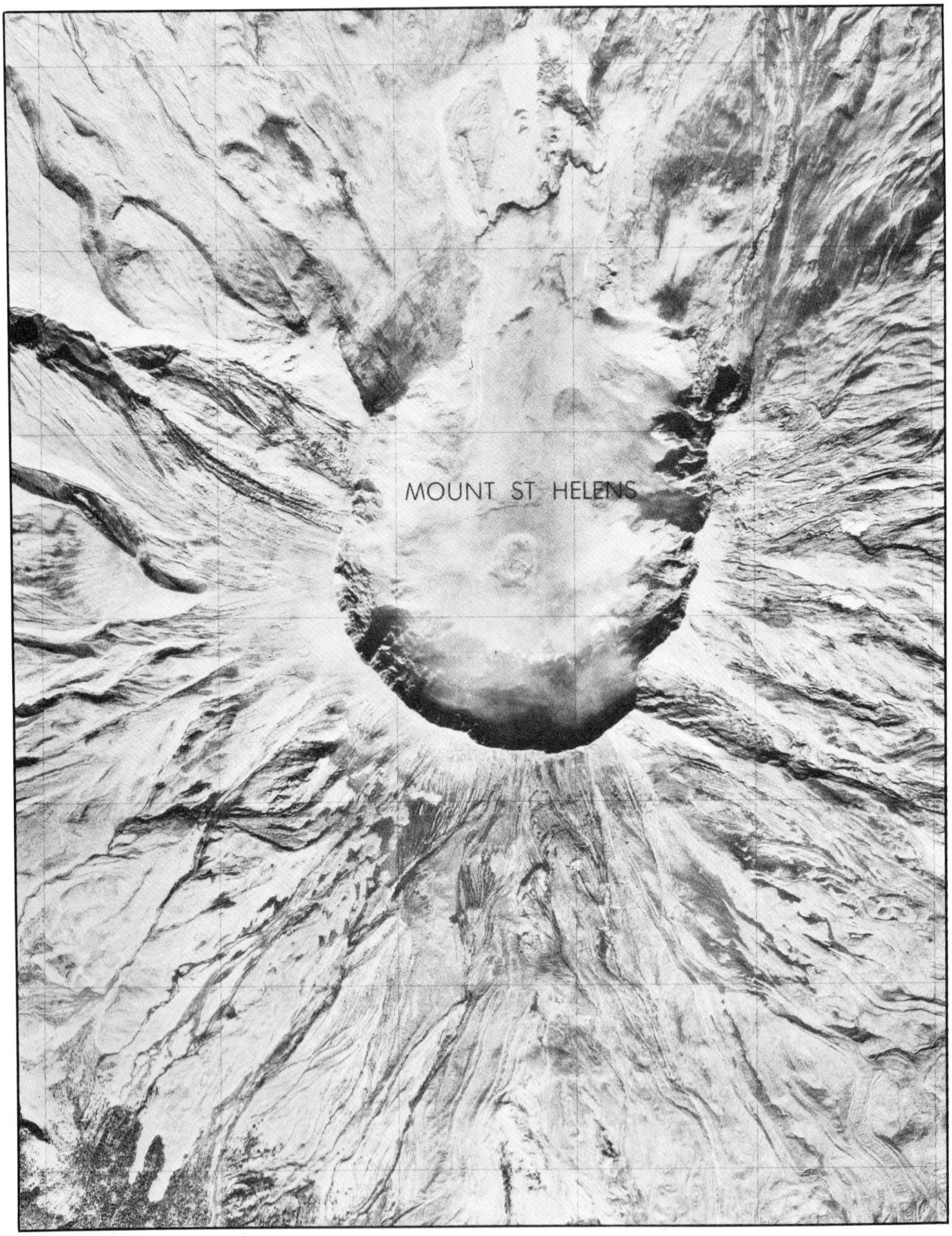

FIGURE 14.27 Part of an orthophotomap of Mt. St. Helens after its eruption of May 18, 1980 (courtesy of U.S. Geological Survey)

and regional developments, engineering feasibility studies, damage and hazard assessments, and coastal wetland studies.

Orthophotomaps are orthophotographs that have undergone cartographic treatments and on which are superimposed regular map information such as contours, place and street names, and map symbols (see Figure 14.27). In the case of relatively featureless terrain such as a broad expanse of desert, the conventional line map is inadequate for portraying relief. However, by applying color to the orthophotographs, ground features can become more recognizable. This is accomplished by using a variety of tones of green, blue, and brown to accentuate such details as marshland limits, saltwater encroachment, and certain geologic features.

14.13 GROUND CONTROLS

It was stated in Section 14.7 that the minimum ground controls needed to perform absolute orientation of a stereo model include: (1) one known ground distance as provided by two horizontal control points for scaling; and (2) three known elevation points for leveling. Thus to map the area covered by the block of 24 photographs (or 21 stereo models) shown in Figure 14.25, about 42 horizontal control points and 63 elevation points are needed. These numbers can be slightly reduced if some of the control points appear in more than one stereo model. However, in practice a much larger number of control points are used to provide the redundancy needed to achieve high quality mapping.

A basic network of horizontal and vertical ground control points must be established by ground survey procedures. Additional control points are determined by a process called phototriangulation (or aerotriangulation), as illustrated in Figure 14.28. In essence, two equations are used to describe each straight line joining a camera position, a photo image point, and its corresponding ground point. All the equations are combined together in a least-squares solution to determine the ground coordinates of the new points. Through this process, the number of ground control points that need to be established by ground survey procedure is significantly reduced.

Figure 14.29 shows the typical distribution of ground controls. For a single strip of photographs as shown in a, two ground control points are needed for each fifth photo. For a block of photographs as shown in b, the ground control points are located near the edge of the stereoscopic coverage area. One ground control point is located near the center of every second photo around the perimeter of the block. In both a and b, both the horizontal positions and elevations of all the ground control points are determined by ground survey procedures.

All ground control points should be covered with proper aerial targets just prior to flying the photography so that they can be positively and

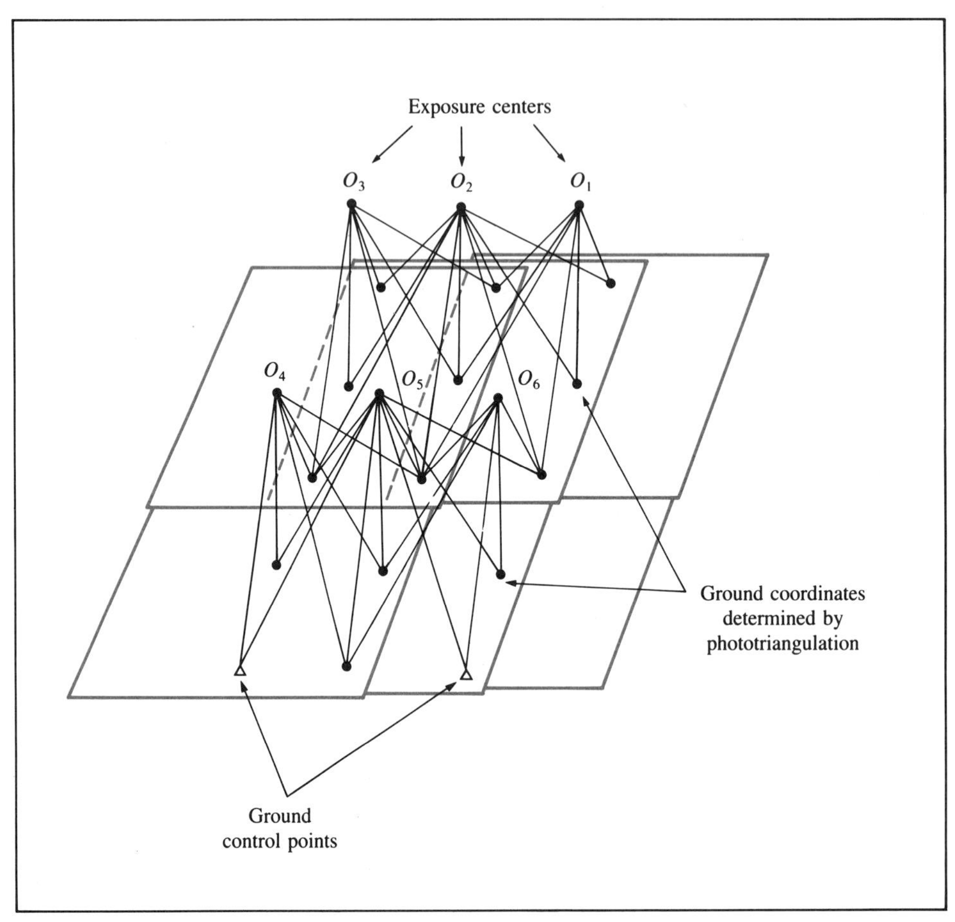

FIGURE 14.28 Phototriangulation

accurately identified in the aerial photographs. Figure 14.30 shows two shapes of targets commonly used for this purpose. They can be made of vinyl or cloth materials.

14.14 USES OF PHOTOGRAMMETRY

Photogrammetric products are used by a broad range of professionals including not only engineers and surveyors, but also planners, foresters, and conservationists.

In some cases considerable information can be obtained from direct viewing of a single aerial photograph. Such an examination may reveal details omitted from existing maps and provide the latest information on changes in the development of an urban area. Stereoscopic viewing of a pair

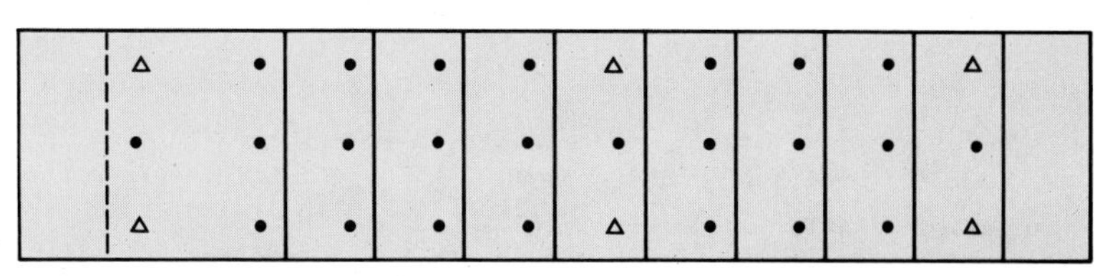

(a) One strip of photographs

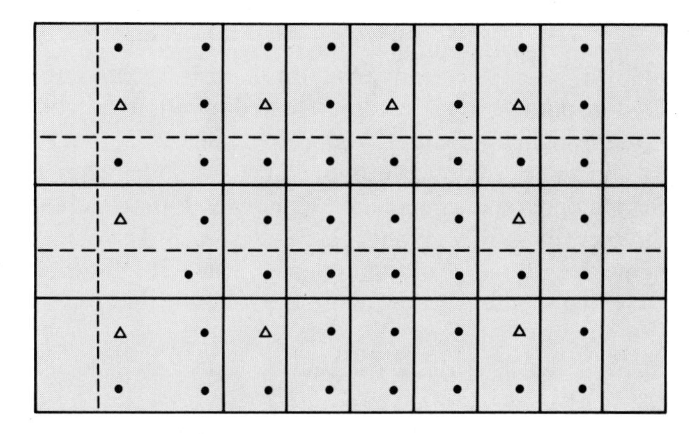

(b) Block of photographs

FIGURE 14.29 Ground control distribution

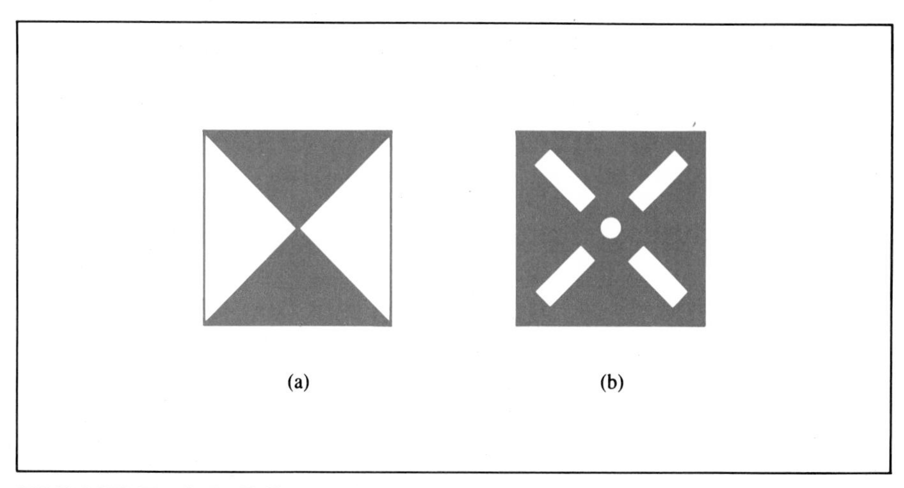

FIGURE 14.30 Aerial targets for ground controls

of overlapping photographs with a simple lens stereoscope will greatly facilitate the study of the terrain. An orthophotograph can be employed as a substitute for a conventional planimetric map and as a base for field investigations in an area where no map is available.

Photogrammetric applications are to be found in the practice of land surveying. Although not readily recognizable to one walking over the ground, old property lines may be visible on the aerial photograph. Mosaics and topographic maps prepared by aerial photogrammetry have been widely utilized by highway planners and designers. Specialized applications of photogrammetry are to be found in traffic accident investigation, hydraulic laboratories, architecture, and many other fields.

The most common use of aerial photogrammetry is in the preparation of topographic maps. Both in the private and public sectors almost all large mapping projects are performed by photogrammetric methods. Some of the principal advantages as compared with ground methods of mapping are greater speed, lower production costs, and greater accuracy of planimetric and topographic details. Note, however, that in areas having heavy vegetative cover such as timber and underbrush it may be necessary to supplement the aerial photogrammetric processes with conventional ground surveying methods to delineate the relief of the terrain.

PROBLEMS

14.1 The scaled distance between the images of two points on a vertical photograph is 4.22 in. The ground distance between these points, which are at substantially the same elevation, is 5,195 ft. What is the approximate scale of the photograph?

14.2 In Problem 14.1 the photo distance is now 128.4 mm and the ground distance is 1,696 m. Find the approximate scale.

14.3 Calculate the average scale of vertical aerial photography taken at an altitude of 7,400 ft above mean sea level with a camera having a focal length of 6.02 in. if the average elevation of the terrain is 525 ft.

14.4 Calculate the average scale of vertical aerial photography taken at a flight altitude of 13,500 ft above mean sea level with a 211.81-mm focal-length camera. The average elevation of the upland prairie is 2,400 ft.

14.5 Vertical photography having an average scale of 1 : 2,400 is desired for highway design purposes. If the focal length of the camera lens is 6 in., at what height above average terrain elevation should the mission be flown?

14.6 A few vertical photographs are to be taken from a helicopter of a highway-railroad grade crossing that has been the scene of several accidents. The ground elevation is 1,200 ft and the desired scale is 1 in. = 100 ft. If the camera has a focal length of 5.98 in., calculate the flight altitude above sea level.

14.7 The image of a radio tower is portrayed on a 1 : 7,500 scale vertical photograph, which was obtained with a 6-in. focal-length camera. The radial distances from the principal point to the images of the bottom and top of the tower are 70.36 mm and 79.87 mm respectively. Calculate the height of the tower to the nearest foot.

14.8 The images of the top and bottom of a smokestack are 4.874 in. and 4.712 in. respectively from the principal point of a vertical photograph. If the flight altitude above the base of the stack is 4,100 ft, calculate the height of the stack.

14.9 The image of the top of a radio tower is measured to be 4.781 in. radially from the principal point of a vertical aerial photograph. The radial distance between top and bottom of the tower is measured to be 0.356 in. The bottom of the tower is known to be situated at an elevation of 550 ft above sea level, and the aerial photograph was flown at an altitude of 10,750 ft above sea level. Determine the height of the tower.

14.10 A stereoscopic pair of aerial photographs was taken with a camera at an altitude of 7,500 ft above average ground elevation. The distance between the two exposure stations (that is, the air base) as measured on the photographs was 4.07 in. The x-distance, measured parallel to the flight line, between the two corresponding images of ground point P was 10.467 in. The x-distance measured between the two corresponding images of ground point G was 10.585 in. The elevation of ground point P is known to be 760.3 ft. The average ground elevation is 740 ft. Compute the elevation of point G on the ground.

14.11 A pair of aerial photographs providing overlapping coverage of an area is known to have a scale of 1/2,400 at a ground elevation of 700 ft. The photographs were taken with an aerial camera that has a focal length of 6 in. The elevation of a ground point g is known to have an elevation of 715.9 ft. The following x-parallaxes were measured for ground points g, m, p, and t according to the procedure illustrated in Figure 14.14:

$$d_g = 10.05''$$
$$d_m = 10.64''$$
$$d_p = 10.31''$$
$$d_t = 9.56''$$

The averaged air base (b) measured on the photographs was computed to be 3.78″.

a. Determine the ground elevations of points m, p, and t.

b. What assumption primarily limits the accuracy of this method of measuring elevation?

14.12 The average height of an aerial camera above the terrain during a photographic mission is 8,200 ft and the air base of a stereopair is 4.27 in. If the difference in parallax of the images of two features is 0.123 in., find the difference in elevation between them.

14.13 Topographic maps of a project site are to be prepared with a contour interval of 2 ft. Determine the maximum allowable flight height above ground if the C-factor of the stereoplotter to be used for the mapping project is:

a. 500

b. 1,000

c. 3,000

14.14 Aerial photographs are to be flown to provide stereoscopic coverage of a rectangular project area that is 11.25 miles long (north-south) and 6.50 miles wide (east-west). The desired average scale of the vertical photography is 1 : 10,000, the focal length of the lens is 6 in., and the picture format is 9 in. × 9 in. A forward overlap of 60% and side overlap of 30% are specified. The average terrain elevation is 1,050 ft. Flight lines are to be directed in a north-south direction. Determine the following:

a. flight altitude above sea level

b. ground area in acres covered by each photograph

c. number of flight lines required

d. number of photographs per flight line

14.15 Aerial photographs are to be taken of a proposed highway corridor with a camera that has a focal length of 6 inches, at a flight altitude of 1,200 ft above the average ground elevation. Aerial photographs will be taken along a single flight path located at the centerline of the corridor (see Figure P14.1). The first photograph will be taken at point A, and successive photos will have 60% forward overlap. The corridor measures 1,500 ft wide and 50,000 ft long. Determine the *minimum* number of 9 in. × 9 in. photographs needed to provide stereoscopic coverage of the entire corridor.

14.16 An aerial survey is to be made of an area 60 mi long by 24 mi wide. The scale is to be 1 in. = 2,000 ft, the photographs are 9 in. × 9 in., forward overlap is 60%, side overlap is 30%, focal length of lens is 6.50 in., and average elevation of terrain is 1,000 ft. Find:

a. flight altitude above mean sea level

b. ground area in acres covered by each photograph

c. minimum number of photographs per flight line

d. number of flight lines

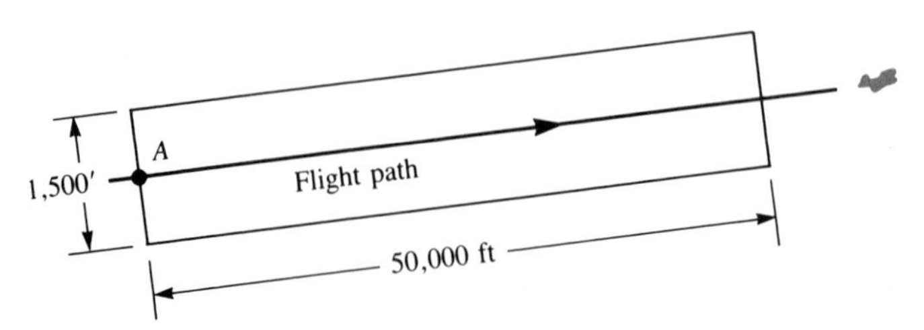

FIGURE P14.1 Highway corridor mapping

REFERENCES

14.1 American Society of Photogrammetry. *Manual of Photogrammetry,* 4th ed., Chester C. Slama, Editor-in-Chief, American Society of Photogrammetry, Falls Church, VA: 1980.

14.2 American Society of Photogrammetry. *Manual of Remote Sensing,* Volumes I and II, 2nd ed., Robert N. Colwell, Editor-in-Chief, American Society of Photogrammetry, Falls Church, VA: 1983.

14.3 U.S. Geological Survey. *Standard Specifications for Aerial Photography for Photogrammetric Mapping,* USGS TD-72-001, January 1972.

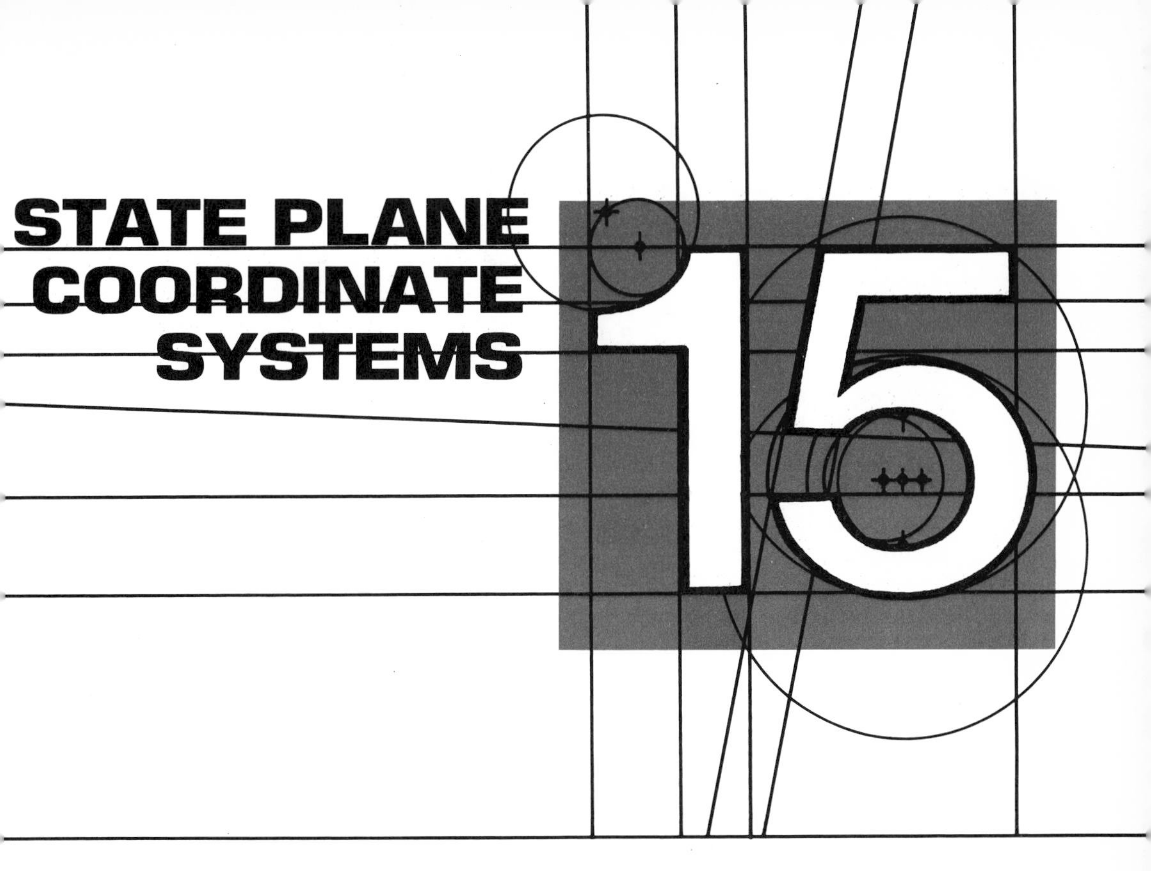

15.1 INTRODUCTION

The position of a point on the surface of the earth is completely defined by stating its latitude and longitude and its elevation above a reference datum.

The necessity of relating the elevations of one survey to those of another has become so common that the desirability of a standard datum for elevations is well recognized. Accordingly, every new survey in the United States involving vertical control of any importance is based upon the National Geodetic Vertical Datum of 1929.

The parallel necessity of relating the horizontal positions of points in one survey to those of another by means of a common *horizontal datum* or a specific system of coordinates has assumed increasing importance in recent years. The number, complexity, and areal extent of modern engineering projects have made urgent the need not only to make more usable existing geodetic position data by reducing them to a convenient system of rectangular coordinates, but also to express the results of future horizontal control surveys in terms of some standard coordinate system. Before explaining the fundamental properties of the state plane coordinate systems, it will be well to trace the steps leading to their evolution, mention the major advantages of such systems as compared with a geographic system, and state the minimum accuracies with which surveys based on state coordinates should be executed.

15.2 NATIONAL NETWORK OF HORIZONTAL CONTROL

Because of the need for perpetuating national and state boundaries, providing control for mapping and charting programs, and serving other purposes, the federal government has long recognized the necessity for an accurate network of horizontal control for the nation. For more than a century the U.S. Coast and Geodetic Survey (now the National Ocean Survey) has engaged in operations that determined the geodetic positions of thousands of well-monumented points in all parts of the country. Supplementing the network of first- and second-order stations of the geodetic component (National Geodetic Survey) of this agency have been the very extensive third-order horizontal control operations of other surveying organizations.

Formerly, it was the practice to publish and make available to the engineers of the country the positions of these survey points in terms of their spherical coordinates only—namely, latitude and longitude. The use of control data in such form required a working knowledge of geodetic surveying formulas and computing methods, which the average surveyor and engineer did not possess. Also, these processes were too slow and tedious when compared with plane surveying calculations to justify their use in common survey practice. Hence for many years the problem besetting the increased use of higher-order horizontal control was that of finding a simple and practicable means for utilizing such position data for detail surveys. Before 1933 this could be done only by setting up local systems of plane coordinates of very limited extent. However, lack of training and experience on the part of engineers and the need for special equipment and mathematical tables prevented any substantial use of geodetic control data by this method. Some cities, notably New York City and Pittsburgh, set up tangent plane coordinate systems based on the national horizontal control system.

15.3 BEGINNINGS OF STATE PLANE COORDINATE SYSTEMS

To make the geodetic data of the national horizontal control system readily available in a form acceptable to engineers and surveyors, the U.S. Coast and Geodetic Survey established in 1935 what are known as the *state plane coordinate systems* (SPCS). Each state has a separate system. The basis of each system is the mathematical projection of the earth's surface upon a surface like that of a cone or cylinder, which can then be developed into a plane. The National Geodetic Survey computes and publishes the plane coordinates, on the appropriate state coordinate system, of all points for which it determines geographic positions. Hence, just as there is but one point on the surface of the earth corresponding to a geographical position expressed by latitude and longitude, so likewise there is but one point corresponding to a given pair of plane coordinates, expressed as x and y, for a particular zone of a given state. The substitution of these simple rectangular plane coordinate systems and their associated position data for the relatively unwieldy and complex geographic coordinates and geodetic computations has led to a substantial increase in the use of national horizontal control data and to a better understanding of the benefits of referencing local surveys to the federal network. Of paramount importance, however, in this connection is the fact that the engineer or surveyor who ties a carefully executed survey to the national net may perform the surveying calculations using the ordinary office procedures of plane surveying.

15.4 ADVANTAGES OF STATE PLANE COORDINATE SYSTEMS

In subsequent discussions of the applications of state coordinates to specific engineering surveys, the detailed advantages of using such a system of coordinates will be explained. It will suffice here to mention three major advantages of a state coordinate system. These are inherited from the national geodetic network on which the state coordinate system depends.

1. Positive checks can be applied to all surveys to prevent the accumulation, beyond permissible limits, of errors in the measurement of angles and directions.

2. Surveys that are initiated at widely separated points, and perhaps for different projects, will have at their junction substantially the same azimuth for a given line and the same coordinates for a given point. If it were not for the errors of field measurement, the agreement in azimuth and coordinates would be exact. Thus, two such surveys can be coordinated and used to supplement each other.

3. Any station whose state coordinates have once been accurately determined may be said to be permanently located. Even though the marker itself is destroyed, its position is perpetuated by the record of coordinates and the station can be restored by careful measurements from the nearest recovered stations in the system.

15.5 QUALITY OF SURVEYS

The field procedures associated with the subsequent computation of state coordinates are the simple methods of plane surveying. It should be emphasized, however, that wholly reliable results will be obtained only if the local survey is tied to first- or second-order control and executed in a careful manner so that third-order or better accuracy is obtained.

To clarify some misconceptions, it is worth mentioning that there is no special operation of surveying that is called *state coordinate surveying*. Any suitably accurate method of carrying distance and azimuth from a point of unquestioned state plane coordinate position will serve to make possible the calculation of the position of other points. Furthermore, there is no survey point that can be properly described as a state plane coordinate station. Defining the position of a station in the nation's horizontal control network by state coordinates is merely utilizing a mode of expressing position that is parallel to that of using geographic coordinates, because for every pair of state plane coordinates there is a pair of unique geographic coordinates and vice versa.

15.6 LOCAL PLANE COORDINATES

Engineers and surveyors have used for many years in their work a variety of unrelated and arbitrary coordinate systems. Frequently, different systems of this type are utilized in the same community or even at the same industrial site. Such coordinate systems are defined by assigning x and y values to a chosen survey point and taking either an assumed meridian or the true meridian through the initial point as grid north (see Section 7.3). These systems lack official recognition, cannot be correlated with other surveys, and pose problems associated with convergency of the meridians when extended over a large area. An isolated system is still too frequently merely an expedient choice. When ties to a recognized system, such as the state coordinate system, are easily feasible, the continued use of local systems is hardly defensible.

15.7 MAP PROJECTIONS

The mathematical theory of map projections is too complex to be treated here, but it is thought desirable to discuss some elementary principles and present the general properties of two kinds of projections.

The surface of a sphere cannot be developed into a plane without distortion; hence, if a considerable portion of the earth's surface is to be shown on a map, the dimensions must be distorted one way or another. The character of the distortion can be controlled if the points on the earth's surface are mathematically projected upon a plane, or upon a surface (cone or cylinder) that can be developed into a plane. After such projection and development, the points will represent in the plane (that is, on the map) with a minimum of scale distortion the correct relative positions of the corresponding points on the earth's surface.

Distortions on a map projection are negligible if they are too small to be plotted at the scale of the map. However, distortions entering into the use of rectangular plane coordinates are negligible only if they are so small as to fall within the usual permissible limits of accidental errors in the field measurements.

In general, a plane coordinate system that is to have maximum engineering utility should have the following features:

1. The y- and x-coordinates of a survey point in the plane rectangular system should be readily obtainable from its latitude and longitude. Also, the reverse process should be equally feasible.

2. The forward and back azimuths of a line in the rectangular grid system should differ by exactly 180°.

3. The length of a survey line as calculated from the grid coordinates of its termini should be equal to the ground distance or a means be available to readily effect a transformation from the grid length to its equivalent ground length.

4. The grid azimuth of a line should be simply related to the true or astronomic azimuth of the line.

Several projections satisfying the requirements of a rectangular plane coordinate system have been devised, but only three will be described here. They may be classed according to area limitations as local systems of coordinates or as regional systems. To the first class belongs the *tangent-plane* projection and to the second belong the *Lambert* and *Transverse Mercator* projections.

15.8 TANGENT PLANE COORDINATES

The only simple way by which the results of the highly accurate geodetic surveys can be easily and readily used for controlling ordinary surveying operations is by transforming the geographic coordinates to some system of plane coordinates.

For a relatively small area, such as an average-sized city, a projection on a tangent plane gives an accuracy that is satisfactory for survey purposes. Such a projection represented the first attempt to utilize existing higher-order horizontal control by transforming geographic to plane coordinates. The projection consisted basically of the representation of points projected radially from the center of the earth to a plane tangent to the earth at a point in the general vicinity of the center of the city.

The use of such a projection was made quite simple by U.S. Coast and Geodetic Survey Special Publication No. 71, entitled "Relation Between Plane Rectangular Coordinates and Geographic Positions." This publication (Reference 15.4) contains tables by means of which one can reduce geographic coordinates to rectangular coordinates on a tangent plane at sea level. This projection is limited to relatively small areas, however, because, as the distance from the origin increases, the difference between the measured length of a line and the length computed from the plane coordinates of its termini increases rather rapidly. At 40 miles from the origin, the scale error is 1 part in 20,000 and at 80 miles it is 1 part in 5,000.

Despite its limitations, the tangent plane coordinate system served very well the needs of several large cities, among them Pittsburgh, Pennsylvania. However, ever since the establishment of the state coordinate systems, which have broad regional coverage, there is no longer any need to establish new tangent plane systems. Hence it is not conceivable that any more projections of this type should be devised for American engineering practice, because they are distinctly inferior to a state plane coordinate system.

It may be added that where a tangent plane system of coordinates has been adequate and satisfactory there is no need to discard the data associated with it. The procedure for transforming coordinates on a local system to a state plane coordinate system is explained in Section 15.21.

15.9 LAMBERT PROJECTION

The Lambert projection has different forms, but the one adopted for state plane coordinate systems is a modified conic projection whose general characteristics may be described by referring to Figure 15.1. It consists of an imaginary cone whose axis *OP* is assumed to be coincident with the axis of the earth, and whose elements, one of which is *PBD*, cut the earth's surface at two points C_1 and C_2. A partial frustrum of the cone is shown as *ABDE*. Two small circles, called *standard parallels,* are generated where the conical surface cuts the earth's surface.

The frustrum *ABDE* is developed into a plane surface. A central meridian will have a known longitude, and the longitude of any point, as *A* or *B*, will be given with respect to the central meridian by adding (or subtracting) the angle θ, at *P*, between the central meridian and the element through the given point. The latitudes of the two circles C_1 and C_2 are known, and the latitude of any point as *A*, not on these circles, can be found from its known distance along its terrestrial meridian, north or south of the standard parallel.

The projections from the earth's surface onto the cone are made along radii from the earth's center *O*.

From the conditions stated above, it is evident that this projection has these following characteristics: (1) Since the projection cone cuts the earth's surface along the standard parallels, the longitude scale along these circles will be exact. (2) The conical surface is so nearly coincident with the earth's surface that the projection may be said to be *conformal;* that is, both the latitude and longitude scales are so nearly exact that angles between lines on the projection are very nearly the same as the angles between the corre-

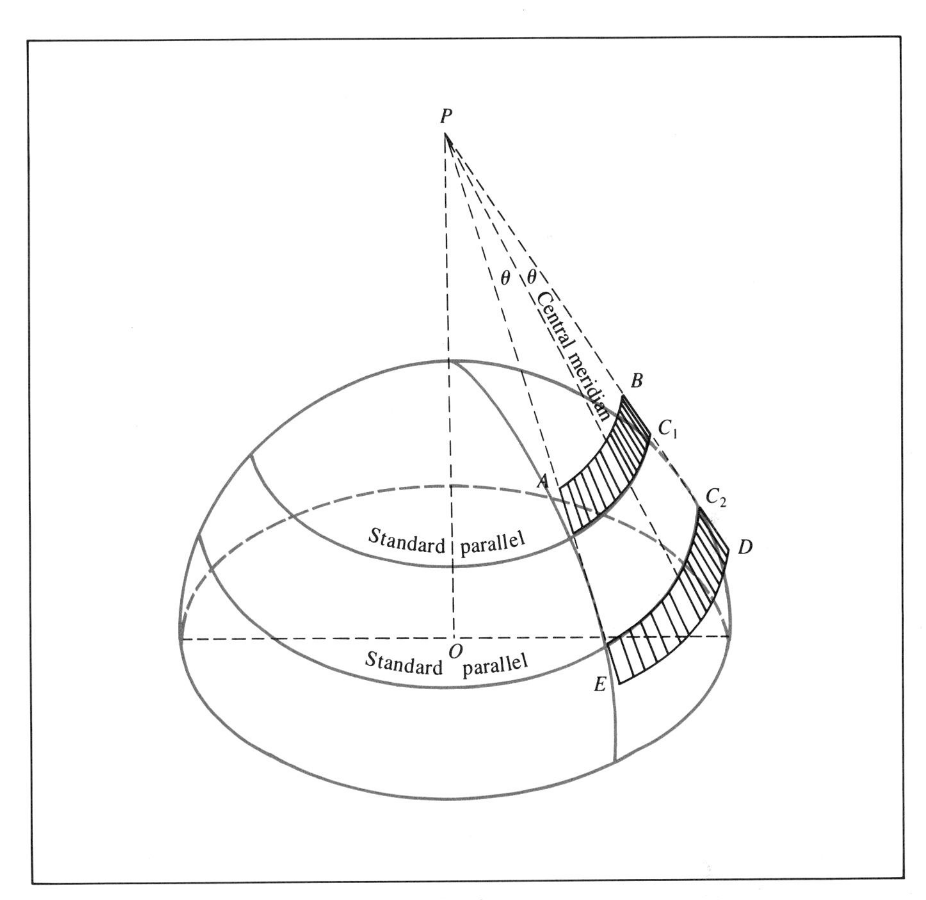

FIGURE 15.1 Lambert projection

sponding lines on the earth's surface. (3) For the zone between the standard parallels, the scale of the projection will be too small; and likewise, for the zones north and south of the middle zone, the scale will be too large. (4) The projection can be extended indefinitely in an east-west direction without affecting the accuracy of the projection; but as the projection is extended in a north-south direction, the scales are modified and change in a rapidly increasing ratio as this dimension is increased.

15.10 THE MERCATOR PROJECTION

The Mercator projection consists of an imaginary cylinder with its axis coincident with the earth's axis and its surface tangent to the earth's surface at the equator. The meridians on this projection are straight lines perpendicular to the baseline, or equator, and hence are parallel with each other. Since the earth's meridians converge at the pole, it is evident that the scale of the projection for east-west dimensions increases as the latitude increases.

At the equator, however, the scale is exact in all directions and the projection is therefore conformal. As the projection is extended northward (or southward) from the equator, the scale of the meridians is changed to correspond with the change in the scale of the parallels, so that this projection possesses the unique quality that a *rhumb line* (that is, a line of constant bearing) is always a straight line. Hence this projection is much used in navigation.

15.11 TRANSVERSE MERCATOR PROJECTION

The Transverse Mercator projection, which is a modification of the Mercator projection, has been designed to meet the requirements of state plane coordinate systems for those states whose greatest dimension lies in a north-south direction. The projection may be described by reference to Figure 15.2.

This projection consists of a cylinder turned 90° from that of the Mercator projection and having its radius slightly reduced, so that instead of

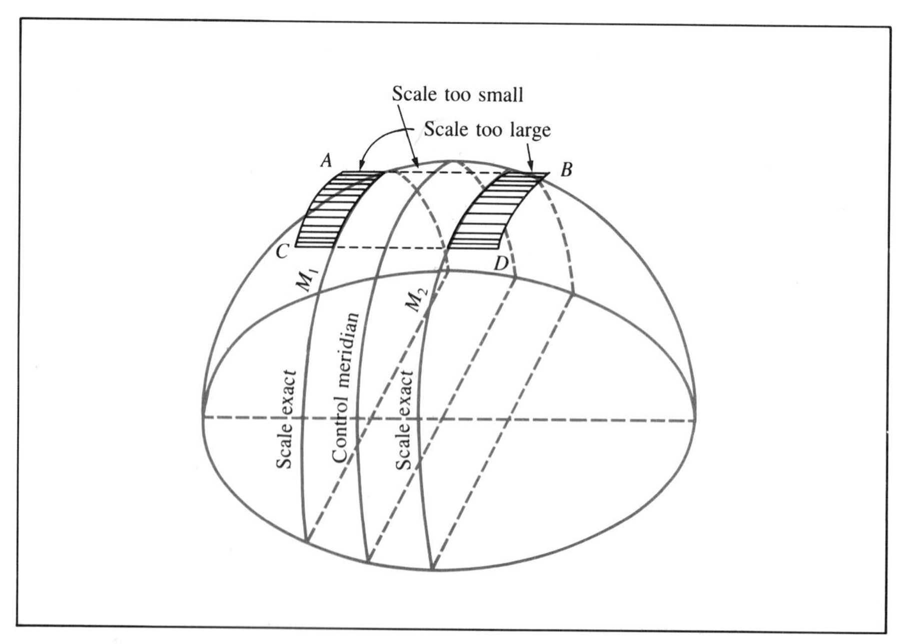

FIGURE 15.2 Transverse mercator projection

being tangent to the earth's surface along a central meridian, it cuts the sea level surface along two parallels as M_1 and M_2. A portion of such a projection is shown as *ABCD*.

This projection has the following characteristics: (1) The scale of the projection is exact along the two parallels M_1 and M_2. (2) The scale is too large for the zones outside of the two intersecting parallels, and it is too small for the zone included between them. (3) This projection can be extended indefinitely in a north-south direction without changing the scale relations, but these relations change rapidly as the length of the projection is extended in the east-west direction.

15.12 THE STATE PLANE COORDINATE SYSTEMS

As mentioned in a previous section, the errors involved in using a system of tangent plane coordinates increase rapidly as the distance from the point of tangency increases. This is a significant weakness of a coordinate system to be used for a statewide project, such as a network of superhighways, or to connect surveys extending over several large counties. The urgency of creating state plane coordinate systems that could utilize existing geodetic data over an entire state without involving anything more complicated than plane surveying procedures led to the establishment of the state coordinate systems.

These systems are based on the Lambert and Transverse Mercator projections. The former is used for those states whose greatest dimension lies in an east-west direction, and the latter is employed for those states whose greatest dimension lies in a north-south direction. In three states, New York, Alaska, and Florida, both projections are used. Almost all the states are divided into several belts or zones, each zone having its own origin and reference meridian.

State coordinate systems based on the Lambert and the Transverse Mercator projections are commonly designated as Lambert and Transverse Mercator grids, respectively.

The states and their grid systems are tabulated in Figure 15.3.

Although all the state grid systems have been mathematically defined and grid coordinates have been computed and published by the National Geodetic Survey for all its horizontal control points, not all states have enacted the necessary legislation to officially recognize the appropriate system. In general, some 35 states have passed legislation that establishes the legal status of the state's grid system and imposes reasonable standards of accuracy in the use of the system when coordinates are incorporated in public records. The legislation usually provides that the use of the system be permissive and not mandatory.

LAMBERT SYSTEM	
Arkansas	North Dakota
California	Ohio
Colorado	Oklahoma
Connecticut	Oregon
Iowa	Pennsylvania
Kansas	South Carolina
Kentucky	South Dakota
Louisiana	Tennessee
Maryland	Texas
Massachusetts	Utah
Michigan	Virginia
Minnesota	Washington
Montana	West Virginia
Nebraska	Wisconsin
North Carolina	
TRANSVERSE MERCATOR SYSTEM	
Alabama	Mississippi
Arizona	Missouri
Delaware	Nevada
Georgia	New Hampshire
Hawaii	New Jersey
Idaho	New Mexico
Illinois	Rhode Island
Indiana	Vermont
Maine	Wyoming
BOTH SYSTEMS	
Florida	New York
Alaska	

FIGURE 15.3 State grid systems

15.13 COMPUTATION OF PLANE COORDINATES ON LAMBERT GRID

Before describing the calculation of a traverse based on state plane coordinates, it will be desirable to show the manner in which the geodetic coordinates of a survey point can be transformed into plane coordinates. Only a limited discussion of the theory underlying the computational procedure for the Lambert grid and none for the Transverse Mercator grid will be presented. However, it should not be difficult for any engineer to perform the calculations shown in this and the following section. Furthermore, since it is the practice of the National Geodetic Survey to compute and publish the coordinates on the appropriate state system of all points for which it determines the geographic positions, it is possible for the interested engineer to make request for such data.

The transformation computation for a Lambert grid will be illustrated by reference to the Minnesota state coordinate system. As shown in Figure 15.4, this state is covered by three overlapping zones, the North, Central, and South zones. Each zone has different axes for x and y, although all y-axes, passing through the center of the respective zone, are given an x value of 2,000,000 ft. The x-axis is placed well below the southern limit of each belt and has a value of zero feet. The geographic coordinates of a point in the North zone will be transformed into plane coordinates.

The North zone has for a central parallel the parallel of latitude 47°50′. Along this line the scale ratio (error of the projection) is 1 part in 10,300 parts too small. The standard parallels along which the scale is true are 47°02′ and 48°38′. The meridian of longitude, 93°06′, is the y-axis and a line perpendicular to it at latitude 46°30′ defines the x-axis. Hence the origin of coordinates for the North zone is a point on the x-axis situated 2,000,000 ft west of longitude 93°06′.

Figure 15.5 shows the fundamental basis for transforming the geographic coordinates of a point P into its Lambert grid coordinates. Point O is the origin of coordinates, and AB is the central meridian of the system. The value of the x-coordinate of the central meridian is designated as C. The point A represents the apex of the cone on which the area is projected, and the arcs PE and DB represent portions of parallels of latitude through point P and the lower extremity of the y-axis. The distance R_b is the largest latitude radius of the zone and is a constant for the zone. It is to be noted that the y-coordinate of point A, the apex of the cone, is equal to R_b. The angle θ is the angle of convergency between the central meridian and the meridian through the point P. The values of R for each whole minute of latitude, and the values of θ for each whole minute of longitude are given in U.S. Coast and Geodetic Survey Special Publication No. 264, entitled "Plane Coordinate Projection Tables—Minnesota." The angle θ is considered positive if P is east of the central meridian and negative if P is west of the central meridian. It is to be noted that *grid north* and *geodetic north* are identical along the *central meridian.*

The expressions for calculating the Lambert coordinates of a point are then as follows:

$$x = R \sin \theta + C$$

$$y = R_b - R \cos \theta$$

A typical transformation computation will now be demonstrated. The pertinent portions of Special Publication No. 264 are shown in Tables X and XI in Appendix B.

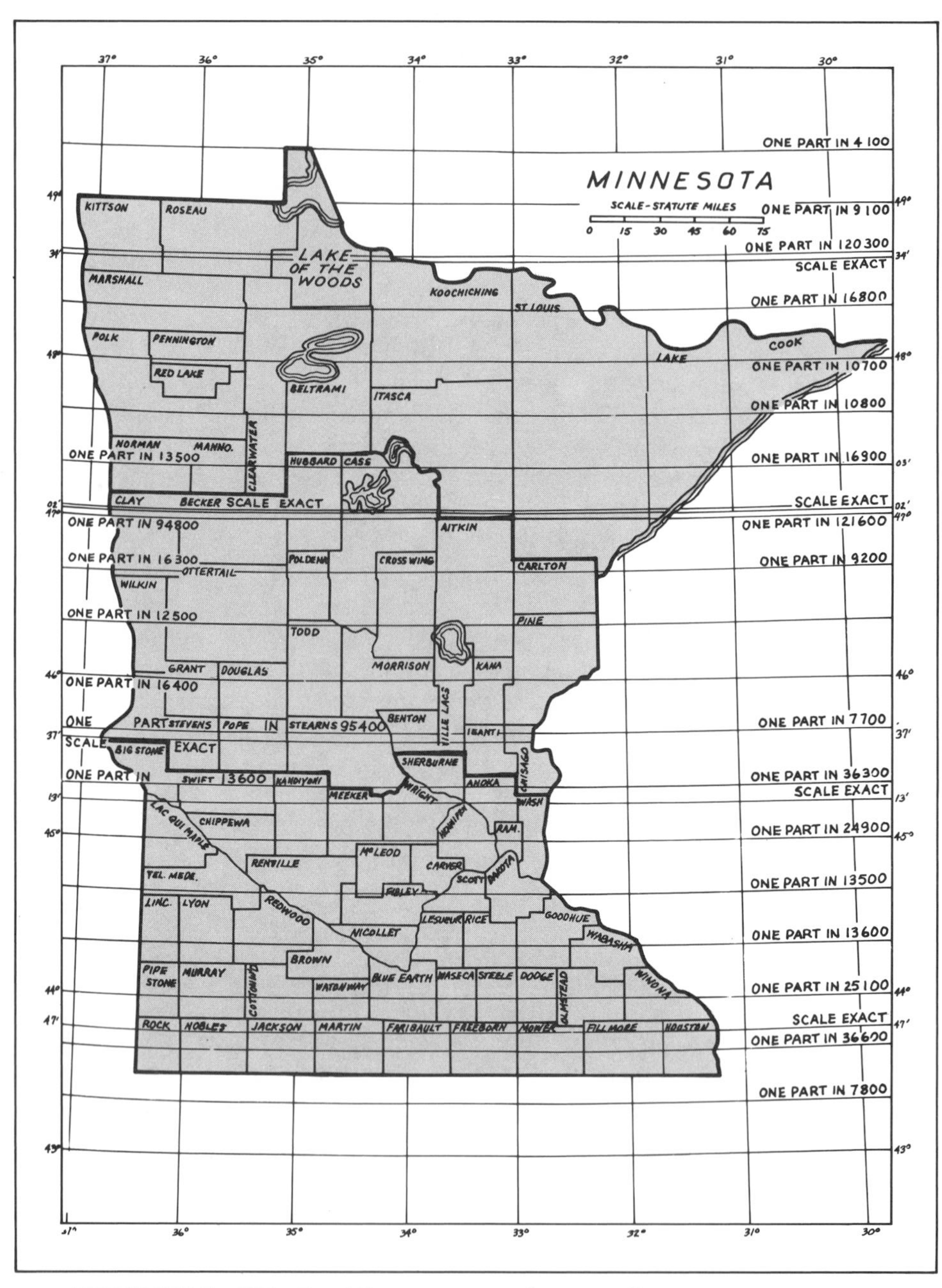

FIGURE 15.4 Minnesota state plane coordinate zones (National Geodetic Survey)

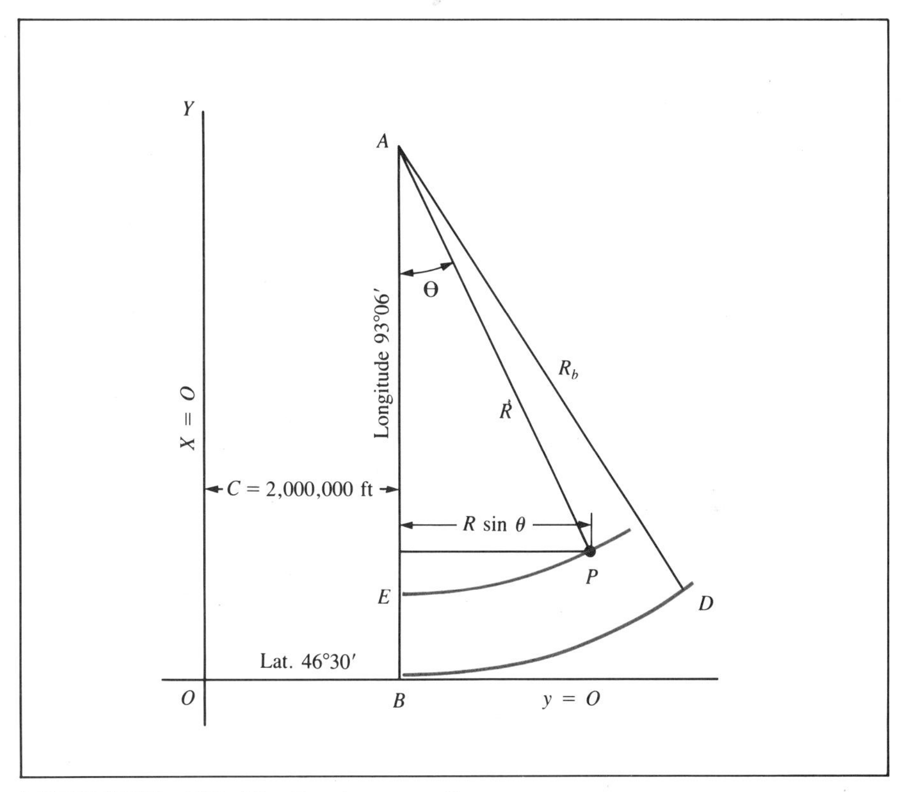

FIGURE 15.5 Lambert coordinates

EXAMPLE 15.1

Given: Station "Blackduck Tank"
Latitude 47°43′50.270″
Longitude 94°32′58.240″
State—Minnesota; Zone—North
C = 2,000,000 ft R_b = 19,471,398.75

Find Lambert coordinates x and y.

SOLUTION

$$R = 19{,}022{,}539.81 \text{ (Table X)}$$

$$\theta = 1°04'27.8621'' \text{ (Table XI)}$$

$$\sin \theta = -0.0187508257$$

$$\cos \theta = +0.9998241876$$

Then,

$$x = R \sin \theta + C$$

$$x = -(19{,}022{,}539.81)(0.0187508257) + C$$

$$x = 1{,}643{,}311.67 \text{ ft}$$

and

$$y = R_b - R \cos \theta$$

$$y = 19{,}471{,}398.75 - (19{,}022{,}539.81) \times (0.9998241876)$$

$$y = 452{,}203.34 \text{ ft}$$

Note that sin θ is negative in this problem. The term $R \cos \theta$ is subtractive in all cases. Also, the preceding calculation makes quite obvious the need for either a 10-digit electronic calculator capable of generating trigonometric values or a 10-bank electric calculator, which would be used with an expanded table of natural sines and cosines. Special Publication No. 246 (Reference 15.4) of the U.S. Coast and Geodetic Survey contains tables of these functions to ten decimal places. This and similar surveying publications are available at nominal prices from the Superintendent of Documents, U.S. Government Printing Office, Washington, D.C. 20402.

15.14 COMPUTATION OF PLANE COORDINATES ON TRANSVERSE MERCATOR GRID

The transformation computation for a Transverse Mercator grid will be illustrated by reference to the Illinois state coordinate system. As will be recalled from a previous article, the Transverse Mercator projection is often described by stating that the axis of the tangent cylinder lies in the plane of the earth's equator. This projection may be illustrated by a cylinder cutting the surface of the spheroid along two small ellipses equidistant from the central meridian of the zone. The cylinder is then considered to be cut along an element and developed into a plane.

As shown in Figure 15.6, the State of Illinois is covered by two overlapping Transverse Mercator zones, the East zone and the West zone. Each zone has its own axis for y, although both axes, passing through the centers of the respective zones, are given an x value of 500,000 ft. Both zones use the same x-axis, which is situated well below the southern limit of the state and has a value of zero feet. The geographic coordinates of a point in the East zone will be transformed into plane coordinates.

The central meridian of the East zone is 88°20′ west longitude. Along this line the scale of the projection is 1 part in 40,000 parts too small. The lines of exact scale are parallel to the central meridian and situated approximately 28 miles (147,900 ft) east and west of it. To the east and west of these lines, respectively, the scale is too large. A straight line perpendicular to the y-axis at latitude 36°40′ defines the x-axis. Hence the origin of coordinates for the East zone is a point on the x-axis situated 500,000 ft west of longitude 88°20′.

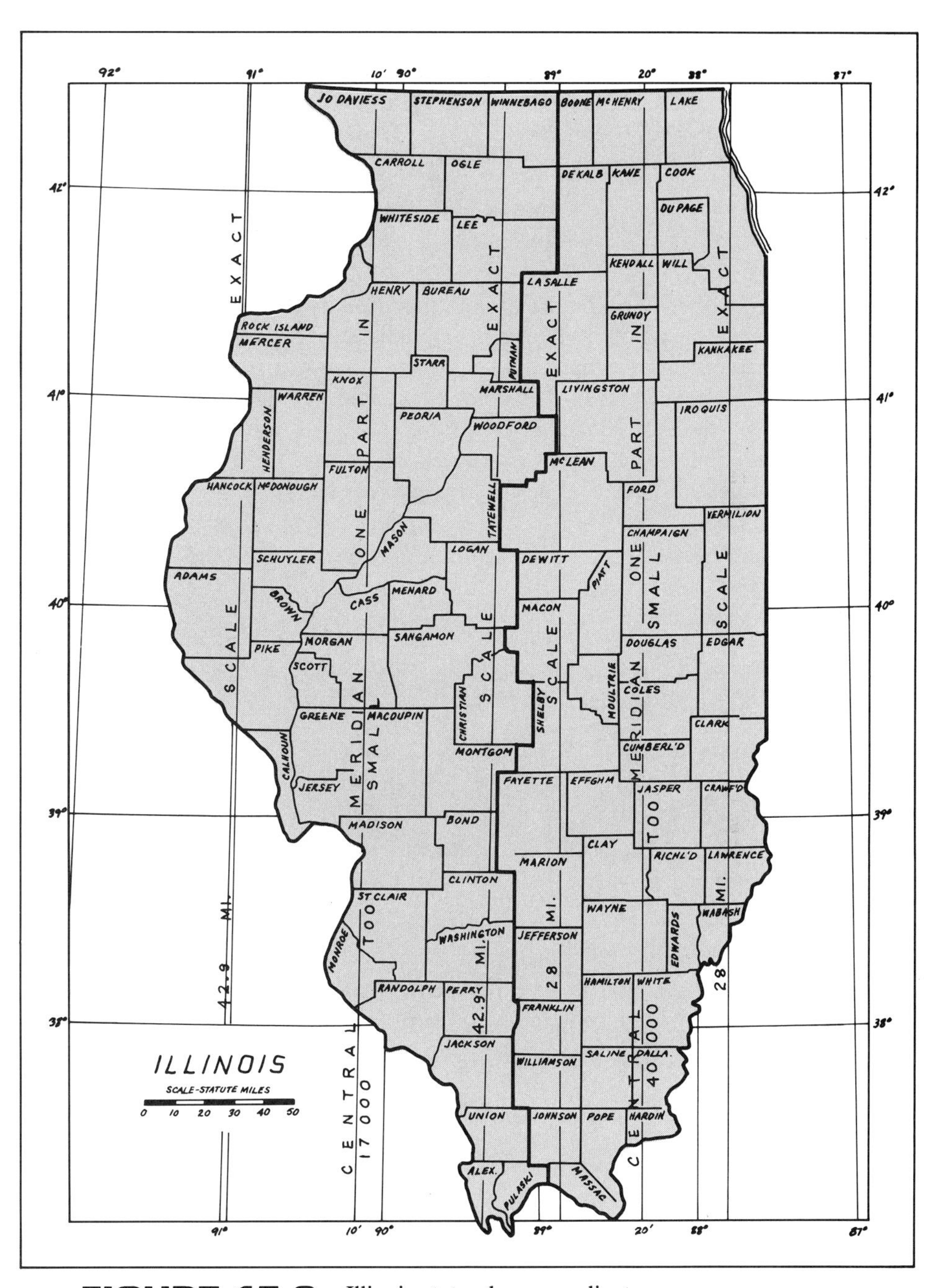

FIGURE 15.6 Illinois state plane coordinate zones

The expressions for calculating the Transverse Mercator coordinates of a point are as follows:

$$x = x' + 500{,}000$$
$$x' = H \cdot \Delta\lambda'' \pm ab$$
$$y = y_o + V\left(\frac{\Delta\lambda''}{100}\right)^2 \pm c$$

where y_0, H, V, and a are quantities based on the geographic latitude, and b and c are based on $\Delta\lambda''$; also x' is the distance the point is either east or west of the central meridian.

A typical transformation computation will now be shown. The pertinent portions of Special Publication No. 303 of the U.S. Coast and Geodetic Survey, entitled "Plane Coordinate Projection Tables—Illinois," are shown in Tables XII, XIII, and XIV in Appendix B.

EXAMPLE 15.2

Given: Station "King"
Latitude 40°43′37.202″
Longitude 88°41′35.208″
State—Illinois; Zone—East
Central Meridian: 88°20′00″

Find Transverse Mercator coordinates x and y.

SOLUTION

$$\Delta\lambda = -0°21'35.208''$$
$$\Delta\lambda'' = -1{,}295.208$$
$$\left(\frac{\Delta\lambda''}{100}\right)^2 = 167.756$$

$H = 76.992654$ (Table XII)
$V = 1.217932$ (Table XII)
$a = -0.492$ (Table XII)
$b = +2.545$ (Table XIII)

$y_0 = 1{,}478{,}725.73$ (Table XII)
$V\left(\frac{\Delta\lambda''}{100}\right)^2 = 204.32$
$c = -0.04$ (Table XIII)
$y = 1{,}478{,}930.01$ ft

$H \cdot \Delta\lambda'' = -99{,}721.50$
$ab = -1.25$
$x' = H \cdot \Delta\lambda \pm ab = -99{,}720.25$*
$x = x' + 500{,}000 = 400{,}279.75$ ft

---

* When ab is negative, decrease $H \cdot \Delta\lambda''$ numerically. If ab is positive, increase $H \cdot \Delta\lambda''$ numerically. Note also that since $\Delta\lambda''$ is negative because the station is west of the central meridian, x' is also negative.

15.15 GRID AZIMUTHS

The projection lines of a state plane coordinate system, whether Lambert or Transverse Mercator, comprise what is called a *grid* because all north-south lines are parallel with the central meridian and perpendicular to all east-west lines. Because of the convergency of the true or geographic meridians, it is obvious that the grid azimuth of a line will be the same as the geodetic azimuth only when the station at which the azimuth is expressed is situated on the central meridian. For all other lines in both systems, the grid azimuth will differ from the geodetic azimuth of the line. This difference becomes greater with increasing distance of the survey station from the central meridian and is substantially equal to the angular convergency between the central meridian and the true meridian passing through the station.

Figure 15.7 shows that for stations west of the central meridian, the grid azimuth is greater than the geodetic azimuth and, for stations east of the central meridian, the grid azimuth is less than the geodetic azimuth.

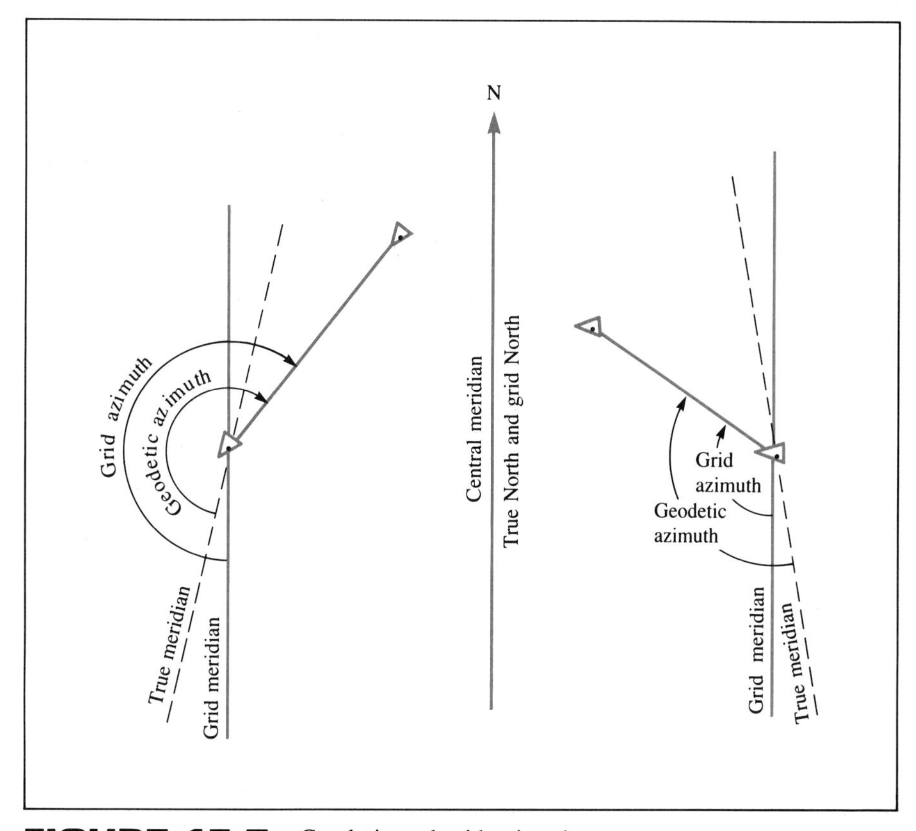

FIGURE 15.7 Geodetic and grid azimuths

The difference between the grid and geodetic azimuths is termed θ in the Lambert systems and $\Delta\alpha$ in the Transverse Mercator systems. Sometimes it is called the *mapping* or *convergence angle*. Its value is positive at points east of the central meridian and negative at points west of the central meridian.

In a state coordinate system, the forward and back grid azimuths of any line differ by exactly 180°. Grid azimuths are frequently reckoned from the south in keeping with the common practice in geodetic surveying.

15.16 COMPUTATION OF GRID AZIMUTHS

When the grid coordinates of two stations are known, the grid azimuth of the connecting line is found by a simple computation in which the tangent of the azimuth angle is equal to the difference in the x-coordinates divided by the difference in the y-coordinates. Thus, $\tan a = \Delta x/\Delta y$. The usual sign conventions apply to this computation; and the magnitude of the azimuth (that is, the quadrant in which it falls) can be readily ascertained by the use of a sketch.

Before the plane coordinates of a traverse can be computed, the grid azimuth of a reference line at the beginning station must be known so that the grid azimuths of each course can be determined. In some cases a distant triangulation or traverse station of known position can be sighted and the traverse initially oriented by this line of known grid azimuth. In other situations a sight can be taken on the azimuth mark, which is usually located at a distance of one-fourth to one-half mile from the triangulation station. The grid azimuth of the line to the azimuth mark is usually published. If it is not, it will be necessary to convert the geodetic azimuth, which is always available, to a grid azimuth by one of the following expressions:

LAMBERT SYSTEMS

$$\textit{grid azimuth} = \textit{geodetic azimuth} - \theta + \textit{second term} \tag{15.1}$$

where θ is the familiar angle of convergency between the central meridian of the projection and the true meridian through the given station, and must be used with careful regard always to sign. The second term is small and may be neglected for all situations involving third-order traverse except those in which the orientation sight is substantially greater than 5 miles.

If no azimuth mark is available, it will be necessary to determine the astronomic azimuth of the first line by solar or stellar observation and convert the astronomic azimuth to a grid azimuth by the foregoing expression. The geodetic azimuth may be considered to be equal to the astronomic azimuth.

TRANSVERSE MERCATOR SYSTEMS

$$\textit{grid azimuth} = \textit{geodetic azimuth} - \Delta\alpha - \textit{second term} \qquad (15.2)$$

where

$$\Delta\alpha'' = \Delta\lambda'' \sin \phi + g \qquad (15.3)$$

The second term is again negligible for most circumstances. The quantity g is obtained from Table XIV and should always be applied to $\Delta\lambda'' \sin \phi$ to increase it numerically.

15.17 DETERMINATION OF GEODETIC DISTANCE FROM GROUND DISTANCE

Before the grid coordinates of the traverse stations of any survey can be computed, it is necessary (1) to reduce all ground distances to mean sea level to determine their equivalent geodetic distances, and (2) to convert these geodetic distances to grid distances on the plane of the state projection.

It is presumed, first of all, that the proper corrections, including those for temperature, inclination, and error in absolute length of tape, and those pertinent to EDM operations have been applied to the observed field distances to obtain the best ground distances. These, then, are to be reduced to sea level.

This reduction is facilitated by the use of Tables VIII and IX. Table VIII lists the factors by which the ground distance at various elevations is to be multiplied in order to obtain the geodetic distance. *Elevation factors* for intermediate elevations can be found by interpolation. Table IX lists the corrections to be subtracted from a ground distance of 1,000 ft at various elevations in order to obtain the sea level distance. In the computation of both tables the mean radius of the earth was taken as 20,908,000 ft.

To illustrate the use of Tables VIII and IX, consider a ground distance of 2,165.87 ft at an elevation of 2,000 ft. From Table VIII the sea level distance is (2,165.87)(0.9999043) or 2,165.66. From Table IX the correction is (2.166)(0.0957) or 0.21. This quantity subtracted from the ground distance gives the sea level distance of 2,165.66 ft.

For most third-order traverse and for higher-order traverse at elevations under 500 ft, the sea level correction is considered insignificant. Furthermore, approximate elevations as obtained from a map are entirely satisfactory for calculating the correction. An error of 500 ft in the elevation of a traverse course will cause a proportional error in the sea level length of the course of only 1 part in 41,800 parts.

15.18 DETERMINATION OF GRID DISTANCE FROM GEODETIC DISTANCE

The ratio of the plane or grid distance of any line in a state plane coordinate system to the geodetic or sea level length of the line is known as the *scale factor*. The scale factor may be found for any line by referring to the appropriate part of the U.S. Coast and Geodetic Survey projection tables (see Tables X and XV) for the given state. For the Lambert systems, the table of scale factors is entered with the latitude. For the Transverse Mercator systems, the entering argument is the quantity x', which is the distance from the central meridian.

The scale factors are expressed both as a ratio and as the correction, in units of the seventh place of logarithms, to the sea level length of the line. The sign of the logarithm and the ratio are those to be used for changing a geodetic distance to a grid distance. In the event the conversion is made in the opposite direction, the sign should be changed and the reciprocal of the indicated scale ratio used.

Note that, when the scale factor is less than 1, it is frequently more convenient to subtract the tabular value of the scale factor from 1 and apply a subtractive correction to the geodetic length. Thus, if the scale factor is 0.9999374 and the geodetic length is 5,280.00 ft, the grid distance is equal to

$$(5{,}280.00) - (5{,}280.00)(0.0000626)$$

or

$$5{,}280.00 - 0.33 = 5{,}279.67 \text{ ft}$$

To determine the scale factors that should be used in the calculation of a traverse on either the Lambert or Transverse Mercator grid, it is essential to plot the traverse to some convenient scale, such as 1 in. equals 4,000 ft. The data for plotting the traverse will be the known grid coordinates of the initial station and the field lengths and azimuths of the various courses.

If the Lambert system is used, it will be necessary to determine the mean latitude of each traverse course. This can be done with sufficient accuracy by scaling the distance, measured parallel to grid north, from the initial station, whose latitude is known, to the middle of each course. This distance in feet can be converted with adequate accuracy to seconds of latitude by dividing by 100. The resulting latitude difference, in seconds of arc, can then be applied to the latitude of the beginning traverse station in order to determine a latitude for entering Table X and finding the scale factor by interpolation.

If the Transverse Mercator System is used, a similar plot can be employed in order to determine the mean value of x' for each traverse course. It should be recalled that if the x-coordinate of the initial traverse station is 520,000.00 ft in the Illinois system—East zone, the value of x' is $+20{,}000.00$ ft. The scaling of east-west distances from the traverse station to the middle of each course will permit the determination of satisfactory values of x' for entering Table XV. Since U.S. Geological Survey topographic maps contain marginal ticks for drawing the grid lines of the state coordinate system, such maps will be found to provide a convenient base for making a plotting of the traverse.

15.19 GRID FACTOR

Since the elevation factors of Table VIII and the scale factors of Tables X and XV are quantities by which the ground distance and geodetic distance, respectively, are multiplied to determine the grid distance, it is helpful to use the product of these factors as a means for more directly calculating the grid distance. This combined factor is called the *grid factor*.

15.20 TRAVERSE COMPUTATIONS

The use of the principles explained in the preceding articles will now be illustrated by the calculation of a traverse. After giving the essential data controlling the traverse, the major steps in the computation and adjustment of the traverse will be explained.

1. *Control Data*. The traverse shown in Figure 15.8 was executed between two fixed points of higher-order horizontal control in western Illinois. The average

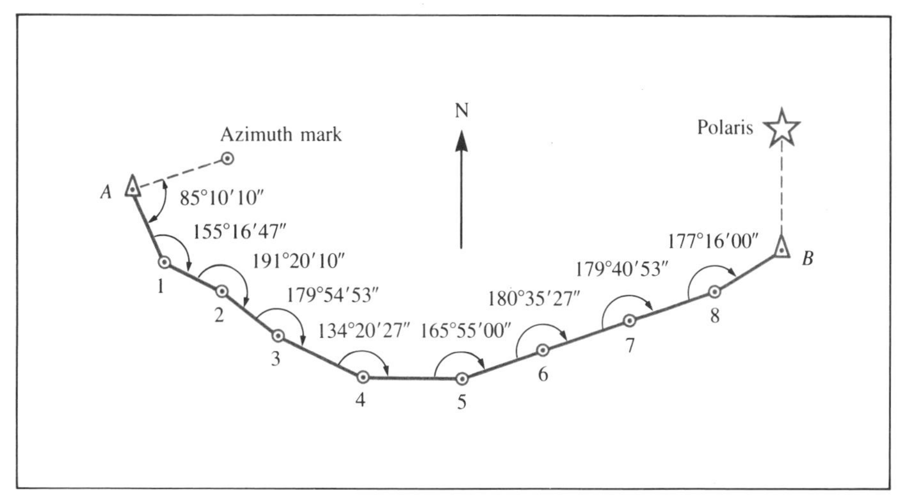

FIGURE 15.8 Traverse

elevation of the traverse above sea level is 750 ft, and the Illinois (West zone) grid coordinates of the initial and terminal points are as follows:

	x	y
A	461,577.68	1,556,540.53
B	474,718.81	1,554,910.45

The grid azimuth from *A* to the azimuth mark is 250°05′52″. At station *B* an observation was made on Polaris for azimuth. The $\Delta\alpha$ angle was applied to the astronomic azimuth as explained in Section 15.16, and the resulting grid azimuth for the line *B* to 8 was found to be 79°35′57″. These azimuth values (reckoned from south) are held fixed. The measured angles-to-right are shown in Figure 15.8, and the distances are tabulated in Figure 15.9.

2. *Computation of Grid Coordinates.* The computation process may be divided into six steps as follows: (a) computation of reference azimuths with respect to the north direction; (b) computation of grid azimuths; (c) reduction of field distances to grid distances; (d) computation of latitudes and departures; (e) computation of preliminary grid coordinates; and (f) computation of adjusted grid coordinates.

a. *Computation of reference azimuths with respect to the north direction.*

azimuth from *A* to azimuth mark	=	250°05′52″	(from south)
		−180°	
	=	70°05′52″	(from north)
azimuth from *B* to station 8	=	79°35′57″	(from south)
		+180°	
	=	259°35′57″	(from north)
azimuth from station 8 to *B*	=	79°35′57″	(from north)

b. *Computation of grid azimuths.* The results are summarized in Figure 15.10. The closure error in the preliminary azimuth is −18″. For a detailed explanation of the computation process, see Section 8.6.

Line		Corrected Field Distance (ft)	Elevation Factor (*a*)	Scale Factor (*b*)	Grid Factor (*a*) × (*b*)	Grid Distance (ft)
From Sta.	To Sta.					
A	1	754.25	0.9999642	0.9999424	0.9999066	754.18
1	2	517.12				517.07
2	3	808.11				808.03
3	4	1,617.63				1,617.48
4	5	982.61				982.52
5	6	3,165.07				3,164.77
6	7	2,354.55				2,354.33
7	8	3,296.43				3,296.12
8	*B*	1,241.74	0.9999642	0.9999424	0.9999066	1,241.62

FIGURE 15.9 Computation of grid distances

Line	Preliminary Azimuth	Correction	Corrected Azimuth
A-Azimuth mark	70°05′52″		
A-1	155°16′02″	+2″	155°16′04″
1-2	130°32′49″	+4″	130°32′53″
2-3	141°52′59″	+6″	141°53′05″
3-4	141°47′52″	+8″	141°48′00″
4-5	96°08′19″	+10″	96°08′29″
5-6	82°03′19″	+12″	82°03′31″
6-7	82°38′46″	+14″	82°39′00″
7-8	82°19′39″	+16″	82°19′55″
8-*B*	79°35′39″	+18″	79°35′57″
	−79°35′57″		

Closure error = −18″

Correction per angle $= +\dfrac{18''}{9} = +2''$

FIGURE 15.10 Summary of preliminary and corrected azimuths

c. *Reduction of field distances to grid distances.* These calculations are shown in Figure 15.9. The elevation factor is obtained by interpolation from Table VIII. The scale factor is obtained from a table for the West zone (not shown) that is similar to Table XV for a mean x' of 31,852 ft (500,000–468,148). The scale factor experiences no substantial change in the entire length of the traverse and an average value is used.
d. *Computation of latitudes and departures.* These quantities are shown in the appropriately headed columns of Figure 15.11.
e. *Computation of preliminary grid coordinates.* These are shown in Figure 15.11 as the first values of the coordinates for each station. From the closure of −0.46 ft in the x-coordinate and +0.62 ft in the y-coordinate, the linear error of closure, 0.77 ft, and the relative error of closure, 1 : 19,000, are calculated.
f. *Computation of adjusted grid coordinates.* By the use of the compass rule, the preliminary x- and y-coordinates are adjusted. The cumulative lengths of the traverse are indicated in parentheses in the column of grid distances, and the coordinate corrections are tabulated in the grid coordinate columns.

Although the traverse computation is essentially completed, it is to be mentioned that the original lengths and adjusted azimuths of all lines were inherently affected by the adjustment of the coordinates of the traverse stations. Hence, if final lengths and azimuths are to be strictly compatible with the final coordinates, they should be determined by inverse calculations, as illustrated in Example 15.3.

EXAMPLE 15.3

Find the azimuth and length of the line 7-8 from the adjusted coordinates of its termini, which are shown in Figure 15.11.

Station	Grid Azimuth from North	Grid Distance d (ft)	Dept. = $d \sin \alpha$	Lat. = $d \cos \alpha$ (Fixed)	Grid Coordinates	
					x	y
A					461,577.68	1,556,540.53
1	155°16′04″	754.18 (754)	+315.53	−685.00	461,893.21 −0.02 461,893.19	1,555,855.53 +0.03 1,555,855.56
2	130°32′53″	517.07 (1,271)	+392.90	−336.14	462,286.11 −0.04 462,286.07	1,555,519.39 +0.05 1,555,519.44
3	141°53′05″	808.03 (2,079)	+498.75	−635.73	462,784.86 −0.06 462,784.80	1,554,883.66 +0.09 1,554,883.75
4	141°48′00″	1,617.48 (3,696)	+1,000.26	−1,271.11	463,785.12 −0.12 463,785.00	1,553,612.55 +0.16 1,553,612.71
5	96°08′29″	982.52 (4,679)	+976.88	−105.11	464,762.00 −0.15 464,761.85	1,553,507.44 +0.20 1,553,507.64
6	82°03′31″	3,164.77 (7,844)	+3,134.42	+437.24	467,896.42 −0.24 467,896.18	1,553,944.68 +0.33 1,553,945.01
7	82°39′00″	2,354.33 (10,198)	+2,334.98	+301.19	470,231.40 −0.32 470,231.08	1,554,245.87 +0.43 1,554,246.30
8	82°19′55″	3,296.12 (13,494)	+3,266.65	+439.81	473,498.05 −0.42 473,497.63	1,554,685.68 +0.57 1,554,686.25
B	79°35′57″	1,241.62 (14,736)	+1,221.22	+224.15 (Fixed)	474,719.27 −0.46 474,718.81	1,554,909.83 +0.62 1,554,910.45

Linear error of closure = $\sqrt{(0.46)^2 + (0.62)^2} = 0.77$ Closure error in $x = -0.46$

Relative error of closure = $\dfrac{1}{19{,}000}$ Closure error in $y = +0.62$

FIGURE 15.11 Computation of coordinates

SOLUTION

$$\begin{array}{ll} x_8 = 473{,}497.63 \text{ ft} & y_8 = 1{,}554{,}686.25 \text{ ft} \\ \underline{x_7 = 470{,}231.08} & \underline{y_7 = 1{,}554{,}246.30} \\ \Delta x = 3{,}266.55 & \Delta y = 439.95 \end{array}$$

$$\tan \alpha = \frac{\Delta x}{\Delta y} = \frac{3{,}266.55}{439.95} = 7.4248210$$

$$\alpha = 82°19'46'' \text{ (from north)}$$

$$\text{azimuth} = 262°19'46'' \text{ (from south)}$$

$$\text{length} = \sqrt{\Delta x^2 + \Delta y^2} = 3{,}296.04 \text{ ft}$$

15.21 INTERCONVERSION OF PLANE COORDINATES

Increasing use of state plane coordinates has led to situations in which it has become necessary to devise a simple procedure for converting plane coordinates on a local system or a tangent plane system to the statewide system that has supplanted them. Other circumstances may make necessary the conversion of state coordinates from one zone to another for a point in the marginal area common to both zones of the same state, or the conversion from Lambert coordinates to Transverse Mercator coordinates or vice versa, in state boundary regions.

If the two rectangular coordinate systems, between which conversions are to be made, are both related to a common geographic system of coordinates, the transformation can be made rigidly by first making an inverse computation to the geodetic coordinates of latitude and longitude and then proceeding to convert these values to plane coordinates on the other system. Such a procedure, while affording the best possible results, is, however, a time-consuming computation. Furthermore, it may not be necessary for results of ordinary accuracy when the area involved is of such a size that significant changes of scale do not take place in the two grid systems between which the conversions are desired. When the latter condition exists, the basic transformation computation can be effected by the method of *translation and rotation.*

In Figure 15.12 point A is a survey station whose geographic position is known as are its rectangular coordinates in grid systems *1* and *2* whose axes are designated with the appropriate subscripts. The rectangular coordinates of point P in system *1* are known.

It is desired to derive expressions for converting the coordinates of P in the first system to those of the second system.

The angle of skew between the two rectangular systems is designated as β.

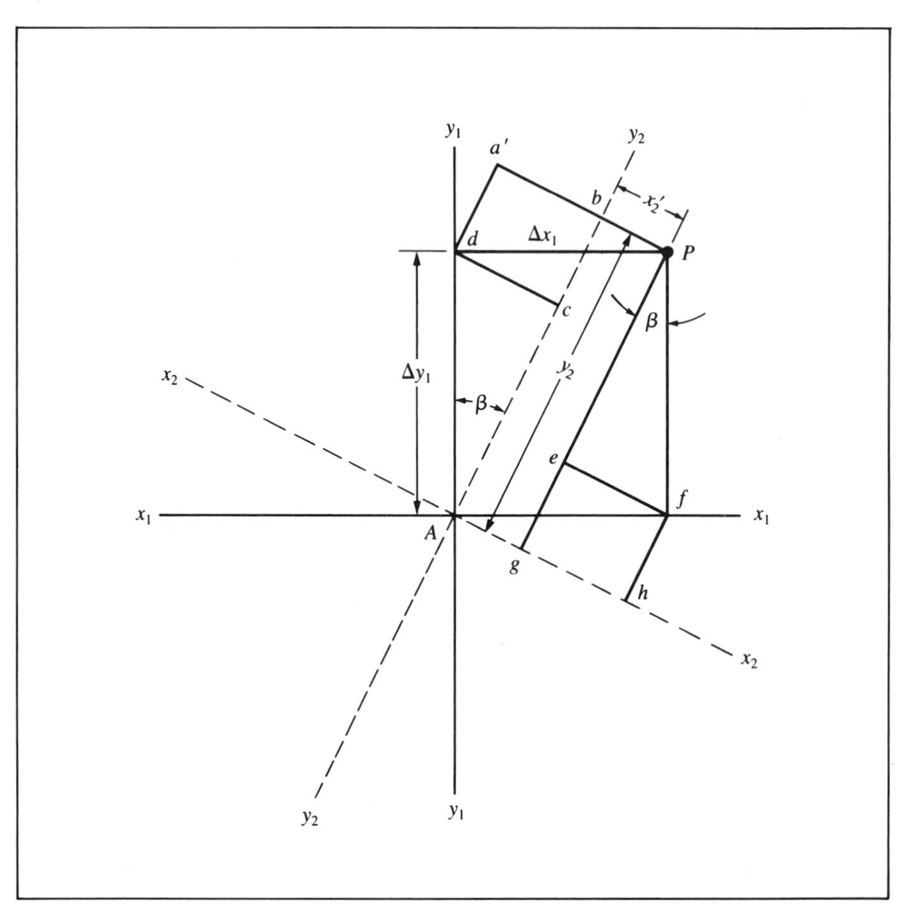

FIGURE 15.12 Interconversion of plane coordinates

The quantities x_2' and y_2' are evaluated as follows:

$$x_2' = aP - ab = aP - dc$$
$$y_2' = eP + eg = eP + fh$$

but,

$$aP = \Delta x_1 \cos \beta$$
$$dc = \Delta y_1 \sin \beta$$
$$eP = \Delta y_1 \cos \beta$$
$$fh = \Delta x_1 \sin \beta$$

where Δx_1 and Δy_1 are the differences in the x- and y-coordinates of points A and P in the first system.

Then, letting the coordinates of A in the second system be indicated by C_x and C_y and substituting for aP and dc their equivalents, we have

$$X_P = C_x + x_2'$$
$$X_P = C_x + \Delta x_1 \cos \beta - \Delta y_1 \sin \beta \qquad (15.4)$$

also,

$$Y_P = C_y + y_2'$$
$$Y_P = C_y + \Delta y_1 \cos \beta + \Delta x_1 \sin \beta \qquad (15.5)$$

Eqs. (15.4) and (15.5) can be generally applied to any of the situations mentioned in the preceding article provided the following conventions are satisfied:

1. The signs of Δx_1 and Δy_1 are those obtained by subtracting the coordinates of point A from those of point P.
2. The skew angle, β, is the θ or $\Delta\alpha$ angle of system 2 at point A minus the θ or $\Delta\alpha$ angle of system 1 at point A.
3. The quantities $\sin \beta$ and $\cos \beta$ must be multiplied by the ratio, (scale factor at A of system 2)/(scale factor at A of system 1), in order to introduce the effect of different scale factors in the two systems on the grid distances.
4. The signs of all quantities must be carefully considered.

EXAMPLE 15.4

Conversion from a Local System of Coordinates to the Lambert System

Given: At the University of Illinois Surveying Camp in northern Minnesota a local system of rectangular coordinates was originally defined by assigning to station ILLINI the arbitrary coordinates x = 10,000 ft and y = 10,000 ft. The y-axis was taken as the true meridian through station ILLINI, and the coordinates of various control stations were determined by traverse calculations using ground distances.

The geographic coordinates of ILLINI are: latitude 47°38′28.30″ and longitude 94°33′06.40″. Its Lambert coordinates in the Minnesota, North zone, are x = 1,642,141.00 and y = 419,598.00. The average elevation of the terrain is 1,350 ft above sea level.

From Table VIII the elevation factor is 0.9999354.

From Table X the scale factor is 0.9999090 and from Table XI the θ angle is −1°04′34″. Station ILLINI is west of the central meridian of the North zone; therefore, θ is negative.

Find the Lambert coordinates of station ALSIP whose local coordinates are: x = 9,114.10 ft and y = 11,485.60 ft.

SOLUTION

1. The grid factor = (0.9999354)(0.9999090)
   = 0.9998444

2. $\sin \beta = -0.01878058$
   $\cos \beta = +0.99982363$
   corrected $\sin \beta = -0.01877766$
   corrected $\cos \beta = +0.99966806$
3. $\Delta x_1 = 9{,}114.10 - 10{,}000 = -885.90$
   $\Delta y_1 = 11{,}485.60 - 10{,}000 = +1{,}485.60$
4. $\Delta x_1 \cos \beta = (-885.90)(0.99966806) = -885.61$
   $\Delta x_1 \sin \beta = (-885.90)(-0.01877766) = +16.64$
   $\Delta y_1 \cos \beta = (1{,}485.60)(0.99966806) = +1{,}485.11$
   $\Delta y_1 \sin \beta = (1{,}485.60)(-0.01877766) = -27.90$
5. $X_2 = C_x + x_2' = C_x + \Delta x_1 \cos \beta - \Delta y_1 \sin \beta$
   $= 1{,}642{,}141.00 - 885.61 + 27.90$
   $X_2 = 1{,}641{,}283.29$
   $Y_2 = C_y + y_2' = C_y + \Delta y_1 \cos \beta + \Delta x_1 \sin \beta$
   $= 419{,}598.00 + 1{,}485.11 + 16.64$
   $Y_2 = 421{,}099.74$ ft

EXAMPLE 15.5

Conversion from East Zone Coordinates to West Zone Coordinates of the Illinois Transverse Mercator System

Given: Traverse station BOWIE is situated in LaSalle County, Illinois, in the marginal area common to both the East and West zones. the coordinates of this station are as follows:

Geodetic: latitude 41°20′41.324″
longitude 89°05′46.141″

East zone: $x = 290{,}536.87$ ft
$y = 1{,}704{,}738.42$ ft

West zone: $x = 793{,}945.96$ ft
$y = 1{,}705{,}573.86$ ft

The East zone coordinates for traverse station JOY are $x = 209{,}514.65$ and $y = 1{,}702{,}533.25$.

The $\Delta\alpha$ angles have been computed with reference to the central meridians of the East and West zones, which are in longitude 88°20′ and 90°10′, respectively. The $\Delta\alpha$ angle for the East zone is −0°30′17.3″; for the West zone it is +0°42′23.2″.

From Special Publication No. 303, U.S. Coast and Geodetic Survey, the scale factors at BOWIE in the East and West zones are 1.0000252 and 1.0000400, respectively.

Find the West zone coordinates of station JOY.

SOLUTION

1. $\beta = \Delta\alpha_2 - \Delta\alpha_1 = +0°42'23.2'' - (-0°30'17.3'')$
   $= +1°12'40.5''$
2. $\dfrac{\text{Scale factor on zone 2 (west)}}{\text{Scale factor on zone 1 (east)}} = \dfrac{1.0000400}{1.0000252}$
   $= 1.0000148$

3. $\sin \beta = +0.02113872$
   $\cos \beta = +0.9997765$
   Corrected $\sin \beta = +0.02113903$
   Corrected $\cos \beta = +0.99979135$
4. $\Delta x_1 = 290{,}514.65 - 290{,}536.87 = -22.22$
   $\Delta y_1 = 1{,}702{,}533.25 - 1{,}704{,}738.42 = -2{,}205.17$
5. $X_2 = C_x + \Delta x_1 \cos \beta - \Delta y_1 \sin \beta$
   $= 793{,}945.96 - 22.22 + 46.62$
   $X_2 = 793{,}970.36$ ft
   $Y_2 = C_y + \Delta y_1 \cos \beta + \Delta x_1 \sin \beta$
   $= 1{,}705{,}573.86 - 2{,}204.71 - 0.47$
   $Y_2 = 1{,}703{,}368.68$ ft

Although the West zone coordinates just computed for station JOY agree exactly with the values derived from a transformation (not shown) from geographic coordinates, such exact agreement is not always to be expected. The error in the coordinates will depend somewhat on the distance to the second point and on the relative change in scale factors between the first and second point.

15.22 SOME USES OF STATE PLANE COORDINATE SYSTEMS

The use of state plane coordinates provides all the advantages of geodetic position data from which they were derived and to which they are permanently related without introducing any of the difficulties associated with geodetic computations. State plane coordinates are conveniently determined through the application of little more than the ordinary field and office procedures of plane surveying. Yet they serve as immutable expressions of horizontal position which can be used for a wide variety of engineering surveys. Some specific uses of state plane coordinates will be described briefly.

The centerline surveys for long route projects, such as a toll highway, are frequently based on state plane coordinates. The use of such horizontal control makes possible the utilization of several survey parties working in widely separated portions of the project with the positive assurance that all survey data can be correlated.

The use of state plane coordinates in extensive city and county mapping projects permits the execution of separate and detached surveys with the knowledge that, when component surveys meet, they will do so harmoniously without overlaps, gaps, or offsets at the junction points.

State plane coordinates provide a means for constructing a simple rectangular projection for the preparation of maps covering large areas and for plotting data thereon. Examples of such maps are those prepared by aerial mapping organizations for state highway departments in connection with extensive highway projects. Figure 15.13 shows a portion of a table (U.S. Coast and Geodetic Survey Special Publication No. 332, Reference 15.4) that gives values of state plane coordinates at $2\frac{1}{2}$ minute intervals of latitude and longitude.

The exchange and use of survey and map information is greatly facilitated by the use of state plane coordinates. However, a multiplicity of local coordinate systems, like an excess of level datum planes, prevents the widest possible utilization of control information.

The use of state plane coordinates makes possible the certain recovery of any lost or destroyed survey point, whether it be a property corner, an important construction control monument, or a triangulation station. The field recovery procedure consists of running a traverse from the nearest point of known position to a temporary marker in the vicinity of the lost

Longitude	Latitude 37 45 00 *x*	Latitude 37 45 00 *y*	Latitude 37 47 30 *x*	Latitude 37 47 30 *y*
86 00 00	403 620.96	91 201.00	403 675.02	106 373.01
02 30	391 573.54	91 246.59	391 634.36	106 418.62
05 00	379 526.10	91 297.55	379 593.68	106 469.60
07 30	367 478.65	91 353.87	367 552.99	106 525.94
86 10 00	355 431.18	91 415.56	355 512.28	106 587.65
12 30	343 383.69	91 482.61	343 471.55	106 654.73
15 00	331 336.18	91 555.02	331 430.80	106 727.17
17 30	319 288.65	91 632.80	319 390.03	106 804.98
86 20 00	307 241.10	91 715.95	307 349.23	106 888.16
22 30	295 193.51	91 804.46	295 308.41	106 976.70
25 00	283 145.91	91 898.33	283 267.56	107 070.61
27 30	271 098.27	91 997.57	271 226.68	107 169.89
86 30 00	259 050.60	92 102.18	259 185.78	107 274.53
32 30	247 002.90	92 212.15	247 144.83	107 384.55
35 00	234 955.17	92 327.49	235 103.86	107 499.93
37 30	222 907.40	92 448.20	223 062.85	107 620.69
86 40 00	210 859.59	92 574.28	211 021.81	107 746.81
42 30	198 811.74	92 705.72	198 980.72	107 878.30
45 00	186 763.86	92 842.53	186 939.60	108 015.16
47 30	174 715.93	92 984.70	174 898.43	108 157.39
86 50 00	162 667.96	93 132.24	162 857.22	108 304.98
52 30	150 619.94	93 285.15	150 815.96	108 457.95
55 00	138 571.88	93 443.44	138 774.66	108 616.29

FIGURE 15.13 Grid intersection coordinates (National Geodetic Survey)

station. A comparison of the computed coordinates of the temporary marker and those of the lost station provide a means for calculating the length and azimuth of the connecting line.

The use of state plane coordinates can eliminate the need for a random line in land surveys. A line between two stations that are not intervisible but can be tied to the state coordinate system can be run out directly with a minimum of clearing in brushy country.

State plane coordinates can be used to express the positions of important engineering structures, such as oil wells, including those in tidal waters, and transmission towers; and they are becoming more widely utilized in pipeline location and description as well as in the planning, construction, and expansion of large industrial plants. Here, good vertical control is likewise very important in the protection of underground utilities and process interconnections.

The number of engineering agencies that use the various state plane coordinate systems as the basis for all control surveys is increasing. These coordinate systems provide the computing and plotting bases for a wide variety of special surveys and all kinds of planimetric and topographic maps.

Occasionally, misunderstandings may develop concerning the requirements of the state enabling acts establishing and recognizing the legality of such coordinate systems. Such legislation usually makes mandatory the use of accurate surveying methods in extending horizontal control from a triangulation station or a traverse point. This provision, however, applies only to court recognition of the use of state plane coordinates in deeds and other legal instruments. Highway and construction surveys, regardless of their quality, can be based on a state plane coordinate system. Naturally, their usefulness will be increased if they are, at least, of third-order accuracy. Plane coordinate projection tables are now available for every state in the nation and can be procured from the Superintendent of Documents, Government Printing Office.

15.23 SOME PRACTICAL CONSIDERATIONS IN USING STATE PLANE COORDINATES

1. *Conflict between Grid and Ground Distances.* In the selection of the projection and the design of the state plane coordinate system based upon it, a major consideration has been to limit the difference between the sea level and grid distances to 1/10,000 part of the length of a line. The effect of high elevation of the ground surface above mean sea level could further accentuate this difference. Thus, an anomalous condition is developed because the actual ground distance cannot be the same as the distance resulting from an inverse calculation between the grid coordinates of two control points in a state plane coordinate system.

This conflict between the ground distance and grid distance may be inconsequential or of some significance depending on a number of circumstances, such as elevation of the terrain, location of the survey line in the state zone as affecting the value of the scale factor, and the accuracy requirements of the engineering study or project layout operations.

As an elementary example consider the situation for a highway centerline survey as follows:

EXAMPLE 15.6

The station numbers on the same tangent of two *POT*'s (points on tangent) were determined by careful EDM measurements to be 231 + 17.63 and 329 + 81.42. The average sea level elevation of the line is 4,570 ft and the scale factor is 0.9999752. Calculate the difference between the ground and grid lengths.

SOLUTION

$$\begin{array}{r} 329 + 81.42 \\ -231 + 17.63 \\ \hline 98 + 63.79 \end{array}$$

Ground distance = 9,863.79 ft

From Table VIII, the elevation factor is 0.9997814.

Grid factor = 0.9997814 × 0.9999752 = 0.9997567

Grid distance = 9,863.79 × 0.9997567 = 9,861.39

Difference = 9,863.79 − 9,861.39 = 2.40 ft

This example, although somewhat extreme, can serve, also, to indicate the discrepancy that would develop if an inverse calculation of the distance between two survey points of known state plane coordinate positions were made and the resulting grid distance compared with the measured ground or project distance.

When the results of a specialized survey of high quality such as for a long, multi-span bridge, are expressed in terms of state plane coordinates and grid lengths and these are portrayed on an engineering plan, it may be advisable to indicate by a suitable note thereon that all displayed grid distances and those calculated from coordinates should be divided by a designated grid factor in order to obtain field lengths. Hence, whenever project dimensions and positions are defined with respect to a system of state plane coordinates, it is essential to recognize the inherent differences between field distances and grid distances and take appropriate steps to reconcile the two.

It should be emphasized that all geodetic data published by the National Geodetic Survey are referred to sea level as are all the state grid systems except the Lambert projection for Michigan. In an effort to reduce the difference between grid and ground distances, the state coordinate system in Michigan is related to the surface of a spheroid 800 ft above sea level, which is the approximate mean elevation for the state.

2. *Calculation of Areas from Grid Coordinates.* This topic was treated in Section 10.2. Probably only in rare cases would this modified area calculation be of major significance.

3. *Survey Points in Overlap Areas.* The civil engineer and land surveyor may sometimes initiate or close horizontal control surveys on a monumented point in the overlap areas between two zones in a given state. It is extremely important that the coordinates be correctly identified with respect to zone and that the proper value of the mapping or convergence angle be used.

In some cases surveys may be conducted along state borders, and state grid systems based on different projections will be involved. Depicted in Figure 15.14 are the horizontal control data for a point along the Arizona–California Boundary. Note that three pairs of coordinates and three values of the mapping angle are shown. They are for the West zone of the Arizona system, East zone of the Nevada system, and Zone V of the California system. The first two relate to Transverse Mercator projections and the third to a Lambert projection.

HORIZONTAL CONTROL DATA
by the
Coast and Geodetic Survey
NORTH AMERICAN 1927 DATUM

PUBLISHED AND PRINTED BY:
U.S. DEPARTMENT OF COMMERCE
COAST AND GEODETIC SURVEY
WASHINGTON D.C.

U.S. DEPARTMENT OF COMMERCE
COAST AND GEODETIC SURVEY

DESCRIPTION OF TRIANGULATION STATION

NAME OF STATION BDRY. REF. PT. NO. 1B   STATE Arizona   COUNTY Mohave
CHIEF OF PARTY L.G. Burdine   YEAR 1964   DESCRIBED BY D. R. Tomlinson

NOTE: HEIGHT OF TELESCOPE ABOVE STATION MARK 1.35 METERS † HEIGHT OF LIGHT ABOVE STATION MARK METERS

1a SURFACE-STATION MARK
7b UNDERGROUND-STATION MARK

DISTANCES AND DIRECTIONS TO AZIMUTH MARK, REFERENCE MARKS AND PROMINENT OBJECTS WHICH CAN BE SEEN FROM THE GROUND AT THE STATION

NOTE	OBJECT	BEARING	DISTANCE FEET	DISTANCE METERS	DIRECTION ° ′ ″
	BDRY. REF. PT, NO. 1A				0 00 00.0
11b	R.M. 1	NNE′	95.56′	29.126′	62 54 43
11b	R.M. 2	ESE′	61.59′	18.772′	159 17 04
	SOTO		14.97	4.563	74 08 44.7

The station is about 11 miles north of Needles, California, near the state boundaries of California, Nevada and Arizona, and on the east bank of the Colorado River. This station was established to determine fixed point number one.

To reach the station from the intersection of Front and H Streets at the northwest corner of the city hall in Needles, California, go north on H Street, crossing the Santa Fe Railroad, for 0.15 mile to a T-road and a golf course on the north side of the intersection. Turn right and go easterly on paved road for 0.4 mile to a T-road. Turn right, south and follow along levee road for 1.0 mile to a bridge over the Colorado River. Continue ahead, crossing the bridge for 0.1 mile to a T-road. Turn left and go northwest on paved road 1.0 mile to a road fork. Take the right fork and go north on the paved road for 8.2 miles to a crossroad and sign "OATMAN–DAVIS DAM" on the right. Turn left and go west on a gravel road for 1.6 miles to a fork. Take the right fork and go northwest on a gravel road for 0.45 mile to a levee road. Turn right and go north on the levee road for 1.8 miles to a side road on the left. Turn left and go west for 0.1 mile to a T-road on the east river bank. Turn right and go north on the east bank river road for 0.1 mile to a turn-out and the station on the right.

Station marks are standard disks stamped POINT NO 1 B 1964. The surface disk is set in the top of a round concrete post projecting 2 inches. It is 34 feet east of the centerline of the river road and 4 feet north of the south edge of the fill of the turn-out. The underground disk is set in an irregular mass of concrete 38 inches below the ground surface.

Reference mark 1 is a standard disk, stamped SOTO NO 1 1964, set in the top of a round concrete post projecting 6 inches. It is 84 feet east of the centerline of the road and about 4 feet lower than the station.

Reference mark 2 is a standard disk, stamped SOTO NO 2 1964, set in the top of a round concrete post projecting 6 inches. It is 68 feet east of the centerline of the road and about 4 feet lower than the station.

A traverse connection was made to triangulation station SOTO. The distance being 4,563 meters or 14.97 feet, north.

The geodetic azimuth and distance to BDRY. PT. NO. 1 are:

Azimuth	Distance meters	Distance feet
135° 09′ 51.″8	273.999	898.95

Detailed description

ADJUSTED HORIZONTAL CONTROL DATA

NAME OF STATION BDRY REF PT NO 18
STATE ARIZONA   SECOND ORDER
LOCALITY ARIZONA-CALIFORNIA BOUNDARY
SOURCE G-13386   FIELD SKETCH ARIZ 52-1

GEODETIC LATITUDE	35° 00′ 00.″12984	ELEVATION	149.5 METERS
GEODETIC LONGITUDE	114 37 48.04945		490 FEET

STATE COORDINATES (*Feet*)

STATE & ZONE	CODE	X	Y	θ (OR Δα) ANGLE
ARIZ. W.	0203	236,442.79	1,456,470.95	− 00° 30′ 17″
CALIF. V	0405	3,009,079.90	562,807.45	+ 01 55 15
NEV. E.	2701	785,502.87	92,356.51	+ 00 32 49

TO STATION OR OBJECT	GEODETIC AZIMUTH (*From south*)	PLANE AZIMUTH (*From south*)	CODE
BDRY REF PT NO 1A	134° 28′ 29.″5	134° 58′ 47″	0203
BDRY REF PT NO 1A	134 28 29.5	132 33 15	0405
BDRY REF PT NO 1A	134 28 29.5	133 55 41	2701

FIGURE 15.14 Horizontal control data (National Geodetic Survey)

15.24 FUTURE DEVELOPMENTS

A readjustment of the national horizontal control network is expected to be completed in late 1984 (see Section 11.18). The readjustment will cause the following changes:

1. A *new* national survey datum, which has been designated as the North American Datum of 1983 (NAD 1983). This datum will replace the North American Datum of 1927 (NAD 1927).
2. *Revised* geographic positions (latitude and longitude) for all the horizontal control points (about 250,000 points) included in the national control network.
3. *Revised* state plane coordinates for some of the horizontal control points in the national network.
4. *Revised* formulas for the calculation of some state plane coordinates.
5. *Revised* definitions of some zones of the state plane coordinate systems.

Thus, although many of the concepts discussed in this chapter will remain valid after 1984, the computation formulas and tables may be superseded by new ones in the near future.

Computer programs for the transformation of state plane coordinates to longitude and latitude or vice versa, and for conversion from one zone to another, have been widely available for many years. New program softwares will undoubtedly be available shortly after the completion of the readjustment project.

PROBLEMS

15.1 A second-order triangulation station in northern Minnesota is situated in latitude 47°35′26.870″ and longitude 94°34′01.505″. Determine (a) the θ angle and (b) the scale factor at this station.

15.2 Compute the Lambert grid coordinates of the triangulation station of Problem 15.1.

15.3 A second-order traverse station in east-central Illinois is situated in latitude 40°46′12.695″ and longitude 88°45′26.359″. Compute the Transverse Mercator (East zone) coordinates of this point.

15.4 For the station of Problem 15.3 determine (a) the $\Delta\alpha$ angle (to the nearest second) and (b) the scale factor.

15.5 The geodetic (sea level) distance between stations Madison (elev. 219.530 m) and Tonica (elev. 205.340 m) on the first-order traverse line of the NGS in central Illinois is 7,337.680 m. Calculate the field (ground) distance between these points.

15.6 The state plane coordinates, in feet, of two survey stations, *A* and *B*, are as follows:

	x	y
A	527,946.70	1,250,583.24
B	528,198.77	1,300,758.10

Compute (a) the grid azimuth (reckoned from south) to the nearest second of the line from station *A* to station *B*, and (b) the grid distance between these points.

15.7 If the geodetic azimuth to the azimuth mark of Problem 15.1 is 27°18′11″, what is the grid azimuth?

15.8 If the geodetic azimuth to the azimuth mark of Problem 15.4 is 329°14′56″, what is the grid azimuth?

15.9 The grid length of a line as calculated from the state plane coordinates of its termini is 4,010.56 ft. If the scale factor is 0.9999800 and the line is situated at an elevation of 4,500 ft, find the ground length.

15.10 Using the adjusted coordinates of the points 4 and 5 of Figure 15.11, compute the final grid bearing and grid length of this line.

15.11 The state plane coordinates of two *POT*'s situated along the same tangent (centerline) of a proposed high-speed railroad were determined by careful triangulation connections to nearby first-order horizontal control markers. The resulting coordinates were as follows:

Point	x(ft)	y(ft)
1	637,427.81	1,281,658.43
2	723,402.14	1,270,400.65

The station numbers of the two points are 637 + 14.22 and 1,504 + 36.33. The average sea level elevation of the line is 1,840 ft and the mean scale factor is 0.9999052.

Calculate the discrepancy between the project (ground) length of the line as determined from the stationing and the ground length deduced from the grid distance as modified by scale and elevation effects. Also, express the discrepancy as a fraction (like 1/16,400) of the distance.

15.12 The area of a tract of valuable land in northern Minnesota was determined from the state plane coordinates of points on its perimeter to be 111.327 acres. The average latitude of the tract is 47°48′ and the elevation is 1,125 ft. Calculate the actual ground area.

REFERENCES

The following publications are available from National Geodetic Information Center, National Ocean Survey, National Oceanic and Atmospheric Administration, Rockville, MD 20852:

15.1 Dracup, Joseph F. *Fundamentals of the State Plane Coordinate Systems,* National Oceanic and Atmospheric Administration geodetic publication, 60 pp., 1974.

15.2 Holdahl, Jeanne H., and Dubester, Dorothy E. *A Computer Program to Adjust a State Plane Coordinate Traverse by the Method of Least Squares,* a National Oceanic and Atmospheric Administration geodetic publication, 232 pp., 1972.

15.3 Simmons, Lansing G. *Survey of the Boundary between Arizona and California,* U.S. Coast and Geodetic Survey Technical Bulletin No. 27, 1965.

15.4 U.S. Coast and Geodetic Survey Special Publications:

No. 71 *Relation Between Plane Rectangular Coordinates and Geographic Positions,* 1932.

No. 235 *The State Coordinate Systems,* 1974.

No. 246 *Sines, Cosines, and Tangents, Ten Decimal Places with Ten-Second Intervals,* 0°–6°, 1971.

No. 264 *Plane Coordinate Projection Tables—Minnesota (Lambert),* 1952.

No. 303 *Plane Coordinate Projection Tables—Illinois (Transverse Mercator),* 1953.

No. 332 *Plane Coordinate Intersection Tables (2½-Minute) Indiana,* 1955.

ASTRONOMICAL DETERMINATION OF DIRECTION

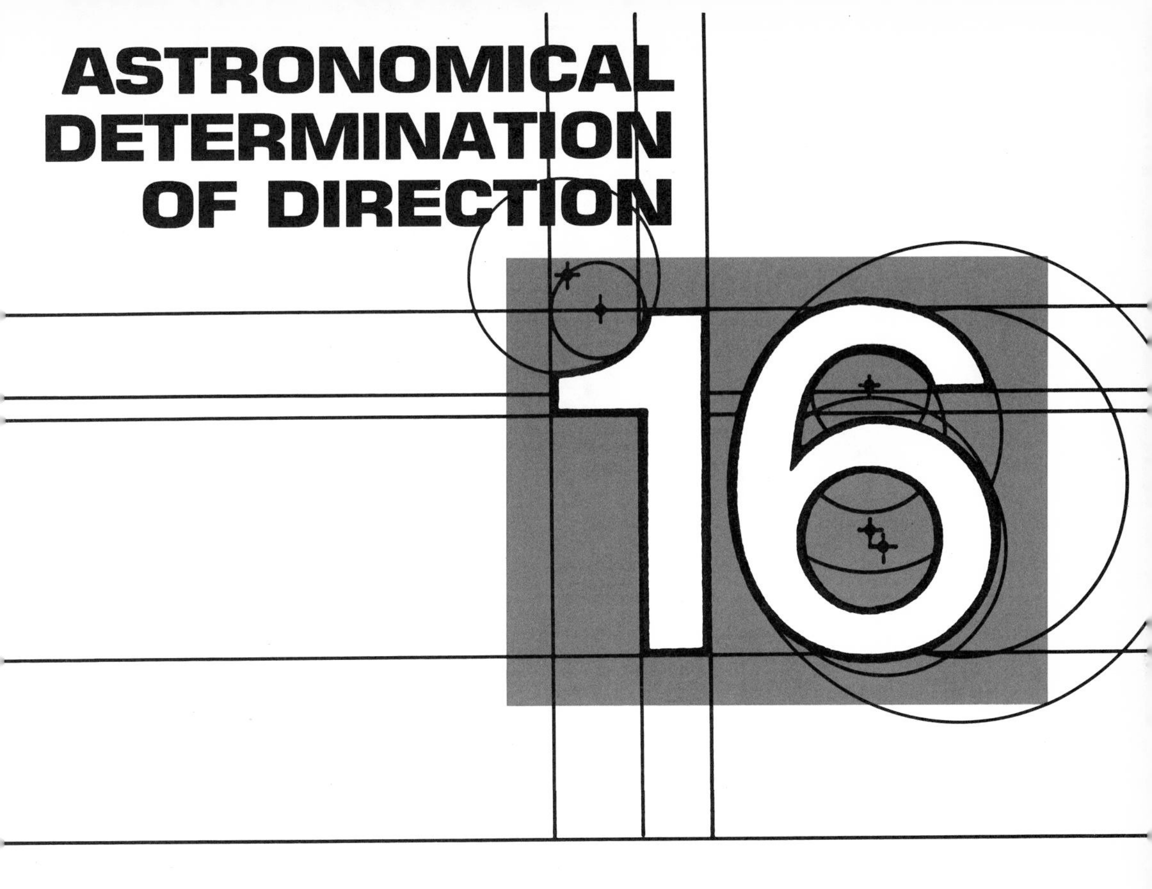

16.1 INTRODUCTION

The direction of a line can be determined by: (1) traversing from a line of known azimuth; (2) using a magnetic compass; (3) using a gyrotheodolite; or (4) making astronomical observations. The first three methods have already been discussed in previous chapters. This chapter will deal with astronomical methods for the determination of directions.

Astronomical determinations of direction can be made with varying degrees of accuracy, depending on the quality of the equipment, the techniques employed, and the skill of the observer.

Following an introduction to the fundamental concepts of engineering astronomy, the field and office procedures for determining the true direction of a line by astronomical observations will be presented. Although both stellar and solar sightings can be taken to fix direction, primary emphasis will be on the method involving the North Star, Polaris. It is presumed an optical transit or a theodolite is available and that an uncertainty of not more than $\pm 10''$ in the derived azimuth of a line is satisfactory. To attain greater accuracy, procedures that are more refined than those described here must be used.

Astronomic observations for azimuth are made to prevent the accumulation of errors in triangulation and traverse, ascertain the true directions of land boundaries, orient radar antennae, and for numerous other purposes.

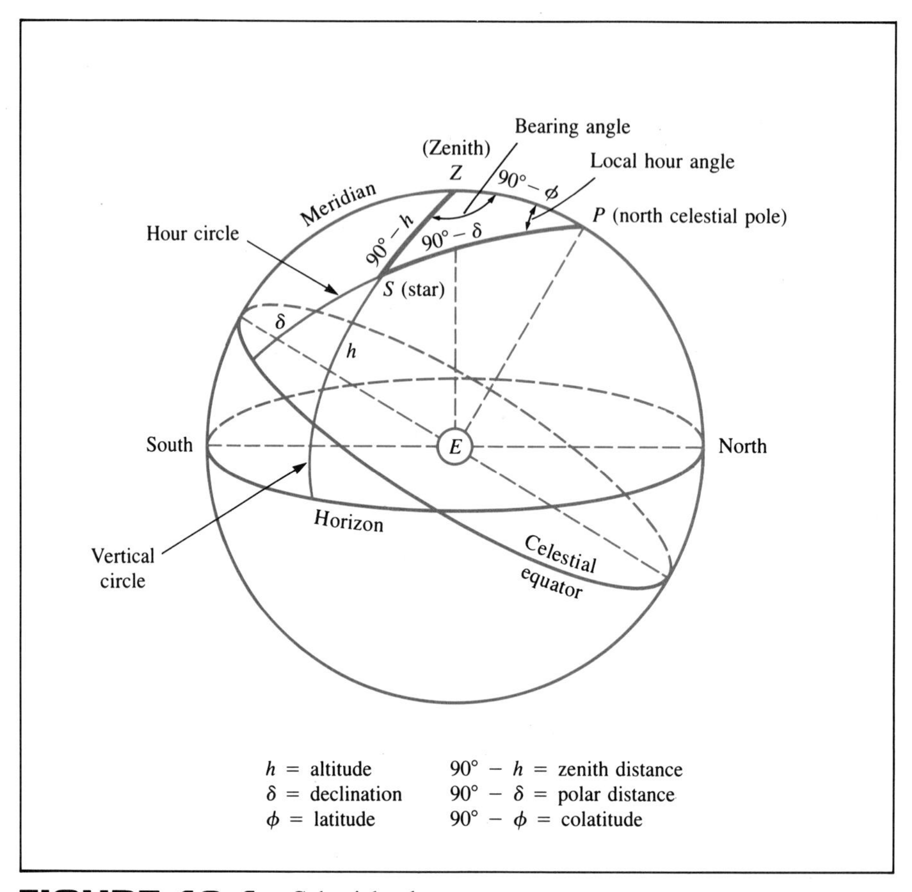

FIGURE 16.1 Celestial sphere

It is occasionally necessary to make an astronomic determination of the azimuth of a line before its state plane or grid azimuth can be calculated.

16.2 CELESTIAL SPHERE

The procedures of practical field astronomy are greatly simplified and an understanding of space relationships more easily achieved by the concept of the celestial sphere. The stars, which are at enormous distances from the earth, are considered to be fixed on the inner surface of a sphere of infinite radius whose center is the earth. It is immaterial in this discussion whether the observer's position on the earth's surface, or the center of the earth, is considered as the center of the celestial sphere, since the radius of the earth is utterly insignificant when compared with the distance to the stars.

Figure 16.1 depicts the celestial sphere whose essential elements are defined as follows:

1. *Great Circle.* A great circle of a sphere is the trace on its surface of the intersection of a plane passing through the center of the sphere.

2. *Celestial Poles.* The north and south geographic poles are at the extremities of the polar axis of the earth. The celestial poles are situated at the points where the earth's polar axis extended intersects the celestial sphere.

3. *Celestial Equator.* The celestial equator is a great circle whose plane is perpendicular to the celestial polar axis.

4. *Zenith.* The zenith is that point on the celestial sphere where a plumb line extended upward intersects it. The opposite point is termed the *nadir.*

5. *Horizon.* The horizon is the great circle that is situated halfway between the observer's zenith and nadir points and whose plane is perpendicular to the plumb line. Azimuth is measured in the plane of the horizon.

6. *Vertical Circle.* A vertical circle is the great circle passing through the observer's zenith and any celestial object. Obviously it must be perpendicular to the horizon. The altitude of a heavenly body above the horizon is measured along a vertical circle.

7. *Hour Circle.* An hour circle is the great circle joining the celestial poles and passing through any celestial body. Thus there is an hour circle for each object in the sky. Of necessity, an hour circle must be perpendicular to the celestial equator.

8. *Meridian.* The observer's celestial meridian is both an hour circle and a vertical circle, since it passes through the celestial poles as well as the zenith and nadir points.

The various angular quantities and the *PZS* astronomical triangle will be explained in subsequent sections.

16.3 APPARENT MOTION OF CELESTIAL SPHERE

The earth is a planet that completes a rotation from west to east about its polar axis in approximately 24 hours and effects a circuit about the sun in approximately 365 days.

For the purposes of astronomical observations, the earth is regarded as stationary and the celestial sphere is assumed to rotate about it from east to west. This concept of motion of the celestial sphere will be helpful in understanding the apparent movement from east to west of the celestial bodies and particularly in making clear the apparent daily rotation of the pole star, Polaris.

16.4 TIME

Time and the conversion of one kind of time to another are of great significance in all aspects of astronomy, since the daily motion of the stars and their positions at any instant are intimately related to time.

If the movement of the sun from east to west is considered, it is recognized that its crossing of each meridian signalizes noon for all points on that meridian and that meridians to the east have already experienced noon, while those to the west have yet to experience noon.

Figure 16.2 shows the standard time zones that have been established in the United States. It will be noticed that the central meridians of the time belts differ by 15° of longitude, or 1 hour of time, since the sun completes an apparent revolution about the earth every 24 hours. Although the instant of noon (12 o'clock) on any observer's meridian is nominally that corresponding to the appearance of the sun exactly over the meridian, it is obvious that the use of a multiplicity of local times would result in endless confusion. Hence, all timepieces within a given time zone are set to keep the same kind of time as that which pertains to the zone's central meridian only. This kind of time is termed *standard time*.

Since the sun apparently moves from east to west, a given moment such as noon is experienced earlier in the zones to the east. For example, 12 o'clock Central Standard Time (CST) is 1 P.M. Eastern Standard Time (EST) and 11 A.M. Mountain Standard Time (MST).

The prime meridian for reckoning world longitude passes through the observatory at Greenwich, England. Thus the longitude of the central meridian of the Greenwich standard time belt is 0° or 0^h. The standard time

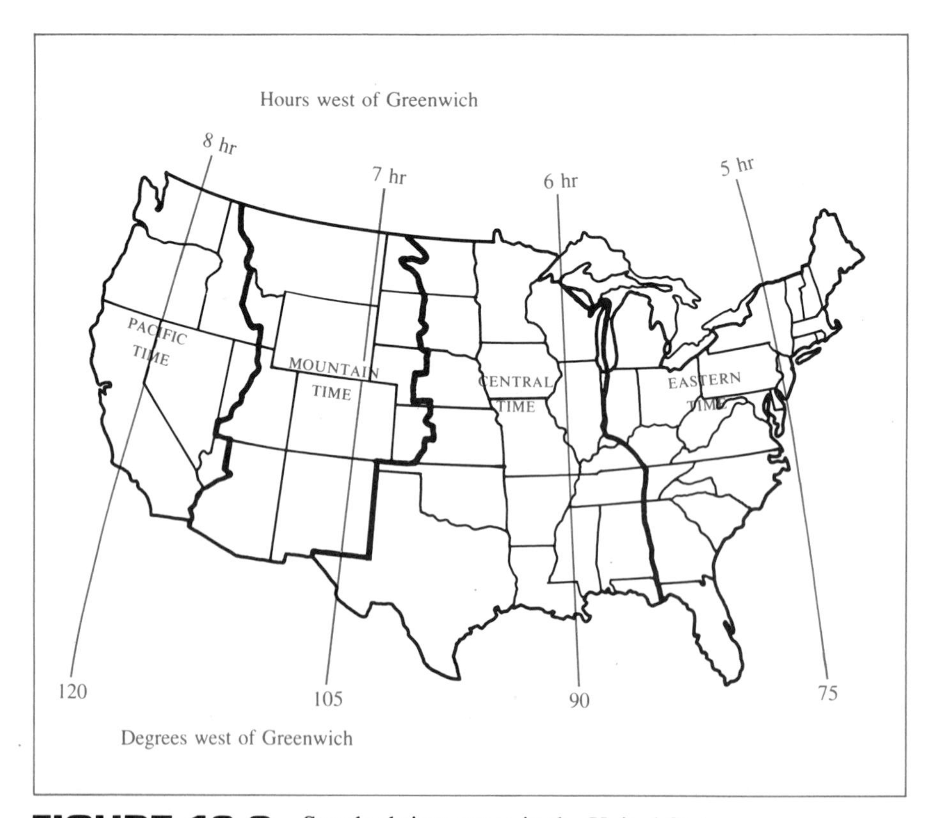

FIGURE 16.2 Standard time zones in the United States

for this zone is called *Greenwich Civil Time* (GCT) (also Greenwich Mean Time or Universal Time) and is widely used in astronomical work. GCT is obtained merely by adding to the standard time of any zone in the United States the appropriate number of hours of longitude that its central meridian is removed from Greenwich.

Daylight Time in any zone is equivalent to standard time in the zone next removed to the east; for example, 9 A.M. Central Daylight Time (CDT) is the same instant as 9 A.M. EST.

In astronomical calculations the hours of the day are numbered consecutively from 0 to 24 beginning at midnight. Thus, an event that occurs at 3:00 A.M., September 1, would be recorded as 3^h00^m September 1, and an event 12 hours later would be recorded as 15^h00^m September 1. The designations of A.M. and P.M. are unnecessary and are frequently productive of serious errors in the expression of time.

EXAMPLE 16.1

Find the GCT of an observation made at 9:50 P.M., CST.

SOLUTION

```
  9:50 P.M., CST
+ 12
  ─────────────
  21h50m CST
+  6
  ─────────────
  3h50m GCT (on the following day)
```

16.5 TIME SIGNAL SERVICE

The ease with which correct standard time, or the error of one's timepiece, can be obtained has contributed materially to the simplicity with which the results of astronomic observations for azimuth can be calculated. Extreme accuracy in time is not required if azimuth values deduced from Polaris observations are to be correct within $\pm 10''$, but it is very essential for the reduction of solar sights. Hence, it is recommended as a principle of uniformity that time service be utilized to secure the watch correction to the nearest second. Any good quality short-wave radio set is suitable for the reception of time signals.

The Time and Frequency Division of the National Bureau of Standards (NBS) broadcasts continuous time signals from its high-frequency radio station, WWV, at Fort Collins, Colorado. The frequencies are 2.5, 5, 10, 15, 20, and 25 MHz. Except during times of severe magnetic disturbances, listeners should be able to receive the signals on at least one of these broadcast frequencies. Generally, frequencies above 10 MHz provide the best daytime reception while the lower frequencies are optimum for nighttime reception.

Voice announcements are made once every minute beginning $7\frac{1}{2}$ seconds before the minute pulse. The time referred to in the announcements is Coordinated Universal Time, which may be considered equivalent to GCT.

The National Research Council of Canada broadcasts radio time signals continuously from Ottawa over station CHU. Transmission is made on three frequencies—namely, 3.330, 7.335, and 14.670 MHz), but reception is most likely on the stronger power output at the intermediate frequency. A voice recording occurs each minute in the 10-second gap between the 50th and 60th second. Announcements alternate between English and French.

16.6 LATITUDE AND LONGITUDE

For the purpose of computing directions from astronomical observations, it is usually sufficiently accurate to scale the latitude and longitude of the observer's position from a good map. The maps or charts published by various government organizations are best for this purpose. Those most widely available are the standard quadrangle topographic maps of the U.S. Geological Survey. It is possible to obtain the latitude and longitude of a survey point with an uncertainty of not more than one second of arc from both the 1/24,000 and 1/62,500 scale maps. Also, some highway maps are now being published with geographic grids. If no reputable map is available, an inquiry for the observer's latitude and longitude may be addressed to the Director, National Geodetic Survey, NOAA, Rockville, Maryland 20852. The location of the observer can be expressed by giving the name of the nearest post office or, in sectionized areas, the number of the section as well as the designation of township.

Longitude can be expressed either in terms of arc or time measure. Thus the longitude of a point in Urbana, Illinois, is $5^h52^m54^s$ West or $88°13'30''$ West. Note that a careful distinction must be maintained between arc and time expressions of the same longitude because a minute of arc is not equivalent to a minute of time. The following tabulation offers a useful summary of time and arc relations:

$$24^h = 360° \qquad 1° = 4^m$$
$$1^h = 15° \qquad 1' = 4^s$$
$$1^m = 15'$$
$$1^s = 15''$$

16.7 POSITION OF A CELESTIAL BODY

In much the same manner that the geographical coordinates, latitude and longitude, are employed to define the position of an observer on the earth, astronomical coordinates are used to indicate the position of a heavenly body on the celestial sphere. These coordinates are *Greenwich Hour Angle* (GHA) and *Declination*.

The GHA of a star is the angular distance, measured westward, from the Greenwich meridian to the hour circle passing through the star. The GHA can be expressed in time measure (0^h to 24^h) or in arc measure (0° to 360°). Since the motion of the celestial sphere is from east to west, while the earth presumably remains stationary, it is apparent that the GHA of all stars constantly increases with time. This rate of increase is somewhat more than 15° per hour of time, because the celestial sphere actually completes slightly more than one complete rotation every 24 hours.

The declination of a star is the angular distance it is above or below the celestial equator. For Polaris, which has a declination of about 89°, it is more convenient to use the term *polar distance,* which is the angular distance the star is from the North Pole. This is approximately 1°. The declination of the sun varies between $23\frac{1}{2}$° South and $23\frac{1}{2}$° North. Although the sun is also a star, the latter designation is infrequently applied to it.

Figure 16.3 shows, for a given moment, the GHA and declination of a star. Several publications of astronomic data have tabulations of GHA for

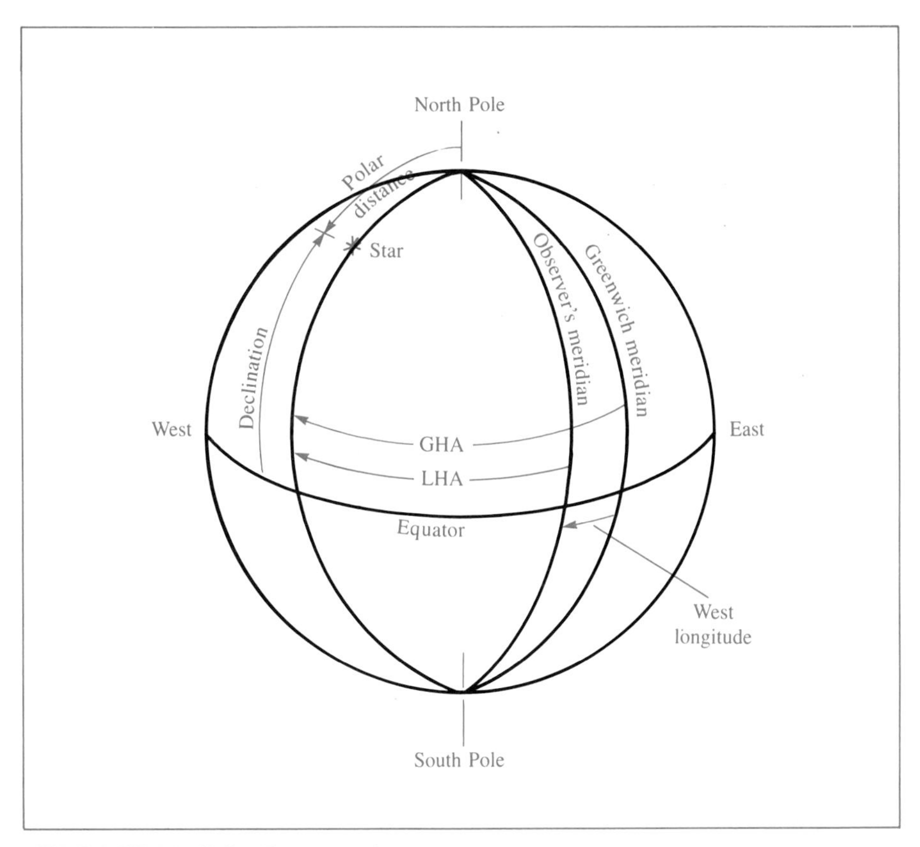

FIGURE 16.3 Coordinates of a star

certain members of the heavenly bodies at the moment of 0^h GCT (Greenwich midnight) for each day in the year. This makes the problem of finding GHA relatively easy. It is merely necessary to convert the standard time of observation to GCT by adding the appropriate number of whole hours (the zone correction) to standard time; then, to the tabulated GHA of the star at 0^h GCT is added the increase in the GHA occurring in the elapsed time interval between 0^h GCT and the GCT of the observation.

EXAMPLE 16.2

An observation was made on Polaris at Pittsburgh, Pennsylvania, at $22^h30^m10^s$ EST on June 17, 1983. What is the GHA of Polaris?

SOLUTION

$$\begin{array}{rl} & 22^h30^m10^s \text{ EST} \\ + & \underline{5^h} \\ & 27^h30^m10^s \end{array}$$

or $3^h30^m10^s$ GCT, June 18

$$\begin{array}{rll} & 232°15.5' & \text{GHA, } 0^h \text{ GCT June 18 (Table I)} \\ + & \underline{52°41.1'} & \text{Increase for elapsed time (Table II)} \\ & 284°56.6' & \text{GHA of Polaris at moment of observation} \end{array}$$

16.8 LOCAL HOUR ANGLE

The *Local Hour Angle* (LHA) of a star is the distance, measured westward, from the observer's meridian to the hour circle through the heavenly body. LHA can be expressed either in time units (0^h to 24^h) or arc measure (0° to 360°). LHA is obtained from GHA merely by subtracting from it the west longitude or adding the east longitude of the local meridian, depending on the position of the observer. Thus,

$$\text{LHA} = \text{GHA} - \text{west longitude} \tag{16.1}$$

or

$$\text{LHA} = \text{GHA} + \text{east longitude} \tag{16.2}$$

Under certain circumstances it will be necessary to add 360° (or 24^h) to the GHA of Eq. (16.1) to perform the required subtraction.

In connection with azimuth observation on Polaris (see Figure 16.4), note that, when the LHA is 0° to 180° (0^h to 12^h), the star is west of north; and when the LHA is 180° to 360° (12^h to 24^h), it is east of north.

A convenient expression of the hour angle position of a star is the *meridian angle* (t). It is reckoned both westward and eastward from the

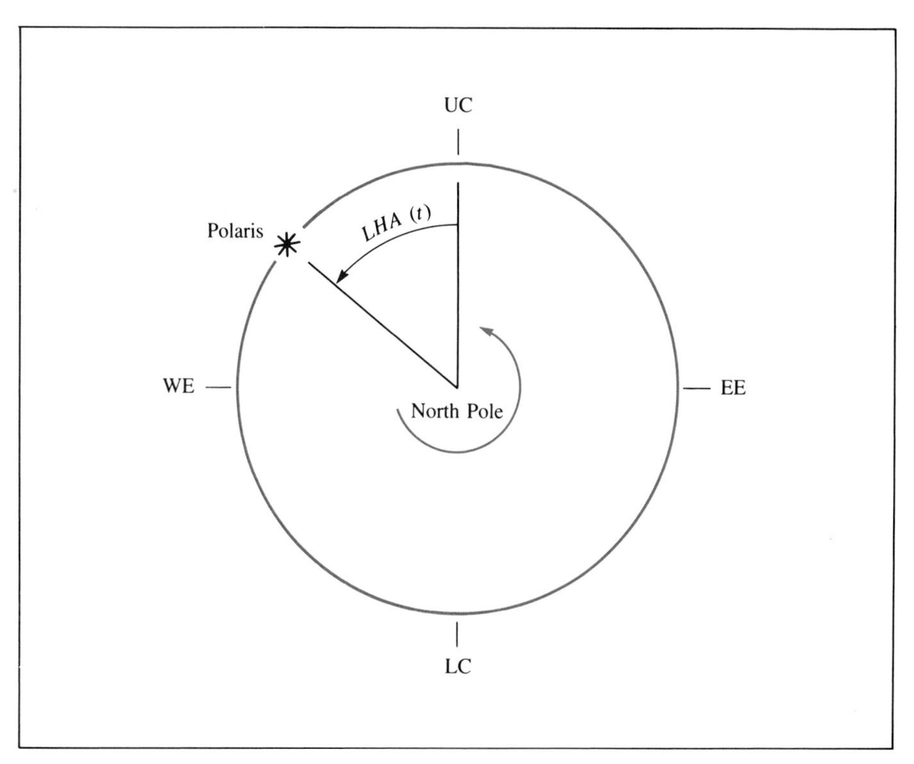

FIGURE 16.4 Motion of polaris (as viewed from the earth)

observer's meridian up to a maximum value of 180°. Thus, when LHA is less than 180°, t is numerically equal to LHA and has the suffix "west." When LHA is greater than 180°, t is equal to 360° less LHA and is termed the meridian angle "east."

For example,

When LHA = 37°10′ then t = 37°10′W.

When LHA = 210°35′ then t = 149°25′E.

16.9 THE ASTRONOMICAL TRIANGLE

A spherical triangle is the figure formed by joining any three points on the surface of a sphere by arcs of great circles. A particular spherical triangle having as its vertices a celestial pole, the observer's zenith, and a heavenly body is called the "astronomical triangle," and it is of much importance in engineering astronomy. See Figure 16.1.

Figure 16.5 shows the *PZS* (pole, zenith, star) triangle as viewed

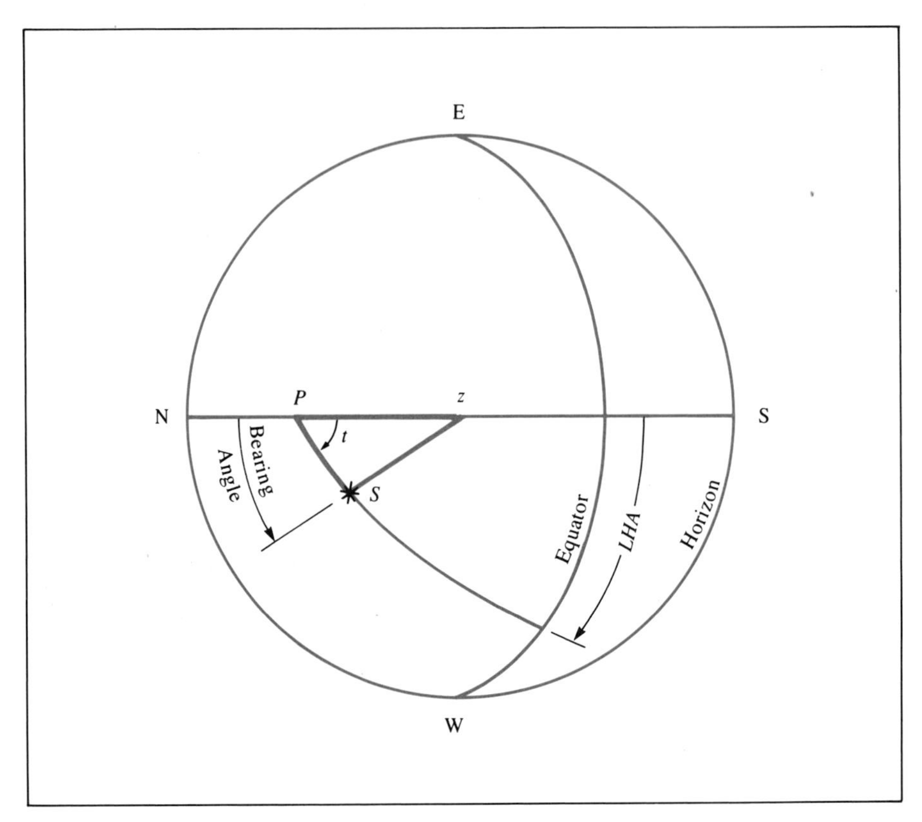

FIGURE 16.5 The astronomical triangle

from the observer's zenith. In the typical azimuth problem, the known quantities in the triangle are sides PZ (90° − latitude) and PS (90° − declination), and the included angle t at the pole. Side ZS equals (90° − altitude). It is frequently called the *zenith distance*. Through application of the spherical trigonometry, the required angle at Z can be calculated. This angle is commonly termed the *bearing angle of the star* and is reckoned both east and west from north. In Figure 16.5 the star is depicted west of the meridian. Note that t is west and is numerically the same as LHA. If the star were east of the meridian, LHA would exceed 180°, t would become east, and the angle at Z would become directly the azimuth of the star.

16.10 POLARIS

Polaris is a celestial body of great importance to the civil engineer and surveyor. It is a fairly bright star (magnitude 2.1) situated about 1° from the north celestial pole. It can be easily identified by prolonging the line connecting the two stars in the bowl of the Big Dipper on the side most remote from the handle. Remember that the altitude of Polaris will be always within 1° of the observer's latitude and that there are no other stars near Polaris that are likely to be confused with it.

As viewed from a position on earth (Figure 16.4), Polaris rotates in a counterclockwise direction about the north celestial pole. It makes a complete revolution approximately every 24 hours.

The instant that Polaris crosses the upper branch of the observer's meridian is called time of Upper Culmination (UC). At that moment LHA of Polaris is zero. Approximately 6 hours later Polaris reaches Western Elongation (WE). Polaris passes over the meridian again at Lower Culmination (LC) and attains its most easterly position at Eastern Elongation (EE).

From this description of the motion of Polaris it is apparent that, at the moments of elongation, the star's motion will be nearly vertical and that its bearing will change but slightly with time. At culmination, however, the rate of change of azimuth of the star with respect to time will be a maximum.

The certainty with which Polaris can be identified, its proximity to the north pole, the fact that it is constantly above the horizon, and the ready availability of specially prepared tables to facilitate the solution of the azimuth problem make it a particularly useful astronomical body.

16.11 THE AZIMUTH PROBLEM

The typical azimuth problem consists of two basic operations regardless of whether the celestial body is the Sun or Polaris. They are as follows:

1. The observation of the celestial body at any time in order to measure the horizontal angle between the star and a terrestrial signal called the "mark." At the moment a pointing is made on the star, time is noted.

2. The computation of the bearing angle of the star at the instant of observation. This calculated angle, when appropriately combined with the measured field angle, will yield the required astronomic azimuth or true bearing of the line.

For any azimuth observation the following data in addition to the measured horizontal angles and the observed times are essential: (1) the date of the observation, (2) the name or identity of the instrument station, (3) the name of the mark, (4) the latitude and longitude of the instrument station as scaled from a map or determined from any other reliable source, (5) the watch error, if any, and (6) the kind of time kept by the watch (MST, CDT, and so on). A good sketch portraying the general position of the terrestrial line with respect to astronomic north is very desirable.

16.12 FIELD PROCEDURE

This treatment emphasizes the inherent advantage of observing Polaris for azimuth at any local hour angle. The facility with which the computations can be made and results of good quality obtained make the hour angle method an entirely satisfactory procedure. It is preferred to the elongation method, even for the most precise azimuth determinations of geodetic surveys.

It can be easily verified that, at the most unfavorable moment (that is, Polaris at culmination) an error of as much as 1 minute in time causes, at a latitude of 40°, a bearing angle error of only 0.3′ of arc. Furthermore, the recommended method permits the observation to be made at any moment convenient to the observer. It is not necessary to wait for the precalculated instant of elongation, which may take place at a most unsuitable time or when the star happens to be obscured by clouds.

The field procedure begins with setting up the instrument over the station, performing the usual centering and leveling, and taking an initial sight at the mark. It will be necessary to illuminate the cross-wires of an ordinary engineer's transit by shining a flashlight obliquely into the objective end of the telescope. Optical transits, however, are wired for internal illumination. It is best that the terrestrial line be at least 1,000 ft long so that the objective lens does not require refocusing when sighting the star. An illuminated traverse target provides the most satisfactory signal over the mark. When pointing upon the star, it is very essential that Polaris appear as a tiny pinpoint of light rather than as a disc. Difficulty will be experienced in finding the star unless the objective lens is correctly focused. This can be accom-

Azimuth by Polaris
Wisconsin River Improvement Project
Traverse from Athens to Wausau

Object	Tel	Time	Angle	Mean angle	Mean time
Mark	D		0°00.0′		
Star		9:10:16	42°17.3′		
Mark	R				
Star		9:11:22	84°35.2′	42°17.6′	9:10:49

Wild T-1, No. 6253
Light Breeze
Temp. 25° F

June 10, 1983
S.L. Finne - Observer
B.O. Cohen - Notes

Station occupied: 26K
Lat. 44°58′16″
Long. 89°42′27″
Mark: 27K

Watch is 16 seconds fast

Time is P.M., Central Standard Time

FIGURE 16.6 Field notes for third-order azimuth observation on Polaris

plished by taking a preliminary sight on any other prominent heavenly body or upon a distant terrestrial light.

The use of either a standard 30″ engineer's transit or a theodolite will provide satisfactory azimuths for most local control work. With both instruments, the horizontal angles are doubled. The field procedure is as follows:

1. Set up the instrument and level it carefully.
2. Set the horizontal circle at 0° and sight the mark.
3. Sight the star and note the time.
4. Read and record the single angle.
5. Release the lower motion, invert the telescope, and sight the mark a second time.
6. Sight the star and note the time.
7. Read and record the double angle.

Typical field notes are shown in Figure 16.6. They constitute a single set of direct and reverse (*D & R*) observations.

16.13 EPHEMERIDES

The astronomical tables in the appendix are available in more complete form in the various ephemerides published by several foreign and domestic governmental agencies and scientific organizations. All are issued annually, and a copy for the year in which field observations are being conducted is indispensable for making the necessary reduction computations.

Abridged ephemerides published by some American surveying equipment manufacturers are available for a nominal charge. Portions of the 1983 Solar Ephemeris of the Keuffel & Esser Company are reproduced in the appendix. Another useful and inexpensive ephemeris for the surveyor is the "Ephemeris of the Sun, Polaris, and other Selected Stars," which is prepared by the Nautical Almanac Office, U.S. Naval Observatory. It can be purchased from the Government Printing Office, Washington, D.C. 20402.

16.14 REDUCTION PROCEDURES

Three reduction procedures for calculating the azimuth of Polaris at any hour angle will be presented.

1. A formula of wide and general application for computing the precise azimuth of any celestial body is as follows:

$$\tan Z = \frac{\sin t}{\cos \phi \tan \delta - \sin \phi \cos t} \tag{16.3}$$

where

ϕ = latitude
δ = declination
t = hour angle
Z = azimuth of star

The quantity Z is reckoned east or west from true north.

2. For a less precise value of the azimuth a more simple formula can be used. It can be derived directly from the law of sines for spherical trigonometry as applied to the *PZS* triangle of Figure 16.5. Hence it is seen that

$$\sin Z = \frac{\sin t \sin (90^\circ - \delta)}{\sin (90^\circ - h)}$$

or

$$\sin Z = \frac{\sin t \sin p}{\cos h} \tag{16.4}$$

where

p = polar distance, $90^\circ - \delta$
h = altitude

For a circumpolar star like Polaris, angles Z and p will always be small. Therefore, we can substitute the values of the angles in seconds (or minutes) for their sines and obtain

$$Z = \frac{p \sin t}{\cos h} \tag{16.5}$$

where Z and p are both in seconds or in minutes of arc.

Note that Eq. (16.5) requires the measurement of the altitude h. This is the value after instrumental corrections and refraction are applied to the observed altitude. If a reliable measurement of altitude seems unlikely and the latitude, on the other hand, has been well established, the following formula can be used:

$$Z = \frac{p \sin t}{\cos (\phi + p \cos t)} \tag{16.6}$$

The quantity $p \cos t$ can be obtained from prepared tables like Table VI. This quantity is the difference between the altitude of Polaris and the North Pole at any hour angle t. In the use of Eq. (16.6) the quantity $p \cos t$ taken from Table VI is either added to or subtracted from the known latitude, depending upon whether the star is above or below the pole at the moment of observation. The position of Polaris is given by its hour angle as shown in Figure 16.4.

3. A convenient solution of the *PZS* triangle for the azimuth of Polaris can be made with certain prepared tables. Such a solution is presented in detail in the following section.

16.15 AZIMUTH FROM POLARIS AT ANY TIME

A typical reduction computation is presented for an azimuth observation made on Polaris on June 1, 1983, at Urbana, Illinois. The given time and horizontal angle represent the mean values from a single set, direct and reverse, of observations on the star.

EXAMPLE 16.3

Station occupied: Illinois
Latitude: 40°06′18″ N
Longitude: 88°13′30″ W
Observed time: $8^h23^m20^s$ P.M., CST
Angle: mark to star (clockwise) 46°17.5′

Mark: Sta. 25
Date: June 1, 1983
Observer: J.L.M.
Watch 35^s slow

Find the true bearing of the line.

SOLUTION

1. *Calculation of GCT*

Watch time (P.M.)	$8^h23^m20^s$ CST
Correction	+ 35^s
Corrected time	$8^h23^m55^s$
Time (24-hr basis)	$20^h23^m55^s$
Zone correction	$+6^h$
GCT	$2^h23^m55^s$
Greenwich date	June 2

2. *Calculation of LHA*

GHA at 0^h GCT	216°34.0′ (Table I)
Increase in GHA	36°04.7′ (Table II)
GHA	252°38.7′
Less west long.	88°13.5′
LHA	164°25.2′
t	164°25.2′ (west)

3. *Preliminary Bearing Angle of Polaris.* A double interpolation is made from Table III, which is entered with the known latitude and previously calculated LHA. This interpolation effects a mechanical solution of the *PZS* triangle for an average value, 0°48.5′, of the polar distance of Polaris for the year.

	Latitude		
LHA	40°	40°06.3′	42°
160°	0°21.4′	0°21.4′	0°22.1′
164°25.2′	—	0°16.8′	—
165°	0°16.2′	0°16.2′	0°16.7′

Thus, the preliminary bearing angle of Polaris is 0°16.8′.

4. *Final Bearing Angle of Polaris*. The actual polar distance of Polaris, 0°48.95′, is obtained from Table IV and affords, together with the preliminary bearing angle of Polaris, a means for entering Table V to obtain a supplementary correction. As a glance at Table V discloses, this correction is usually very small.

Preliminary bearing angle	0°16.8′
Correction (Table V)	+ 0.2′
Final Bearing Angle of Polaris	0°17.0′

5. *True Bearing of Line*. The final bearing of Polaris is combined with the measured field angle to obtain the bearing of the line. Since the LHA of Polaris is between 0° and 180°, the star is west of north. A good sketch (Figure 16.7) depicting the position of the star in relation to the meridian and the terrestrial line always should be drawn.

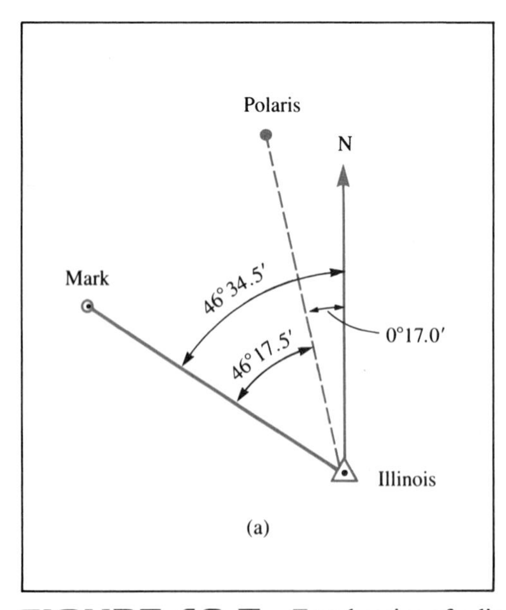

FIGURE 16.7 True bearing of a line by Polaris

The true bearing of the line is thus found to be N46°34.5′ W.

Alternate solution. An alternate solution of the problem will now be effected with the use of Eq. (16.6) as follows:

$$Z = \frac{p \sin t}{\cos (\phi + p \cos t)}$$

where

$$\begin{aligned} p &= 48.95' && \text{(from Table IV)} \\ t &= 164°25.2' && \text{(as previously calculated)} \\ p \cos t &= -46.7' && \text{(from Table VI)} \\ \phi &= 40°06.3' \text{ N} \end{aligned}$$

hence,

$$Z = \frac{(48.95) \sin 164°25.2'}{\cos 39°19.6'} = 0°17.0'$$

16.16 QUALITY OF POLARIS AZIMUTH DETERMINATION

The quality of an astronomic determination of azimuth is dependent on the accuracy of the measured field angle, as well as on the accuracy of the computed bearing angle of the star.

Consideration will be given to the effect of an error in time upon the quality of the calculated bearing angle of Polaris. Assume that an observation was made on June 1, 1983, in latitude 40°10′ N when the hour angle of Polaris was 5°. The polar distance for the date is 0°48.95′. Assume uncertainty in time is ±5 seconds.

Differentiating Eq. (16.5), we have

$$dz = \frac{p \cos t \, dt}{\cos h} \tag{16.7}$$

where p and dz are in minutes of arc and dt is in radians.

$$h = \phi + p \cos t = 40°10' + 48.76' = 40°58.76'$$

Recalling that 1 second of time equals 15″ of arc and 1″ of arc equals 0.00000485 radian, we have

$$dt = (5)(15)(0.00000485) = \pm 0.00036 \text{ radian}$$

Hence,

$$dz = \frac{(48.95 \cos 5°)(0.00036)}{\cos 40°58.76'} = \pm 0.023' = \pm 1.4''$$

Since the preceding observation was made near the moment of upper culmination when the azimuth of the star is changing most rapidly, it is obvious that time is not critical for azimuth observations of ordinary accuracy on Polaris.

16.17 ALTITUDE OF POLARIS AT ANY TIME

Figure 16.8 portrays a section along the observer's meridian through the portion of the celestial sphere above the horizon. It is readily seen that, since the polar axis is perpendicular to the equator and the plumb line is normal to the plane of the horizon, the altitude of the pole equals the latitude. At culmination Polaris is above or below the pole by an amount of arc equal to its polar distance. At elongation it is directly opposite the pole and its altitude equals the latitude. For other intermediate positions, Table VI tabulates the angular corrections which, for various values of LHA, can be applied to the latitude to obtain the true altitude of Polaris. The calculated

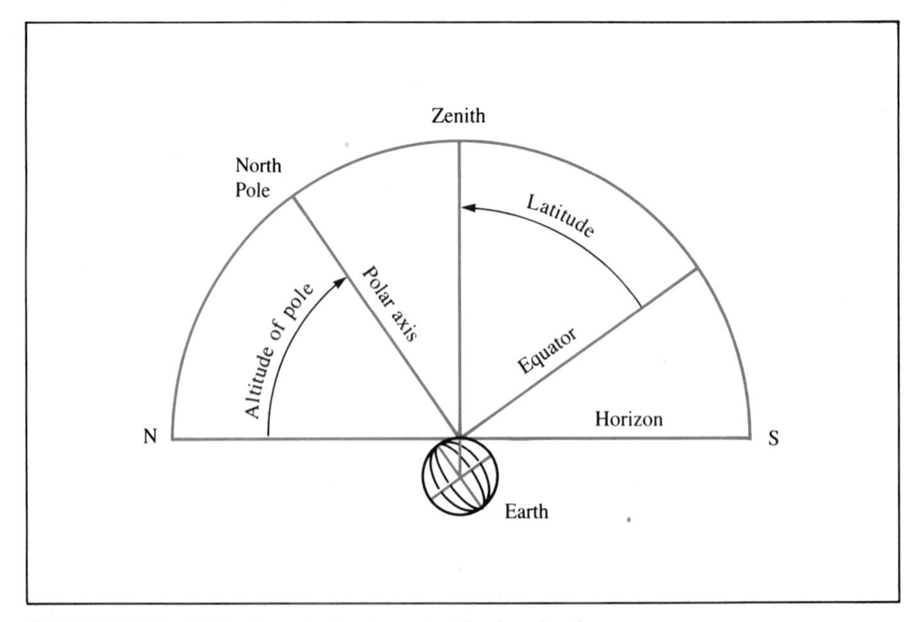

FIGURE 16.8 Latitude and altitude of pole

altitude can be of assistance in locating the star when preparing to make sightings on it for azimuth.

EXAMPLE 16.4

The azimuth observations of Section 16.15 were scheduled to begin at 8:15 P.M., CST. Calculate the zenith distance of Polaris for that moment.

SOLUTION

CST	$20^h15^m00^s$	
Zone correction +	6^h	
GCT	$2^h15^m00^s$	
Date	June 2	
GHA of Polaris at 0^h GCT	216°34.0′	(June 2)
Increase in GHA	33°50.5′	
GHA	250°24.5′	
Less west long.	88°13.5′	
LHA	162°11.0′	
or *t*	162°11.0′	West
Lat	40°06.3′	
Correction	−46.2′	(Table VI)
Altitude	39°20.1′	
or Z.D.	50°40′	

16.18 SOLAR OBSERVATIONS

The same fundamental field and office procedures that were used to obtain an azimuth by observations of Polaris can be employed to determine direction by sights on the sun. Generally, the results of solar observations are not as reliable as those made upon Polaris. The pole star has a relatively slow motion and appears as a tiny pin point of light, which assures pointing accuracy. The sun moves rapidly and its image, which has a diameter of 32′, is difficult to bisect. Furthermore, it is necessary to interpose a darkening lens in order to safely view the sun. If an attachment, known as the *Roeloef's solar prism,* is available and mounted over the transit's objective lens, the quality of a solar azimuth is very appreciably enhanced. This prism enables the instrument operator to make an accurate pointing on the sun's center. Although the field technique of observing Polaris is simpler than that of observing the sun, even with a solar prism, and the reduction calculations are less involved than for solar work, it may be still desirable to determine azimuth by solar observations. This would be the case where the lines are so difficult to reach that night trips for Polaris observations are unwarranted. Of greater significance in the case of local control surveys would be the economic disadvantage of returning the party to the field after darkness sets in. The additional cost entailed by this action might make it mandatory to rely on solar azimuths. Further, if the accuracy requirements are not too demanding, such azimuths can be entirely satisfactory.

Although the same astronomic terms and concepts apply to solar observations as well as to Polaris sights, a broader treatment of the principles of time is essential.

16.19 PRINCIPLES OF TIME

The basis of time measurement is the uniform rate of rotation of the earth about its axis. This rotation causes the sun and stars to appear to cross the sky from east to west. Every celestial body crosses the plane of the observer's meridian twice during each rotation of the earth, or each apparent revolution of the celestial sphere.

That moment when any object on the celestial sphere appears on the meridian of the observer is called the *time of transit* or culmination of that object over that meridian. When the object is on the upper branch of the meridian, the moment is termed *upper transit;* when it is on the lower branch, it is termed *lower transit.*

All of the various kinds of time are defined fundamentally by the hour angle of a celestial object as measured from a particular meridian. For example, Greenwich time is reckoned with respect to the Greenwich Meridian; local time is reckoned with respect to the local meridian. Also, the time will be dependent upon the selection of the celestial object. This can be (1) the *apparent* or real sun, or (2) a fictitious or *mean sun.*

16.20 APPARENT SOLAR TIME

Apparent solar time is reckoned with respect to the real sun. When the real sun is on the lower branch of the observer's meridian, the apparent solar day is just beginning. When the apparent solar time is 12^h, the real sun is at upper transit. Hence, *local apparent time,* LAT, is the hour angle of the real sun plus 12 hours. *Greenwich apparent time* is indicated by GAT.

Despite the fact that the real sun seems an obvious choice of a celestial body for reckoning time, it is not a good time indicator because the length of the apparent solar day, or the time interval between two successive lower transits of the real sun, is not constant. The two basic reasons for this irregularity are the elliptical path of the apparent sun across the sky and the nonuniform velocity of travel of the apparent sun.

16.21 MEAN SOLAR TIME

To overcome the difficulties of using apparent solar time, astronomers have devised a fictitious body, called the *mean sun,* which is assumed to move along the celestial equator at the average or mean rate of motion of the real sun. The time provided by the mean sun is such that every day is of exactly the same duration and equal to the average length of an apparent solar day. Hence, *local mean time,* also called *local civil time, LCT,* is the hour angle of the mean sun plus 12 hours.

16.22 EQUATION OF TIME

Any of the ephemerides mentioned in Section 16.13 provides values of the time interval between the transits of the mean and apparent suns as they cross the Greenwich Meridian at 0^h GCT each day. This interval is called the *equation of time.* It is used to make conversions from one kind of solar time to the other and is defined by the following equation:

$$\text{Equation of time} = \text{apparent solar time} - \text{mean solar time}$$

The equation of time can be positive (real sun ahead of mean sun) or negative (mean sun ahead of real sun) and may amount to as much as 16 minutes.

16.23 STANDARD TIME

Since the sun does not rise and set in different parts of the country at the same time, it is evident that mean solar time varies from place to place. In order to reduce confusion, standard time zones have been adopted. These are depicted in Figure 16.2.

Standard time is essentially mean solar time for a particular meridian—that is, the central meridian of the zone.

Daylight time in any zone is equivalent to standard time in the zone next removed to the east.

16.24 SOLAR AZIMUTH OBSERVATION

The use of a Roeloef's Solar Prism depicted in Figure 16.9 is essential if sun azimuths are to be of good quality. This device fits on the objective end of the theodolite, although in some cases an adapter may be required. Its darkening lens is hinged so that it can be closed when viewing the sun and opened when sighting the terrestrial mark. The attachment is optically so constructed that it makes possible a direct pointing on the sun's center when the intersection of the cross-wires is placed upon the small black square in the center of the field of view.

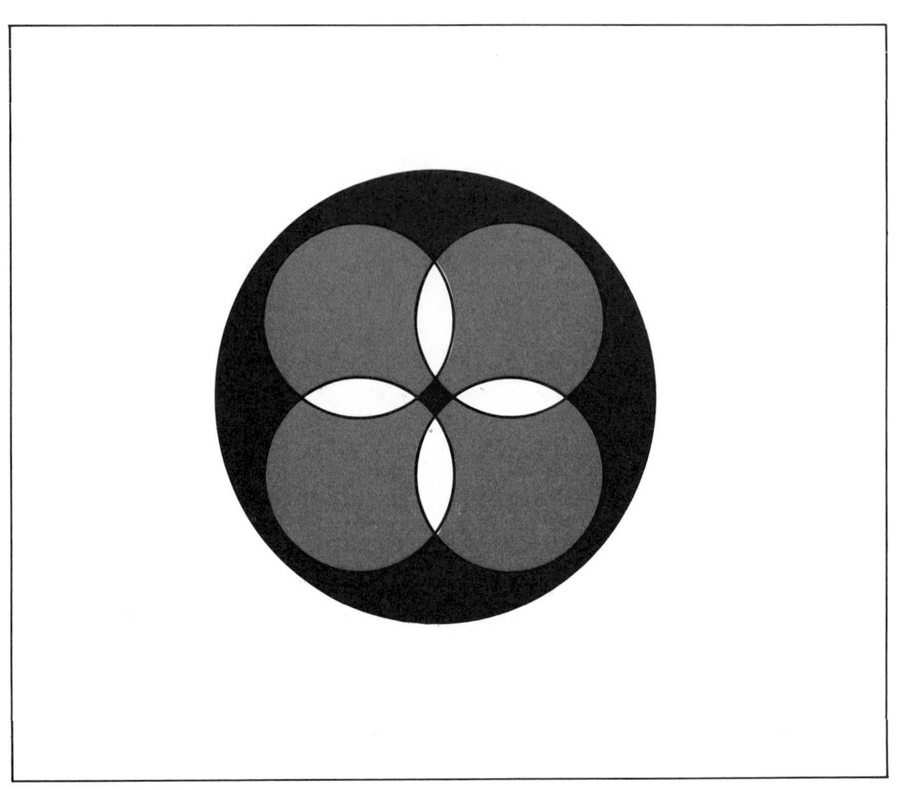

FIGURE 16.9 Sun's image with Roeloef's Solar Prism

High accuracy in the timing of solar pointings is extremely important. Time control should be maintained with a reliable watch or chronometer that has been compared before and after returning from the field with radio time signals. A battery-powered radio set taken into the field would be better.

The form of field record could be essentially the same as that shown in Figure 16.6.

The reduction calculation is effected with the use of Eq. (16.3), which requires that a new approach be taken with respect to the evaluation of the meridian angle t. Basically, this involves changing the zone or standard time of observation to GCT and converting this to GAT with the equation of time. The GAT is then changed to LAT by means of the known longitude of the survey point and the LHA or t is found from the basic equation

$$\text{LAT} = \text{LHA} + 12^{\text{h}} \tag{16.8}$$

A complete reduction computation is shown in the following example.

EXAMPLE 16.5

Station occupied: Illinois
Latitude: 40°06′18″N
Longitude: 88°13′30″W ($5^h52^m54^s$)
Locality: Champaign, IL
Date: June 4, 1983
Mark: TT 64
Observed time: $3^h20^m10^s$ P.M., CDT
Watch 11^s slow
Angle: Mark to sun (clockwise) = 122°16′06″
Find the bearing of the line.

SOLUTION

1. Calculation of meridian angle t

Observed watch time (P.M.)	$3^h20^m10^s$
Correction +	11^s
Corrected time (CDT)	$3^h20^m21^s$
Time (24-hr basis)	$15^h20^m21^s$
Zone correction +	5^h
GCT	$20^h20^m21^s$

Equation of time at 0^h GCT (June 4)	$1^m55.6^s$ (Table I)
Change (20.3×0.42^s) −	8.5^s
Corrected equation of time	$1^m47.1^s$

GCT	$20^h20^m21^s$
(plus) Equation of time +	01^m47^s
GAT	$20^h22^m08^s$
(minus) West longitude	$5^h52^m54^s$
LAT	$14^h29^m14^s$
t	$2^h29^m14^s$W
or t	37°18′30″W

2. Calculation of declination

Declination at 0^h GCT (June 4)	N22°20.3′ (Table I)
Change (20.3 × 0.30′) +	06.1′
Corrected declination	N22°26.4′

3. Calculation of bearing of sun

$$\tan Z = \frac{\sin t}{\cos \phi \cdot \tan \delta - \sin \phi \cdot \cos t}$$

$\tan \delta = 0.412987$ $\quad \sin t = 0.606104$
$\sin \phi = 0.644190$ $\quad \cos t = 0.795385$
$\cos \phi = 0.764865$

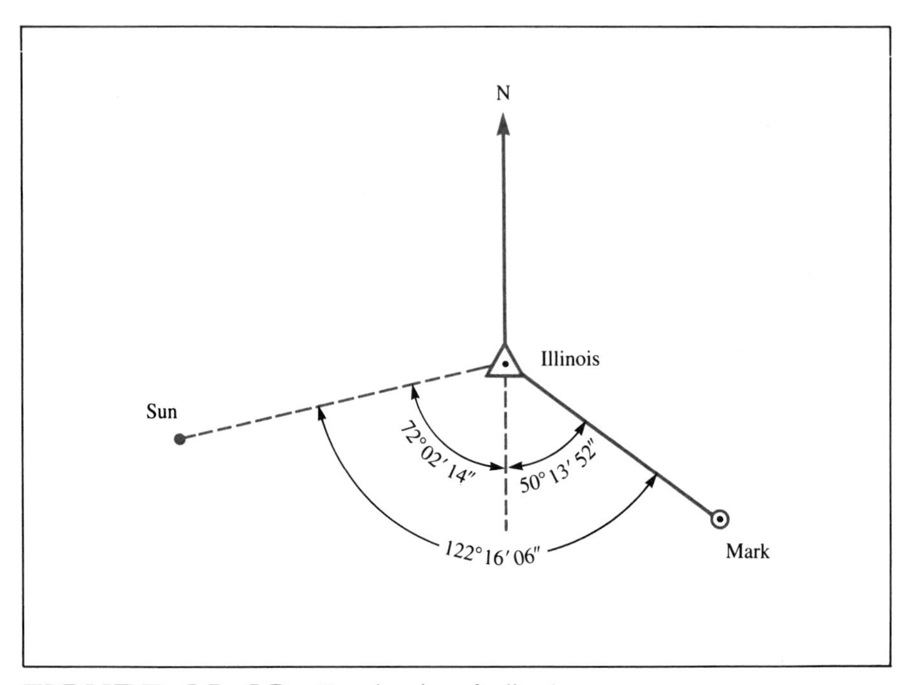

FIGURE 16.10 True bearing of a line by sun

$$\tan Z = \frac{0.606104}{0.315879 - 0.512379} = \frac{0.606104}{-0.196500}$$

$$= -3.084498$$

$$Z = 72°02'14''$$

$$\text{Bearing of sun} = \text{S}72°02'14''\ \text{W}$$

Care must be exercised in the interpretation of the sign of tan Z. When it is positive, Z is reckoned from north; when negative, from south. When the sun is observed east of the meridian, Z is measured toward the east; in the afternoon it is measured toward the west. Also, when the declination is south, tan δ is negative.

4. Bearing of line

The bearing of the line is combined with the measured field angle as shown in Figure 16.10. Hence, bearing of line is S 50°13′52″ E.

PROBLEMS

16.1 An azimuth observation was made upon Polaris on June 20, 1983, at Pittsburgh, Pennsylvania. The latitude of the station is 41°40.0′N, and the longitude is 83°36.5′W. The mean angle, measured clockwise, from the mark to the star is 23°14½′ and the corresponding observed time is $10^h43^m25^s$ P.M., EST. The watch is known to be 30^s fast. Find the true bearing of the line.

16.2 An azimuth observation was made on Polaris on June 28, 1983, at a third-order traverse station near Milwaukee, Wisconsin. As scaled from a map, the latitude of the station is 43°07′20″N, and the longitude is 87°58′50″W. The mean of the angles, measured clockwise, from the azimuth mark to the star with a Wild T-1 transit is 56°42.3′, and the corresponding observed time is $9^h16^m15^s$ P.M., CDT. The watch is known to be 10^s slow. Find the astronomic azimuth of the line (from south).

16.3 Reduce the field notes of Figure 16.6 to find the azimuth (from south) of the line.

16.4 As an aid in locating Polaris in the field situation of Problem 16.1, calculate the star's altitude at 10:30 P.M., EST.

16.5 Determine the zenith distance of Polaris at the traverse station of Problem 10.2 at 9:00 P.M., CDT.

16.6 An azimuth observation was made on the sun on June 11, 1983, at a section corner (Lat. 39°57′20″N, Long. 88°17′15″W) near Sadoris, Illinois. The corrected time of the solar pointing was $4^h27^m34^s$ P.M., CDT and the clockwise angle, mark to sun, was 62°11′12″. Calculate the azimuth (from south) to the mark.

16.7 A series of 12 astronomic observations were made of the azimuth of a baseline. The calculated directions were as follows:

Observation	Azimuth	Observation	Azimuth
1	143°16′44″	7	143°16′38″
2	38″	8	46″
3	40″	9	40″
4	41″	10	44″
5	38″	11	42″
6	45″	12	38″

Calculate the mean azimuth and its standard error.

REFERENCES

16.1 Hosmer, G. L., and J. M. Robbins. *Practical Astronomy,* 4th ed., Wiley, New York: 1956.

16.2 *K + E Solar Ephemeris*, Keuffel & Esser Company, Morristown, N.J.: (Published annually).

16.3 Nassau, J. J. *Practical Astronomy,* 2nd ed., McGraw-Hill, New York: 1948.

HYDROGRAPHIC SURVEYING

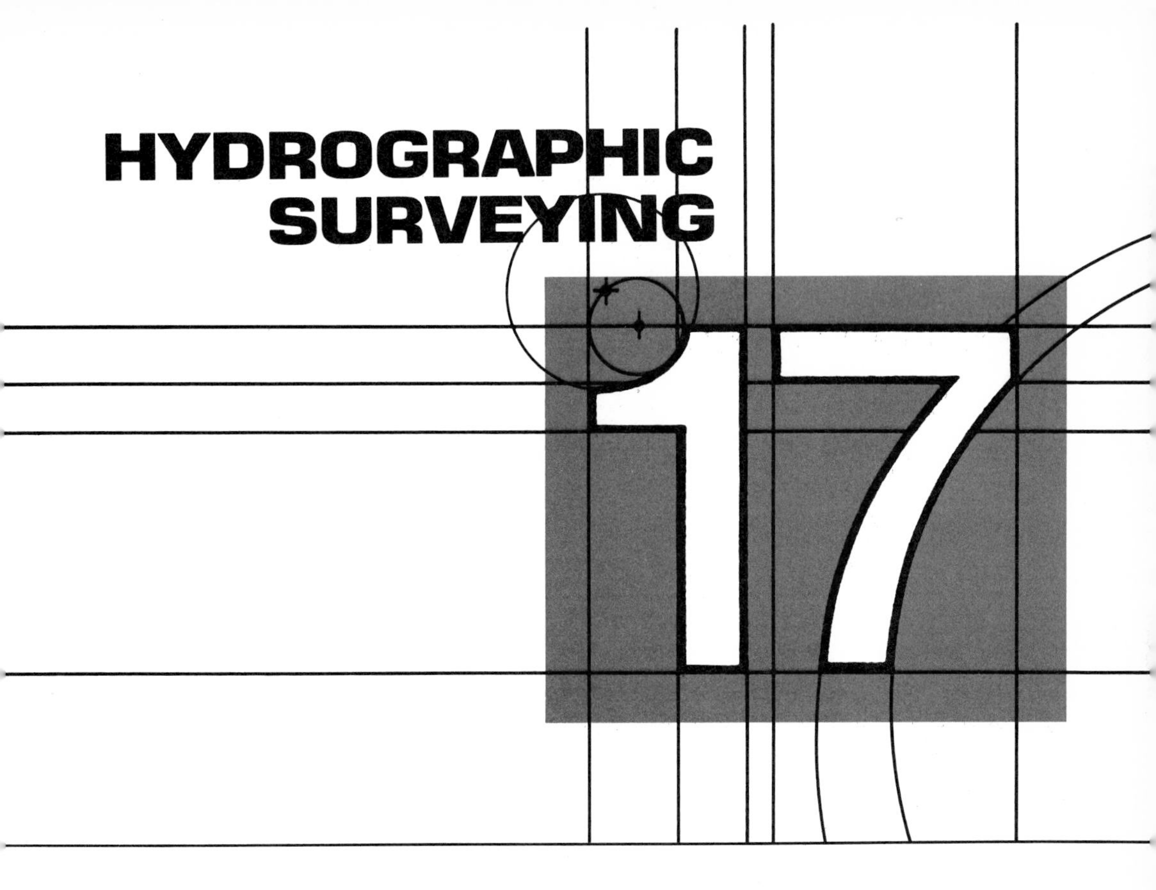

17.1 INTRODUCTION

A *hydrographic survey* is one whose principal purpose is to secure information concerning the physical features of water areas. Such information is essential for the preparation of modern nautical charts, on which are shown available depths, improved channels, breakwaters, piers, aids to navigation, harbor facilities, shoals, menaces to navigation, magnetic declinations, sailing courses, and other details of concern to mariners. Also, a hydrographic survey may deal with various subaqueous investigations that are conducted to secure information needed for the construction, development, and improvement of port facilities; to obtain data necessary for the design of piers and other subaqueous structures; to determine the loss in capacity of lakes or reservoirs because of silting; and to ascertain the quantities of dredged material.

Since the 1960s there has been a tremendous increase in the exploitation of natural resources, particularly oil and gas, beyond the ocean shorelines. Drilling platforms, underwater cables and pipelines, and construction barges must be located and positioned. These activities have greatly accelerated developments in *marine surveying*.

The fundamental principles of conducting the hydrographic survey of a harbor or an inland lake are substantially the same as those employed in executing a comprehensive survey of a large tidal estuary or in sounding a vast oceanic area like the Gulf of Mexico, but there are marked differences in the sounding vessels, instrumental equipment, and the surveying techniques. This section deals mainly with the basic procedures for executing a hydrographic survey of limited scope and with the relationship of such surveys to the practice of civil engineering in such aspects as waterway improvement, dock and harbor construction, beach erosion control, and sewage disposal.

17.2 THE HYDROGRAPHIC SURVEY

A hydrographic survey is distinguished by the measurements and observations that are made to determine and subsequently portray the submarine or underwater topography, as well as to locate various marine features of interest to the navigator. The chief elements of a hydrographic surveying project will now be briefly outlined. The detailed treatment of such operations as those of making and locating soundings will be presented in subsequent sections.

1. *Reconnaissance.* Although the principal operation of the hydrographic survey is the obtaining of hydrography or making the soundings, this phase of the project cannot be performed until certain preliminary steps are undertaken. The first of these is a careful reconnaissance of the area to be surveyed in order to select the most expeditious manner of performing the survey and to plan all operations so that the project mission is satisfactorily completed in accordance with the general instructions and specifications governing such work. The use of aerial photography can be of considerable assistance in this preliminary study.

2. *Horizontal Control.* The next step is the establishment of horizontal control or the framework by which land and marine features are held in their true relationship to each other. Triangulation and, to a lesser extent, traverse are most commonly executed to provide horizontal control. It is not practicable to make a general statement regarding the accuracy of such control. In the case of original surveys over large bodies of water, second- or third-order triangulation may be required. For detached surveys of small and isolated reservoirs it may be entirely satisfactory to develop a control system by a combination of stadia and graphical triangulation procedures with the plane table.

Previously established control is a very important asset in any hydrographic survey. Every effort should be made to obtain and utilize data from earlier surveys in the area. Sometimes a sufficient number of former survey stations can be recovered to satisfy the requirements of a revision survey and no new horizontal control will be necessary.

3. *Vertical Control.* Before sounding operations are begun, it is essential to execute the vertical control in order that the stage or elevation of the water surface can be known when the soundings are obtained. Vertical control data are also needed for the limited topography shown on all nautical charts. When surveys are conducted in tidal bodies of water whose low water level is not known, it is necessary to establish a tide station and begin observations of the tidal fluctuations so as to define a plane of reference for the soundings. This datum is then tied to one or more nearby benchmarks by leveling.

4. *Topographic Survey.* Topographic surveys are conducted of the chart area back of the shoreline. Since the navigator's only interest in this area is in any prominent landmarks that it may contain, only a relatively narrow fringe of topography is shown on the chart.

5. *Hydrography.* The measurement of water depths is the most important operation in nautical charting or in hydrographic surveys that are related to civil engineering.

6. *Preparation of the Hydrographic Map or Nautical Chart.* This (Figure 17.1) usually represents the final product of the hydrographic survey. In the case of subsurface surveys for engineering purposes, the end result may be the calculation of quantities of silt, dredged material, or the preparation of underwater profiles needed for subaqueous construction.

17.3 SOUNDING DATUMS

In topographic surveying it is necessary to determine the position of various points on land and also their elevations above some datum, usually mean sea level. In hydrographic surveying the depth of water with respect to some particular *stage* or height of the water surface is required, as well as the horizontal position of the sounding. It should seem obvious, therefore, that observed soundings must be reduced to some plane of reference if any uniformity in the record of soundings is to be obtained.

In the case of nautical charting surveys it is common practice to reduce observed depths to the *plane of low water*. These reduced depths are then portrayed on the chart and will indicate to the navigator the least depth that will exist at any time. The *Low Water Datum* for Lake Michigan as used by the *Lake Survey Center* of NOS is defined as the surface of the lake when it is at elevation 578.5 ft above mean sea level. This corresponds quite closely with the lowest recorded water surface elevation for that lake. To illustrate the reduction of a measured depth to a particular plane of reference, consider soundings made in Chicago Harbor during a day on which the stage of the lake read 5.8 on the gage (Figure 17.2) and it was known that the zero of the gage was at elevation 576.0. All measured depths must be reduced by $(576.0 + 5.8) - 578.5$ or 3.3 ft in order to refer them to the Low Water Datum.

In tidal bodies of water it is the usual practice to refer soundings to *Mean Low Water* (MLW) or *Mean Lower Low Water* (MLLW), depending upon the character of the rise and fall of the sea.

In addition to the nautical applications of tidal data, there are some interesting and important engineering aspects. The harbor engineer must consider carefully tidal data when planning or designing waterfront structures. During construction the tidal range will influence the freeboard requirements of cofferdams and caissons, the type of dock construction, and the cutoff elevations for timber piles. The predicted times of high and low

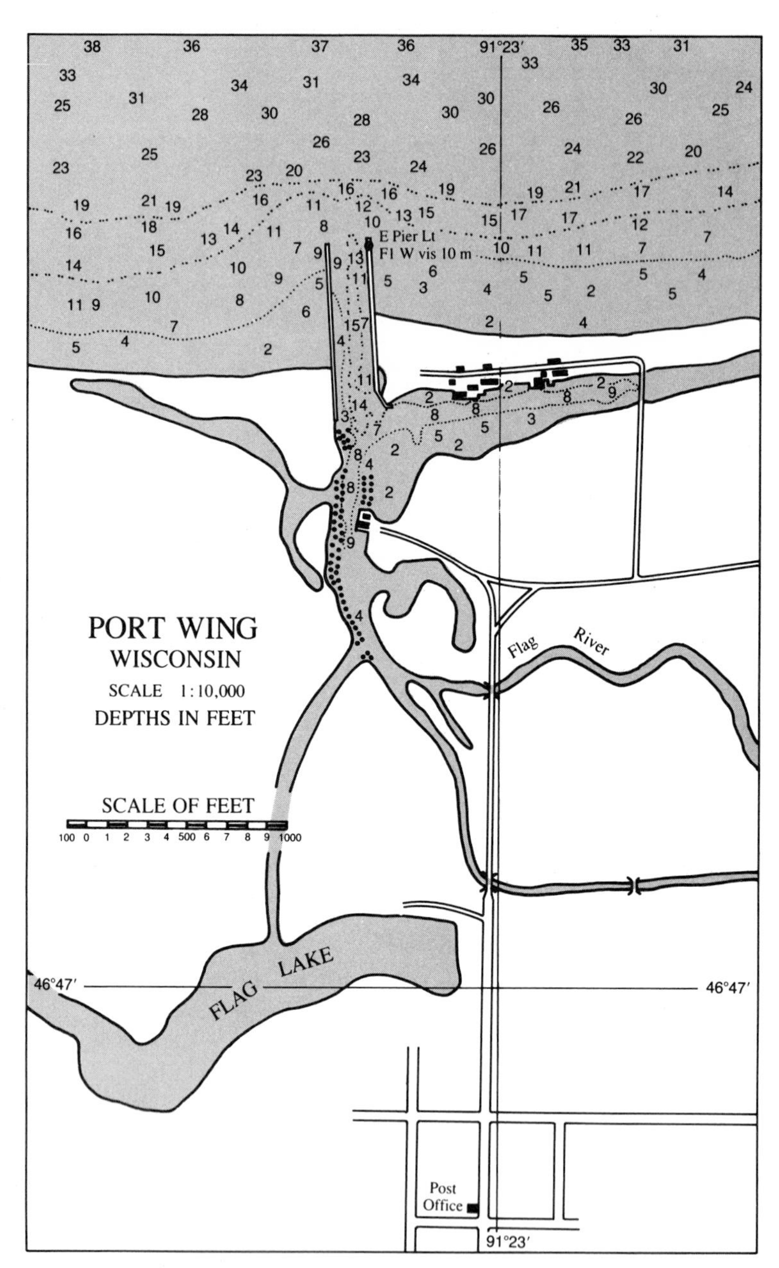

FIGURE 17.1 Harbor chart (courtesy of Lake Survey Center)

FIGURE 17.2 Recording lake gage (courtesy of Lake Survey Center)

water will affect the scheduling of certain construction operations. Soundings that are made to determine depths in tidewater reaches of rivers and in sea harbors for pipeline crossings, submarine cables, sewer outfalls, and other construction must be reduced to the appropriate tidal datum plane.

17.4 TOPOGRAPHIC AND SHORELINE SURVEYS

Although the depths in the water area are most important data on a nautical chart, the topographic features of a seacoast or the shore of a lake are indispensable in providing orientation for the mariner and improving the appearance of the sheet. Topographic surveys provide this information. In the past most of such surveys were made with the plane table and to some extent by the transit-stadia method. Today such ground survey methods are used only for hydrographic surveys of limited extent or for securing the information necessary to make periodic cultural revisions in the shore area of a harbor chart.

For new and extensive hydrographic survey projects the aerial photogrammetric method has very largely superseded ground mapping procedures because of the pronounced economies in time and cost which are made possible by aerial mapping. Mapping the extent of reefs and shoals is also made possible by the examination of aerial photographs by one skilled in their interpretation.

17.5 EQUIPMENT FOR HYDROGRAPHY

The modern survey ships of the National Ocean Survey and the U.S. Naval Oceanographic Office are complete, mobile, and self-sustaining surveying and chart producing plants. They carry the equipment and personnel needed for carrying out the entire hydrographic charting mission, including the procurement of aerial photography, the execution of horizontal and vertical control, the development of hydrography, and the reproduction of the finished chart. The remarks that follow are confined to brief descriptions of the kind of equipment that would be used in conducting sounding operations in a lake, reservoir, or harbor with a small boat or launch rather than from a survey ship operating miles off the seacoast.

BOATS

Various types of launches and small boats are used in hydrographic surveying. Most fishing or working-type launches are satisfactory, because they are seaworthy, have reliable motor performance at low speeds for sounding work, and can be adapted to the various operations associated with hydrographic surveying.

For limited surveys in protected areas small boats like the dinghy can be used. This boat has a rounded bottom and sufficient keel to aid in maintaining a steady course. It can be powered with an outboard motor. Figure 17.3 depicts the type of boat used by the U.S. Corps of Engineers for conducting hydrographic surveys on the Mississippi River. It is equipped with the latest electronic equipment to facilitate position determination and depth measurement.

SOUNDING POLE

For measuring water depths up to 12 ft the soundings can be more easily obtained with a sounding pole. Such a pole can be made of a 15-ft length of 1½-in. rounded lumber, with painted gradations at foot or half-foot intervals and a metal shoe at each end to hasten sinking.

FIGURE 17.3 Hydrographic survey boat (courtesy of Corps of Engineers)

LEADLINE

A leadline or sounding line consists of a suitable length of good quality sash cord, at the end of which a weight or sounding lead is attached. The leadline can be graduated by fathom or foot marks in various distinctive patterns so that no difficulty is experienced in reading the required water depth. In sounding operations the lead is lowered until it touches the bottom; and at the moment that the line is vertical and taut, the depth is determined from the markings on the leadline.

Even a well-seasoned leadline will change its length as the result of normal use. At regular intervals its length should be verified by comparison with a steel tape and, if necessary, suitable corrections should be applied to the observed depths. The following notes show the data from a comparison test made at the end of a day's sounding operations.

Test of Leadline

July 27, 1984 — by Hooper and Kinch

Leadline	Tape (ft)	Leadline	Tape (ft)	Leadline	Tape (ft)
5	4.9	50	47.5	95	91.3
10	9.7	55	52.5	100	96.2
15	14.5	60	57.4	105	100.9
20	19.3	65	62.3	110	105.9
25	23.9	70	67.1	115	110.6
30	28.6	75	71.9	120	115.4
35	33.4	80	76.7	125	120.1
40	38.2	85	81.5	130	124.9
45	42.8	90	86.4	135	129.8

From a plotting of the leadline readings against the computed leadline corrections, a table of corrections was prepared. This table indicates the applicable correction, to the nearest full foot, that is to be applied to soundings in various ranges as follows:

Leadline Corrections

Observed Depth (ft)	Correction (ft) (to be subtracted)
0–14	0
15–29	1
30–49	2
50–84	3
85–119	4
120–140	5

FATHOMETER

On modern hydrographic surveys of any importance or extent, measurements of water depth are made with an echo sounding instrument called a *fathometer*. *Echo sounding* is a method for obtaining water depths by determining the time required for sound waves to travel from a point near the surface of the water to the bottom and back. A fathometer is designed to produce a signal, transmit it downward, receive and amplify the echo, measure the intervening time interval, and automatically convert this interval into feet or fathoms of depth. The fathometer may indicate the depth visually or record it graphically on a roll of specially prepared chart paper. Hence every line of echo soundings provides a virtual profile of the lake or harbor bottom beneath the course of the survey launch even though the boat proceeds at full speed. Water depths can be easily scaled from the resulting *fathogram*. If the instrument (Figure 17.4) is set to operate in the 0 to 70-ft range, the appropriate depth scale reads 47 ft at the bottom of the trench.

SEXTANT

The sextant (see Figure 17.5) is a portable instrument that is used for measuring horizontal and vertical angles from a ship. In celestial navigation the sextant is used to measure the altitude of various prominent stars above the horizon. In hydrographic surveying the sextant is used to measure the horizontal angles between designated objects on land. This makes possible the subsequent solution of the resection problem for determining the position of the sounding vessel at various selected times.

THREE-ARM PROTRACTOR

The three-arm protractor is used extensively in hydrographic surveying to plot boat positions during sounding operations. A transparent

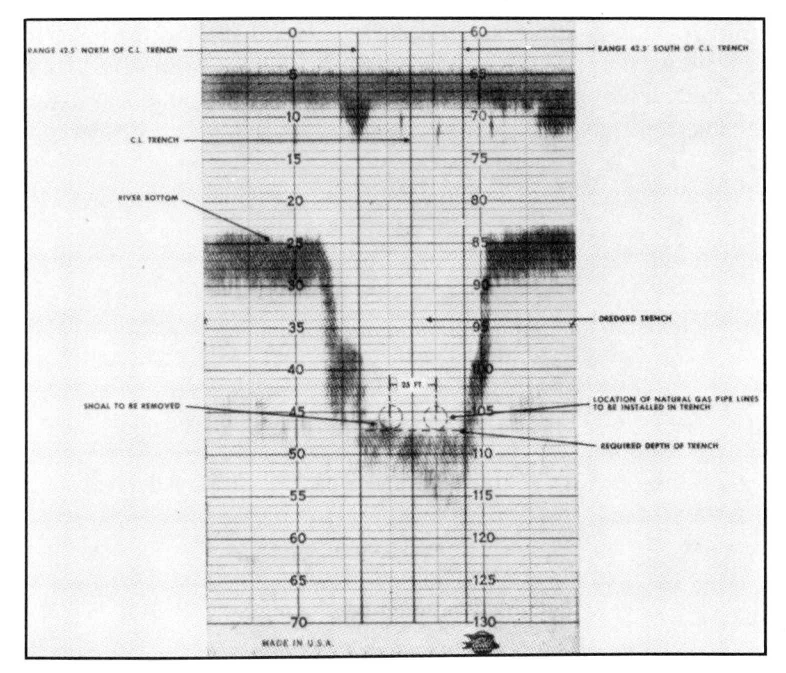

FIGURE 17.4 Fathogram (courtesy of Edo Corporation)

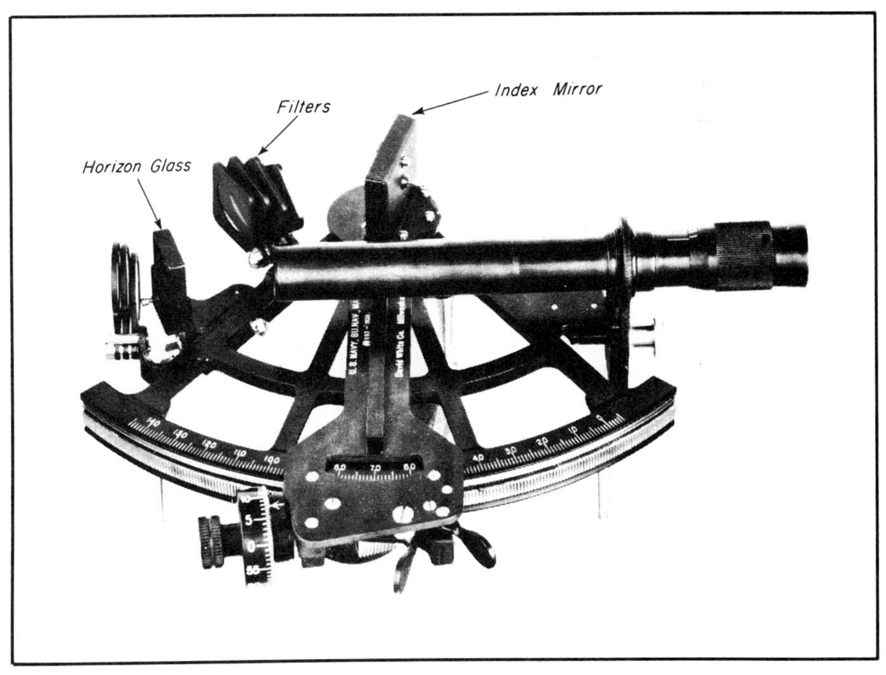

FIGURE 17.5 Sextant (courtesy of David White Corporation)

protractor constructed of celluloid is very satisfactory for all but the most accurate plottings. This protractor consists of a solid disc about 12 in. in diameter containing a circle graduated in degrees, and one fixed and two movable arms extending about 13 in. beyond the edge of the disc. Each of the movable arms has a vernier permitting angles to be set with an accuracy of 2′.

17.6 SOUNDING OPERATIONS

Sounding operations constitute the basic element of the hydrographic survey. However, the determination of depth is useless unless the horizontal position of the point at which the depth was measured is simultaneously obtained.

Although a variety of methods can be employed to locate the soundings, only three principal ones will be mentioned.

1. *By Range Line and One Angle from Shore*. Figure 17.6 shows a common method for locating soundings on small lakes. The sounding craft is kept on a range line as defined by shore signals, and at various intervals a sounding is obtained at the same moment that the bow of the boat or any other appropriate part is "cut in" by a sight with a plane table at the shore station, *A*. The plane table operator and the boat party must synchronize their watches before beginning operations and note and record the time for each sounding. Only in this way is it possible to identify the position of each sounding when the depths are subsequently entered on the plane table sheet.

Although a transit can be utilized for locating soundings by this method, the plane table is much more suitable because it can be used to delineate the shoreline,

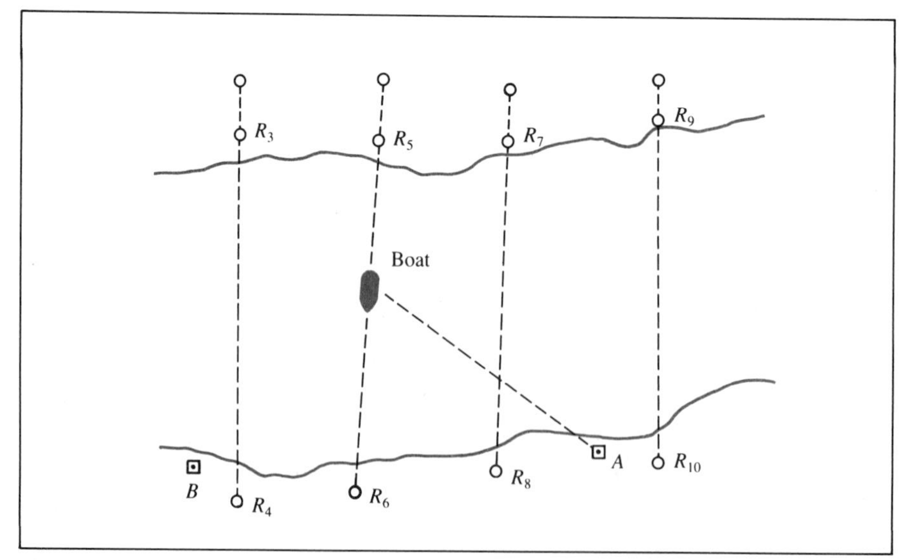

FIGURE 17.6 Range line and one angle from shore

FIGURE 17.7 Locating soundings (courtesy of Tennessee Valley Authority)

obtain the topography, and secure other related information needed to complete the hydrographic survey.

In Figure 17.7 a sounding rig working in a large reservoir is being kept on range by radio telephone from the transit operator on shore while a plane table operator periodically measures the distance to the boat by stadia.

SOUNDINGS							
No.	Time h m	Obser. depth (ft)	Lead-line corr.	Datum corr.	Total corr.	Corr. depth (ft)	Lake Benjamin Windy 72° F Beltrami, County Minn. July 29, 1984
47	10 07	22	−1	−1	−2	20	Thuma-Chief
48	10 09	26	1	1	2	24	Talbot-Leadline
49	10 11	30	2	1	3	27	Boyer-Notes
50	10 13	32	2	1	3	29	Clark-Oars
51	10 15	40	2	1	3	37	Leadline No. 3
52	10 17	47	2	1	3	44	Staff gage: 1.42 ft
53	10 20	52	3	1	4	48	Elev. zero of staff gage = 1347.79 (msl)
54	10 22	62	3	1	4	58	Datum for soundings = Elev. 1348.0
55	10 24	60	3	1	4	56	
56	10 27	58	3	1	4	54	
57	10 31	49	2	1	3	46	
58	10 33	40	2	1	3	37	
59	10 36	30	2	1	3	27	
60	10 38	22	1	1	2	20	
61	10 42	18	1	1	2	16	
62	10 45	15	1	1	2	13	
63	10 47	12	0	1	1	11	
64	10 49	7	0	1	1	6	
65	10 51	3	0	1	1	2	(Near shore by Sta. G)

FIGURE 17.8 Sounding notes (two angles from shore method)

2. *By Two Angles from Shore*. Where it is impracticable to establish ranges because of steep or heavily wooded shores, or where river currents make it difficult to keep the boat on range, the soundings may be located by angles read simultaneously from two transit positions on shore. At a given signal from the boat party both transit operators sight some definite object on the sounding craft and read the horizontal angle. Figures 17.8 and 17.9 show the records of soundings and the transit observations at one shore station, respectively.

3. *By Two Angles from the Boat*. An important and widely used method for locating a sounding is that known as the sextant three-point fix. This procedure involves the simultaneous measurement on board the sounding vessel of the two horizontal angles between three selected shore signals (Figure 17.10) whose positions are known. The angles are measured with sextants at the same moment that the sounding is obtained. The position of the vessel is then immediately determined by using a three-arm protractor, which effects a graphical solution of the problem. The advantages of this method are that all the hydrographic surveying operations are performed on board the survey vessel, the frequency of soundings and their coverage of the area are conveniently discernible, and the boat may be directed to those parts of the lake, river, or harbor where additional development of hydrography seems necessary. Figure 17.11 shows the record of soundings and sextant angles when this method is employed.

The three preceding methods are equally applicable to leadline and echo-sounding operations. However, if a fathometer is used, the sextant three-point fix is particularly useful for locating soundings. At the precise moment the sextant angles

LOCATION OF SOUNDINGS

Station Occupied – ATHENS　　　Station Sighted – CASEY (0°-00′)

Location of soundings					Lake Benjamin K & E No. 5149 Cloudy 92° F	Beltrami County, Minn. Aug. 12, 1984
No.	Time h	m	Azimuth °	′		
82	13	10	299	10	Hooper – transit	
83	13	12	302	15	Chamberlin – notes	
84	13	14	304	05		
85	13	17	308	55		
86	13	20	314	00		
87	13	22	316	50		
88	13	24	320	25		
89	13	27	323	15		
90	13	31	328	10		
91	13	33	330	25		
92	13	34	332	10		
93	13	37	336	30		
94	13	40	338	05		
95	13	43	340	55		
96	13	46	343	10		
97	13	50	348	15		
98	13	52	350	25		

FIGURE 17.9 Transit notes (two angles from shore method)

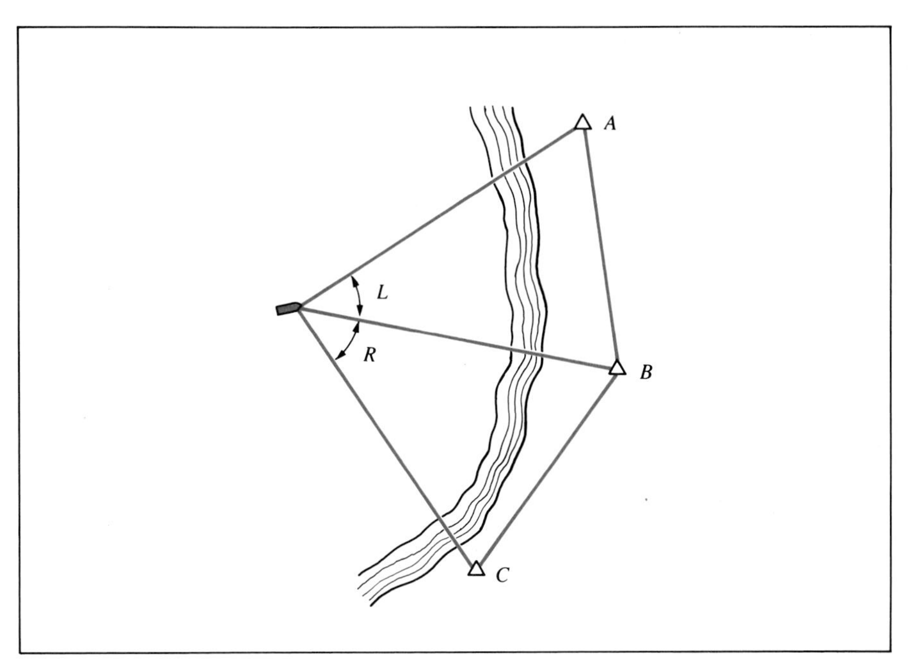

FIGURE 17.10 Two angles from boat

HYDROGRAPHY								Lake Benjamin Calm 88°F
Angles		Signals			Obser. depth (ft)	Total corr.	Corr. depth (ft)	Beltrami County, Minn. June 22, 1984
Left ° ′	Right ° ′	Left	Center	Right				
29 16	73 20	A	B	F	6	−1	5	D. White Sextant No. 7120
29 36	74 08	A	B	F	11	1	10	Leadline No. 3
29 15	77 41	A	B	F	20	2	18	Eck–Chief
31 10	85 25	A	B	F	35	3	32	Easley–notes
107 50	33 27	G	I	J	55	4	51	Stokes–right angle
97 02	30 55	G	I	J	81	4	77	Hursh–left angle
90 43	29 00	G	I	J	96	5	91	Rees–leadline
82 36	28 18	G	I	J	85	5	80	Schnoor–oars
74 00	27 03	G	I	J	66	4	62	Staff gage: 1.27 ft
65 48	27 03	G	I	J	30	3	27	Elev. zero of gage = 1347.79 (msl)
62 16	26 38	G	I	J	20	2	18	Datum for soundings = 1348.0
59 00	26 44	G	I	J	6	1	5	Note: Total sounding correction
57 02	26 39	G	I	J	0	(shore)	0	equals leadline correction
58 02	22 10	G	I	J	12	1	11	minus one foot for datum
60 59	22 26	G	I	J	24	2	22	correction
65 01	22 43	G	I	J	36	3	33	
72 59	22 25	A	C	F	71	4	67	
82 58	18 54	A	C	F	128	6	122	
84 21	17 19	A	C	F	135	6	129	

FIGURE 17.11 Record of soundings and sextant angles (three-point method)

are read, a button is pressed by the fathometer attendant and a mark is thereby registered on the fathogram. A pencil notation is then immediately made on the fathogram identifying the number of the position determination. This makes possible the subsequent scaling of intermediate depths from the fathogram between the consecutive sextant fixes.

17.7 REDUCTION OF SOUNDINGS

The observed soundings must be reduced or converted to values that would have been obtained if the water surface had been coincident with the selected datum or plane of reference. In tidal waters the stage or gage height of the sea above the datum is subtracted from the measured depth. The rapid fluctuation of the tide may make necessary the application of several different reduction factors in the course of one day's sounding operations.

In the case of waters not subjected to tidal influences, a similar but less variable correction must be applied. It is equal to the difference in elevation between the actual water surface and the selected datum and is usually negative. Figure 17.8 indicates the application of this correction to the soundings made in a small inland lake. It is to be noted that this correction is combined with the leadline correction of Section 17.5 and the total correction is subtracted from the observed depths. Since depths were measured to the nearest foot only, all corrections were likewise expressed to the nearest foot.

17.8 PLOTTING SOUNDINGS

The plotting of boat or sounding positions requires no explanation except in the case of a sextant three-point fix (two angles from the boat).

It is necessary to call attention to the use of the boat sheet without which a sextant fix could not be made on board the survey vessel as the hydrographic survey is conducted. The *boat sheet* is the work sheet used by the hydrographer to plot the soundings as soon as they are made. Its use enables the hydrographer to cover an area with properly spaced lines of soundings, to judge the completeness of the survey, and to determine where additional soundings or investigations are needed.

To permit the fixing of the successive positions of the sounding craft the boat sheet must contain the plotted positions of all shore signals. When the left and right sextant angles are simultaneously measured between the objects *A, B,* and *C* (Figure 17.10), these angles are recorded and immediately set off on a three-arm protractor. The protractor is then moved over the boat sheet until the three arms pass through the plotted positions of the appropriate signals. When this is effected, a pencil mark is made at the center of the protractor. This indicates the position of the boat or the sounding.

The hydrographer plots the successive positions of the sounding craft as each new fix is obtained so as to determine whether the area is being systematically covered. Also, each day the depths from the fathogram are entered on the boat sheet.

The depth curves commonly drawn on nautical charts are depicted in Figure 17.1. There they are the one-, two-, and three-fathom curves. Such a chart is usually multicolor.

17.9 CHART PROCUREMENT

Nautical charts and other hydrographic publications relating to United States coastal and intra-coastal waters including the Great Lakes may be purchased from the National Ocean Survey, Riverdale, Maryland 20840. Various other governmental agencies such as certain U.S. Army Engineer Districts, the Tennessee Valley Authority, and the Canadian Hydrographic Service issue maps and charts for their respective areas of jurisdiction.

REFERENCES

17.1 Ciani, John B., and Palmer, Harold D. "Acoustic Techniques for Offshore Positioning," *Journal of the Surveying and Mapping Division,* American Society of Civil Engineers, Vol. 108, No. SU1 (April 1982) 34–45.
17.2 Collins, James. "Coastal Mapping Handbook," *Journal of the Surveying and Mapping Division,* American Society of Civil Engineers, Vol. 106, No. SU1 (November 1980) 23–25.
17.3 Collins, James. "Satellite Positioning at Sea," *Journal of the Surveying and Mapping Division,* American Society of Civil Engineers, Vol. 107, No. SU1 (November 1981) 1–9.
17.4 Jones, George E. "Industries' Role in Marine Surveying and Mapping," *Proceedings of the 31st Annual Meeting,* American Congress on Surveying and Mapping (March 1971) 628–640.
17.5 Lyddan, Robert H. "The National Mapping Program of the United States," *Proceedings of the 36th Annual Meeting,* American Congress on Surveying and Mapping (February 1976) 342–351.
17.6 O'Brien, Thomas J. "Orthophotomapping for Prudhoe Bay Development," *Journal of the Surveying and Mapping Division,* American Society of Civil Engineers (November 1971) 199–204.

APPENDIX A DERIVATIONS

A.1 LAW OF PROPAGATION OF RANDOM ERRORS

Suppose that the value of a parameter, Y, can be calculated from the measured values of n other parameters, say $X_1, X_2, X_3, \ldots, $ and X_n. Let Y be related to the n parameters by a continuous function $F(\ )$; that is,

$$Y = F(X_1, X_2, X_3, \ldots, X_n) \tag{A.1}$$

Furthermore, let $\hat{\sigma}_{X_i}$ be the estimated standard error of parameter X_i, and $\hat{\sigma}_Y$ be the estimated standard error of Y. The law of propagation of random errors states that

$$\hat{\sigma}_Y^2 = \left(\frac{\partial F}{\partial X_1}\right)^2 \hat{\sigma}_{X_1}^2 + \left(\frac{\partial F}{\partial X_2}\right)^2 \hat{\sigma}_{X_2}^2 + \cdots + \left(\frac{\partial F}{\partial X_n}\right)^2 \hat{\sigma}_{X_n}^2 \tag{A.2}$$

where $\partial F/\partial X_i$ is the partial derivative of the function $F(\ )$ with respect to X_i. Equation (A.2) thus relates the estimated standard error of the computed parameter to the estimated standard errors of the measured parameters.

DERIVATION

Let x_i be a measurement of the parameter X_i and let its unknown error be denoted by v_i. Furthermore, let y denote a value of Y computed from one set of measurements: $x_1, x_2, x_3, \ldots, x_n$, and let v_y denote the unknown error in y. Then, according to Eq. (A.1),

$$y - v_y = F(x_1 - v_1, x_2 - v_2, \ldots, x_n - v_n) \tag{A.3}$$

The right-hand side of this expression can be approximated by using Newton's first-order approximation as follows:

$$y - v_y = F(x_1, x_2, \ldots, x_n) - \left(\frac{\partial F}{\partial X_1}\right) v_1 - \left(\frac{\partial F}{\partial X_2}\right) v_2 \cdots - \left(\frac{\partial F}{\partial X_n}\right) v_n \tag{A.4}$$

The partial derivatives in Eq. (A.4) are computed using the measured values $x_1, x_2, \ldots,$ and x_n.

Let

$$\alpha_i = \left(\frac{\partial F}{\partial X_i}\right) \tag{A.5}$$

then Eq. (A.4) can be expressed as follows:

$$y - v_y = F(x_1, x_2, \ldots, x_n) - \alpha_1 v_1 - \alpha_2 v_2 \cdots - \alpha_n v_n \tag{A.6}$$

Since

$$y = F(x_1, x_2, \ldots, x_n),$$
$$v_y = \alpha_1 v_1 + \alpha_2 v_2 + \cdots + \alpha_n v_n \tag{A.7}$$

Suppose that there are m independent sets of measured values (x_1, x_2, . . . , and x_n). A value of y can be computed for each set of ($x_1, x_2, \ldots,$ and x_n). By introducing a second subscript k to denote the quantities associated with the kth set of measurements, Eq. (A.7) may be written as follows:

$$v_{yk} = \alpha_1 v_{1k} + \alpha_2 v_{2k} + \cdots + \alpha_n v_{nk} \tag{A.8}$$

The differences in the values of the coefficient α_j for the k different sets of ($x_1, x_2, \ldots,$ and x_n) are very small and can be neglected.

Squaring both sides of Eq. (A.8) yields

$$\begin{aligned} v_{yk}^2 = {} & \alpha_1^2 v_{1k}^2 + \alpha_2^2 v_{2k}^2 + \cdots + \alpha_n^2 v_{nk}^2 \\ & + 2\,\alpha_1 v_{1k}[\alpha_2 v_{2k} + \alpha_3 v_{3k} + \cdots + \alpha_n v_{nk}] \\ & + 2\,\alpha_2 v_{2k}[\alpha_3 v_{3k} + \alpha_4 v_{4k} + \cdots + \alpha_n v_{nk}] \\ & \;\cdot \\ & \;\cdot \\ & \;\cdot \\ & + 2\,\alpha_{n-1} v_{n-1,k}[\alpha_n v_{nk}] \end{aligned} \tag{A.9}$$

Since

$$\hat{\sigma}_Y = \pm \sqrt{\frac{\sum_{k=1}^{m} v_{yk}^2}{m-1}}$$

Then

$$\hat{\sigma}_Y^2 = \pm \frac{\sum_{k=1}^{m} v_{yk}^2}{m-1} = \frac{\alpha_1^2 \sum_{k=1}^{m} v_{1k}^2}{m-1} + \frac{\alpha_2^2 \sum_{k=1}^{m} v_{2k}^2}{m-1} + \cdots + \frac{\alpha_n^2 \sum_{k=1}^{m} v_{nk}^2}{m-1}$$

$$+ \frac{2\ \alpha_1 \sum_{k=1}^{m} v_{1k}(\alpha_2 v_{2k} + \alpha_3 v_{3k} + \cdots + \alpha_n v_{nk})}{m - 1}$$

$$+ \frac{2\ \alpha_2 \sum_{k=1}^{m} v_{2k}(\alpha_3 v_{3k} + \alpha_4 v_{4k} + \cdots + \alpha_n v_{nk})}{m - 1}$$

$$\vdots$$

$$+ \frac{2\ \alpha_{n-1}\alpha_n \sum_{k=1}^{m} v_{n-1,k} v_{nk}}{m - 1} \tag{A.10}$$

Since it is the basic characteristic of random errors that

$$\sum_{k=1}^{m} v_{ik} = 0$$

it must also be true that

$$\sum_{k=1}^{m} v_{ik} v_{jk} = 0 \tag{A.11}$$

when i is not equal to j. Furthermore,

$$\hat{\sigma}_{X_i} = \pm \sqrt{\frac{\sum_{k=1}^{m} v_{ik}^2}{m - 1}} \tag{A.12}$$

Therefore, Eq. (A.10) can be simplified to the following form:

$$\hat{\sigma}_Y^2 = \alpha_1^2 \hat{\sigma}_{X_1}^2 + \alpha_2^2 \hat{\sigma}_{X_2}^2 + \cdots + \alpha_n^2 \hat{\sigma}_{X_n}^2 \tag{A.13}$$

From Eq. (A.5),

$$\alpha_i = \left(\frac{\partial F}{\partial X_i}\right)$$

Therefore, Eq. (A.13) may be written as follows:

$$\hat{\sigma}_Y^2 = \left(\frac{\partial F}{\partial X_1}\right)^2 \hat{\sigma}_{X_1}^2 + \left(\frac{\partial F}{\partial X_2}\right)^2 \hat{\sigma}_{X_2}^2 + \cdots + \left(\frac{\partial F}{\partial X_n}\right)^2 \hat{\sigma}_{X_n}^2$$

which is the same as Eq. (A.2).

A.2 THE LEAST-SQUARES CONDITION

The method of least squares is based on the following assumptions:

1. There are more measurements than the minimum number needed to determine the unknown parameters.
2. The measurements contain only random errors.
3. The measurements are made independently.

Let $x_1, x_2, \ldots,$ and x_n be the measured values of n parameters, and let $v_1, v_2, \ldots,$ and v_n be the errors in these n measured values. Furthermore, let σ_i be the standard error of the measurement x_i. Then, the most probable solution to the unknown survey parameters is that which results in satisfying the following condition:

$$\sum_{i=1}^{n}\left(\frac{\sigma_o v_i}{\sigma_i}\right)^2 = \text{minimum} \tag{A.14}$$

where σ_o is a constant that can be assigned any value. Hence the solution is called the least-squares solution. The derivation for this least-squares condition follows below.

DERIVATION

Since all the measurement errors are random, the distribution function of error v_i is represented by the following expression:

$$F(v_i) = \frac{1}{\sqrt{2\pi}\,\sigma_i}\, e^{-1/2(v_i/\sigma_i)^2} \tag{A.15}$$

Furthermore, since all the measurements are mutually independent, the joint distribution function of the n errors ($v_1, v_2, \ldots,$ and v_n) is equal to the product of their individual distribution function; that is,

$$F(v_1, v_2, \ldots, v_n) =$$

$$\frac{1}{\sqrt{2\pi}\,\sigma_1}\, e^{-1/2(v_1/\sigma_1)^2}\; \frac{1}{\sqrt{2\pi}\,\sigma_2}\, e^{-1/2(v_2/\sigma_2)^2} \cdots \frac{1}{\sqrt{2\pi}\,\sigma_n}\, e^{-1/2(v_n/\sigma_n)^2}$$

$$F(v_1, v_2, \ldots, v_n) = \left(\frac{1}{\sqrt{2\pi}}\right)^n \left(\frac{1}{\sigma_1}\cdot\frac{1}{\sigma_2}\cdots\frac{1}{\sigma_n}\right) e^{-1/2 \sum_{i=1}^{n} (v_i/\sigma_i)^2} \tag{A.16}$$

The probability that the random errors take on a specific set of values ($v_1, v_2, \ldots,$ and v_n) is then given by the following expression:

$$P(v_1, v_2, \ldots, v_n) = \left(\frac{1}{\sqrt{2\pi}}\right)^n \left(\frac{1}{\sigma_1} \cdot \frac{1}{\sigma_2} \cdots \frac{1}{\sigma_n}\right) e^{-1/2 \sum_{i=1}^{n} (v_i/\sigma_i)^2} d\varepsilon_1 d\varepsilon_2 \cdots d\varepsilon_n \quad \text{(A.17)}$$

where $d\varepsilon_1, d\varepsilon_2, \ldots,$ and $d\varepsilon_n$ are some infinitely small values. The probability $P(v_1, v_2, \ldots, v_n)$ is maximized if

$$\sum_{i=1}^{n} \left(\frac{v_i}{\sigma_i}\right)^2 = \text{minimum}$$

Multiplying the left-hand side by a constant σ_o^2 yields

$$\sum_{i=1}^{n} \left(\frac{\sigma_o v_i}{\sigma_i}\right)^2 = \text{minimum}$$

which is the condition stated in Eq. (A.14) for a least-squares solution.

A.3 THE WEIGHTED MEAN AND ITS ESTIMATED STANDARD ERROR

Let $x_1, x_2, \ldots,$ and x_n be n independent measurements of a quantity and let $\sigma_1, \sigma_2, \ldots,$ and σ_n be the corresponding standard errors of the n measurements. Thus the measurements are assumed to be made with different precision. It can be shown that the most probable value ($\hat{x}$) of the quantity is given by the weighted mean of the n measurements; that is,

$$\hat{x} = \frac{w_1 x_1 + w_2 x_2 + \cdots + w_n x_n}{w_1 + w_2 + \cdots + w_n} \quad \text{(A.18)}$$

where w_i is the weight of x_i as defined by Eq. (2.14). Furthermore, an estimate of the standard error of the weighted mean ($\hat{\sigma}_{\hat{x}}$) can be computed from the following expression:

$$\sigma_{\hat{x}} = \frac{\sigma_o}{\sqrt{\sum_{i=1}^{n} w_i}} \quad \text{(A.19)}$$

DERIVATION

Let v_i denote the error in measurement x_i. Then,

$$v_i = x_i - \hat{x} \tag{A.20}$$

According to the method of least squares, the most probable value, $\hat{x}$, is that which results in

$$\sum_{i=1}^{n} \frac{\sigma_o^2 v_i^2}{\sigma_i^2} = \text{minimum} \tag{A.21}$$

Substituting Eq. (A.20) into Eq. (A.21) yields the following condition:

$$\sum_{i=1}^{n} \frac{\sigma_o^2}{\sigma_i^2} (x_i - \hat{x})^2 = \text{minimum} \tag{A.22}$$

To find the value of $\hat{x}$ that satisfies the least-squares condition, the left-hand side of Eq. (A.22) can be differentiated with respect to $\hat{x}$ and then the results equated to zero. That is,

$$\frac{d}{d\hat{x}} \sum_{i=1}^{n} \frac{\sigma_o^2}{\sigma_i^2} (x_i - \hat{x})^2 = 0$$

Therefore,

$$\sum_{i=1}^{n} \frac{\sigma_o^2}{\sigma_i^2} 2(x_i - \hat{x})(-1) = 0$$

which can be simplified to the following:

$$\sum_{i=1}^{n} \frac{\sigma_o^2}{\sigma_i^2} (x_i - \hat{x}) = 0$$

Since, from Eq. (2.14) in Section 2.11,

$$w_i = \frac{\sigma_o^2}{\sigma_i^2}$$

Then

$$\sum_{i=1}^{n} w_i x_i - \hat{x} \sum_{i=1}^{n} w_i = 0$$

Therefore,

$$\hat{x} = \frac{\sum_{i=1}^{n} w_i x_i}{\sum_{i=1}^{n} w_i} = \frac{w_1 x_1 + w_2 x_2 + \cdots + w_n x_n}{w_1 + w_2 + \cdots + w_n}$$

which is the same as Eq. (A.18).

According to the law of propagation of random errors,

$$\hat{\sigma}_{\hat{x}}^2 = \left(\frac{\partial \hat{x}}{\partial x_1}\right)^2 \sigma_1^2 + \left(\frac{\partial \hat{x}}{\partial x_2}\right)^2 \sigma_2^2 + \cdots + \left(\frac{\partial \hat{x}}{\partial x_n}\right)^2 \sigma_n^2$$

But,

$$\begin{aligned}
\left(\frac{\partial \hat{x}}{\partial x_k}\right)^2 \sigma_k^2 &= \left(\frac{w_k}{\sum_{i=1}^{n} w_i}\right)^2 \sigma_k^2 \\
&= \frac{1}{\left(\sum_{i=1}^{n} w_i\right)^2} \cdot \left(\frac{\sigma_o^2}{\sigma_k^2}\right)^2 \cdot \sigma_k^2 \\
&= \frac{\sigma_o^2 w_k}{\left(\sum_{i=1}^{n} w_i\right)^2}
\end{aligned}$$

Hence,

$$\hat{\sigma}_{\hat{x}}^2 = \frac{\sigma_o^2 \sum_{i=1}^{n} w_i}{\left(\sum_{i=1}^{n} w_i\right)^2} = \frac{\sigma_o^2}{\sum_{i=1}^{n} w_i}$$

which is the same as Eq. (A.19).

APPENDIX B

TABLE I Solar Ephemeris, June, 1983 For 0^h Universal Time or Greenwich Civil Time (from the 1983 Solar Ephemeris, Keuffel & Esser Company)

Day of Month & Week		The Sun's Apparent Declination	Diff. in Declin. for 1 hour	Equation of Time: True Sol. Time = LCT + Eq. of Time	Equation of Time: Differ. for 1 hour	GHA of Polaris
		° ′	′	m s	s	° ′
1	W	N21 56.7	0.34	+02 23.6	0.37	215 35.0
2	TH	N22 04.9	0.33	+02 14.7	0.39	216 34.0
3	FR	N22 12.8	0.31	+02 05.4	0.41	217 32.9
4	SA	N22 20.3	0.30	+01 55.6	0.42	218 31.8
5	SU	N22 27.4	0.28	+01 45.5	0.44	219 30.7
6	M	N22 34.1	0.26	+01 35.0	0.45	220 29.6
7	TU	N22 40.4	0.25	+01 24.2	0.46	221 28.4
8	W	N22 46.4	0.23	+01 13.1	0.48	222 27.3
9	TH	N22 51.9	0.21	+01 01.6	0.49	223 26.1
10	FR	N22 57.0	0.20	+00 49.9	0.50	224 24.9
11	SA	N23 01.7	0.18	+00 38.0	0.51	225 23.6
12	SU	N23 06.0	0.16	+00 25.8	0.52	226 22.4
13	M	N23 09.9	0.15	+00 13.4	0.52	227 21.2
14	TU	N23 13.4	0.13	+00 00.8	0.53	228 20.0
15	W	N23 16.5	0.11	−00 11.8	0.53	229 18.9
16	TH	N23 19.2	0.09	−00 24.6	0.54	230 17.7
17	FR	N23 21.4	0.08	−00 37.5	0.54	231 16.6
18	SA	N23 23.3	0.06	−00 50.5	0.54	232 15.5
19	SU	N23 24.7	0.04	−01 03.5	0.54	233 14.3
20	M	N23 25.7	0.03	−01 16.5	0.54	234 13.1
21	TU	N23 26.3	0.01	−01 29.5	0.54	235 11.9
22	W	N23 26.5	0.01	−01 42.5	0.54	236 10.6
23	TH	N23 26.3	0.03	−01 55.4	0.54	237 09.4
24	FR	N23 25.6	0.04	−02 08.3	0.53	238 08.1
25	SA	N23 24.6	0.06	−02 21.1	0.53	239 06.9
26	SU	N23 23.1	0.08	−02 33.8	0.52	240 05.6
27	M	N23 21.3	0.10	−02 46.4	0.52	241 04.4
28	TU	N23 19.0	0.11	−02 58.8	0.51	242 03.2
29	W	N23 16.3	0.13	−03 11.0	0.50	243 02.0
30	TH	N23 13.2	0.15	−03 23.1	0.49	244 00.8
31	FR	N23 09.7		−03 35.0		244 59.7

Hourly differences in declination and equation of time are for the 24-hours following 0-hours of date in left column.

TABLE II Increase in GHA for Elapsed Time (from the 1983 Solar Ephemeris, Keuffel & Esser Company)

Min.	Hours of Greenwich Civil Time 0h		1h		2h		3h		Sec.	Corr.
	°	′	°	′	°	′	°	′		′
0	0	0.0	15	2.5	30	4.9	45	7.4	0	0.0
1	0	15.0	15	17.5	30	19.9	45	22.4	1	0.3
2	0	30.1	15	32.5	30	35.0	45	37.5	2	0.5
3	0	45.1	15	47.6	30	50.1	45	52.5	3	0.8
4	1	0.2	16	2.6	31	5.1	46	7.6	4	1.0
5	1	15.2	16	17.7	31	20.1	46	22.6	5	1.3
6	1	30.2	16	32.7	31	35.2	46	37.6	6	1.5
7	1	45.3	16	47.8	31	50.2	46	52.7	7	1.8
8	2	0.3	17	2.8	32	5.3	47	7.7	8	2.0
9	2	15.4	17	17.8	32	20.3	47	22.8	9	2.3
10	2	30.4	17	32.9	32	35.3	47	37.8	10	2.5
11	2	45.5	17	47.9	32	50.4	47	52.8	11	2.8
12	3	0.5	18	3.0	33	5.4	48	7.9	12	3.0
13	3	15.5	18	18.0	33	20.5	48	22.9	13	3.3
14	3	30.6	18	33.0	33	35.5	48	38.0	14	3.5
15	3	45.6	18	48.1	33	50.5	48	53.0	15	3.8
16	4	0.7	19	3.1	34	5.6	49	8.0	16	4.0
17	4	15.7	19	18.2	34	20.6	49	23.1	17	4.3
18	4	30.7	19	33.2	34	35.7	49	38.1	18	4.5
19	4	45.8	19	48.2	34	50.7	49	53.2	19	4.8
20	5	0.8	20	3.3	35	5.7	50	8.2	20	5.0
21	5	15.9	20	18.3	35	20.8	50	23.3	21	5.3
22	5	30.9	20	33.4	35	35.8	50	38.3	22	5.5
23	5	45.9	20	48.4	35	50.9	50	53.3	23	5.8
24	6	1.0	21	3.4	36	5.9	51	8.4	24	6.0
25	6	16.0	21	18.5	36	21.0	51	23.4	25	6.3
26	6	31.1	21	33.5	36	36.0	51	38.5	26	6.5
27	6	46.1	21	48.6	36	51.0	51	53.5	27	6.8
28	7	1.1	22	3.6	37	6.1	52	8.5	28	7.0
29	7	16.2	22	18.7	37	21.1	52	23.6	29	7.3
30	7	31.2	22	33.7	37	36.2	52	38.6	30	7.5
31	7	46.3	22	48.7	37	51.2	52	53.7	31	7.8
32	8	1.3	23	3.8	38	6.2	53	8.7	32	8.0
33	8	16.4	23	18.8	38	21.3	53	23.7	33	8.3
34	8	31.4	23	33.9	38	36.3	53	38.8	34	8.5
35	8	46.4	23	48.9	38	51.4	53	53.8	35	8.8
36	9	1.5	24	3.9	39	6.4	54	8.9	36	9.0
37	9	16.5	24	19.0	39	21.4	54	23.9	37	9.3

TABLE II *(continued)*

Min.	0h °	′	1h °	′	2h °	′	3h °	′	Sec.	Corr. ′
38	9	31.6	24	34.0	39	36.5	54	39.0	38	9.5
39	9	46.6	24	49.1	39	51.5	54	54.0	39	9.8
40	10	1.6	25	4.1	40	6.6	55	9.0	40	10.0
41	10	16.7	25	19.1	40	21.6	55	24.1	41	10.3
42	10	31.7	25	34.2	40	36.7	55	39.1	42	10.5
43	10	46.8	25	49.2	40	51.7	55	54.2	43	10.8
44	11	1.8	26	4.3	41	6.7	56	9.2	44	11.0
45	11	16.8	26	19.3	41	21.8	56	24.2	45	11.3
46	11	31.9	26	34.4	41	36.8	56	39.3	46	11.5
47	11	46.9	26	49.4	41	51.9	56	54.3	47	11.8
48	12	2.0	27	4.4	42	6.9	57	9.4	48	12.0
49	12	17.0	27	19.5	42	21.9	57	24.4	49	12.3
50	12	32.1	27	34.5	42	37.0	57	39.4	50	12.5
51	12	47.1	27	49.6	42	52.0	57	54.5	51	12.8
52	13	2.1	28	4.6	43	7.1	58	9.5	52	13.0
53	13	17.2	28	19.6	43	22.1	58	24.6	53	13.3
54	13	32.2	28	34.7	43	37.1	58	39.6	54	13.5
55	13	47.3	28	49.7	43	52.2	58	54.6	55	13.8
56	14	2.3	29	4.8	44	7.2	59	9.7	56	14.0
57	14	17.3	29	19.8	44	22.3	59	24.7	57	14.3
58	14	32.4	29	34.8	44	37.3	59	39.8	58	14.5
59	14	47.4	29	49.9	44	52.4	59	54.8	59	14.8
60	15	2.5	30	4.9	45	7.4	60	9.9	60	15.0

Hours of Greenwich Civil Time

TABLE II *(continued)*

Min.	4h °	′	5h °	′	6h °	′	7h °	′	8h °	′	9h °	′
0	60	9.9	75	12.3	90	14.8	105	17.2	120	19.7	135	22.2
1	60	24.9	75	27.4	90	29.8	105	32.3	120	34.8	135	37.2
2	60	39.9	75	42.4	90	44.9	105	47.3	120	49.8	135	52.3
3	60	55.0	75	57.4	90	59.9	106	2.4	121	4.8	136	7.3
4	61	10.0	76	12.5	91	14.9	106	17.4	121	19.9	136	22.3
5	61	25.1	76	27.5	91	30.0	106	32.5	121	34.9	136	37.4
6	61	40.1	76	42.6	91	45.0	106	47.5	121	50.0	136	52.4

Hours of Greenwich Civil Time

TABLE II *(continued)*

	Hours of Greenwich Civil Time											
Min.	4^h		5^h		6^h		7^h		8^h		9^h	
	°	′	°	′	°	′	°	′	°	′	°	′
7	61	55.1	76	57.6	92	0.1	107	2.5	122	5.0	137	7.5
8	62	10.2	77	12.6	92	15.1	107	17.6	122	20.0	137	22.5
9	62	25.2	77	27.7	92	30.2	107	32.6	122	35.1	137	37.5
10	62	40.3	77	42.7	92	45.2	107	47.7	122	50.1	137	52.6
11	62	55.3	77	57.8	93	0.2	108	2.7	123	5.2	138	7.6
12	63	10.3	78	12.8	93	15.3	108	17.7	123	20.2	138	22.7
13	63	25.4	78	27.9	93	30.3	108	32.8	123	35.2	138	37.7
14	63	40.4	78	42.9	93	45.4	108	47.8	123	50.3	138	52.8
15	63	55.5	78	57.9	94	0.4	109	2.9	124	5.3	139	7.8
16	64	10.5	79	13.0	94	15.4	109	17.9	124	20.4	139	22.8
17	64	25.6	79	28.0	94	30.5	109	32.9	124	35.4	139	37.9
18	64	40.6	79	43.1	94	45.5	109	48.0	124	50.4	139	52.9
19	64	55.6	79	58.1	95	0.6	110	3.0	125	5.5	140	8.0
20	65	10.7	80	13.1	95	15.6	110	18.1	125	20.5	140	23.0
21	65	25.7	80	28.2	95	30.6	110	33.1	125	35.6	140	38.0
22	65	40.8	80	43.2	95	45.7	110	48.2	125	50.6	140	53.1
23	65	55.8	80	58.3	96	0.7	111	3.2	126	5.7	141	8.1
24	66	10.8	81	13.3	96	15.8	111	18.2	126	20.7	141	23.2
25	66	25.9	81	28.3	96	30.8	111	33.3	126	35.7	141	38.2
26	66	40.9	81	43.4	96	45.8	111	48.3	126	50.8	141	53.2
27	66	56.0	81	58.4	97	0.9	112	3.4	127	5.8	142	8.3
28	67	11.0	82	13.5	97	15.9	112	18.4	127	20.9	142	23.3
29	67	26.0	82	28.5	97	31.0	112	33.4	127	35.9	142	38.4
30	67	41.1	82	43.6	97	46.0	112	48.5	127	50.9	142	53.4
31	67	56.1	82	58.6	98	1.1	113	3.5	128	6.0	143	8.4
32	68	11.2	83	13.6	98	16.1	113	18.6	128	21.0	143	23.5
33	68	26.2	83	28.7	98	31.1	113	33.6	128	36.1	143	38.5
34	68	41.2	83	43.7	98	46.2	113	48.6	128	51.1	143	53.6
35	68	56.3	83	58.8	99	1.2	114	3.7	129	6.1	144	8.6
36	69	11.3	84	13.8	99	16.3	114	18.7	129	21.2	144	23.7
37	69	26.4	84	28.8	99	31.3	114	33.8	129	36.2	144	38.7
38	69	41.4	84	43.9	99	46.3	114	48.8	129	51.3	144	53.7
39	69	56.5	84	58.9	100	1.4	115	3.9	130	6.3	145	8.8
40	70	11.5	85	14.0	100	16.4	115	18.9	130	21.4	145	23.8
41	70	26.5	85	29.0	100	31.5	115	33.9	130	36.4	145	38.9
42	70	41.6	85	44.0	100	46.5	115	49.0	130	51.4	145	53.9
43	70	56.6	85	59.1	101	1.5	116	4.0	131	6.5	146	8.9
44	71	11.7	86	14.1	101	16.6	116	19.1	131	21.5	146	24.0
45	71	26.7	86	29.2	101	31.6	116	34.1	131	36.6	146	39.0
46	71	41.7	86	44.2	101	46.7	116	49.1	131	51.6	146	54.1

TABLE II *(continued)*

Min.	Hours of Greenwich Civil Time 4h	5h	6h	7h	8h	9h
	° ′	° ′	° ′	° ′	° ′	° ′
47	71 56.8	86 59.2	102 1.7	117 4.2	132 6.6	147 9.1
48	72 11.8	87 14.3	102 16.8	117 19.2	132 21.7	147 24.1
49	72 26.9	87 29.3	102 31.8	117 34.3	132 36.7	147 39.2
50	72 41.9	87 44.4	102 46.8	117 49.3	132 51.8	147 54.2
51	72 56.9	87 59.4	103 1.9	118 4.3	133 6.8	148 9.3
52	73 12.0	88 14.5	103 16.9	118 19.4	133 21.8	148 24.3
53	73 27.0	88 29.5	103 32.0	118 34.4	133 36.9	148 39.3
54	73 42.1	88 44.5	103 47.0	118 49.5	133 51.9	148 54.4
55	73 57.1	88 59.6	104 2.0	119 4.5	134 7.0	149 9.4
56	74 12.2	89 14.6	104 17.1	119 19.5	134 22.0	149 24.5
57	74 27.2	89 29.7	104 32.1	119 34.6	134 37.0	149 39.5
58	74 42.2	89 44.7	104 47.2	119 49.6	134 52.1	149 54.6
59	74 57.3	89 59.7	105 2.2	120 4.7	135 7.1	150 9.6
60	75 12.3	90 14.8	105 17.2	120 19.7	135 22.2	150 24.6

TABLE III Bearing of Polaris at All Local Hour Angles, 1983 (from 1983 Solar Ephemeris, Keuffel & Esser Company)

1983 computed for a Polar distance of 0°48.50′ for local hour angles 0° to 180° the star is West of North and from 180° to 360° it is East of North.

Lat.	10°	20°	26°	30°	32°	34°	36°	38°	Lat.
LHA	° ′	° ′	° ′	° ′	° ′	° ′	° ′	° ′	LHA
0	0 00.0	0 00.0	0 00.0	0 00.0	0 00.0	0 00.0	0 00.0	0 00.0	360
5	0 04.3	0 04.5	0 04.7	0 04.9	0 05.0	0 05.1	0 05.3	0 05.4	355
10	0 08.6	0 09.0	0 09.4	0 09.8	0 10.0	0 10.3	0 10.5	0 10.8	350
15	0 12.8	0 13.4	0 14.1	0 14.6	0 14.9	0 15.3	0 15.7	0 16.1	345
20	0 16.9	0 17.7	0 18.6	0 19.3	0 19.7	0 20.2	0 20.7	0 21.3	340
25	0 20.9	0 21.9	0 22.9	0 23.8	0 24.4	0 24.9	0 25.6	0 26.3	335
30	0 24.7	0 25.9	0 27.1	0 28.2	0 28.8	0 29.5	0 30.2	0 31.1	330
35	0 28.3	0 29.7	0 31.1	0 32.3	0 33.0	0 33.8	0 34.7	0 35.6	325
40	0 31.7	0 33.3	0 34.9	0 36.2	0 37.0	0 37.9	0 38.8	0 39.9	320
45	0 34.9	0 36.6	0 38.3	0 39.8	0 40.7	0 41.6	0 42.7	0 43.9	315
50	0 37.8	0 39.7	0 41.5	0 43.1	0 44.1	0 45.1	0 46.2	0 47.5	310
55	0 40.4	0 42.4	0 44.4	0 46.1	0 47.1	0 48.2	0 49.4	0 50.7	305
60	0 42.7	0 44.8	0 46.9	0 48.7	0 49.7	0 50.9	0 52.2	0 53.6	300

TABLE III *(continued)*

1983 computed for a Polar distance of 0°48.50′
for local hour angles 0° to 180° the star is West of North
and from 180° to 360° it is East of North.

Lat.	10°	20°	26°	30°	32°	34°	36°	38°	Lat.
LHA	° ′	° ′	° ′	° ′	° ′	° ′	° ′	° ′	LHA
65	0 44.7	0 46.9	0 49.0	0 50.9	0 52.0	0 53.2	0 54.6	0 56.0	295
70	0 46.3	0 48.6	0 50.8	0 52.8	0 53.9	0 55.2	0 56.5	0 58.1	290
75	0 47.6	0 49.9	0 52.2	0 54.2	0 55.4	0 56.6	0 58.1	0 59.6	285
80	0 48.5	0 50.9	0 53.2	0 55.2	0 56.4	0 57.7	0 59.1	1 00.7	280
85	0 49.1	0 51.4	0 53.8	0 55.8	0 57.0	0 58.3	0 59.8	1 01.4	275
90	0 49.2	0 51.6	0 54.0	0 56.0	0 57.2	0 58.5	0 59.9	1 01.5	270
95	0 49.1	0 51.4	0 53.7	0 55.7	0 56.9	0 58.2	0 59.7	1 01.3	265
100	0 48.5	0 50.8	0 53.1	0 55.1	0 56.2	0 57.5	0 58.9	1 00.5	260
105	0 47.5	0 49.8	0 52.0	0 54.0	0 55.1	0 56.4	0 57.8	0 59.3	255
110	0 46.2	0 48.4	0 50.6	0 52.5	0 53.6	0 54.8	0 56.1	0 57.6	250
115	0 44.6	0 46.7	0 48.8	0 50.6	0 51.6	0 52.8	0 54.1	0 55.5	245
120	0 42.6	0 44.6	0 46.6	0 48.3	0 49.3	0 50.4	0 51.7	0 53.0	240
125	0 40.3	0 42.2	0 44.0	0 45.7	0 46.6	0 47.7	0 48.8	0 50.1	235
130	0 37.7	0 39.4	0 41.2	0 42.7	0 43.6	0 44.5	0 45.6	0 46.8	230
135	0 34.8	0 36.4	0 38.0	0 39.4	0 40.2	0 41.1	0 42.1	0 43.2	225
140	0 31.6	0 33.0	0 34.5	0 35.8	0 36.5	0 37.3	0 38.2	0 39.2	220
145	0 28.2	0 29.5	0 30.8	0 31.9	0 32.6	0 33.3	0 34.1	0 35.0	215
150	0 24.6	0 25.7	0 26.8	0 27.8	0 28.4	0 29.0	0 29.7	0 30.5	210
155	0 20.8	0 21.7	0 22.7	0 23.5	0 24.0	0 24.5	0 25.1	0 25.8	205
160	0 16.8	0 17.6	0 18.3	0 19.0	0 19.4	0 19.8	0 20.3	0 20.8	200
165	0 12.7	0 13.3	0 13.9	0 14.4	0 14.7	0 15.0	0 15.4	0 15.8	195
170	0 08.5	0 08.9	0 09.3	0 09.6	0 09.8	0 10.1	0 10.3	0 10.6	190
175	0 04.3	0 04.5	0 04.7	0 04.8	0 04.9	0 05.1	0 05.2	0 05.3	185
180	0 00.0	0 00.0	0 00.0	0 00.0	0 00.0	0 00.0	0 00.0	0 00.0	180

Lat.	40°	42°	44°	46°	48°	50°	60°	70°	Lat.
LHA	° ′	° ′	° ′	° ′	° ′	° ′	° ′	° ′	LHA
0	0 00.0	0 00.0	0 00.0	0 00.0	0 00.0	0 00.0	0 00.0	0 00.0	360
5	0 05.6	0 05.8	0 06.0	0 06.2	0 06.4	0 06.7	0 08.7	0 12.9	355
10	0 11.1	0 11.5	0 11.9	0 12.3	0 12.8	0 13.3	0 17.3	0 25.6	350
15	0 16.6	0 17.1	0 17.7	0 18.3	0 19.0	0 19.9	0 25.7	0 38.1	345
20	0 21.9	0 22.6	0 23.4	0 24.2	0 25.2	0 26.2	0 34.0	0 50.3	340
25	0 27.0	0 27.9	0 28.9	0 29.9	0 31.1	0 32.4	0 41.9	1 02.1	335
30	0 32.0	0 33.0	0 34.1	0 35.4	0 36.7	0 38.3	0 49.5	1 13.4	330
35	0 36.7	0 37.8	0 39.1	0 40.5	0 42.1	0 43.9	0 56.8	1 24.0	325

TABLE III (*continued*)

1983 computed for a Polar distance of 0°48.50′
for local hour angles 0° to 180° the star is West of North
and from 180° to 360° it is East of North.

Lat.	40°	42°	44°	46°	48°	50°	60°	70°	Lat.
LHA	° ′	° ′	° ′	° ′	° ′	° ′	° ′	° ′	LHA
40	0 41.1	0 42.4	0 43.8	0 45.4	0 47.2	0 49.1	1 03.5	1 33.9	320
45	0 45.1	0 46.6	0 48.1	0 49.9	0 51.8	0 54.0	1 09.8	1 43.1	315
50	0 48.9	0 50.4	0 52.1	0 54.0	0 56.1	0 58.4	1 15.5	1 51.4	310
55	0 52.2	0 53.9	0 55.7	0 57.7	0 59.9	1 02.4	1 20.6	1 58.8	305
60	0 55.2	0 56.9	0 58.8	1 00.9	1 03.3	1 05.9	1 25.0	2 05.2	300
65	0 57.7	0 59.5	1 01.5	1 03.7	1 06.1	1 08.9	1 28.8	2 10.6	295
70	0 59.7	1 01.6	1 03.7	1 05.9	1 08.5	1 11.3	1 31.9	2 15.0	290
75	1 01.3	1 03.2	1 05.4	1 07.7	1 10.3	1 13.2	1 34.3	2 18.3	285
80	1 02.5	1 04.4	1 06.6	1 08.9	1 11.6	1 14.5	1 35.9	2 20.5	280
85	1 03.1	1 05.1	1 07.2	1 09.6	1 12.3	1 15.3	1 36.8	2 21.7	275
90	1 03.3	1 05.3	1 07.4	1 09.8	1 12.5	1 15.4	1 37.0	2 21.7	270
95	1 03.0	1 04.9	1 07.1	1 09.5	1 12.1	1 15.0	1 36.4	2 20.7	265
100	1 02.2	1 04.1	1 06.2	1 08.6	1 11.2	1 14.1	1 35.1	2 18.7	260
105	1 01.0	1 02.8	1 04.9	1 07.2	1 09.7	1 12.6	1 33.1	2 15.6	255
110	0 59.3	1 01.1	1 03.1	1 05.3	1 07.7	1 10.5	1 30.4	2 11.5	250
115	0 57.1	0 58.8	1 00.8	1 02.9	1 05.3	1 07.9	1 27.0	2 06.4	245
120	0 54.5	0 56.2	0 58.0	1 00.0	1 02.3	1 04.8	1 23.0	2 00.4	240
125	0 51.5	0 53.1	0 54.8	0 56.7	0 58.8	1 01.2	1 18.4	1 53.6	235
130	0 48.1	0 49.6	0 51.2	0 53.0	0 55.0	0 57.2	1 13.2	1 46.0	230
135	0 44.4	0 45.7	0 47.2	0 48.9	0 50.7	0 52.7	1 07.4	1 37.6	225
140	0 40.3	0 41.5	0 42.9	0 44.4	0 46.0	0 47.9	1 01.2	1 28.5	220
145	0 36.0	0 37.0	0 38.2	0 39.6	0 41.0	0 42.7	0 54.5	1 18.8	215
150	0 31.3	0 32.3	0 33.3	0 34.5	0 35.8	0 37.2	0 47.5	1 08.6	210
155	0 26.5	0 27.3	0 28.1	0 29.1	0 30.2	0 31.4	0 40.1	0 57.9	205
160	0 21.4	0 22.1	0 22.8	0 23.6	0 24.4	0 25.4	0 32.4	0 46.8	200
165	0 16.2	0 16.7	0 17.2	0 17.8	0 18.5	0 19.2	0 24.5	0 35.4	195
170	0 10.9	0 11.2	0 11.6	0 12.0	0 12.4	0 12.9	0 16.4	0 23.7	190
175	0 05.5	0 05.6	0 05.8	0 06.0	0 06.2	0 06.5	0 08.3	0 11.9	185
180	0 00.0	0 00.0	0 00.0	0 00.0	0 00.0	0 00.0	0 00.0	0 00.0	180

TABLE IV Polar Distance of Polaris, 1983 for 0^h Universal Time or Greenwich Civil Time (from 1983 Solar Ephemeris, Keuffel & Esser Company)

Polar Distance			Polar Distance		
1983	Angle ° ′	Cotan	1983	Angle ° ′	Cotan
Jan. 1	0 48.55	70.80	July 10	0 49.02	70.12
11	0 48.52	70.85	20	0 49.02	70.12
21	0 48.50	70.88	30	0 49.00	70.15
31	0 48.50	70.88			
			Aug. 9	0 48.98	70.18
Feb. 10	0 48.50	70.88	19	0 48.96	70.21
20	0 48.52	70.85	29	0 48.92	70.27
Mar. 2	0 48.54	70.82	Sep. 8	0 48.88	70.33
12	0 48.57	70.77	18	0 48.83	70.40
22	0 48.62	70.70	28	0 48.78	70.47
Apr. 1	0 48.66	70.64	Oct. 8	0 48.72	70.56
11	0 48.71	70.57	18	0 48.65	70.66
21	0 48.76	70.50	28	0 48.59	70.75
May 1	0 48.81	70.43	Nov. 7	0 48.53	70.83
11	0 48.86	70.35	17	0 48.47	70.92
21	0 48.90	70.30	27	0 48.40	71.02
31	0 48.94	70.24			
			Dec. 7	0 48.35	71.10
June 10	0 48.98	70.18	17	0 48.30	71.17
20	0 49.00	70.15	27	0 48.26	71.23
30	0 49.01	70.14			

Declination = 90° − Polar Distance

TABLE V Corrections to Preliminary Bearings of Polaris, 1983 (from 1983 Solar Ephemeris, Keuffel & Esser Company)

Bearing	0°	0°20′	0°40′	1°00′	1°20′	1°40′	2°00′	2°20′	2°40′
Polar Dist.									
° ′	′	′	′	′	′	′	′	′	′
0 49.00	0.0	+0.2	+0.4	+0.6	+0.8	+1.0	+1.2	+1.4	+1.6
0 48.90	0.0	+0.2	+0.3	+0.5	+0.7	+0.8	+1.0	+1.2	+1.3
0 48.80	0.0	+0.1	+0.2	+0.4	+0.5	+0.6	+0.7	+0.9	+1.0
0 48.70	0.0	+0.1	+0.2	+0.2	+0.3	+0.4	+0.5	+0.6	+0.7
0 48.60	0.0	+0.0	+0.1	+0.1	+0.2	+0.2	+0.2	+0.3	+0.3
0 48.50	0.0	0.0	0.0	0.0	0.0	0.0	0.0	0.0	0.0
0 48.40	0.0	−0.0	−0.1	−0.1	−0.2	−0.2	−0.2	−0.3	−0.3
0 48.30	0.0	−0.1	−0.2	−0.2	−0.3	−0.4	−0.5	−0.6	−0.7
0 48.20	0.0	−0.1	−0.2	−0.4	−0.5	−0.6	−0.7	−0.9	−1.0
0 48.10	0.0	−0.2	−0.3	−0.5	−0.7	−0.8	−1.0	−1.2	−1.3
0 48.00	0.0	−0.2	−0.4	−0.6	−0.8	−1.0	−1.2	−1.4	−1.6

TABLE VI Corrections to Be Applied to Latitude to Obtain the True Altitude of Polaris, 1983 (from 1983 Solar Ephemeris, Keuffel & Esser Company)

t	Corr.	t	Corr.	t	Corr.	t	Corr.
°	′	°	′	°	′	°	′
000	+48.5	045	+34.1	090	−00.3	135	−34.5
001	+48.5	046	+33.5	091	−01.2	136	−35.1
002	+48.5	047	+32.9	092	−02.0	137	−35.6
003	+48.4	048	+32.3	093	−02.9	138	−36.2
004	+48.4	049	+31.6	094	−03.7	139	−36.7
005	+48.3	050	+31.0	095	−04.6	140	−37.3
006	+48.2	051	+30.3	096	−05.4	141	−37.8
007	+48.1	052	+29.6	097	−06.2	142	−38.3
008	+48.0	053	+29.0	098	−07.1	143	−38.9
009	+47.9	054	+28.3	099	−07.9	144	−39.4
010	+47.8	055	+27.6	100	−08.8	145	−39.8
011	+47.6	056	+26.9	101	−09.6	146	−40.3
012	+47.4	057	+26.2	102	−10.4	147	−40.8
013	+47.2	058	+25.5	103	−11.2	148	−41.2
014	+47.0	059	+24.7	104	−12.1	149	−41.7

TABLE VI (*continued*)

t	Corr.	t	Corr.	t	Corr.	t	Corr.
°	′	°	′	°	′	°	′
015	+46.3	060	+24.0	105	−12.9	150	−42.1
016	+46.6	061	+23.2	106	−13.7	151	−42.5
017	+46.4	062	+22.5	107	−14.5	152	−42.9
018	+46.1	063	+21.7	108	−15.3	153	−43.3
019	+45.8	064	+21.0	109	−16.1	154	−43.7
020	+45.5	065	+20.2	110	−16.9	155	−44.0
021	+45.2	066	+19.4	111	−17.7	156	−44.4
022	+44.9	067	+18.7	112	−18.5	157	−44.7
023	+44.6	068	+17.9	113	−19.2	158	−45.0
024	+44.2	069	+17.1	114	−20.0	159	−45.3
025	+43.9	070	+16.3	115	−20.8	160	−45.6
026	+43.5	071	+15.5	116	−21.5	161	−45.9
027	+43.1	072	+14.7	117	−22.3	162	−46.2
028	+42.7	073	+13.9	118	−23.0	163	−46.4
029	+42.3	074	+13.1	119	−23.8	164	−46.6
030	+41.9	075	+12.2	120	−24.5	165	−46.9
031	+41.5	076	+11.4	121	−25.2	166	−47.1
032	+41.0	077	+10.6	122	−25.9	167	−47.3
033	+40.6	078	+09.8	123	−26.7	168	−47.5
034	+40.1	079	+08.9	124	−27.4	169	−47.6
035	+39.6	080	+08.1	125	−28.0	170	−47.8
036	+39.1	081	+07.3	126	−28.7	171	−47.9
037	+38.6	082	+06.4	127	−29.4	172	−48.0
038	+38.1	083	+05.6	128	−30.1	173	−48.1
039	+37.6	084	+04.7	129	−30.7	174	−48.2
040	+37.0	085	+03.9	130	−31.4	175	−48.3
041	+36.5	086	+03.0	131	−32.0	176	−48.4
042	+35.9	087	+02.2	132	−32.6	177	−48.4
043	+35.3	088	+01.4	133	−33.3	178	−48.5
044	+34.7	089	+00.5	134	−33.9	179	−48.5
045	+34.1	090	−00.3	135	−34.5	180	−48.5

This table is computed for latitude 45°00′00″N and declination of 89°11′30″N.

TABLE VII Stadia Reductions

	0°		1°		2°		3°	
Minutes	Hor. Dist.	Diff. Elev.	Hor. Dist.	Diff. Elev.	Hor. Dist.	Diff. Elev.	Hor. Dist.	Diff. Elev.
0	100.00	.00	99.97	1.74	99.88	3.49	99.73	5.23
2	100.00	.06	99.97	1.80	99.87	3.55	99.72	5.28
4	100.00	.12	99.97	1.86	99.87	3.60	99.71	5.34
6	100.00	.17	99.96	1.92	99.87	3.66	99.71	5.40
8	100.00	.23	99.96	1.98	99.86	3.72	99.70	5.46
10	100.00	.29	99.96	2.04	99.86	3.78	99.69	5.52
12	100.00	.35	99.96	2.09	99.85	3.84	99.69	5.57
14	100.00	.41	99.95	2.15	99.85	3.89	99.68	5.63
16	100.00	.47	99.95	2.21	99.84	3.95	99.68	5.69
18	100.00	.52	99.95	2.27	99.84	4.01	99.67	5.75
20	100.00	.58	99.95	2.33	99.83	4.07	99.66	5.80
22	100.00	.64	99.94	2.38	99.83	4.13	99.66	5.86
24	100.00	.70	99.94	2.44	99.82	4.18	99.65	5.92
26	99.99	.76	99.94	2.50	99.82	4.24	99.64	5.98
28	99.99	.81	99.93	2.56	99.81	4.30	99.63	6.04
30	99.99	.87	99.93	2.62	99.81	4.36	99.63	6.09
32	99.99	.93	99.93	2.67	99.80	4.42	99.62	6.15
34	99.99	.99	99.93	2.73	99.80	4.47	99.61	6.21
36	99.99	1.05	99.92	2.79	99.79	4.53	99.61	6.27
38	99.99	1.11	99.92	2.85	99.79	4.59	99.60	6.32
40	99.99	1.16	99.92	2.91	99.78	4.65	99.59	6.38
42	99.99	1.22	99.91	2.97	99.78	4.71	99.58	6.44
44	99.98	1.28	99.91	3.02	99.77	4.76	99.58	6.50
46	99.98	1.34	99.90	3.08	99.77	4.82	99.57	6.56
48	99.98	1.40	99.90	3.14	99.76	4.88	99.56	6.61
50	99.98	1.45	99.90	3.20	99.76	4.94	99.55	6.67
52	99.98	1.51	99.89	3.26	99.75	4.99	99.55	6.73
54	99.98	1.57	99.89	3.31	99.74	5.05	99.54	6.79
56	99.97	1.63	99.89	3.37	99.74	5.11	99.53	6.84
58	99.97	1.69	99.88	3.43	99.73	5.17	99.52	6.90
60	99.97	1.74	99.88	3.49	99.73	5.23	99.51	6.96

TABLE VII *(continued)*

	4°		5°		6°		7°	
Minutes	Hor. Dist.	Diff. Elev.	Hor. Dist.	Diff. Elev.	Hor. Dist.	Diff. Elev.	Hor. Dist.	Diff. Elev.
0	99.51	6.96	99.24	8.68	98.91	10.40	98.51	12.10
2	99.51	7.02	99.23	8.74	98.90	10.45	98.50	12.15
4	99.50	7.07	99.22	8.80	98.88	10.51	98.49	12.21
6	99.49	7.13	99.21	8.85	98.87	10.57	98.47	12.27
8	99.48	7.19	99.20	8.91	98.86	10.62	98.46	12.32
10	99.47	7.25	99.19	8.97	98.85	10.68	98.44	12.38
12	99.46	7.30	99.18	9.03	98.83	10.74	98.43	12.43
14	99.46	7.36	99.17	9.08	98.82	10.79	98.41	12.49
16	99.45	7.42	99.16	9.14	98.81	10.85	98.40	12.55
18	99.44	7.48	99.15	9.20	98.80	10.91	98.39	12.60
20	99.43	7.53	99.14	9.25	98.78	10.96	98.37	12.66
22	99.42	7.59	99.13	9.31	98.77	11.02	98.36	12.72
24	99.41	7.65	99.11	9.37	98.76	11.08	98.34	12.77
26	99.40	7.71	99.10	9.43	98.74	11.13	98.33	12.83
28	99.39	7.76	99.09	9.48	98.73	11.19	98.31	12.88
30	99.38	7.82	99.08	9.54	98.72	11.25	98.30	12.94
32	99.38	7.88	99.07	9.60	98.71	11.30	98.28	13.00
34	99.37	7.94	99.06	9.65	98.69	11.36	98.27	13.05
36	99.36	7.99	99.05	9.71	98.68	11.42	98.25	13.11
38	99.35	8.05	99.04	9.77	98.67	11.47	98.24	13.17
40	99.34	8.11	99.03	9.83	98.65	11.53	98.22	13.22
42	99.33	8.17	99.01	9.88	98.64	11.59	98.20	13.28
44	99.32	8.22	99.00	9.94	98.63	11.64	98.19	13.33
46	99.31	8.28	98.99	10.00	98.61	11.70	98.17	13.39
48	99.30	8.34	98.98	10.05	98.60	11.76	98.16	13.45
50	99.29	8.40	98.97	10.11	98.58	11.81	98.14	13.50
52	99.28	8.45	98.96	10.17	98.57	11.87	98.13	13.56
54	99.27	8.51	98.94	10.22	98.56	11.93	98.11	13.61
56	99.26	8.57	98.93	10.28	98.54	11.98	98.10	13.67
58	99.25	8.63	98.92	10.34	98.53	12.04	98.08	13.73
60	99.24	8.68	98.91	10.40	98.51	12.10	98.06	13.78

TABLE VII (*continued*)

	8°		9°		10°		11°	
Minutes	Hor. Dist.	Diff. Elev.	Hor. Dist.	Diff. Elev.	Hor. Dist.	Diff. Elev.	Hor. Dist.	Diff. Elev.
0	98.06	13.78	97.55	15.45	96.98	17.10	96.36	18.73
2	98.05	13.84	97.53	15.51	96.96	17.16	96.34	18.78
4	98.03	13.89	97.52	15.56	96.94	17.21	96.32	18.84
6	98.01	13.95	97.50	15.62	96.92	17.26	96.29	18.89
8	98.00	14.01	97.48	15.67	96.90	17.32	96.27	18.95
10	97.98	14.06	97.46	15.73	96.88	17.37	96.25	19.00
12	97.97	14.12	97.44	15.78	96.86	17.43	96.23	19.05
14	97.95	14.17	97.43	15.84	96.84	17.48	96.21	19.11
16	97.93	14.23	97.41	15.89	96.82	17.54	96.18	19.16
18	97.92	14.28	97.39	15.95	96.80	17.59	96.16	19.21
20	97.90	14.34	97.37	16.00	96.78	17.65	96.14	19.27
22	97.88	14.40	97.35	16.06	96.76	17.70	96.12	19.32
24	97.87	14.45	97.33	16.11	96.74	17.76	96.09	19.38
26	97.85	14.51	97.31	16.17	96.72	17.81	96.07	19.43
28	97.83	14.56	97.29	16.22	96.70	17.86	96.05	19.48
30	97.82	14.62	97.28	16.28	96.68	17.92	96.03	19.54
32	97.80	14.67	97.26	16.33	96.66	17.97	96.00	19.59
34	97.78	14.73	97.24	16.39	96.64	18.03	95.98	19.64
36	97.76	14.79	97.22	16.44	96.62	18.08	95.96	19.70
38	97.75	14.84	97.20	16.50	96.60	18.14	95.93	19.75
40	97.73	14.90	97.18	16.55	96.57	18.19	95.91	19.80
42	97.71	14.95	97.16	16.61	96.55	18.24	95.89	19.86
44	97.69	15.01	97.14	16.66	96.53	18.30	95.86	19.91
46	97.68	15.06	97.12	16.72	96.51	18.35	95.84	19.96
48	97.66	15.12	97.10	16.77	96.49	18.41	95.82	20.02
50	97.64	15.17	97.08	16.83	96.47	18.46	95.79	20.07
52	97.62	15.23	97.06	16.88	96.45	18.51	95.77	20.12
54	97.61	15.28	97.04	16.94	96.42	18.57	95.75	20.18
56	97.59	15.34	97.02	16.99	96.40	18.62	95.72	20.23
58	97.57	15.40	97.00	17.05	96.38	18.68	95.70	20.28
60	97.55	15.45	96.98	17.10	96.36	18.73	95.68	20.34

TABLE VII *(continued)*

Minutes	12° Hor. Dist.	12° Diff. Elev.	13° Hor. Dist.	13° Diff. Elev.	14° Hor. Dist.	14° Diff. Elev.	15° Hor. Dist.	15° Diff. Elev.
0	95.68	20.34	94.94	21.92	94.15	23.47	93.30	25.00
2	95.65	20.39	94.91	21.97	94.12	23.52	93.27	25.05
4	95.63	20.44	94.89	22.02	94.09	23.58	93.24	25.10
6	95.61	20.50	94.86	22.08	94.07	23.63	93.21	25.15
8	95.58	20.55	94.84	22.13	94.04	23.68	93.18	25.20
10	95.56	20.60	94.81	22.18	94.01	23.73	93.16	25.25
12	95.53	20.66	94.79	22.23	93.98	23.78	93.13	25.30
14	95.51	20.71	94.76	22.28	93.95	23.83	93.10	25.35
16	95.49	20.76	94.73	22.34	93.93	23.88	93.07	25.40
18	95.46	20.81	94.71	22.39	93.90	23.93	93.04	25.45
20	95.44	20.87	94.68	22.44	93.87	23.99	93.01	25.50
22	95.41	20.92	94.66	22.49	93.84	24.04	92.98	25.55
24	95.39	20.97	94.63	22.54	93.82	24.09	92.95	25.60
26	95.36	21.03	94.60	22.60	93.79	24.14	92.92	25.65
28	95.34	21.08	94.58	22.65	93.76	24.19	92.89	25.70
30	95.32	21.13	94.55	22.70	93.73	24.24	92.86	25.75
32	95.29	21.18	94.52	22.75	93.70	24.29	92.83	25.80
34	95.27	21.24	94.50	22.80	93.67	24.34	92.80	25.85
36	95.24	21.29	94.47	22.85	93.65	24.39	92.77	25.90
38	95.22	21.34	94.44	22.91	93.62	24.44	92.74	25.95
40	95.19	21.39	94.42	22.96	93.59	24.49	92.71	26.00
42	95.17	21.45	94.39	23.01	93.56	24.55	92.68	26.05
44	95.14	21.50	94.36	23.06	93.53	24.60	92.65	26.10
46	95.12	21.55	94.34	23.11	93.50	24.65	92.62	26.15
48	95.09	21.60	94.31	23.16	93.47	24.70	92.59	26.20
50	95.07	21.66	94.28	23.22	93.45	24.75	92.56	26.25
52	95.04	21.71	94.26	23.27	93.42	24.80	92.53	26.30
54	95.02	21.76	94.23	23.32	93.39	24.85	92.49	26.35
56	94.99	21.81	94.20	23.37	93.36	24.90	92.46	26.40
58	94.97	21.87	94.17	23.42	93.33	24.95	92.43	26.45
60	94.94	21.92	94.15	23.47	93.30	25.00	92.40	26.50

TABLE VIII Elevation Factors

Elevation (ft)	Elevation Factor	Elevation (ft)	Elevation Factor
Sea level	1.0000000	3,000	0.9998565
500	0.9999761	3,500	0.9998326
1,000	0.9999522	4,000	0.9998087
1,500	0.9999283	4,500	0.9997848
2,000	0.9999043	5,000	0.9997609
2,500	0.9998804	5,500	0.9997370

TABLE IX Elevation Corrections, in Feet, per 1,000 ft

Elevation (ft)	Correction	Elevation (ft)	Correction
Sea level	0.00	3,000	0.1435
500	0.0239	3,500	0.1674
1,000	0.0478	4,000	0.1913
1,500	0.0717	4,500	0.2152
2,000	0.0957	5,000	0.2391
2,500	0.1196	5,500	0.2630

TABLE X Values of R, y', and Scale Factors—Minnesota North Zone

Lambert Projection for Minnesota—North Zone:

Lat.	R (ft)	y' y Value on Central Meridian (ft)	Tabular Difference for 1″ of Lat. (ft)	Scale in Units of 7th Place of Logs	Scale Expressed as a Ratio
47° 31′	19,100,580.81	370,817.94	101.31550	−355.2	0.9999182
32	19,095,501.88	376,896.87	101.31550	−362.0	0.9999166
33	19,088,422.95	382,975.80	101.31567	−368.4	0.9999152
34	19,082,344.01	389,054.74	101.31583	−374.5	0.9999138
35	19,076,265.06	395,133.69	101.31600	−380.2	0.9999125
47° 36′	19,070,186.10	401,212.65	101.31617	−385.6	0.9999112
37	19,064,107.13	407,291.62	101.31633	−390.6	0.9999101
38	19,058,028.15	413,370.60	101.31650	−395.2	0.9999090
39	19,051,949.16	419,449.59	101.31683	−399.4	0.9999080
40	19,045,870.15	425,528.60	101.31700	−403.3	0.9999071
47° 41′	19,039,791.13	431,607.62	101.31717	−406.8	0.9999063
42	19,033,712.10	437,686.65	101.31750	−410.0	0.9999056
43	19,027,633.05	443,765.70	101.31767	−412.8	0.9999050
44	19,021,553.99	449,844.76	101.31783	−415.2	0.9999044
45	19,015,474.92	455,923.83	101.31817	−417.3	0.9999039
47° 46′	19,009,395.83	462,002.92	101.31850	−419.0	0.9999035
47	19,003,316.72	468,082.03	101.31867	−420.3	0.9999032
48	18,997,237.60	474,161.15	101.31900	−421.2	0.9999030
49	18,991,158.46	480,240.29	101.31933	−421.8	0.9999029
50	18,985,079.30	486,319.45	101.31950	−422.1	0.9999028

TABLE XI Values of θ—Minnesota North Zone

Lambert Projection for Minnesota—North Zone
1″ of long. = 0″.7412196637 of θ:

Long.	θ	Long.	θ
94° 21′	−0 55 35.4885	94° 41′	−1 10 24.9521
22	−0 56 19.9617	42	−1 11 09.4253
23	−0 57 04.4348	43	−1 11 53.8984
24	−0 57 48.9080	44	−1 12 38.3716
25	−0 58 33.3812	45	−1 13 22.8448
94° 26′	−0 59 17.8543	94° 46′	−1 14 07.3180
27	−1 00 02.3276	47	−1 14 51.7912
28	−1 00 46.8007	48	−1 15 36.2643
29	−1 01 31.2739	49	−1 16 20.7375
30	−1 02 15.7471	50	−1 17 05.2107
94° 31′	−1 03 00.2202	94° 51′	−1 17 49.6839
32	−1 03 44.6935	52	−1 18 34.1571
33	−1 04 29.1666	53	−1 19 18.6302
34	−1 05 13.6398	54	−1 20 03.1034
35	−1 05 58.1130	55	−1 20 47.5766
94° 36′	−1 06 42.5862	94° 56′	−1 21 32.0498
37	−1 07 27.0594	57	−1 22 16.5230
38	−1 08 11.5325	58	−1 23 00.9961
39	−1 08 56.0057	59	−1 23 45.4693
40	−1 09 40.4789	95° 00′	−1 24 29.9425

TABLE XII Values of H and V—Illinois East Zone

Transverse Mercator Projection—Illinois, East Zone:

Lat.	Y_0 (ft)	ΔY_0 per second	H	ΔH per second	V	ΔV per second	a
40° 35′	1,426,385.98	101.19683	77.158010	319.28	1.216989	1.85	−0.509
40 36	1,432,457.79	101.19700	77.138853	319.42	1.217100	1.85	−0.507
40 37	1,438,529.61	101.19733	77.119688	319.52	1.217211	1.83	−0.505
40 38	1,444,601.45	101.19767	77.100517	319.62	1.217321	1.83	−0.503
40 39	1,450,673.31	101.19800	77.081340	319.73	1.217431	1.82	−0.501
40 40	1,456,745.19	101.19817	77.062156	319.85	1.217540	1.82	−0.499
40 41	1,462,817.08	101.19850	77.042965	319.95	1.217649	1.80	−0.497
40 42	1,468,888.99	101.19883	77.023768	320.05	1.217757	1.80	−0.495
40 43	1,474,960.92	101.19917	77.004565	320.18	1.217865	1.80	−0.493
40 44	1,481,032.87	101.19950	76.985354	320.27	1.217973	1.78	−0.491
40 45	1,487,104.84	101.19967	76.966138	320.40	1.218080	1.78	−0.489
40 46	1,493,176.82	101.20000	76.946914	320.48	1.218187	1.77	−0.487
40 47	1,499,248.82	101.20033	76.927685	320.62	1.218293	1.77	−0.485
40 48	1,505,320.84	101.20067	76.908448	320.72	1.218399	1.77	−0.483
40 49	1,511,392.88	101.20083	76.889205	320.82	1.218505	1.75	−0.481
40 50	1,517,464.93	101.20133	76.869956	320.93	1.218610	1.75	−0.479
40 51	1,523,537.01	101.20150	76.850700	321.05	1.218715	1.73	−0.477
40 52	1,529,609.10	101.20167	76.831437	321.15	1.218819	1.73	−0.475
40 53	1,535,681.20	101.20217	76.812168	321.25	1.218923	1.73	−0.473
40 54	1,541,753.33	101.20233	76.792893	321.38	1.219027	1.72	−0.471

TABLE XIII Values of b and c—Illinois Zones

Transverse Mercator Projection—Illinois, Both Zones:			
$\Delta\lambda''$	b	Δb	c
0	0.000	+0.212	0.000
100	+0.212	+0.212	0.000
200	+0.424	+0.210	−0.001
300	+0.634	+0.208	−0.002
400	+0.842	+0.207	−0.003
500	+1.049	+0.203	−0.005
600	+1.252	+0.201	−0.007
700	+1.453	+0.196	−0.010
800	+1.649	+0.192	−0.014
900	+1.841	+0.187	−0.018
1000	+2.028	+0.181	−0.022
1100	+2.209	+0.175	−0.027
1200	+2.384	+0.169	−0.032
1300	+2.553	+0.162	−0.038
1400	+2.715	+0.153	−0.043
1500	+2.868	+0.146	−0.049
1600	+3.014	+0.137	−0.055
1700	+3.151	+0.128	−0.061
1800	+3.279	+0.118	−0.067
1900	+3.397	+0.107	−0.073
2000	+3.504	+0.097	−0.079

TABLE XIV Values of g—Illinois Zones

Transverse Mercator Projection—Values of g:							
	$\Delta\lambda''$						
Latitude	0″	1000″	2000″	3000″	4000″	5000″	6000″
36°	0.00	0.00	0.02	0.08	0.19	0.38	0.65
37	0	0	0.02	0.08	0.19	0.38	0.65
38	0	0	0.02	0.08	0.19	0.38	0.65
39	0	0	0.02	0.08	0.19	0.37	0.64
40	0	0	0.02	0.08	0.19	0.37	0.64
41	0	0	0.02	0.08	0.19	0.37	0.63
42	0	0	0.02	0.08	0.18	0.36	0.63
43	0	0	0.02	0.08	0.18	0.36	0.62
44	0	0	0.02	0.08	0.18	0.35	0.61
45	0	0	0.02	0.08	0.18	0.35	0.60

TABLE XV Values of Scale Factors—Illinois East Zone

Transverse Mercator Projection—Illinois, East Zone:					
x' (ft)	Scale in Units of 7th Place of Logs	Scale Expressed as a Ratio	x' (ft)	Scale in Units of 7th Place of Logs	Scale Expressed as a Ratio
0	−108.6	0.9999750	75,000	−80.7	0.9999814
5,000	−108.5	0.9999750	80,000	−76.8	0.9999823
10,000	−108.1	0.9999751	85,000	−72.7	0.9999833
15,000	−107.5	0.9999752	90,000	−68.4	0.9999843
20,000	−106.6	0.9999755	95,000	−63.8	0.9999853
25,000	−105.5	0.9999757	100,000	−58.9	0.9999864
30,000	−104.1	0.9999760	105,000	−53.9	0.9999876
35,000	−102.5	0.9999764	110,000	−48.5	0.9999888
40,000	−100.7	0.9999768	115,000	−42.9	0.9999901
45,000	−98.5	0.9999773	120,000	−37.1	0.9999915
50,000	−96.2	0.9999778	125,000	−31.0	0.9999929
55,000	−93.6	0.9999784	130,000	−24.7	0.9999943
60,000	−90.7	0.9999791	135,000	−18.1	0.9999958
65,000	−87.6	0.9999798	140,000	−11.3	0.9999974
70,000	−84.3	0.9999806	145,000	−4.2	0.9999990

TABLE XVI Greek Letters

Form		English Equivalent	Name
Α	α	a	Alpha
Β	β	b	Beta
Γ	γ	g	Gamma
Δ	δ	d	Delta
Ε	ϵ	e (short)	Epsilon
Ζ	ζ	z	Zeta
Η	η	e (long)	Eta
Θ	θ	th	Theta
Ι	ι	i	Iota
Κ	κ	k, c	Kappa
Λ	λ	l	Lambda
Μ	μ	m	Mu
Ν	ν	n	Nu
Ξ	ξ	x	Xi
Ο	ο	o	Omicron
Π	π	p	Pi
Ρ	ρ	r	Rho
Σ	σ	s	Sigma
Τ	τ	t	Tau
Υ	υ	u	Upsilon
Φ	ϕ	ph	Phi
Χ	χ	ch	Chi
Ψ	ψ	ps	Psi
Ω	ω	o	Omega

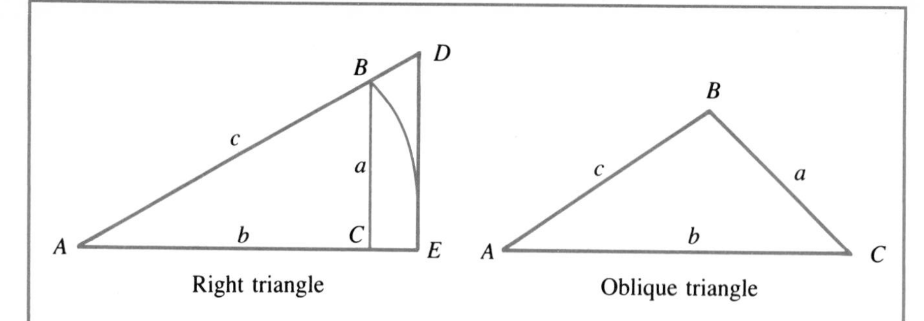

Right triangle

Oblique triangle

Right Triangles

$$\sin A = \frac{a}{c} = \cos B \qquad \sec A = \frac{c}{b} = \operatorname{cosec} B$$

$$\cos A = \frac{b}{c} = \sin B \qquad \operatorname{cosec} A = \frac{c}{a} = \sec B$$

$$\tan A = \frac{a}{b} = \cot B \qquad \cot A = \frac{b}{a} = \tan B$$

$$a = c \sin A = c \cos B = b \tan A = b \cot B = \sqrt{c^2 - b^2}$$

$$b = c \cos A = c \sin B = a \cot A = a \tan B = \sqrt{c^2 - a^2}$$

$$c = \frac{a}{\sin A} = \frac{a}{\cos B} = \frac{b}{\sin B} = \frac{b}{\cos A}$$

Oblique Triangles

Given	Sought	Formulas
A, B, a	b, c	$b = \frac{a}{\sin A} \cdot \sin B \qquad c = \frac{a}{\sin A} \cdot \sin(A + B)$
A, a, b	B, c	$\sin B = \frac{\sin A}{a} \cdot b \qquad c = \frac{a}{\sin A} \cdot \sin C$
C, a, b	$\frac{1}{2}(A + B)$	$\frac{1}{2}(A + B) = 90° - \frac{1}{2}C$
	$\frac{1}{2}(A - B)$	$\text{Tan } \frac{1}{2}(A - B) = \frac{a - b}{a + b} \cdot \tan \frac{1}{2}(A + B)$
a, b, c	A	Given $s = \frac{1}{2}(a + b + c)$, then $\sin \frac{1}{2} A = \sqrt{\frac{(s - b)(s - c)}{bc}}$ $\cos \frac{1}{2}A = \sqrt{\frac{s(s - a)}{bc}}$ $\tan \frac{1}{2}A = \sqrt{\frac{(s - b)(s - c)}{s(s - a)}}$ $\sin A = 2\, \frac{\sqrt{s(s - a)(s - b)(s - c)}}{bc}$
	Area	$\text{Area} = \sqrt{s(s - a)(s - b)(s - c)}$
C, a, b	Area	$\text{Area} = \frac{1}{2} ab \sin C$

TABLE XVIII Commonly Used Conversion Factors

a. *Length relationships*

1 ft = 0.3048006096 m (U.S. Survey Foot)
1 m = 3.280833333 ft
1 cm = 0.01 m
1 mm = 0.001 m
1 micrometer (μm) or micron (μ) = 0.001 mm = 10^{-6} m
1 millimicron (mμ) = 0.001 μm = 10^{-9} m

1 yard = 3 ft
1 mile = 5,280 ft
1 Nautical mile = 6,076.10 ft
1 chain = 100 links
1 chain = 66 ft
1 chain = 4 rods
1 rod = 16.5 ft

b. *Area relationships*

1 acre = 43,560 sq ft
1 acre = 10 sq chains
1 hectare = 2.47104 acres

c. *Angular relationships*

$400^g = 360°$
$1^g = 0.9° = 0.01570797$ rad
$.01^g = 1^c = 0.009° = 0°00'32.4''$
$0.0001^g = 1^{cc} = 0.00009° = 0°00'0.324''$

1 mil = 0.05625° = 3.37500′
1° = 17.77778 mils
1′ = 0.29630 mils

1 rad = 57.2957795° = 57°17′44.806″
1 rad = 3437.4677′ = 206,264.806″
1° = 0.0174533 rad
1′ = 0.0002909 rad
1″ = 0.00000485 rad

When θ is a small angle,

$\sin\theta = \theta$ (in radians) — $\tan\theta = \theta$ (in radians)
$\sin\theta = \theta'' \sin 1''$ — $\tan\theta = \theta'' \tan 1''$
$\sin 1'' = 0.00000485$ — $\tan 1'' = 0.00000485$
θ (radians) $= \theta'' \times .00000485$

d. *Other useful relationships*

$\pi = 3.141592654$
°F = 9/5° C + 32
°C = 5/9(° F − 32)
1 kg = 2.205 lb

INDEX